# Big Data in Omics and Imaging

Integrated Analysis and Causal Inference

# CHAPMAN & HALL/CRC
## Mathematical and Computational Biology Series

**Aims and scope:**
This series aims to capture new developments and summarize what is known over the entire spectrum of mathematical and computational biology and medicine. It seeks to encourage the integration of mathematical, statistical, and computational methods into biology by publishing a broad range of textbooks, reference works, and handbooks. The titles included in the series are meant to appeal to students, researchers, and professionals in the mathematical, statistical and computational sciences, fundamental biology and bioengineering, as well as interdisciplinary researchers involved in the field. The inclusion of concrete examples and applications, and programming techniques and examples, is highly encouraged.

**Series Editors**

N. F. Britton
*Department of Mathematical Sciences*
*University of Bath*

Xihong Lin
*Department of Biostatistics*
*Harvard University*

Nicola Mulder
*University of Cape Town*
*South Africa*

Maria Victoria Schneider
*European Bioinformatics Institute*

Mona Singh
*Department of Computer Science*
*Princeton University*

Proposals for the series should be submitted to one of the series editors above or directly to.
**CRC Press, Taylor & Francis Group**
3 Park Square, Milton Park
Abingdon, Oxfordshire OX14 4RN
UK

# Published Titles

**An Introduction to Systems Biology: Design Principles of Biological Circuits**
*Uri Alon*

**Glycome Informatics: Methods and Applications**
*Kiyoko F. Aoki-Kinoshita*

**Computational Systems Biology of Cancer**
*Emmanuel Barillot, Laurence Calzone, Philippe Hupé, Jean-Philippe Vert, and Andrei Zinovyev*

**Python for Bioinformatics, Second Edition**
*Sebastian Bassi*

**Quantitative Biology: From Molecular to Cellular Systems**
*Sebastian Bassi*

**Methods in Medical Informatics: Fundamentals of Healthcare Programming in Perl, Python, and Ruby**
*Jules J. Berman*

**Chromatin: Structure, Dynamics, Regulation**
*Ralf Blossey*

**Computational Biology: A Statistical Mechanics Perspective**
*Ralf Blossey*

**Game-Theoretical Models in Biology**
*Mark Broom and Jan Rychtář*

**Computational and Visualization Techniques for Structural Bioinformatics Using Chimera**
*Forbes J. Burkowski*

**Structural Bioinformatics: An Algorithmic Approach**
*Forbes J. Burkowski*

**Spatial Ecology**
*Stephen Cantrell, Chris Cosner, and Shigui Ruan*

**Cell Mechanics: From Single Scale-Based Models to Multiscale Modeling**
*Arnaud Chauvière, Luigi Preziosi, and Claude Verdier*

**Bayesian Phylogenetics: Methods, Algorithms, and Applications**
*Ming-Hui Chen, Lynn Kuo, and Paul O. Lewis*

**Statistical Methods for QTL Mapping**
*Zehua Chen*

**An Introduction to Physical Oncology: How Mechanistic Mathematical Modeling Can Improve Cancer Therapy Outcomes**
*Vittorio Cristini, Eugene J. Koay, and Zhihui Wang*

**Normal Mode Analysis: Theory and Applications to Biological and Chemical Systems**
*Qiang Cui and Ivet Bahar*

**Kinetic Modelling in Systems Biology**
*Oleg Demin and Igor Goryanin*

**Data Analysis Tools for DNA Microarrays**
*Sorin Draghici*

**Statistics and Data Analysis for Microarrays Using R and Bioconductor, Second Edition**
*Sorin Drăghici*

**Computational Neuroscience: A Comprehensive Approach**
*Jianfeng Feng*

**Mathematical Models of Plant-Herbivore Interactions**
*Zhilan Feng and Donald L. DeAngelis*

**Biological Sequence Analysis Using the SeqAn C++ Library**
*Andreas Gogol-Döring and Knut Reinert*

**Gene Expression Studies Using Affymetrix Microarrays**
*Hinrich Göhlmann and Willem Talloen*

**Handbook of Hidden Markov Models in Bioinformatics**
*Martin Gollery*

**Meta-Analysis and Combining Information in Genetics and Genomics**
*Rudy Guerra and Darlene R. Goldstein*

# Published Titles (continued)

# Published Titles (continued)

**Cancer Systems Biology**
*Edwin Wang*

**Stochastic Modelling for Systems Biology, Second Edition**
*Darren J. Wilkinson*

**Big Data in Omics and Imaging: Association Analysis**
*Momiao Xiong*

**Big Data Analysis for Bioinformatics and Biomedical Discoveries**
*Shui Qing Ye*

**Bioinformatics: A Practical Approach**
*Shui Qing Ye*

**Introduction to Computational Proteomics**
*Golan Yona*

**Big Data in Omics and Imaging: Integrated Analysis and Causal Inference**
*Momiao Xiong*

# Big Data in Omics and Imaging
## Integrated Analysis and Causal Inference

Momiao Xiong

**CRC Press**
Taylor & Francis Group
Boca Raton  London  New York

CRC Press is an imprint of the
Taylor & Francis Group, an **informa** business

A CHAPMAN & HALL BOOK

CRC Press
Taylor & Francis Group
6000 Broken Sound Parkway NW, Suite 300
Boca Raton, FL 33487-2742

First issued in paperback 2021

© 2018 by Taylor & Francis Group, LLC
CRC Press is an imprint of Taylor & Francis Group, an Informa business

No claim to original U.S. Government works

ISBN-13: 978-1-03-209523-3 (pbk)
ISBN-13: 978-0-8153-8710-7 (hbk)

*To Ping*

# Contents

# *Preface*

Despite significant progress in dissecting the genetic architecture of complex diseases by association analysis, understanding the etiology and mechanism of complex diseases remains elusive. It is known that significant findings of association analysis have lacked consistency and often proved to be controversial. The current approach to genomic analysis lacks breadth (number of variables analyzed at a time) and depth (the number of steps which are taken by the genetic variants to reach the clinical outcomes across genomic and molecular levels) and its paradigm of analysis is association and correlation analysis. Next generation genomic, epigenomic, sensing, and image technologies are producing ever deeper multiple omic, physiological, imaging, environmental, and phenotypic data, the causal inference of which is a cornerstone of scientific discovery and an essential component for discovery of mechanism of diseases. It is time to shift the current paradigm of genetic analysis from shallow association analysis to deep causal inference and from genetic analysis alone to integrated genomic, epigenomic, imaging and phenotypic data analysis for unraveling the mechanism of psychiatric disorders.

This book is a natural extension of the book *Big Data in Omics and Imaging: Association Analysis*. The focus of this book is integrated genomic, epigenomic, and imaging data analysis and causal inference. To make the paradigm shift feasible, this book will (1) develop novel or apply existing causal inference methods for genome-wide and epigenome-wide causal studies of complex diseases; (2) develop unified frameworks for systematic casual analysis of integrated genomic, epigenomic, image, and clinical phenotype data analysis, and inferring multilevel omic and image causal networks which lead to discovery of paths of genetic variants to the disease via multiple omic and image causal networks; (3) develop novel and apply existing methods for gene expression and methylation deconvolution, and develop novel methods for inferring cell specific multiple omic causal networks; and (4) introduce deep learning for genomic, epigenomic, and imaging data analysis and develop methods for combining deep learning with causal inference.

This book is organized into seven chapters. The following is a description of each chapter. Chapter 1, "Genotype–Phenotype Network Analysis," studies directed and undirected genotype–phenotype networks, which are major topics of causal inference. Efficient genetic analysis consists of two major parts: (1) breadth (the number of phenotypes which the genetic variants affect) and (2) depth (the number of steps which are taken by the genetic variants to reach the clinical outcomes). Causal inference theory and chain graph models provide an innovative analytic platform for deep and precise multilevel hybrid causal genotype–disease network analysis. Very few genetic and epigenetic textbooks cover causal inference theory in depth; therefore, Chapter 1 and Chapter 2 will provide solid knowledge and efficient tools for causal inference in genomic and epigenomic analysis. Chapter 1 includes (1) undirected graphs for genotype network, (2) alternating direction method of multipliers for estimation of Gaussian graphical model, (3) coordinate descent algorithm and graphical Lasso, (4) multiple graphical models, (5) directed graphs and structural equation models for networks, (6) sparse linear structural equations, (7) functional structural equation models for genotype–phenotype networks with next-generation sequencing data, and (8) effect decomposition and estimation.

Chapter 2, "Causal Analysis and Network Biology," covers (1) Bayesian networks as a general framework for causal inference, (2) structural equations and score metrics for continuous causal networks, (3) network penalized logistic regression for learning hybrid Bayesian networks, (4) statistical methods for pedigree-based causal inference, (5) nonlinear structural equation models, (6) mixed linear and nonlinear structural equation models, (7) jointly interventional and observational data for causal inference, and (8) integer programming for causal structure leaning.

Chapter 3, "Wearable Computing and Genetic Analysis of Function-Valued Traits," studies the genetics of function-valued traits. Early detection of diseases and health monitoring are primary goals of health care and disease management. Physiological traits such as ECG, EEG, SCG, EMG, MEG, and oxygen saturation levels provide important information on the health status of humans and can be used to monitor and diagnose diseases. Wearable sensors with a capacity of noninvasive and continuous personal health monitoring will not only measure health parameters of individuals at rest, but also generate signals of transient events that may be of profound prognostic or therapeutic importance. These physiological traits are a function-valued trait. Analysis of genomic and space-temporal physiological data can provide the holistic genetic structure of disease, but also poses great methodological and computational challenges. There is a lack of statistical methods for genetic analysis of function-valued traits in the literature. In this chapter, we propose novel statistical methods for genetic analysis of physiological traits. Chapter 3 covers wearable computing for automated disease diagnosis and real time health care monitoring, deep learning for physiological time series data analysis, functional linear models with both functional response and functional predictors for association analysis of physiological traits with next-generation sequencing data, mixed functional linear models with functional response for family-based genetic analysis of physiological traits, functional regression models with both functional response and functional predictors for gene gene interaction analysis, and functional canonical correlation analysis for association studies of physiological traits.

Chapter 4, "RNA-Seq Data Analysis," covers (1) data normalization and preprocessing, (2) functional principal component analysis test for differential expression analysis with RNA-seq or miRNA-seq data, (3) multivariate functional principal component analysis for allele-specific expression analysis, (4) eQTL and eQTL epistasis analysis with RNA-seq data, (5) co-expression networks, (6) linear and nonlinear regulatory networks, (7) gene expression imputation, and (8) genotype–expression regulatory networks, (9) dynamic Bayesian networks and longitudinal expression data analysis, and (10) single cell RNA-seq data analysis, gene expression deconvolution, and genetic screening.

Chapter 5, "Methylation Data Analysis," discusses methylation data analysis. The statistical methods for differential gene expression, eQTL analysis, and genotype–expression regulatory networks can be easily extended to methylation data analysis. Epigenome-wide causal studies, a new concept for epigenetic analysis, will be first introduced in this chapter. In addition to these analyses, Chapter 5 will put emphasis on inference on whole genome methylation and expression causal networks. Since both gene expression and methylation data involve more than 20,000 genes, it is impossible to construct a causal network with more than 40,000 nodes. Therefore, multiple level methylation-expression networks should be designed. Chapter 5 addresses three essential issues in the estimation of multiple level methylation expression networks: (1) low rank model for representation of either gene expression or methylation in a pathway or a cluster, (2) construction of methylation and expression networks using low rank model representation of methylation and gene expression in the pathways or clusters, and (3) construction of methylation and gene expression causal networks using original methylation and gene expression values in the local connected pathways or clusters. Chapter 5 also investigates the methylations in what cells regulate what cell gene expression. This chapter presents several novel approaches to methylation and gene expression analysis.

Chapter 6, "Imaging and Genomics," focuses on imaging signal processing, automatic image diagnosis, and genetic-imaging data analysis. There is increasing interest in statistical methods and computational algorithms to analyze high dimensional, space-correlated, and complex imaging data, and clinical and genetic data for disease diagnosis, management, and disease mechanism research. This chapter covers (1) deep learning for medical image semantic segmentation, (2) three-dimensional functional principal component analysis for imaging signal extraction, (3) imaging network construction and connectivity analysis, (4) causal machine learning for automated imaging diagnosis of disease, (5) multiple functional linear models for imaging genetics analysis with next-generation sequencing data, (6) quadratically regularized functional canonical correlation analysis for imaging genetics or imaging RNA-seq data analysis, (7) causal analysis for imaging genetics and imaging RNA-seq data analysis, (8) time series structural equation models for integrated causal analysis of fMRI and genomic data, and (9) causal machine learning.

Chapter 7, "From Association Analysis to Integrated Causal Inference," will develop novel statistical methods for genome-wide causal studies and investigate integrated genomic, epigenomic, imaging, and multiple phenotype data analysis. Chapter 7 presents mathematical formulation of causal analysis and discusses principles underlying causation. The criterions for distinguishing causation tests from association tests are also introduced in Chapter 7. In genomic and epigenomic data analysis, we usually consider four types of associations: association of discrete variables with continuous variables, continuous variables with continuous variables, discrete variables with binary trait, and continuous variables with binary trait (disease status). These four types of association analyses are extended to four types of causation analyses in this chapter. Chapter 7 also covers several powerful tools, including additive noise models, information geometry, trace methods, and Haar measure and distance correlation, for casual inference. There are multiple steps between genes and phenotypes. Only broadly and deeply searching enormous path space connecting genetic variants to the clinical outcomes allows us to uncover the mechanism of diseases. Precision medicine demands deep, systematic, comprehensive, and precise analysis of genotype–phenotype – "and the deeper you go, the more you know." Chapter 7 proposes to use causal inference theory to develop an innovative analytic platform for deep and precise multilevel hybrid causal genotype–disease network analysis, which integrates gene association subnetworks, environment subnetworks, gene regulatory subnetworks, causal genetic-methylation subnetworks, methylation-gene expression networks, genotype–gene expression-imaging subnetworks, the intermediate phenotype subnetworks, and multiple disease subnetworks into a single connected multilevel genotype–disease network to reveal the deep causal chain of mechanisms underlying the disease. In addition, Chapter 7 also covers causal inference with confounders.

Overall, this book introduces state-of-the-art studies and practice achievements in causal inference, deep learning, genomic, epigenomic, imaging, and multiple phenotype data analysis. This book sets the basis and analytic platforms for further research in this challenging and rapidly changing field. The expectation is that the presented concepts, statistical methods, computational algorithms and analytic platforms in the book will facilitate training next-generation statistical geneticists, bioinformaticians, and computational biologists.

I would like to thank Sara A. Barton for editing the book. I am deeply grateful to my colleagues and collaborators Li Jin, Eric Boerwinkle, and others whom I have worked with for many years. I would especially like to thank my former and current students and postdoctoral fellows for their strong dedication to the research and scientific contributions to the book: Jinying Zhao, Li Luo, Shenying Fang, Nan Lin, Rong Jiao, Zixin Hu, Panpan Wang, Kelin Xu, Dan Xie, Xiangzhong Fang, Jun Li, Shicheng Guo, Shengjun Hong, Pengfei Hu, Tao Xu, Wenjia Peng, Xueson Wu, Yun Zhu, Dung-Yang Lee, Lerong Li, Getie A. Zewdie, Long Ma, Hua Dong, Futao Zhang, and Hoicheong Siu. Finally, I must thank my editor, David Grubbs, for his encouragement and patience during the process of creating this book.

MATLAB® is a registered trademark of The MathWorks, Inc. For product information, please contact:

The MathWorks, Inc.
3 Apple Hill Drive
Natick, MA 01760-2098 USA
Tel: 508-647-7000
Fax: 508-647-7001
Email: info@mathworks.com
Web: www.mathworks.com

# *Author*

**Momiao Xiong**, is a professor in the Department of Biostatistics and Data Science, University of Texas School of Public Health; a regular member in the Genetics & Epigenetics (G&E) Graduate Program at The University of Texas MD Anderson Cancer, UTHealth Graduate School of Biomedical Science; and a distinguished professor in the school of Life Science, Fudan University, China.

# 1

## Genotype–Phenotype Network Analysis

## 1.1 Undirected Graphs for Genotype Network

### 1.1.1 Gaussian Graphic Model

Genetic variants are correlated. Dependence relationship among genetic variants are classically measured using linkage disequilibrium. Linkage disequilibrium is widely used to quantify pair-wise correlation between genetic variants. However, high order linkage disequilibrium is complicated and difficult to use in practice. How to characterize the dependence relationship among many genetic variants is an open question.

Undirected graph is a widely used tool to infer conditional dependence among many variables. It is a natural approach to characterizing dependence among genetic variants (Mohan et al. 2014; Zhou et al. 2011). A graph $G = (V, E)$ consists of nodes $V = \{1,\dots,p\}$ and undirected edges. Each node in the graph represents a single variable (genetic variant) and the edge connecting two nodes implies the dependence relationships between two variables. A simple example for an undirected graph representing a genotype network is shown in Figure 1.1. If a node in the graph is used to denote a gene, then a statistic summarizing genetic information across the gene will be used to represent the gene.

Undirected graph characterizes joint distribution of the variables. Under normal assumption of the joint distribution, the structure of the graph clearly encodes the dependence relationships among variables. Absence of an edge in the graph indicates that two variables connected by the edge are independent, given all the remaining variables. Undirected graphs are determined by zeros in the concentration matrix (Appendix 1.A).

**Example 1.1**

To illustrate the relationship between the undirected graph and concentration matrix, we present a sample example. We consider a graph with four nodes. The covariance matrix is assumed as

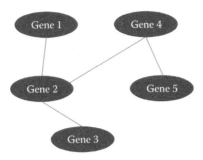

**FIGURE 1.1**
Gene network.

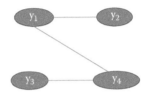

**FIGURE 1.2**
Graph for Example 1.1.

$$V = \begin{bmatrix} 0 & 1.25 & 0 & 0 \\ 1.25 & 0 & 0.6944 & 0 \\ 0 & 0.6944 & 0 & 1.111 \\ 0 & 0 & 1.111 & 0 \end{bmatrix}.$$

The concentration matrix is

$$V^{-1} = \begin{bmatrix} 0 & 0.8 & 0 & -0.5 \\ 0.8 & 0 & 0 & 0 \\ 0 & 0 & 0 & 0.9 \\ -0.5 & 0 & 0.9 & 0 \end{bmatrix}.$$

The corresponding graph is shown in Figure 1.2. The node $y_1$ is connected with the nodes $y_2$ and $y_4$. The node $y_3$ is connected with the node $y_4$.

### 1.1.2 Alternating Direction Method of Multipliers for Estimation of Gaussian Graphical Model

Next, we study how to estimate the structure of the graph and parameters in the Gaussian graphic model. It is well known that learning a single Gaussian graphic model from sampled $N$ individuals with the normal distribution $N(\mu, \Sigma_p)$ is equivalent to learning the sparse inverse matrix $\Sigma_p^{-1}$. Suppose that $N$ individuals are sampled. Let $Y_i = [y_{i1}, ..., y_{ip}]^T$ and $Y = [Y_1, ..., Y_N]^T$. The log-likelihood of the data $Y$ is given by

$$
\begin{aligned}
l &= -\frac{1}{2}pN\log(2\pi) - \frac{1}{2}N\log|\Sigma_p| - \frac{1}{2}\sum_{i=1}^{N}(Y_i - \mu)^T\Sigma_p^{-1}(Y_i - \mu) \\
&= \frac{1}{2}pN\log(2\pi) + \frac{1}{2}N\log|\Sigma_p^{-1}| - \frac{N}{2}Tr\left(\Sigma_p^{-1}S\right),
\end{aligned}
\tag{1.1}
$$

where $S = \frac{1}{N}\sum_{i=1}^{N}(Y_i - \mu)(Y_i - \mu)^T$ is a sampling matrix.

When the sample size is larger than the number of variables ($N > p$), the inverse matrix $\Sigma_p^{-1}$ is estimated by maximizing the log-likelihood. However, when the number of nodes $p$ is larger than the sample size $N$, the covariance matrix is singular. The maximization likelihood estimation method for estimation of the inverse of the covariance matrix is infeasible. The concentration matrix is then induced by imposing a graphical lasso penalty on $\Sigma p$ to shrink the parameters in the penalized maximum likelihood estimates of $\Sigma_p^{-1}$ (Mohan et al. 2014):

$$
\min_{\Theta} \quad Tr(S\Theta) - \log|\Theta| + \lambda\|\Theta\|_1,
\tag{1.2}
$$

where $\lambda$ is a penalty parameter r controlling the number of elements (or edges) of zeros in the matrix and $\|\cdot\|_1$ is the $L_1$ norm of the matrix and defined as the sum of the absolute values of the elements of the matrix. Many methods such as the coordinate descent algorithm (Friedman et al. 2008; http://www-stat.stanford.edu/~tibs/glasso) and the alternating direction method of multipliers (ADDM) (Boyd et al. 2011) for solving the optimization problem (1.2) have been developed. Here, we mainly introduce ADDM. The optimization problem (1.2) is a nonsmooth optimization problem that involves both smooth and nonsmooth functions. Basic tools for solving the optimization problem are calculus. Since derivatives cannot directly be applied to nonsmooth functions, the great barrier in solving the optimization problem (1.2) lies in dealing with nonsmooth functions. The strategies of the ADDM algorithms are to separate the smooth part from the nonsmooth part in the objective function of the optimization problem by imposing constraints and to use dual decomposition for developing decentralized algorithms. Let $f(\Theta) = Tr(S\Theta) - \log|\Theta|)$ and $g(Z) = \lambda\|Z\|_1$. The optimization problem (1.2) can be transformed to

$$
\begin{aligned}
&\min \quad f(\Theta) + g(Z) \\
&\text{subject to } \Theta - Z = 0
\end{aligned}
\tag{1.3}
$$

We form the augmented Lagrangian:

$$L_\rho(\Theta, Z, U) = f(\Theta) + g(Z) + \frac{\rho}{2} \| \Theta - Z + U \|_2^2. \tag{1.4}$$

ADMM consists of the iterations to decompose an optimization problem into two separate smooth and nonsmooth optimization problems:

$$\Theta^{(k+1)} := \arg\min_{\Theta} \ L_\rho\left(\Theta, Z^{(k)}, U^{(k)}\right) \tag{1.5}$$

$$Z^{(k+1)} := \arg\min_{Z} \ L_\rho\left(\Theta^{(k+1)}, Z, U^{(k)}\right) \tag{1.6}$$

$$U^{(k+1)} := U^{(k)} + \rho\left(\Theta^{(k+1)} - Z^{(k+1)}\right). \tag{1.7}$$

Specifically, the iteration procedures (1.5–1.7) can be rewritten as

$$\Theta^{(k+1)} := \underset{\Theta}{\mathrm{argmin}} \ \left(\mathrm{Tr}(S\Theta) - \log|\Theta| + \frac{\rho}{2} \| \Theta - Z^{(k)} + U^{(k)} \|_F^2\right) \tag{1.8}$$

$$Z^{(k+1)} := \underset{\Theta}{\mathrm{argmin}}\left(\lambda \| Z \|_1 + \frac{\rho}{2} \| \Theta^{(k+1)} - Z + U^{(k)} \|_F^2\right) \tag{1.9}$$

$$U^{(k+1)} := U^{(k)} + \Theta^{(k+1)} - Z^{(k+1)}, \tag{1.10}$$

where $\|\cdot\|_F$ is the Frobenius norm of the matrix, defined as the square root of the sum of the squares of the elements of the matrix.

To solve the problem (1.8), we need two formulae from matrix calculus:

$$\frac{\partial(\mathrm{Tr}(AB))}{\partial B} = A^T \tag{1.11}$$

$$\frac{\partial \log|A|}{\partial A} = A^{-T}, \tag{1.12}$$

where $A$ and $B$ are matrices.

We first solve the problem (1.8). Let $W = \mathrm{Tr}(S\Theta) - \log|\Theta| + \frac{\rho}{2} \| \Theta - Z^{(k)} + U^{(k)} \|_F^2$. Using formulae (1.11) and (1.12) we obtain

$$\frac{\partial W}{\partial \Theta} = S - \Theta^{-1} + \rho\left(\Theta - Z^{(k)} + U^{(k)}\right) = 0, \tag{1.13}$$

which can be rewritten as

$$\rho\Theta - \Theta^{-1} = \rho\left(Z^{(k)} - U^{(k)}\right) - S. \tag{1.14}$$

Using singular value decomposition (SVD), the matrix $\rho(Z^{(k)} - U^{(k)}) - S$ can be decomposed as

$$\rho\left(Z^{(k)} - U^{(k)}\right) - S = Q\Lambda Q^T,$$

where $Q$ is an orthonormal matrix: $Q^T Q = QQ^T = I$ and $\Lambda = diag(\lambda_1,...,\lambda_p)$. Multiplying Equation 1.14 by $Q^T$ on the left and by $Q$ on the right, we obtain

$$\rho Q^T \Theta Q - Q^T \Theta^{-1} Q = \Lambda \quad \text{or}$$

$$\rho \tilde{\Theta} - \tilde{\Theta}^{-1} = \Lambda. \tag{1.15}$$

Since $\Lambda$ is a diagonal matrix, $\tilde{\Theta}$ should have the form: $\tilde{\Theta} = diag(\tilde{\Theta}_{11},...,\tilde{\Theta}_{pp})$. Equation 1.15 is then reduced to

$$\rho \tilde{\Theta}_{ii} - \frac{1}{\tilde{\Theta}_{ii}} = \lambda_i, \quad i = 1,...,p. \tag{1.16}$$

Solution of Equation 1.16 is

$$\tilde{\Theta}_{ii} = \frac{\lambda_i + \sqrt{\lambda_i^2 + 4\rho}}{2\rho}, i = 1,...,p.$$

Thus, the solution to Equation 1.14 is $\Theta^{(k+1)} := Q\tilde{\Theta}Q^T$.

Next, we solve the nonsmooth optimization problem (1.1). By definition, $\|Z\|_1$ is $\|Z\|_1 = \sum_{i=1}^{p} \sum_{j=1}^{p} |z_{ij}|$. The generalized derivative of $\|Z\|_1$ is given by

$$\frac{\partial \| Z \|_1}{\partial z_{ij}} = \delta_{ij} = \begin{cases} 1 & z_{ij} > 0 \\ [-1,1] & z_{ij} = 0 \\ -1 & z_{ij} < 0 \end{cases} \tag{1.17}$$

Let $D = \lambda \| Z \|_1 + \frac{\rho}{2} \| \Theta^{(k+1)} - Z + U^{(k)} \|_F^2$. The generalized derivative of the function $D$ with respect to $z_{ij}$ is defined as

$$\frac{\partial D}{\partial z_{ij}} = \lambda \delta_{ij} - \rho \left( \Theta_{ij}^{(k+1)} - z_{ij} + U_{ij}^{(k)} \right) = 0, \tag{1.18}$$

which can be reduced to

$$z_{ij} = \Theta_{ij}^{(k+1)} + U_{ij}^{(k)} - \frac{\lambda}{\rho} \delta_{ij}. \tag{1.19}$$

We can show that if $\Theta_{ij}^{(k+1)} + U_{ij}^{(k)} > 0$, then $Z_{ij} \geq 0$. Otherwise, if $z_{ij} < 0$, then $\delta_{ij} = -1$. Thus, $\Theta_{ij}^{(k+1)} + U_{ij}^{(k)} + \frac{\lambda}{\rho} > 0$ which contradicts our assumption that $Z_{ij} < 0$. Similarly, we can show that if $\Theta_{ij}^{(k+1)} + U_{ij}^{(k)} < 0$ then $Z_{ij} \leq 0$. Therefore, the solution is

$$z_{ij}^{(k+1)} = sign\left(\Theta_{ij}^{(k+1)} + U_{ij}^{(k)}\right)\left(\left|\Theta_{ij}^{(k+1)} + U_{ij}^{(k)}\right| - \frac{\lambda}{\rho}\right)_+, \qquad (1.20)$$

where sign (x) is a sign function and

$$f(x)_+ = \begin{cases} x & x \geq 0 \\ 0 & x < 0. \end{cases}$$

In summary, the ADDM algorithm for estimating the Gaussian graphic model is given as follows:

Step 1: Initialization ($k = 0$).

Using SVD, we obtain $-S = Q\Lambda Q^T$. Calculate $\tilde{\Theta}_{ii} = \dfrac{\lambda_i + \sqrt{\lambda_i^2 + 4\rho}}{2\rho}$,

$i = 1, \ldots, p$. Define the matrix $\tilde{\Theta} = diag(\tilde{\Theta}_{11}, \ldots, \tilde{\Theta}_{pp})$ and $\Theta^{(1)} := Q\tilde{\Theta}Q^T$.

Calculate $z_{ij}^{(1)} = sign\left(\Theta_{ij}^{(1)}\right)\left(|\Theta_{ij}^{(k+1)}| - \dfrac{\lambda}{\rho}\right)_+$ and $U^{(1)} := \Theta^{(1)} - Z^{(1)}$.

Iterate between step 2 and step 4 until convergence occurs.

Step 2: Update matrix $\Theta^{(k+1)}$.

i. Using SVD, we obtain the orthogonal eigenvalue decomposition: $\rho(Z^{(k)} - U^{(k)}) - S = Q\Lambda Q^T$.

ii. Calculate $\tilde{\Theta}_{ii} = \dfrac{\lambda_i + \sqrt{\lambda_i^2 + 4\rho}}{2\rho}, i = 1, \ldots, p$.

iii. Update the matrix: $\Theta^{(k+1)} := Q\tilde{\Theta}Q^T$.

Step 3: Update matrix $Z^{(k+1)}$: $z_{ij}^{(k+1)} = sign\left(\Theta_{ij}^{(k+1)} + U_{ij}^{(k)}\right)\left(|\Theta_{ij}^{(k+1)} + U_{ij}^{(k)}| - \dfrac{\lambda}{\rho}\right)_+, i = 1, \ldots, p, j = 1, \ldots, p.$

Step 4: Update matrix $U^{(k+1)}$: $U^{(k+1)} := U^{(k)} + \Theta^{(k+1)} - Z^{(k+1)}$.

### 1.1.3 Coordinate Descent Algorithm and Graphical Lasso

The coordinated descent algorithm was developed to solve the Graphical Lasso (Glasso) problem (Friedman et al. 2007; Mazumder and Hastie, 2012). Using Equations 1.11 and 1.12, we can obtain the optimality conditions for a solution to the nonsmooth optimization problem (1.2):

$$S - \Theta^{-1} + \lambda\Delta = 0, \qquad (1.21)$$

where

$$\Delta_{ij} = \begin{cases} 1 & \Theta_{ij} > 0 \\ [-1,1] & \Theta_{ij} = 0 \\ -1 & \Theta_{ij} < 0. \end{cases}$$

We use a block-coordinate method to solve the nonlinear matrix equation (1.21). Suppose that the matrices $\Theta$ and $\Delta$ are partitioned as

$$\Theta = \begin{bmatrix} \Theta_{11} & \Theta_{12} \\ \Theta_{21} & \theta_{22} \end{bmatrix}, \Delta = \begin{bmatrix} \Delta_{11} & \Delta_{12} \\ \Delta_{21} & \Delta_{22} \end{bmatrix} \text{ and } H = \Theta^{-1} = \begin{bmatrix} H_{11} & H_{12} \\ H_{21} & h_{22} \end{bmatrix}.$$

Similar arguments are used to prove Theorem 1.A.1, we have

$$\Theta^{-1} = \begin{bmatrix} \left(\Theta_{11} - \dfrac{\Theta_{12}\Theta_{21}}{\theta_{22}}\right)^{-1} & -H_{11}\dfrac{\Theta_{12}}{\theta_{22}} \\ H_{11}\dfrac{\Theta_{21}}{\theta_{22}} & \dfrac{1}{\theta_{22}} + \dfrac{\Theta_{21}H_{11}\Theta_{12}}{\theta_{22}^2} \end{bmatrix} \tag{1.22}$$

Using Theorem A.3.4. from Anderson (1984), we obtain

$$\left(\Theta_{11} - \frac{\Theta_{12}\Theta_{21}}{\theta_{22}}\right)^{-1} = \Theta_{11}^{-1} + \frac{\Theta_{11}^{-1}\Theta_{12}\Theta_{21}\Theta_{11}^{-1}}{\theta_{22} - \Theta_{21}\Theta_{11}^{-1}\Theta_{12}}. \tag{1.23}$$

It is easy to check that $(\Theta_{12}\Theta_{21}\Theta_{11}^{-1})\Theta_{12} = (\Theta_{21}\Theta_{11}^{-1}\Theta_{12})\Theta_{12}$, which implies that

$$\Theta_{12}\Theta_{21}\Theta_{11}^{-1} = \Theta_{21}\Theta_{11}^{-1}\Theta_{12}. \tag{1.24}$$

Using Equation 1.24 we have

$$H_{11} = \Theta_{11}^{-1} + \frac{\Theta_{11}^{-1}\Theta_{12}\Theta_{21}\Theta_{11}^{-1}}{\theta_{22} - \Theta_{21}\Theta_{11}^{-1}\Theta_{12}} = \Theta_{11}^{-1}\left[I + \frac{\Theta_{12}\Theta_{21}\Theta_{11}^{-1}}{\theta_{22} - \Theta_{21}\Theta_{11}^{-1}\Theta_{12}}\right]$$

$$= \Theta_{11}^{-1}\frac{\theta_{22}}{\theta_{22} - \Theta_{21}\Theta_{11}^{-1}\Theta_{12}}. \tag{1.25}$$

Therefore, combining Equations (1.22) and (1.25) we have

$$- H_{11}\frac{\Theta_{12}}{\theta_{22}} = -\frac{\Theta_{11}^{-1}\Theta_{12}}{\theta_{22} - \Theta_{21}\Theta_{11}^{-1}\Theta_{12}} \tag{1.26}$$

and

$$\frac{1}{\theta_{22}} + \frac{\Theta_{21}H_{11}\Theta_{12}}{\theta_{22}^2} = \frac{1}{\theta_{22}} + \frac{\Theta_{21}\Theta_{11}^{-1}\Theta_{12}}{\theta_{22}(\theta_{22} - \Theta_{21}\Theta_{11}^{-1}\Theta_{12})} = \frac{1}{\theta_{22} - \Theta_{21}\Theta_{11}^{-1}\Theta_{12}}. \tag{1.27}$$

Combing Equations 1.22, 1.26, and 1.27 results in

$$H = \Theta^{-1} = \begin{bmatrix} \Theta_{11}^{-1} + \dfrac{\Theta_{11}^{-1}\Theta_{12}\Theta_{21}\Theta_{11}^{-1}}{\theta_{22} - \Theta_{21}\Theta_{11}^{-1}\Theta_{12}} & -\dfrac{\Theta_{11}^{-1}\Theta_{12}}{\theta_{22} - \Theta_{21}\Theta_{11}^{-1}\Theta_{12}} \\[4mm] \dfrac{\Theta_{21}\Theta_{11}^{-1}}{\theta_{22} - \Theta_{21}\Theta_{11}^{-1}\Theta_{12}} & \dfrac{1}{\theta_{22} - \Theta_{21}\Theta_{11}^{-1}\Theta_{12}} \end{bmatrix}. \tag{1.28}$$

Glasso algorithms iteratively solve Equation 1.21 for a column at a time as shown below:

$$\begin{bmatrix} H_{11}^{(k)} & H_{12}^{(k+1)} \\[2mm] \left(H_{12}^{(k+1)}\right)^{T} & h_{22}^{(k+1)} \end{bmatrix}$$

In other words, we assume that $H_{11}^{(k)}$ is known and seek to calculate the vector $H_{12}^{(k+1)}$ and the number $h_{22}^{(k+1)}$. The last column of Equation 1.2 is

$$-H_{12} + S_{12} + \lambda\Delta_{12} = 0. \tag{1.29}$$

It follows from Equation (1.22) that

$$H_{12} = -\frac{H_{11}\Theta_{12}}{\theta_{22}}. \tag{1.30}$$

Substituting Equation 1.30 into Equation 1.29, we obtain

$$H_{11}^{(k)}\frac{\Theta_{12}}{\theta_{22}} + S_{12} + \lambda\Delta_{12} = 0. \tag{1.31}$$

Let $\beta = \dfrac{\Theta_{12}}{\theta_{22}}$. Equation (1.31) is reduced to

$$H_{11}^{(k)}\beta + S_{12} + \lambda\Delta_{12} = 0. \tag{1.32}$$

Equation 1.21 requires that $\theta_{ii} > 0$ which implies that

$$h_{ii} = s_{ii} + \lambda > 0, i = 1, ..., p \tag{1.33}$$

and signs of the parameters $\beta$, $\theta_{12}$, and $\Delta_{12}$ are the same. For convenience of discussion, we let $V = H_{11}^{(k)}$, $u = S_{12}$ and $\delta = \Delta_{12}$. Equation 1.21 can be reduced to

$$V\beta + u + \lambda\delta = 0. \tag{1.34}$$

The *j*-th component of Equation 1.34 is given by

$$V_{jj}\beta_j + \sum_{k \neq j} V_{jk}\beta_{k=0} + \lambda\delta_j = 0, \tag{1.35}$$

which is equivalent to the stationary equation of a lasso regression. Evoking lasso techniques for regression, we obtain the solution:

$$\beta_j = \frac{sign\left(u_j + \sum_{k \neq j} V_{jk}\beta_k\right)\left(|u_j + \sum_{k \neq j} V_{jk}\beta_k| - \lambda\right)_+}{V_{jj}}, j = 1, ..., p-1, \tag{1.36}$$

where sign is a sign function.

Combining Equations 1.30 and 1.36, we can estimate

$$\hat{H}_{12} = H_{11}^{(k)}\beta$$

The element in the last row and column $h_{pp}$ can be estimated by Equation 1.33

$$\hat{h}_{pp} = S_{pp} + \lambda.$$

The element $h_{22}$ is also denoted by $\hat{h}_{pp}$.

Recall from Equation 1.22 that we have

$$\frac{1}{\theta_{22}} = h_{22} - \hat{\beta}^T \hat{H}_{12}.$$

The element $\hat{\theta}_{22}$ of the covariance matrix can be estimated by

$$\hat{\theta}_{22} = \frac{1}{h_{22} - \hat{\beta}^T \hat{H}_{12}}. \tag{1.37}$$

The vector $\hat{\theta}_{12}$ of the covariance matrix is estimated by

$$\hat{\Theta}_{12} = \hat{\theta}_{22}\hat{\beta}.$$

In summary, we have Algorithm 1.1.

**Algorithm 1.1**

Input: The sampling covariance matrix $S$ and penalty parameter $\lambda$

1. Initialize $H^{(1)} = S + \lambda I$.
2. Cycle with the columns repeatedly and perform the following steps until convergence:
   a. Rearrange the columns and rows so that the last column is the target column.

b.  Let $V = H_{11}^{(k)}$, $u = S_{12}$. Calculate

$$\beta_j = \frac{sign\left(u_j + \sum_{k \neq j} V_{jk}\beta_k\right)\left(\left|u_j + \sum_{k \neq j} V_{jk}\beta_k\right| - \lambda\right)_+}{V_{jj}}, j = 1, ..., p-1$$

$$\hat{\beta} = \left[\beta_1, ..., \beta_p\right]^T$$

c.  Calculate $\hat{H}_{12} = H_{11}^{(k)}\beta$.
d.  Calculate $\hat{h}_{22} = S_{22} + \lambda(\hat{h}_{pp} = S_{pp} + \lambda)$.
e.  Calculate $\hat{\theta}_{22} = \dfrac{1}{\hat{h}_{22} - \hat{\beta}^T \hat{H}_{12}}$

f.  $\hat{\Theta}_{12} = \hat{\theta}_{22}\hat{\beta}$.
g.  Save $\hat{\beta}$ for this column in the matrix $B$.

3.  Calculate $h_{jj} = S_{jj} + \lambda$ and use Equation 1.37 to calculate $\theta_{jj}$. Use $\hat{\Theta}_{12} = \hat{\theta}_{22}\hat{\beta}$ to convert the $B$ matrix to $\Theta$.

## 1.1.4 Multiple Graphical Models

In many applications we need to consider multiple graphical models. For example, we have disease and normal samples. Disease samples can be further divided into multiple disease subtypes. Accordingly, genotype networks can be divided into genotype subnetworks for various disease subtypes and normal samples. These subnetworks will share the same nodes, but differ in their dependence structures (Danaher et al. 2014; Guo et al. 2011). The subnetworks for disease samples and the normal samples have some edges common across all subnetworks and other edges unique to each subnetwork. If the graphical models for the disease and normal samples are estimated separately, the substantial structure similarity between multiple graphical models will not be explored. This will reduce the power to identify the true graphical models from the observational data.

To utilize all available information in the data, methods for joint estimation of multiple graphical models in which the estimates for multiple graphical models explore structural similarity, while allowing for some structure differences should be developed. Two types of approaches to joint estimation of multiple graphical models: edge-based and node-based approaches are widely used. We first briefly introduce the edge-based approach, then focus on the node-based approach to the joint estimation of multiple graphical models.

### 1.1.4.1 Edge-Based Joint Estimation of Multiple Graphical Models

We consider $K$ classes in the data. We assume that $X_1^{(k)}, ..., X_{n_k}^{(k)} \in R^P$ are independent and identically distributed from a normal $N(\mu_k, \Sigma_k)$ distribution,

where $n_k$ is the number of samples in the $k$-th class. Let $S^{(k)} = \frac{1}{n_k} \sum_{i=1}^{n_k} (X_i -$
$\bar{X}^{(k)})(X_i - \bar{X}^{(k)})^T$ be the sampling covariance matrix for the $k$-th class. Let $\Theta^{(k)} = \Sigma_k^{(-1)}$ be the precision matrix for the $k$-th class. The negative log likelihood function is defined as

$$L\left(\Theta^{(1)}, .., \Theta^{(K)}\right) = \sum_{k=1}^{K} n_k \left(-\log\left|\Theta^{(k)}\right| + \mathrm{Tr}\left(S^{(k)}\Theta^{(k)}\right)\right). \qquad (1.38)$$

The traditional approach is to maximize Equation 1.38 with respect to $\Theta^{(1)},...,\Theta^{(k)}$ which results in the estimate $(S^{(1)})^{-1}, ..., (S^{(K)})^{-1}$. However, when the number of variables is larger than the number of samples, the covariance matrix $S^{(k)}$ is singular and its inverse $(S^{(k)})^{-1}$ does not exist.

Penalization methods will be explored. When we collect multiple heterogeneous datasets, we wish to explore information to simultaneously estimate the multiple precision matrices, rather than estimating each precision matrix separately. To achieve this, we seek $\hat{\Theta}^{(k)}, k = 1, ..., K$ to minimize the following penalized negative log likelihood:

$$\min_{\Theta} \quad L\left(\Theta^{(1)}, ..., \Theta^{(K)}\right) + \lambda_1 \sum_{k=1}^{K} \|\Theta^{(k)}\|_1$$
$$+\lambda_2 \sum_{i=1}^{p} \sum_{j=1, j\neq i}^{p} P\left(\Theta_{ij}^{(1)}, ..., \Theta_{ij}^{(K)}\right), \qquad (1.39)$$

where $\lambda_1$ and $\lambda_2$ are penalty parameters. $P(\Theta_{ij}^{(1)}, ..., \Theta_{ij}^{(K)})$ is a convex penalty function applied to edges to encourage similarity among them. We can consider two specific penalty functions: a fused lasso penalty on the differences between pairs of network edges and a group lasso penalty on the edges themselves (Tibshirani et al. 2005; Yuan and Lin 2007; Mohan et al. 2014; Danaher et al. 2014):

$$P\left(\Theta_{ij}^{(1)}, ..., \Theta_{ij}^{(K)}\right) = \sum_{k<k'} \left|\Theta_{ij}^{(k)} - \Theta_{ij}^{(k')}\right|, \qquad \text{and} \qquad (1.40)$$

$$P\left(\Theta_{ij}^{(1)}, ..., \Theta_{ij}^{(K)}\right) = \sqrt{\sum_{k=1}^{K} \left(\Theta_{ij}^{(k)}\right)^2}. \qquad (1.41)$$

Since joint estimation of multiple graphical models using Equation 1.39 borrow all available information in estimating each subnetwork, it can offer more accurate estimation of multiple graphical models than estimating each subnetwork separately. ADMM algorithms can be used to solve the problem (1.39) with constraints (1.40) or (1.41) (Exercise 1.4).

### 1.1.4.2 Node-Based Joint Estimation of Multiple Graphical Models

Now we study the node-based joint estimation of multiple graphical models. The structural differences in the network are due to particular nodes that are perturbed across different conditions. For example, some SNPs may be mutated. This will cause variation in connectivity patterns of the nodes in the network.

An essential issue for node-based joint estimation of multiple graphical models is how to define the penalty function. Due to the symmetry of $\Theta^{(1)}$ and $\Theta^{(2)}$, there is large overlap among the p columns. It is obvious that the $(i, j)$ th element of $\Theta^{(1)} - \Theta^{(2)}$ is included in both the $i$-th and $j$-th columns of the matrices $\Theta^{(1)}$ and $\Theta^{(2)}$. The traditional group lasso (Yuan and Lin, 2007) penalty to the columns of the precision matrices $\sum_{j=1}^{p} \| \Theta_j^{(1)} - \Theta_j^{(2)} \|_2$ will lead to the estimates whose support is the intersections of the columns and rows, rather the union of the columns and rows. To overcome this limitation, Mohan et al. (2014) developed the row-column overlap norm (RCON) for creating penalty function.

**Definition 1.1**

Define a $Kp \times p$ matrix $V = \begin{bmatrix} V^{(1)} \\ \vdots \\ V^{(K)} \end{bmatrix}$, where $V^{(k)}$ is a $p \times p$ matrix. The row-column overlap norm induced by a matrix norm $\|\cdot\|$ is define as the following minimization problem:

$$\Omega\left(\Theta^{(1)}, ..., \Omega^{(k)}\right) = \min_{\Omega^{(1)},...,\Omega^{(K)}} \| V \| \tag{1.42}$$

$$\text{subject to} \quad \Omega^{(k)} = V^{(k)} + \left(V^{(k)}\right)^T, k = 1, .., K.$$

If $\| V^{(k)} \| = \| (V^{(k)})^T \|$, then Equation 1.42 is reduced to

$$\Omega\left(\Theta^{(1)}, ..., \Omega^{(k)}\right) = \frac{1}{2} \| \Omega \|, \tag{1.43}$$

where $\Omega = \begin{bmatrix} \Omega^{(1)} \\ \vdots \\ \Omega^{(K)} \end{bmatrix}$.

This definition attempts to interpret the precision matrix $\Omega^{(k)}$ as a set of columns plus a set of rows. The norm $\|\cdot\|$ is often an $l_1/l_q$ norm, defined as $\| V \| = \sum_{j=1}^{p} \| V_j \|_q$. For example, an $l_1/l_2$ norm is $\| V \| = \sum_{j=1}^{p} \| V_j \|_2$.

Now the node-based joint estimation of multiple graphical models can be defined as

$$\min_{\Theta^{(1)},...,\Theta^{(k)}} \left\{ L\left(\Theta^{(1)}, ..., \Theta^{(k)}\right) + \lambda_1 \sum_{k=1}^{K} \| \Theta^{(k)} \|_1 \right.$$

$$\left. + \lambda_2 \sum_{k<k'} \Omega_q\left(\Theta^{(k)} \quad \Theta^{(k')}\right) \right\}. \tag{1.44}$$

The nonsmooth convex optimization problem (1.44) can be solved by the ADMM algorithm. For simplicity, we consider $K = 2$ and $q = 2$. In this case, the problem (1.44) can be rewritten as

$$\min_{\Theta^{(1)},\dots,\Theta^{(k)}} \quad \left\{ L\left(\Theta^{(1)},\Theta^{(2)}\right) + \lambda_1 \| \Theta^{(1)} \|_1 + \lambda_1 \| \Theta^{(2)} \|_1 + \lambda_2 \sum_{j=1}^{p} \| V_j \|_2 \right\}$$

(1.45)

$$\text{subject to} \quad \Theta^{(1)} - \Theta^{(2)} = V + V^T.$$

Now we introduce some artificial variables and constraints to decouple the first term with the remaining terms in the objective function of the problem (1.45) that are not easy to solve jointly. We impose the constraints: $\Theta^{(1)} = Z^{(1)}$, $\Theta^{(2)} = Z^{(2)}$, and $W = V^T$. The optimization problem (1.45) is then transformed to

$$\min_{\Theta^{(1)},\dots,\Theta^{(k)}} \quad \left\{ L\left(\Theta^{(1)},\Theta^{(2)}\right) + \lambda_1 \| \Theta^{(1)} \|_1 + \lambda_1 \| \Theta^{(2)} \|_1 + \lambda_2 \sum_{j=1}^{p} \| V_j \|_2 \right\}$$

$$\text{s.t.} \quad \Theta^{(1)} - \Theta^{(2)} = V + W, V = W^T, \Theta^{(1)} = Z^{(1)}, \Theta^{(2)} = Z^{(2)}. \tag{1.46}$$

Using the Lagrangian multipliers, we can transform (1.46) to the following unconstrained optimization problem:

$$\min \quad L\left(\Theta^{(1)},\Theta^{(2)}\right) + \lambda_1 \| Z^{(1)} \|_1 + \lambda_1 \| Z \|_2 + \lambda_2 \sum_{j=1}^{p} \| V_j \|_2$$

$$+ \frac{\rho}{2} \| \Theta^{(1)} - \Theta^{(2)} - (V + W) + U_1 \|_2^2 + \tag{1.47}$$

$$\frac{\rho}{2} \| V - W^T + U_2 \|_2^2 + \frac{\rho}{2} \| \Theta^{(1)} - Z^{(1)} + U_3 \|_2^2 + \frac{\rho}{2} \| \Theta^{(2)} - Z^{(2)} + U_4 \|_2^2.$$

The ADMM algorithm consists of three types of iterations:

(1): $\Theta^{(1)(l+1)} \leftarrow \arg\min_{\Theta^{(1)}}$

$$L_\rho\left(\Theta^{(1)},\Theta^{(2)(l)},Z^{(1)(l)},Z^{(2)(l)},W^{(l)},V^{(l)},U_1^{(l)},U_2^{(l)},U_3^{(l)},U_4^{(l)}\right)$$

$\Theta^{(2)(l+1)} \leftarrow \arg\min_{\Theta^{(2)}}$

$$L_\rho\left(\Theta^{(1)(l+1)},\Theta^{(2)},Z^{(1)(l)},Z^{(2)(l)},W^{(l)},V^{(l)},U_1^{(l)},U_2^{(l)},U_3^{(l)},U_4^{(l)}\right)$$

$W^{(l+1)} \leftarrow \arg\min_{\Theta^{(2)}}$

$$L_\rho\left(\Theta^{(1)(l+1)},\Theta^{(2)(l+1)},Z^{(1)(l)},Z^{(2)(l)},W,V^{(l)},U_1^{(l)},U_2^{(l)},U_3^{(l)},U_4^{(l)}\right)$$

(2): $Z^{(1)(l+1)} \leftarrow \arg\min_{\Theta^{(2)}}$

$$L_\rho\left(\Theta^{(1)(l+1)},\Theta^{(2)(l+1)},Z^{(1)},Z^{(2)(l)},W^{(l+1)},V^{(l)},U_1^{(l)},U_2^{(l)},U_3^{(l)},U_4^{(l)}\right)$$

$Z^{(2)(l+1)} \leftarrow \arg\min_{\Theta^{(2)}}$

$$L_\rho\left(\Theta^{(1)(l+1)},\Theta^{(2)(l+1)},Z^{(1)(l+1)},Z^{(2)},W^{(l+1)},V^{(l)},U_1^{(l)},U_2^{(l)},U_3^{(l)},U_4^{(l)}\right)$$

$V^{(l+1)} \leftarrow \arg\min_{\Theta^{(2)}}$

$$L_\rho\left(\Theta^{(1)(l+1)},\Theta^{(2)(l+1)},Z^{(1)(l+1)},Z^{(2)(l+1)},W^{(l+1)},V,U_1^{(l)},U_2^{(l)},U_3^{(l)},U_4^{(l)}\right)$$

(3):  $\quad U_1^{(l+1)} = U_1^{(l)} + \Theta^{(1)(l+1)} - \Theta^{(2)(l+1)} - V^{(l+1)} - W^{(l+1)}$

$\qquad U_2^{(l+1)} = U_2^{(l)} + V^{(l+1)} - \left(W^{(l+1)}\right)^{(T)}$

$\qquad U_3^{(l+1)} = U +^3 \Theta^{(1)(l+1)} - Z^{(1)(l+1)}$

$\qquad U_4^{(l+1)} = U_4^{(l)} + \Theta^{(2)(l+1)} - Z^{(2)(l+1)}.$

The algorithm is summarized below (Appendix 1.B).

**Algorithm 1.2**

Input sampling covariance matrices: $S^{(1)}$, $S^{(2)}$.

Step 1: Initialization ($l = 0$)

Define $B_1 = -\dfrac{n_1}{2\rho} S^{(1)}$. Take the orthogonal eigenvalue decomposition of

the matrix $B_1$:

$B_1 = Q_1 \Lambda_1 Q_1^T$, where $\Lambda_1 = diag(\lambda_1^{(1)}, ..., \lambda_p^{(1)})$.

Define $\tilde{X}_{ii}^{(1)} = \dfrac{\lambda_i^{(1)} + \sqrt{(\lambda_i^{(1)})^2 + \dfrac{2n_1}{\rho}}}{2}$ and $\tilde{X}^{(1)} = diag(\tilde{X}_{11}^{(1)}, ..., \tilde{X}_{pp}^{(1)})$.

Calculate $\hat{\Theta}^{(1)(0)} = Q_1 \tilde{X}^{(1)} Q_1^T$.

Define $B_2 = \dfrac{1}{2}[\Theta^{(1)(0)}] - \dfrac{n_2}{2\rho} S^{(2)}$. Take the orthogonal eigenvalue decom-

position of the matrix $B_2$:

$B_2 = Q_2 \Lambda_2 Q_2^T$, where $\Lambda_2 = diag(\lambda_1^{(2)}, ..., \lambda_p^{(2)})$.

Define $\tilde{X}_{ii}^{(2)} = \dfrac{\lambda_i^{(2)} + \sqrt{(\lambda_i^{(2)})^2 + \dfrac{2n_2}{\rho}}}{2}$ and $\tilde{X}^{(2)} = diag(\tilde{X}_{11}^{(2)}, ..., \tilde{X}_{pp}^{(2)})$.

Calculate $\hat{\Theta}^{(2)(0)} = Q_2 \tilde{X}^{(2)} Q_2^T$.

$$Z^{(1)(0)} = sign\left(\Theta^{(1)(0)}\right)\left(\left|\Theta^{(1)(0)}\right| - \frac{\lambda_1}{\rho}\right)_+$$

$$Z^{(2)(0)} = sign\left(\Theta^{(2)(0)}\right)\left(\left|\Theta^{(2)(0)}\right| - \frac{\lambda_2}{\rho}\right)_+$$

For $j = 1, ..., p$ do

Define $G_j = \dfrac{1}{2}[\Theta^{(1)(0)} - \Theta^{(2)(0)}]_{(j)}$.

$$V_j^{(0)} = \left(1 - \frac{\lambda_2}{2\rho\|G_j\|_2}\right)_+ G_j, \quad V^{(0)} = \left[V_1^{(0)} \cdots V_p^{(0)}\right]$$

End

$$W^{(0)} = \frac{1}{2}\left[\Theta^{(1)(0)} - \Theta^{(2)(0)} + \left(V^{(0)}\right)^T - V^{(0)}\right].$$

$$U_1^{(0)} = \Theta^{(1)(0)} - \Theta^{(2)(0)} - V^{(0)} - W^{(0)}$$

$$U_2^{(0)} = V^{(0)} - \left(W^{(0)}\right)^{(T)}$$

$$U_3^{(0)} = \Theta^{(1)(0)} - Z^{(1)(0)}$$

$$U_4^{(0)} = \Theta^{(2)(0)} - Z^{(2)(0)}.$$

Repeat Steps 2, 3, and 4 until convergence.

Step 2:

Define $B_1 = \frac{1}{2}\left[\Theta^{(2)(l)} + V^{(l)} + W^{(l)} + Z^{(1)(l)} - U_1^{(l)} - U_3^{(l)}\right] - \frac{n_1}{2\rho}S^{(1)}.$

Take the orthogonal eigenvalue decomposition of the matrix $B_1$:

$B_1 = Q_1\Lambda_1Q_1^T$, where $\Lambda_1 = diag(\lambda_1^{(1)}, ..., \lambda_p^{(1)})$.

Define $\tilde{X}_{ii}^{(1)} = \dfrac{\lambda_i^{(1)} + \sqrt{(\lambda_i^{(1)})^2 + \dfrac{2n_1}{\rho}}}{2}$ and $\tilde{X}^{(1)} = diag(\tilde{X}_{11}^{(1)}, ..., \tilde{X}_{pp}^{(1)}).$

Calculate $\hat{\Theta}^{(1)(l+1)} = Q_1\tilde{X}^{(1)}Q_1^T.$

Define $B_2 = \frac{1}{2}\left[\Theta^{(1)(l+1)} - V^{(l)} - W^{(l)} + Z^{(2)(l)} + U_1^{(l)} - U_4^{(l)}\right] - \frac{n_2}{2\rho}S^{(2)}.$

Take the orthogonal eigenvalue decomposition of the matrix $B_2$:

$B_2 = Q_2\Lambda_2Q_2^T$, where $\Lambda_2 = diag(\lambda_1^{(2)}, ..., \lambda_p^{(2)})$.

Define $\tilde{X}_{ii}^{(2)} = \dfrac{\lambda_i^{(2)} + \sqrt{(\lambda_i^{(2)})^2 + \dfrac{2n_2}{\rho}}}{2}$ and $\tilde{X}^{(2)} = diag(\tilde{X}_{11}^{(2)}, ..., \tilde{X}_{pp}^{(2)}).$

Calculate $\hat{\Theta}^{(2)(l+1)} = Q_2\tilde{X}^{(2)}Q_2^T.$

Step 3:

$$Z^{(1)(l+1)} = sign\left(\Theta^{(1)(l+1)} + U_3^{(l)}\right)\left(\left|\Theta^{(1)(l+1)} + U_3^{(l)}\right| - \frac{\lambda_1}{\rho}\right)_+,$$

$$Z^{(2)(l+1)} = sign\left(\Theta^{(2)(l+1)} + U_4^{(l)}\right)\left(\left|\Theta^{(2)(l+1)} + U_4^{(l)}\right| - \frac{\lambda_2}{\rho}\right)_+,$$

For $j = 1, ..., p$ do

Define $G_j = \frac{1}{2}\left[\Theta^{(1)(l+1)} - \Theta^{(2)(l+1)} - W^{(l)} + U_1^{(l)} + \left(W^{(l)}\right)^T - U_2^{(l)}\right]_{(j)}.$

$$V_j^{(l+1)} = \left(1 - \frac{\lambda_2}{2\rho\|G_j\|_2}\right)_+ G_j, \quad V^{(l+1)} = \left[V_1^{(l+1)} \quad ... \quad V_p^{(l+1)}\right]$$

End

$$W^{(l+1)} = \frac{1}{2}\left[\Theta^{(1)(l+1)} - \Theta^{(2)(l+1)} + \left(V^{(l+1)}\right)^T - V^{(l+1)} + U_1^{(l+1)} + \left(U_2^{(l+1)}\right)^T\right].$$

Step 4:

$$U_1^{(l+1)} = U_1^{(l)} + \Theta^{(1)(l+1)} - \Theta^{(2)(l+1)} - V^{(l+1)} - W^{(l+1)}$$

$$U_2^{(l+1)} = U_2^{(l)} + V^{(l+1)} - \left(W^{(l+1)}\right)^{(T)}$$

$$U_3^{(l+1)} = U_3^{(1)} \Theta^{(1)(l+1)} - Z^{(1)(l+1)}$$

$$U_4^{(l+1)} = U_4^{(l)} + \Theta^{(2)(l+1)} - Z^{(2)(l+1)}.$$

## 1.2 Directed Graphs and Structural Equation Models for Networks

### 1.2.1 Directed Acyclic Graphs

Most genetic analyses of quantitative traits have focused on a single trait association analysis, analyzing each phenotype independently (Stephens 2013). However, multiple phenotypes are correlated. For example, metabolism of lipoproteins involves cholesterol, triglycerides, very low-density lipoproteins (VLDL), low-density lipoproteins, and high-density lipoproteins. These multiple traits are dependent. The integrative analysis of correlated phenotypes often increases statistical power to identify genetic associations (Aschard et al. 2014). The association analysis of multiple phenotypes is expected to become popular in the near future (Zhou and Stephens 2014). Three major approaches are commonly used to test association of genetic variants with multiple correlated phenotypes: multiple regression methods, integration of p values of univariate analysis, and dimension reduction methods. All these methods can only estimate the effect of the genetic variant on the phenotype. However, the genetic effects can be classified into three types of effects: direct, indirect, and total effects. These methods are unable to reveal causal mechanisms underlying the genetic structures of multiple phenotype association analysis.

Directed graphical models and structural equations can be used as a tool to model the complex structures among phenotypes, risk factors, and genotypes, which are referred to as the genotype–phenotype networks (Bollen 1981; Rosa et al. 2011; Mi et al. 2010).

A graphical model consists of nodes and edges. The nodes represent variables and edges represent the dependence structures among variables. A directed graphic model is defined as the graph in which all the inter-node connections have a direction visually denoted by an arrowhead. Directed acyclic graphics (DAGs) are defined as directed graphics with no cycles. In other words, we can never start at a node $X$, travel edges in the directions of

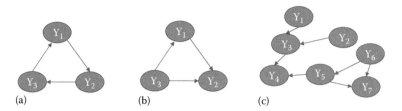

**FIGURE 1.3**
(a) Directed cyclic graph. (b, c) Directed acyclic graph.

the arrows, and get back to the node $X$. Figure 1.3 is an example of directed cyclic graphs and a directed acyclic graph.

A DAG with nodes encodes conditional dependence structure of the variables $Y_1,...Y_n$. We define the parents of a node as the nodes pointing directly to it. For example, the parents of the node $Y_7$ are the nodes $Y_5$ and $Y_6$. The concept of parents provides an easy way to read off conditional independence from DAGs. We associate "correlation" with "connectedness" in the graph and independence with "separation." Any node $Y_i$ is conditionally independent of its non-descendents given its parents. This assumption is often referred to as the causal Markov assumption. For example, given $Y_5$ and $Y_6$, the variable $Y_7$ is independent of the variables $Y_1, Y_2, Y_3$, and $Y_4$. The joint distribution can be simply factorized by the conditional independence. We can use DAGs to write the joint probability distribution:

$$P(Y_1,...,Y_n) = \prod_{i=1}^{n} P(Y_i|\text{parent}_i), \tag{1.48}$$

where $\text{parent}_i$ represents the parent nodes of the node $Y_i$. If the $\text{parent}_i$ is empty, then $P(Y_i|\text{parent}_i = \phi) = P(Y_i)$.

For example, we can write the joint distribution for Figure 1.3c as

$$P(Y_1,...,Y_7) = P(Y_1)P(Y_2)P(Y_6)P(Y_3|Y_1,Y_2)P(Y_4|Y_3,Y_5)P(Y_5|Y_6)P(Y_7|Y_5,Y_6).$$

## 1.2.2 Linear Structural Equation Models

Traditional regressions describe one-way or unidirectional relationships among variables in which the variables on the left sides of the equations are dependent variables and the variables on the right sides of the equations are explanatory variables or independent variables. The explanatory variables are used to predict the outcomes of the dependent variables. However, in many cases, there are two ways, or simultaneous relationships between the variables. Variables in some equations are response variables, but will be predictors in other equations. The variables in equations may influence each other. It is difficult to distinguish dependent variables and explanatory variables. The phenotype data are generated by the existing biological systems and should be described as a system of simultaneous relations among the random phenotype variables. The structural equation models (SEMs) are a powerful mathematic tool to describe such data generating mechanisms and infer causal relationships among the variables (Chen and Pearl 2014).

The SEMs classify variables into two class variables: endogenous and exogenous variables. The jointly dependent variables that are determined in the model are called endogenous variables. The explanatory variables that are determined outside the model or predetermined are called exogenous variables. In the genotype–phenotype networks, the phenotype variables such as BMI, blood pressure, high-density lipoprotein and low-density lipoprotein, are endogenous variables, age, sex, race, environments, and genotypes are exogenous variables.

We consider $M$ phenotypes that are referred to as endogenous variables. Assume that $n$ individuals are sampled. We denote the $n$ observations on the $M$ endogenous variables by the matrix $Y = [y_1, y_2,..., y_M]$, where $y_i = [y_{1i},..., y_{ni}]^T$ is a vector of collecting $n$ observation of the endogenous variable $i$. Covariates, genetic variants as exogenous or predetermined variables are denoted by $X = [x_1,...x_K]$ where $x_i = [x_{1i},...,x_{ni}]^T$. Similarly, random errors are denoted by $E = [e_1,...,e_M]$, where we assume $E[e_i] = 0$ and $E[e_ie_i^T] = \sigma_i^2 I_n$ for $i = 1,...,M$. The linear structural equations for modeling relationships among phenotypes and genotypes can be written as (Judge et al. 1182):

$$y_1\gamma_{11} + y_2\gamma_{21} + ... + y_M\gamma_{M1} + x_1\beta_{11} + x_2\beta_{21} + ... + x_K\beta_{K1} + e_1 = 0$$

$$\vdots \qquad\qquad\qquad\qquad\qquad \vdots \qquad\qquad (1.49)$$

$$y_1\gamma_{1M} + y_2\gamma_{2M} + ... + y_M\gamma_{MM} + x_1\beta_{1M} + x_2\beta_{2M} + ... + x_K\beta_{KM} + e_M = 0$$

where the $\gamma$'s and $\beta$'s are the structural parameters of the system that are unknown. Variables in the SEMs can be classified into two basic types of variables: observed variables that can be measured and the residual error variables that cannot be measured and represent all other unmodeled causes of the variables. Most observed variables (e.g., phenotypes) are random. Some observed variables may be nonrandom or control variables (e.g., genotypes, drug dosages) whose values remain the same in repeated random sampling or might be manipulated by the experimenter. The observed variables will be further classified into exogenous variables (e.g., genotypes), which lie outside the model, and endogenous variables (e.g., phenotypes), whose values are determined through joint interaction with other variables within the system. All nonrandom variables can be viewed as exogenous variables. Phenotypes are viewed as endogenous variables. The terms exogenous and endogenous are model specific. It may be that an exogenous variable in one model is endogenous in another.

The sample values of the joint endogenous and exogenous variables can be denoted by an $n \times M$ matrix

$$
Y = \begin{bmatrix} y_{11} & y_{12} & \cdots & y_{1M} \\ y_{21} & y_{22} & \cdots & y_{2M} \\ \vdots & \vdots & \ddots & \vdots \\ y_{n1} & y_{n2} & \cdots & y_{nM} \end{bmatrix} = [Y_1 \cdots Y_M]
$$

and an $n \times K$ matrix

$$
X = \begin{bmatrix} x_{11} & x_{12} & \cdots & x_{1K} \\ x_{21} & x_{22} & \cdots & x_{2K} \\ \vdots & \vdots & \ddots & \vdots \\ x_{n1} & x_{n2} & \cdots & x_{nK} \end{bmatrix} = [X_1 \cdots X_K], \text{respectively.}
$$

Let

$$
E = \begin{bmatrix} e_{11} & e_{12} & \cdots & e_{1M} \\ e_{21} & e_{22} & \cdots & e_{2M} \\ \vdots & \vdots & \ddots & \vdots \\ e_{n1} & e_{n2} & \cdots & e_{nM} \end{bmatrix} = [e_1 \cdots e_M]
$$

be an $n \times M$ matrix of unobservable values of the random error vectors. We define an $M \times M$ matrix of coefficients of the endogenous variables as

$$
\Gamma = \begin{bmatrix} \gamma_{11} & \gamma_{12} & \cdots & \gamma_{1M} \\ \gamma_{21} & \gamma_{22} & \cdots & \gamma_{2M} \\ \vdots & \vdots & \ddots & \vdots \\ \gamma_{M1} & \gamma_{M2} & \cdots & \gamma_{MM} \end{bmatrix} = [\Gamma_1 \cdots \Gamma_M]
$$

and a $K \times M$ matrix of coefficients of the exogenous variables as

$$
B = \begin{bmatrix} \beta_{11} & \beta_{12} & \cdots & \beta_{1M} \\ \beta_{21} & \beta_{22} & \cdots & \beta_{2M} \\ \vdots & \vdots & \ddots & \vdots \\ \beta_{K1} & \beta_{K2} & \cdots & \beta_{KM} \end{bmatrix} = [B_1 \cdots B_M].
$$

In matrix notation the SEMs (1.49) can be rewritten as

$$Y\Gamma + XB + E = 0, \tag{1.50}$$

In a matrix form, the $i$-the equation in (1.49) can be written as

$$Y\Gamma_i + XB_i + e_i = 0. \tag{1.51}$$

Traditionally, we often select one endogenous variable to appear on the left-hand side of the equation. Specifically, the $i$-th equation is

$$y_1\gamma_{1i} + \cdots + y_{i-1}\gamma_{i-1i} + y_i\gamma_{ii} + y_{i+1}\gamma_{i+1i} + \cdots + y_M\gamma_{Mi} + x_{1i}\beta_{1i} + \cdots + x_{Ki}\beta_{Ki} + e_i = 0$$

Dividing both sides of the above equation by $-\gamma_{ii}$ and replacing $-\dfrac{\gamma_{ji}}{\gamma_{ii}}$ and $-\dfrac{\beta_{ji}}{\gamma_{ii}}$ by $\gamma_{ji}$ and $\beta_{ji}$, respectively, we obtain

$$y_i = y_1\gamma_{1i} + \cdots + y_{i-1}\gamma_{i-1i} + y_{i+1}\gamma_{i+1i} + \cdots + y_M\gamma_{Mi} + x_{1i}\beta_{1i} + \cdots + x_{Ki}\beta_{Ki} + e_i, \tag{1.52}$$

where $\gamma_{ji}$ is a path coefficient that measures the strength of the causal relationship from $y_j$ to $y_i$, $\beta_{ki}$ is a path coefficient from the exogenous variable to the endogenous variable which measures the causal effect of the exogenous variable $x_{ki}$ on the endogenous variable $y_i$. The coefficients $\gamma_{ji} = 0$ and $\beta_{ki} = 0$ imply the zero direct influence of $Y_j$ and $x_{ki}$ on $Y_i$, respectively, and are usually omitted from the equation. Therefore, Equation 1.52 is reduced to

$$\begin{aligned} y_i &= Y_{-i}\gamma_i + X_i\beta_i + +e_i \\ &= W_i\Delta_i + e_i \end{aligned} \tag{1.53}$$

where $Y_{-i}$ is a vector of the endogenous variables after removing variable $y_i$, $\gamma_i$ is a vector of the path coefficients associated with $Y_{-i}$, and

$$W_i = [Y_{-i}\ X_i], \Delta_i = \begin{bmatrix} \gamma_i \\ \beta_i \end{bmatrix}.$$

Typically, we have two scenarios: (1) the structure of the equation is known and (2) the structure of the equation is unknown. When the structure is known from biological knowledge, what we need is to estimate the parameters in the equation and to make causal inference. When the structure is unknown we need to estimate both the structure and parameters. Now we focus on the first scenario with the known structure.

The structure of the SEMs can be described by DAGs. The variables are represented by nodes. Straight single-headed arrows represent causal relationships between the variables connected by the arrows. Each arrow or directed edge is associated with a coefficient in the SEM, which is often referred to as a path coefficient. There may be intermediate steps between the

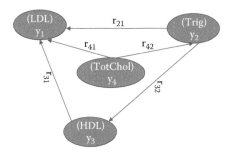

**FIGURE 1.4**
Phenotype network.

nodes, which are unknown and omitted. A curved two-headed arrow or an undirected edge indicates an association between two variables. The variables may be associated for any of a number of reasons. The association may be due to both variables depending on some other variable(s), or the variables may have a causal relationship, but this remains unspecified or indicates that the two error terms are independent.

Figure 1.4 is a DAG for modeling the network of four phenotypes: high-density lipoprotein cholesterol (HDL), low-density lipoprotein cholesterol (LDL), triglyceride (Trig), and total cholesterol (TotChol). The SEM for the DAG in Figure 1.4 is given by

$$
\begin{aligned}
y_1 &= e_1 \\
y_2 &= \gamma_{12} y_2 + e_2 \\
y_3 &= \gamma_{23} y_2 + e_3 \\
y_4 &= \gamma_{24} y_2 + \gamma_{34} y_3 + e_4.
\end{aligned}
\tag{1.54}
$$

To further describe the relations between the DAG and SEM, we introduce additional terminologies. If there exists a sequence of arrows all of which are directed from $y_j$ to $y_i$, then $y_j$ is defined as an ancestor of $y_i$ or $y_i$ is a descendant of $y_j$. The set of nodes connected to $y_i$ by undirected edges are defined as the siblings of $y_i$. A sequence of edges connecting two nodes $y_j$ and $y_i$ is defined as a path between $y_j$ and $y_i$. A path consisting only of arrows from $y_j$ to $y_i$ is defined as a directed path from $y_j$ to $y_i$.

### 1.2.3 Estimation Methods

Estimation of the parameters in the structural equations is rather complex. It involves many different estimation methods with varying statistical properties. In this chapter we introduce three popular methods: maximum likelihood method, two-stage least squares (2SLS) method, and three-stage squares (3SLS) method.

### 1.2.3.1 *Maximum Likelihood (ML) Estimation*

One observation on all $M$ structural equations can be written as (Judge et al. 1182)

$$y_i^T \Gamma + x_i^T B + e_i^T = 0, \qquad i = 1, \ldots, n, \qquad (1.55)$$

where $y_i^T$, $x_i^T$, and $e_i^T$ are the $i^{th}$ row of matrices $Y$, $X$, and $E$, respectively. Let

$$e_i = \begin{bmatrix} e_{i1} \\ \vdots \\ e_{iM} \end{bmatrix}.$$

The probability density function of $e_i$ is

$$p(e_i) = (2\pi)^{-\frac{M}{2}} |\Sigma|^{-\frac{1}{2}} \exp\left\{ -\frac{1}{2} e_i^T \Sigma^{-1} e_i \right\}.$$

To derive the probability density of the variables $y_i$, we make a change of variables. The Jacobian of the transformation is

$$\frac{\partial y_i}{\partial e_i^T} = -\Gamma^{-1}.$$

The joint probability density function for $y$ is (Judge et al. 1982)

$$f(y) = (2\pi)^{-\frac{nM}{2}} |\Sigma|^{-\frac{n}{2}} \| \Gamma \|^n \exp\left\{ -\frac{1}{2} \sum_{i=1}^{n} (y_i^T \Gamma + x_i^T B) \Sigma^{-1} (\Gamma^T y_i + B^T x_i) \right\}, \quad (1.56)$$

where $\|\Gamma\|$ is the absolute value of the determinant of the matrix $\Gamma$.

The log likelihood function is defined as

$$\log l(\Delta, \Sigma | y, W) == -\frac{nM}{2} \log(2\pi) + \frac{n}{2} \log |\Sigma|^{-1} + n$$

$$\times \log \| \Gamma \| - \frac{1}{2} \sum_{i=1}^{n} (y_i^T \Gamma + x_i^T B) \Sigma^{-1} (\Gamma^T y_i + B^T x_i). \quad (1.57)$$

The estimators $\hat{\Delta}$, $\hat{\Sigma}$ of the parameters in the structural equations can be found by maximizing the log likelihood function in Equation 1.57.

### 1.2.3.2 *Two-Stage Least Squares Method*

The previous maximum likelihood method needs to assume the normal distribution of the endogenous variables. However, in practice, normality assumptions on these variables will not hold. Therefore, we need to develop estimation methods that do not assume distributions of the endogenous and exogenous variables. Theoretic analysis indicates that the ordinary least square methods for estimation of the parameters in the structural equations, in general, will lead to inconsistent estimators. The statistical methods without assuming distribution of the endogenous and exogenous variables are either single-equation methods, which can be applied to each equation of the structural equations, or complete systems methods, which are applied to all structural equations. Here, we introduce the most important and widely used single equation method: two-stage least squares (2SLS) method.

Multiplying by the matrix $X^T$ on both sides of Equation 1.53, we obtain

$$X^T y_i = X^T W_i \Delta_i + X^T e_i. \tag{1.58}$$

It is known that

$$\text{cov}\left(X^T e_i, X^T e_i\right) = X^T X \sigma_i^2.$$

The generalized least square estimate $\hat{\Delta}_i$ is given by

$$
\begin{aligned}
\hat{\Delta}_i &= \left[W_i^T X \left(\sigma_i^2 X^T X\right)^{-1} X^T W_i\right]^{-1} W_i^T X \left(\sigma_i^2 X^T X\right)^{-1} X^T y_i \\
&= \left[W_i^T X \left(X^T X\right)^{-1} X^T W_i\right]^{-1} W_i^T X \left(X^T X\right)^{-1} X^T y_i.
\end{aligned}
\tag{1.59}
$$

The generalized least square estimate $\hat{\Delta}_i$ can be interpreted as a two-stage least square estimate (Judge et al. 1182).

Suppose that in the first stage, $Y_{-i}$ is regressed on $X$ to obtain

$$\hat{\Pi}_i = \left(X^T X\right)^{-1} X^T Y_{-i} \text{ and } \hat{Y}_{-i} = X \hat{\Pi}_i. \tag{1.60}$$

Then,

$$
\begin{aligned}
\hat{W}_i &= \left[\hat{Y}_{-i} \ X\right] \\
&= \left[X\left(X^T X\right)^{-1} X^T Y_{-i} \ X\right] \\
&= \left[X\left(X^T X\right)^{-1} X^T Y_{-i} \ X\left(X^T X\right)^{-1} X^T X\right] \\
&= X\left(X^T X\right)^{-1} X^T \left[Y_{-i} \ X\right] \\
&= X\left(X^T X\right)^{-1} X^T W_i.
\end{aligned}
\tag{1.61}
$$

Substituting Equation 1.61 into Equation 1.59, we obtain

$$\hat{\Delta}_i = \left(\hat{W}_i^T \hat{W}_i\right)^{-1} \hat{W}_i^T y_i. \tag{1.62}$$

Therefore, if $W_i$ in Equation 1.53 is replaced by $\hat{W}_i$, Equation 1.62 can be interpreted as that in the second stage, $y_i$ is regressed on $\hat{Y}_i$ and $X$ to obtain estimate $\hat{\Delta}_i$.

The variance-covariance matrix of the estimate $\hat{\Delta}_i$ is given by

$$\Lambda_i = \mathrm{var}\left(\hat{\Delta}_i\right) = n\hat{\sigma}_i^2 \left[W_i^T X (X^T X)^{-1} X^T W_i\right]^{-1}, \tag{1.63}$$

where $\hat{\sigma}_i^2$ is estimated by

$$\hat{\sigma}_i^2 = \frac{1}{n - m_i + 1 - k_i} (y_i - W_i\Delta_i)^T (y_i - W_i\Delta_i), \tag{1.64}$$

with $m_i$ and $k_i$ be the number of endogenous and exogenous variables present in the $i$th equation, respectively.

Under fairly general conditions, $\sqrt{n}(\hat{\Delta}_i - \Delta_i)$ is asymptotically distributed as a normal distribution $N(0,\Lambda_i)$.

The presence of the directed edge can be tested. Suppose that we want to test whether the phenotype $y_j$ causes the changes of the phenotype $y_i$. In other words, we want to test whether the parameter $\gamma_{ji}$ is equal to zero. Let $\Lambda_{ji}$ be the element of the variance-covariance matrix $\Lambda_i$ which corresponds to the variance of the estimator $\hat{\gamma}_{ji}$. If $j < i$, then $\Lambda_{ji}$ is the $j$ th diagonal element of the matrix $\Lambda i$ and if $j > i$, then $\Lambda_{ji}$ is the $j - 1$ th diagonal element of the matrix $\Lambda_i$. Define the test statistics as

$$T_\gamma = n \frac{\hat{\gamma}_{ji}^2}{\Lambda_{ji}}. \tag{1.65}$$

Under the null hypothesis of no directed connection from $y_j$ to $y_i$, the statistic $T_\gamma$ is asymptotically distributed as a central $\chi_{(1)}^2$ distribution.

### 1.2.3.3 Three-Stage Least Squares Method

To fully explore information in the structural equations and jointly estimate all structural equations, we consider the whole system of M structural equations:

$$\begin{bmatrix} X^T y_1 \\ X^T y_2 \\ \vdots \\ X^T y_M \end{bmatrix} = \begin{bmatrix} X^T W_1 & & & \\ & X^T W_2 & & \\ & & \ddots & \\ & & & X^T W_M \end{bmatrix} \begin{bmatrix} \Delta_1 \\ \Delta_2 \\ \vdots \\ \Delta_M \end{bmatrix} + \begin{bmatrix} X^T e_1 \\ X^T e_2 \\ \vdots \\ X^T e_M \end{bmatrix} \tag{1.66}$$

which can be compactly written as

$$[I \otimes X^T] Y = [I \otimes X^T] W\Delta + (I \otimes X^T) e, \tag{1.67}$$

where

$$W = \begin{bmatrix} W_1 & & & \\ & W_2 & & \\ & & \ddots & \\ & & & W_M \end{bmatrix}_{TM \times ([M(M-1)]}, \quad Y = \begin{bmatrix} y_1 \\ y_2 \\ \vdots \\ y_M \end{bmatrix}_{TM}, \quad \Delta = \begin{bmatrix} \Delta_1 \\ \Delta_2 \\ \vdots \\ \Delta_M \end{bmatrix}_M \quad \text{and} \quad e = \begin{bmatrix} e_1 \\ e_2 \\ \vdots \\ e_M \end{bmatrix}_M.$$

Recall that

$$\mathrm{cov}\left(\left(I_M \otimes X^T\right)e\right) = \left(I_M \otimes X^T\right)\left(\Sigma \otimes I_n\right)\left(I_M \otimes X\right)$$
$$= \Sigma \otimes X^T X. \tag{1.68}$$

It follows from Equation 1.67 and 1.68 that the generalized least square estimator of $\Delta$ is

$$\hat{\Delta} = \left\{\left[W^T\left(I_M \otimes X\right)\right]\left[\Sigma^{-1} \otimes \left(X^T X\right)^{-1}\right]\left[\left(I_M \otimes X^T\right)W\right]\right\}^{-1}\left[W^T\left(I_M \otimes X\right)\right]$$
$$\times \left[\Sigma^{-1} \otimes \left(X^T X\right)^{-1}\right]\left(I_M \otimes X^T\right)Y \tag{1.69}$$
$$= \left\{W^T\left[\Sigma^{-1} \otimes X\left(X^T X\right)^{-1}X^T\right]W\right\}^{-1}W^T\left[\Sigma^{-1} \otimes X\left(X^T X\right)^{-1}X^T\right]Y.$$

Using the 2SLS results for the estimators $\hat{\Delta}_i$, we can estimate the covariance matrix $\Sigma$ by

$$\hat{\sigma}_{ij} = \frac{\left(y_i - W_i\hat{\Delta}_i\right)^T\left(y_i - W_i\hat{\Delta}_i\right)}{\tau_{ij}}, \tag{1.70}$$

where $\tau_{ij} = \sqrt{(n - m_i + 1 - k_i)(n - m_j + 1 - k_j)}$.

The variance-covariance matrix of the estimator $\hat{\Delta}$ is given by

$$Var\left(\hat{\Delta}\right) = \left\{W^T\left[\Sigma^{-1} \otimes X\left(X^T X\right)^{-1}X^T\right]W\right\}^{-1}W^T\left[\Sigma^{-1} \otimes X\left(X^T X\right)^{-1}X^T\right]\left[\Sigma \otimes I_N\right]$$
$$\times \left[\Sigma^{-1} \otimes X\left(X^T X\right)^{-1}X^T\right]W\left\{W^T\left[\Sigma^{-1} \otimes X\left(X^T X\right)^{-1}X^T\right]W\right\}^{-1}$$
$$= \left\{W^T\left[\Sigma^{-1} \otimes X\left(X^T X\right)^{-1}X^T\right]W\right\}^{-1}$$
$$W^T\left[\Sigma^{-1}\Sigma\Sigma^{-1} \otimes X\left(X^T X\right)^{-1}X^T X\left(X^T X\right)^{-1}X^T\right]W$$
$$\times \left\{W^T\left[\Sigma^{-1} \otimes X\left(X^T X\right)^{-1}X^T\right]W\right\}^{-1}$$
$$= \left\{W^T\left[\Sigma^{-1} \otimes X\left(X^T X\right)^{-1}X^T\right]W\right\}^{-1}\left\{W^T\left[\Sigma^{-1} \otimes X\left(X^T X\right)^{-1}X^T\right]W\right\}$$
$$\left\{W^T\left[\Sigma^{-1} \otimes X\left(X^T X\right)^{-1}X^T\right]W\right\}^{-1}$$
$$= \left\{W^T\left[\Sigma^{-1} \otimes X\left(X^T X\right)^{-1}X^T\right]W\right\}^{-1} \tag{1.71}$$

Therefore, $\hat{\Delta}$ is asymptotically distributed as a normal distribution $N(\Delta, \Lambda)$, where

$$\Lambda = \left\{ W^T \left[ \Sigma^{-1} \otimes X \left( X^T X \right)^{-1} X^T \right] W \right\}^{-1}. \tag{1.72}$$

All path coefficients are included in the vector $\Delta$. Total elements in the vector $\Delta$ is

$$L = \sum_{i=1}^{M} (m_i + k_i - 1).$$

A causal inference can be made by testing the element in the vector $\Delta$ of path coefficients. Let $\Lambda_{ll}$ denote the $l$-th diagonal element of the covariance matrix $\Lambda$. The null hypothesis is defined as

$$H_0 : \Delta_l = 0.$$

We define the statistic for testing the nonzero element of the vector $\Delta$:

$$T_l = \frac{\Delta_l^2}{\Lambda_{ll}}. \tag{1.73}$$

Under the null hypothesis $H_0 : \Delta_l = 0$, the statistic $T_l$ is asymptotically distributed as a central $\chi^2_{(1)}$ distribution with one degree of freedom.

To investigate pleiotropic causal effects or study causal relationships among a set of variables, we can develop a statistic to test a set of path coefficients. Specifically, we consider a set of parameters $\Delta_S$ which define the causal relationships among the set of variables $S$. Let $\Delta_S$ be a submatrix of $\Lambda$ which corresponds to $\Delta_S$. Suppose that the null hypothesis is $H_0 : \Delta_S = 0$. Define the statistic:

$$T_S = \hat{\Delta}_S^T \hat{\Lambda}_S^{-1} \hat{\Delta}_S. \tag{1.74}$$

Under the null hypothesis $H_0 : \Delta_S = 0$, $T_S$ is asymptotically distributed as a central $\chi^2_{(p)}$ with $p$ degrees of freedom where $p$ is the number of parameters being tested in the set $\Delta_S$.

## 1.3 Sparse Linear Structural Equations

In general, the genotype–phenotype networks are sparse. Therefore, $\Gamma$ and $B$ are sparse matrices. To obtain sparse estimates of $\Gamma$ and $B$, the natural approach is the $l_1$-norm penalization.

### 1.3.1 $L_1$-Penalized Maximum Likelihood Estimation

In order to obtain sparse estimates of $\Gamma$ and $B$, a natural approach is to maximize the log likelihood regularized by the $l_1$-norm terms $\|\Gamma\|_1$ and $\|B\|_1$. The proposed $l_1$-penalized ML estimation approach is to maximize

$$\max_{\Gamma, B} \quad \frac{n}{2} \log |\Sigma|^{-1} + n \log \|\Gamma\| - \frac{1}{2} \sum_{i=1}^{n} (y_i^T \Gamma + x_i^T B) \Sigma^{-1} (\Gamma^T y_i + B^T x_i)$$

$$- \lambda_1 \|\Gamma\|_1 - \lambda_2 \|B\|_1. \tag{1.75}$$

The optimization problem (1.75) can be solved by the ADMM algorithm. To separate the smooth optimization part from the nonsmooth optimization part, we introduce new matrix variables $Z_1$ and $Z_2$, and form the following constrained optimization problem:

$$\min f(\Gamma, B) + \lambda_1 \|Z_1\|_1 + \lambda_2 \|Z_2\|_1$$

$$\text{subject to } \Gamma - Z_1 = 0 \tag{1.76}$$

$$B - Z_2 = 0,$$

where $f(\Gamma, B) = -n \log \|\Gamma\| + \frac{1}{2} \sum_{i=1}^{n} (y_i^T \Gamma + x_i^T B) \Sigma^{-1} (\Gamma^T y_i + B^T x_i)$.

To solve the optimization problem (1.76), we form the augmented Lagrangian

$$L_\rho(\Gamma, B, Z, \mu) = f(\Gamma, B) + \lambda_1 \|Z_1\|_1 + \lambda_2 \|Z_2\|_1 + tr\left(\mu_1^T(\Gamma - Z_1)\right) + tr\left(\mu_2^T(B - Z_2)\right)$$

$$+ \frac{\rho}{2} \left(\|\Gamma - Z_1\|_F^2 + \|B - Z_2\|_F^2\right) \tag{1.77}$$

where $\|\cdot\|_F$ denotes the Frobenius norm of a matrix.

Note that

$$\frac{\rho}{2} \left(\|\Gamma - Z_1\|_2^2 + \|B - Z_2\|_2^2\right) + tr\left(\mu_1^T(\Gamma - Z_1)\right) + tr\left(\mu_2^T(B - Z_2)\right)$$

$$= \frac{\rho}{2} \left\{\|\Gamma - Z_1 + \frac{\mu_1}{\rho}\|_F^2 + \|B - Z_2 + \frac{\mu_2}{\rho}\|_F^2\right\} - \frac{\|\mu_1\|_F^2}{2\rho} - \frac{\|\mu_2\|_F^2}{2\rho}. \tag{1.78}$$

Using Equation 1.78, Equation 1.77 can be transformed to

$$L_\rho(\Gamma, B, Z, \mu) = f(\Gamma, B) + \lambda_1 \|Z_1\|_1 + \lambda_2 \|Z_2\|_1$$

$$+ \frac{\rho}{2} \left\{\|\Gamma - Z_1 + \frac{\mu_1}{\rho}\|_F^2 + \|B - Z_2 + \frac{\mu_2}{\rho}\|_F^2\right\} - \frac{\|\mu_1\|_F^2}{2\rho} - \frac{\|\mu_2\|_F^2}{2\rho}. \tag{1.79}$$

Define the matrices $U_1 = \dfrac{\mu_1}{\rho}$ and $U_2 = \dfrac{\mu_2}{\rho}$. Equation 1.79 can be further reduced to

$$L_\rho(\Gamma, B, Z, \mu) = f(\Gamma, B) + \lambda_1 \| Z_1 \|_1 + \lambda_2 \| Z_2 \|_1$$

$$+ \frac{\rho}{2} \left\{ \| \Gamma - Z_1 + U_1 \|_F^2 + \| B - Z_2 + U_2 \|_F^2 \right\} - \| U_1 \|_F^2 - \| U_2 \|_F^2.$$

$$(1.80)$$

The alternating direction method of multipliers (ADMM) consists of the iterations:

$$\Gamma^{(k+1)} == \underset{\Gamma}{\arg\min} \left( f\left(\Gamma, B^{(k)}\right) + \frac{\rho}{2} \| \Gamma - Z_1^{(k)} + U_1^{(k)} \|_F^2 \right) \qquad (1.81)$$

$$B^{(k+1)} == \underset{B}{\arg\min} \left( f\left(\Gamma^{(k+1)}, B\right) + \frac{\rho}{2} \| B - Z_2^{(k)} + U_2^{(k)} \|_F^2 \right) \qquad (1.82)$$

$$Z_1^{(k+1)} : == \underset{Z_1}{\arg\min} \left( \lambda \| Z_1 \|_1 + \frac{\rho}{2} \| \Gamma^{(k+1)} - Z_1 + U_1^{(k)} \|_F^2 \right) \qquad (1.83)$$

$$Z_2^{(k+1)} : == \underset{Z_2}{\arg\min} \left( \lambda \| Z_2 \|_1 + \frac{\rho}{2} \| B^{(k+1)} - Z_2 + U_2^{(k)} \|_F^2 \right) \qquad (1.84)$$

$$U_1^{(k+1)} : == U_1^{(k)} + \left( \Gamma^{(k+1)} - Z_1^{(k+1)} \right). \qquad (1.85)$$

$$U_2^{(k+1)} : == U_2^{(k)} + \left( B^{(k+1)} - Z_2^{(k+1)} \right). \qquad (1.86)$$

### 1.3.2 $L_1$-Penalized Two Stage Least Square Estimation

Using weighted least square and $l_1$-norm penalization of Equation 1.58, we can form the following optimization problem:

$$\min_{\Delta_i} \ f(\Delta_i) + \lambda \| \Delta_i \|_1$$

$$\text{where } f(\Delta_i) = \left( X^T y_i - X^T W_i \Delta_i \right)^T \left( X^T X \right)^{-1} \left( X^T y_i - X^T W_i \Delta_i \right).$$

$$(1.87)$$

To separate the smooth optimization $\min f(\Delta_i)$ from nonsmooth optimization $\| \Delta_i \|_1$, we can introduce constraints $\Delta_i - Z_i = 0$ and transform the unconstrained optimization problem (1.87) into the constrained optimization problem:

$$\min_{\Delta_i} \ f(\Delta_i) + \lambda \| Z_i \|_1$$

$$\text{subject to } \Delta_i - Z_i = 0$$

$$(1.88)$$

To solve the optimization problem (1.88), we form the augmented Lagrangian

$$L_\rho(\Delta_i, Z_i, \mu) = f(\Delta_i)/2 + \lambda \|Z_i\|_1 + \mu^T(\Delta_i - Z_i) + \frac{\rho}{2}\|\Delta_i - Z_i\|_2^2. \tag{1.89}$$

Note that

$$\mu^T(\Delta_i - Z_i) + \frac{\rho}{2}\|\Delta_i - Z_i\|_2^2 = \frac{\rho}{2}\|\Delta_i - Z_i + \frac{\mu}{\rho}\|_2^2 - \frac{\|\mu\|_2^2}{2\rho}.$$

Equation (1.89) can be transformed to

$$L_\rho(\Delta_i, Z_i, \mu) = f(\Delta_i)/2 + \lambda\|Z_i\|_1 + \frac{\rho}{2}\|\Delta_i - Z_i + \frac{\mu}{\rho}\|_2^2 - \frac{\|\mu\|_2^2}{2\rho}. \tag{1.90}$$

Let $u = \frac{\mu}{\rho}$. Equation 1.90 can be further reduced to

$$L_\rho(\Delta_i, Z_i, \mu) = f(\Delta_i)/2 + \lambda\|Z_i\|_1 + \frac{\rho}{2}\|\Delta_i - Z_i + u\|_2^2 - \frac{1}{2}\|u\|_2^2. \tag{1.91}$$

The alternating direction method of multipliers (ADMM) consists of the iterations:

$$\Delta_i^{(k+1)} := == \arg\min_{\Delta_i} \left( f(\Delta_i)/2 + \frac{\rho}{2}\|\Delta_i - Z_i^{(k)} + u^{(k)}\|_2^2 \right) \tag{1.92}$$

$$Z_i^{(k+1)} := == \arg\min_{Z_i} \left( \lambda\|Z_i\|_1 + \frac{\rho}{2}\|\Delta_i^{(k+1)} - Z_i + u^{(k)}\|_2^2 \right) \tag{1.93}$$

$$u^{(k+)} := == u^{(k)} + \left( \Delta_i^{(k+1)} - Z_i^{(k+1)} \right). \tag{1.94}$$

For some special forms of the function $f(\Delta_i)$, the problem (1.92) can be solved in a closed form. To solve the minimization problem (1.92), we first need to calculate the following derivatives:

$$\frac{\partial f}{\partial \Delta_i} = -2W_i^T X (X^T X)^{-1} (X^T y_i - X^T W_i \Delta_i) \text{ and}$$

$$\frac{\partial \left[ \rho\|\Delta_i - Z_i^{(k)} + u^{(k)}\|_2^2/2 \right]}{\partial \Delta_i} = \rho \left( \Delta_i - Z_i^{(k)} + u^{(k)} \right). \tag{1.95}$$

Setting $\dfrac{\partial L_\rho(\Delta_i, Z_i, u)}{\partial \Delta_i} = 0$ and using Equation 1.14, we obtain

$$-W_i^T X (X^T X)^{-1} (X^T y_i - X^T W_i \Delta_i) + \rho \left( \Delta_i - Z_i^{(k)} + u^{(k)} \right) = 0. \tag{1.96}$$

Solving Equation 1.15 yields

$$\Delta_i^{(k+1)} = \left[ W_i^T X (X^T X)^{-1} X^T W_i + \rho I \right]^{-1} \left[ W_i^T X (X^T X)^{-1} X^T y_i + \rho \left( Z_i^k - u^k \right) \right], \quad (1.97)$$

which can be reduced to

$$\Delta_i^{(k+1)} = \left[ \frac{1}{\rho} I - \frac{1}{\rho} W_i^T X \left( \rho X^T X + X^T W_i W_i^T X \right)^{-1} X^T W_i \right]$$
$$\left[ W_i^T X (X^T X)^{-1} X^T y_i + \rho \left( Z_i^k - u^k \right) \right].$$

The optimization problem (1.11) is non-differentiable. Although the first term in (1.93) is not differentiable, we still can obtain a simple closed-form solution to the problem (1.93) using subdiffenrential calculus (Boyd et al. 2011). Let $\Gamma j$ be a generalized derivative of the $j$-th component $Z_i^j$ of the vector $Z_i$ and $\Gamma = [\Gamma_1,...,\Gamma_{M+K-1}]^T$ where

$$\Gamma_j = \begin{cases} 1 & Z_i^j > 0 \\ [-1,1] & Z_i^j = 0 \\ -1 & Z_i^j < 0. \end{cases}$$

Then, we have

$$\frac{\lambda}{\rho} \Gamma + Z_i = \Delta_i^{k+1} + u^k,$$

which implies that

$$Z_i^{(k+1)} = \mathrm{sgn}\left( \Delta_i^{k+1} + u^k \right) \left( \left| \Delta_i^{k+1} + u^k \right| - \frac{\lambda}{\rho} \right)_+, \quad (1.98)$$

where

$$(x)_+ = \begin{cases} x & x \geq 0 \\ 0 & x < 0. \end{cases}$$

In summary, the algorithms are given below.

**Algorithm 1.3**

For $i = 1,...,M$

Step 1. Initialization

$$u^0 := 0$$
$$\Delta_i^0 := \left[ W_i^T X (X^T X)^{-1} X^T W_i + \rho I \right]^{-1} W_i^T X (X^T X)^{-1} X^T y_i$$
$$Z_i^0 := \Delta_i^0.$$

Carry out Steps 2, 3, and 4 until convergence.

Step 2.

$$\Delta_i^{(k+1)} := \left[ \frac{1}{\rho} I - \frac{1}{\rho} W_i^T X \left( \rho X^T X + X^T W_i W_i^T X \right)^{-1} X^T W_i \right]$$
$$\left[ W_i^T X \left( X^T X \right)^{-1} X^T y_i + \rho \left( Z_i^k - u^k \right) \right].$$

Step 3.

$$Z_i^{(k+1)} := \text{sgn}\left( \Delta_i^{k+1} + u^k \right) \left( \left| \Delta_i^{k+1} + u^k \right| - \frac{\lambda}{\rho} \right)_+,$$

where

$$|x|_+ = \begin{cases} x & x \geq 0 \\ 0 & x < 0. \end{cases}$$

Step 4.

$$u^{(k+)} := = u^{(k)} + \left( \Delta_i^{(k+1)} - Z_i^{(k+1)} \right).$$

Most of the elements of matrices $\Gamma$ and $B$ are equal to zero. The $l_1$–regularized Lasso for the two-stage least squares approach and ADMM algorithms are expected to shrink most of the coefficient matrices $\Gamma$ and $B$ toward zero, yielding sparse network structures. The sparsity-controlling parameter $\lambda$ will be estimated via cross validation or set by users to get reasonable results.

### 1.3.3 $L_1$-Penalized Three-Stage Least Square Estimation

Recall that the whole system of M structural equations is given by

$$\begin{bmatrix} X^T y_1 \\ X^T y_2 \\ \vdots \\ X^T y_M \end{bmatrix} = \begin{bmatrix} X^T W_1 & & & \\ & X^T W_2 & & \\ & & \ddots & \\ & & & X^T W_M \end{bmatrix} \begin{bmatrix} \Delta_1 \\ \Delta_2 \\ \vdots \\ \Delta_M \end{bmatrix} + \begin{bmatrix} X^T e_1 \\ X^T e_2 \\ \vdots \\ X^T e_M \end{bmatrix}. \tag{1.99}$$

which can be compactly written as

$$[I \otimes X^T] Y = [I \otimes X^T] W \Delta + (I \otimes X^T) e, \tag{1.100}$$

where

$$W = \begin{bmatrix} W_1 & & & \\ & W_2 & & \\ & & \ddots & \\ & & & W_M \end{bmatrix}, Y = \begin{bmatrix} y_1 \\ y_2 \\ \vdots \\ y_M \end{bmatrix}, \Delta = \begin{bmatrix} \Delta_1 \\ \Delta_2 \\ \vdots \\ \Delta_M \end{bmatrix} \text{ and } e = \begin{bmatrix} e_1 \\ e_2 \\ \vdots \\ e_M \end{bmatrix}.$$

Recall that the covariance matrix of $(I \otimes X^T)e$ is given by $\Sigma \otimes X^T X$. Then, we can define the objective function by the weighted least square methods:

$$f(\Delta) = \left[ (I \otimes X^T) Y - (I \otimes X^T) W\Delta \right]^T \left[ \Sigma^{-1} \otimes (X^T X)^{-1} \right] \left[ (I \otimes X^T) Y - (I \otimes X^T) W\Delta \right].$$

The covariance matrix $\Sigma$ can be estimated by

$$\hat{\sigma}_{ij} = \frac{(y_i - W_i \Delta_i)^T (y_i - W_i \Delta_i)}{\sqrt{(n - m_i + 1 - k_i)(n - m_j + 1 - k_j)}}. \tag{1.101}$$

The ADMM for the three-stage least squares estimator is defined as

$$\min_{\Delta} \quad f(\Delta) + \lambda \| \Delta \|_1 \tag{1.102}$$

To separate smooth optimization problem which can be solved in a closed form from the nonsmooth optimization problem, we formulate the optimization problem (1.102) as

$$\min_{\Delta, Z} \quad f(\Delta) + \lambda \| Z \|_1$$
$$\text{subject to } \Delta - Z = 0. \tag{1.103}$$

The ADMM procedure for solving the optimization problem (1.103) is given by the following:

Step 1. Initialization

$$u^0 = 0$$
$$\Delta^0 = \left[ W^T \left( \Sigma^{-1} \otimes X(X^T X)^{-1} X^T \right) W \right]^{-1} W^T \left[ \Sigma^{-1} \otimes X(X^T X)^{-1} X^T \right] Y$$
$$Z^0 = \Delta^0.$$

Step 2.

$$\Delta^{k+1} = \arg\min_{\Delta} \left( f(\Delta) + \frac{\rho}{2} \| \Delta - Z^k + u^k \|_2^2 \right). \text{ Solving this, we have}$$

$$\Delta^{k+1} = \left\{ W^T \left[ \Sigma^{-1} \otimes X(X^T X)^{-1} X^T \right] W + \rho I \right\}^{-1}$$
$$\left\{ W^T \left[ \Sigma^{-1} \otimes X(X^T X)^{-1} X^T \right] Y + \rho \left( Z^k - u^k \right) \right\}.$$

Step 3.

$$Z^{k+1} = \arg\min_{Z} \left( \lambda \| Z \|_1 + \frac{\rho}{2} \| \Delta - Z^k + u^k \|_2^2 \right).$$

Note that by soft thresholding, we have

$$x_i^+ := \arg\min_{x_i} \left( \lambda |x_i| + \frac{\rho}{2} (x_i - v_i)^2 \right) = S_{\lambda/\rho}(v_i),$$

where

$$S_d(a) = \begin{cases} a - d & a > d \\ 0 & |a| \le d \\ a + d & a < -d. \end{cases}$$

For the j-th component of the vector $Z$, we have

$$Z_j^{k+1} = S_{\lambda/\rho} \left( \Delta_j^{k+1} + u_j^k \right), j = 1, 2, \dots, M(M-1).$$

Step 4.

$$u^{k+1} = u^k + \Delta^{k+1} - Z^{k+1}.$$

Update Steps 2, 3, and 4 until convergence, that is,

$$\| \Delta^{k+1} - \Delta^k \|_2 < \varepsilon.$$

## 1.4 Functional Structural Equation Models for Genotype–Phenotype Networks

In the previous section, the SEMs carry out variant by variant analysis. However, the power of the traditional variant-by-variant analytical tools for association analysis of rare variants with the phenotypes will be limited. Large simulations have shown that combining information across multiple variants in a genomic region of analysis will greatly enhance the power to detect association of rare variants. To utilize multi-locus genetic information, we propose to use a genomic region or a gene as a unit in multiple trait association analysis and develop sparse structural functional equation models (SFEMs) for construction and analysis of the phenotype and genotype networks.

### 1.4.1 Functional Structural Equation Models

We first define a genotype function. Let $t$ be a genomic position. Define a genotype function $x_i(t)$ of the $i$-th individual as

$$x_i(t) = \begin{cases} 2P_q(t), & QQ \\ P_q(t)\text{-}P_Q(t), & Qq \\ -2P_Q(t), & qq \end{cases}$$

where $Q$ and $q$ are two alleles of the marker at the genomic position $t$, $P_Q(t)$ and $P_q(t)$ are the frequencies of the alleles $Q$ and $q$, respectively. Suppose that we are interested in $k$ genomic regions or genes $[a_j, b_j]$, denoted as $T_j, j = 1,2,\ldots, k$. We consider the following functional structural equation models (FSEMs):

$$y_1\gamma_{11} + y_2\gamma_{21} + \ldots + y_M\gamma_{M1} + \int_{T_1} x_1(t)\beta_{11}(t)dt + \ldots + \int_{T_k} x_k(t)\beta_{k1}(t)dt + e_1 = 0$$

$$y_1\gamma_{12} + y_2\gamma_{22} + \ldots + y_M\gamma_{M2} + \int_{T_1} x_1(t)\beta_{12}(t)dt + \ldots + \int_{T_k} x_k(t)\beta_{k2}(t)dt + e_2 = 0$$

$$\vdots \qquad\qquad \vdots \qquad\qquad \vdots$$

$$y_1\gamma_{1M} + y_2\gamma_{2M} + \ldots + y_M\gamma_{MM} + \int_{T_1} x_1(t)\beta_{1M}(t)dt + \ldots + \int_{T_k} x_k(t)\beta_{kM}(t)dt + e_M = 0$$

$$(1.104)$$

where $\beta_{ji}(t), j = 1,\ldots,k, i = 1,\ldots,M$ are genetic effect functions.

Functional principal components (FPCs) are efficient summary statistics. The FPCs simultaneously employ genetic information of the individual variants and correlation information (LD) among all variants. The FPCs view the genetic variation across the genomic region as a function of its genomic location and uses intrinsic functional dependence structure of the data and all available genetic information of the variants in the genomic region. The neighboring genetic variants are linked. The genotypes at one SNP are dependent on the genotypes at nearby SNPs. The FPCs account for the space-ordering of the genetic variation data. Expanding the genotype functions in terms of a few orthogonal FPCs will substantially reduce the dimensions of the genetic variation data while preserving the intrinsic correlation structure and the space-ordering of the data. Specifically, for each genomic region or gene, we use functional principal component analysis to calculate principal component function. Let $N$ be the number of sampled individuals. We expand $x_{nj}(t), n = 1,\ldots,N$, $j = 1,2,\ldots,k$ in each genomic region in terms of orthogonal principal component functions:

$$x_{nj}(t) = \sum_{l=1}^{L_j} \eta_{njl}\phi_{jl}(t), j = 1,\ldots,k,$$

where $\phi_{jl}(t)$, $j = 1,\ldots,k, l = 1,\ldots,L_j$ are the $l$-th principal component function in the $j$-th genomic region or gene and $\eta_{njl}$ are the functional principal component scores of the $n$-th individual. Using the functional principal component expansion of $x_{nj}(t)$, we obtain

$$\int_{T_j} x_{nj}(t)\beta_{ji}(t)dt = \int_{T_j} \sum_{l=1}^{L_j} \eta_{njl}\phi_{jl}(t)\beta_{ji}(t)dt = \sum_{l=1}^{L_j} \eta_{njl}b_{jl}^{(i)}, n = 1,\ldots,N, j$$

$$= 1,\ldots,k, i = 1,\ldots,M, \tag{1.105}$$

where $b_{jl}^{(i)} = \int_{T_j} \phi_{jl}(t)\beta_{ji}(t)dt$.

Let $x_j(t) = [x_{1j}(t),\ldots,x_{nj}(t)]^T$, $\eta_{jl} = [\eta_{1jl},\ldots,\eta_{Njl}]^T$. Substituting Equation 1.105 into Equation 1.104, we obtain

$$y_1 r_{11} + y_2 r_{21} + \cdots + y_M r_{M1} + \sum_{l=1}^{L_1} \eta_{1l}b_{1l}^{(1)} + \cdots + \sum_{l=1}^{L_k} \eta_{kl}b_{kl}^{(1)} + e_1 = 0$$

$$\vdots \qquad\qquad \vdots \qquad\qquad \vdots \tag{1.106}$$

$$y_1 r_{1M} + y_2 r_{2M} + \cdots + y_M r_{MM} + \sum_{l=1}^{L_1} \eta_{1l}b_{1l}^{(M)} + \cdots + \sum_{l=1}^{L_k} \eta_{kl}b_{kl}^{(M)} + e_M = 0.$$

Let $\eta = [\eta_{11}, ..., \eta_{1L_1,...}, \eta_{k1}, ..., \eta_{kL_k}]$, $B = \begin{bmatrix} b_{11}^{(1)} & \cdots & b_{11}^{(M)} \\ \vdots & \vdots & \vdots \\ b_{1L_1}^{(1)} & \cdots & b_{1L_1}^{(M)} \\ \vdots & \vdots & \vdots \\ b_{k1}^{(1)} & \cdots & b_{k1}^{(M)} \\ \vdots & \vdots & \vdots \\ b_{kL_k}^{(1)} & \vdots & b_{kL_k}^{(M)} \end{bmatrix} = [B_1 \cdots B_M]$ and $Y, \Gamma$

and $E$ be defined as before.

In matrix form, Equation 1.106 can be rewritten as

$$Y\Gamma + \eta B + E = 0. \tag{1.107}$$

If we consider only one genomic region or gene, the matrices $\eta$ and $B$ will be reduced to

$$\eta = [\eta_1, ..., \eta_L] \text{ and } B = \begin{bmatrix} b_{11} & \cdots & b_{1M} \\ \vdots & \vdots & \vdots \\ b_{L1} & \cdots & b_{LM} \end{bmatrix}.$$

If we take functional principal component scores as predictors, the models and algorithms for the network structure and parameter estimation will be similar to that discussed in the previous section. Specifically, the $i$-th equation is given by

$$Y\Gamma_i + \eta B_i + e_i = 0,$$

which can be rewritten as

$$y_i = W_i \Delta_i + e_i, \tag{1.108}$$

where $W_i = [Y_{-i} \ \eta], \Delta_i = [\gamma_{-i} \ B_i]$.

Then, the sparse FSEMs are transformed to

$$\min_{\Delta_i} \quad f(\Delta_i) + \lambda \| \Delta_i \|_1$$
$$\text{where } f(\Delta_i) = (\eta^T y_i - \eta^T W_i \Delta_i)^T (\eta^T \eta)^{-1} (\eta^T y_i - \eta^T W_i \Delta_i). \tag{1.109}$$

Finally, ADMM algorithms are given by

**Algorithm 1.4**

For $i = 1,\ldots,M$

Step 1. Initialization

$$u^0 := 0$$

$$\Delta_i^0 := \left[W_i^T \eta(\eta^T \eta)^{-1} \eta^T W_i\right]^{-1} W_i^T \eta(\eta^T \eta)^{-1} \eta^T y_i$$

$$Z_i^0 := \Delta_i^0.$$

Carry out Steps 2, 3, and 4 until convergence.

Step 2.

$$\Delta_i^{(k+1)} := \left[\frac{1}{\rho}I - \frac{1}{\rho}W_i^T \eta(\rho\eta^T \eta + \eta^T W_i W_i^T \eta)^{-1} \eta^T W_i\right]$$

$$\left[W_i^T \eta(\eta^T \eta)^{-1} \eta^T y_i + \rho\left(Z_i^k - u^k\right)\right].$$

Step 3.

$$Z_i^{(k+1)} := \mathrm{sgn}\left(\Delta_i^{k+1} + u^k\right)\left(\left|\Delta_i^{k+1} + u^k\right| - \frac{\lambda}{\rho}\right)_+.$$

Step 4.

$$u^{(k+)} := == u^{(k)} + \left(\Delta_i^{(k+1)} - Z_i^{(k+1)}\right).$$

## 1.4.2 Group Lasso and ADMM for Parameter Estimation in the Functional Structural Equation Models

Expansion of genotype function often has a number of functional principal components in a genomic region or a gene. However, in the genotype–phenotype network, a gene is allowed to have a directed edge from the gene to a phenotype. Therefore, we need to treat all functional principal components of genotype function expansion in a gene as a group and use group lasso to penalize all functional principal components in the gene or genomic region.

Assume that $\Delta_i = \begin{bmatrix} \gamma_i \\ B_i \end{bmatrix} = \begin{bmatrix} \gamma_i \\ \Delta_{1i} \\ \vdots \\ \Delta_{Ki} \end{bmatrix}$, $\Delta_{ki} = \begin{bmatrix} \Delta_{k1}^{(i)} \\ \vdots \\ \Delta_{kL_k}^{(i)} \end{bmatrix}$ $\Delta_{kl}^{(i)} = b_{kl}^{(i)}, l = 1, \ldots, L_k$. Let $Z_{\gamma_i} =$

$\gamma_i$, $Z_{ki} = \Delta_{ki}$, $k = 1,\ldots, K$, $Z_{ki} = \begin{bmatrix} Z_{k1}^{(i)} \\ \vdots \\ Z_{kL_k}^{(i)} \end{bmatrix} = \begin{bmatrix} \Delta_{k1}^{(i)} \\ \vdots \\ \Delta_{kL_k}^{(i)} \end{bmatrix}$, $Z_i = \begin{bmatrix} Z_{\gamma i} \\ Z_{1i} \\ \vdots \\ Z_{Ki} \end{bmatrix} = \Delta_i$. Sparse functional SEMs attempts to solve the following optimization problem:

$$\min \ f(\Delta_i) + \lambda \left[ \| Z_{\gamma i} \|_1 + \sum_{k=1}^{K} \sqrt{L_k} \| Z_{ki} \|_2 \right] \tag{1.110}$$

subject to $\Delta_i - Z_i = 0$.

To solve the optimization problem (1.110), we form the augmented Lagrangian

$$L_\rho(\Delta_i, Z_i, \mu) = f(\Delta_i) + \lambda \| Z_{\gamma i} \|_1 + \lambda \sum_{k=1}^{K} \sqrt{L_k} \| Z_{ki} \|_2 + \mu^T(\Delta_i - Z_i)$$

$$+ \rho/2 \| \Delta_i - Z_i \|_2^2$$

$$= f(\Delta_i) + \lambda \| Z_{\gamma i} \|_1 + \lambda \sum_{k=1}^{K} \sqrt{L_k} \| Z_{ki} \|_2$$

$$+ \rho/2 \| \Delta_i - Z_i + \mu \|_2^2 - \rho/2 \| \mu \|_2^2. \tag{1.111}$$

The ADMM algorithm consists of the iterations:

$$\Delta_i^{(l+1)} := \arg\min_{\Delta_i} L_\rho \left( \Delta_i, Z_i^{(l)}, \mu^{(l)} \right) \tag{1.112}$$

$$Z_i^{(l+1)} := \arg\min_{Z_i} L_\rho \left( \Delta_i^{(l+1)}, Z_i, \mu^{(l)} \right) \tag{1.113}$$

$$\mu^{(l+1)} := \mu^{(l)} + \Delta_i^{(l+1)} - Z_i^{(l+1)}. \tag{1.114}$$

or

$$\Delta_i^{(l+1)} := \arg\min_{\Delta_i} \left( f(\Delta_i) + \frac{\rho}{2} \| \Delta_i - Z_i^{(l)} + u^{(l)} \|_2^2 \right) \tag{1.115}$$

$$Z_i^{(l+1)} := \arg\min_{Z_i} \left( \lambda \| Z_{\gamma i} \|_1 + \lambda \sum_{k=1}^{K} \sqrt{L_k} \| Z_{ki} \|_2 + \rho/2 \| \Delta_i^{(l+1)} - Z_i + \mu^{(l)} \|_2^2 \right) \tag{1.116}$$

$$\mu^{(l+1)} := \mu^{(l)} + \Delta_i^{(l+1)} - Z_i^{(l+1)}. \tag{1.117}$$

First, we solve the minimization problem (1.112). Let $\mu = \begin{bmatrix} \mu_\gamma \\ \mu_{1i} \\ \vdots \\ \mu_{Ki} \end{bmatrix}$, $\mu_{ki} = \begin{bmatrix} \mu_{k1}^{(i)} \\ \vdots \\ \mu_{kL_k}^{(i)} \end{bmatrix}$.

Setting the partial derivative of $L_\rho$ to zero yields

$$\frac{\partial L_\rho}{\partial \Delta_i} = -W_i^T \eta [\eta^T \eta]^{-1} (\eta^T y_i - \eta^T W \Delta_i) + \rho \left( \Delta_i - Z_i^{(l)} + \mu^{(l)} \right) = 0, \qquad (1.118)$$

which implies that

$$\Delta_i^{(l+1)} = \left[ W_i^T \eta (\eta^T \eta)^{-1} \eta^T W_i + \rho I \right]^{-1} \left[ W_i^T \eta (\eta^T \eta)^{-1} \eta^T y_i + \rho \left( Z_i^{(l)} - \mu^{(l)} \right) \right]. \quad (1.119)$$

The optimization problem (1.116) is non-differentiable. Although the first two terms in (1.116) are not differentiable, we still can obtain simple closed-form solutions to the problem (1.116) using subdiffenrential calculus. We first consider the generalized gradient of $\|Z_{\gamma i}\|_1$. Let $\Phi_i^{(m)}$ be a generalized derivative of the $m$-th component of the vector $\|Z_{\gamma i}\|_1$:

$$\Phi_i^{(m)} = \begin{cases} 1 & Z_{\gamma i}^{(m)} = \gamma_{im} > 0 \\ [-1, 1] & Z_{\gamma i}^{(m)} = \gamma_{im} = 0 \\ -1 & Z_{\gamma i}^{(m)} = \gamma_{im} < 0. \end{cases}$$

Let $\Phi_i = \begin{bmatrix} \Phi_i^{(1)} \\ \vdots \\ \Phi_i^{(M)} \end{bmatrix}$.

Then, we have

$$\frac{\lambda}{\rho} \Phi_i + Z_{\gamma i} = \Delta_{\gamma i}^{(l+1)} + \mu_\gamma^{(l)},$$

which implies that

$$Z_{\gamma i}^{(l+1)} = \text{sgn}\left( \Delta_{\gamma i}^{(l+1)} + \mu_\gamma^{(l)} \right) \left( \left| \Delta_{\gamma i}^{(l+1)} + \mu_\gamma^{(l)} \right| - \frac{\lambda}{\rho} \right)_+, \qquad (1.120)$$

where

$$|x|_+ = \begin{cases} x & x \geq 0 \\ 0 & x < 0. \end{cases}$$

Next, we consider group lasso. The generalized gradient $\dfrac{\partial L_\rho}{\partial Z_{ki}}$ is given by

$$\frac{\partial L_\rho}{\partial Z_{ki}} = \lambda \sqrt{L_k} s + \rho \left( Z_{ki} - \left( \Delta_{ki}^{(l+1)} - \mu_{ki}^{(l)} \right) \right) = 0, \qquad (1.121)$$

where

$$s = \begin{cases} 0 & \| \Delta_{ki}^{(l+1)} - \mu_{ki}^{(l)} \|_2 < \dfrac{\lambda \sqrt{L_k}}{\rho} \\[3mm] \dfrac{Z_{ki}}{\|Z_{ki}\|_2} & \| \Delta_{ki}^{(l+1)} - \mu_{ki}^{(l)} \|_2 \geq \dfrac{\lambda \sqrt{L_k}}{\rho}. \end{cases}$$

Equation (1.121) implies

$$\left( 1 + \frac{\lambda \sqrt{L_k}}{\rho \| Z_{ki} \|_2} \right) Z_{ki} = \Delta_{ki}^{(l+1)} - \mu_{ki}^{(l)} \tag{1.122}$$

or

$$\left( 1 + \frac{\lambda \sqrt{L_k}}{\rho \| Z_{ki} \|_2} \right) \| Z_{ki} \|_2 = \| \Delta_{ki}^{(l+1)} - \mu_{ki}^{(l)} \|_2. \tag{1.123}$$

Solving Equation (1.123), we obtain

$$\| Z_{ki} \|_2 = \| \Delta_{ki}^{(l+1)} - \mu_{ki}^{(l)} \|_2 - \frac{\lambda \sqrt{L_k}}{\rho}.$$

Thus, we have

$$\left( 1 + \frac{\lambda \sqrt{L_k}}{\rho \| Z_{ki} \|_2} \right) = \frac{1}{1 - \dfrac{\lambda \sqrt{L_k}}{\rho \| \Delta_{ki}^{(l+1)} - \mu_{ki}^{(l)} \|_2}} \tag{1.124}$$

Combining Equations 1.122 and 1.124, we obtain

$$Z_{ki}^{(l+1)} = \left( 1 - \frac{\lambda \sqrt{L_k}}{\rho \| \Delta_{ki}^{(l+1)} - \mu_{ki}^{(l)} \|_2} \right)_+ \left( \Delta_{ki}^{(l+1)} - \mu_{ki}^{(l)} \right), k = 1, \ldots, K. \tag{1.125}$$

The ADMM algorithm for solving FSEMs is summarized as follows. For $i = 1, \ldots, M$

Step 1. Initialization

$$\mu^0 := 0$$
$$\Delta_i^0 := \left[ W_i^T \eta (\eta^T \eta)^{-1} \eta^T W_i + \rho I \right]^{-1} W_i^T \eta (\eta^T \eta)^{-1} \eta^T y_i$$
$$Z_i^0 := \Delta_i^0.$$

Carry out Steps 2, 3, and 4 until convergence.

Step 2.

$$\Delta_i^{(l+1)} = \left[ W_i^T \eta (\eta^T \eta)^{-1} \eta^T W_i + \rho I \right]^{-1} \left[ W_i^T \eta (\eta^T \eta)^{-1} \eta^T y_i + \rho \left( Z_i^{(l)} - \mu^{(l)} \right) \right].$$

Step 3.

$$Z_{yi}^{(l+1)} = \text{sgn}\left(\Delta_{yi}^{(l+1)} + \mu_{\gamma}^{(l)}\right)\left(\left|\Delta_{yi}^{(l+1)} + \mu_{\gamma}^{(l)}\right| - \frac{\lambda}{\rho}\right)_{+},$$

where

$$|x|_{+} = \begin{cases} x & x \geq 0 \\ 0 & x < 0. \end{cases}$$

$$Z_{ki}^{(l+1)} = \left(-\frac{\lambda\sqrt{L_k}}{\rho\|\Delta_{ki}^{(l+1)} - \mu_{ki}^{(l)}\|_2}\right)_{+}\left(\Delta_{ki}^{(l+1)} - \mu_{ki}^{(l)}\right), k = 1, ..., K.$$

Let

Step 4.

$$\mu_i^{(l+1)} = \mu_i^{(l)} + \left(\Delta_i^{(l+1)} - Z_i^{(l+1)}\right).$$

## 1.5 Causal Calculus

### 1.5.1 Effect Decomposition and Estimation

In the genotype–phenotype networks we are interested in the estimation of effects of genetic variants on phenotypes, which is referred to as genetic effects and effects of treatment on phenotypes. All genetic effects and treatment effects can be decomposed as total (causal), direct effects, and indirect effects. Distinction between total, direct, and indirect effects are of great practical importance in genetic analysis (Pearl 2001). The total effect measures the changes of response variable $Y$ would take on the value $y$ when variable $X$ is assigned to $x$ by external intervention. The action of setting a variable $X$ to the value $x$ is represented by removing the structural equation for $X$ and replacing it with the equality $X = x$. Direct effect is defined as sensitivity of $Y$ to changes in $X$ while all other variables in the model are held fixed. The indirect effect is to measure the portion of the effect which can be explained by mediation alone, while inhibiting the capacity of $Y$ to respond to $X$. The total effect is equal to the summation of direct and indirect effects.

Given a DAG model $G$, one can compute total effects using intervention calculus (Maathuis et al. 2001). Suppose that the expected value of a response variable $Y$, after $X$ is assigned value $x$ by intervention is denoted by $E[Y \mid do (X = x)]$. The total effect is defined as

$$\frac{\partial}{\partial x} E[Y|do(X = x)]. \tag{1.126}$$

Note $X_j$ is called a parent of $X$ in $G$ if there is a directed edge $X_j \rightarrow X$. Let $pa_x$ denote the set of all parents of $X$ in $G$. In the linear SEMs, $E[Y \mid X, pa_x]$ is linear in $X$ and $pa_x$:

$$E[Y|X, pa_x] = \alpha + \beta X + \gamma^T pa_x. \tag{1.127}$$

Then,

$$\frac{\partial}{\partial x} E[Y|do(X = x)] = \beta.$$

We can also write $\beta$ as $\beta_{YX \cdot pa_x}$.

When a DAG is given, it is easy to calculate the total effect. Assume that there are $k$ directed paths from $X$ to $Y$ and $p_i$ are the product of the path coefficients along the $i$-th path. The total effect of $X$ on $Y$ is then defined as $\sum_{i=1}^{k} p_i$. As shown in Figure 1.5, the total effect of $X$ on $Y$ is $ag + bdh + cdh$. By its definition, the direct effect measures the sensitivity of $Y$ to changes in $X$ while all other variables in the model are held fixed. In other words, all links from $X$ to $Y$ other than the direct link will be blocked. Consequently, the direct effect is equal to the path coefficient from $X$ to $Y$. In the linear SEMs, the indirect effect of $X$ on $Y$ mediated by $M$ is equal to the sum of the products associated with directed paths from $X$ to $Y$ through $M$. In Figure 1.6, there is no direct effect from $X$ to $Y$. The indirect effect $X$ on $Y$ is mediated by $B$ and $D$ is equal to $bdh$.

In the SEMs for genotype–phenotype networks, since all SNPs only form undirected graph and there are no directed links between SNPs although we can observe linkage disease (or correlation) between SNPs, SNPs in the genotype–phenotype networks do not have parents. The total effect of SNP $X$ on $Y$ is the regression coefficient $\beta$ of the following linear regression:

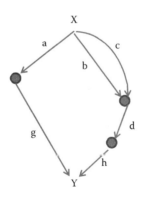

**FIGURE 1.5**
Total effect.

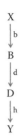

**FIGURE 1.6**
Indirect effect.

$$E[Y|do\ (X = x)] = \alpha + \beta x,$$

which is a simple regression of $Y$ on $X$. This indicates that the traditional simple regression for association studies captures the total effect of a genetic variant on a phenotype.

If we include environments and risk factors such as smoking and obesity in the model and want to evaluate the effects of the environments and risk factors on the phenotype, these variables play mediation roles and will also be taken as phenotypes. We denote these mediation phenotypes by $Y_{ME}$. Since genetic variants and other risk factors and phenotypes will affect the mediation phenotypes, the mediation phenotypes in the graphics may have parents. Their parents are denoted by $S$. Total effect of the mediation phenotype on the target phenotype is calculated by

$$E[Y|do\ (Y_{ME} = y_{ME}, X_{pa} = x_{pa})] = \alpha + \beta y_{ME} + \gamma^T x_{pa}, \tag{1.128}$$

where $\beta$ is the total effect of the mediation phenotype $Y_{ME}$ on the target phenotype $Y$. In this case, a simple regression of $Y$ on $Y_{ME}$ can no longer be used to measure the total effect of the mediation phenotype $Y_{ME}$ on the target phenotype $Y$. To see this, we simulated 1000 individuals with the SEM as shown in Figure 1.7. Each variable has a noise term distributed as $N(0,1)$.

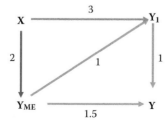

**FIGURE 1.7**
Network structure of mediation phenotype, parent of mediation phenotype and target phenotype.

The total effect of the mediation phenotype $Y_{ME}$ on the target phenotype $Y$ is 3.5. We obtain the simple regression:

$$Y = 1.39 + 5.85Y_{ME}.$$

It is clear that the coefficient of the simple regression is 5.85. This value is far away from the total effect 3.5. However, using Equation 1.13 we obtain

$$Y = 3.54Y_{ME} + 5.85X,$$

where the regression coefficient 3.54 measured the total effect of the mediation phenotype $Y_{ME}$ on the target phenotype $Y$.

### 1.5.2 Graphical Tools for Causal Inference in Linear SEMs

#### 1.5.2.1 Basics

After linear SEMs are learned from the data, our main focus is to make causal inferences. In this section we introduce graphical tools as a major tool for causal inference in linear SEMs in which many materials are from Chen and Pearl's (2015) works. The results are complimentary to an algebraic approach and underlie genetic epidemiology and therapy evaluation (Shrout et al. 2010).

Observed variables and directed graph were discussed in the previous section. Now we study how the hidden variables are represented in the graphical model. We consider three observed variables $X(LDL), Y(HDL)$, and $Z$ (systolic BP), and two hidden variables $H_1$ (environment 1) and $H_2$ (environment 2) which are not directly measurable (Figure 1.8). Observed variables

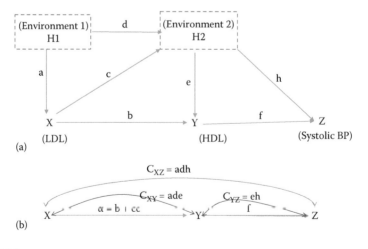

**FIGURE 1.8**
(a) Model with hidden variables ($H_1$ and $H_2$) shown explicitly (b) with hidden variables summarized.

are denoted by nodes and hidden variables are denoted by dashed boxes. Error terms are typically not displayed in the graph, unless they have correlations. The arrows in the graph represent the direction of causation. The corresponding structural coefficients are associated with arrows. Model 1 explicitly summarizes the relationships among the variables including both observed and hidden variables.

**Model 1**

$$H_1 = U_1$$
$$X = aH_1 + U_2$$
$$H_2 = cX + aH_1 + U_3$$
$$Y = bX + eH_2 + U_4$$
$$Z = fY + hH_2 + U_5,$$

where $U_1, U_2, U_3, U_4,$ and $U_5$ are noises.

Figure 1.8 demonstrates causal graphs. Figure 1.8a displays the causal relations among the variables with the hidden variables $H_1$ and $H_2$ shown explicitly. However, the hidden variables cannot be measured. The hidden variables are incorporated into the analysis by considering correlations the hidden variables induce on the error terms. Figure 1.8b shows the model with the hidden variables summarized. Now the effect of the variable $X$ on $Y$ is summarized by the coefficient $a = b + ce$ in Figure 1.8b. The correlation terms $C_{XY} = ade$ between $X$ and $Y$, $C_{YZ} = eh$ between $Y$ and $Z$, and $C_{XZ} = adh$ between $X$ and $Z$ are represented by their corresponding bidirected arrows. The model for the graph in Figure 1.8b is summarized in Model 2.

**Model 2**

$$X = U_X$$
$$Y = \alpha X + U_Y$$
$$Z = fY + U_Z.$$

Before studying how to calculate the causal effects from the DAG, we introduce some basic definition in a graph.

**Definition 1.2**

**Edge:** An edge is a connection between two nodes. The edge can be either an arrow or a bidirected arc.
**Path:** A path between nodes $X$ and $Y$ is defined as a sequence of edges, connecting two nodes. A path may be along or against the direction of directed edges. A directed path from $X$ to $Y$ is defined as a path consisting of only directed edges pointed toward $Y$. A back-door path from $X$ to $Y$ is defined as a path starting with an arrow pointing to $X$ and ends with an

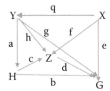

**FIGURE 1.9**
Model illustrating total effect calculation.

arrow pointing to $Y$. A collider is defined as a node in which two arrowheads meet. An acyclic model without correlated error terms is referred to as Markovian. A model with correlated error terms is referred to as non-Markovian model and an acyclic non-Markovian model is referred to as a semi-Markovian model.

Now we study how to calculate the causal effects from the graphs.

**Result 1.1**

Assume that $\Pi = \{\pi_1,...,\pi_m\}$ is the set of directed paths from $X$ to $Y$ and that $p_i$ is the product of the structural coefficients (coefficients in the SMEs) along path $\pi_i$. The total causal effect of $X$ on $Y$ is the summation of all path effects $\alpha_T = \sum_{i=1}^{m} p_i$.

**Example 1.2**

A graph is shown in Figure 1.9. Node $G$ is a collider. There are four directed paths from $Y$ to $G$: $Y \to H \to G, Y \to Z \to G, Y \to G, Y \to H \to Z \to G$ with $p_1 = ab, p_2 = hd, p_3 = g$ and $p_4 = acd$. The total effect of $Y$ on $G$ is $ab + hd + g + acd$.

### 1.5.2.2 Wright's Rules of Tracing and Path Analysis

Path analysis is one of major graphical tools for causal analysis and Wright's rules of tracing is an essential component of path analysis (Wright 1921). The models for causal analysis are divided into two categories: recursive and non-recursive models. In a recursive model, causation flows in only one direction. In a non-recursive model, causation may flow in more than one direction. The Wright's rules of tracing are designed for models of graphs that are non-recursive. For the convenience of presentation, we assume that all variables are standardized to mean zero and unit variance. We consider a set of paths between $X$ and $Y$: $\Pi = \{\pi_1, \pi_2, ..., \pi_m\}$. These paths do not trace a collider and follow three rules:

1. No loops are allowed. The same variable in the path does not enter more than once.
   In Figure 1.10a, consider the paths between $X_4$ and $X_5$. The path $X_4 \leftarrow X_3 \to X_5$ is valid, but $X_4 \leftarrow X_3 \leftarrow X_1 \leftrightarrow X_2 \to X_3 \to X_5$ is not valid.

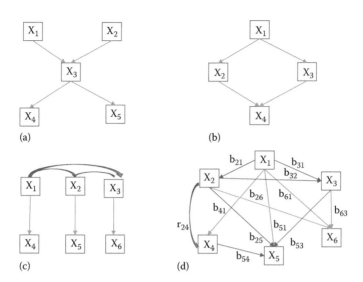

**FIGURE 1.10**
Illustration of Wright's Rules of Tracing. (a) The same variable in the path does not enter more than once. (b) Going forward and then backward is not permitted. (c) No more than one bidirectional edge in a path are allowed. (d) A causal graph for Example 1.3.

2. Going forward and then backward is not permitted. You can go backward up a directed edge and then forward along the next edge or go forward from one variable to the other, but never forward and then backward.

    As Figure 1.10b showed, considering the paths between $X_2$ and $X_3$, the path $X_2 \leftarrow X_1 \rightarrow X_3$ is valid, and $X_2 \rightarrow X_4 \leftarrow X_3$ is not valid.

3. No more than one bidirectional edge in a path is allowed.

Consider the paths between $X_4$ and $X_6$. The path $X_4 \leftarrow X_1 \leftrightarrow X_3 \rightarrow X_6$ is valid, but $X_4 \leftarrow X_1 \leftrightarrow X_2 \leftrightarrow X_3 \rightarrow X_6$ is not valid.

**Result 1.2**

Let $p_i$ denote the product of path coefficients along path $\pi_i$. Then, the covariance between any pair of variables, $X$ and $Y$, is equal to the sum of products of the path coefficients and error covariance along certain paths between $X$ and $Y$: $\sigma_{XY} = \text{cov}(X, Y) = \sum_{i=1}^{m} p_i$.

**Example 1.3**

A causal graph is shown in Figure 1.10d. We want to calculate the covariance between $X_2$ and $X_5$. The set of valid paths for calculation of the covariance between $X_2$ and $X_5$ is

$$X_2 \rightarrow X_5$$
$$X_2 \leftrightarrow X_4 \rightarrow X_5$$
$$X_2 \rightarrow X_3 \rightarrow X_5$$
$$X_2 \leftarrow X_1 \rightarrow X_3 \rightarrow X_5 \,.$$

The covariance between $X_2$ and $X_5$ is

$$\sigma_{25} = b_{25} + r_{24}b_{54} + b_{32}b_{53} + b_{21}b_{31}b_{51}\,.$$

### 1.5.2.3 Partial Correlation, Regression, and Path Analysis

Partial correlation and regression coefficients can be expressed in terms of path coefficients. Consider a covariance matrix

$$\Sigma_{xyz} = \begin{bmatrix} \sigma_{xx} & \sigma_{xy} & \sigma_{xz} \\ \sigma_{yx} & \sigma_{yy} & \sigma_{yz} \\ \sigma_{zx} & \sigma zy & \sigma_{zz} \end{bmatrix}.$$

The conditional variances $\sigma_{xx.z}$ and $\sigma_{yy.z}$ are, respectively, given by

$$\sigma_{xx.z} = \sigma_{xx} - \frac{\sigma_{xz}^2}{\sigma_{zz}} \quad \text{and} \tag{1.129}$$

$$\sigma_{yy.z} = \sigma_{yy} - \frac{\sigma_{yz}^2}{\sigma_{zz}}. \tag{1.130}$$

The conditional covariance matrix between $X$ and $Y$, given a scalar variable is

$$\Sigma_{xy\,z} = \begin{bmatrix} \sigma_{xx} & \sigma_{xy} \\ \sigma_{yx} & \sigma_{yy} \end{bmatrix} - \begin{bmatrix} \sigma_{xz} \\ \sigma_{yz} \end{bmatrix} \sigma_{zz}^{-1} \begin{bmatrix} \sigma_{zx} & \sigma_{zy} \end{bmatrix}, \tag{1.131}$$

which implies the following results.

**Result 1.3**

$$\sigma_{yx.z} = \bar{\sigma}_{yx} - \frac{\sigma_{yz}\sigma_{zx}}{\sigma_{zz}} \tag{1.132}$$

$$\rho_{yx.z} = \frac{\rho_{yx} - \rho_{yz}\rho_{xz}}{\left[(1 - \rho_{yz}^2)(1 - \rho_{xz}^2)\right]^{\frac{1}{2}}} \tag{1.133}$$

$$\beta_{yx.z} = \frac{\sqrt{\sigma_{yy}}}{\sqrt{\sigma_{xx}}} \frac{\rho_{yx} - \rho_{yz}\rho_{zx}}{1 - \rho_{xz}^2},$$ (1.134)

where $\beta_{yx.z}$ is a regression coefficient, of $Y$ on $X$, given $Z$. In general, regression of $Y$ on $X$ and $Z$ can be written as

$$Y = \beta_{yx.z}X + \beta_{yz.x}Z + \varepsilon.$$ (1.135)

If we assume that $Z$ is a scalar variable and $S$ is a set of variables then using Equation 1.132 we can define the conditional covariance between $Y$ and $X$, given $S$:

$$\sigma_{yx.s} = \sigma_{yx} - \sigma_{ys}\sigma_{ss}^{-1}\sigma_{xs}.$$ (1.136)

The conditional covariance $\sigma_{yx.zs}$ between $Y$ and $X$, given $ZS$ can be viewed as the conditional variance of the random variable $\sigma_{yx.s}$, given $Z$:

$$\sigma_{yx.zs} = \sigma_{yx.s} - \frac{\sigma_{yz.s}\sigma_{xz.s}}{\sigma_{z.s}^2}.$$ (1.137)

Similarly, we have

$$\rho_{yx.zs} = \frac{\rho_{yx.s} - \rho_{yz.s}\rho_{xz.s}}{\sqrt{(1 - \rho_{yz.s}^2)(1 - \rho_{xz.s}^2)}} \quad \text{and}$$ (1.138)

$$\beta_{yx.zs} = \frac{\sqrt{\sigma_{yy.s}}}{\sqrt{\sigma_{xx.s}}} \frac{\rho_{yx.s} - \rho_{yz.s}\rho_{zx.s}}{1 - \rho_{xz.s}^2}.$$ (1.139)

Heritability or the phenotype variation can be explained by the variation of predictor variables in the regression and can be calculated by Wright's tracing rules.

**Example 1.4**

Consider the directed graph in Figure 1.11. The correlations that can be calculated by Wright's tracing rules are

$$\rho_{31} = b_{31} + r_{12}b_{32} \quad \text{and} \quad \rho_{32} = b_{32} + r_{12}b_{31}.$$

The regression coefficients using formula (1.134) are

$$\beta_{31.2} = \frac{\rho_{31} - \rho_{32}\rho_{12}}{1 - r_{12}^2} \quad \text{and}$$

$$\beta_{32.1} = \frac{\rho_{32} - \rho_{31}\rho_{21}}{1 - r_{12}^2}.$$

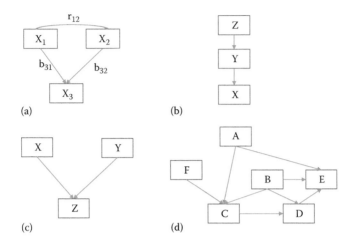

**FIGURE 1.11**

Illustrate some examples. (a) Directed graph for Example 9.5.3. (b) Conditionally independent in a Bayesian network. (c) Assess all independences between variables. (d) Path diagrams for Example 9.5.4.

### 1.5.2.4 Conditional Independence and D-Separation

A Bayesian network assumes that each variable is conditionally independent of its non-descendants, given its parents. For example, Figure 1.11b shows that since we know the value of all of $X$'s parents (namely,$Y$), and $Y$ is not a descendant of $X$, $X$ is conditionally independent of $Z$. However, this rule is not sufficient to assess all independences between variables. Consider Figure 1.11c. Since $X$ has no parents, values of its parents are trivially known and $Z$ is not a descendent of $X$, the variables $X$ and $Y$ are marginally independent. However, what will happen if we know the values of $Z$?

To assess conditional independence in a more general case, we introduce a useful concept, D-separation which associates "correlation" with "connectedness" and independence with "separation." To account for the orientation of directed graph, we use the terms "d-separation."

#### Definition 1.3: D-Separation

Two sets of variables $X$ and $Y$ are d-separated by a third set of variables $Z$ if and only if every undirected path from $X$ to $Y$ is blocked, where a path is "blocked" if only one or more of the following conditions are true:

1. There exists a variable $V \in Z$ such that the arcs putting $V$ in the path are "tail-to-tail" ($\leftarrow V \rightarrow$).
2. There exists a variable $V \in Z$ such that the arcs putting $V$ in the path are "tail-to-head" ($\rightarrow V \rightarrow$).
3. There exists a variable $V$ in the path which is not in the set $Z$ or its descendants. In addition, $V$ is not a collider.

If $X$ and $Y$ are d-separated by $Z$, then $X$ and $Y$ are conditionally independent, given variables $Z$. The following procedures are often used to compute d-separation:

1. Draw the ancestral graph.
   Construct the "ancestral graph" of all variables of interests. This is a reduced version of the original directed graph, consisting only of the variables being studied and all of their ancestors (parents, parents' parents.)
2. "Moralize" the ancestral graph by "marrying" the parents.
   For each pair of variables with a common child, draw an undirected edge between them. (If a variable has more than two parents, draw edges between every pair of parents.)
3. Replace the directed edges with undirected edges.
4. Delete the conditioned nodes and their edges.

After these four procedures, we can directly assess the independence from the graphs. If the variables are disconnected in this graph, they are ensured to be independent. If the variables are connected in this graph, they are not ensured to be independent.

**Example 1.5**

Consider path diagrams shown in Figure 1.11d. We are interested in conditional independence between $F$ and $E$, given $A$ and $C$. We first construct the ancestral graph that consists of nodes $F$, $E$, $A$, $C$ and their parents as shown in Figure 1.12a. Then, we moralize the ancestral graph

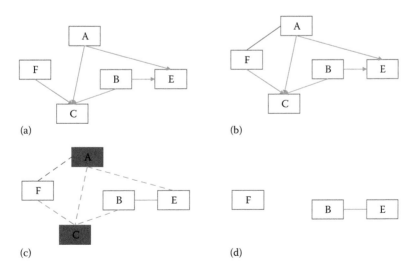

**FIGURE 1.12**

Illustration for procedures to assess d-separation. (a) Ancestral graph for Example 1.5. (b) Moralize the ancestral graph. (c) Replace all directed edges by undirected edges. (d) The final graph after deleting the set of conditional nodes $A$ and $C$, and their connected edges from Figure 1.12c.

by marring nodes $F$ and $A$ which have a common child $C$ (Figure 1.12b). Replace all directed edges by undirected edges (Figure 1.12c). Finally, delete the set of conditional nodes $A$ and $C$, and their connected edges $F - A, F - C, A - C, C - B$, and $A - E$, leading to the graph in Figure 1.12d. It is clear that $F$ is d-separated from $E$ by $A$ and $C$. In other words, $F$ and $E$ are conditionally independent, given $A$ and $C$.

### 1.5.3 Identification and Single-Door Criterion

**Definition 1.4**

Identification consists of parameter identification and model identification. If a model parameter can be uniquely determined from the probability distribution over the model variables, it is referred to as identified. If every parameter in the model is identified, then the model is referred to as identified (Chen and Pearl 2015).

In SEMs, a parameter can be identified if it can be uniquely expressed in terms of the covariance matrix. Some software is available for checking the identifiability of the model. Here, we focus on using graphical criteria for assessing the identifiability of the parameters and model.

Consider a path diagram in Figure 1.13a and a structural equation model:

$$Y = \beta_1 X + \beta_2 Z + \varepsilon_Y. \tag{1.140}$$

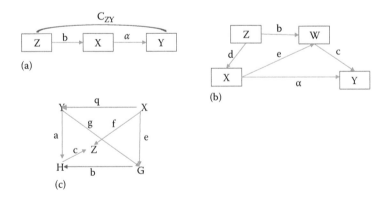

**FIGURE 1.13**
D-separation and identifiability. (a) A path diagram. (b) A path diagram for Example 9.5.5. (c) A path diagram for Example 9.5.6.

Parameters are identified if the error $\varepsilon_Y$ is independent of $X$ and $Z$. However, if we use the regression model:

$$Y = \beta X + U_Y \tag{1.141}$$

since $U_Y$ is correlated with $X$, then the estimator $\hat{\beta}$ is biased.

To overcome the limitations of simple regression, we add the variable $Z$ to the regression (1.141). The causal effect $\alpha$ can be estimated by the regression coefficient, $\beta_{YX.Z}$, of $Y$ on $X$, given $Z$. In fact, using Equation 1.134

$$
\begin{aligned}
\beta_{yx.z} &= \frac{\sqrt{\sigma_{yy}}}{\sqrt{\sigma_{xx}}} \frac{\rho_{yx} - \rho_{yz}\rho_{zx}}{1 - \rho_{xz}^2} \\
&= \frac{1}{1} \frac{\alpha + C_{zy}b - \left(\alpha b + C_{zy}\right)b}{1 - b^2} \\
&= \frac{\alpha - \alpha b^2}{1 - b^2} \\
&= \alpha.
\end{aligned}
$$

Adding a set of variables $Z$ to a regression to estimate a path coefficient is referred to as adjusting for $Z$. Now the question is how to select and determine if a set of variables is adequate for adjustment to ensure that the parameters and models are identifiable. Next, we introduce graphical identifiability conditions that allow to assess parameter identifiability by inspection of the path diagram (Pearl 2001; Chen and Pearl 2015).

## Theorem 1.1: Single-Door Criterion

Let $G$ be an acyclic directed graph in which $\alpha$ is the path coefficient associated with directed edge $X \rightarrow Y$, and $G_\alpha$ be the directed graph that results when $X \rightarrow Y$ is removed from the original graph $G$. The path coefficient $\alpha$ is identifiable if a set of variables $Z$ exists such that (1) no descendent of $Y$ is present in the set $Z$ and (2) $X$ and $Y$ are d-separated in $G_\alpha$ by $Z$. The path coefficient $\alpha$ is then estimated by the regression coefficient $\beta_{YX.Z}$.

### Example 1.6

Consider the path diagram in Figure 1.13b. We want to estimate $\alpha$. In $G_\alpha$, $X$ and $Y$ are d-separated by $W$. Therefore, $\alpha$ is identified and equal to $\beta_{YX.W}$. Using Wright's path tracing rules, we obtain

$$
\begin{aligned}
\rho_{YX} &= \alpha + ec + dbc \\
\rho_{YW} &= c + e\alpha + bd\alpha \\
\rho_{WX} &= e + db.
\end{aligned}
$$

Using Equation 1.134, we obtain

$$\beta_{YX.W} = \frac{\sqrt{\sigma_{YY}}}{\sqrt{\sigma_{XX}}} \frac{\rho_{YX} - \rho_{YW}\rho_{XW}}{1 - \rho_{XW}^2}$$

$$= \frac{\sqrt{1}}{\sqrt{1}} \frac{\alpha + ec + dbc - (c + e\alpha + bd\alpha)(e + db)}{1 - (e + db)^2}$$

$$= \alpha \frac{1 - (e + db)^2}{1 - (e + db)^2}$$

$$= \alpha.$$

Using Theorem 1.1 we can show that any Markovian (acyclic DAG without correlated error terms) model can be identified. Consider a path coefficient $\alpha$ associated with directed edge $X \rightarrow Y$. Its structural equation is

$$Y = \beta_1 X + \beta_2 Pa(Y),$$

where $Pa(Y)$ is the set of parents of $Y$. It is clear that $Y$ is d-separated from $X$ by the set of parents of $Y$. The path coefficient $\alpha$ can then be unbiasedly estimated:

$$\hat{\alpha} = \beta_{YX.Pa(Y)}.$$

In general, we have the following result.

**Result 1.4**

Any Markovian (acyclic directed graph without correlated error terms) model can be identified equation by equation using regression.

**Example 1.7**

Consider Figure 1.13c and the path coefficient $b$ associated with the directed edge $G \rightarrow H$. The parent of $H$ is $Y$. The graph in Figure 1.13c is Markovian. The path coefficient $b$ is identified and can be estimated by

$$b = \beta_{HG.Y}$$

$$= \frac{\sqrt{\sigma_{HH}}}{\sqrt{\sigma_{GG}}} \frac{\rho_{HG} - \rho_{HY}\rho_{GY}}{1 - \rho_{GY}^2}$$

$$- \frac{\sqrt{1}}{\sqrt{1}} \frac{b + eqa + ga - (a + qeb + gb)(g + qe)}{1 - (g + qe)^2}$$

$$= \frac{b[1 - (g + qe)^2]}{1 - (g + qe)^2}$$

$$= b.$$

### 1.5.4 Instrument Variables

A single-door admissible set may represent confounding factors. Some confounding factors may be unmeasured. Single-door creation cannot be applied. Consider Figure 1.14a which shows the directed acyclic graph for the relationship between an instrument variable $Z$, a factor $D$ causing changes in outcome $Y$, and unmeasured confounders $U$. Due to the presence of unmeasured confounding, unbiased estimation of the causal effect of $D$ on $Y$ is difficult. To overcome this limitation, instrumental variables (IVs) are used to control for unmeasured confounding and measurement error in causal inferences with observational data. The basic idea of the IV method is to find instrument variables that are correlated with causal factors, but are independent of unmeasured confounders and have no direct effects on the outcome except through their effects on causal factors (Baiocchi et al. 2014). Formally, we present the following definition (Bollen 2012; Chen and Pearl 2015):

#### Definition 1.5

Consider a structural equation, $Y = \beta_1 X_1 + \ldots + \beta_m X_m + U_Y$, $Z$ is an instrumental variable if the following conditions are satisfied:

1. $Z$ is correlated with $X = \{X_1, \ldots, X_m\}$ and
2. $Z$ is uncorrelated with $U_Y$.

We used the correlation between $Y$ and $X$ to model the unmeasured confounder $U$ (Figure 1.14b). Using Wright's path tracing rules, we obtain

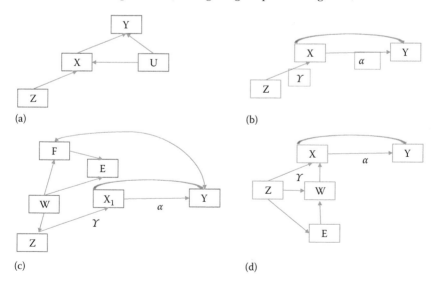

**FIGURE 1.14**

Instrument variables. (a) Directed acyclic graph. (b) Illustration of definition of instrument variable. (c) Conditioning on some sets of variables, the potential instrument variables become d-separated from outcome or response $Y$. (d) Illustration of definition of conditional instrumental variable.

$\sigma_{YZ} = \gamma\alpha$ and $\sigma_{XZ} = \gamma$. Thus, the parameter $\alpha$ can be estimated by $\hat{\alpha} = \dfrac{\sigma_{YZ}}{\sigma_{XZ}}$. Without instrument variable $Z$, using Wright's path tracing rules we only have $\sigma_{YX} = \alpha + U_{YX}$, which implies that $\alpha$ is not identifiable.

Definition 1.5 defines an IV for an equation. However, using IVs, the equation, as a whole, is identified, but some parameters in the equation may still not be identified. The next definition developed by Pearl (2001) specifies IVs for the parameters of interests (Chen and Pearl 2015).

### Definition 1.6

Let $G_\alpha$ be the subgraph reduced from the original graph $G$ by removing the directed edge $X \rightarrow Y$ where the path coefficient $\alpha$ is associated with. If a variable $Z$ satisfies two conditions:

1. $Z$ is d-separated from $Y$ in the reduced graph $G_\alpha$ and
2. $Z$ is not d-separated from $X$ in $G_\alpha$.

For example, in Figure 1.14b, $Z$ is d-separated from $Y$, but not d-separated from $X$ in $G_\alpha$ where $X \rightarrow Y$ is removed. Therefore, $Z$ is an IV for the parameter $\alpha$.

In practice, many variables that can be potential IVs are not d-separated from outcome $Y$. However, conditioning on some sets of variables, these potential IVs become d-separated from outcome or response $Y$. For example, in Figure 1.14c, $Z$ is not an IV for estimation of the parameter $\alpha$ associated with the directed edge $X \rightarrow y$ because $Z$ is not d-separated from $Y$ in the subgraph $G_\alpha$. However, conditioning on $F$ or $W$, the connection between $Z$ and $Y$ is blocked and, hence, $Z$ can be selected as an IV. Combining conditioning and instruments, the parameters can be identified. Therefore, we introduce the following definition for conditioning instrument variables to expand the power of instrument variables for causal inference.

### Definition 1.7

A variable $Z$ is a conditional instrument variable given a conditional set $W$ for a path coefficient $\alpha$ associated with the directed edge $X \rightarrow Y$ if the following three conditions are satisfied:

1. 1 $W$ contains no descendants of $Y$.
2. $Z$ is d-separated from $Y$ by the set $W$ in the subgraph $G_\alpha$.
3. $W$ does not d-separate $Z$ from $X$ in the subgraph $G_\alpha$.

In this case, the parameter is identified and can be estimated by $\hat{\alpha} = \dfrac{\beta_{YZ.W}}{\beta_{XZ.W}}$. To illustrate a definition of the conditional instrumental variable, we inspect Figure 1.14d. Since $W$ d-separates $Z$ from $Y$ given $W$ in subgraph $G_\alpha$, $Z$ is a conditional instrument variable for $\alpha$ given $W$.

The concept of a single instrument variable can be extended to a set of instrument variables as follows.

**Definition 1.8**

Consider $m$ path coefficients $\alpha_1,\ldots,\alpha_m$ associated with directed edges $X_1 \to Y,\ldots,X_m \to Y$. Let $\pi_h[V_i\ldots V_j]$ denote the subpath that begins with $V_i$ and ends with $V_j$. If the following conditions are satisfied, then $\{Z_1,\ldots,Z_m\}$ is an instrument set for the coefficients $\alpha_1,\ldots,\alpha_m$.

1. Let $G_m$ be the subgraph induced from $G$ by removing directed edges $X_1 \to Y,\ldots,X_m \to Y$. Then, $Z_i$ is d-separated from $Y$ in $G_m$ for all $i = 1,\ldots,m$.
2. Paths $\pi_1,\ldots,\pi_m$ where $\pi_i$ is a path from $Z_i$ to $Y$ including directed edge $X_i \to Y$ are present. If paths $\pi_i$ and $\pi_j$ have a common node, then we have paths either

   a. $\pi_i[Z_i\ldots V]_i : Z_i\ldots \to V$ and $\pi_j\,[V\ldots Y] : V \leftarrow \ldots Y$ or
   b. $\pi_j[Z_j\ldots V]_i : Z_j\ldots \to V$ and $\pi_i[V\ldots Y] : V \leftarrow \ldots Y$ for all $i$, $j = 1,2,\ldots,m, i \neq j$.

The condition indicates that the common node must be a collider.

With the aid of the instrument set, the path coefficients can be estimated by solving a system of linear equations that can be obtained by covariance analysis (Brito and Pearl 2002).

**Theorem 1.2**

Let $\{Z_1,\ldots,Z_m\}$ be an instrumental set for the path coefficients $\alpha_1,\ldots,\alpha_m$ associated with directed edges $X_1 \to Y_1,\ldots,X_m \to Y$. Then, the following system of linear equations:

$$A_{ZX}\alpha = b_{ZY},\qquad(1.142)$$

where

$$A_{ZX} = \begin{bmatrix} \sigma_{Z_1 X_1} & \cdots & \sigma_{Z_1 X_m} \\ \vdots & \vdots & \vdots \\ \sigma_{Z_m X_1} & \cdots & \sigma_{Z_m X_m} \end{bmatrix}, \alpha = \begin{bmatrix} \alpha_1 \\ \vdots \\ \alpha_m \end{bmatrix} \text{ and } b_{ZY} = \begin{bmatrix} \sigma_{Z_1 Y} \\ \vdots \\ \sigma_{Z_m Y} \end{bmatrix}$$

are linearly independent for almost all the parameterization of the model. In other words, the matrix $A_{ZX}$ is nonsingular and the system of linear equations has a single solution. The path coefficients $\alpha_1,\ldots,\alpha_m$ can be estimated using Equation 1.142.

**Example 1.8**

Consider a directed acyclic graph in Figure 1.15a. It is clear that $Z_1,Z_2$, and $Z_3$ are d-separated from $Y$ in $G_m$. Path $\pi_1 : Z_1 \to Z_2 \to X_1 \to Y$ and path $\pi_2 : Z_2 \leftrightarrow X_2 \to Y$ have a common node $Z_2$. It is clear that $Z_2$ is a collider. Therefore, by Definition 1.8, $\{Z_1,Z_2,Z_3\}$ is an instrument set. Now we calculate the matrix $A_{ZX}$. Using Wright's rule, we obtain

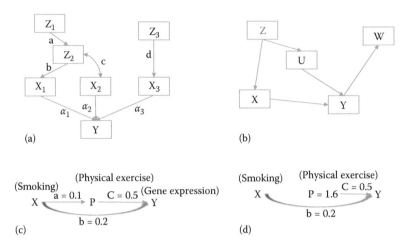

**FIGURE 1.15**
Backdoor and counterfactuals. (a) A directed acyclic graph for Example 9.5.7. (b) Back-door condition for Example 9.5.8. (c) Counterfactuals and linear SEMs. (d) The modified model.

$$\sigma_{Z_1 X_1} = ab, \sigma_{Z_1 X_2} = 0, \sigma_{Z_1 X_3} = 0$$
$$\sigma_{Z_2 X_1} = b, \sigma_{Z_2 X_2} = c, \sigma_{Z_2 X_3} = 0$$
$$\sigma_{Z_3 X_1} = 0, \sigma_{Z_3 X_2} = 0, \sigma_{Z_3 X_3} = d.$$

Therefore, we have

$$A_{ZX} = \begin{bmatrix} \sigma_{Z_1 X_1} & 0 & 0 \\ \sigma_{Z_2 X_1} & \sigma_{Z_2 X_2} & 0 \\ 0 & 0 & \sigma_{Z_3 X_3} \end{bmatrix} = \begin{bmatrix} ab & 0 & 0 \\ b & c & 0 \\ 0 & 0 & d \end{bmatrix} \text{ and}$$

$$A_{ZX}^{-1} = \begin{bmatrix} \dfrac{1}{ab} & 0 & 0 \\ -\dfrac{1}{ac} & \dfrac{1}{c} & 0 \\ 0 & 0 & \dfrac{1}{d} \end{bmatrix}.$$

Using Equation 1.142, we obtain the total effects and backdoor criterion.

## 1.5.5 Total Effects and Backdoor Criterion

In the previous discussion, it is known that even when the model as a whole is not identified, some coefficients may be identifiable. We also observe that when the total effect is identifiable, it is not necessary to identify all coefficients along a causal path that is a part of the paths to calculate the total effect. In practice, we often observe that treatment and response variables are not observed, but we can measure some surrogate variables effected by observed variables. Through these surrogate variables, we can still estimate the causal effect of the treatment on the outcomes or response variables. Here, we introduce the back-door criterion that is a sufficient condition for the identification of the total effect due to Pearl (Pearl 2001; Chen and Pearl 2015).

**Theorem 1.3: Back-Door Criterion**

For any two variables $X$ and $Y$ in a causal diagram, the total of effects of $X$ on $Y$ is identifiable if there exists a set of variables $Z$ such that

1. No node in $Z$ is a descendent of $X$ and
2. $Z$ d-separates (blocks) every backdoor path between $X$ and $Y$ that contains an arrow into $X$.

Then the total effect $T_{YX}$ of $X$ on $Y$ is estimated by $\beta_{YX.Z}$.

> **Example 1.9**
>
> Consider Figure 1.15b. $Z$ is not a descendent of $X$ and blocks the backdoor path $X \leftarrow Z \rightarrow U \rightarrow Y$. Therefore, $Z$ satisfies the backdoor condition and the total effect $T_{YX}$ of $X$ on $Y$ is identifiable and can be estimated by $\beta_{YX.Z}$.
>
> Backdoor paths begin with the directed edge $\rightarrow X$. All backdoor paths should be blocked by the set $Z$. The set $Z$ may be empty. In this case, $\beta_{YX.Z} = \beta_{XY}$.

## 1.5.6 Counterfactuals and Linear SEMs

Graphical models encode scientific assumptions and counterfactuals emerge as a general framework to formally formularize research questions of interest (Pear 2013; Chen and Pearl 2015). If the model is identified, a counterfactual query will be raised to predict values of the variables in the model using the available information. For example, given that we observe the evidence, $E = e$ for a given individual, what would we expect the value of the variable $B$ for the individual if the variable $A$ had been $a$? Consider Figure 1.15c where $X$ represents the number of cigarettes smoked each day by the individual, $P$ the measure of physical exercise, and $Y$ the overall expression level of some gene. The value of each variable is given as the number of standard deviations above the mean. Therefore, the data are standardized to mean 0 and variance 1. The model for Figure 1.15c is given below.

**Model**

$$
\begin{aligned}
X &= U_X \\
P &= aX + U_P \\
Y &= bX + cP + U_Y,
\end{aligned}
\tag{1.143}
$$

where $a = 0.1$, $b = 0.2$, and $c = 0.5$.

Suppose that measurements of another individual are given by $X = 1$, $P = 0.8$, and $Y = 1.2$. The question is what would be his or her gene expression level $Y$ if his or her physical exercise activities are doubled? To use SEMs to solve this problem, we first need to use the observed evidence data for the individual to obtain the subject specific values of the random variables $U_X, U_P, U_Y$ for this individual. Using the model (1.143) and observed evidence, we obtain

$$U_X = 1,$$
$$U_P = P - aX = 0.8 - 0.1^*1 = 0.7$$
$$U_Y = Y - bX - cP = 1.2 - 0.2^*1 - 0.5^*0.8 = 0.6.$$

Next, we simulate the action of doubling the individual physical exercise by replace the equation $P = 0.1X + U_P$ by $P = 1.6$. The modified model is shown in Figure 1.15d. If physical activities of the individual are doubled, then the gene expression level can be predicted from the third equation in the model (1.143):

$$Y = 0.2^*1 + 0.5^*1.6 + 0.6 = 1.6.$$

In general, we fit the SEMs to the population data and derive the SEMs:

$$Y = \Gamma Y + BX + U. \tag{1.144}$$

**Counterfactual analysis consists of three steps:**

1. Abduction. By the evidence $(Y_*, X_*)$ of observed new endogenous and exogenous variables for the individual, we estimate the random vector $U_*$:

$$U_* = Y_* - \Gamma Y_* - BX_*. \tag{1.145}$$

2. Action. Assume that $A$ consists of many variables in $Y$ and $X$. Set $A = a$ and replace equations in the model (1.144) involving $A$ by $A = a$ and obtain the modified model:

$$Y_A = \Gamma Y_A + BX_A + U_*. \tag{1.146}$$

3. Prediction. Use the modified model (1.146) to compute the expectation of $B$.

## 1.6 Simulations and Real Data Analysis

### 1.6.1 Simulations for Model Evaluation

The performance of the sparse SEM approach for genotype–phenotype network analysis can be evaluated by large-scale simulation studies

(Wang et al. 2016). Two types of simulations: SNP-based simulations and gene-based simulations were considered. The simulations were carried out for common variants (MAF ≤5%), rare variants (MAF <5%), and half common and half rare variants. The genotype data were selected from the NHLBI's Exome Sequencing Project (ESP) with 3248 individuals of European origin, which were then used to generate a population of 1,000,000 individuals.

We first introduce the SNP-based simulations. The genotype–phenotype network consisted of two parts. The first part was the phenotype network that was modeled by a DAG. The second part was the connections between the genotypes and phenotypes in which the genotypes were directed to the phenotypes. We randomly generated a genotype–phenotype network structure. On the average, each phenotype node was connected with three other phenotype nodes. The parameters $\Gamma_{ij}$ in the SEMs for modeling the phenotype sub-network were generated from a uniformly distributed random variable over the interval (0.5, 1) or (–1, –0.5) if an edge from node $j$ to node $i$ was presented in the phenotype sub-network; otherwise $\Gamma_{ij} = 0$. Similarly, the parameters $B_{ij}$ in the SEMs for modeling the direction from the genotype (SNP) node $j$ to the phenotype node $i$ were generated from a uniformly distributed random variable over the interval (0, 1) or (–1,0) if an edge from node $j$ to node $i$ was presented in the network, otherwise $B_{ij} = 0$. The SNPs are coded by their number of minor alleles. Using the randomly generated network structure and parameters in the structural equations we produced the phenotypes by the model: $Y = - XB\Gamma^{-1} + \varepsilon\Gamma^{-1}$, where $\varepsilon \sim N(0, 0.01 \times I)$, $I$ is an identity matrix, and $X$ is a matrix of indicator variables for coding genotypes. For the randomly generated phenotype network, the expected number of degrees per node is 3. Simulations were repeated 100 times. Fivefold cross validation was used to determine the penalty parameter $\lambda$ that was then employed for the sparse SEM calculations. Two measures: the power of detection (PD) and the false discovery rate (FDR) were used to evaluate the performance of the algorithms for identification of the network inference. Specifically, let $N_t$ be the total number of edges among 1000 replicates of the network and $\hat{N}_t$ be the total number of edges detected by the inference algorithm, $N_{true}$ will be the total number of true edges detected among simulated network and $N_{false}$ be the false edges detected among $\hat{N}_t$.

Now, the power of detection (PD) is defined by $\dfrac{N_{True}}{N_t}$ and false discovery rate (FDR) is defined by $\dfrac{N_{False}}{\hat{N}_t}$.

In the SNP-based simulations we first compared the sparse two-stage SEM (S2SEM) with ADMM algorithms with the sparse maximum likelihood SEMs (SML) with coordinate ascent algorithms (Cai et al. 2013). The SML method assumes that each phenotype has one a priori known QTL, and only focuses on the inference of the phenotype network. So, in this comparison we only calculate PD and FDR for the phenotype network.

We first compare the power and FDR of the S2SEM and SML under the assumption that each QTL had only connection with one phenotype, and no pleiotropic effects were present. We considered two scenarios: 10 phenotypes and 10 SNPs, 30 phenotypes and 30 SNPs. Three types of SNPs: common, rare, and all variants were considered. Results were shown in Figure 1.16. We observed that if the variants were common variants, the SML had a higher power and a lower FDR than the S2SEM. However, once rare variants were included, the SML substantially lost power and increased the FDR. The S2SEM was much better than the SML. The power and FDR of the S2SEM and SML under the assumption of the presence of pleiotropic genetic effects were similar.

The S2SEM requires much less computation time than the SML for network inference and can construct genotype–phenotype networks. It was reported that the power of S2SEM for detecting the structure of the genotype (common variants, rare variants, and both common and rare variants)–phenotype network with 100 phenotypes and 1000 SNPs as a function of sample size was still high and their FDR was low (Wang et al. 2016).

The functional SEMs (FSEMs) for network analysis using a genomic region or gene as a unit of analysis can reduce data dimensions and increase the power. In all three cases: common, rare, and both common and rare variants, the gene-based FSEM had a much higher power and smaller FDR than the SNP-based SEMs. It is interesting to observe that even for the rare variants the gene-based method can reach the power as high as 85% when sample sizes were larger than 3000 (Wang et al. 2016).

### 1.6.2 Application to Real Data Examples

To illustrate its performance, the sparse functional SEMs with a gene as a unit of analysis were applied to two samples of the UK-10K dataset with 765 individuals and the Taizhou (China) cohort study dataset with 21,501 individuals. A total of 13 common phenotypes was summarized in Table 1.1. Sparse SEMs were used to construct phenotype-networks from each dataset. For each dataset, a phenotype network with 13 nodes and 54 edges was identified (Wang 2016). It was observed that 33 edges were shared between two phenotype networks derived from the UK-10K dataset and Taizhou dataset. The replication rate was 61%. This demonstrated that the SEMs for inferring networks were robust. Two phenotype networks were shown in Figure 1.17 where edges in green color represent common edges between two networks, edges in pink color represent edges that were presented only in the network derived from the UK-10K dataset, edges in black color represent edges only presented in the network derived from Taizhou cohort dataset, and edges in pepper color represent the same edge with a different direction. From Figure 1.17 we observed that most esti mated edges had two directions and only a small number of edges had only one direction. The methods for converting hybrid phenotype networks with

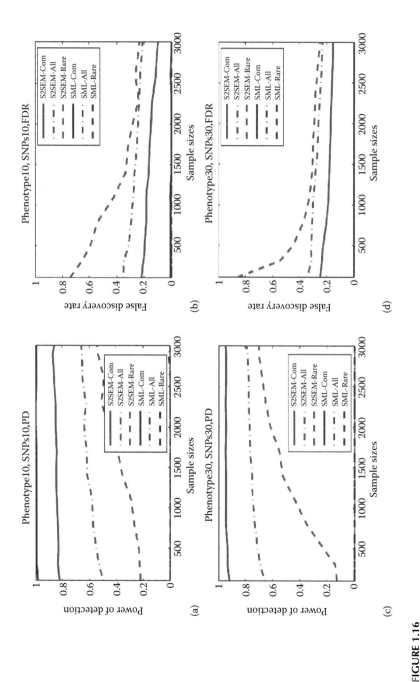

**FIGURE 1.16**

Performance of S2SEM and SML for phenotype network inference. The power and FDR of the two methods for inference of phenotype networks when the phenotype and genotype number is 10 (a and b) and 30 (c and d), respectively.

**TABLE 1.1**

A List of 13 Common Phenotypes Shared between UK-10K and Taizhou Cohort

| Phenotypes | TWINSUK, 756 Samples | | | Taizhou, 21, 501 Samples | | |
|---|---|---|---|---|---|---|
| | Mean | Median | Standard Deviation | Mean | Median | Standard Deviation |
| Glucose (mmol/l) | 5.04 | 4.90 | 0.86 | 5.43 | 5.16 | 1.55 |
| Total Cholesterol (mmol/l) | 5.55 | 5.50 | 1.03 | 5.01 | 4.94 | 1.04 |
| HDL (mmol/l) | 1.85 | 1.81 | 0.48 | 1.51 | 1.46 | 1.11 |
| LDL (mmol/l) | 3.17 | 3.09 | 0.99 | 2.41 | 2.49 | 0.84 |
| Triglycerides (mmol/l) | 1.16 | 0.99 | 0.64 | 1.68 | 1.22 | 1.79 |
| Creatinine (umol/l) | 75.63 | 74.00 | 30.20 | 68.74 | 64.80 | 26.99 |
| Urea (mmol/l) | 5.02 | 4.80 | 1.43 | 6.02 | 5.80 | 2.48 |
| Bilirubin (umol/l) | 8.94 | 8.00 | 3.51 | 14.11 | 13.40 | 5.86 |
| BMI (Kg/m2) | 26.50 | 25.81 | 4.72 | 24.56 | 24.35 | 3.18 |
| WHR | 0.79 | 0.78 | 0.05 | 0.87 | 0.87 | 0.07 |
| ECG_HeartRate (Beats/min) | 66.58 | 66.00 | 10.29 | 76.75 | 76.00 | 10.61 |
| Systolic_BP (mm/Hg) | 130.17 | 129.00 | 16.88 | 136.04 | 134.50 | 20.00 |
| Diastolic_BP (mm/Hg) | 78.55 | 78.00 | 9.77 | 81.51 | 81.00 | 12.27 |

mixed directed edges and undirected edges to directed graphics will be discussed in Chapter 2.

## Appendix 1.A

Consider $p$ genes. We can use a summary statistic to represent the genomic content across the gene. For example, for the $i$-th gene, its genomic content can be summarized either by $y_i = \dfrac{1}{T_i} \int_{T_i} g_i(t) dt$, where $g_i(t)$ is a genotype function and $T_i$ is the genomic region of the $i$-th gene or by the functional principal component scores which are defined in the previous section. In general, these summarizing statistics asymptotically follow normal distributions. Therefore, we assume that a vector $Y = [y_1, ..., y_p]^T$ follows a normal distribution with a mean of zero and covariance matrix $\Sigma = (\sigma_{ij})$. Let the concentration matrix be denoted by $\Omega = (o^{ij}) = \Sigma^{-1}$. Let $y_{-(i,j)}$ be the random vector containing all elements in $y$ after removing the variables $y_i$ and $y_j$. The partial correlation between the variables $y_i$ and $y_j$ is defined as the conditional correlation between $y_i$ and $y_j$, given the remaining variables in the graph: $\rho^{ij} = Corr(y_i, y_j | y_{-(i,j)})$. The structure of the graph is determined by the concentration

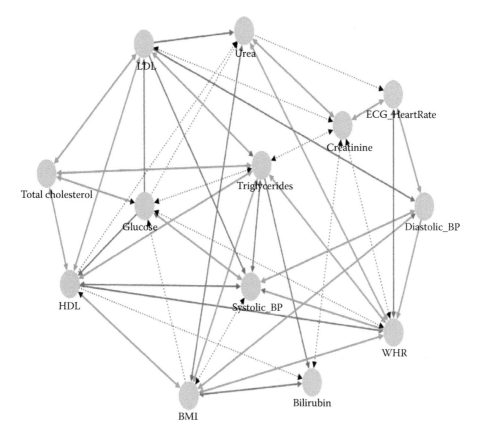

**FIGURE 1.17**
Two phenotype networks with 13 nodes and 54 edges identified from UK-10K and Taizhou cohort datasets.

matrix. Absence of an edge between the variables $y_i$ and $y_j$ is equivalent to the element of the concentration matrix $\sigma^{ij} = 0$.

Theoretically, we have Theorem 1.A.1.

## Theorem 1.A.1

Absence of an edge between the variables $y_i$ and $y_j$ if and only if the variables $y_i$ and $y_j$ are conditionally independent, given all remaining variables, or if and only if the partial correlation $\rho^{ij} = 0$.

Proof. Let the covariance matrix be partitioned as $\Sigma = \begin{bmatrix} \Sigma_{(ij)} & \Sigma_{(ij,-ij)} \\ \Sigma_{(ij,-ij)} & \Sigma_{(-ij)} \end{bmatrix}$ and

$\Sigma_{ij.(-ij)} = \begin{bmatrix} \sigma_{ii.(-ij)} & \sigma_{ij.(-ij)} \\ \sigma_{ji.(-ij)} & \sigma_{jj.(-ij)} \end{bmatrix} = \Sigma_{ij} - \Sigma_{(ij,-ij)}\Sigma_{(-jj)}^{-1}\Sigma_{(ij,-ij)}$. Theorem 1.4 states that $\sigma^{ij} = 0$ if and only if the variables $y_i$ and $y_j$ are conditionally independent, given all remaining variables, which are equivalent to $\sigma_{(ij).(-ij)} = 0$. First, we establish the relationship between $\Sigma^{-1}$ and $\Sigma_{ij.(-ij)}$. Define

$$
E_1 = \begin{bmatrix} I & 0 \\ 0 & \Sigma^{-1}_{(-ij)} \end{bmatrix}, \ E_2 = \begin{bmatrix} I & -\Sigma_{(ij,-ij)} \\ 0 & I \end{bmatrix}, \ E_3 = \begin{bmatrix} \Sigma^{-1}_{ij.(-ij)} & 0 \\ 0 & I \end{bmatrix}, \text{ and } E_4 = \begin{bmatrix} I & 0 \\ -\Sigma^{-1}_{(-ij)}\Sigma_{(ij,-ij)} & I \end{bmatrix}.
$$

Then, we have

$E_4E_3E_2E_1\Sigma = I$. Therefore, $\Sigma^{-1} = \begin{bmatrix} \Sigma^{-1}_{ij.(-ij)} & * \\ * & * \end{bmatrix}$. By definition of the inverse matrix, we obtain

$$
\Sigma^{-1}_{ij(-ij)} = \frac{1}{\Delta} \begin{bmatrix} \sigma_{jj(-ij)} & -\sigma_{ij(-ij)} \\ -\sigma_{ji(-ij)} & \sigma_{ii(-ij)} \end{bmatrix}, \tag{1.A.1}
$$

where $\Delta = \sigma_{ii.(-ij)}\sigma_{jj.(-ij)} - \sigma_{ij.(-ij)}\sigma_{ji.(-ij)}$, which implies that

$$
\sigma^{ij} = \frac{-\sigma_{ij(-ij)}}{\Delta}. \tag{1.A.2}
$$

Equation 1.A.2 indicates that $\sigma^{ij} = 0$ if and only if $\sigma_{ij.(-ij)} = 0$, which implies that the variables $y_i$ and $y_j$ are conditionally independent, given all remaining variables if the joint distribution of the variables in the graph follows a normal distribution. Next, we show that this is equivalent to $\rho^{ij} = 0$.

In fact, by the definition, $\rho^{ij} = \dfrac{\sigma_{ij.(-ij)}}{\sqrt{\sigma_{ii.(-ij)}\sigma_{jj.(-ij)}}}$. It follows from Equation 1.A.1 that

$$
\sigma^{ii} = \frac{\sigma_{jj(-ij)}}{\Delta}, \ \sigma^{jj} = \frac{\sigma_{ii(-ij)}}{\Delta} \tag{1.A.3}
$$

Combining Equations 1.A.2 and 1.A.3 yields

$$
\rho^{ij} = \frac{\sigma_{ij(-ij)}}{\sqrt{\sigma_{ii(-ij)}\sigma_{jj(-ij)}}} = -\frac{\Delta\sigma^{ij}}{\sqrt{\Delta^2\sigma^{ii}\sigma^{jj}}} = -\frac{\sigma^{ij}}{\sqrt{\sigma^{ii}\sigma^{jj}}}, \tag{1.A.4}
$$

which implies that $\rho^{ij} = 0$ if and only if $\sigma^{ij} = 0$ or if and only if the variables $y_i$ and $y_j$ are conditionally independent, given all remaining variables.

The partial correlation coefficients are closely related to the coefficients of the multiple regression as Theorem 1.A.2 shows.

**Theorem 1.A.2**

Assume that a multiple regression is given by $y_i = \beta_{ij}y_j + \sum\limits_{k \neq i,j} \beta_{ik}y_k + \varepsilon_i$. Then, we have $\beta_{ij} = \rho^{ij}\sqrt{\dfrac{\sigma^{jj}}{\sigma^{ii}}}$.

Proof

The coefficient $\beta_{ij}$ is the regression coefficient of $y_i$ on $y_j$ in the multiple regression equation of $y_i$ on $y_j$, $y_k$, $k \neq i, j$. Thus, we have

$$\beta_{ij} = \frac{\text{cov}\left(y_i, y_j \big| Y_{(-ij)}\right)}{\text{var}\left(y_j \big| Y_{(-ij)}\right)} = \frac{\sigma_{ij\,(-ij)}}{\sigma_{jj\,(-ij)}} = \rho^{ij}\sqrt{\frac{\sigma_{ii\,(-ij)}}{\sigma_{jj\,(-ij)}}} = \rho^{ij}\sqrt{\frac{\sigma^{jj}}{\sigma^{ii}}}.$$

The multiple regression can be rewritten as

$$y_i = \beta_{ij\,(-ij)}y_j + \sum_{k \neq i,j} \beta_{ik\,(-ik)}y_k + \varepsilon_i. \tag{1.A.5}$$

Consider the regression of $y_i$ on $y_j$: $y_i = \beta_{ij}y_j + e_i$, which implies that

$$\beta_{ij}y_j + e_i = \beta_{ij\,(-ij)}y_j + \sum_{k \neq i,j} \beta_{ik\,(-ik)}y_k + \varepsilon_i. \tag{1.A.6}$$

On both sides of Equation 1.A.6, we make the covariance with $y_j$, we obtain

$$\beta_{ij}\sigma_{jj} = \beta_{ij\,(-ij)}\sigma_{jj} + \sum_{k \neq i,j} \beta_{ik\,(-ik)}\sigma_{kj}. \tag{1.A.7}$$

Dividing on both sides of Equation 1.A.7 by $\sigma_{jj}$, we obtain

$$\beta_{ij} = \beta_{ij\,(-ij)} + \sum_{k \neq i,j} \beta_{ik\,(-ik)}\beta_{kj}. \tag{1.A.8}$$

---

## Appendix 1.B

Now we derive Algorithm 1.2. We first consider the $\Theta$-minimization. Taking the derivative of the Lagrangian function $L_\rho(\Theta^{(1)})$ with respect to $\Theta^{(1)}$ and setting it to be zero, we obtain

$$\begin{aligned}
\frac{\partial L_\rho}{\partial \Theta^{(1)}} &= n_1 S^{(1)} - n_1 \left(\Theta^{(1)}\right)^{(-1)} + \rho(\Theta^{(1)} - \Theta^{(2)} - (V + W) + U_1 \\
&\quad + \Theta^{(1)} - Z^{(1)} + U_3) = 0,
\end{aligned} \tag{1.B.1}$$

which implies that

$$\Theta^{(1)} - \frac{n_1}{2\rho} \left( \Theta^{(1)} \right)^{-1} = \frac{1}{2} \left( \Theta^{(2)} + V + W - U_1 - U_3 + Z^{(1)} \right) - \frac{n_1}{2\rho} S^{(1)}. \qquad (1.B.2)$$

Let

$$B_1 = \frac{1}{2} \left( \Theta^{(2)} + V + W - U_1 - U_3 + Z^{(1)} \right) - \frac{n_1}{2\rho} S^{(1)}.$$

To solve the matrix equation 1.B.2, we take the orthogonal eigenvalue decomposition of the matrix $B_1$:

$$B_1 = Q_1 \Lambda_1 Q_1^T, \qquad (1.B.3)$$

where $\Lambda_1 = \text{diag} \,(\lambda_1^{(1)}, ..., \lambda_p^{(1)})$ and $QQ^T = Q^TQ = I$.
   Multiplying Equation 1.B.2 by the orthonormal matrix $Q^T$ on its left side and by $Q$ on its right side, we obtain

$$\tilde{X}_1 - \frac{n_1}{2\rho} \tilde{X}_1^{-1} = \Lambda_1, \qquad (1.B.4)$$

where $\tilde{X}_1 = Q_1^T \Theta^{(1)} Q_1$ is a diagonal matrix.
   Therefore, Equation 1.B.4 implies the following linear algebra equation:

$$\tilde{X}_{ii} - \frac{n_1}{2\rho \tilde{X}_{ii}} = \lambda_i, i = 1, ..., p. \qquad (1.B.5)$$

Solving Equation 1.B.5 and taking a positive root, we have

$$\tilde{X}_{ii} = \frac{\lambda_i + \sqrt{\lambda_i^2 + \frac{2n_1}{\rho}}}{2}, i = 1, ..., p.$$

The precision matrix for the first subnetwork is estimated by

$$\hat{\Theta}^{(1)(l+1)} = Q_1 \tilde{X}_1 Q_1^T, \qquad (1.B.6)$$

where $\tilde{X}_1 = \text{diag} \,(\tilde{X}_{11}, ..., \tilde{X}_{pp})$.
   For $\Theta^{(2)}$, we similarly have

$$\frac{\partial L_\rho}{\partial \Theta^{(2)}} = n_2 S^{(2)} - n_2 \left( \Theta^{(2)} \right)^{-1} - \rho \left[ \Theta^{(1)(l+1)} - \Theta^{(2)} - \left( V^{(l)} + W^{(l)} \right) + U_1^{(l)} \right]$$

$$+ \rho (\Theta^{(2)} - Z^{(2)(l)} + U_4^{(l)}) = 0,$$

which leads to the matrix equation

$$\Theta^{(2)} - \frac{n_2}{2\rho}\left(\Theta^{(2)}\right)^{-1} = B_2, \tag{1.B.7}$$

where $B_2 = \frac{1}{2}\left[\Theta^{(1)(l+1)} - (V^{(l)} + W^{(l)}) + U_1^{(l)} + Z^{(2)(l)} - U_4^{(l)}\right] - \frac{n_2}{2\rho}S^{(2)}$.

Taking the orthogonal eigenvalue decomposition of the matrix $B_2$, we obtain

$$B_2 = Q_2\Lambda_2 Q_2^T. \tag{1.B.8}$$

Combining Equations 1.B.7 and 1.B.8 yields

$$\tilde{X}_2 - \frac{n_2}{2\rho}\left(\tilde{X}_2\right)^{-1} = \Lambda_2, \tag{1.B.9}$$

where $\tilde{X}_2 = Q_2^T\Theta^{(2)}Q_2$, $\Lambda_2 = \text{diag}\,(\lambda_i^{(2)},...,\lambda_p^{(2)})$.

Solving the matrix equation 1.B.9 we obtain the solution:

$$\tilde{X}_{ii} = \frac{\lambda_i^{(2)} + \sqrt{\left(\lambda_i^{(2)}\right)^2 + \frac{2n_2}{\rho}}}{2} \quad \text{and} \quad \tilde{X}_2 = \text{diag}\,\left(\tilde{X}_{11},...,\tilde{X}_{pp}\right).$$

Therefore, we have

$$\Theta^{(2)(l+1)} = Q_2\tilde{X}_2 Q_2^T. \tag{1.B.10}$$

Setting $\frac{\partial L_\rho}{\partial W} = 0$, we obtain

$$W^{(l+1)} = \frac{1}{2}\left[\Theta^{(1)(l+1)} - \Theta^{(2)(l+1)} + \left(V^{(l)}\right)^T - V^{(l)} + U_1^{(l)} + \left(U_2^{(l)}\right)^T\right]. \tag{1.B.11}$$

Now we use the generalized gradient or subgradient to obtain $Z^{(1)}, Z^{(2)}$ and $V$.

Taking a generalized gradient of $L_\rho$ with respect to $Z^{(1)}$, we obtain

$$\frac{\partial L_\rho}{\partial Z^{(1)}} = \lambda_1\Gamma_1 - \rho\left(\Theta^{(1)(l+1)} - Z^{(1)} + U_3^{(l)}\right) = 0, \tag{1.B.12}$$

where $\Gamma_1 = [\gamma_1,...,\gamma_p]^T$ and

$$\gamma_j = \begin{cases} 1 & Z_j^{(1)} > 0 \\ [-1,1] & Z_j^{(1)} = 0 \\ -1 & Z_j^{(1)} < 0 \end{cases}$$

Solving Equation 1.B.12, we obtain

$$Z^{(1)(l+1)} = sign\left(\Theta^{(1)(l+1)} + U_3^{(l)}\right)\left(\left|\Theta^{(1)(l+1)} + U_3^{(l)}\right| - \frac{\lambda_1}{\rho}\right)_+ . \qquad (1.B.13)$$

Similarly, we have

$$Z^{(2)(l+1)} = sign\left(\Theta^{(2)(l+1)} + U_4^{(l)}\right)\left(\left|\Theta^{(2)(l+1)} + U_4^{(l)}\right| - \frac{\lambda_1}{\rho}\right)_+ . \qquad (1.B.14)$$

Now we consider the group lasso for $V$. Taking a generalized gradient of $L_\rho$ with respect to $V_j$, we obtain

$$\frac{\partial L_\rho}{\partial V_j} = \lambda_2 h - \rho\left[\Theta^{(1)(l+1)} - \Theta^{(2)(l+1)} - \left(V + W^{(l)}\right) + U_1^{(l)}\right]_{(j)}$$
$$+ \rho\left(V - \left(W^{(l+1)}\right)^T + U_2^{(l)}\right)_{(j)} = 0, \qquad (1.B.15)$$

where

$$h = \begin{cases} \dfrac{V_j}{\|V_j\|_2} & V_j \neq 0 \\[2mm] \|h\|_2 < 1 & V_j = 0. \end{cases}$$

Using algebra manipulation, Equation 1B.15 can be reduced when $V_j \neq 0$ to

$$\left[1 + \frac{\lambda_2}{2\rho\|V_j\|_2}\right]V_j = G_j, \qquad (1.B.16)$$

where $G_j = \dfrac{1}{2}\left[\Theta^{(1)(l+1)} - \Theta^{(2)(l+1)} - W^{(l+1)} + U_1^{(l)} + (W^{(l+1)})^T - U_2^{(l)}\right]$.

Taking norm $\|.\|_2$ on both sides of Equation 1.B.16, we obtain

$$\left[1 + \frac{\lambda_2}{2\rho\|V_j\|_2}\right]\|V_j\|_2 = \|G_j\|_2. \qquad (1.B.17)$$

Solving Equation 1.B.17 for $\|V_j\|_2$, we obtain

$$\|V_j\|_2 = \|G_j\|_2 - \frac{\lambda_2}{2\rho}. \qquad (1.B.18)$$

Substituting Equation 1.B.18 into Equation 1.B.16 for $V_j \neq 0$ yields

$$V_j = \left(1 - \frac{\lambda_2}{2\rho\|G_j\|_2}\right)G_j. \qquad (1.B.19)$$

Combining Equation 1.B.19 with the definition of subgradient of $L_\rho$ at $V_j = 0$ we obtain the solution:

$$V_j = \left(1 - \frac{\lambda_2}{2\rho \| G_j \|_2}\right)_+ G_j, j = 1, ..., p. \qquad (1.B.20)$$

## Exercises

Exercise 1. Let $\Sigma_x$ be a covariance matrix of $X$. We define the standardized predictors as $Z = \Sigma_x^{-1/2}(X - E(X))$. Prove that var$(Z) = 1$.

Exercise 2. Show $\frac{1}{2}\sum_{i=1}^{N}(Y_i - \mu)^T \Sigma_p^{-1}(Y_i - \mu) = \frac{N}{2} Tr(\Sigma_p^{-1}S)$.

Exercise 3. Let $A$ and $B$ be matrices. Show that

$$\frac{\partial(\text{Tr}(AB))}{\partial B} = A^T \text{ and } \frac{\partial \log|A|}{\partial A} = A^{-T}.$$

Exercise 4. Derive ADMM algorithm for solving the problem (1.39) with constraint (1.32) assuming $K = 2$.

Exercise 5. Derive ADMM algorithm for three-stage least square estimation problem (1.102).

Exercise 6. Show that the partial coefficient $\rho_{yx.z}$ is given by

$$\rho_{yx.z} = \frac{\rho_{yx} - \rho_{yz}\rho_{xz}}{\left[(1 - \rho_{yz}^2)(1 - \rho_{xz}^2)\right]^{\frac{1}{2}}}.$$

Exercise 7. Define regression coefficient, $\beta_{yx.z}$, of $Y$ on $X$, given $Z$ as $\beta_{yx.z} = \frac{\sigma_{yx.z}}{\sigma_{xx.z}}$. Show

$$\beta_{yx.z} = \frac{\sqrt{\sigma_{yy}}}{\sqrt{\sigma_{xx}}} \frac{\rho_{yx} - \rho_{yz}\rho_{zx}}{1 - \rho_{xz}^2}.$$

Exercise 8. In Figure 1.8b, calculate the regression coefficient $\beta_{yz.x}$ in terms of path coefficients.

Exercise 9. Consider Figure 1.11b. Assess independence between $C$ and $E$, given $A$, $D$, and $B$.

Exercise 10. Consider Figure 1.13c. Use Result 1.4 to estimate the path coefficient $a$.

# 2

# Causal Analysis and Network Biology

Causal inference is fundamental in genetic, epigenetic, and image data analysis. The major aim of causal inference is to estimate the causal dependencies from observational and interventional data. Observational data are commonly referred to as the data that are generated from the system of variables under consideration without external manipulations.

Similar to other statistical problems, likelihood function is a general framework for modeling causal networks. However, unlike other statistical problems, likelihood function for a causal network can be factorized according to the structure of the causal networks. The structures of causal networks and their parameters are unknown. The major aim of learning causal networks is to search a causal network with the maximum likelihood. Learning causal networks from data is an NP-hard problem (de Campos 2000).

Learning causal networks consists of two parts: a scoring metric and a search procedure. There are two types of variables: discrete variables and continuous variables, and three types of network connections: connections between discrete variables, connections between continuous variables, and connections between discrete variables and continuous variables. The widely used score metrics for connections between discrete variables are Bayesian Dirichlet equivalent uniform (BDeu) or conditional probability tables (Barlett and Cussens 2013; Scanagatta et al. 2014). The score metrics for connections between continuous variables will be defined using structural equations and the score metrics for connections between discrete and continuous variables are defined using a network penalized logistic regression. The structural equations are introduced in Chapter 1. The network penalized logistic regressions are investigated in Chapter 8 in the book *Big Data in Omics and Imaging: Association Analysis*. In this chapter, we will introduce BDeu.

There are two approaches to learning the structure of causal networks: constraint-based and score-based learning. The constraint-based learning methods are to learn the causal networks by testing conditional independence from the data (Cheng et al. 2002). However, the constrained-based learning methods are sensitive to noise. The score-based methods formulize the learning of causal networks as a combinatorial optimization problem by assigning score function of the network and searching for networks with the best score. The score-search algorithms for learning causal networks from observed data include dynamic programming (Koivisto et al. 2004), A* search (Fan et al. 2014), and integer programming (Barlett and Cussens 2013). Since integer programming can solve large causal networks, we will mainly introduce integer programming for searching causal network structure.

## 2.1 Bayesian Networks as a General Framework for Causal Inference

We consider $p$ random variables $Y = (Y_1, ..., Y_p)$ with joint distribution $P(y_1, ..., y_p)$ which follows a Markov condition with respect to an underlying directed acyclic graph (DAG). We assume that a DAG $G = (V, E)$ has $p$ nodes $V = \{1, 2, ..., p\}$. Each node $i$ represents a variable $y_i$, $i = 1, ..., p$ and each edge is directed. A directed edge from the nodes $i$ to $j$ is denoted by $i \rightarrow j$. The node $i$ is referred to as a parent of node $j$ and node $j$ is referred to as a child of node $i$. We denote the set of parents of a node $j$ by $pa_D(j) = \{i \mid i \rightarrow j \text{ in } G\}$ and its set of children by $ch(j)$. A cycle is defined as a path that starts and ends at the same node. A directed acyclic graph (DAG) is defined as a directed graph that has no cycles.

The joint probability $P(y_1, y_2, ..., y_p)$ can be factorized as

$$P(y_1, y_2, ..., y_p) = P(y_1)P(y_2|y_1)...P(y_p \mid y_1, y_2, ...y_{p-1})$$
$$= \prod_{i=}^{p} P(y_i|y_1, ..., y_{i-1}) \qquad (2.1)$$
$$= \prod_{i=1}^{p} P(y_i|pa_i),$$

where $P(y_i \mid pa_i) = P(y_i)$ when the set of parents of node $i$ is empty.

Suppose that a Bayesian network is given in Figure 2.1. The probability function $P(y_1, y_2, ..., y_7)$ can be factorized as

$$P(y_1, y_2, y_3, y_4, y_5, y_6, y_7) = P(y_1)P(y_2)P(y_3|y_1, y_4)P(y_4|y_2)$$
$$*P(y_5|y_2)P(y_6|y_3, y_4)P(y_7|y_4, y_5).$$

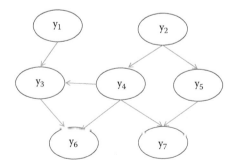

**FIGURE 2.1**
Probability factorization for Bayesian networks.

In a Bayesian network, every variable is conditionally independent of its non-descendant non-parent variables given its parent variables (Markov condition).

---

## 2.2 Parameter Estimation and Bayesian Dirichlet Equivalent Uniform Score for Discrete Bayesian Networks

The first task for learning Bayesian networks is to estimate the parameters given the structure of the Bayesian networks. Consider $p$ variables $Y_1, Y_2, \ldots, Y_p$ which form a DAG. Assume that the variable $Y_i$ can take $r_i$ values $\{y_{i1}, y_{i2}, \ldots, y_{ir_i}\}$. Its collection of all possible values will be denoted by $\Omega_{Y_i} = \{y_{i1}, y_{i2}, \ldots, y_{ir_i}\}$. Let $r_{pa_i} = \prod_{t \in pa_i} r_t$ be the number of all possible values which the parents of the variable $Y_i$ can take. Let $\pi_{ij}$ ($i \in \{1, \ldots, p\}, j \in \{1, \ldots, r_{pa_i}\}$) be the value which the parent of the node $i$ takes. Define $\theta_{ijk} = P(Y_{ik} \mid \pi_{ij})$ to be the conditional probability of observing $y_{ik}$, given the value $\pi_{ij}$ which the parent set takes. Let $\theta = (\theta_{ijk})_{\forall ijk}$ be the entire vector of parameters $\theta_{ijk}$.

Let $\chi$ be a space of all random variables $\{Y_1, \ldots, Y_p\}$ and bold letters denote the sets or vectors. Define $\Omega_Y = \bigcup_{Y \in Y} \Omega_Y$ for all $Y \subseteq \chi$ as all instantiations or values of the variables in $Y$. Assume that $n$ subjects are sampled. Consider a complete dataset $D = \{D_1, \ldots, D_n\}$. The parameters $\theta_{ijk}$ can be used to capture the joint distributions of $\{Y_1, \ldots, Y_p\}$. The score metric can be defined as the posterior distribution of the parameters $\theta_{ijk}$. The prior distribution for the parameter $\theta_{ijk}$ is a Dirichlet. Its density function is given by

$$P\left(\theta_{ijk} \mid pa_i\right) = \Gamma\left(\sum_{k=1}^{r_i} \alpha_{ijk}\right) \prod_{k=1}^{r_i} \frac{\theta_{ijk}^{\alpha_{ijk}-1}}{\Gamma\left(\alpha_{ijk}\right)}. \tag{2.2}$$

Let $P(\theta \mid G)$ be a priori of $\theta$ for a given $G$. It follows from Equation 2.2 that Dirichlet prior of $\theta$ is

$$P(\theta \mid G) = \prod_{i=1}^{p} \prod_{j=1}^{r_{pa_i}} \Gamma\left(\sum_{k=1}^{r_i} \alpha_{ijk}\right) \prod_{k=1}^{r_i} \frac{\theta_{ijk}^{\alpha_{ijk}-1}}{\Gamma\left(\alpha_{ijk}\right)}. \tag{2.3}$$

Let $n_{ijk}$ be the number of subjects who contain both $y_{ik}$ and $\pi_{ij}$. Define $n_{ij} = \sum_{k=1}^{r_i} n_{ijk}$ and $n_i = \sum_{j=1}^{r_{pa_i}} n_{ij}$. The observed number $n_{ijk}$ follows a multinomial distribution:

$$P\left(n_{ijk} \mid pa_i, \theta_{ijk}\right) = \frac{n_i!}{\prod_{j=1}^{r_{pa_i}} \prod_{k=1}^{r_i} n_{ijk}!} \theta_{ijk}^{n_{iijk}}. \tag{2.4}$$

Let $P(D \mid G, \theta)$ be the probability distribution of observing the data, given the structure $G$ of the DAG. Then, using Equation 2.4, we obtain

$$P(D|G, \theta) = \prod_{i=1}^{p}\prod_{j=1}^{r_{pa_i}}\prod_{k=1}^{r_i} \frac{n_i!}{n_{ijk}!}\theta_{ijk}^{n_{ijk}}. \tag{2.5}$$

The probability of observing the data $D$, given the structure of the DAG is

$$P(D|G) = \int P(D|G, \theta)P(\theta|G)d\theta. \tag{2.6}$$

Recall that

$$\int \prod_{k=1}^{r_i} \theta_{ijk}^{\alpha_{ijk}+n_{ijk}-1} d\theta_{ijk} = \frac{\prod_{k=1}^{r_i}\Gamma\left(\alpha_{ijk}+n_{ijk}\right)}{\Gamma\left(\displaystyle\sum_{k=1}^{r_i}(\alpha_{ijk}+n_{ijk})\right)}. \tag{2.7}$$

Substituting Equations 2.3, 2.5 and 2.7 into Equation 2.6 gives

$$P(D|G) = \prod_{i=1}^{p}\prod_{j=1}^{r_{pa_i}}\Gamma\left(\sum_{k=1}^{r_i}\alpha_{ijk}\right)\prod_{k=1}^{r_i}\frac{1}{\Gamma\left(\alpha_{ijk}\right)}\int \theta_{ijk}^{n_{ijk}+\alpha_{ijk}-1}d\theta_{ijk}$$

$$= \prod_{i=1}^{p}\prod_{j=1}^{r_{pa_i}}\frac{\Gamma\left(\sum_{k=1}^{r_i}\alpha_{ijk}\right)}{\Gamma\left(\sum_{k=1}^{r_i}(\alpha_{ijk}+n_{ijk})\right)}\prod_{k=1}^{r_i}\frac{\Gamma\left(\alpha_{ijk}+n_{ijk}\right)}{\Gamma\left(\alpha_{ijk}\right)}. \tag{2.8}$$

The score metric for the DAG is defined as

$$Score_D(G) = \log P(D|G)$$

$$= \sum_{i=1}^{p}\sum_{j=1}^{r_{pa_i}}\left(\log\frac{\Gamma\left(\sum_{k=1}^{r_i}\alpha_{ijk}\right)}{\Gamma\left(\sum_{k=1}^{r_i}(\alpha_{ijk}+n_{ijk})\right)} + \sum_{k=1}^{r_i}\log\frac{\Gamma\left(\alpha_{ijk}+n_{ijk}\right)}{\Gamma\left(\alpha_{ijk}\right)}\right). \tag{2.9}$$

Equation 2.9 shows that the score metric is decomposable and can be written in terms of the local nodes of the DAG:

$$Score_D(G) = \sum_{i=1}^{p}Score_i(pa_i), \tag{2.10}$$

where

$$Score_i(pa_i) = \sum_{j=1}^{r_{pa_i}}\left(\log\frac{\Gamma\left(\sum_{k=1}^{r_i}\alpha_{ijk}\right)}{\Gamma\left(\sum_{k=1}^{r_i}(\alpha_{ijk}+n_{ijk})\right)} + \sum_{k=1}^{r_i}\log\frac{\Gamma\left(\alpha_{ijk}+n_{ijk}\right)}{\Gamma\left(\alpha_{ijk}\right)}\right). \tag{2.11}$$

The score metric $Score_i(pa_i)$ is referred to as the score metric of the node $i$.

## Example 2.1

Consider a simple data example of three variables summarized in Table 2.1. Assume that the network structure is shown in Figure 2.2. Then, the probability distribution of the variables $x_1$, $x_2$, and $x_3$ can be calculated by

$$P(x_1, x_2, x_3) = P(x_1)P(x_2|x_1)P(x_3|x_2). \qquad (2.12)$$

Now we calculate conditional probability as follows.

$$P(x_1 = 0) = \frac{1}{2}, P(x_1 = 1) = \frac{1}{2},$$

$$P(x_2 = 0|x_1 = 0) = \frac{3}{4}, P(x_2 = 1|x_1 = 0) = \frac{1}{4},$$

$$P(x_2 = 0|x_1 = 1) = \frac{1}{5}, P(x_2 = 1|x_1 = 1) = \frac{4}{5},$$

$$P(x_3 = 0|x_2 = 0) = \frac{4}{5}, P(x_3 = 1|x_2 = 0) = \frac{1}{5},$$

$$P(x_3 = 0|x_2 = 1) = 0, P(x_3 = 1|x_2 = 1) = 1.$$

Using these conditional probabilities, we can calculate the joint probabilities. For example, the probability $P(x_1 = 0, x_2 = 0, x_3 = 0)$ is given by

**TABLE 2.1**

An Example of Three Variables

| Subject | $x_1$ | $x_2$ | $x_3$ |
|---------|-------|-------|-------|
| 1 | 1 | 0 | 0 |
| 2 | 1 | 1 | 1 |
| 3 | 0 | 0 | 1 |
| 4 | 1 | 1 | 1 |
| 5 | 0 | 0 | 0 |
| 6 | 0 | 1 | 1 |
| 7 | 1 | 1 | 1 |
| 8 | 0 | 0 | 0 |
| 9 | 1 | 1 | 1 |
| 10 | 0 | 0 | 0 |

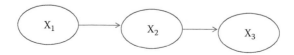

**FIGURE 2.2**
An example of a causal network.

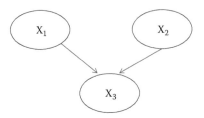

**FIGURE 2.3**
An example of a causal network.

$$P(x_1 = 0, x_2 = 0, x_3 = 0) = P(x_1 = 0)P(x_2 = 0|x_1 = 0)P(x_3 = 0|x_2 = 0)$$

$$= \frac{1}{2}\frac{3}{4}\frac{4}{5} = \frac{3}{10}. \qquad (2.13)$$

Now suppose that the network structure is changed to Figure 2.3. We calculate the probabilities:

$$P(x_2 = 0) = \frac{1}{2} \text{ and } x_3 = 0|x_1 = 0, x_2 = 0) = \frac{3}{4}. \text{ Then, we have}$$

$$P(x_1 = 0, x_2 = 0, x_3 = 0) = \frac{1}{2}\frac{1}{2}\frac{3}{4} = \frac{3}{16}.$$

This example shows that the joint probabilities depend on the network structures. Although the observed data are the same, joint probabilities under different network structures may be different.

## 2.3 Structural Equations and Score Metrics for Continuous Causal Networks

### 2.3.1 Multivariate SEMs for Generating Node Core Metrics

BDeu and other score metrics for discrete nodes cannot be applied to the nodes with continuous variables. SEMs studied in Chapter 1 offer a powerful tool for score metric selection when the node variables are continuous. Assume that the target node and its parent nodes consist of $M$ continuous endogenous variables, and $K$ exogenous variables that can be either continuous variables or discrete variables. We denote the $n$ observations on the $M$ endogenous variables by the matrix $y = [y_1,...,y_M]$ and on the exogenous variables by $X = [x_1,...,x_K]$. Recall that the linear structural equations for modeling relationships among these variables can be written as

$$y_1\gamma_{1i} + ... + y_M\gamma_{Mi} + x_1\beta_{1i} + ... + x_K\beta_{Ki} + e_i = 0, i = 1, ..., M, \qquad (2.14)$$

where the $\gamma$'s and $\beta$'s are the structural parameters of the system that are unknown. In matrix notation, the $ith$ equation can be rewritten as

$$Y\Gamma_i + XB_i + E_i = 0, \tag{2.15}$$

where $\Gamma_i$, $B_i$, $E_i$ are corresponding vectors. Let $y_i$ be the vector of observations of the variable $i$. Let $Y_{-i}$ be the observation matrix $Y$ after removing $y_i$ and $\gamma_{-1}$ be the parameter vector $\Gamma_i$ after removing the parameter $\gamma_{ii}$. The $i$th equation can be written as

$$y_i = W_i \Delta_i + e_i, \tag{2.16}$$

where $W_i = [\, Y_{-i}, X \,]$, $\Delta_i = [\, \gamma_{-i}, B_i \,]$. The estimator investigated in Chapter 1 is

$$\hat{\Delta}_i = \left( \hat{W}_i^T \hat{W}_i \right)^{-1} \hat{W}_i^T y_i, \tag{2.17}$$

where $\hat{W}_i = X(X^T X)^{-1} X^T W_i$. The squared $l_2$-loss

$$Score_i = ||y_i - W_i \hat{\Delta}||_2^2 \tag{2.18}$$

is taken as a score metric of the node $y_i$.

## 2.3.2 Mixed SEMs for Pedigree-Based Causal Inference

Population-based sample design is the current major study design for causal inference. As an alternative to the population-based causal inference, family-based designs have several remarkable features over the population-based causal inference. Family data convey more information than population data. Family data not only include genetic information across the genome, but also contain correlation between individuals. The segregation of rare variants in families offers information on multiple copies of the segregated rare variants. Family data provide rich information on transmission of genetic variants from generation o generation which will improve accuracy for imputation of rare variants.

In addition, population substructures are often present. Population substructures may create spurious causal structures. Similar to family-based association analysis, in this chapter we introduce mixed structural equation models (MSEMs) for deriving score metrics for continuous variables and for family-based causal inference.

### 2.3.2.1 Mixed SEMs

Consider a MSEM:

$$
\begin{aligned}
y_1\gamma_{11} + \dots + y_M\gamma_{M1} + x_1\beta_{11} + \dots + x_K\beta_{K1} + g_1 u_{11} + \dots + g_q u_{q1} + e_1 &= 0 \\
y_1\gamma_{12} + \dots + y_M\gamma_{M2} + x_1\beta_{12} + \dots + x_K\beta_{K2} + g_1 u_{12} + \dots + g_q u_{q2} + e_2 &= 0 \\
\vdots \qquad\qquad \vdots \qquad\qquad \vdots \qquad\qquad& \\
y_1\gamma_{1M} + \dots + y_{MM}\gamma_{MM} + x_1\beta_{1M} + \dots + x_K\beta_{KM} + g_1 u_{1M} + \dots + g_q u_{qM} + e_M &= 0
\end{aligned}
\tag{2.19}
$$

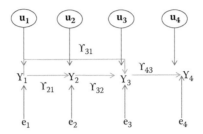

**FIGURE 2.4**
Causal structure of four phenotypes.

where $y_1,..., y_M$ are $M$ endogenous variables (traits), $x_1,..., x_K$ are the $K$ exogenous and predetermined variables (covariates), $g_1,..., g_q$ are the $q$ standardized genetic variables (Figure 2.4). As defined in Chapter 6 in the book *Big Data in Omics and Imaging: Association Analysis*. The $\gamma'$ s and the $\beta'$ s are structure parameters, $u_{11}, ..., u_{qM}$ are random effects, and $e_1, ..., e_M$ are M random variables.

Define vectors of random effects as

$$u_1 = \begin{bmatrix} u_{11} \\ \vdots \\ u_{1M} \end{bmatrix}, ..., u_q = \begin{bmatrix} u_{q1} \\ \vdots \\ u_{qM} \end{bmatrix}.$$

The structural equation model (2.19) can be written in a matrix form:

$$Y\Gamma + XB + GU + E = 0, \tag{2.20}$$

where

$$Y = \begin{bmatrix} y_{11} & \cdots & y_{1M} \\ \vdots & \vdots & \vdots \\ y_{n1} & \cdots & y_{nM} \end{bmatrix} = [Y_1, ..., Y_M],$$

$$\Gamma = \begin{bmatrix} \gamma_{11} & \cdots & \gamma_{1M} \\ \vdots & \vdots & \vdots \\ \gamma_{M1} & \cdots & \gamma_{MM} \end{bmatrix} = [\Gamma_1, ..., \Gamma_M],$$

$$X = \begin{bmatrix} x_{11} & \cdots & x_{1p} \\ \vdots & \vdots & \vdots \\ x_{n1} & \cdots & x_{np} \end{bmatrix} = [X_1, ..., X_p],$$

$$B = \begin{bmatrix} \beta_{11} & \cdots & \beta_{1M} \\ \vdots & \vdots & \vdots \\ \beta_{p1} & \cdots & \beta_{pM} \end{bmatrix} = [B_1, \ldots, B_M] \text{ is a } p \times M \text{ dimensional genetic additive matrix,}$$

$$G = \begin{bmatrix} g_{11} & \cdots & g_{1q} \\ \vdots & \vdots & \vdots \\ g_{n1} & \cdots & g_{nq} \end{bmatrix} = \left[ G_1, \ldots, G_q \right],$$

$$U = \begin{bmatrix} u_{11} & \cdots & u_{1M} \\ \vdots & \vdots & \vdots \\ u_{q1} & \cdots & u_{qM} \end{bmatrix} = [u_1, \ldots, u_M] \text{ is a } q \times M \text{ dimensional random effect matrix,}$$

and

$$\varepsilon = \begin{bmatrix} \varepsilon_{11} & \cdots & \varepsilon_{1M} \\ \vdots & \vdots & \vdots \\ \varepsilon_{n1} & \cdots & \varepsilon_{nM} \end{bmatrix} = [\varepsilon_1, \ldots, \varepsilon_M].$$

We assume that the joint distribution of the random effects and residuals

$$\begin{bmatrix} vec(U) \\ vec(\varepsilon) \end{bmatrix} \sim N\left( \begin{bmatrix} 0 \\ 0 \end{bmatrix} \begin{bmatrix} \Sigma_0 \otimes I_q & 0 \\ 0 & R_0 \otimes I_n \end{bmatrix} \right), \tag{2.21}$$

where

$$\Sigma_0 = \begin{bmatrix} \sigma_{11}^u & \cdots & \sigma_{1M}^u \\ \vdots & \vdots & \vdots \\ \sigma_{M1}^u & \cdots & \sigma_{MM}^u \end{bmatrix} \text{ and } R_0 = \begin{bmatrix} \sigma_{11}^e & \cdots & \sigma_{1M}^e \\ \vdots & \vdots & \vdots \\ \sigma_{M1}^e & \cdots & \sigma_{MM}^e \end{bmatrix}.$$

Next, we consider the distribution of $ZU + \varepsilon$. Let $W = GU + \varepsilon$. Then, we have

$$\begin{aligned} vec(W) &= vec(GU) + vec(\varepsilon) \\ &= (I \otimes G)vec(U) + vec(\varepsilon). \end{aligned} \tag{2.22}$$

It follows from Equations (2.24) and (2.25) that

$$\begin{aligned} \Lambda = cov(vec(W)) &= (I_M \otimes G)\left( \Sigma_0 \otimes I_q \right)\left( I_M \otimes G^T \right) + R_0 \otimes I_n \\ &= \left( \Sigma_0 \otimes (GG^T) \right) + R_0 \otimes I_n. \end{aligned} \tag{2.23}$$

### 2.3.2.2 Two-Stage Estimate for the Fixed Effects in the Mixed SEMs

Consider the *ith* equation for the model (2.19):

$$Y\Gamma_i + XB_i + Gu_i + e_i = 0. \tag{2.24}$$

To use the regression approach, we often select one endogenous variable to appear on the left-hand side of the equation. Specifically, the *ith* equation is

$$y_1\gamma_{1i} + \dots + y_{i-1}\gamma_{i-1i} + y_i\gamma_{ii} + y_{i+1}\gamma_{i+1i} + \dots + y_M\gamma_{Mi} + x_1\beta_{1i} + \dots + x_K\beta_{Ki} + g_1u_{1i} + \dots$$
$$+ g_qu_{qi} + e_i = 0.$$

Dividing both sides of the above equation by $-\gamma_{ii}$ and replacing $-\dfrac{\gamma_{ij}}{\gamma_{ii}}, -\dfrac{\beta_{ji}}{\gamma_{ii}}$ and $-\dfrac{u_{ji}}{\gamma_{ii}}$ by $\gamma_{ji}, \beta_{ji}$ and $u_{ji}$, respectively, we obtain

$$y_i = y_1\gamma_{1i} + \dots + y_{i-1}\gamma_{i-1i} + y_{i+1}\gamma_{i+1i} + \dots + y_M\gamma_{Mi} + x_1\beta_{1i} + \dots + x_K\beta_{Ki}$$
$$+ g_1u_{1i} + \dots + g_qu_{qi} + e_i. \tag{2.25}$$

Some coefficients of $\Gamma_i$ and $B_i$ may be zero. Assume

$$\Gamma_i = \begin{bmatrix} -1 \\ \gamma_i \\ \gamma_i^* \end{bmatrix} = \begin{bmatrix} -1 \\ \gamma_i \\ 0 \end{bmatrix}, B_i = \begin{bmatrix} \beta_i \\ \beta_i^* \end{bmatrix} = \begin{bmatrix} \beta_i \\ 0 \end{bmatrix}, Y = [y_i\ Y_i\ Y_i^*] \text{ and } X = [X_i\ X_i^*].$$

Equation (2.25) can then be rearranged as

$$y_i = Y_i\gamma_i + Y_i^*\gamma_i^* + X_i\beta_i + X_i^*\beta_i^* + Gu_i + \phi_i$$
$$= Y_i\gamma_i + X_i\beta_i + Gu_i + \varepsilon_i$$
$$= [Y_i\ X_i]\begin{bmatrix} \gamma_i \\ \beta_i \end{bmatrix} + Gu_i + \varepsilon_i \tag{2.26}$$
$$= Z_i\delta_i + Gu_i + \varepsilon_i$$
$$= Z_i\delta_i + W_i,$$

where

$$Z_i = [Y_i\ X_i], \delta_i = \begin{bmatrix} \gamma_i \\ \beta_i \end{bmatrix} \text{ and } W_i = Gu_i + \varepsilon_i.$$

Recall from equation (2.21) that

$$\text{var}(u_i) = \sigma_{ii}^u I_q,$$

which implies that

$$W_i \sim N\left(0\ \sigma_{ii}^u GG^T + \sigma_{ii}^e I_n\right).$$ (2.27)

Pre-multiplying Equation 2.26 by $X^T$, we obtain

$$X^T y_i = X^T Z_i \delta_i + X^T W_i.$$ (2.28)

The covariance matrix of $X^T W_i$ is

$$H = X^T \left(\sigma_{ii}^u GG^T + \sigma_{ii}^e I_n\right) X.$$ (2.29)

A generalized least squares estimator of the parameters for equation (2.28) is given by

$$\hat{\delta}_i = \left[\left(X^T Z_i\right)^T \left(X^T \left(\sigma_{ii}^u GG^T + \sigma_{ii}^e I_n\right) X\right)^{-1} X^T Z_i\right]^{-1}$$
$$\left(X^T Z_i\right)^T \left(X^T \left(\sigma_{ii}^u GG^T + \sigma_{ii}^e I_n\right) X\right)^{-1} X^T y_i.$$ (2.30)

The variance of the estimator is

$$\operatorname{var}\left(\hat{\delta}_i\right) = \left[\left(X^T Z_i\right)^T \left(X^T \left(\sigma_{ii}^u GG^T + \sigma_{ii}^e I_n\right) X\right)^{-1} X^T Z_i\right]^{-1}.$$ (2.31)

### 2.3.2.3 Three-Stage Estimate for the Fixed Effects in the Mixed SEMs

Similar to the classical SEMs, the fixed effects can be estimated using three-stage least square methods. Equation 1.65 in Chapter 1 can be modified as

$$\begin{bmatrix} X^T y_1 \\ X^T y_2 \\ \vdots \\ X^T y_M \end{bmatrix} = \begin{bmatrix} X^T Z_1 & & & \\ & X^T Z_2 & & \\ & & \ddots & \\ & & & X^T Z_M \end{bmatrix} \begin{bmatrix} \delta_1 \\ \delta_2 \\ \vdots \\ \delta_M \end{bmatrix} + \begin{bmatrix} X^T W_1 \\ X^T W_2 \\ \vdots \\ X^T W_M \end{bmatrix},$$ (2.32)

where $W_i = Gu_i + e_i$, $i = 1, 2,\ldots,M$.
   Again, Equation 2.32 can be written in a compact matrix form:

$$\left[I \otimes X^T\right] y = \left[I \otimes X^T\right] Z\delta + \left(I \otimes X^T\right) W,$$ (2.33)

where

$$Z = \begin{bmatrix} Z_1 & & & \\ & Z_2 & & \\ & & \ddots & \\ & & & Z_M \end{bmatrix}_{TM \times ([M(M-1)],}$$

$$y = \begin{bmatrix} y_1 \\ y_2 \\ \vdots \\ y_M \end{bmatrix}_{TM}, \delta = \begin{bmatrix} \delta_1 \\ \delta_2 \\ \vdots \\ \delta_M \end{bmatrix}_M \text{ and } W = \begin{bmatrix} W_1 \\ W_2 \\ \vdots \\ W_M \end{bmatrix}_M.$$

Recall that the covariance of the residual term $(I \otimes X^T) W$ is

$$\begin{aligned} \Lambda_0 &= \mathrm{cov}\big( (I_M \otimes X^T) W \big) = (I_M \otimes X^T) \big[ \Sigma_0 \otimes GG^T + R_0 \otimes I_n \big] (I_M \otimes X) \\ &= \Sigma_0 \otimes (X^T GG^T X) + R_0 \otimes (X^T X). \end{aligned} \tag{2.34}$$

It follows from Equations 2.33 and 2.34 that the generalized least square estimator of $\delta$ is

$$\hat{\delta} = \big\{ [Z^T (I_M \otimes X)] \Lambda_0^{-1} [(I_M \otimes X^T) Z] \big\}^{-1} [Z^T (I_M \otimes X)] \Lambda_0^{-1} (I_M \otimes X^T) y. \tag{2.35}$$

Recall that the covariance matrix of $y$ is given by

$$\mathrm{cov}(y) = \Sigma_0 \otimes GG^T + R_0 \otimes I_n,$$

which implies

$$(I_M \otimes X^T) \mathrm{cov}(y)(I_M \otimes X) = \Lambda_0. \tag{2.36}$$

Using Equations 2.35 and 2.36, we can obtain the variance of the estimator $\hat{\delta}$:

$$\begin{aligned} \mathrm{var}\big( \hat{\delta} \big) &= \big\{ [Z^T (I_M \otimes X)] \Lambda_0^{-1} [(I_M \otimes X^T) Z] \big\}^{-1} \\ &\quad [Z^T (I_M \otimes X)] \Lambda_0^{-1} \Lambda_0 \Lambda_0^{-1} (I_M \otimes X^T) Z \\ &\quad * \big\{ [Z^T (I_M \otimes X)] \Lambda_0^{-1} [(I_M \otimes X^T) Z] \big\}^{-1} \\ &= \big\{ [Z^T (I_M \otimes X)] \Lambda_0^{-1} [(I_M \otimes X^T) Z] \big\}^{-1} \end{aligned} \tag{2.37}$$

### 2.3.2.4 The Full Information Maximum Likelihood Method

Both two-stage and three-stage estimation methods require the assumption that the variances of random effects and residuals are known. However, in practice, the variances of random effects and residuals are unknown and need to be estimated. The full information maximum likelihood methods will provide likelihood functions that can be used to estimate the variance components. Assume that both random genetic effects and structural errors are normally distributed. Consider the model:

$$
\begin{bmatrix} y_1 \\ \vdots \\ y_M \end{bmatrix} = \begin{bmatrix} Z_1 & \cdots & 0 \\ \vdots & \vdots & \vdots \\ 0 & \cdots & Z_M \end{bmatrix} \begin{bmatrix} \delta_1 \\ \vdots \\ \delta_M \end{bmatrix} + \begin{bmatrix} G & \cdots & 0 \\ \vdots & \vdots & \vdots \\ 0 & \cdots & G \end{bmatrix} \begin{bmatrix} u_1 \\ \vdots \\ u_M \end{bmatrix} + \begin{bmatrix} \varepsilon_1 \\ \vdots \\ \varepsilon_M \end{bmatrix}.
\tag{2.38}
$$

Let

$$
Z = \begin{bmatrix} Z_1 & & & \\ & Z_2 & & \\ & & \ddots & \\ & & & Z_M \end{bmatrix}_{TM \times ([M(M-1)],}
$$

$$
y = \begin{bmatrix} y_1 \\ y_2 \\ \vdots \\ y_M \end{bmatrix}_{TM}, \delta = \begin{bmatrix} \delta_1 \\ \delta_2 \\ \vdots \\ \delta_M \end{bmatrix}_M, u_0 = \begin{bmatrix} u_1 \\ \vdots \\ u_M \end{bmatrix} \text{ and } \varepsilon_0 = \begin{bmatrix} \varepsilon_1 \\ \varepsilon_2 \\ \vdots \\ \varepsilon_M \end{bmatrix}_M.
$$

Then, equation (2.38) can be written in a matrix form:

$$
\begin{aligned}
y &= Z\delta + (I \otimes G)u_0 + \varepsilon_0 \\
&= Z\delta + W_0,
\end{aligned}
\tag{2.39}
$$

where $W_0 = (I \otimes G) u_0 + \varepsilon_0$.

Next, we derive the distribution of the vector $W_0$. Recall that the joint distribution of $u_0$ and $\varepsilon_0$ is

$$
\begin{bmatrix} u_0 \\ \varepsilon_0 \end{bmatrix} \sim N \left( \begin{bmatrix} 0 \\ 0 \end{bmatrix}, \begin{bmatrix} \Sigma_0 \otimes I_q & 0 \\ 0 & R_0 \otimes I_n \end{bmatrix} \right),
\tag{2.40}
$$

which implies

$$
W_0 \sim N\left(0, \Sigma_0 \otimes (GG^T) + R_0 \otimes I_n\right).
\tag{2.41}
$$

The joint probability density function for $W_0$ is

$$
f(W_0) = \frac{1}{(2\pi)^{\frac{nM}{2}} |\Sigma_0 \otimes (GG^T) + R_0 \otimes I_n|^{\frac{1}{2}}}
$$
$$
\exp\left\{ -\frac{1}{2} W_0^T \left[\Sigma_0 \otimes (GG^T) + R_0 \otimes I_n\right]^{-1} W_0 \right\}.
\tag{2.42}
$$

To find the joint probability density function for $y$, we make a change of variables. Taking vector operation of the matrices in Equation 2.20, we obtain

$$\left(\Gamma^T \otimes I_n\right) vec(Y) + \left(B^Y \otimes I_n\right) vec(X) + vec(W) = 0 \quad \text{or}$$

$$W_0 = -\left(\Gamma^T \otimes I_n\right) y - \left(B^Y \otimes I_n\right) vec(X). \tag{2.43}$$

The Jacobian matrix $\left|\dfrac{\partial y}{\partial W_0^T}\right|$ is

$$\left|\frac{\partial y}{\partial W_0^T}\right|^{-1} = \left|\frac{\partial W_0}{\partial y^T}\right| = \left|\Gamma^T \otimes I_n\right| = |\Gamma|^n. \tag{2.44}$$

Using change of variables, we obtain the following results for the joint probability density function for $y$:

**Result 2.1**

The joint probability density function for $y$ is

$$f(y) = \frac{1}{(2\pi)^{\frac{nM}{2}} |\Sigma_0 \otimes (GG^T) + R_0 \otimes I_n|^{\frac{1}{2}}} |\Gamma|^n$$

$$\exp\left\{-\frac{1}{2}(y - Z\delta)^T \left[\Sigma_0 \otimes (GG^T) + R_0 \otimes I_n\right]^{-1}(y - Z\delta)\right\} \tag{2.45}$$

The log likelihood function is

$$l(\delta, \Sigma_0, R_0 | y) = -\frac{nM}{2} \log(2\pi) - \frac{1}{2} \log |\Sigma_0 \otimes (GG^T) + R_0 \otimes I_n| + n \log |\Gamma|$$

$$-\frac{1}{2}(y - Z\delta)^T \left[\Sigma_0 \otimes (GG^T) + R_0 \otimes I_n\right]^{-1}(y - Z\delta). \tag{2.46}$$

Equation 2.46 is the basis for the maximum likelihood estimation of both fixed effects, random effects, and variance components. It uses the full information of the data.

### 2.3.2.5  Reduced Form Representation of the Mixed SEMs

The first proposed and widely used method for the mixed SEMs is the reduced form representation of the SEMs (Gianola and Sorensen 2004; Rosa et al. 2011; Valente et al. 2013).

Consider a mixed structural equation model:

$$\Lambda y_i = X_i \beta + u_i + \varepsilon_i, \tag{2.47}$$

where

$$
y_i = \begin{bmatrix} y_{i1} \\ \vdots \\ y_{iM} \end{bmatrix}, \Lambda = \begin{bmatrix} \gamma_{11} & \cdots & \gamma_{1M} \\ \vdots & \vdots & \vdots \\ \gamma_{M1} & \cdots & \gamma_{MM} \end{bmatrix}, X_i = \begin{bmatrix} x_{i1}^T & \cdots & 0 \\ \vdots & \vdots & \vdots \\ 0 & \cdots & x_{iM}^T \end{bmatrix}, x_{ij} = \begin{bmatrix} x_{ij1} \\ \vdots \\ x_{ijp_j} \end{bmatrix}, \beta = \begin{bmatrix} \beta_1 \\ \vdots \\ \beta_M \end{bmatrix},
$$

$$
\beta_j = \begin{bmatrix} \beta_{j1} \\ \vdots \\ \beta_{jp_j} \end{bmatrix},
$$

$$
u_i = \begin{bmatrix} u_{i1} \\ \vdots \\ u_{iM} \end{bmatrix}, \varepsilon_i = \begin{bmatrix} \varepsilon_{i1} \\ \vdots \\ \varepsilon_{iM} \end{bmatrix}.
$$

We assume that

$$
u_i \sim N(0, \Sigma_0) \text{ and } e_i \sim N(0, R_0).
$$

If $\Lambda$ has full rank, then multiplying $\Lambda^{-1}$ on both sides of Equation 2.47, we obtain the reduced form:

$$
y_i = \Lambda^{-1} X_i \beta + \Lambda^{-1} u_i + \Lambda^{-1} \varepsilon_i. \tag{2.48}
$$

The marginal distribution of the $M$ endogenous variables $y_i$ follows a normal distribution:

$$
y_i \sim N\big(\Lambda^{-1} X_i \beta, \Lambda^{-1} (\Sigma_0 + R_0) \Lambda^{-T}\big). \tag{2.49}
$$

From Equation 2.48 we can obtain the following covariance matrices:

$$
\text{cov}(u_i, y_i) = \Sigma_0 \Lambda^{-T}. \tag{2.50}
$$

The joint distribution of $u_i$ and $y_i$ is

$$
\begin{pmatrix} u_i \\ y_i \end{pmatrix} \sim N\left( \begin{bmatrix} 0 \\ \Lambda^{-1} X_i \beta \end{bmatrix} \begin{bmatrix} \Sigma_0 & \Sigma_0 \Lambda^{-T} \\ \Lambda^{-1} \Sigma_0 & \Lambda^{-1}(\Sigma_0 + R_0) \Lambda^{-T} \end{bmatrix} \right). \tag{2.51}
$$

The random effects can be estimated by the conditional mean of $u_i$, given the observed endogenous variables $y_i$:

$$
\begin{aligned}
\hat{u}_i = E[u_i | y_i] &= 0 + \Sigma_0 \Lambda^{-T} \big[ \Lambda^{-1}(\Sigma_0 + R_0) \Lambda^{-T} \big]^{-1} \big( y_i - \Lambda^{-1} X_i \beta \big) \\
&= \Sigma_0 (\Sigma_0 + R_0)^{-1} \Lambda \big( y_i - \Lambda^{-1} X_i \beta \big).
\end{aligned} \tag{2.52}
$$

The model (2.47) can be extended to the entire dataset:

$$
\begin{bmatrix} \Lambda y_1 \\ \vdots \\ \Lambda y_n \end{bmatrix} = \begin{bmatrix} X_1 \\ \vdots \\ X_n \end{bmatrix} \beta + \begin{bmatrix} u_1 \\ \vdots \\ u_n \end{bmatrix} + \begin{bmatrix} \varepsilon_1 \\ \vdots \\ \varepsilon_n \end{bmatrix} \tag{2.53}
$$

$$
= X\beta + u + \varepsilon,
$$

or

$$
(I_n \otimes \Lambda)y = X\beta + u + \varepsilon.
$$

We assume that the vector of genetic random effects and residual errors follow a normal distribution:

$$
\begin{bmatrix} u \\ \varepsilon \end{bmatrix} \sim N \left( \begin{bmatrix} 0 \\ 0 \end{bmatrix} \begin{bmatrix} A \otimes \Sigma_0 & 0 \\ 0 & I_n \otimes R_0 \end{bmatrix} \right), \tag{2.54}
$$

where $A$ is a genetic relationship matrix.

Let $w = u + \varepsilon$ and $\Sigma = \mathrm{cov}(w) = A \otimes \Sigma_0 + I_n \otimes R_0$. The probability density function for $w$ is

$$
f(w) = \frac{1}{(2\pi)^{\frac{nM}{2}} |\Sigma|^{\frac{1}{2}}} \exp\left\{ -\frac{1}{2} w^T \Sigma^{-1} w \right\}. \tag{2.55}
$$

To find the probability density function for $y$, we make the following change of variables:

$$
w = (I_n \otimes \Lambda)y - X\beta. \tag{2.56}
$$

The Jacobian matrix for the change of variables is

$$
J = \frac{\partial w}{\partial y^T} = I_n \otimes \Lambda. \tag{2.57}
$$

Using Equations 2.55, 2.56, and 2.57 and a theorem for change of variables, we obtain the probability density function for $y$:

$$
f(y|\Lambda, \beta, \Sigma) = \frac{1}{(2\pi)^{\frac{nM}{2}} |\Sigma|^{\frac{1}{2}}} |\Lambda|^n
$$

$$
\exp\left\{ -\frac{1}{2} ((I_n \otimes \Lambda)y - X\beta)^T \Sigma^{-1} ((I_n \otimes \Lambda)y - X\beta) \right\}. \tag{2.58}
$$

## 2.4 Bayesian Networks with Discrete and Continuous Variables

Many genetic and epigenetic networks contain discrete and continuous variables. For example, phenotype-disease networks contain discrete disease variables and continuous phenotype networks. Calculation of the score metrics with discrete variables and calculation of the score metrics with continuous variables were investigated in Sections 2.2 and 2.3, respectively. In this section, we need to consider hybrid causal networks with both discrete and continuous variables. Unfortunately, the statistical methods and computational algorithms for learning causal networks with hybrid variables have not been well developed. A widely used approach to learning causal networks with hybrid variables is to discretize all the variables (Kozlov 1997). However, the discretization approach cannot ensure accurate estimation and fast computation. In this section, the network penalized logistic regression and variational approximation approach will be introduced for learning hybrid causal networks.

### 2.4.1 Two-Class Network Penalized Logistic Regression for Learning Hybrid Bayesian Networks

Consider a Bayesian network of binary discrete nodes with both continuous and discrete parents. Assume that the parent nodes form subnetworks denoted by $G = \{V, E\}$ where $V$ denotes a set of nodes and $E$ denotes a set of directed edges (Figure 2.5). Assume that the graph $G$ consists of three subnetworks: the subnetwork with continuous nodes $G_1 = (V_1, E_1)$, the subnetwork both continuous and discrete nodes $G_2 = (V_2, E_2)$, and the subnetwork $G_3 = (V_3, E_3)$ that connects the network $G_2$ with the network $G_1$.

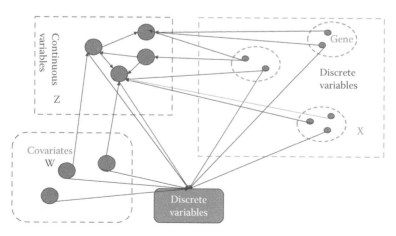

**FIGURE 2.5**
Hybrid Bayesian network (discrete nodes with hybrid parents).

Assume that $|V_1| = K$, $|V_2| = L + K - 1$ and $|V_3| = m + K$. The structure of the whole network is characterized by the adjacent matrix. The elements of the adjacent matrix are defined as follows.

$$S_{uv} = \begin{cases} \bar{s}_{uv} \ (u,v) \in E_1 \\ 0 \ (u,v) \notin E_1 \end{cases}, \quad S_{ul} = \begin{cases} \bar{s}_{ul} \ (u,l) \in E_2 \\ 0 \ (u,l) \notin E_2 \end{cases}, \quad S_{uj} = \begin{cases} \bar{s}_{uj} \ (u,j) \in E_3 \\ 0 \ (u,j) \in E_3 \end{cases}.$$

The adjacency matrix is then defined as

$$S_{\delta\delta} = \begin{bmatrix} 0 & s_{u_1 u_2} & \cdots & s_{u_1 u_K} \\ \vdots & \vdots & \vdots & \vdots \\ s_{u_K u_1} & s_{u_K u_2} & \cdots & 0 \end{bmatrix}_{K \times K}, S_{\delta\eta} = \begin{bmatrix} s_{u_1 l_1} & \cdots & s_{u_1 l_L} \\ \vdots & \vdots & \vdots \\ s_{u_K l_1} & \cdots & s_{u_K l_L} \end{bmatrix}_{K \times L},$$

$$S_{\delta\alpha} = \begin{bmatrix} s_{u_1 j_1} & \cdots & s_{u_1 j_m} \\ \vdots & \vdots & \vdots \\ s_{u_K j_1} & \cdots & s_{u_K j_m} \end{bmatrix}_{K \times m} \quad \text{and } S = \begin{bmatrix} S_{\delta\delta} & S_{\delta\eta} & S_{\delta\alpha} \\ S_{\delta\eta}^T & 0 & 0 \\ S_{\delta\alpha}^T & 0 & 0 \end{bmatrix}_{(K+L+m) \times (K+L+m)}.$$

Let $\mathbf{1}$ be a $(K + L + m)$ dimensional vector with all elements of one. Define a $(K + L + m)$ dimensional degree vector: $d = S\mathbf{1}$ and a degree matrix $D = \text{diag}$ $(d_1, d_2,..., d_{(K+L+m)})$. The Laplacian matrix associated with the whole network is $D - S$.

Each discrete node is modeled by a logistic regression. Consider a discrete node. The binary node value of *ith* individual will be denoted by $y_i \in \{-1, 1\}$. Recall that the logistic model is given by

$$\pi_i = p(y_i = 1 | H_i, \beta) = \frac{e^{\xi_i}}{1 + e^{\xi_i}}, \tag{2.59}$$

where $\xi_i = H_i\beta$, $H_i = [z_i \ w_i \ x_i]$ and $\beta = \begin{bmatrix} \delta \\ \eta \\ \alpha \end{bmatrix}$. Specifically, the vector of continuous variables (phenotypes) $z$ and their coefficients are, respectively, denoted by $z_i = [z_{i1},..., z_{iK}]$ and $\delta = [\delta_1, \delta_2, ..., \delta_K]^T$. The discrete variables (genotype indicator variables) or continuous variables (functional principal component scores) are denoted by

$$x_i = \left[x_i^{(1)}, ..., x_i^{(G)}\right], x_i^{(g)} = \left[x_{i1}^{(g)}, ..., x_{ik_g}^{(g)}\right], g = 1, ..., G.$$

Correspondingly, we denote the genetic effects by

$$\alpha = \begin{bmatrix} \alpha_1 \\ \vdots \\ \alpha_G \end{bmatrix}, \quad \alpha_g = \begin{bmatrix} \alpha_{g1} \\ \vdots \\ \alpha_{gk_g} \end{bmatrix}, \quad g = 1, ..., G, m = \sum_{g=1}^{G} k_g.$$

We can simply write $\alpha = [\alpha_1, \alpha_2, ..., \alpha_m]^T$. The covariates $w_i$ and the vector of their logistic regression coefficients $\eta$ are hybrid variables (discrete or continuous variables).

For each sample, we have a vector of observed variables (covariates and genotypes $H_i$) in parents and an observed class $y_i$. The class variable $y_i$ follows a Bernoulli distribution with the conditional probability $p(Y_i \mid H_i, \beta)$ as its parameter. The likelihood and log-likelihood, are, respectively, given by

$$L(\beta) = \prod_{i=1}^{n} p(y_i|H_i, \beta)^{y_i}(1 - p(Y_i|H_i, \beta))^{1-y_i},\tag{2.60}$$

and

$$
\begin{aligned}
l(\beta) &= \sum_{i=1}^{n}[y_i \log p(y_i|H_i, \beta) + (1 - y_i) \log(1 - p(y_i|H_i, \beta))]\\
&= \sum_{i=1}^{n}\left[y_i \log \frac{p(y_i|H_i, \beta)}{1 - p(y_i|H_i, \beta)} + \log(1 - p(y_i \mid H_i, \beta))\right]\\
&= \sum_{i=1}^{n}\left[y_i(H_i\beta) - \log\left(1 + e^{H_i\beta}\right)\right].
\end{aligned}\tag{2.61}
$$

Now we consider the first type of penalty on the path coefficients associated with continuous and discrete variables in the parents (phenotype coefficients $\delta$ and covariate coefficients $\eta$). The $L_1$ norm regularization on $\delta$ and $\eta$ is defined as

$$J_1 = \sum_{u=1}^{K}||\delta_u||_1 + \sum_{l=2}^{L}||\eta_l||_1.\tag{2.62}$$

The second type of penalty is group LASSO for the genotype variables in the parents and is defined as

$$J_2(\alpha) = \sum_{g=1}^{G}\phi_g||\alpha_g||_2.\tag{2.63}$$

The third type of penalty is a network penalty in which constraints are posed on graphs. For the subnetwork we use edge penalty to penalize the network. We penalize the difference between the variables at adjacent nodes. First, we consider the edges in the phenotype subnetworks (that may contain continuous variables). Consider the edge between nodes $u$ and $v$ with weight $s_{uv}$. The constraint posed on the edge is $\bar{s}_{uv}||\delta_u - \delta_v||_2^2$. Thus, the penalty for the phenotype subnetwork is

$$\sum_{(u,v)\in E_1} \bar{s}_{uv}||\delta_u - \delta_v||_2^2.\tag{2.64}$$

Then, consider the penalty for the environment-phenotype subnetwork. The penalty for the edge connecting a covariate $\eta_l$ and a phenotype $\delta_u$ is $\bar{s}_{ul}||\eta_l - \delta_u||_2^2$. The penalty for

$$\sum_{(l,u)\in E_2} \bar{s}_{ul}||\eta_l - \delta_u||_2^2.\tag{2.65}$$

Finally, consider the penalty for the genotype–phenotype connect subnetwork that contains both discrete and continuous variables. The penalty for the edge connecting a SNP $\alpha_j$ and a phenotype $\delta_u$ is $\bar{s}_{uj}||\alpha_j - \delta_u||_2^2$. The penalty for the genotype–phenotype connect subnetwork is

$$\sum_{(j,u)\in E_3} \bar{s}_{uj}||\alpha_j - \delta_u||_2^2. \tag{2.66}$$

Combining Equations 2.64, 2.65, and 2.66 gives the penalty for the whole network:

$$J_3(\beta) = \sum_{(u,v)\in E_1} \bar{s}_{uv}||\delta_u - \delta_v||_2^2 + \sum_{(l,u)\in E_2} \bar{s}_{ul}||\eta_l - \delta_u||_2^2$$
$$+ \sum_{(j,u)\in E_3} \bar{s}_{uj}||\alpha_j - \delta_u||_2^2. \tag{2.67}$$

Let **1** be a $(K + L + m)$ dimensional vector with all elements of one. Define a $(K + L + m)$ dimensional degree vector: $d = S\mathbf{1}$ and a degree matrix $D = \text{diag}$ $(d_1, d_2, ..., d_{(K+L+m)})$. The Laplacian matrix associated with the whole network is $D - S$. We can use the Laplacian matrix to rewrite equation in a matrix form:

$$J_3(\beta) = \beta^T (D - S)\beta. \tag{2.68}$$

Therefore, the penalized loglikelihood function is defined as

$$l_p(\beta) = -l(\beta) + \lambda_1 \sum_{u=1}^{K}||\delta_u||_1 + \lambda_2 \sum_{l=2}^{L}||\eta_l||_1] + \lambda_3 \sum_{g=1}^{G}\phi_g||\alpha_g||_2$$
$$+ \lambda_4 \beta^T (D - S)\beta, \tag{2.69}$$

where $l(\beta) = \sum_{i=1}^{n}[y_i\xi_i - \log(1 + e^{\xi_i})]$, $\xi_i = H_i\beta$, $H_i = [z_i \; w_i \; x_i]$ and $\beta = \begin{bmatrix} \delta \\ \eta \\ \alpha \end{bmatrix}$.

Parameters can be estimated by proximate methods. The minimum of a penalized log-likelihood is defined as a score metric for the discrete node with continuous variables in the parents.

### 2.4.2 Multiple Network Penalized Functional Logistic Regression Models for NGS Data

The majority of variants in the NGS data are rare variants. Multivariate logistic regression is difficult to apply to NGS data. We need to extend the multivariate network penalized logistic regression model to the network penalized functional logistic regression. Let $g_i = x_i\alpha$. Replacing $g_i = x_i\alpha$ by $g_i = \sum_{j=1}^{J}\int_{\tau_j} x_i(\tau)\alpha$

$(\tau)d\tau$, we extend the logistic regression model in Section 2.4.1 to the functional logistic regression. The genotype function $x_i(\tau)$ is expanded in terms of their eigenfunction $\phi_m(\tau)$:

$$x_i(\tau) = \sum_{m=1}^{M_j} \xi_{ijm}\phi_m(\tau). \tag{2.70}$$

Using expansion (2.70) gives

$$g_i = \sum_{j=1}^{J}\sum_{m=1}^{M_j} \xi_{ijm} b_{jm}, \tag{2.71}$$

where $b_{jm} = \int_{\tau_j} \alpha(\tau)\phi_m(\tau)d\tau$.

Thus, the functional logistic regression is transformed into the multivariate logistic regression and the score function problem is reduced to the network penalized multivariate logistic regression analysis.

### 2.4.3 Multi-Class Network Penalized Logistic Regression for Learning Hybrid Bayesian Networks

When a discrete node is a categorical variable and takes multiple values, the multi-class logistic regression model can be used to quantify the score of the discrete node with categorical variables. In Section 8.1.6, we discuss the network penalized multi-class logistic regression. Assume that the number of classes for the categorical variable is $C$. As we discussed in Section 2.4.2, predictors consist of three parts: covariates including environments, genotypes, and phenotypes (or gene expressions). Assume $L$ covariates including 1, $G$ genes, and $K$ phenotypes and environments. The log-likelihood is defined as

$$l_p(\beta) = \sum_{i=1}^{n}\sum_{c=1}^{C-1}\left[I(y_i = c)\log\frac{e^{H_i\beta^{(c)}}}{1 + \sum_{c=1}^{C-1}e^{H_i\beta^{(c)}}}\right]$$

$$= \sum_{i=1}^{n}\sum_{c=1}^{C-1}\left[I(y_i = c)\left(H_i\beta^{(c)} - \log\left(1 + \sum_{c=1}^{C-1}e^{H_i\beta^{(c)}}\right)\right)\right], \tag{2.72}$$

where $I(y_i = c) = \begin{cases} 1 & y_i = c \\ 0 & y_i \neq c \end{cases}$.

The data vector $H_i$ and the parameter vector $\beta^{(c)}$ are defined as

$H_i = \begin{bmatrix} z_i & w_i & x_i \end{bmatrix}$ and

$$\beta^{(c)} = \begin{bmatrix} \delta^{(c)} \\ \eta^{(c)} \\ \alpha^{(c)} \end{bmatrix}, \; \delta^{(c)} = \begin{bmatrix} \delta_1^{(c)} \\ \vdots \\ \delta_K^{(c)} \end{bmatrix}, \; \eta^{(c)} = \begin{bmatrix} \eta_1^{(c)} \\ \vdots \\ \eta_L^{(c)} \end{bmatrix}, \; \alpha^{(c)} = \begin{bmatrix} \alpha_1^{(c)} \\ \vdots \\ \alpha_G^{(c)} \end{bmatrix}, \; \alpha_g^{(c)} = \begin{bmatrix} \alpha_{g1}^{(c)} \\ \vdots \\ \alpha_{gk_g}^{(c)} \end{bmatrix},$$

$$c = 1, \ldots, C - 1$$

$$\tag{2.73}$$

The goal is to minimize

$$\min_{\beta} \quad l_k(\beta). \tag{2.74}$$

The proximal algorithms summarized in Result 8.4 in the book *Big Data in Omics and Imaging: Association Analysis* can be used to solve the optimization problem. The score function of the node $k$ with multi-class is defined as

$$\text{Score}_D(k) = l_k\left(\hat{\beta}\right). \tag{2.75}$$

Finally, the integer programming that will be discussed in Section 2.6 will be used to estimate the causal genotype–phenotype networks.

---

## 2.5 Other Statistical Models for Quantifying Node Score Function

### 2.5.1 Nonlinear Structural Equation Models

SEMS are often used for modelling causal networks between some given variables, where each variable is expressed as a function of some other variables (its causes) as well as some noise (Nowzohour and Bühlmann 2016). The model consists of three essential components: (1) causal structure, (2) the functional dependencies among causal and effect variables, and (3) the joint distribution of the noises. We often assume that (1) there are no unobserved variables and hence that the noise terms are independent and (2) the difference between the effect variable and some noise term is a deterministic function of the causal variables. The classical approach is to parametrize the model and assume that the functional dependency is linear, and the noise is Gaussian. In Chapter 1, we study linear SEMs. However, as Nowzohour and Bühlmann (2016) pointed out that when the linearity or the Gaussianity assumptions in the model is removed, the model becomes fully identifiable. To overcome this limitation, in this section, we introduce a nonparametric approach to causal inference where we assume functional causal models to represent the effect as a function of direct causes and some unmeasured noises (Zhang et al. 2016) and develop score-based methods for parameter estimation.

### 2.5.1.1 Nonlinear Additive Noise Models for Bivariate Causal Discovery

In this section, we focus on the bivariate causal discovery (Mooij et al. 2016). It is well known that linear regression analysis cannot distinguish the causal relationships between two variables. Only nonlinear functional models can aid in identifying causal relationships between two variables from observed data.

Assume two variables $X$ and $Y$ with joint distribution $P_{X,Y}$. If an external intervention cause changes in some aspect of the system, it will also result in changes in the joint distribution $P_{X,Y}$. Assume no confounding, no selection bias, and no feedback. Our task is to distinguish cause from effect, that is, determine whether $X$ causes $Y$ or $Y$ causes $X$ by using only purely observational data.

Consider a bivariate additive noise model $X \to Y$ where $Y$ is a nonlinear function of $X$ and independent additive noise $E_Y$:

$$Y = f_Y(X) + E_Y$$
$$X \sim P_X, \quad E_Y \sim P_{E_Y},$$
(2.76)

where $X$ and $E_Y$ are independent. The density $P_{XY}$ is said to be induced by the additive noise model $X \to Y$ (Figure 2.6a). If the density $P_{XY}$ is induced by the additive noise model $X \to Y$, but not by the additive noise model $Y \to X$, then the additive noise model is identifiable.

The alternative additive noise model between $X$ and $Y$ is the additive noise model $Y \to X$:

$$X = f_X(Y) + E_X$$
$$Y \sim P_Y, \quad E_X \sim P_{E_X},$$
(2.77)

where $Y$ and $E_X$ are independent.

Now we discuss how to determine cause direction. Recall that $X$ causes $Y$ if $P_{Y \mid do(x)} \neq P_{Y \mid do(x')}$ for some $x, x'$ where $P_{Y \mid do(x)}$ is the resulting distribution of $Y$ when the variable $X$ is enforced to have the value $x$ (do $(X = x)$). Figure 2.6 shows two causal structures: $X$ causes $Y$ (Figure 2.6a) and $Y$ causes $X$. When $X$ causes $Y$, changing $X$ will cause changes in the distribution of $Y$, the interventional distribution $P_{Y \mid do(x)}$ will be equal to the conditional distribution $P_{Y \mid X}$, but will not be equal to the marginal distribution $P_Y$. Also, we

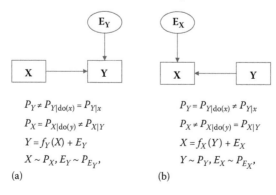

$$P_Y \neq P_{Y \mid do(x)} = P_{Y \mid x}$$
$$P_X = P_{X \mid do(y)} \neq P_{X \mid Y}$$
$$Y = f_Y(X) + E_Y$$
$$X \sim P_X, E_Y \sim P_{E_Y},$$
(a)

$$P_Y = P_{Y \mid do(x)} \neq P_{Y \mid x}$$
$$P_X \neq P_{X \mid do(y)} = P_{X \mid Y}$$
$$X = f_X(Y) + E_X$$
$$Y \sim P_Y, E_X \sim P_{E_X},$$
(b)

**FIGURE 2.6**
Two additive noise models correspond to two causal structures. (a) The additive noise model $X \to Y$. (b) The additive noise model $Y \to X$.

know that when $X$ causes $Y$, enforcing $Y$ to take a specific value $y$ will not affect the distribution of $X$. Therefore, in this case, the intervention distribution $P_{X|do(Y)}$ will be equal to the marginal distribution $P_X$, but will not be equal to the conditional distribution $P_{X|y}$. The same arguments apply to the case where $Y$ causes $X$ (Figure 2.6b).

Empirically, if the additive noise model $X \rightarrow Y$ fits the data, then we infer that $X$ causes $Y$, or if the additive noise model $Y \rightarrow X$ fits the data, then $Y$ causes $X$ will be concluded. Although this statement cannot be rigorously proved, in practice, this principle will provide the basis for bivariate cause discovery (Mooij et al. 2016). To implement this principal, we need to develop statistical methods for assessing whether the additive noise model fits the data. The following result offers a rigorous statistical method for assessing the model fitting by testing the independence between residuals and causal variable (Mooij et al. 2016).

### Result 2.2

Let $P(x, y)$ be a joint density of two random variables $X$ and $Y$. Assume that the conditional expectation $E(Y | X)$ exists. Then, $P(x, y)$ is induced by a bivariate additive noise model $X \rightarrow Y$ if and only if the residuals $E_Y = Y - E(Y | X)$ have a finite mean and are independent of $X$.

Result 2.2 can be easily proved. For the self-contained of the book, we rephrase the proof of Mooij et al. (2016) here. In fact, if the density $P(x, y)$ is induced by the model (2.76), then we have

$$Y = f(x) + e, \tag{2.78}$$

where $e \perp\!\!\!\perp X$.

Taking conditional expectation on both sides of Equation 2.78 gives the conditional expectation:

$$E(Y|X) = f(x) + E(e), \tag{2.79}$$

Therefore, we obtain $Y - E(Y | X) = e - E(e) = E_Y$. Since $e \perp\!\!\!\perp X$, then $e - E(e) \perp\!\!\!\perp X$, and hence $E_Y \perp\!\!\!\perp X$.

Then, conversely, we denote $f_Y(x) = E(Y | X)$. Then, we have

$$Y = f_Y(x) + E_Y, \tag{2.80}$$

where $E_Y \perp\!\!\!\perp X$. Model 2.80 is exactly the model 2.76. The density $P(x,y) = P(x)p(E_Y)$ is induced by the bivariate additive noise model $X \rightarrow Y$.

Result 2.2 provides a general procedure for bivariate causal discovery. Nonparametric regression can be used to approximate the conditional expectation

$f_Y(x) = E(Y \mid X)$ and the Hilbert–Schmidt independence criterion (HSIC) can be used to test the independence between the residuals and causal variable.

Divide the dataset into a training dataset $D_{train} = \{Y_n, X_n\}$, $Y_n = [y_1, \ldots, y_n]^T$, $X_n = [x_1, \ldots, x_n]^T$ for fitting the model and a test dataset $D_{test} = \{\tilde{Y}_m, \tilde{X}_m\}$, $\tilde{Y}_m = [\tilde{y}_1, \ldots, \tilde{y}_m]^T$, $\tilde{X}_m = [\tilde{x}_1, \ldots, \tilde{x}_m]^T$ for testing the independence, where $n$ is not necessarily equal to $m$. There are many non-parametric methods that can be used to regress $Y$ on $X$ or regress $X$ on $Y$. In this section, we primarily introduce smoothing spline regression methods (Appendix 2.A) and consider model (2.76). The model (2.77) can be similarly dealt with. Let $L$ be a continuous linear functional defined on a RKHS. For example, $L$ can be an evaluational functional $L_x$ defined as

$$L_x f = f(x).$$

We consider a general smoothing spline regression model:

$$y_i = L_i f + \varepsilon_i, \tag{2.81}$$

$$i = 1, 2, \ldots, n,$$

where $y_i$ is the observed value of the variable $Y$ of the $i$th individual, $L_i$ is a continuous functional defined on $H$, $f$ represents a general unknown function $f_Y$, and $\varepsilon_i$ represents $E_Y$ and are zero-mean independent random errors with a common variance $\sigma_e^2$. If $L_i$ is an evaluational functional, then Equation 2.81 is reduced to

$$y_i = f(x_i) + \varepsilon_i,$$

$$i = 1, 2, \ldots, n.$$

The algorithm for smoothing spline regression is defined as follows.

Step 1: Select the penalty parameter $\lambda$.

Step 2: Compute the matrices

$$T = \begin{bmatrix} L_{1(x)}\varphi_1(x) & \cdots & L_{1(x)}\varphi_p(x) \\ \vdots & \vdots & \vdots \\ L_{n(x)}\varphi_1(x) & \cdots & L_{n(x)}\varphi_p(x) \end{bmatrix} \text{ and}$$

$$\Sigma = \begin{bmatrix} L_{1(x)}L_{1(z)}R_1(x,z) & \cdots & L_{1(x)}L_{n(z)}R_1(x,z) \\ \vdots & \vdots & \vdots \\ L_{n(x)}L_{1(z)}R_1(x,z) & \cdots & L_{n(x)}L_{n(z)}R_1(x,z) \end{bmatrix}.$$

If the functional is evaluational functional, then the matrices $T$ and $\Sigma$ are

$$T = \begin{bmatrix} \varphi_1(x_1) & \cdots & \varphi_p(x_1) \\ \vdots & \vdots & \vdots \\ \varphi_1(x_n) & \cdots & \varphi_p(x_n) \end{bmatrix} \text{ and } \Sigma = \begin{bmatrix} R_1(x_1, z_1) & \cdots & R_1(x_1, z_n) \\ \vdots & \vdots & \vdots \\ R_1(x_n, z_1) & \cdots & R_1(x_n, z_n) \end{bmatrix},$$

where $\{z_n\} = \{x_n\}$ are observed values of $X$.
Step 3: Perform QR decomposition of the matrix $T$:

$$T = [Q_1 \ Q_2] \begin{bmatrix} R \\ 0 \end{bmatrix}.$$

Step 4: Compute coefficients of the smoothing spline regression

$$\hat{a} = R^{-1}Q_1^T \left[ I - MQ_2 (Q_2^T M Q_2)^{-1} Q_2^T \right] Y \text{ and } \hat{b} = Q_2 (Q_2^T M Q_2)^{-1} Q_2^T Y,$$

where $M = \Sigma + n\lambda I$.
Step 5: Compute the smoothing spline regression function

$$\hat{f}(x) = \sum_{j=1}^p \hat{a}_j \varphi_j(x) + \sum_{i=1}^n \hat{b}_i \xi_i(x),$$

where $\xi_i(x) = L_{i(z)}R_1(x, z)$.
Step 6. Compute the fitted value:

$$\hat{f} = H(\lambda)Y,$$

where $H(\lambda) = I - n\lambda Q_2 (Q_2^T M Q_2)^{-1} Q_2^T$.

The basis functions and RKs $R_1(x, z)$ of the cubic spline based on the classic polynomials are

$$\phi_1(x) = 1, \ \phi_2(x) = x \quad \text{and} \quad (x \wedge z)^2 \frac{3(x \vee z) - x \wedge z}{6},$$

where $x \wedge z = \min(x, z)$ and $x \vee z = \max(x, z)$.
The basis functions and RKs $R_1(x, z)$ of the cubic spline based on Bernoulli polynomials are, respectively, given by

$$\phi_1(x) = 1, \ \phi_2(x) = k_1(x), \text{ and}$$
$$R_1(x, z) = k_2(x)k_2(z) - k_4(|x - z|),$$

where

$$k_0(x) = 1,$$

$$k_1(x) = x - \frac{1}{2},$$

$$k_2(x) = \frac{1}{2}\left\{ k_1^2(x) - \frac{1}{12} \right\},$$

$$k_4(x) = \frac{1}{24}\left\{ k_1^4(x) - \frac{1}{2}k_1^2(x) + \frac{7}{240} \right\}.$$

In Section 5.3.2, we discuss covariance operator and dependence measures. A covariance operator can measure the magnitude of dependence and is a useful tool for assessing dependence between variables. Specifically, we will use the norm of the Hilbert–Schmidt norm of the cross-covariance operator or its approximation, the Hilbert–Schmidt independence criterion (HSIC) to measure the degree of dependence between the residuals and potential causal variable.

Now we calculate the HSIC. It consists of the following steps.

Step 1: Using test data set to compute the following:

Compute the smoothing spline regression function

$$\hat{f}(x_i) = \sum_{j=1}^{p} \hat{a}_j \varphi_j(x_i) + \sum_{i=1}^{n} \hat{b}_i \xi_i(x_i), i = 1, ..., m.$$

where $\xi_i(x) = L_{i(z)} R_1(x, z)$.

Step 2: Compute the residuals:
Compute the residuals:

$$\varepsilon = E_Y = Y - \hat{f},$$                                       (2.82)

where $\varepsilon_i = E_Y(i) = y_i - \hat{f}_i,\ \ i = 1, ..., m$.

Step 3: Select two kernel functions $k_E(\varepsilon_i, \varepsilon_j)$ and $k_x(x_1, x_2)$. Compute the Kernel matrices.

$$K_{E_Y} = \begin{bmatrix} k_E(\varepsilon_1, \varepsilon_1) & \cdots & K_E(\varepsilon_1, \varepsilon_m) \\ \vdots & \vdots & \vdots \\ K_E(\varepsilon_m, \varepsilon_1) & \cdots & K_E(\varepsilon_m, \varepsilon_m) \end{bmatrix},$$

$$K_X = \begin{bmatrix} k_x(x_1, x_1) & \cdots & k_x(x_1, x_m) \\ \vdots & \vdots & \vdots \\ k_x(x_m, x_1) & \cdots & k_x(x_m, x_m) \end{bmatrix}.$$

Step 4: Compute the Hilbert–Schmidt independence criterion for measuring dependence between the residuals and potential causal variable.

$$HSIC^2(E_Y, X) = \frac{1}{m^2} \mathrm{Tr}\ (K_{E_Y} H K_X H),$$

where $H = I - \frac{1}{m} \mathbf{1}_m \mathbf{1}_m^T$, $\mathbf{1}_m = [1, 1, ..., 1]^T$ and $\mathrm{Tr}$ denotes trace of the matrix.

The general procedure for bivariate causal discovery is

1. Divide a data set into a training data set and a test data set;
2. Use the training data set to estimate the regression functions in both directions: $X \rightarrow Y$ or $Y \rightarrow X$;

3. Compute the corresponding residuals;
4. Estimate the dependence measure $HSIC^2$ ($E_Y$, $X$) and $HSIC^2$ ($E_X$, $Y$);
5. Infer the direction that has the lowest dependence measure as causal direction.

The specific algorithms for bivariate causal discovery are summarized as follows (Mooij et al. 2016):

Step 1: Divide a data set into a training data set $D_{train} = \{Y_n, X_n\}$ for fitting the model and a test data set $D_{test} = \{\tilde{Y}_m, \tilde{X}_m\}$ for testing the independence.

Step 2: Use the training data set and smoothing spline.

    a. to regress $Y$ on $X$ : $Y = f_Y(x) + E_Y$ and
    b. to regress $X$ on $Y$ : $X = f_X(y) + E_X$.

Step 3: Use the test data set and estimated smoothing spline regressions to predict residuals:

    a. $\hat{E}_{Y_X} = \tilde{Y} - \hat{f}_Y(\tilde{X})$
    b. $\hat{E}_{X_Y} = \tilde{X} - \hat{f}_X(\tilde{Y})$.

Step 4: Calculate the dependence measures $HSIC^2$ ($E_Y$, $X$) and $HSIC^2$ ($E_X$, $Y$).
Step 5: Infer causal direction:

$$X \rightarrow Y \ \text{if} \ HSIC^2(E_Y, X) < HSIC^2(E_X, Y);$$

$$Y \rightarrow X \ \text{if} \ HSIC^2(E_Y, X) > HSIC^2(E_X, Y).$$

If $HSIC^2$ ($E_Y$, $X$) = $HSIC^2$ ($E_X$, $Y$), then causal direction is undecided.

### 2.5.1.2 Nonlinear Structural Equations for Causal Network Discovery

The functional models, also known as structural causal models, or non-parametric-structural equation models (Mooij et al. 2016) discussed in Section 2.5.1.1 can be extended to multivariate (Nowzohour and Buhlman 2016). Consider $M$ continuous endogenous variables $Y_1,..., Y_M$ and $K$ exogenous variables $X_1,..., X_K$. Let $pa_D(d)$ be the parent set of the node $d$ including both endogenous and exogenous variables. Consider a nonlinear structural equation model:

$$Y_d = f_d\left(Y_i \in pa_D(d), X_j \in pa_D(d)\right) + \varepsilon_d, d = 1, ..., M, \tag{2.83}$$

where $f_d$ is a nonlinear function whose forms are in general unknown and the errors $\varepsilon_d$ are independent and follow distribution $P_{\varepsilon_d}$. The model (2.83) is also called an additive noise model (ANM).

The joint density or the likelihood function of the ANM (2.83) with respect to a DAG $D$ is then given by

$$L = \prod_{d=1}^{M} P_{\varepsilon_d}\left(y_d - f_d\left(y_i \in pa_D(d), x_j \in pa_D(d)\right)\right). \tag{2.84}$$

The likelihood function $L$ depends on the parent sets of the node or the structure of the DAG. The nonlinear functions can be either parametric functions or nonparametric functions. For the convenience of discussion, we denote $W_l = (Y_i, X_j) \in pa_D(d)$, that is, $W_l$ is the set of parents of the node $d$ including both endogenous and exogenous variables. Equation (2.83) is then reduced to

$$Y_d = f_d(W_l \in pa_D(d)) + \varepsilon_d. \tag{2.85}$$

Since the functional form $f_d$ is, in general, unknown, we often take a non-parametric approach. Here, we mainly introduce smoothing spline to approximate the function $f_d$ (Wang 2011).

The multivariate function $f_d(W)$ is defined indirectly through a linear functional (Appendix 2.A). Consider a RKHS $H$. Let $L$ be a continuous linear functional defined on the RKHS $H$. Consider a general multivariate nonlinear function model:

$$Y_d^i = L_i f_d + \varepsilon_i, \tag{2.86}$$

$$i = 1, \ldots, n,$$

where $Y_d^i$ is the observed $Y_d$ in the $i$th sample, $L_i$ are continuous functionals, and $\varepsilon_i$ are zero-mean independent random errors with common variance $\sigma_e^2$. Consider $p$ variables $\{w_1, \ldots, w_p\}$ and the tensor product $M = H = H^{(1)} \otimes H^{(2)} \otimes \ldots \otimes H^{(p)}$ on domain $\chi = \chi_1 \times \chi_2 \times \ldots \times \chi_p$. Each $H^{(k)}$ can be decomposed into

$$H^{(k)} = H_0^{(k)} \oplus H_1^{(k)} \oplus \ldots \oplus H_{r_k-1}^{(k)} \oplus H_{*1}^{(k)}, \quad k = 1, \ldots, p.$$

Now we group the model space $M$ into two subspaces:

$$H_0^* = H^0 \text{ and } H_1^* = H^1 \oplus \ldots \oplus H^q,$$

where

$$H^0 = \sum_{j_1=0}^{r_1-1} \ldots \sum_{j_p=0}^{r_p-1} H_{j_1}^{(1)} \otimes \ldots \otimes H_{j_p}^{(p)}$$

which is a finite dimensional space including all functions that will not be penalized and $H^1, \ldots, H^q$ are orthogonal RKHS's with RKs $R^1, \ldots, R^q$. The inner product and RK defined on $H_0^0$ will be used for $H_0^*$. Now define the inner product and RK for the RKHS $H_1^*$. The RK for the $H_1^*$ is defined as

$$R_1^* = \sum_{j=1}^{q} \theta_j R^j,$$

where $R^j$ is the RK for the $H^j$.

Let $\varphi_0^{(k)}, \ldots, \varphi_{r_k-1}^{(k)}$ be the set of basis functions for the space $H_0^{(k)}$. Consider all possible combinations of basis functions for $H^0$:

$$\left\{ \varphi_1^{(1)} \oplus \ldots \oplus \varphi_{r_1}^{(1)} \right\} \ldots \left\{ \varphi_1^{(p)} \oplus \ldots \oplus \varphi_{r_p}^{(p)} \right\} = \sum_{j_1=0}^{r_1-1} \ldots \sum_{j_p=0}^{r_p-1} \varphi_{j_1}^{(1)} \ldots \varphi_{j_p}^{(p)} \tag{2.87}$$

$$= \phi_1 + .. + \phi_r.$$

where $r = r_1 \ldots r_p$ and $\phi_v, v = 1, \ldots, r$ is one of elements $\varphi_{j_1}^{(1)} \ldots \varphi_{j_p}^{(p)}$. If $r_1 = \ldots = r_p = 2$ then $r = 2^p$.

In this section, for simplicity we consider only cubic spline. In this scenario, the tensor product space is

$$H = \left\{ H_0^{(1)} \oplus H_1^{((1))} \oplus H_2^{(1)} \right\} \otimes \left\{ H_0^{(2)} \oplus H_1^{((2))} \oplus H_2^{(2)} \right\} \otimes \ldots$$

$$\otimes \left\{ H_0^{(p)} \oplus H_1^{((p))} \oplus H_2^{(p)} \right\}$$

$$= \left\{ H_0^{(1)} \otimes H_0^{(2)} \otimes \ldots \otimes H_0^{(p)} \right\} \oplus \left\{ H_1^{(1)} \otimes H_0^{(2)} \otimes \ldots \otimes H_0^{(p)} \right\} \oplus \ldots$$

$$\oplus \left\{ H_1^{(1)} \otimes H_1^{(2)} \otimes \ldots \otimes H_1^{(p-1)} \otimes H_0^{(p)} \right\}$$

$$\oplus \left\{ H_2^{(1)} \otimes H_0^{(2)} \otimes \ldots \otimes H_0^{(p)} \right\} \oplus .. \oplus \left\{ H_2^{(1)} \otimes H_2^{(2)} \otimes \ldots H_2^{(p)} \right\}$$

$$= H^{(0)} \oplus H^{(1)} \oplus \ldots H^{(q)},$$

where

$$H^{(0)} = \left\{ H_0^{(1)} \otimes H_0^{(2)} \otimes \ldots \otimes H_0^{(p)} \right\} \oplus \left\{ H_1^{(1)} \otimes H_0^{(2)} \otimes \ldots \otimes H_0^{(p)} \right\} \oplus \ldots$$

$$\oplus \left\{ H_1^{(1)} \otimes H_1^{(2)} \otimes \ldots \otimes H_1^{(p-1)} \otimes H_1^{(p)} \right\}, q = 3^p - 2^p.$$

Consider $p = 3$, cubic spline and Bernoulli polynomials. Then, the basis functions are given by

$$\phi_1(x_1, x_2, x_3) = 1, \quad \phi_2(x_1, x_2, x_3) = x_1 - 0.5, \phi_3(x_1, x_2, x_3)$$

$$= x_2 - 0.5, \quad \phi_4(x_1, x_2, x_3) = x_3 - 0.5,$$

$$\phi_5(x_1, x_2, x_3) = (x_1 - 0.5)(x_2 - 0.5), \quad \phi_6(x_1, x_2, x_3)$$

$$= (x_1 - 0.5)(x_3 - 0.5), \quad \phi_7(x_1, x_2, x_3) = (x_2 - 0.5)(x_3 - 0.5),$$

$$\phi_3(x_1, x_2, x_3) = (x_1 - 0.5)(x_2 - 0.5)(x_3 - 0.5).$$

Next consider RKs $R^j$. RKs $R^j$ is the product of individual RKs of $H_{j_1}^{(1)}, \ldots, H_{j_p}^{(p)}$, where $R_0^{(l)} = 1$, $R_1^{(l)} = k_1(x_l)k_1(z_l)$ and $R_2^{(l)} = k_2(x_l)k_2(z_l) - k_4(|x_l - z_l|)$ are RKs corresponding the RKHSs $H_0^{(l)}$, $H_1^{(l)}$ and $H_2^{(l)}$, respectively. For example, $R^1$ of $H^1 = H_2^{(1)} \otimes H_0^{(2)} \otimes H_0^{(3)}$ is $K_2(x_1)k_2(z_1) - k_4(|x_1 - z_1|)$, $R^4$ of $H^4 = H_2^{(1)} \otimes H_1^{(2)} \otimes H_0^{(3)}$ is

$$(K_2(x_1)k_2(z_1) - k_4(|x_1 - z_1|))k_1(x_2)k_1(z_2).$$

To estimate smoothing splines in regression (2.86), we minimize

$$\min_{f \in M} \ \frac{1}{n}\sum_{i=1}^{n}\left(y_d^i - L_i f\right)^2 + \lambda \|P_1^* f\|_*^2, \tag{2.88}$$

where $\lambda \|P_1^* f\|_*^2 = \sum_{j=1}^{q} \lambda_j \|P_j f\|^2$, $\lambda_j = \dfrac{\lambda}{\theta_j}$ and $P_j$ is the orthogonal projection of the function onto the RKHS $H^j$, $j = 0, 1, \ldots, q$.

The algorithm for solving optimization problem (2.88) is

Step 1: Select the penalty parameter $\lambda$. Define $Y_d = [Y_d^1, \ldots, Y_d^n]^T$.

Step 2: Compute the matrices

$$T = \begin{bmatrix} L_{1(x)}\phi_1(x) & \cdots & L_{1(x)}\phi_r(x) \\ \vdots & \vdots & \vdots \\ L_{n(x)}\phi_1(x) & \cdots & L_{n(x)}\phi_r(x) \end{bmatrix},$$

$$\Sigma_j = \begin{bmatrix} L_{1(x)}L_{1(z)}R^j(x,z) & \cdots & L_{1(x)}L_{n(z)}R^j(x,z) \\ \vdots & \vdots & \vdots \\ L_{n(x)}L_{1(z)}R^j(x,z) & \cdots & L_{n(x)}L_{n(z)}R^j(x,z) \end{bmatrix} \quad \text{and}$$

$\Sigma_\theta = \theta_1 \Sigma_1 + \ldots + \theta_q \Sigma_q$, where $\theta_1, \ldots, \theta_q$ are pre-determined weights. For the evaluational functional, $T$ and $\Sigma_j$ can be calculated by

$$T = \begin{bmatrix} \phi_1(x_1) & \cdots & \phi_r(x_1) \\ \vdots & \vdots & \vdots \\ \phi_1(x_n) & \cdots & \phi_r(x_n) \end{bmatrix} \text{and } \Sigma_j = \begin{bmatrix} R^j(x_1,z_1) & \cdots & R^j(x_1,z_n) \\ \vdots & \vdots & \vdots \\ R^j(x_n,z_1) & \cdots & R^j(x_n,z_n) \end{bmatrix}.$$

Step 3: Perform QR decomposition of the matrix $T$:

$$T = [Q_1 \ Q_2] \begin{bmatrix} R \\ 0 \end{bmatrix}.$$

Step 4: Compute coefficients of the smoothing spline regression

$$\hat{a} = R^{-1}Q_1^T\left[I - MQ_2\left(Q_2^T MQ_2\right)^{-1}Q_2^T\right]Y_d \text{ and } \hat{b} = Q_2\left(Q_2^T MQ_2\right)^{-1}Q_2^T Y_d,$$

where $M = \Sigma + n\lambda I$.

Step 5: Compute the smoothing spline regression function

$$\hat{f}(\mathbf{x}) = \sum_{j=1}^{r} \hat{a}_j \phi_j(\mathbf{x}) + \sum_{v=1}^{n} \hat{b}_v \sum_{j=1}^{q} \theta_j L_{v(z)} R^j(\mathbf{x,z}).$$

Step 6: Compute the fitted value:

$$\hat{f} = H(\lambda)Y_d,$$

where $H(\lambda) = I - n\lambda Q_2 (Q_2^T M Q_2)^{-1} Q_2^T$.

Step 7: Calculate the score of the node $d$:

$$\text{Score}_d = \frac{1}{n}||Y_d - T\hat{a} - \Sigma_\theta \hat{b}||^2 + \hat{b}^T \Sigma_\theta \hat{b}. \qquad (2.89)$$

The total score of the nodes for a causal network with $M$ nodes is

$$\text{Score}(D) = \sum_{d=1}^{M} \text{Score}_d. \qquad (2.90)$$

Consider $N$ candidate GAGs $D_1,...,D_N$. The true causal graph $D_*$ will be found by minimizing the total score:

$$D_* = \underset{D_1,...,D_N}{\arg\min} \ \text{Score}(D_i) \qquad (2.91)$$

or by integer programming that will be discussed in Section 2.6.

### 2.5.2 Mixed Linear and Nonlinear Structural Equation Models

In previous sections, we discussed linear structural equation models and nonlinear structural equation models. In this section, we will introduce linear structural equation models with Gaussian noise or mixed linear and nonlinear structural equation models (Ernest et al. 2016). In general, some edge functions in the network may be linear, but other edge functions may be nonlinear. The mixed linear and nonlinear structural equation models are the most generous causal models. For example, Figure 2.7 shows the mixed linear and nonlinear structural equation model where solid lines represent the nonlinear edges and dashed lines represent the linear edges.

The general nonlinear SEMs in Equation 2.77 can be written as (Ernest et al. 2016)

$$Y_j = \mu_j + \sum_{i \in pa_D(j)} f_{ji}(W_i) + \varepsilon_j, \qquad (2.92)$$

where $\mu_j \in R$, $f_{ji}$ is twice differentiable, $E[f_{ji}(W_i)] = 0$, and $\varepsilon_j \sim N(0, \sigma_j^2)$, $\sigma_j^2 > 0$, $j = 1, 2, ..., p$. The functions $f_{ji}$ that are associated with the directed edge $i \to j$ in the graph $D$ can be either linear or nonlinear. If the function $f_{ji}$ is linear, then

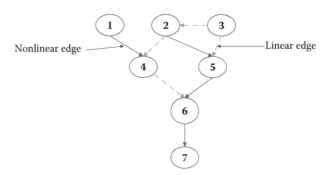

**FIGURE 2.7**
Illustration of mixed linear and nonlinear structural equation models.

the edge $i \rightarrow j$ is called a linear edge, otherwise, it is called a nonlinear edge. As Ernest et al. (2016) pointed out that the (non-) linearity of an edge is defined with respect to a specific DAG $D$.

The total causal effects can be decomposed into linear causal effects and nonlinear causal effects. The model (2.92) can be rewritten as

$$y_j = \mu_j + \sum_{l=1}^{l_j} \alpha_{jl} w_l + \sum_{i=1}^{i_j} f_{ji}(w_i) + \varepsilon_j, \tag{2.93}$$

where $l_j$ denotes the number of parents of the node $j$ with linear edge connections and $i_j$ denotes the number of parents of the node $j$ with nonlinear edge connections. For example, the functional model of the node 4 is

$$y_4 = \mu_4 + \alpha_{41} y_1 + f_{42}(y_2) + \varepsilon_4.$$

In general, the functions $f_{ji}$ are unknown and will be approximated by nonparametric functions. Again, in this section smoothing splines are used to represent the nonlinear functions $f_{ji}$. Consider the smoothing spline regression model (Appendix 2.A):

$$y_j = \mu_j + \sum_{l=1}^{l_j} \alpha_{jl} w_l + \sum_{i=1}^{i_j} L_{ji} f_{ji} + \varepsilon_j, \tag{2.94}$$

where $L_{ji}$ is a continuous functional defined on RKHS $H^i$, $f_{ji}$ represents a general unknown function, and $\varepsilon_i$ are zero-mean independent random errors with a common variance $\sigma_j^2$. The *Hilbert space* $H^i$ can be decomposed into two subspaces (Wang 2011):

$$H^i = H_0^i \oplus H_1^i,$$

where $H_0^i$ is a finite dimensional space with orthonormal basis functions $\varphi_1^i(x), \ldots, \varphi_p^i(x)$ and its orthogonal complement $H_1^i$ that is an RKHS with reproducing kernel (RK) $R_1^i(x, z)$.

To estimate the regression coefficients $\alpha_{jl}$ function $f_{ji}$ in Equation 2.94, we should make both the errors between the observations $Y_j$ and the estimators $\hat{\alpha}_{ji}$ and $\hat{f}_{ji}$, and departure from the smoothness as small as possible. Let the projection $P_1 f_{ji}$ of the function $f_{ji}$ on the RKHS $H_1$ be the penalty functions. Therefore, to achieve this goal, the objective function for the regression coefficient estimators and smoothing spline estimators of the function $f_{ji}$ is given by

$$\frac{1}{n} \sum_{v=1}^{n} \left( y_j^v - \sum_{l=1}^{l_j} \alpha_{jl} w_l^v - \sum_{i=1}^{i_j} L_v f_{ji} \right)^2 + \lambda \left\| \sum_{i=1}^{i_j} P_1 f_{ji} \right\|^2, \qquad (2.95)$$

where $\lambda$ is a smoothing penalty parameter to balance the goodness of fit and the smoothness. Let $R^i(w, z)$ be a reproducing kernel associated with $f_{ji}$, which can be decomposed into

$$R^i(w, z) = R_0^i(w, z) + R_1^i(w, z),$$

where

$$R_0^i(w, z) = \sum_{m=1}^{p} \varphi_m^i(w) \varphi_m^i(z). \qquad (2.96)$$

Assume that the estimator $\hat{f}_{ji}$ is expressed as

$$\hat{f}_{ji}(w) = \sum_{m=1}^{p} a_m^i \varphi_m^i(w) + \sum_{u=1}^{n} b_u^i \xi_u^i(w), \qquad (2.97)$$

where $\xi_u^i(w) = L_{u(z)} R_1^i(w, z)$.

Therefore, we have

$$L_{v(w)} \hat{f}_{ji}(w) = \sum_{m=1}^{p} a_m^i L_{v(w)} \varphi_m^i(w) + \sum_{u=1}^{n} b_u^i L_{v(w)} L_{u(z)} R_1^i(w, z) \qquad (2.98)$$

or

$$\begin{bmatrix} L_{1(w)} \hat{f}_{ji}(w) \\ \vdots \\ L_{n(x)} \hat{f}(x) \end{bmatrix} = \begin{bmatrix} L_{1(w)} \varphi_1^i(w) & \cdots & L_{1(w)} \varphi_p^i(w) \\ \vdots & \vdots & \vdots \\ L_{n(w)} \varphi_1^i(w) & \cdots & L_{n(w)} \varphi_p^i(w) \end{bmatrix} \begin{bmatrix} a_1^i \\ \vdots \\ a_p^i \end{bmatrix}$$

$$+ \begin{bmatrix} L_{1(w)} L_{1(z)} R_1^i(x, z) & \cdots & L_{1(w)} L_{n(z)} R_1^i(x, z) \\ \vdots & \vdots & \vdots \\ L_{n(w)} L_{1(z)} R_1^i(x, z) & \cdots & L_{n(w)} L_{n(z)} R_1^i(x, z) \end{bmatrix} \begin{bmatrix} b_1^i \\ \vdots \\ b_n^i \end{bmatrix}$$

$$= T^i a^i + \Sigma^i b^i,$$

where

$$T^i = \begin{bmatrix} L_{1(w)}\varphi_1^i(w) & \cdots & L_{1(w)}\varphi_p^i(w) \\ \vdots & \vdots & \vdots \\ L_{n(w)}\varphi_1^i(w) & \cdots & L_{n(w)}\varphi_p^i(w) \end{bmatrix},$$

$$a^i = \begin{bmatrix} a_1^i \\ \vdots \\ a_p^i \end{bmatrix}, \Sigma^i = \begin{bmatrix} L_{1(w)}L_{1(z)}R_1^i(x,z) & \cdots & L_{1(w)}L_{n(z)}R_1^i(x,z) \\ \vdots & \vdots & \vdots \\ L_{n(w)}L_{1(z)}R_1^i(x,z) & \cdots & L_{n(w)}L_{n(z)}R_1^i(x,z) \end{bmatrix} \text{ and } b^i = \begin{bmatrix} b_1^i \\ \vdots \\ b_n^i \end{bmatrix}.$$

Let

$$Y_j = \begin{bmatrix} y_j^1 \\ \vdots \\ y_j^n \end{bmatrix}, W = \begin{bmatrix} w_1^1 & \cdots & w_{l_j}^1 \\ \vdots & \vdots & \vdots \\ w_1^n & \cdots & w_{l_j}^n \end{bmatrix}, \alpha^j = \begin{bmatrix} \alpha_{j1} \\ \vdots \\ \alpha_{jl_j} \end{bmatrix}, a = \begin{bmatrix} a^1 \\ \vdots \\ a^{i_j} \end{bmatrix}, b = \begin{bmatrix} b^1 \\ \vdots \\ b^{i_j} \end{bmatrix},$$

$$\gamma = \begin{bmatrix} \alpha \\ a \end{bmatrix}, T = \begin{bmatrix} T^1 & \cdots & T^{i_j} \end{bmatrix}, A = \begin{bmatrix} W & T \end{bmatrix} \text{ and } \Sigma = \begin{bmatrix} \Sigma^1 & \cdots & \Sigma^{i_j} \end{bmatrix}.$$

It can also be shown that

$$|\sum_{i=1}^{i_j} P_1 f_{ji}|^2 = b^T \Sigma b. \tag{2.99}$$

In a matrix form, Equation 2.95 can be reduced as

$$\min_{\gamma,b} \quad \frac{1}{n}||Y_j - A\gamma - \Sigma b||^2 + \lambda b^T \Sigma b. \tag{2.100}$$

The solution to the optimization problem (2.100) is (Appendix 2.A)

$$A\gamma + (\Sigma + n\lambda I)b = Y_j$$
$$A^T b = 0 \tag{2.101}$$

Perform QR decomposition of the matrix $A$:

$$T = \begin{bmatrix} Q_1 & Q_2 \end{bmatrix} \begin{bmatrix} R \\ 0 \end{bmatrix}.$$

Finally, we obtain the coefficients of the smoothing spline regression

$$\hat{\gamma} = R^{-1}Q_1^T \left[ I - MQ_2(Q_2^T MQ_2)^{-1}Q_2^T \right] Y_j \text{ and } \hat{b} = Q_2(Q_2^T MQ_2)^{-1}Q_2^T Y_j, \tag{2.102}$$

where $M = \Sigma + n\lambda I$.

Now we define the node score for the parent set with linear and nonlinear edge connections. Suppose that the number of parents of the node $j$ is $m_j$. Consider the directed edge $i \rightarrow j$. Define the indicator variable for the linear edge:

$$\pi_i = \begin{cases} 1 & \text{edge } i \rightarrow j \text{ is linear} \\ 0 & \text{edge } i \rightarrow j \text{ is nonlinear.} \end{cases} \tag{2.103}$$

The functional model (2.93) can be rewritten as

$$y_j = \mu_j + \sum_{li=1}^{m_j} \pi_i \alpha_{jl} w_l + \sum_{i=1}^{im_j}(1 - \pi_i)f_{ji}(w_i) + \varepsilon_j, \tag{2.104}$$

To fit the model (2.104) to the data, we need to minimize the following penalized objective function:

$$\frac{1}{n}\sum_{v=1}^{n}\left(y_j^v - \sum_{i=1}^{m_j}\pi_i \alpha_{jl} w_l^v - \sum_{i=1}^{m_j}(1 - \pi_i)L_v f_{ji}\right)^2 + \lambda\left\|\sum_{i=1}^{m_j}(1 - \pi_i)P_1 f_{ji}\right\|^2. \tag{2.105}$$

Expanding the continuous functional $L_v$, Equation 2.105 can be further reduced to

$$\frac{1}{n}\sum_{v=1}^{n}\left\{y_j^v - \sum_{i=1}^{m_j}\pi_i \alpha_{ji} w_i^v - \sum_{i=1}^{m_j}(1 - \pi_i)\right.$$
$$\left.\left[\sum_{q=1}^{p}a_q^i L_{v(w)}\,\varphi_q^i(w) + \sum_{u=1}^{n}b_u^i L_{v(w)} L_{v(z)} R_1^i(w,z)\right]\right\}^2 \tag{2.106}$$
$$+\lambda\left\|\sum_{i=1}^{m_j}(1 - \pi_i)P_1 f_{ji}\right\|^2.$$

In a matrix form, the optimization problem (2.106) can be rewritten as

$$\min_{\pi,\alpha,a,b}\ \frac{1}{n}\|Y_j - W\pi\alpha - T(I - \pi)a - \Sigma(I - \pi)b\|^2 + \lambda b^T(I - \pi)\Sigma(I - \pi)b, \tag{2.107}$$

where $\pi = \text{diag}\,(\pi_1, \dots, \pi_{m_j\backslash})$ and $I$ is an identity matrix, others are defined as before. In the optimization problem (2.107), the elements of the diagonal matrix $\pi$ take values of 0 and 1, the vectors $\alpha$, $a$, and $b$ take real numbers. The optimization problem (2.107) are mixed integer programming problems. Its solution will be discussed in Section 2.6. After the optimization problem is solved, the score function of the node $j$ is defined as

$$\text{Score}\,(j,\ pa_j) = \frac{1}{n}\|Y_j - W\pi\alpha - T(I - \pi)a - \Sigma(I - \pi)b\|^2$$
$$+ \lambda b^T(I - \pi)\Sigma(I - \pi)b. \tag{2.108}$$

### 2.5.3 Jointly Interventional and Observational Data for Causal Inference

Causal network reconstruction uses two types of data: observational and experimental data. In many cases, the randomized experiments are expensive, unethical, or technically infeasible, we often estimate the Markov equivalence class of DAGs that all follow the same conditional independence from the observational data. However, the gold standard for causal inference is interventional experiments. An intervention is to force the value of one or several random variables of the system to designed values. For example, consider two variables. If one variable $X$ directly causes the change of another variable $Y$ and intervention on $X$ is forced such that $X$ and $Y$ are associated while other variables in the system are held fixed (Eberhardt and Scheines 2007). In many molecular biology studies, both observational and interventional data are available. Causal inference from joint interventional and observation data is particularly useful for precision medicine.

In this section, we primarily discuss the causal estimation from both observational and interventional data. Interventions include two types of interventions: structural and parametric. Causal discovery from both observational and interventional data depends on the types of interventions and assumptions one can make about the models. We assume that the observational distribution is Markovian and faithful to true underlying DAG $D_0$ to indicate that no conditional independence relations other than those entailed by the Marko property are present (Ernest et al. 2016; Hauser and Bühlmann 2015). The different interventional distributions are assumed to be linked to the DAG $D_0$. We further assume that latent confounders are present and graphs are acyclic.

Statistical methods to incorporate interventional data to learn causal models include the Bayesian procedures (Cooper and Yoo 1999; Eaton and Murphy 2007), active learning (He and Geng, 2008; Eberhardt, 2008), and Greedy Interventional Equivalent Search (GIES) (Hauser and Bühlmann 2012). Here, we propose a novel algorithm that combines structural equation models with integer programming for causal discovery from both observational and interventional data.

#### 2.5.3.1 Structural Equation Model for Interventional and Observational Data

Again, consider $p$ endogenous variables $y_1, \ldots y_p$ (phenotypes, gene expressions, methylations, and imaging signals) and $q$ exogenous variables $x_1, \ldots, x_q$ (genotypes, environments, and other covariates). Assume that $n_{obs}$ samples of observational data and $n_{int}$ samples of interventional data are available. Let $n = n_{obs} + n_{int}$. Define an observational data set: $Y^{(j)} = [y_1^{(j)}, \ldots, y_p^{(j)}]^T$, $X^{(j)} = [x_1^{(j)}, \ldots, x_q^{(j)}]$, $j = 1, \ldots, n_{obs}$, $Y^{(obs)} = [Y^{(1)}, \ldots Y^{(n_{obs})}]$, and $X^{(obs)} = [X^{(1)}, \ldots, X^{(n_{obs})}]$, and interventional data set: $Y^{(i)} = [y_1^{(i)}, \ldots, y_p^{(i)}]^T$, $X^{(i)} = [x_1^{(i)}, \ldots, x_q^{(i)}]^T$, $i = 1, \ldots, n_{int}$, $Y^{(int)} = [Y^{(1)}, \ldots, Y^{(n_{int})}]$ and $X^{(int)} = [X^{(1)}, \ldots, X^{(n_{int})}]$. Given a DAG $D$, the intervention DAD $D_I$ is defined as $D$, but delete all directed edges which point into $i \in I$, for all $i \in I$.

For the observation data set, we define the linear structural equation model for a DAD $D$:

$$y_{obs,k}^{(j)} = \sum_{l=1}^{p} \beta_{kl} y_{obs,l}^{(j)} + \sum_{m=1}^{q} \alpha_{km} x_{obs,m}^{(j)} + \varepsilon_k^{(j)}, k = 1, ..., p, \; j = 1, ..., n_{obs}, \quad (2.109)$$

where $\beta_{kl} = 0$, if $l \notin pa_D(k)$, $\alpha_{km} = 0$, $m \notin pa_D(k)$, $\varepsilon_k^{(j)} \sim N(0, \sigma_k^2)$, $\varepsilon_1^{(j)}, ..., \varepsilon_p^{(j)}$ are independent, and are independent of $y_k^{(j)}, x_m^{(j)}, \varepsilon_{obs}^{(j)} = [\varepsilon_{obs,1}^{(j)}, ..., \varepsilon_{obs,p}^{(j)}]^T, \varepsilon_{obs}^{(1)}, ..., \varepsilon_{obs}^{(n_{net})}$ are independent.

Before describing the model for intervention, we first clarify the concept of intervention. Intervention is interpreted as forcing the intervened variables to take prespecified values. Let $I \subseteq \{1, ..., p\}$ be the set of intervention target variables. Intervention is often represented by do calculus (Pearl 2000). The interventions are divided into two types of interventions: deterministic intervention and stochastic intervention. Deterministic intervention do $X_I = u_I$ is defined as doing an intervention at the set of variables $X_I$ by setting them to the values $u_I$. Stochastic intervention is defined as setting the intervened variables $X_I$ to the values of a random vector $U_I \sim \prod_{j \in I} f_{U_j}(u_j) du_j$ with independent, but not identically distributed densities $f_{U_j}(.)$ $(j \ni I)$. We often assume that the densities for the intervention variables are Gaussian: $U_j \sim N(\mu_{U_j}, \tau_j^2)$.

Now consider structural equation models for the interventional data. Assume that an interventional target is $T = I$. The endogenous variables $y_{int,k}^{(i)}$ is represented as

$$y_{int,k}^{(i)} = \begin{cases} \sum_{l \notin I} \beta_{kl} y_{int,l}^{(i)} + \sum_{l \ni I} \beta_{kl} U_l^{(i)} + \sum_{m \notin I} \alpha_{km} x_{int,m}^{(i)} + \sum_{m \ni I} \alpha_{km} U_m^{(i)} + \varepsilon_k^{(i)}, & \text{if } k \notin I \\ U_k^{(i)} & , \text{if } k \in I, \end{cases}$$
$$(2.110)$$

where $\beta_{kl}$, $\alpha_{km}$ are defined as that in the model (2.109).

For the simplicity of presentation, models (2.109) and (2.110) can be written in a matrix form. Define the matrices:

$$B \in B(D) = \begin{cases} \beta_{kl}, & l \in pa_D(k), \\ 0, & l \notin pa_D(k), \end{cases} \quad (2.111)$$

$$A \in A(D) = \begin{cases} \alpha_{km}, & m \in pa_D(k), \\ 0, & m \notin pa_D(k) \end{cases} \quad (2.112)$$

The model (2.109) can be rewritten as

$$Y^{(obs)} = BY^{(obs)} + AX^{(obs)} + \varepsilon^{(obs)}, \quad (2.113)$$

where $\varepsilon_{obs}^{(j)}, j = 1, ..., n_{obs}$ are distributed as $N_p(0, \text{diag}(\sigma_1^2, ..., \sigma_p^2))$.

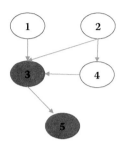

**FIGURE 2.8**
Illustration of intervention.

Next, we start the matrix representation of the model (2.110) by defining the matrices that describe the map from the original data points to the intervention target data points. Easy ways to define these matrices are by examples. Consider five nodes $\{1, 2, 3, 4, 5\}$ (Figure 2.8) and assume that the nodes $I = \{3, 5\}$. The original untargeted data points are rearranged to $\{1, 2, 3\}$. The map from $\{1, 2, 3, 4, 5\}$ to $\{1, 2, 4\}$ is $1 \rightarrow 1; 2 \rightarrow 2; 4 \rightarrow 3$. The original point is represented by a vector with 1 to be the element corresponding to the label of the original point and all other elements will be zeros. For example, the node 3 will be represented by a vector $[\,0\ 0\ 1\ 0\ 0\,]$. Let $I^c = \{1, 2, 4\}$ be the set of untargeted variables. Define $P^{(I)} : R^p \rightarrow R^{p-|I|}$ be a matrix mapping the original data points to the untargeted data points. In this example, the matrix $P^{(I)}$ is

$$P^{(I)} = \begin{bmatrix} 1 & 0 & 0 & 0 & 0 \\ 0 & 1 & 0 & 0 & 0 \\ 0 & 0 & 0 & 1 & 0 \end{bmatrix}_{p \times (p-|I|)}.$$

Similarly, define the matrix $Q^{(I)} : R^p \rightarrow R^{|I|}$ to be a matrix mapping the original data points to the target data points. The intervened variables $\{3, 5\}$ are represented in the original data set are $[\,0\ 0\ 1\ 0\ 0\,]$ and $[\,0\ 0\ 0\ 0\ 1\,]$. The intervened data points $\{3, 5\}$ are arranged to $\{1, 2\}$ in the new intervention data set. The matrix $Q^{(I)}$ in this example is

$$Q^{(I)} = \begin{bmatrix} 0 & 0 & 1 & 0 & 0 \\ 0 & 0 & 0 & 0 & 1 \end{bmatrix}.$$

Finally, we define the matrix $R: R^p \rightarrow R^p$ mapping the original intervened data points from the original data set to the new intervened data set as $R^{(I)} = (P^{(I)})^T P^{(I)}$. The matrix $R^{(I)}$ in this example is

$$R^{(I)} = \begin{bmatrix} 1 & 0 & 0 & 0 & 0 \\ 0 & 1 & 0 & 0 & 0 \\ 0 & 0 & 0 & 0 & 0 \\ 0 & 0 & 0 & 1 & 0 \\ 0 & 0 & 0 & 0 & 0 \end{bmatrix}.$$

The model (2.110) can then be written in the following matrix form:

$$Y_{int}^{(i)} = R^{(I)}\left(BY_{int}^{(i)} + AX_{int}^{(i)} + \varepsilon_{int}^{(i)}\right) + \left(Q^{(I)}\right)^T U^{(i)}, \tag{2.114}$$

or

$$Y^{(int)} = R^{(I)}\left(BY^{(int)} + AX^{(int)} + \varepsilon^{(int)}\right) + \left(Q^{(I)}\right)^T U. \tag{2.115}$$

For the convenience of presentation, we consider the entire data set { $Y^{(i)}$, $X^{(i)}$, $i = 1,..., n$, $n = n_{obs} + n_{int}$ } of observational and interventional data points to unify observational and interventional data points in a common framework. Let $T$ be the sequence of intervention targets $T^{(1)}, ..., T^{(n)}$. If the *ith* sampled data are observational, then we define $T^{(i)} = \phi$, the empty target, and $U^{(i)} = 0$. The complete data set is denoted by $Y = [Y^{(1)} \cdots Y^{(n)}]$, $X = [X^{(1)} \cdots X^{(n)}]$, $U = [U^{(1)} \cdots U^{(n)}]$ and $\varepsilon = [\varepsilon^{(1)} \cdots \varepsilon^{(n)}]$. The structural equation for the entire data set is given by

$$Y = R^{(I)}(BY + AX + \varepsilon) + \left(Q^{(I)}\right)^T U, \tag{2.116}$$

where $X^{(i)}$ are often assumed to be fixed variables $X^{(1)} = ... = X^{(n)} = X_0$.

### 2.5.3.2 Maximum Likelihood Estimation of Structural Equation Models from Interventional and Observational Data

We start to derive likelihood functions with the distributions of the intervened variables. Recall that $U = [U^{(1)} \cdots U^{(n)}]$, where $U^{(i)} = [U_1^{(i)} \cdots U_p^{(i)}]$. Assume that $U_1^{(i)}, ..., U_p^{(i)}$ are independent with distribution

$$U_j^{(i)} = \begin{cases} 0 & i = 1, ..., n_{obs} \\ N\left(\mu_{u_j}^{(i)}, \tau_j^2\right) & i = n_{obs} + 1, ..., n, \, j = 1, ..., p \end{cases} \tag{2.117}$$

Denote

$$\mu_U^{(i)} = \begin{cases} 0 & i = 1, ..., n_{obs} \\ \left[\mu_{u_1}^{(i)}, ..., \mu_{u_p}^{(i)}\right]^T & i = n_{obs} + 1, ..., n. \end{cases} \tag{2.118}$$

Assume that the errors follow a normal distribution $\varepsilon^{(i)} \sim N(0, \Sigma^{(i)})$ where

$$\Sigma^{(i)} = \text{diag}\left(\left(\sigma_1^{(i)}\right)^2, ..., \left(\sigma_p^{(i)}\right)^2\right).$$

It follows from Equation 2.116 that

$$Y = \left(I - R^{(I)}B\right)^{-1} R^{(I)} AX + \left(I - R^{(I)}B\right)^{-1} R^{(I)} \varepsilon$$

$$+ \left(I - R^{(I)}B\right)^{-1} \left(Q^{(I)}\right)^T U, \tag{2.119}$$

or

$$Y^{(i)} = \left(I - R^{(I)}B\right)^{-1} R^{(I)} A X^{(i)} + \left(I - R^{(I)}B\right)^{-1} R^{(I)} \varepsilon^{(i)}$$
$$+ \left(I - R^{(I)}B\right)^{-1} \left(Q^{(I)}\right)^T U^{(i)}. \qquad (2.120)$$

The vector $Y^{(i)}$ follows a normal distribution $N(\mu_Y^{(i)}, \Sigma_Y^{(i)})$, where

$$\mu_Y^{(i)} = \left(I - R^{(I)}B\right)^{-1} R^{(I)} A X^{(i)} + \left(I - R^{(I)}B\right)^{-1} \left(Q^{(I)}\right)^T \mu_U^{(i)}, \qquad (2.121)$$

$$\Sigma_Y^{(i)} = \left(I - R^{(I)}B\right)^{-1} \left[ R^{(I)} \Sigma_\varepsilon^{(i)} R^{(I)} + \left(Q^{(I)}\right)^T \Sigma_U^{(i)} Q^{(I)} \right] \left(I - R^{(I)}B\right)^{-T}. \qquad (2.122)$$

Its density function can be written as

$$f_{Y^{(i)}} = \frac{1}{(2\pi)^{\frac{p}{2}} |\Sigma_Y^{(i)}|^{\frac{1}{2}}} \exp\left\{ -\frac{1}{2} \left(Y^{(i)} - \mu_Y^{(i)}\right)^T \left(\Sigma_Y^{(i)}\right)^{-1} \left(Y^{(i)} - \mu_Y^{(i)}\right) \right\}. \qquad (2.123)$$

Negative log-likelihood is then expressed as

$$-l_D(A, B, T, \Sigma_\varepsilon) = \frac{np}{2} \log(2\pi) + \frac{1}{2} \sum_{i=1}^{n} \left[ \log |\Sigma_Y^{(i)}| \right.$$
$$\left. + \left(Y^{(i)} - \mu_Y^{(i)}\right)^T \left(\Sigma_Y^{(i)}\right)^{-1} \left(Y^{(i)} - \mu_Y^{(i)}\right) \right]. \qquad (2.124)$$

We can show that the negative likelihood in 2.124 can be reduced to (Appendix 2.B)

$$-l_D(A, B, T, \Sigma_\varepsilon) \approx \frac{1}{2} \sum_{i=1}^{n} \left[ \mathrm{Tr}\ (K^{(i)} Y^{(i)} (Y^{(i)})^T) - 2(Y^{(i)})^T (I - B)^T R^{(I)} (\Sigma_\varepsilon^{(i)})^{-1} \right.$$
$$R^{(I)} A X^{(i)}$$
$$\left. + \mathrm{Tr}\ (A^T R^{(I)} \left(\Sigma_\varepsilon^{(i)}\right)^{-1} R^{(I)} A X^{(i)} (X^{(i)})^T - \sum_{j \notin I} \log \sigma_j^{-2} \right] \qquad (2.125)$$

Define $n^{(I)} = |\{i \mid T^{(i)} = I\}|$ to be the number of samples with the set of intervened variables $I$ and its associated sampling matrix $S^{(I)} = \frac{1}{n^{(I)}} \sum_{i : T^{(i)} = I}$

$Y^{(i)}(Y^{(i)})^T$. Removing the terms that do not contain the model parameters in the matrix $K^{(l)}$, Equation 2.125 can be further reduced to

$$-l_D(A,B,T,\Sigma_\varepsilon) \approx \frac{1}{2}\sum_{l\in T}\left\{ n^{(l)}\mathrm{Tr}\left(S^{(l)}(I-B)^T R^{(l)}\left(\Sigma_\varepsilon^{(i)}\right)^{-1}R^{(l)}(I-B)\right)\right.$$

$$-n^{(l)}\sum_{j\notin I}\log\sigma_j^{-2}$$

$$-\sum_{i:T^{(i)}=I}\left[2\left(Y^{(i)}\right)^T(I-B)^T R^{(l)}\left(\Sigma_\varepsilon^{(i)}\right)^{-1}R^{(l)}AX^{(i)}\right.$$

$$\left.\left. -\mathrm{Tr}\left(A^T R^{(l)}\left(\Sigma_\varepsilon^{(i)}\right)^{-1}R^{(l)}AX^{(i)}\left(X^{(i)}\right)^T\right)\right]\right\}.$$

Define $n^{(-k)} = \sum_{l\in T:k\notin I}n^{(l)}$. In the example in Figure 2.8, $n^{(-1)} = n^{(-2)} = n^{(-4)} = n$ and $n^{(-3)}=0$, $n^{(-5)}=0$. Define $S^{(-k)} = \sum_{l\in T:k\in I}\dfrac{n^{(l)}}{n^{(-k)}}S^{(l)}$, where $S^{(-k)}=0$, if $n^{(-k)} = 0$. Define $E_c = \{i \mid$ at least one $a_{ij} \ne 0\}$ which is the set of nodes that are connected with the exogenous variables, $n^{(-l)} = \sum_{l\in T:(l\in E_c)\cap(l\notin I)}n^{(l)}$. The number of samples in which the node $l$ connecting with exogenous variables is not intervened, $S_{XY}^{(l)} = \dfrac{1}{n^{(l)}}\sum_{i:T^{(i)}=I}X^{(i)}(Y^{(i)})^T$, $S_{XX}^{(l)} = \dfrac{1}{n^{(l)}}\sum_{i:T^{(i)}=I}X^{(i)}(X^{(i)})^T$, $S_{XY}^{(-l)} =$

$$\sum_{l\in T:(l\in E_c)\cap(l\notin I),,i:T^{(i)}=I}\frac{n^{(l)}}{n^{(-l)}}S_{XY}^{(l)} \quad \text{and} \quad S_{XY}^{(-l)} = \sum_{l\in T:(l\in E_c)\cap(l\notin I),,i:T^{(i)}=I}\frac{n^{(l)}}{n^{(-l)}}S_{XY}^{(l)}.$$

Then, we obtain the following negative log-likelihood function (Appendix 2.B)

### Result 2.3: Negative Log-Likelihood Function

$$-l_D(A,B,T,\Sigma_\varepsilon) = \sum_{k=1}^{p}l_k\left(A_k,B_k,\sigma_k^2,T,Y,X\right), \tag{2.126}$$

where
when the node has connections with exogenous variables,

$$l_k\left(A_{k.},B_{k.},\sigma_k^2,T,Y,X\right) = \frac{1}{2}n^{(-k)}\left\{\sigma_k^{-2}(I-B)_k.S^{(-k)}((I-B)_k.)^T - \log\sigma_k^{-2}\right.$$

$$\left.-\sigma_k^{-2}\left[2A_{k.}S_{XY}^{(-k)}((I-B)_k.)^T - A_{k.}S_{XX}^{(-k)}(A_{k.})^T\right]\right\}, \tag{2.127}$$

when the node does not have connections with any exogenous variables,

$$l_k\left(B_{k.},\sigma_k^2,T,Y,X\right) = \frac{1}{2}n^{(-k)}\left\{\sigma_k^{-2}(I-B)_k.S^{(-k)}((I-B)_k.)^T - \log\sigma_k^{-2}\right\}. \tag{2.128}$$

Since the negative log-likelihood function can be decomposed into summation of the log-likelihood function of each node, the parameters $A$, $B$ and the variance of the errors can be estimated separately for each node. We first estimate the parameters of the system without exogenous variables. We can show the following (Appendix 2.B):

**Result 2.4: Parameter Estimation Without Exogenous Variables**

$$B_{k.}^T = \left(S^{(-k)}\right)^{-1} S_{.k}^{(-1)}, \tag{2.129}$$

$$B_{k,pa_D(k)} = S_{k,pa_D(k)}^{(-k)} \left(S_{pa_D(k),pa_D(k)}^{(-k)}\right)^{-1}, \tag{2.130}$$

$$\hat{\sigma}_k^2 = \left(I - \hat{B}\right)_{k.} S^{(-k)} \left(\left(I - \hat{B}\right)_{k.}\right)^T, k = 1, \dots, p, \tag{2.131}$$

where $B_{k,pa_D(k)}$ is a row vector consisting of only the entry of the node and its parents in the parameter matrix $B$.

The score of the node $k$ is then defined as

$$\text{Score}_D(k) = \frac{1}{2} n^{(-k)} \left(1 + \log \hat{\sigma}_k^2\right). \tag{2.132}$$

Next, consider the system with both endogenous and exogenous variables. In Appendix 2.B we derive the estimators of the matrices $A$, $B$, variance $\sigma_k^2$, and node score.

**Result 2.5: Parameter Estimation with Exogenous Variables**

$$\begin{bmatrix} (B_{k.})^T \\ (A_{k.})^T \end{bmatrix} = \begin{bmatrix} S^{(-k)} & S_{YX}^{(-k)} \\ S_{XY}^{(-k)} & S_{XX}^{(-k)} \end{bmatrix}^{-1} \begin{bmatrix} \left(S_{k.}^{(-k)}\right)^T \\ \left(S_{XY}^{(-k)}\right)_{.k} \end{bmatrix}, \tag{2.133}$$

$$\hat{\sigma}_k^2 = \left(I - \hat{B}\right)_{k.} S^{(-k)} \left(\left(I - \hat{B}\right)_{k.}\right)^T - 2\hat{A}_{k.} S_{XY}^{(-k)} \left(\left(I - \hat{B}\right)_{k.}\right)^T$$
$$+ \hat{A}_{k.} S_{XX}^{(-k)} \left(\hat{A}_{k.}\right)^T, \tag{2.134}$$

$$\text{Score}_D(k) = \max_{A_{k.}, B_{k.}, \sigma_k^2} l_k \left(A_{k.}, B_{k.}, \sigma_k^2, \mathbf{T}, \mathbf{Y}, \mathbf{X}\right) = \frac{1}{2} n^{(-k)} \left(1 + \log \hat{\sigma}_k^2\right). \tag{2.135}$$

### 2.5.3.3 Sparse Structural Equation Models with Joint Interventional and Observational Data

Causal networks are often sparse. Therefore, the matrices $A$ and $B$ are sparse. To obtain sparse estimates of $A$ and $B$, the natural approach is the $l_1$-norm penalization. Since the matrices depend on the structures of graphs, the matrices $A$ and $B$ are denoted by $A(D)$ and $B(D)$. Consider the general negative log-likelihood:

$$l_k\left(A_{k.}, B_{k.}, \sigma_k^2, T, Y, X\right) = \frac{1}{2}n^{(-k)}\left\{\sigma_k^{-2}(I - B)_{k.}S^{(-k)}((I - B)_{k.})^T - \log\sigma_k^{-2}\right.$$
$$\left. -\sigma_k^{-2}\left[2A_{k.}S_{XY}^{(-k)}((I - B)_{k.})^T - A_{k.}S_{XX}^{(-k)}(A_{k.})^T\right]\right\}.$$

Define *dim* $(D_k)$ as the number of non-zero elements in $\hat{A}_{k.}(D)$ and $\hat{B}_k(D)$. The $L_1$-norm penalized likelihood optimization problem for the sparse SEMs with joint interventional and observation data is defined as

$$\min_{A_{k.}, B_{k.}}(I - B)_{k.}S^{(-k)}((I - B)_{k.})^T - \left[2A_{k.}S_{XY}^{(-k)}((I - B)_{k.})^T - A_{k.}S_{XX}^{(-k)}(A_{k.})^T\right]$$
$$+ \lambda_1\|B_{k.}\|_1 + \lambda_2\|A_{k.}\|_1 \tag{2.136}$$

where $\|B_{k.}\|_1 = \sum_{j=1}^{p}|\beta_{kj}|$ and $\|A_{k.}\|_1 = \sum_{l=1}^{q}|\alpha_{kl}|$, $\lambda_1$ and $\lambda_2$ are penalty parameters and are chosen to balance the fitness of the model fitting the data and sparsity of the networks. If the endogenous variables are RNA-seq data and exogenous variables are NGS data, the constraints for the gene-based network analysis should be

$$\lambda_1\Omega_1(B_{k.}) + \lambda_2\Omega_2(A_{k.}) = \lambda_1\sum_{j=1}^{p}\|B_{kj}\|_2 + \lambda_2\sum_{l=1}^{q}\|A_{kl}\|_2, \tag{2.137}$$

where

$$\|B_{kj}\|_2 = \sqrt{\sum_{i=1}^{l_j}\beta_{kj_i}^2} \text{ and } \|A_{kl}\|_2 = \sqrt{\sum_{m=1}^{M_l}\alpha_{kl_m}^2}. \tag{2.138}$$

Equation 2.136 for the gene-based network analysis will become

$$\min_{A_{k.}, B_{k.}}(I - B)_{k.}S^{(-k)}((I - B)_{k.})^T - \left[2A_{k.}S_{XY}^{(-k)}((I - B)_{k.})^T - A_{k.}S_{XX}^{(-k)}(A_{k.})^T\right]$$
$$+ \lambda_1\Omega_1(B_k) + \lambda_2\Omega_2(A_{k.}). \tag{2.139}$$

The objective function in the optimization problem (2.136) can be decomposed into a differential function part and non-smooth function part. Define the differential function part as

$$f(A_{k.}, B_{k.}) = (I - D)_{k.}S^{(-k)}((I - B)_k)^T - \left[2A_{k.}S_{XY}^{(-k)}((I - B)_{k.})^T - A_{k.}S_{XX}^{(-k)}(A_{k.})^T\right]$$

and non-differential function part as

$$\Omega(A_{k.}, B_{k.}) = \lambda_1\|B_{k.}\|_1 + \lambda_2\|A_{k.}\|_1.$$

Similarly, the differential part of the decomposition of the objective function in the optimization problem (2.139) is the same as that for (2.136), but the non-differential part takes the form:

$$\lambda_1\Omega_1(B_{k.}) + \lambda_2\Omega_2(A_{k.}).$$

The optimization problem (2.136) or (2.139) will be solved by the proximal method discussed in Section 1.3.

Recall that in Equation 1.38 we consider convex optimization problems of the forms:

$$\min_w f(u, v) + \lambda_1\Omega_1(u) + \lambda_2\Omega_2(v),$$

where $f(u, v)$ is a convex differentiable function, and $\Omega_1(u)$ and $\Omega_2(v)$ are nonsmooth functions, typically nonsmooth norms. The proximal operator for solving the optimization problem (2.139) is then defined as

$$\text{Prox}_{\lambda_1\Omega_1 + \lambda_2\Omega_2}(u, v)$$

$$= \underset{w \in \mathbb{R}^p}{\text{argmin}} \left( \Omega(w) + \frac{1}{2\lambda_1} \left|\left| w_1 - u \right|\right|_2^2 + \frac{1}{2\lambda_2} \left|\left| w_2 - v \right|\right|_2^2 \right). \qquad (2.140)$$

The optimization problem (2.139) can be solved by the proximal gradient method. To give a unified form for two types of constraints, we define

$$\Omega(B_{k.}, A_k) = \lambda_1\Omega_1(B_{k.}) + \lambda_2\Omega_2(A_{k.}), \qquad (2.141)$$

where $\Omega_1(B_{k.}) = |B_{k.}|_1$ and $\Omega_2(A_{k.}) = |A_{k.}|_1$.

Using Equations 1.45, 1.25, and 2.141, we obtain

$$\text{Prox}_{\lambda_1|\Omega 1}(u) = \begin{bmatrix} \text{sign}(u_1)(|u_1| - \lambda_1)_+ \\ \vdots \\ \text{sign}(u_p)(|u_p| - \lambda_1)_+ \end{bmatrix} \qquad (2.142)$$

and

$$\text{Prox}_{\lambda_2|\Omega_2}(v) = \begin{bmatrix} \text{sign}(v_1)(|v_1| - \lambda_2)_+ \\ \vdots \\ \text{sign}(v_q)(|v_q| - \lambda_2)_+ \end{bmatrix}. \qquad (2.143)$$

Similarly, using Equations 1.149, 1.45, and 2.137 gives the proximal operator for the group lasso constraints (gene-based network analysis)

$$\text{Prox}_{\lambda_1\Omega_1}(u) = \begin{bmatrix} \left(1 - \dfrac{\lambda_1}{||u_1||_2}\right)_+ u_1 \\ \vdots \\ \left(1 - \dfrac{\lambda_1}{||u_p||_2}\right)_+ u_p \end{bmatrix}, \qquad (2.144)$$

$$\text{Prox}_{\lambda_2 \Omega_2}(v) = \begin{bmatrix} \left(1 - \dfrac{\lambda_1}{||v_1||_2}\right)_+ v_1 \\ \vdots \\ \left(1 - \dfrac{\lambda_1}{||v_q||_2}\right)_+ v_q \end{bmatrix}. \tag{2.145}$$

In general, the proximal gradient method is given by

$$\begin{bmatrix} u^{t+1} \\ v^{t+1} \end{bmatrix} = \begin{bmatrix} \text{Prox}_{\lambda_1 \Omega_1}\left(u^t - \rho^k \dfrac{\partial f}{\partial \beta}(u^t, v^t)\right) \\ \text{Prox}_{\lambda_2 \Omega_2}\left(v^t - \rho^k \dfrac{\partial f(u^t, v^t)}{\partial \alpha}\right) \end{bmatrix}, \tag{2.146}$$

where $\rho^k > 0$ is a step size.

Define

$$\frac{\partial f(u^t, v^t)}{\partial \beta} = \frac{\partial f(u^t, v^t)}{\partial u} = \begin{bmatrix} \dfrac{\partial f}{\partial u_1} \\ \vdots \\ \dfrac{\partial f}{\partial u_p} \end{bmatrix}$$

$$= n^{(-k)} \sigma_k^{(-2)} \left\{ -(I - B)_{k.} S^{(-k)} + A_{k.} S_{XY}^{(-k)} \right\}^T, \tag{2.147}$$

$$\frac{\partial f(u^t, v^t)}{\partial \alpha} = \frac{\partial f(u^t, v^t)}{\partial v} = \begin{bmatrix} \dfrac{\partial f}{\partial v_1} \\ \vdots \\ \dfrac{\partial f}{\partial v_q} \end{bmatrix}$$

$$= n^{(-k)} \sigma_k^{(-2)} \left\{ -(I - B)_{k.} S^{(-k)} + A_{k.} S_{XY}^{(-k)} \right\}^T. \tag{2.148}$$

The proximal gradient method can also take the following form to ensure the convergence of the iterative algorithms.

Define the function:

$$\hat{f}_\lambda(u, v, y_1, y_2) = f(y) + \nabla_u f(y)^T (u - y_1) + \nabla_v f(y)^T (v - y_2) + \frac{1}{2\lambda_1} ||u - y_1||_2^2$$

$$+ \frac{1}{2\lambda_2} ||v - y_2||_2^2,$$

with $\lambda_1 > 0$, $\lambda_2 > 0$.

The alternative algorithm for the proximal gradient method is given as follows:

**Algorithm 2.1**

Step 1: Given $u^k, v^k, \lambda_1^{k-1}, \lambda_2^{k-1}$, and parameter $\delta \in (0, 1)$. Set $\lambda_1 = \lambda_1^{k-1}$, $\lambda_2 = \lambda_2^{k-1}$

Step 2:
 Repeat

 1. Set

 2. $z = \begin{bmatrix} z_1 \\ z_2 \end{bmatrix} = \begin{bmatrix} \text{Prox}_{\lambda_1 \Omega_1}\left(u^t - \lambda_1 \dfrac{\partial f(u^t, v^t)}{\partial u}\right) \\ \text{Prox}_{\lambda_2 \Omega_2}\left(v^t - \lambda_2 \dfrac{\partial f(u^t, v^t)}{\partial v}\right) \end{bmatrix}.$

 3. Break if $f(z) \le \hat{f}_\lambda(z, u^k, v^k)$.
 4. Update $\lambda_1 = \delta_1 \lambda_1, \lambda_2 = \delta_2 \lambda_2$.

Step 3: return $\lambda_1^t = \lambda_1, \lambda_2^t = \lambda_2, u^{t+1} = z_1, v^{t+1} = z_2$.

What Equations 2.142 and 2.143 or 2.144 and 2.145 are used to calculate the proximal operators depends on what norms of the vectors are used to penalize the endogenous and exogenous variables or what gene expression values and genotype values are used to study genotype–expression causal networks. If we use overall expression levels to represent the expression of a gene and an SNP to represent genotype, then Equations 2.142 and 2.143 will be used to calculate the proximal operators. If we use an RNA-seq profile to represent an expression curve of the gene and multiple SNPs or multiple function principal scores to represent genotypes in the gene, then we will use Equations 2.144 and 2.145 to calculate the proximal operators.

After the parameters in the structural equations are estimated, we can calculate the score function of the node:

$$\text{Score}(k) = \left(I - \hat{B}\right)_{k.} S^{(-k)} \left(\left(I - \hat{B}\right)_{k.}\right)^T - \left[2A_{k.} S_{XY}^{(-k)} \left(\left(I - \hat{B}\right)_{k.}\right)^T - \hat{A}_{k.} S_{XX}^{(-k)} \left(\hat{A}_{k.}\right)^T\right]$$
$$+ \lambda_1 \Omega_1 \left(\hat{B}_{k.}\right) + \lambda_2 \Omega_2 \left(\hat{A}_{k.}\right)$$

(2.149)

Integer programming will then be used to find causal networks which will be discussed in the next section.

---

## 2.6 Integer Programming for Causal Structure Leaning

Two basic approaches to causal network or Bayesian network learning: score-based and constrained-based (Cussens et al. 2016). The score-based methods learn causal networks via maximizing the score metrics that characterize the causal networks, while constrained-based methods learn causal networks via testing conditional independence. The score-based approach is the most popular approach to causal network learning. This section will focus on score-based methods for causal network learning.

The score-based causal network learning is a combinatorial optimization problem of searching a causal network structure that optimizes a score metric from the data (observational, or interventional or both observational and interventional data). Learning optimal causal network structure is an NP-hard problem (Chickering 1996). Recently, several computational algorithms including dynamic programming (Yuan and Malone 2013), A* search method (Yuan et al. 2011), and integer linear programming (IP) (Cussens et al. 2016) have been developed for causal structure learning. In this section, we will mainly introduce GOBNILP (Cussens 2011) for DAG (causal network) learning. The GOBNILP combines IP, cutting planes and branch-cut methods to develop efficient computational algorithms for score-based causal network learning.

### 2.6.1 Introduction

Integer linear programming is a widely used method for solving combinatorial optimization problems and can be used for exact DAG learning and causal inference. The integer programming consists of three components: integer variables, constraints, and objective function. These constraints must be linear in the variables and are used to limit values of variables to a feasible region. A standard form of integer linear programming is

$$
\begin{aligned}
\text{Min} \quad & C^T X \\
& AX \le b \\
& X \ge 0 \\
& X \in Z^n,
\end{aligned}
\tag{2.150}
$$

where $C \in R^n$, $b \in R^m$, a matrix $A \in R^{m \times n}$ and $Z = \{0, 1, 2, \dots\}$.

If all variables are restricted to the values from $B = \{0, 1\}$ we have a 0-1-integer linear programming:

$$
\begin{aligned}
\text{Min} \quad & C^T X \\
& AX \le b \\
& X \ge 0 \\
& X \in B^n.
\end{aligned}
\tag{2.151}
$$

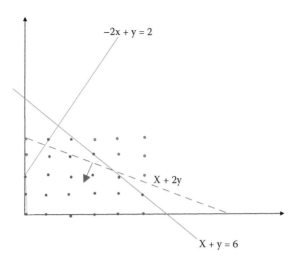

**FIGURE 2.9**
Illustration of Example 2.2.

**Example 2.2**

Consider the following integer linear programming:

$$\text{Min} \qquad x + 2y$$
$$x + y \leq 6$$
$$-2x + y \leq 2$$
$$x \geq 0, y \geq 0.$$

The feasible region is shown in Figure 2.9. It consists of the integral points in red.

## 2.6.2 Integer Linear Programming Formulation of DAG Learning

The "score and search" approach is a popular method for DAG learning (Jaakkola et al. 2002; Barlett and Cussens 2013; Cussens 2011, 2014). In the previous section we discussed how to define score functions for DAGs. Suppose that score functions for DAGs are available. We now study the formulation of DAG learning in terms of 0-1-integer linear programming (ILP). Each candidate DAG has a score measuring how well the DAG fits the data. The task is to search a DAG that optimizes the score via IP. A DAG is defined as $G = (V, E)$, where the set $V$ of nodes represents a set of random variables $Y = \{Y_1, \ldots, Y_p\}$ with $p = |V|$ and $E$ denotes the set of directed edges. The set of parent variables for a variable $v \in V$ is denoted as $W_v$. A DAG can be encoded

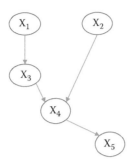

**FIGURE 2.10**
A typical DAG.

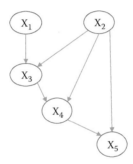

**FIGURE 2.11**
A DAG for exercise.

by the set $W = \{W_1, ..., W_p\}$ of parent variables for all nodes $V$ in the graph $G$. To illustrate how to encode a DAG we plot Figure 2.10 (more example, please see Exercise 16 and Figure 2.11). The sets of parent variables are $W_1 = \{\phi\}$, $W_2 = \{\phi\}$, $W_3 = \{X_1\}$, $W_4 = \{X_2, X_3\}$ and $W_5 = \{X_4\}$. Learning a DAG is to identify the DAG with an optimal score by searching all possible DAGs that are represented by the sets of parent variables.

The global score is the summation of the score of each node. We use $C$ $(v, W_v)$ to denote a score function for the pair of node $v$ and its parent set $W_v$. We assume that all score functions are positive. We first define a general optimization problem for learning DAG. The global score is defined as

$$C(D) = \sum_{i \in V} C(v, W_v).$$

The learning task is to find a DAG that optimizes the global score $C(D)$ over all possible DAGs $D$ or parent sets:

$$\min_{D} \quad \sum_{i \in V, W_v \in D} C(v, W_v).$$

Next, we formulate the DAG learning problem into a specific optimization problem, ILP problem. We define a variable $x$ $(W_v \to v)$ to indicate the presence or absence of the parent set $W_v$ in the DAG. In other words, $x$ $(W_v \to v) = 1$ if and only if it is the parent set for the node $v$. The parent set $W_v$ can be an empty set. The objective function for the ILP formulation of a DAG learning can be defined as

$$\sum_{v=1}^{p} \sum_{j_v=1}^{J_v} C\left(v, W_{j_v}\right) x\left(W_{j_v} \to v\right). \qquad (2.152)$$

For example, we consider a DAG with three nodes $\{X_1, X_2, X_3\}$ that can be denoted by $\{1,2,3\}$ for simplicity. Figure 2.12 presents a DAG with these three variables and Table 2.2 lists an indicator variable encoding for the DAG in Figure 2.12. The objective function for the DAG is defined as

$$C(1, \phi)x(\phi \to 1) + C(1, \{2\})x(\{2\} \to 1) + C(1, \{3\})x(\{3\} \to 1)$$
$$+ C(1, \{2,3\})x(\{2,3\} \to 1) +$$
$$C(2, \phi)x(\phi \to 1) + C(2, \{1\})x(\{1\} \to 2) + C(2, \{3\})x(\{3\} \to 2)$$
$$+ C(2, \{1,3\})x(\{1,3\} \to 2) +$$
$$C(3, \phi)x(\phi \to 3) + C(3, \{1\})x(\{1\} \to 3) + C(3, \{2\})x(\{2\} \to 3)$$
$$+ C(3, \{1,2\})x(\{1,2\} \to 3).$$

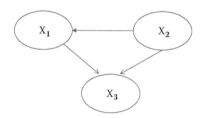

**FIGURE 2.12**
An example of a causal network.

**TABLE 2.2**

Indicator Variable Encoding for Figure 2.12

| $1 \leftarrow \{\}$ | $1 \leftarrow \{2\}$ | $1 \leftarrow \{3\}$ | $1 \leftarrow \{2,3\}$ |
|---|---|---|---|
| 0 | 1 | 0 | 0 |
| $2 \leftarrow \{\}$ | $2 \leftarrow \{1\}$ | $2 \leftarrow \{3\}$ | $2 \leftarrow \{1,3\}$ |
| 1 | 0 | 0 | 0 |
| $3 \leftarrow \{\}$ | $3 \leftarrow \{1\}$ | $3 \leftarrow \{2\}$ | $3 \leftarrow \{1,2\}$ |
| 0 | 0 | 0 | 1 |

Our goal is to find a candidate parent set $W_v$ for each node $v$ by optimizing (minimizing or maximizing) the objective function (2.152). It is clear that every DAG can be encoded by a zero-one indicator variable. However, any set of zero-one numbers may not encode a DAG. A set of linear constraints must be posted to make the set of indicator variables to represent a DAG. Without constraints all indicator variables for the parent sets will be equal to either zero or one. These solutions will not form a DAG. The constraints need to be imposed to ensure that the solutions encode a DAG. All variables in the objective function are to indicate the presence or absence of the sets of parent variables and hence should take values 0 or 1. To ensure that the sets of parent variables encode a valid DAG, each node has exactly one (perhaps empty) parent set. This constraint that is referred to as convexity constraint, can be expressed as

$$\sum_{j_v=1}^{J_v} x\left(W_{j_v} \to v\right) = 1, v = 1, ..., p. \tag{2.153}$$

For example, we consider a DAG with three nodes $\{X_1, X_2, X_3\}$. The parent sets are represented by (Table 2.2)

$$W_{1,1} = \{\phi\}, W_{2,1} = \{X_2\}, W_{3,1} = \{X_3\}, W_{4,1} = \{X_2, X_3\}$$
$$W_{1,2} = \{\phi\}, W_{2,2} = \{X_1\}, W_{3,2} = \{X_3\}, W_{4,2} = \{X_1, X_3\}$$
$$W_{1,3} = \{\phi\}, W_{2,3} = \{X_1\}, W_{3,3} = \{X_2\}, W_{4,3} = \{X_1, X_2\}.$$

The constraints for the DAG with three nodes are

$$x\left(W_{1,1} \to 1\right) + x\left(W_{2,1} \to 1\right) + x\left(W_{3,1} \to 1\right) + x\left(W_{4,1} \to 1\right) = 1$$
$$x\left(W_{1,2} \to 2\right) + x\left(W_{2,2} \to 2\right) + x\left(W_{3,2} \to 2\right) + x\left(W_{4,2} \to 2\right) = 1 \tag{2.154}$$
$$x\left(W_{1,3} \to 3\right) + x\left(W_{2,3} \to 3\right) + x\left(W_{3,3} \to 3\right) + x\left(W_{4,3} \to 3\right) = 1.$$

The convexity constraints (2.153) can define a directed graph. However, the generated directed graph may have cycles. A directed cycle is defined as a directed path (with at least one edge) whose first and last nodes are the same. For example, if we assume that $x(W_{2,1} \to 1) = 1, x(W_{3,2} \to 2) = 1, x(W_{2,3} \to 3) = 1$ and all other variables are equal zeros. These solutions satisfy constraints (2.154), but they form a cycle $X_3 \to X_2 \to X_1 \to X_3$.

To eliminate a cycle, we need to impose other constraints. We observe that any subset $C$ of the nodes $V$ in a DAG must contain at least one node that has no parent in the subset $C$. Mathematically, this constraint is expressed as

$$\forall C \subseteq V: \sum_{v \in C} \sum_{W : W \cap C = \phi} x(W \to v) \geq 1, \tag{2.155}$$

which is referred to as cluster-based constraints.

To illustrate that a directed cycle will violate the constraint (2.155) we consider a cycle $X_1 \to X_2 \to X_3 \to X_4 \to X_1$. It is clear that all parent variables are in the set $C = \{X_1, X_2, X_3, X_4\} \subseteq V$. Therefore, we have $\forall C \subseteq V: \sum_{v \in C} \sum_{W : W \cap C = \phi} x(W \to v) = 0$. The cluster-based constraint (2.155) is violated.

Now we consider $X_1 \rightarrow X_2 \rightarrow X_3 \rightarrow X_4$, $X_1 \rightarrow X_4$, since this will not form a directed cycle, the parent set of the variable $X_1$ is empty. Then, we have $x\,(\phi \rightarrow X_1) = 1$ which implies that the constraint (2.155) is satisfied.

The constraint (2.155) basically claims that a DAG has at least one node whose parents are not in the C. The constraint (2.155) can be extended to more general cases. We consider a node that has exactly one parent in C, that is, $|\, W \cap C\,| = 1 < 2$. In this case, to ensure that a DAG has no directed cycle, we must have $\forall\, C \subseteq V: \sum_{v \in C} \sum_{W:W \cap C=1} x(W \rightarrow v) \geq 2$. For example, we consider a DAG with three nodes: $X_1 \rightarrow X_2 \rightarrow X_3$. There is no cycle in the DAG. We observe that

$$x(\phi \rightarrow 1) = 1$$

$$x(\{1\} \rightarrow 2) + x(\{2\} \rightarrow 3) = 1 + 1 = 2 \geq 2,$$

which implies that $\forall\, C \subseteq V: \sum_{v \in C} \sum_{W:W \cap C=1} x(W \rightarrow v) \geq 2$ holds.

In general, we have

$$\forall\, C \subseteq V: \sum_{v \in C} \sum_{W:W \cap C < k} x(W \rightarrow v) \geq k, \forall\, k, 1 \leq k \leq |C|, \tag{2.156}$$

which is referred to as *k-cluster-based constraints*.

In summary, a DAG learning can be formulated as the following 0-1 integer linear programming:

$$\text{Min} \quad \sum_{v=1}^{p} \sum_{j_v=1}^{J_v} C\left(v, W_{j_v}\right) x\left(W_{j_v} \rightarrow v\right)$$

$$\sum_{j_v=1}^{J_v} x\left(W_{j_v} \rightarrow v\right) = 1, v = 1, ..., p$$

$$\forall\, C \subseteq V: \sum_{v \in C} \sum_{W_{j_v}: W_{j_v} \cap C=\phi, j_v=1,...,J_v} x\left(W_{j_v} \rightarrow v\right) \geq 1$$

$$x\left(W_{j_v} \rightarrow v\right) = 0, \text{ or } 1 \tag{2.157}$$

or

$$\text{Min} \quad \sum_{v=1}^{p} \sum_{j_v=1}^{J_v} C\left(v, W_{j_v}\right) x\left(W_{j_v} \rightarrow v\right)$$

$$\sum_{j_v=1}^{J_v} x\left(W_{j_v} \rightarrow v\right) = 1, v = 1, ..., p$$

$$\forall\, C \subseteq V: \sum_{v \in C} \sum_{W_{j_v}: W_{j_v} \cap C<k, j_v=1,...,J_v} x\left(W_{j_v} \rightarrow v\right) \geq k, \forall\, k, 1 \leq k \leq |C|,$$

$$x\left(W_{j_v} \rightarrow v\right) = 0, \text{ or } 1. \tag{2.158}$$

### 2.6.3 Cutting Plane for Integer Linear Programming

Integer linear programming is an NP-hard problem (Fugenschuh and Martin 2005). The efficient algorithms for solving integer linear programming in polynomial time do not exist. A common approach is to use relaxation methods for solving IP problems. In the initial phase of the algorithms for solving IP, no cluster constraints (2.155) are included in the IP and the integrality conditions are relaxed.

A major part in integer programming which causes difficulty in solving the optimization problem lies in the integrality constraints. To improve efficiency of the algorithms, we need to remove the integrality constraints which requires integer solutions, but retain the original solutions in the feasible region of the relaxed problem. In other words, we add cutting planes as new constraints to the problem such that (1) the set of feasible integer solutions remains the same and (2) the new constraints cut off the current LP relaxation solutions making the feasible region of the new LP-relaxation smaller (Cussens et al. 2016; Bartlett and Cussens 2015).

Figure 2.13 shows a convex region of feasible solutions defined by several constraints. The optimum solution $\hat{x}$ (yellow point) to the LP relaxation problem is not an integer and hence is not a solution to the original integer programming. If we can find a plane to separate the LP relaxation solution $\hat{x}$ from the convex hull of integer points in the original feasible region and remove part of the region, which contains noninteger solutions, we could possibly reach an optimal integer solution. The plane we used to remove part

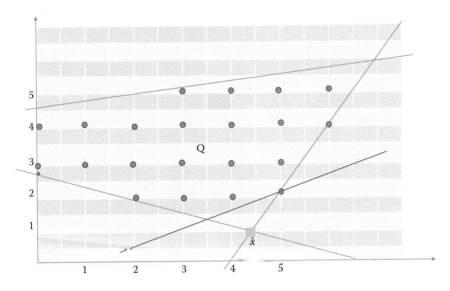

**FIGURE 2.13**
Illustration of cutting plane techniaque.

of the region from the solution search and as additional linear constraints is called a cutting plane. The cutting plane mathematically defined by $a^T x \le b$ and represented by a red line in Figure 2.13 is violated by the current LP relaxation solution $\hat{x}$, but does not cut off feasible integer solutions from the feasible region $Q$.

Many cutting planes exist. Here, we mainly focus on the constraint-based cutting planes. Selecting good cutting planes that reduce error between the true objective and the bound given by the LP relaxation solution as much as possible and cut deep into the LP feasible region is a key to the success of the IP for DAG learning. In this section, we formulate a cutting plane search problem as a sub-IP problem to search for cluster cuts. A popular criterion for cutting plane selection is the efficiency of the cutting planes. The selected cutting planes and the current LP relaxation $\hat{x}$ should be on the same side (outside) of the feasible region. The efficiency of the cut is defined as:

$$d = a^T \hat{x} - b, \tag{2.159}$$

where the cutting plane is defined as $a^T x \le b$.

To reach the large efficacy, we take the constraints of the cluster, say $C$, as a candidate cutting plane. The convexity constraints state that any node has exactly one parent set. Cluster constraints for the cluster $C$ (Exercise 18 and Figure 2.14) state that any subset $C$ of the nodes $V$ in a DAG must contain at least one node that has no parent in the subset $C$. In other words, any subset $C$ of the nodes $V$ in a DAG contains at most $|C| - 1$ nodes that has parents in the subset $C$. Therefore, the constraint (2.151) can be reformulated as

$$\forall C \subseteq V : \sum_{v \in C} \sum_{W : W \cap C \ne \phi} x(W \to v) \le |C| - 1. \tag{2.160}$$

Equation 2.160 implies that not all nodes in the cluster $C$ can have parents in the cluster $C$.

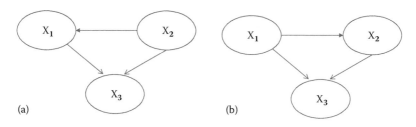

(a)  (b)

**FIGURE 2.14**
(a) Directed acyclic graph, (b) directed cyclic graph.

Take Figure 2.12 as an example. Consider the cluster $C = \{1, 2, 3\}$. The constraints in Equation 2.160 for Figure 2.12 are

$$I(\{2\} \rightarrow 1) + I(\{1,2\} \rightarrow 3) = 2 \leq |3| - 1.$$

Using Equations 2.151 and 2.152, the distance $d$ can be reduced to

$$d = -|C| + \sum_{v \in C} \sum_{W : W \cap C \neq \phi} \hat{x}(W \rightarrow v). \tag{2.161}$$

To cut deep into the feasible region, we wish to find a cluster such that the distance $d$ between the current LP relaxed solution $\hat{x}$ and the cutting plane is as large as possible.

Let the cluster be represented by an indicator function $I\,(v \in C)$ and $J\,(W \rightarrow v)$ be an indicator variable associated with $\hat{x}(W \rightarrow v)$. To make the efficacy the largest, we solve the following optimization problem:

$$\text{Max} \qquad -|C| + \sum_{v} \sum_{W} \hat{x}(W \rightarrow v)J(W \rightarrow v) \tag{2.162}$$

For each $J\,(W \rightarrow v)$:

$$J(W \rightarrow v) \Rightarrow I(v \in C) \tag{2.163}$$

$$J(W \rightarrow v) \Rightarrow \bigvee_{w \in W} I(w \in C) \tag{2.164}$$

$$|C| \geq 2.$$

The constraints are expressed in the language of the first order logic. $\phi \Rightarrow \psi$ indicates that $\neg\,\phi \vee \psi$. The constraint (2.163) states that $J\,(W \rightarrow v) = 1$ implies that $v \in C$ and the constraint (2.164) states that $J\,(W \rightarrow v)$ also implies that at least one member of the parent set $W$ is in the cluster set $C$. Consequently, to satisfy the constraints (2.163) and (2.153), we must have that a non-zero $v$ is in the cluster $C$ and at least one parent is also in the cluster $C$. In any feasible solution, we obtain that

$$-|C| + \sum_{v} \sum_{W} \hat{x}(W \rightarrow v)J(W \rightarrow v) \geq -|C| + |C| = 0 > -1 \quad \text{or}$$

$$-|C| + \sum_{v \in C} \sum_{W : W \cap C \neq \phi} \hat{x}(W \rightarrow v)J(W \rightarrow v) \geq -|C| + |C| = 0 > -1. \tag{2.165}$$

Any feasible solution identifies the cluster whose associated constraint (2.160) is violated by the current LP relaxation solution and defines a cutting plane. The above optimization problem will find the best cutting plane that tightens the relaxation.

An algorithm for finding a cutting plane is given below.

**Algorithm 2.2: Cutting Plane Algorithm**

1. $P_r$ = LP relaxation of ILP
2. $\hat{x}$ = solution of $P_r$;
3. If $P_r$ is unbounded or infeasible then
   stop;
   end
4. $C$ = set of clusters that will possibly contain directed cycles in $\hat{x}$;
5. While $C \neq \phi$ do
6. $c^*$ = most effective cutting plane $\in C$ by solving the optimization problem (2.162, 2.163, 2.164);
7. add the constraint $c^*$ to $P_r$;
8. $\tilde{x}$ = solution of $P_r$;
9. if $P_r$ is infeasible then
   stop
   else

   a. $\hat{x} = \tilde{x}$;
   b. $C$ = set of acyclicity constraint violated in $\hat{x}$;

   end
10. end
11. return $x^* = \hat{x}$.

## 2.6.4 Branch-and-Cut Algorithm for Integer Linear Programming

The branch and bound method is a popular algorithm that ensures finding an optimal solution to the 0-1 ILP problem. It solves the problem by conducting an implicit exhaustive and nonredundant search of the $2^p$ zero-one solutions. When the number of $p$ nodes in the DAG increases the computational time for DAG learning will increase exponentially. The basic idea of the branch and bound method is to successively divide the ILP problem into smaller problems that are easy to solve and reduce search space. We also know that a parent set is not optimal when a subset has a better score. The branch-and-cut algorithm is given below.

**Algorithm 2.3: Branch-and-Cut Algorithm**

Step 1 (Root Node): Initialize subproblem list $L := \{R\}, \hat{c} := \infty$. Calculate the score functions with a number of parent sets for each node. Create convexity constraints for each node.

Step 2: If $L = \phi$, stop and return $x^* = \hat{x}$ and $c^* = \hat{c}$.

Step 3: Node selection. Use a cutting method to check whether there are valid cluster based constraints which are not satisfied by the current LP relaxation solutions $\hat{x}$. Add constraints that are violated by the current solutions $\hat{x}$. Select $Q \in L$ and set $L := L\backslash\{Q\}$.

Step 4: A LP relaxation $Q_{relax}$ of the ILP $Q$ is obtained by removing integrality and cluster based constraints. Solve LP relaxation $Q_{relax}$ by a simplex method. If the solution is infeasible (empty), set $\check{c} := \infty$. Otherwise, let $\check{x}$ be an optimal solution of $Q_{relax}$ and $\check{c}$ its objective value.

Step 5: If $\check{c} \geq \hat{c}$, discard solution $\hat{x}$, cut off subproblems, prune the large parts of the search tree, and go to step 2. Otherwise go to step 6.

Step 6: If solution $\check{x}$ is integer-valued and feasible, then the current problem is solved and set $x^* = \hat{x} := \check{x}$ and $c^* = \hat{c} := \check{c}$. Go to step 2. Otherwise, if solution $\check{x}$ is not integer-valued, go to step 7.

Step 7: Branch on a variable with fractional value in $\hat{x}$, for example $\hat{x}_j \notin Z, j \in \{1, ..., p\}$. Create two subproblems $Q_j^- = Q \cap \{x_j \leq \lfloor \hat{x}_j \rfloor\}$ and $Q_j^+ = Q \cap \{\lceil \hat{x}_j \rceil\}$.

Branching is a key to the branch-and-cut algorithms. A popular strategy for splitting a problem into two subproblems is to branch variables. The basic algorithm for branching variable selection is given below.

**Algorithm 2.4: Branching Variable Selection Algorithm**

1. Let $F = \{k \in \{1, ..., p\} \mid x_k \notin Z\}$ be the set of branching candidate variables.
2. For all candidates $k \in F$, calculate a score value $s_k = \max\{\check{c}_{Q_k^-} - \check{c}, 10^{-6}\}\max\{\check{c}_{Q_k^+} - \check{c}_Q, 10^{-6}\}$.
3. Return an index $j \in F$ with $s_j = \max_{k \in F} \{s_k\}$.

We can branch on variable $x_j$.

### 2.6.5 Sink Finding Primal Heuristic Algorithm

Sink finding is to search for a feasible integer solution that generates a DAG near the solution to the current LP relaxation (Barlett and Cussens 2013). The purpose of a sink finding algorithm is to find a suboptimal feasible solution in the early stage of the branching process to prune the search earlier. A feasible solution ensures that a directed graph is acyclic.

The sink-finding algorithm begins with defining a cost function for selecting the best sink candidate. We hope to find a variable (node) which has estimates $\hat{x}_s(W \rightarrow s)$ close to the current LP relaxation solution. Suppose that the indicator variables for the parent set are arranged in a table. Each node is represented by a row in the table. The total number of rows is $p = |V|$. In each row we arrange the indicator variables for the parent sets according to their associated objective coefficient $C(V, W)$ (Table 2.3). For example, $x(1 \rightarrow 3)$ is the best parent set with the largest objective coefficients for the node 3 and $x(2 \rightarrow 3)$ for the worst.

Let $\hat{x}(W_{v,1} \rightarrow v)$ be the indicator variable for the best parent set with the largest object coefficients of the node $v$. For each indicator variable for the parent set of the node $v$, we define a cost: $f_v = 1 - \hat{x}(W_{v,1} \rightarrow v)$. The sink seeking algorithm is given as follows.

**TABLE 2.3**

(A) Cost Functions That Are Associated with the DAG and (B) the List of Indicator Variables Arranged According to Their Objective Coefficients

A

| Node | 1 | 2 | 3 | 4 |
|------|---|---|---|---|
| 1 | 0 | 0 | 3 | 2 |
| 2 | 0 | 0 | 2 | 0 |
| 3 | 0 | 0 | 0 | 1 |
| 4 | 0 | 0 | 0 | 0 |

B

| Node | 1 | 2 | 3 | 4 |
|------|---|---|---|---|
| 1 | {} | {} | {} | {} |
| 2 | {} | {} | {} | {} |
| 3 | $x(1{\rightarrow}3)$ | $x(2{\rightarrow}3)$ | {} | {} |
| 4 | $x(1{\rightarrow}4)$ | $x(3{\rightarrow}4)$ | {} | {} |

**Algorithm 2.5: Sink Finding Algorithm**

1. Select the sink candidate $V_j = \min_v f_v$.
2. Delete all parent sets that contain the parents of the selected node $V_j$ and generate a new table in which these parent sets are deleted. For example, the selected node is $V_j = 3$. The parent sets containing the parents of the selected node $V_j = 3$ are node {1} and {2}. The transformed table from Table 2.3B after removing the indicator $x$ $(1 \rightarrow 4)$ in which the parent node {1} in $x$ $(1 \rightarrow 4)$ contained the parent node {1} in $x$ $(1 \rightarrow 3)$ is shown in Table 2.4.

Repeat Steps 1 and 2 until all nodes visited.

A sink node is a node that has no children. A DAG must have at least one sink node. The sink finding algorithm seeks parent sets for each node in the DAG. It begins to find the best node $v$ with the smallest cost as a sink node. Then, we remove all parent sets for other nodes which contains $v$ to ensure that the node $v$ has no children. In the second iteration, the sink finding algorithm repeats the same process to seek a sink node for a DAG with the remaining nodes $V \backslash \{v\}$. In the subsequent iteration, the algorithm works similarly until the parent sets for all nodes in the graph are found and a DAG is completely constructed.

**TABLE 2.4**

The Transformed Table from Table 2.3B After Removing the Parent Sets That Contain the Parents of the Selected Sink Node {3}

| Node | 1 | 2 | 3 | 4 |
|------|---|---|---|---|
| 1 | {} | {} | {} | {} |
| 2 | {} | {} | {} | {} |
| 3 | $x$ $(1 \rightarrow 3)$ | $x$ $(2 \rightarrow 3)$ | {} | {} |
| 4 | | $x$ $(3 \rightarrow 4)$ | {} | {} |

## 2.7 Simulations and Real Data Analysis

### 2.7.1 Simulations

The performance of the score-based method for network structure inference was evaluated in simulation studies of a phenotype network. The simulations were carried out for the network with 30 nodes and 190 directed edges. The sparse SEMs were used to model the phenotype network and to calculate the node score metrics. The SEMs for randomly generating the network structures and parameters were

$$Y = \varepsilon \Gamma^{-1}, \tag{2.166}$$

where $\varepsilon \sim N (0, 0.05 \times I)$. The parameters $\Gamma_{ij}$ in the SEMs for modeling phenotype sub-networks were generated from a uniformly distributed random variable over the interval $(0.5, 1)$ or $(-1, -0.5)$ if an edge from node $j$ to node $i$ was presented in the phenotype sub-network; otherwise $\Gamma_{ij} = 0$.

For the randomly generated phenotype network, the expected number of degrees per node is 3. Simulations were repeated 5000 times. Fivefold cross-validations were used to determine the penalty parameter $\lambda$ that was then employed to infer the network while running power simulations. Two measures: the power of detection (PD) and the false discovery rate (FDR) were used to evaluate the performance of the algorithms for identification of the network structures. Specifically, let $N_t$ be the total number of edges among 5000 replicates of the network and $\hat{N}_t$ be the total number of edges detected by the inference algorithm, $N_{true}$ be the total number of true edges detected among simulated network, and $N_{False}$ be the false edges detected among $\hat{N}_t$. Now, the PD is defined by $\dfrac{N_{True}}{N_t}$ and the false discovery rate (FDR) is defined by $\dfrac{N_{False}}{\hat{N}_t}$.

The PD and FDR of five statistical methods: the proposed SEMs combined with IP (SEMIP), the SEM, SEM with maximum likelihood (sparse MLE) (Cai et al. 2013), QTLnet (Neto et al. 2008), and the co-association (correlation) network inferred by the weighted correlation network analysis. Figure 2.15 showed the PD and FDR of five statistics as a function of the sample sizes. We observed two remarkable features. First, the SEMIP outperforms all the other four methods. The SEMIP had the highest PD and the smallest FDR. Second, the PD of all five statistical methods increased, but their FDR decreased, as the

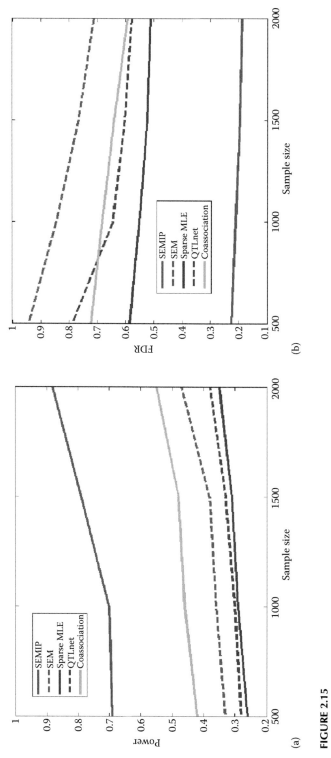

**FIGURE 2.15**

(a) The PD and (b) FDR curves of the three methods for causal phenotype network inference.

sample size increased. Third, the FDR of all four other methods was very high. It was not expected that their FDR was still higher than 60% even when the sample size was 2000. This demonstrated that the many popular statistical methods without combinatorial optimization to search the networks for the best score might not be suitable for causal network analysis.

### 2.7.2 Real Data Analysis

To further evaluate its performance, we applied the proposed SEMIP to the UK-2K dataset. The UK2K Cohorts project used a low read depth whole-genome sequencing (WGS) to assess the contribution of the genetic variants to the 64 different traits (Walter et al. 2015). However, missing phenotypes were found in many individuals. To ensure that there were no missing phenotypes in individuals, we included 765 individuals with 2,240,049 SNPs in 33,746 genes, and shared 39 traits in 13 major phenotypic groups, which covered a wide range of traits in the analysis. We took the rank-based inverse normal transformation of the phenotypes as trait values to ensure that the trait values follow normal distributions.

To study the genotype–phenotype network with rare variants, the gene-based functional SEMs were used to construct a genotype–phenotype network that will be used as the initial network covering the true causal genotype–phenotype network. The procedures for inferring genome-wide genotype–phenotype networks consist of four steps. The first step is to estimate functional principal component scores for each gene. The second step is to identify all genes significantly associated with 39 traits using the quadratically regularized functional canonical correlation analysis (QRFCCA). The third step is to estimate the initial network structure using the FSEMs. The final step is to infer the genotype–phenotype network using the FSME combined with the IP method where the FSEMs are used to calculate the score function.

Applying the QRFCCA to the UK-2K dataset, a total of 79 genes with pure rare variants were identified to be significantly associated with 39 traits. Then, we use 39 traits and 79 genes to construct the genotype–phenotype networks. The largest connected genotype–phenotype causal network with 39 trait nodes, 58 gene nodes, and 323 directed edges was shown in Figure 2.16 where the nodes in red represent the trait nodes and the nodes in green represent the gene nodes, black solid lines represent the causal relations between phenotypes, and blue dotted lines represent the causal relations between the genotype and phenotype (Wang 2016).

We observed directed and indirect causal effects between variables. The directed causal effect means that variation of a variable directly affects changes in another variable and the indirect causal effect indicates that the variation of a variable affects the changes in another variable via mediation of other variables. Through path search, we find 1112 pairs of causal relations in the network. Among them, 916 pairs of causal relations showed indirect causal effects and accounted for 82.4% of the causal relations. Table 2.5 listed

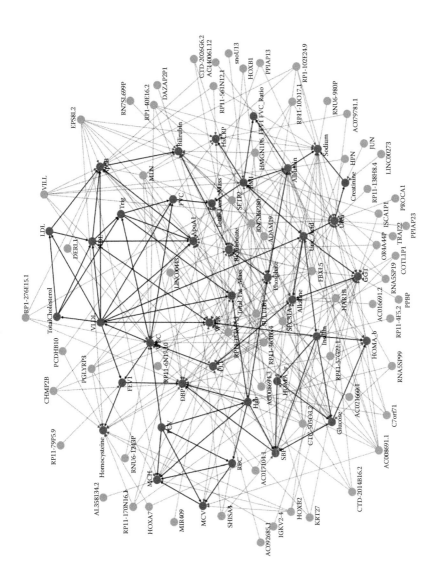

**FIGURE 2.16**

The estimated genotype–phenotype network with 39 phenotype nodes and 58 gene nodes.

**TABLE 2.5**

A List of 20 Pairs of Causal Phenotype–Phenotype and Genotype–Phenotype Relations

| Response | Predictor | Direct Effect | Indirect Effect | Total Effect | Marginal Effect |
|---|---|---|---|---|---|
| Alkaline | PLT | 0.1397 | 0.0053 | 0.1450 | 0.1467 |
| ApoA1 | HDL | 0.8451 | −0.0369 | 0.8082 | 0.7814 |
| ApoA1 | VLDL | 0.0840 | −0.3284 | −0.2444 | −0.1687 |
| ApoB | HDL | −0.1877 | 0.2109 | 0.0233 | −0.1789 |
| FEV1 | WBC | −0.1577 | −0.0260 | −0.1837 | −0.1818 |
| FVC | VLDL | −0.0610 | −0.0131 | −0.0740 | −0.1625 |
| FVC | WBC | −0.1500 | −0.0309 | −0.1809 | −0.1798 |
| Glucose | Insulin | −0.4846 | 0.9631 | 0.4785 | 0.4668 |
| HDL | TotalCholesterol | 2.1633 | −1.9076 | 0.2556 | 0.1939 |
| MCH | Hgb | 1.1621 | 0.6690 | 1.8311 | 0.2664 |
| Total_Fat_Mass | VLDL | 0.1217 | 0.0261 | 0.1477 | 0.2667 |
| Alkaline | AC008694.3 | −0.0901 | 0.0181 | −0.0720 | −0.0634 |
| ApoA1 | RP11-363E6.4 | 0.0534 | 0.0005 | 0.0539 | 0.0553 |
| ApoB | RP1-276E15.1 | −0.0915 | −0.0058 | −0.0973 | −0.0723 |
| Bicarbonate | HOXB1 | −0.0950 | 0.0112 | −0.0839 | −0.0709 |
| Bicarbonate | OR4A44P | −0.0870 | 0.0010 | −0.0860 | −0.0652 |
| Bilirubin | RP11-153M7.3 | −0.0176 | 0.0067 | −0.0109 | −0.0117 |
| Total_Lean_Mass | MLN | −0.0628 | −0.0056 | −0.0684 | −0.0628 |
| Urea | FBXL5 | 0.0661 | 0.0008 | 0.0669 | 0.0632 |
| WBC | RP11-363E6.4 | −0.0946 | 0.0842 | −0.0104 | −0.0133 |

20 pairs of causal phenotype–phenotype relations and genotype–phenotype relations. The direct effect is the path coefficient of the direct edge from the causal node to the outcome end node. The indirect effect is the product of path coefficients of the path from the causal node to the end outcome node through intermediate mediation nodes. The total effect is the summation of direct and indirect effects. In Chapter 1, we show that if all causal paths including direct and indirect effects are discovered, then the marginal regression effect will be equal to the total effect. If the marginal and total effects of a variable act on another variable, then the causal paths between two variables have not been completely discovered. We observed two remarkable features from Table 2.5. The first, all causal phenotypes and genotypes had both direct and indirect effects. The second, the marginal and total effect of the three phenotype-phenotype pairs: PLT Alkaline, WBC -> FEV1 and WBC -> FVC, and one genotype–phenotype pair RP11-363E6.4 -> ApoA1 were approximately equal. However, marginal effects of the remaining 16 causal phenotypes or genotypes were either larger or smaller than the total effects. This showed that many of the confounding factors were unmeasured or not included in the analysis.

## Software Package

The software for exact Bayesian network learning including Bayesian-Dirichlet equivalence (BDe) and minimal description length (MDL) score metrics can be found in the G6G Directory of Omics and Intelligent Software (http://g6g-softwaredirectory.com/bio/cross-omics/pathway-analysis -grns/20697-Univ-Warsaw-BNFinder.php). The bnlearn R is an R package that includes several algorithms for learning the structure of Bayesian networks with either discrete or continuous variables. Both constraint-based and score-based algorithms are implemented (https://arxiv.org/pdf/0908.3817). The "bene" package contains software for constructing the globally optimal Bayesian network structure using decomposable scores AIC, BIC, BDe, fNML, and LOO (https://github.com/tomisilander/bene). R-package pcalg (Methods for Graphical Models and Causal Inference) includes functions for causal structure learning and causal inference using graphical models. The main algorithms for causal structure learning are PC (for observational data without hidden variables), FCI and RFCI (for observational data with hidden variables), and GIES (for a mix of data from observational studies [i.e., observational data], and data from experiments involving interventions [i.e., interventional data] without hidden variables). For causal inference the IDA algorithm, the generalized backdoor criterion (GBC), and the generalized adjustment criterion (GAC) are implemented. Pcalg can be downloaded from https://cran.r-project.org/web/packages/pcalg/index.html. GOBNILP (Globally Optimal Bayesian Network learning using Integer Linear Programming) is a C program which learns Bayesian networks from complete discrete data or from local scores. GOBNILP uses the SCIP framework for constraint integer programming. It can be downloaded from https://www.cs.york.ac.uk/aig /sw/gobnilp/. The COMICN package is a causal inference for OMICs with next-generation sequencing data. It can exactly learn causal networks from both observational and interventional data. The COMICN takes a core-based approach to causal network learning. It combines network penalized logistic regression, structural equations, functional structural equations, and integer programming. The COMICN can be downloaded from http://www.sph.uth .tmc.edu/hgc/faculty/xiong/index.htm.

## Appendix 2.A    Introduction to Smoothing Splines

To make the book as self-contained as possible, in this appendix we introduce the basic theory of smoothing splines. Much of the materials for smoothing splines are from the book written by Wang (2011). For more details, we refer the reader to Wang (2011).

### 2.A.1  Smoothing Spline Regression for a Single Variable

We defined a *Hilbert space H* and RKHS in Section 4.3. Let $L$ be a continuous linear functional defined on a *Hilbert space H*. For example, $L$ can be an evaluational functional $L_x$ defined as

$$L_x f = f(x), \qquad (2.A.1)$$

or an integral functional defined as

$$Lf = \int_0^1 f(x)dx.$$

We consider a general smoothing spline regression model:

$$y_i = L_i f + \varepsilon_i, \qquad (2.A.2)$$

$$i = 1, 2, ..., n,$$

where $y_i$ is the observed response of the *ith* individual, $z_i$ is the design point of the *ith* individual, $L_i$ is a continuous functional defined on $H$, $f$ represents a general unknown function, and $\varepsilon_i$ are zero-mean independent random errors with a common variance $\sigma_e^2$. If $L_i$ is evaluational functional, then Equation 2.A.2 is reduced to

$$y_i = f(x_i) + \varepsilon_i, \ i = 1, 2, .., n. \qquad (2.A.3)$$

The *Hilbert space H* can be decomposed into two subspaces (Wang 2011):

$$H = H_0 \oplus H_1, \qquad (2.A.4)$$

where $H_0$ is a finite dimensional space with orthonormal basis functions $\varphi_1(x)$, ..., $\varphi_p(x)$ and its orthogonal complement $H_1$ that is an RKHS with reproducing kernel (RK) $R_1(x, z)$. Suppose that $H$ is an RKHS with RK $R(x, z)$. Let $R_0(x, z)$ be RK associated with $H_0$. Then, we have

$$R(x, z) = R_0(x, z) + R_1(x, z). \qquad (2.A.5)$$

We can show that RK $R_0(x, z)$ is given by

$$R_0(x, z) = \sum_{j=1}^{p} \varphi_j(x)\varphi_j(z). \qquad (2.A.6)$$

To show equality (2.A.6), by definition of the reproducing kernel, we only need to show that

$$g(x) = (g(.), R_0(x, .) >, \qquad (2.A.7)$$

where $g \in H_0$. In fact, since $g \in H_0$ and $\{\varphi_1(x),\ldots, \varphi_p(x)\}$ are a set of orthonormal bases functions, any function $g \in H_0$ can be expressed as

$$g(x) = \sum_{i=1}^{p} a_i \varphi_i(x). \tag{2.A.8}$$

Thus, we have

$$
\begin{aligned}
< g(.), R_0(x,.) > &= < \sum_{i=1}^{p} a_i \varphi_i(.), \sum_{j=1}^{p} \varphi_j(x)\varphi_j(.) > \\
&= \sum_{i=1}^{p} a_i < \varphi_i(.), \sum_{j=1}^{p} \varphi_j(x)\varphi_j(.) > \\
&= \sum_{i=1}^{p} a_i \sum_{j=1}^{p} < \varphi_i(.), \varphi_j(.) > \varphi_j(x)
\end{aligned}
\tag{2.A.9}
$$

The assumption of the orthonormality of the basis functions implies that

$$< \varphi_i, \varphi_j >= \begin{cases} 1 & i = j \\ 0 & i \neq j \end{cases}. \tag{2.A.10}$$

Substituting Equation 2.A.10 into Equation 2.A.9 gives

$$< g(.), R_0(x,.) >= \sum_{i=1}^{p} a_i \varphi_i(x) = g(x).$$

This shows that $R_0(x,z) = \sum_{j=1}^{p} \varphi_j(x)\varphi_j(z)$ is the RK for the RKHS $H_0$.

Similarly, suppose that $\{\psi_1(x), \ldots, \psi_q(x)\}$ is a set of orthonormal basis functions for the RKHS $H_1$, then the RK $R_1(x, z)$ is given by

$$R_1(x,z) = \sum_{i=1}^{q} \psi_i(x)\psi_i(z). \tag{2.A.11}$$

The function $f$ to be fitted consists of two parts: (1) projection function $f_0$ on $H_0$ and projection function $f_1$ on $H_1$, that is,

$$f = f_0 + f_1. \tag{2.A.12}$$

To estimate the function $f$ in Equation 2.A.2, we should make both the errors between the observations $y_i$ and the estimator function $\hat{f}$ and departure from the null space $H_0 \|f_1\|^2$ small.

The function $f_1$ is the projection $P_1f$ of the function $f$ on the RKHS $H_1$. Therefore, to achieve this goal, the objective function for the smoothing spline estimator of the function $f$ is given by

$$\frac{1}{n} \sum_{i=1}^{n} (y_i - L_i f)^2 + \lambda \| P_1 f \|^2, \tag{2.A.13}$$

where $\lambda$ is a smoothing penalty parameter to balance the goodness of fit and the smoothness. Increasing the penalty $\lambda$ will increase the smoothness, but decrease the prediction accuracy. Our goal is to find $f$ to minimize the objective function in Equation 2.A.13:

$$\min_f \quad \frac{1}{n}\sum_{i=1}^n (y_i - L_i f)^2 + \lambda \|P_1 f\|^2. \tag{2.A.14}$$

Next, we develop algorithms for searching for the optimal solution to 2.A.14. First, we discuss how to use a reproducing kernel to transform the functional optimization problem (2.A.14) into the classical real-valued optimization problem. A key for achieving this is to find the appropriate representation of the general function $f$.

A powerful tool for representation of the function $f$ is the reproducing kernel theory. We assume that $L$ is a continuous functional, by the Riesz representation theorem, there exists a representor $\eta_i(x) \in H$ such that

$$L_i f = < \eta_i, f) >. \tag{2.A.15}$$

Since $H$ is a RKHS, we can find RK $R_x(z) = R(x, z)$ such that

$$\eta_i(x) = < \eta_i(.), R(x,.) >= L_i R(x,.) = L_{i(z)} R(x,z). \tag{2.A.16}$$

Equation 2.A.16 recovered the relationship between the representor $\eta_i(x)$ and the RK and implies that it can be obtained by applying the functional to the RK $R(x, z)$. It is known that the reproducing kernel $R(x, z)$ can be decomposed into

$$R(x,z) = R_0(x,z) + R_1(x,z), \tag{2.A.17}$$

where $R_0(x, z)$ and $R_1(x, z)$ are RKs of $H_0$ and $H_1$, respectively. Therefore, it follows from Equation 2.A.17 that

$$\eta_i(x) = L_{i(z)} R(x,z) = L_{i(z)} R_0(x,z) + L_{i(z)} R_1(x,z). \tag{2.A.18}$$

Next, we calculate $L_{i(z)} R_1(x, z)$. Using Equation 2.A.15 gives

$$L_{i(z)} R_1(x,z) = < \eta_i, R_1(x,z) >. \tag{2.A.19}$$

Note that the RK $R_1(x, z)$ is projection of $R(x, z)$ on to $H_1$:

$$R_1(x,z) = P_1 R(x,z). \tag{2.A.20}$$

Substituting Equation 2.A.20 into Equation 2.A.19 yields

$$L_{i(z)} R_1(x,z) = < \eta_i, P_1 R(x,z) >$$
$$= < P_1 \eta_i, R(x,z) > \tag{2.A.21}$$
$$= P_1 \eta_i.$$

Denote $P_1 \eta_i$ by $\xi_i(x)$. Equation 2.A.19 is reduced to

$$L_{i(z)}R_1(x, z) = \xi_i(x). \tag{2.A.22}$$

Next we calculate $L_{i(z)}R_0(x, z)$. Equation 2.A.6 gives

$$L_{i(z)}R_0(x, z) = \sum_{j=1}^{p} \varphi_j(x) L_{i(z)} \varphi_j(z). \tag{2.A.23}$$

Combing Equations 2.A.18, 2.A.22, and 2.A.23, we obtain

$$\eta_i(x) = \sum_{j=1}^{p} \left( L_{i(z)} \varphi_j(z) \right) \varphi_j(x) + \xi_i(x). \tag{2.A.24}$$

We take $\{\eta_1(x), \ldots, \eta_n(x)\}$ as a set of basis functions in $H$. Function $\hat{f} \in H$ can then be expressed as

$$
\begin{aligned}
\hat{f}(x) &= \sum_{i=1}^{n} b_i \eta_i(x) \\
&= \sum_{j=1}^{p} \left( L_{i(z)} \varphi_j(z) \right) \varphi_j(x) + \sum_{i=1}^{n} b_i \xi_i(x)
\end{aligned}
\tag{2.A.25}
$$

Let $a_j = L_{i(z)} \phi_j(z)$. Then, Equation 2.A.25 is reduced to

$$\hat{f}(x) = \sum_{j=1}^{p} a_j \varphi_j(x) + \sum_{i=1}^{n} b_i \xi_i(x). \tag{2.A.26}$$

Now it is ready to calculate the objective function (2.A.13). Note that Equation 2.A.22 gives

$$L_{i(x)} \xi_j(x) = L_{i(x)} L_{j(z)} R_1(x, z). \tag{2.A.27}$$

Since $P_1$ is a projection operator on $H_1$ and $\varphi_j \in H_0$, $\xi_j \in H_1$, applying the projection operator $P_1$ to $\hat{f}(x)$ yields

$$
\begin{aligned}
P_1 \hat{f}(x) &= \sum_{i=1}^{n} b_i \xi_i(x) \\
&= \xi^T b,
\end{aligned}
\tag{2.A.28}
$$

where

$$\xi = \begin{bmatrix} \xi_1(x) \\ \vdots \\ \xi_n(x) \end{bmatrix} \text{ and } b = \begin{bmatrix} b_1 \\ \vdots \\ b_n \end{bmatrix}.$$

Since $\xi_j \in H_1$ and $\xi_j = P_1 \eta$, then we obtain

$$L_{i(x)} \xi_j = < \xi_i, \xi_j >. \tag{2.A.29}$$

Combining Equations 2.A.27 and 2.A.29, we obtain

$$L_{i(x)}L_{j(z)}R_1(x,z) = <\xi_i, \xi_j>. \tag{2.A.30}$$

Applying the operator $L_{(ix)}$ to Equation 2.A.26, we obtain

$$L_{i(x)}\hat{f}(x) = \sum_{j=1}^{p} a_j L_{i(x)}\varphi_j(x) + \sum_{i=1}^{n} b_i L_{i(x)}\xi_i(x). \tag{2.A.31}$$

Substituting Equation 2.A.27 into Equation 2.A.31 gives

$$L_{i(x)}\hat{f}(x) = \sum_{j=1}^{p} a_j L_{i(x)}\varphi_j(x) + \sum_{i=1}^{n} b_i L_{i(x)}L_{j(z)}R_1(x,z). \tag{2.A.32}$$

Therefore, we have

$$
\begin{bmatrix} L_{1(x)}\hat{f}(x) \\ \vdots \\ L_{n(x)}\hat{f}(x) \end{bmatrix} = \begin{bmatrix} L_{1(x)}\varphi_1(x) & \cdots & L_{1(x)}\varphi_p(x) \\ \vdots & \vdots & \vdots \\ L_{n(x)}\varphi_1(x) & \cdots & L_{n(x)}\varphi_p(x) \end{bmatrix} \begin{bmatrix} a_1 \\ \vdots \\ a_p \end{bmatrix}
$$
$$
+ \begin{bmatrix} L_{1(x)}L_{1(z)}R_1(x,z) & \cdots & L_{1(x)}L_{n(z)}R_1(x,z) \\ \vdots & \vdots & \vdots \\ L_{n(x)}L_{1(z)}R_1(x,z) & \cdots & L_{n(x)}L_{n(z)}R_1(x,z) \end{bmatrix} \begin{bmatrix} b_1 \\ \vdots \\ b_n \end{bmatrix} \tag{2.A.33}
$$
$$
= Ta + \Sigma b,
$$

where

$$
T = \begin{bmatrix} L_{1(x)}\varphi_1(x) & \cdots & L_{1(x)}\varphi_p(x) \\ \vdots & \vdots & \vdots \\ L_{n(x)}\varphi_1(x) & \cdots & L_{n(x)}\varphi_p(x) \end{bmatrix}, a = \begin{bmatrix} a_1 \\ \vdots \\ a_p \end{bmatrix},
$$

$$
\Sigma = \begin{bmatrix} L_{1(x)}L_{1(z)}R_1(x,z) & \cdots & L_{1(x)}L_{n(z)}R_1(x,z) \\ \vdots & \vdots & \vdots \\ L_{n(x)}L_{1(z)}R_1(x,z) & \cdots & L_{n(x)}L_{n(z)}R_1(x,z) \end{bmatrix} \text{ and } b = \begin{bmatrix} b_1 \\ \vdots \\ b_n \end{bmatrix}.
$$

Next, we calculate $\| P_1 f \|^2$. It follows from Equations 2.A.28 and 2.A.30 that

$$\|P_1\hat{f}\|^2 = b^T \xi\xi^T b$$
$$= b^T \Sigma b \tag{2.A.34}$$

Let

$$
Y = \begin{bmatrix} y_1 \\ \vdots \\ y_n \end{bmatrix}.
$$

Using Equations 2.A.33 and 2.A.34, we can transform the optimization problem (2.A.14) to

$$\min_{a,b} \ \frac{1}{n}||Y - Ta - \Sigma b||^2 + \lambda b^T \Sigma b. \qquad (2.A.35)$$

Let $F = \frac{1}{n}||Y - Ta - \Sigma b||^2 + \lambda b^T \Sigma b$. Setting the partial derivatives of the function $F$ with respect to the vectors $a$ and $b$ to zero yields

$$\frac{\partial F}{\partial a} = -\frac{2}{n}T^T(Y - Ta - \Sigma b) = 0 \qquad (2.A.36)$$

$$\frac{\partial F}{\partial b} = -\frac{2}{n}\Sigma(Y - Ta - \Sigma b) + 2\lambda\Sigma b = 0. \qquad (2.A.37)$$

Solving Equations 2.A.36 and 2.A.37, we obtain

$$\Sigma Ta + (\Sigma + n\lambda I)\Sigma b = \Sigma Y, \qquad (2.A.38)$$

$$TT^T a + T^T \Sigma b = T^T Y. \qquad (2.A.39)$$

Multiplying $\Sigma^{-1}$ on both sides of Equation 2.A.38 yields

$$Ta + (\Sigma + n\lambda I)b = Y. \qquad (2.A.40)$$

Again, multiplying $T^T$ on both sides of Equation 2.A.40 gives

$$T^T Ta + T^T \Sigma b + n\lambda IT^T b = T^T Y. \qquad (2.A.41)$$

Substituting Equation 2.A.39 into Equation 2.A.41 yields

$$T^T b = 0. \qquad (2.A.42)$$

In summary, the solution to the optimization problem (2.A.35) satisfies the following two equations:

$$Ta + (\Sigma + n\lambda I)b = Y, \qquad (2.A.43)$$

$$T^T b = 0. \qquad (2.A.44a)$$

Let $M = \Sigma + n\lambda I$. Solving Equation 2.A.43 for $b$, we obtain

$$b = M^{-1}y - M^{-1}Ta. \qquad (2.A.44b)$$

Substituting Equation 2.A.44a into Equation 2.A.44b gives

$$T^T M^{-1}y - T^T M^{-1}Ta = 0. \qquad (2.A.45)$$

Solving Equation 2.A.45 for $a$, we obtain

$$a = (T^T M^{-1} T)^{-1} T^T M^{-1} Y. \tag{2.A.46}$$

Substituting Equation 2.A.46 into Equation 2.A.44 gives the solution:

$$b = M^{-1} \left[ I - T(T^T M^{-1} T)^{-1} T^T M^{-1} \right] Y. \tag{2.A.47}$$

To efficiently compute $a$ and $b$, we use the QR decomposition of $T$:

$$T = [Q_1 \ Q_2] \begin{bmatrix} R \\ 0 \end{bmatrix},$$

where $Q = [Q_1 \ Q_2]$ is an orthogonal matrix and $R$ is an invertible upper triangular matrix. Using Equation 2.A.44, we obtain

$$T^T b = [R^T \ 0] \begin{bmatrix} Q_1^T \\ Q_2^T \end{bmatrix} b = R^T Q_1^T b = 0. \tag{2.A.48}$$

Since $R$ is invertible, Equation 2.A.48 implies

$$Q_1^T b = 0. \tag{2.A.49}$$

By the assumption that $Q$ is an orthogonal matrix, we have

$$QQ^T = I, \text{ which implies}$$

$$b = (QQ^T)b = Q_1 Q_1^T b + Q_2 Q_2^T b. \tag{2.A.50}$$

Substituting Equation 2.A.49 into Equation 2.A.50 leads to

$$b = Q_2 Q_2^T b. \tag{2.A.51}$$

Note that

$$Q_2^T T = [Q_2^T Q_1 \ Q_2^T Q_2] \begin{bmatrix} R \\ 0 \end{bmatrix} = [0 \ I] \begin{bmatrix} R \\ 0 \end{bmatrix} = 0. \tag{2.A.52}$$

Multiplying Equation (2.A.40) by $Q_2^T$ and using Equation (2.A.52) gives

$$Q_2^T M b = Q_2^T Y. \tag{2.A.53}$$

Substituting Equation 2.A.51 into Equation 2.A.53 yields

$$Q_2^T M Q_2 Q_2^T b = Q_2^T Y. \tag{2.A.54}$$

Solving Equation 2.A.54 for $b$, we obtain the solution for $b$:

$$\hat{b} = Q_2 \left( Q_2^T M Q_2 \right)^{-1} Q_2^T Y. \tag{2.A.55}$$

Note that

$$Q_1^T T = \begin{bmatrix} Q_1^T Q_1 & Q_1^T Q_2 \end{bmatrix} \begin{bmatrix} R \\ 0 \end{bmatrix} = \begin{bmatrix} I & 0 \end{bmatrix} \begin{bmatrix} R \\ 0 \end{bmatrix} = R. \tag{2.A.56}$$

Again, multiplying Equation 2.A.40 by $Q_1$ gives

$$Ra = Q_1^T (Y - Mb), \text{ or}$$

$$a = R^{-1} Q_1^T (Y - Mb). \tag{2.A57}$$

Substituting Equation 2.A.55 into Equation 2.A.57 gives

$$\hat{a} = R^{-1} Q_1^T \left[ I - M Q_2 \left( Q_2^T M Q_2 \right)^{-1} Q_2^T \right] Y. \tag{2.A.58}$$

The fitted value $\hat{f}$ is defined as

$$\hat{f} = T\hat{a} + \Sigma \hat{b}. \tag{2.A.59}$$

Substituting Equation 2.A.59 into Equation 2.A.40 and using Equation 2.A.55, we obtain the formula for computing the fitted value:

$$\hat{f} = Y - n\lambda b = \left[ I - n\lambda Q_2 \left( Q_2^T M Q_2 \right)^{-1} Q_2^T \right] Y = H(\lambda)Y, \tag{2.A.60}$$

where

$$H(\lambda) = I - n\lambda Q_2 \left( Q_2^T M Q_2 \right)^{-1} Q_2^T. \tag{2.A.61}$$

In summary, the algorithm for smoothing spline regression is

Step 1: Select the penalty parameter $\lambda$.

Step 2: Compute the matrices

$$T = \begin{bmatrix} L_{1(x)} \varphi_1(x) & \cdots & L_{1(x)} \varphi_p(x) \\ \vdots & \vdots & \vdots \\ L_{n(x)} \varphi_1(x) & \cdots & L_{n(x)} \varphi_p(x) \end{bmatrix} \text{ and } \Sigma$$

$$= \begin{bmatrix} L_{1(x)} L_{1(z)} R_1(x,z) & \cdots & L_{1(x)} L_{n(z)} R_1(x,z) \\ \vdots & \vdots & \vdots \\ L_{n(x)} L_{1(z)} R_1(x,z) & \cdots & L_{n(x)} L_{n(z)} R_1(x,z) \end{bmatrix}$$

Step 3: Perform QR decomposition of the matrix $T$:

$$T = [Q_1 \ Q_2] \begin{bmatrix} R \\ 0 \end{bmatrix}.$$

Step 4: Compute coefficients of the smoothing spline regression

$$\hat{a} = R^{-1}Q_1^T \left[ I - MQ_2 \left( Q_2^T MQ_2 \right)^{-1} Q_2^T \right] Y \text{ and } \hat{b} = Q_2 \left( Q_2^T MQ_2 \right)^{-1} Q_2^T Y,$$

where $M = \Sigma + n\lambda I$.
Step 5: Compute the smoothing spline regression function

$$\hat{f}(x) = \sum_{j=1}^{p} \hat{a}_j \varphi_j(x) + \sum_{i=1}^{n} \hat{b}_i \xi_i(x),$$

where $\xi_i(x) = L_{i(z)} R_1(x, z)$.
Step 6. Compute the fitted value:

$$\hat{f} = H(\lambda)Y,$$

where $H(\lambda) = I - n\lambda Q_2 (Q_2^T MQ_2)^{-1} Q_2^T$.

**Example 2.A.1: Linear and Cubic Splines**

First, we consider a linear spline. The RKHS $H_0$ and RKHS $H_1$ are formed, respectively, by

$$H_0 = span\{1\},$$

$$H_1 = \left\{ f : f = 0, \int_a^b \left( f' \right)^2 dx < \infty \right\},$$

with corresponding RKs:

$$R_0(x, z) = 1 \ ,$$

$$R_1(x, z) = \int_0^{\min(x,z)} du = \min(x, z).$$

Now we consider a cubic spline. Its $H_0$ and $H_1$ are given by

$$H_0 = span\{1, x\},$$

$$H_1 = \left\{ f : f = 0, f'(a) = 0, \int_a^b \left( f'' \right)^2 dx < \infty \right\}.$$

Their corresponding RKs are, respectively, given by

$$R_0(x,z) = 1 + xz,$$

$$R_1(x,z) = \int_0^{x \wedge z} (x \wedge z - u)(x \vee z - u)du$$

$$= \int_0^{x \wedge z} \left[ (x \wedge z)(x \vee z) - (x \wedge z + x \vee z)u + u^2 \right] du$$

$$= (x \vee z)(x \wedge z)^2 - (x \wedge z + x \vee z)\frac{(x \wedge z)^2}{2} + \frac{(x \wedge z)^3}{3}$$

$$= (x \wedge z)^2 \left[ x \vee z - \frac{x \wedge z + x \vee z}{2} + \frac{x \vee z}{3} \right]$$

$$= (x \wedge z)^2 \frac{3(x \vee z) - x \wedge z}{6}$$

**Example 2.A.2: Linear and Cubic Splines Based on Bernoulli Polynomials**

First, we define Bernoulli polynomials as

$$k_r(x) = \frac{B_r(x)}{r!}, \tag{2.A.62}$$

where $B_r(x)$ are recursively defined as

$$B_0(x) = 1,$$
$$B_r'(x) = rB_{r-1}(x),$$
$$\int_0^1 B_r(x)dx = 0$$

Now we derive the first four-scaled Bernoulli polynomials. It is clear that $B_0(x) = 1$.

$$B_1(x) = \int B_0(x)dx = \int dx = x + c, \text{ but } B_1(x) \text{ must satisfy}$$

$$\int_0^1 B_1(x)dx = \frac{1}{2} + c = 0$$

which implies $c = -\frac{1}{2}$. Therefore, $B_1(x)$ is

$$B_1(x) = x - \frac{1}{2}.$$

Next we compute $B_2(x)$. By the recursive formula, we obtain

$$B_2(x) = 2\int B_1(x)dx = 2\left[\frac{x^2}{2} - \frac{x}{2} + c\right] = x^2 - x + c \text{ and}$$

$$\int_0^1 B_2(x)dx = \frac{1}{3} - \frac{1}{2} + c = 0.$$

Solving the above equation for $c$ gives $c = \frac{1}{6}$. Thus, $B_2(x)$ is

$$B_2(x) = x^2 - x + \frac{1}{6}.$$

Similarly, we obtain

$$B_3(x) = x^3 - \frac{3x^2}{2} + \frac{x}{2} \text{ and}$$

$$B_4(x) = x^4 - 2x^3 + x^2 - \frac{1}{30}.$$

It follows from Equation 2.A.62 that the first four-scaled Bernoulli polynomials are

$$k_0(x) = 1,$$

$$k_1(x) = x - \frac{1}{2},$$

$$k_2(x) = \frac{1}{2}\left\{k_1^2(x) - \frac{1}{12}\right\},$$

$$k_4(x) = \frac{1}{24}\left\{k_1^4(x) - \frac{1}{2}k_1^2(x) + \frac{7}{240}\right\}$$

We define the *Sobolev* space $W_2^m[a, b]$ as a set of functions with absolutely continuous $f, f', ..., f^{(m-1)}$ and $\int_a^b (f^{(m)})^2 dx < \infty$.

We define an inner product in the *Sobolev* space $W_2^m[0, 1]$:

$$<f, g> = \sum_{j=0}^{m-1}\left(\int_0^1 f^{(j)}dx\right)\left(\int_0^1 g^{(j)}dx\right) + \int_0^1 f^{(m)}g^{(m)}dx.$$

It can be shown (Wang 2011) that $W_2^m[0, 1] = H_0 \oplus H_1$ where

$$H_0 = \text{span}\{k_0(x), k_1(x), ..., k_{m-1}(x)\} \text{ and}$$

$$H_1 = \left\{f : \int_0^1 f^{(j)}dx = 0, j = 0, 1, ..., m-1, \int_0^1 \left(f^{(m)}\right)^2 dx < \infty\right\}.$$

We can show that $\{\, k_0(x)\,, k_1(x)\,, \ldots, k_{m-1}\,(x)\,\}$ are a set of orthonormal basis functions. In fact,

$$\int_0^1 B_j^{(v)} dx = \begin{cases} 1 & v = j \\ 0 & v \neq j \end{cases},$$

which implies

$$\int_0^1 B_j^{(v)} dx \int_0^1 B_i^{(v)} dx = \begin{cases} 1 & i = j = v \\ 0 & \text{otherwise} \end{cases}.$$

Therefore, we have

for $i = j$

$$< k_j(x), k_i(x) >= \sum_{v=0}^{i} \left( \int_0^1 \left( k_j^{(v)} dx \right) \right) \left( \int_0^1 \left( k_i^{(v)} dx \right) \right)$$

$$= \sum_{v=0}^{i} \int_0^1 \frac{B_j^{(v)}}{j!} dx \int_0^1 \frac{B_i^{(v)}}{i!} dx = 1,$$

for $j < i$,

$$< k_j(x), k_i(x) >= \sum_{v=0}^{i} \left( \int_0^1 \left( k_j^{(v)} dx \right) \right) \left( \int_0^1 \left( k_i^{(v)} dx \right) \right)$$

$$= \sum_{v=0}^{i} \int_0^1 \frac{B_j^{(v)}}{j!} dx \int_0^1 \frac{B_i^{(v)}}{i!} dx = 0.$$

$H_0$ and $H_1$ are RKHS's with corresponding RKs where

$$R_0(x, z) = \sum_{j=0}^{m-1} k_j(x)k_j(z),$$

$$R_1(x, z) = k_m(x)k_m(z) + (-1)^{m-1}k_{2m}(|x - z|).$$

Consider a linear spline. The basis function is $\{1\}$ and

$$R_0(x, z) = 1, R_1(x, z) = k_1(x)k^1(z) + k_2(|x - z|).$$

If a spline is cubic, then the set of base functions is $\{1, k_1(x)\}$. The RKs are

$$R_0(x, z) = 1 + k_1(x)k_1(z),$$

$$R_1(x, z) = k_2(x)k_2(z) - k_4(|x - z|).$$

## 2.A.2 Smoothing Spline Regression for Multiple Variables

Consider a multivariate nonlinear regression model:

$$y_i = f(\mathbf{x_i}) + \varepsilon_i, \, i = 1, ..., n, \tag{2.A.63}$$

where $\mathbf{x_i} = (x_{i1}, ..., x_{ip})$ and $\varepsilon_i$ are zero-mean independent errors with variance $\sigma_e^2$. Usually, the function forms are unknown, again smoothing splines are used to approximate the function $f$. To extend a single variate smoothing spline, we start with the tensor product of the marginal spaces for each independent variable.

Consider a function $f: \chi = \chi_1 \times \chi_2 \times ... \times \chi_p$. Assume that $H^{(j)}$ is RKHS defined on the domain $\chi_j$ with RK $R^{(j)}$. Then, the tensor product reproducing kernel Hilbert spaces are defined as

$$H = H^{(1)} \otimes H^{(2)} \otimes ... \otimes H^{(p)} \tag{2.A.64}$$

and associated RK is defined as

$$R(\mathbf{x}, \mathbf{z}) = R^{(1)}(x_1, z_1) R^{(2)}(x_2, z_2) ... R^{(p)}\left(x_p, z_p\right). \tag{2.A.65}$$

In Section 2.A.1, the function $f$ is decomposed into two parts: a parametric $f_0$ and a smooth component $f_1$, that is,

$$f = f_0 + f_1. \tag{2.A.66}$$

Smoothing spline analysis for the multivariate variables is also based on the decomposition of the tensor product. Decomposition is a key concept for the smoothing spline analysis. Therefore, we first extend the decomposition of the marginal space for a single variable to a general case.

The general decomposition is based on averaging operators. An operator $A$ is called an averaging operator if $A^2 = A$. The averaging operator is also called an idempotent operator. The commonly used averaging operators are projection operators.

Assume that the model space $H$ is decomposed into $H = H_0 \oplus H_1$, where $H_0$ is a finite dimensional space with a set of orthonormal base functions $\varphi_1(x)$, ..., $\varphi_p(x)$. Assume that $A_j$ is the projection operator onto the subspace $\varphi_j(x)$, $j = 1, ..., p$, and $A_{p+1}$ is the projection operator onto $H_1$. Let $A = A_1 + ... + A_p + A_{p+1}$. Then, the function $f$ is decomposed into

$$f = Af = \left(A_1 + ... + A_p + A_{p+1}\right)f = \left(A_1 + ... + A_p\right)f + A_{p+1}f$$

$$= f_{01} + ... + f_{0p} + f_1 = f_0 + f_1, \tag{2.A.67}$$

where $f_0 = f_{01} + ... + f_{0p}$.

Equation 2.A.67 indicates that the function $f$ is decomposed into a parametric component and a smooth component.

**Example 2.A.3: Decomposition of $W_2^m[a, b]$**

The basis functions in $H_0$ are $\dfrac{(x-a)^{j-1}}{(j-1)!}$, $j = 1, ..., m$. The projection of the function $f$ onto the basis function $\dfrac{(x-a)^{j-1}}{(j-1)!}$, denoted by $A_j f$ is

$$A_j f = <f, \frac{(x-a)^{j-1}}{(j-1)!}> \frac{(x-a)^{j-1}}{(j-1)!}, \qquad (2.A.68)$$

$$j = 1, ..., m.$$

Recall that the inner product $<f, g>$ is defined as

$$<f, g> = \sum_{v=0}^{m-1} f^{(v)}(a) g^{(v)}(a) + \int_a^b f^{(m)} g^{(m)} dx. \qquad (2.A.69)$$

Using Equation 2.A.68, we obtain

$$<f, \frac{(x-a)^{j-1}}{(j-1)!}> = \sum_{v=0}^{m-1} f^{(v)}(a) \left[ \frac{(x-a)^{j-v-1}}{(j-v-1)!} \right]\bigg|_a \qquad (2.A.70)$$

$$= f^{(j-1)}(a)$$

Substituting Equation 2.A.70 into Equation 2.A.68 gives

$$A_j f = f^{(j-1)}(a) \frac{(x-a)^{j-1}}{(j-1)!}. \qquad (2.A.71)$$

It is easy to see that

$$A_j^2 f = A_j \left( A_j f \right) = A_j \left( f^{(j-1)}(a) \frac{(x-a)^{j-1}}{(j-1)!} \right)$$

$$= f^{(j-1)}(a) A_j \left( \frac{(x-a)^{j-1}}{(j-1)!} \right) = f^{(j-1)}(a) \frac{(x-a)^{j-1}}{(j-1)!}$$

$$= A_j f,$$

which implies that $A_j$ is an averaging operator. Define

$$A_{m+1} = I - A_1 - ... - A_m.$$

The decomposition

$$f = A_1 f + \ldots + A_m f + A_{m+1} f$$

$$= \sum_{j=0}^{m-1} f^{(j-1)}(a) \frac{(x-a)^{j-1}}{(j-1)!} + \int_a^b f^{(m)}(u) \frac{(x-u)^{m-1}}{(m-1)!} du$$

Corresponds to the Taylor expansion. The Sobolev space $W_2^m[a,b]$ is then decomposed into

$$W_2^m[a,b] = \{1\} \oplus \{x-a\} \oplus \ldots \oplus \frac{(x-a)^{m-1}}{(m-1)!} \oplus H_1, \qquad (2.A.72)$$

where $H_1 = \{f : f^{(j)}(a) = 0, j = 0, 1, \ldots, m-1, \int_a^b (f^m(x))^2 dx < \infty\}$.

### Example 2.A.4: Decomposition of $W_2^m[a,b]$ Based on Bernoulli Polynomials

Under the Bernoulli polynomial construction of $W_2^m[a,b]$, the inner product for the Hilbert space is defined as

$$<f,g> = \sum_{v=0}^{m-1} \left( \int_0^1 f^{(v)}(x) dx \right) \left( \int_0^1 g^{(v)}(x) dx \right)$$

$$+ \int_0^1 f^{(m)}(x) g^{(m)}(x) dx. \qquad (2.A.73)$$

We first compute the projection of the function $f \in H$ onto $k_j(x)$. Note that

$$\int_0^1 \left( k_j(x) \right)^{(v)} dx = \int_0^1 \left( \frac{B_j(x)}{j!} \right)^{(v)} dx = \int_0^1 (B_0(x))^{v-j} dx = \begin{cases} 1 & v = j \\ 0 & v \neq j \end{cases}. \qquad (2.A.74)$$

Using Equations 2.A.73 and 2.A.74 gives

$$<f, k_j(x)> = \int_0^1 f^{(j)}(x) dx, \text{ which implies}$$

$$A_j f = \left\{ \int_0^1 f^{(j-1)}(x) dx \right\} k_{j-1}(x), \qquad j = 1, \ldots, m. \qquad (2.A.75)$$

Note that

$$A_j k_{j-1}(x) = \int_0^1 k_{j-1}^{(j-1)}(x) dx k_{j-1}(x)$$

$$= \int_0^1 \frac{B_j^{(j-1)}(x)}{(j-1)!} dx k_{j-1}(x) \qquad (2.A.76)$$

$$= \int_0^1 B_0(x) dx k_{j-1}(x)$$

$$= k_{j-1}(x).$$

Using Equation 2.A.76, we obtain

$$
\begin{aligned}
A_j^2 f &= A_j \left( A_j f \right) \\
&= A_j \left( \left\{ \int_0^1 f^{(j-1)}(x)dx \right\} k_{j-1}(x) \right) \\
&= \int_0^1 f^{(j-1)}(x)dx A_j(k_{j-1}(x)) \\
&= \int_0^1 f^{(j-1)}(x)dx k_{j-1}(x), \qquad j = 1,\ldots,m.
\end{aligned}
\tag{2.A.77}
$$

Again, we show that $A_j$ is an averaging operator. Specifically, $A_1 f = \int_0^1 f(x)dx$ is the average of the function over interval $[0, 1]$.

Let $A_{m+1} = I - A_1 - \ldots - A_m$. We obtain the decomposition: $f = A_1 f + \ldots + A_m f + A_{m+1} f$, which corresponds to the decomposition of Sobolev space $W_2^m[a, b]$ into

$$
W_2^m[a, b] = \{k_0(x)\} \oplus \{k_1(x)\} \oplus \ldots \oplus \{k_{m-1}(x)\} \oplus H_1,
\tag{2.A.78}
$$

where $H_1 = \{f : f^{(j)}(a) = 0, j = 0, 1, \ldots, m - 1, \int_a^b (f^m(x))^2 dx < \infty\}$.

Next, we discuss tense product decomposition. Consider $p$ dependent variables $\{x_1, \ldots, x_p\}$ and the tensor product $H = H^{(1)} \otimes H^{(2)} \otimes \ldots \otimes H^{(p)}$ on domain $\chi = \chi_1 \times \chi_2 \times \ldots \times \chi_p$. For each function of the variable, we define the set of averaging operators $A_1^{(k)}, \ldots, A_{r_k}^{(k)}$ and one-way decomposition of the function:

$$
f = A_1^{(k)} f + \ldots + A_{r_k}^{(k)},
\tag{2.A.79}
$$

$$
k = 1, \ldots, p,
$$

where $A_1^{(k)} + \ldots + A_{r_k}^{(k)} = I$. The decomposition of the function $f(x_1, \ldots, x_p)$ onto the tense product is given by

$$
\begin{aligned}
f &= \left\{ A_1^{(1)} + \ldots + A_{r_1}^{(1)} \right\} \ldots \left\{ A_1^{(p)} + \ldots + A_{r_p}^{(p)} \right\} f \\
&= \sum_{j_1}^{r_1} \ldots \sum_{j_p}^{r_p} A_{j_1}^{(1)} \ldots A_{j_p}^{(p)} f.
\end{aligned}
\tag{2.A.80}
$$

Assume that the RKHS $H^{(k)}$ is decomposed into

$$
H^{(k)} = H_1^{(k)} \oplus \ldots \oplus H_{r_k}^{(k)},
\tag{2.A.81}
$$

where $H_{j_k}^{(k)}$ is RKHS with RK $R_{j_k}^{(k)}$. Then, the decomposition of the RKHS $H$ is

$$H = H^{(1)} \otimes \ldots \otimes H^{(p)}$$
$$= \left( H_1^{(1)} \oplus \ldots \oplus H_{r_1}^{(1)} \right) \otimes \ldots \otimes \left( H_1^{(p)} \oplus \ldots \oplus H_{r_p}^{(p)} \right) \quad (2.A.82)$$
$$= \sum_{j_1=1}^{r_1} \ldots \sum_{j_p=1}^{r_p} H_{j_1}^{(1)} \otimes \ldots \otimes H_{j_p}^{(p)}$$

Next, we discuss the model selection and parameter selection for the multivariate smoothing splines. Consider a model space:

$$M = H^0 \oplus H^1 \oplus \ldots \oplus H^q, \quad (2.A.83)$$

where $H^0$ is a finite dimensional space including all functions that will not be penalized and $H^1, \ldots, H^q$ are orthogonal RKHS's with RKs $R^1, \ldots, R^q$. Let $P_j$ be the orthogonal projection of the function onto the RKHS $H^j, j = 0, 1, \ldots, q$. We define the norm of the function $f$ as

$$\|f\|^2 = \|P_0 f\|^2 + \sum_{j=1}^{q} \|P_j f\|^2. \quad (2.A.84)$$

Similar to a single variate smoothing spline, the multivariate smoothing splines also consist of two parts: (1) functions from the finite dimensional space and (2) smoothing components to be penalized. Consider the multivariate smoothing spline regression model (Wang 2011):

$$y_i = L_i f + \varepsilon_i, \quad (2.A.85)$$

$$i = 1, \ldots, n,$$

where $L_i$ are continuous functionals and $\varepsilon_i$ are zero-mean independent random errors with common variance $\sigma_e^2$.

Now we group the model space $M$ into two subspaces:

$$H_0^* = H^0 \text{ and } H_1^* = H^1 \oplus \ldots \oplus H^q. \quad (2.A.86)$$

The inner product and RK defined on $H^0$ will be used for $H_0^*$. Now define the inner product and RK for the RKHS $H_1^*$. For any $f \in H_1^*$, we have

$$f(x) = f_1(x) + \ldots + f_q(x), \quad f_j \in H^j, \ j = 1, \ldots, q.$$

The inner product for $H_1^*$ is defined as the weighted inner products defined on the individual $H^j, j = 1, \ldots, q$:

$$<f, g>_* = \sum_{j=1}^{q} \theta_j^{-1} <f_j, g_j>, \quad (2.A.87)$$

where $\theta_j$, $j = 1, \ldots, q$ are parameters. Then, the norm of the function $f$ is

$$||f||_*^2 = \sum_{j=1}^{q} \theta_j^{-1} ||f_j||^2.$$

Define $P_1^* = \sum_{j=1}^{q} P_j$ as the orthogonal projection in $M$ onto $H_1^*$. Then, we have

$$||P_1^* f||_*^2 = \sum_{j=1}^{q} \theta_j^{-1} ||P_j f||^2. \tag{2.A.88}$$

The RK for the $H_1^*$ is defined as

$$R_1^* = \sum_{j=1}^{q} \theta_j R^j, \tag{2.A.89}$$

where $R^j$ is the RK for the $H^j$.

Now we show that $R_1^*$ is indeed the RK for the $H_1^*$. By the definition of the RK, we have

$$< R^j(x, .) f_j(.) >= f_j(x). \tag{2.A.90}$$

It is clear that for any $f \in H_1^*$, using Equations 2.A.87, 2.A.89, and 2.A.90, we have

$$< R_1^*(\mathbf{x}, .), f(.)>_* = \; < \sum_{j=1}^{q} \theta_j^{-1} < \theta_j R^j(\mathbf{x}, .), f_j(.) >$$

$$= \sum_{j=1}^{q} < R^j(\mathbf{x}, .), f_j(.) >$$

$$= \sum_{j=1}^{q} f_j(\mathbf{x}) = f(\mathbf{x}).$$

Let $\varphi_1^{(k)}, \ldots, \varphi_{r_k}^{(k)}$ be the set of base functions for the space $H_0^{(k)}$. Consider all possible combinations of base functions for $H^0$:

$$\left\{ \varphi_1^{(1)} \oplus .. \oplus \varphi_{r_1}^{(1)} \right\} .. \left\{ \varphi_1^{(p)} \oplus .. \oplus \varphi_{r_p}^{(p)} \right\} = \sum_{j_1=1}^{r_1} \cdots \sum_{j_p=1}^{r_p} \varphi_{j_1}^{(1)} \cdots \varphi_{j_p}^{(p)}$$

$$= \phi_1 + \ldots + \phi_r, \tag{2.A.91}$$

where $r = r_1 \ldots r_p$ and $\phi_v$, $v = 1, \ldots, r$ is one of the elements $\varphi_{j_1}^{(1)} \ldots \varphi_{j_p}^{(p)}$.

To estimate smoothing splines in the regression (2.A.85), we minimize

$$\min_{f \in M} \; \frac{1}{n} \sum_{i=1}^{n} (y_i - L_i f)^2 + \lambda ||P_1^* f||_*^2, \tag{2.A.92}$$

where $\lambda ||P_1^* f||_*^2 = \sum_{j=1}^{q} \lambda_j ||P_j f||^2$, $\lambda_j = \dfrac{\lambda}{\theta_j}$.

Equation 2.A91 has the same form as 2.A.14 if $H_1$ and $P_1$ are replaced by $H_1^*$ and $P_1^*$, respectively. To solve the problem, we first need to study the representation of the function $f(x_1, \ldots, x_p)$. Similar to Equation 2.A.26, we need to find the base functions and projections to $H_1^*$. The set of base functions are specified in Equation 2.A.91. Now we find the projection of the function onto $H_1^*$. Similar to Equation 2.A.22, for $v = 1, \ldots, n$, we have

$$\xi_v(x) = L_{v(z)} R_1^*(\mathbf{x}, \mathbf{z})$$

$$= L_{v(z)} \sum_{j=1}^{q} \theta_j R^j(\mathbf{x}, \mathbf{z}) \tag{2.A.93}$$

$$= \sum_{j=1}^{q} \theta_j L_{v(z)} R^j(\mathbf{x}, \mathbf{z}).$$

The general solution to the optimization problem (2.A.92) is

$$f(\mathbf{x}) = \sum_{j=1}^{r} a_j \phi_j(\mathbf{x}) + \sum_{v=1}^{n} b_v \xi_v(x) \text{ or}$$

$$f(\mathbf{x}) = \sum_{j=1}^{r} a_j \phi_j(\mathbf{x}) + \sum_{v=1}^{n} b_v \sum_{j=1}^{q} \theta_j L_{v(z)} R^j(\mathbf{x}, \mathbf{z}). \tag{2.A.94}$$

Similar to Equation 2.A.32, we have

$$L_{i(x)} f(\mathbf{x}) = \sum_{j=1}^{r} a_j L_{i(x)} \phi_j(\mathbf{x}) + \sum_{v=1}^{n} b_v \sum_{j=1}^{q} \theta_j L_{i(x)} L_{v(z)} R^j(\mathbf{x}, \mathbf{z}) \text{ ,i}$$

$$= 1, \ldots, n. \tag{2.A.95}$$

Again, define the matrix $T$:

$$T = \begin{bmatrix} L_{1(x)} \phi_1(\mathbf{x}) & \cdots & L_{1(x)} \phi_r(\mathbf{x}) \\ \vdots & \vdots & \vdots \\ L_{n(x)} \phi_1(\mathbf{x}) & \cdots & L_{n(x)} \phi_r(\mathbf{x}) \end{bmatrix},$$

the matrix $\Sigma_j$:

$$\Sigma_j = \begin{bmatrix} L_{1(x)} L_{1(z)} R^j(\mathbf{x}, \mathbf{z}) & \cdots & L_{1(x)} L_{n(z)} R^j(\mathbf{x}, \mathbf{z}) \\ \vdots & \vdots & \vdots \\ L_{n(x)} L_{1(z)} R^j(\mathbf{x}, \mathbf{z}) & \cdots & L_{n(x)} L_{n(z)} R^j(\mathbf{x}, \mathbf{z}) \end{bmatrix} \text{ and } \Sigma_\theta = \theta_1 \Sigma_1 + \ldots + \theta_q \Sigma_q.$$

Let

$$L\hat{f} = \begin{bmatrix} L_{1(x)} \hat{f}(\mathbf{x}) \\ \vdots \\ L_{n(x)} \hat{f}(\mathbf{x}) \end{bmatrix}.$$

Equation 2.A.94 can be written in a matrix form:

$$L\hat{f} = Ta + \Sigma b, \tag{2.A.96}$$

where $a = [a_1, \ldots, a_r]^T$ and $b = [b_1, \ldots, b_n]^T$.

Now we calculate $||P_1^* f||_*^2$. It follows from Equation 2.A.94 that

$$P_1^* f = \sum_{v=1}^n b_v \sum_{j=1}^q \theta_j L_{v(\mathbf{z})} R^j(\mathbf{x},\mathbf{z})$$
$$= \sum_{j=1}^q \theta_j \sum_{v=1}^n b_v L_{v(\mathbf{z})} R^j(\mathbf{x},\mathbf{z}) \qquad (2.A.97)$$

By the definition of inner product in the $H_1^*$ space (2.A.87), we obtain

$$||P_1^* f||_*^2 = \sum_{j=1}^q \frac{1}{\theta_j} < \theta_j \sum_{v=1}^n b_v L_{v(\mathbf{z})} R^j(\mathbf{x},\mathbf{z}), \theta_j \sum_{u=1}^n b_u L_{u(\mathbf{z})} R^j(\mathbf{x},\mathbf{z}) >$$
$$= \sum_{j=1}^q \theta_j \sum_{v=1}^n \sum_{u=1}^n b_v b_u < L_{v(\mathbf{z})} R^j(\mathbf{x},\mathbf{z}), L_{u(\mathbf{z})} R^j(\mathbf{x},\mathbf{z}) >$$
$$(2.A.98)$$

It follows from Equations 2.A.22, 2.A.27, and 2.A.29 that

$$< L_{v(\mathbf{z})} R^j(\mathbf{x},\mathbf{z}), L_{u(\mathbf{z})} R^j(\mathbf{x},\mathbf{z}) > = L_{v(\mathbf{z}) L_{u(\mathbf{z})} R^j(\mathbf{x},\mathbf{z})}. \qquad (2.A.99)$$

Substituting Equation 2.A.98 into Equation 2.A.97 gives

$$||P_1^* f||_*^2 = \sum_{j=1}^q \theta_j \sum_{v=1}^n \sum_{u=1}^n b_v b_u L_{v(\mathbf{z})} L_{u(\mathbf{z})} R^j(\mathbf{x},\mathbf{z})$$
$$= \sum_{j=1}^q \theta_j b^T \Sigma_j b \qquad (2.A.100)$$
$$= b^T \Sigma_\theta b$$

Let $Y = [y_1, \ldots, y_n]^T$. Then, substituting Equations 2.A.96 and 2.A.100 into Equation 2.A.91, we obtain

$$\min_{a,b} \quad \frac{1}{n} ||Y - Ta - \Sigma_\theta b||^2 + b^T \Sigma_\theta b. \qquad (2.A.101)$$

The form of the optimization problem (2.A.100) is exactly the same as that of the optimization problem (2.A.35) if the matrix $\Sigma_\theta$ is replaced by the matrix $\Sigma$. Therefore, the vectors $a$ and $b$ are the solutions to

$$Ta + (\Sigma_b + n\lambda I)b = Y, \qquad (2.A.102)$$

$$T^T b = 0. \qquad (2.A.103)$$

Similarly, using QR decomposition of the matrix $T$ (Wang 2011):

$$T = [Q_1 \ Q_2] \begin{bmatrix} R \\ 0 \end{bmatrix},$$

we can obtain the solutions:

$$\hat{a} = R^{-1} Q_1^T \left[ I - MQ_2 (Q_2^T MQ_2)^{-1} Q_2^T \right] Y \text{ and } \hat{b}$$
$$= Q_2 (Q_2^T MQ_2)^{-1} Q_2^T Y, \qquad (2.A.104)$$

where $M = \Sigma + n\lambda I$.

The smoothing spline regression function is given by

$$\hat{f}(\mathbf{x}) = \sum_{j=1}^{r} \hat{a}_j \phi_j(x) + \sum_{i=1}^{n} \hat{b}_i \xi_i(x), \qquad (2.A.105)$$

where $\xi_i(x) = L_{i(z)} R_1(x, z)$.
   Finally, we obtain the fitted value:

$$\hat{f} = H(\lambda) Y, \qquad (2.A.106)$$

where $H(\lambda) = I - n\lambda Q_2 (Q_2^T M Q_2)^{-1} Q_2^T$.


### Example 2.A.5: Smoothing Splines on $W_2^m[a, b] \otimes W_2^m[a, b]$

Consider two continuous variables $x_1$ and $x_2$ in the interval $[0, 1]$. The model spaces for the variables $x_1$ and $x_2$ are assumed to be the tensor product $W_2^m[a, b] \otimes W_2^m[a, b]$. Assume $m_1 = m_2 = 2$, that is, cubic splines are considered.
   Define four averaging operators:

$$A_1^{(k)} f = \int_0^1 f dx_k$$

$$A_2^{(k)} f = \int_0^1 \frac{\partial f(\mathbf{x})}{\partial x_k} dx_k (x_k - 0.5), \qquad k = 1, 2.$$

Therefore, we have

$$A_1^{(1)} A_1^{(2)} f = A_1^{(1)} \left( \int_0^1 f(x_1, x_2) dx_2 \right) = \int_0^1 \int_0^1 f(x_1, x_2) dx_1 dx_2 = \mu,$$

$$A_2^{(1)} A_1^{(2)} f = A_2^{(1)} \left( \int_0^1 f(x_1, x_2) dx_2 \right) = \int_0^1 \int_0^1 \frac{\partial f(x_1, x_2)}{\partial x_1} dx_1 dx_2 (x_1 - 0.5)$$

$$= \beta_1 (x_1 - 0.5),$$

$$A_1^{(1)} A_2^{(2)} f = A_1^{(1)} \left( \int_0^1 \frac{\partial f(\mathbf{x})}{\partial x_2} dx_2 (x_2 - 0.5) \right) = \int_0^1 \int_0^1 \frac{\partial f(\mathbf{x})}{\partial x_2} dx_1 dx_2 (x_2 - 0.5)$$

$$= \beta_2 (x_2 - 0.5),$$

$$A_2^{(1)} A_2^{(2)} f = A_2^{(1)} \left( \int_0^1 \frac{\partial f(\mathbf{x})}{\partial x_2} dx_2 (x_2 - 0.5) \right)$$

$$= \int_0^1 \int_0^1 \frac{\partial^2 f(\mathbf{x})}{\partial x_1 \partial x_2} dx_1 dx_2 (x_1 - 0.5)(x_2 - 0.5)$$

$$= \gamma_{12} (x_1 - 0.5)(x_2 - 0.5),$$

$$(2.A.107)$$

where

$$\mu = \int_0^1 \int_0^1 f(x_1, x_2) dx_1 dx_2, \ \beta_1 = \int_0^1 \int_0^1 \frac{\partial f(x_1, x_2)}{\partial x_1} dx_1 dx_2, \ \beta_2$$

$$= \int_0^1 \int_0^1 \frac{\partial f(\mathbf{x})}{\partial x_2} dx_1 dx_2, \ \gamma_{12} = \int_0^1 \int_0^1 \frac{\partial^2 f(\mathbf{x})}{\partial x_1 \, \partial x_2} dx_1 dx_2.$$

Define $A_3^{(k)} = I - A_1^{(k)} - A_2^{(k)}$, k = 1,2. Then,

$$A_3^{(k)} f = f(\mathbf{x}) - \int_0^1 f dx_k - \int_0^1 \frac{\partial f(\mathbf{x})}{\partial x_k} dx_k (x_k - 0.5),$$

$$A_1^{(1)} A_3^{(2)} f = \int_0^1 f(x_1, x_2) dx_1 - \int_0^1 \int_0^1 f(x_1, x_2) dx_1 dx_2$$

$$- \int_0^1 \int_0^1 \frac{\partial f(x_1, x_2)}{\partial x_2} dx_1 dx_2 (x_2 - 0.5)$$

$$= g_2(x_2) - \mu - \beta_2 (x_2 - 0.5),$$

$$A_2^{(1)} A_3^{(2)} f = \int_0^1 \frac{\partial f(x_1, x_2)}{\partial x_1} dx_1 (x_1 - 0.5) - \int_0^1 \int_0^1 \frac{\partial f(x_1, x_2)}{\partial x_1} dx_1 dx_2 (x_1 - 0.5)$$

$$- \int_0^1 \int_0^1 \frac{\partial^2 f(x_1, x_2)}{\partial x_1 \, \partial x_2} dx_1 dx_2 (x_1 - 0.5)(x_2 - 0.5)$$

$$= \alpha_1 (x_1 - 0.5) - \beta_1 (x_1 - 0.5) - \gamma_{12} (x_1 - 0.5)(x_2 - 0.5),$$

$$A_3^{(1)} A_1^{(2)} f = \int_0^1 f(x_1, x_2) dx_2 - \int_0^1 \int_0^1 f(x_1, x_2) dx_1 dx_2$$

$$- \int_0^1 \int_0^1 \frac{\partial f(x_1, x_2)}{\partial x_1} dx_1 dx_2 (x_1 - 0.5)$$

$$= g_1(x_1) - \mu - \beta_1 (x_1 - 0.5),$$

$$A_3^{(1)} A_2^{(2)} f = \int_0^1 \frac{\partial f(x_1, x_2)}{\partial x_2} dx_2 (x_2 - 0.5) - \int_0^1 \int_0^1 \frac{\partial f(x_1, x_2)}{\partial x_2} dx_1 dx_2 (x_2 - 0.5)$$

$$- \int_0^1 \int_0^1 \frac{\partial^2 f(x_1, x_2)}{\partial x_1 \, \partial x_2} dx_1 dx_2 (x_1 - 0.5)(x_2 - 0.5)$$

$$= \alpha_2 (x_2 - 0.5) - \beta_2 (x_2 - 0.5) - \gamma_{12} (x_1 - 0.5)(x_2 - 0.5),$$

$$A_3^{(1)}A_3^{(2)}f = f(x_1, x_2) + \int_0^1\int_0^1 f(x_1, x_2)dx_1dx_2 - \int_0^1 f(x_1, x_2)dx_1 - \int_0^1 f(x_1, x_2)dx_2$$

$$+ \left[\int_0^1\int_0^1 \frac{\partial f(x_1, x_2)}{\partial x_1}dx_1dx_2 - \int_0^1 \frac{\partial f(x_1, x_2)}{\partial x_1}\right](x_1 - 0.5)$$

$$+ \left[\int_0^1\int_0^1 \frac{\partial f(x_1, x_2)}{\partial x_2}dx_1dx_2\right.$$

$$- \int_0^1 \frac{\partial f(x_1, x_2)}{\partial x_2}dx_2(x_2 - 0.5)$$

$$+ \int_0^1\int_0^1 \frac{\partial^2 f(x_1, x_2)}{\partial x_1 \partial x_2}dx_1dx_2(x_1 - 0.5)(x_2 - 0.5)$$

$$= f(x_1, x_2) + \mu - g_1(x_1) - g_2(x_2) + (\beta_1 - \alpha_1)(x_1 - 0.5)$$

$$+ (\beta_2 - \alpha_2)(x_2 - 0.5)$$

$$+ \gamma_{12}(x_1 - 0.5)(x_2 - 0.5),$$

where $g_1(x_1) = \int_0^1 f(x_1, x_2)dx_2, g_2(x_2) = \int_0^1 f(x_1, x_2)dx_1, \ \alpha_1 = \int_0^1 \frac{\partial f(x_1, x_2)}{\partial x_1}dx_1,$

$\alpha_2 = \int_0^1 \frac{\partial f(x_1, x_2)}{\partial x_2}dx_2.$

Using Equation 2.A.80, the decomposition of the function $f(x_1, x_2)$ onto the tense product space $W_2^m[a, b] \otimes W_2^m[a, b]$ is

$$f = \left\{A_1^{(1)} + A_2^{(1)} + A_3^{(1)}\right\}\left\{A_1^{(2)} + A_2^{(2)} + A_3^{(2)}\right\}f$$

$$= A_1^{(1)}A_1^{(2)}f + A_1^{(1)}A_2^{(2)}f + A_1^{(1)}A_3^{(2)}f + A_2^{(1)}A_1^{(2)}f + A_2^{(1)}A_2^{(2)}f$$

$$+ A_2^{(1)}A_3^{(2)}f + A_3^{(1)}A_1^{(2)}f + A_3^{(1)}A_2^{(2)}f + A_3^{(1)}A_3^{(2)}f$$

$$= \mu + \beta_2(x_2 - 0.5) + f_2^s(x_2) + \beta_1(x_1 - 0.5) + \gamma_{12}(x_1 - 0.5)(x_2 - 0.5)$$

$$+ f_{12}^{ls}(x_1, x_2) + f_1^s(x_1) + f_{12}^{sl}(x_1, x_2) + f_{12}^{ss}(x_1, x_2),$$

where

$$f_1^s(x_1) = A_3^{(1)}A_1^{(2)}f = g_1(x_1) - \mu - \beta_1(x_1 - 0.5),$$

$$f_2^s(x_2) = A_1^{(1)}A_3^{(2)}f = g_2(x_2) - \mu - \beta_2(x_2 - 0.5),$$

$$f_{12}^{ls}(x_1, x_2) = A_2^{(1)}A_3^{(2)}f = \alpha_1(x_1 - 0.5) - \beta_1(x_1 - 0.5) - \gamma_{12}(x_1 - 0.5)(x_2 - 0.5),$$

$$f_{12}^{sl}(x_1, x_2) = A_3^{(1)}A_2^{(2)}f = \alpha_2(x_2 - 0.5) - \beta_2(x_2 - 0.5) - \gamma_{12}(x_1 - 0.5)(x_2 - 0.5),$$

$$f_{12}^{ss}(x_1, x_2) = A_3^{(1)}A_3^{(2)}f = f(x_1, x_2) + \mu - g_1(x_1) - g_2(x_2) + (\beta_1 - \alpha_1)(x_1 - 0.5)$$

$$+ (\beta_2 - \alpha_2)(x_2 - 0.5) + \gamma_{12}(x_1 - 0.5)(x_2 - 0.5),$$

$f_1^s(x_1)$ and $f_2^s(x_2)$ measure the main effects of $x_1$ and $x_2$, respectively, $f_{12}^{ls}$ $(x_1, x_2)$, $f_{12}^{sl}(x_1, x_2)$, and $f_{12}^{ss}(x_1, x_2)$ are linear-smooth, smooth-linear, and smooth-smooth interaction functions of $x_1$ and $x_2$. Equation 2.A.27 indicates that the function can be decomposed into a linear function and smoothing nonlinear interactions.

Next, we study the tense product decomposition of model space $W_2^m$ $[0,1] \otimes W_2^m[0,1]$. Suppose that the Sobolev space $W_2^m[0,1]$ is decomposed as a finite dimensional base function subspace with two base functions $H_0^{(k)} = \{1\}$ and $H_1^{(k)} = \{x_k - 0.5\}$, and an infinite dimensional function space

$$H_2^{(k)} = \{ f \in W_2^m[0,1] : \int_0^1 f(x_1, x_2)dx_k = 0, \int_0^1 \frac{\partial f(\mathbf{x})}{\partial x_k}dx_k = 0 \}, \ k = 1,2.$$ The tensor product space of $W_2^m[0,1] \otimes W_2^m[0,1]$ can be decomposed into

$$W_2^m[0,1] \otimes W_2^m[0,1] = \left\{ H_0^{(1)} \oplus H_1^{(1)} \oplus H_2^{(1)} \right\}\left\{ H_0^{(2)} \oplus H_1^{(2)} \oplus H_2^{(2)} \right\}$$

$$= \left\{ H_0^{(1)} \otimes H_0^{(2)} \right\} \oplus \left\{ H_0^{(1)} \otimes H_1^{(2)} \right\} \oplus \left\{ H_0^{(1)} \otimes H_2^{(2)} \right\}$$

$$\oplus \left\{ H_1^{(1)} \otimes H_0^{(2)} \right\} \oplus \left\{ H_1^{(1)} \otimes H_1^{(2)} \right\} \oplus \left\{ H_1^{(1)} \otimes H_2^{(2)} \right\}$$

$$\oplus \left\{ H_2^{(1)} \otimes H_0^{(2)} \right\} \oplus \left\{ H_2^{(1)} \otimes H_1^{(2)} \right\} \oplus \left\{ H_2^{(1)} \otimes H_2^{(2)} \right\}$$

$$\tag{2.A.108}$$

Equation 2.A.28 can be regrouped into

$$W_2^m[0,1] \otimes W_2^m[0,1] = H^{(0)} \oplus H^{(1)} \oplus H^{(2)} \oplus H^{(3)} \oplus H^{(4)} \oplus H^{(5)}, \tag{2.A.109}$$

where

$$H^{(0)} = \left\{ H_0^{(1)} \otimes H_0^{(2)} \right\} \oplus \left\{ H_1^{(1)} \otimes H_0^{(2)} \right\} \oplus \left\{ H_0^{(1)} \otimes H_1^{(2)} \right\} \oplus \left\{ H_1^{(1)} \otimes H_1^{(2)} \right\}$$

$$= \{\phi_1(\mathbf{x}) = 1\} \oplus \{\phi_2(\mathbf{x}) = x_1 - 0.5\} \oplus \{\phi_3(\mathbf{x}) = x_2 - 0.5\}$$

$$\oplus \{\phi_4(\mathbf{x}) = (x_1 - 0.5)(x_2 - 0.5)\},$$

$$H^1 = H_2^{(1)} \otimes H_0^{(2)}, \ H^2 = H_2^{(1)} \otimes H_1^{(2)}, \ H^3 = H_0^{(1)} \otimes H_2^2, \ H^4$$

$$= H_1^{(1)} \otimes H_2^{(2)}, \ H^5 = H_2^{(1)} \otimes H_2^{(2)}.$$

The RKs $R^1$, $R^2$, $R^3$, $R^4$ and $R^5$ of $H^1$, $H^2$, $H^3$, $H^4$, and $H^5$ can be, respectively, calculated as follows:

$$R^1 = k_2(x_1)k_2(z_1) - k_4(|x_1 - z_1|),$$
$$R^2 = (k_2(x_1)k_2(z_1) - k_4(|x_1 - z_1|))k_1(x_2)k_1(z_2),$$
$$R^3 = k_2(x_2)k_2(z_2) - k_4(|x_2 - z_2|), \tag{2.A.110}$$
$$R^4 = k_1(x_1)k_1(z_1)(k_2(x_2)k_2(z_2) - k_4(|x_2 - z_2|)),$$
$$R^5 = (k_2(x_1)k_2(z_1) - k_4(|x_1 - z_1|))(k_2(x_2)k_2(z_2) - k_4(|x_2 - z_2|)).$$

Assume that the continuous functional is an evaluation functional. The matrix $T$ is

$$T = \begin{bmatrix} 1 & x_1^{(1)} - 0.5 & x_2^{(1)} - 0.5 & \left(x_1^{(1)} - 0.5\right)\left(x_2^{(1)} - 0.5\right) \\ 1 & x_1^{(2)} - 0.5 & x_2^{(2)} - 0.5 & \left(x_1^{(2)} - 0.5\right)\left(x_2^{(2)} - 0.5\right) \\ \vdots & \vdots & \vdots & \vdots \\ 1 & x_1^{(n)} - 0.5 & x_2^{(n)} - 0.5 & \left(x_1^{(n)} - 0.5\right)\left(x_2^{(n)} - 0.5\right) \end{bmatrix}, \quad (2.A.111)$$

where $\mathbf{x}_1 = [x_1^{(1)}, \ldots, x_1^{(n)}]^T$ and $\mathbf{x}_2 = [x_2^{(1)}, \ldots, x_2^{(n)}]^T$ are two vectors of observed independent variables.

Next we calculate the matrix $\Sigma_j$:

$$\Sigma_1 = \begin{bmatrix} R^1\left(x^{(1)}, x^{(1)}\right) & \cdots & R^1\left(x^{(1)}, x^{(n)}\right) \\ \vdots & \vdots & \vdots \\ R^1\left(x^{(n)}, x^{(1)}\right) & \cdots & R^1\left(x^{(n)}, x^{(n)}\right) \end{bmatrix}$$

$$= \begin{bmatrix} \left[k_2\left(x_1^{(1)}\right)\right]^2 - k_4(0) & \cdots & k_2\left(x_1^{(1)}\right)k_2\left(x_1^{(n)}\right) - k_4\left(\left|x_1^{(1)} - x_1^{(n)}\right|\right) \\ \vdots & \vdots & \vdots \\ k_2\left(x_1^{(n)}\right)k_2\left(x_1^{(1)}\right) - k_4(|x_1^{(n)} - x_1^{(1)}|) & \cdots & \left[k_2\left(x_1^{(n)}\right)\right]^2 - k_4(0) \end{bmatrix}$$

Similarly, using Equation 2.A.12, we can calculate the matrices $\Sigma_2, \Sigma_3, \Sigma_4$, and $\Sigma_5$. Finally, we calculate $\Sigma_\theta = \theta_1\Sigma_1 + \theta_2\Sigma_2 + \theta_3\Sigma_3 + \theta_4\Sigma_4 + \theta_5\Sigma_5$.

---

## Appendix 2.B    Penalized Likelihood Function for Jointly Observational and Interventional Data

This appendix extends the structural equation model of Hauser and Buhlmann (2012) for both interventional and observation data to include exogenous variables. Let $K^{(i)} = (\Sigma_Y^{(i)})^{-1}$ and $\xi^{(i)} = K^{(i)}\mu_Y^{(i)}$. Note that

$$\left(Y^{(i)} - \mu_Y^{(i)}\right)^T \left(\Sigma_Y^{(i)}\right)^{-1} \left(Y^{(i)} - \mu_Y^{(i)}\right) = \left(Y^{(i)} - \mu_Y^{(i)}\right)^T K^{(i)} \left(Y^{(i)} - \mu_Y^{(i)}\right)$$

$$= \mathrm{Tr}\left(\left(Y^{(i)}\right)^T K^{(i)} Y^{(i)}\right) - 2\left(Y^{(i)}\right)^T K^{(i)}\mu_Y^{(i)}$$

$$+ \left(\mu_Y^{(i)}\right)^T K^{(i)}\mu_Y^{(i)}$$

$$= \mathrm{Tr}\left(K^{(i)}Y^{(i)}\left(Y^{(i)}\right)^T\right) - 2\left(Y^{(i)}\right)^T \xi^{(i)}$$

$$+ \left(\xi^{(i)}\right)^T \left(K^{(i)}\right)^{-1} \xi^{(i)}.$$

$$(2.B.1)$$

Using Equation 2.B.1 and ignoring the constant term, Equation 2.124 can be reduced to

$$-l_D(A, B, T, \Sigma_\varepsilon) \approx \frac{1}{2} \sum_{i=1}^{n} \left[ \mathrm{Tr}\, \left( K^{(i)} Y^{(i)} \left( Y^{(i)} \right)^T \right) - 2 \left( Y^{(i)} \right)^T \xi^{(i)} \right.$$

$$\left. + \left( \xi^{(i)} \right)^T \left( K^{(i)} \right)^{-1} \xi^{(i)} - \log |K_{Y|}^{(2)}| \right]. \tag{2.B.2}$$

The likelihood function is used as a tool to estimate the parameters in the model. Next, we will show that the terms of $(Y^{(i)})^T \xi^{(i)}$ and $(\xi^{(i)})^T (K^{(i)})^{-1} \xi^{(i)}$ involve no model parameters $A$, $B$ and $\Sigma_\varepsilon$. We first show that

$$\left[ R^{(l)} \Sigma_\varepsilon^{(i)} R^{(l)} + \left( Q^{(l)} \right)^T \Sigma_U^{(i)} Q^{(l)} \right]^{-1}$$

$$= R^{(l)} \left( \Sigma_\varepsilon^{(i)} \right)^{-1} R^{(l)} + \left( Q^{(l)} \right)^T \left( \Sigma_U^{(i)} \right)^{-1} Q^{(l)}. \tag{2.B.3}$$

Consider the example in Figure 2.8. The matrices $R^{(l)}$ and $Q^{(l)}$ are

$$R^{(l)} = \begin{bmatrix} 1 & 0 & 0 & 0 & 0 \\ 0 & 1 & 0 & 0 & 0 \\ 0 & 0 & 0 & 0 & 0 \\ 0 & 0 & 0 & 1 & 0 \\ 0 & 0 & 0 & 0 & 0 \end{bmatrix} \text{ and } Q^{(l)} = \begin{bmatrix} 0 & 0 & 1 & 0 & 0 \\ 0 & 0 & 0 & 0 & 1 \end{bmatrix}.$$

It is clear that the operation of $R^{(l)} \Sigma_\varepsilon^{(i)} R^{(l)}$ in this example is to keep the variances of the un-intervened nodes in the diagonal matrix $\Sigma_\varepsilon$ and the operation of $(Q^{(l)})^T \Sigma_U^{(i)} Q^{(l)}$ is to keep the variance of the intervened variables in the diagonals of the matrix $\Sigma_U^{(i)}$. In this example, we have

$$R^{(l)} \Sigma_\varepsilon^{(i)} R^{(l)} + \left( Q^{(l)} \right)^T \Sigma_U^{(i)} Q^{(l)} = \mathrm{diag}\, (\sigma_1^2, \sigma_2^2, \tau_1^2, \sigma_4^2, \tau_2^2),$$

which implies

$$\left[ R^{(l)} \Sigma_\varepsilon^{(i)} R^{(l)} + \left( Q^{(l)} \right)^T \Sigma_U^{(i)} Q^{(l)} \right]^{-1} = \mathrm{diag}\, (\sigma_1^{-2}, \sigma_2^{-2}, \tau_1^{-2}, \sigma_4^{-2}, \tau_2^{-2})$$

$$= R^{(l)} \left( \Sigma_\varepsilon^{(i)} \right)^{-1} R^{(l)} + \left( Q^{(l)} \right)^T \left( \Sigma_U^{(i)} \right)^{-1} Q^{(l)}$$

Using $R^{(l)}$ and $Q^{(l)}$ in Figure 2.8, we can easily check if the following equalities hold:

$$Q^{(I)}\left(Q^{(I)}\right)^T = I, \ Q^{(I)}R^{(I)} = 0, \text{ and } R^{(I)}R^{(I)} = R^{(I)}.$$

Using these equalities and equality (2.B.3), we obtain

$$K^{(i)} = \left(I - R^{(I)}B\right)^T\left[R^{(I)}\left(\Sigma_\varepsilon^{(i)}\right)^{-1}R^{(I)} + \left(Q^{(I)}\right)^T\left(\Sigma_U^{(i)}\right)^{-1}Q^{(I)}\right]\left(I - R^{(I)}B\right)$$

$$= \left(I - R^{(I)}B\right)^T R^{(I)}\left(\Sigma_\varepsilon^{(i)}\right)^{-1}R^{(I)}\left(I - R^{(I)}B\right) + \left(I - R^{(I)}B\right)^T\left(Q^{(I)}\right)^T$$

$$\left(\Sigma_U^{(i)}\right)^{-1}Q^{(I)}\left(I - R^{(I)}B\right)$$

$$= (I - B)^T R^{(I)}\left(\Sigma_\varepsilon^{(i)}\right)^{-1}R^{(I)}(I - B) + \left(Q^{(I)}\right)^T\left(\Sigma_U^{(i)}\right)^{-1}Q^{(I)}. \tag{2.B.4}$$

Now we calculate $\xi^{(i)} = K^{(i)}\mu_Y^{(i)}$. Using Equation 2.B.4 and Equation 2.121 gives

$$\xi^{(i)} = \left(I - R^{(I)}B\right)^T\left[R^{(I)}\left(\Sigma_\varepsilon^{(i)}\right)^{-1}R^{(I)} + \left(Q^{(I)}\right)^T\left(\Sigma_U^{(i)}\right)^{-1}Q^{(I)}\right]\left(I - R^{(I)}B\right)$$

$$*\left[\left(I - R^{(I)}B\right)^{-1}R^{(I)}AX^{(i)} + \left(I - R^{(I)}B\right)^{-1}\left(Q^{(I)}\right)^T\mu_U^{(i)}\right]$$

$$= \left(I - R^{(I)}B\right)^T R^{(I)}\left(\Sigma_\varepsilon^{(i)}\right)^{-1}R^{(I)}AX^{(i)} + \left(I - R^{(I)}B\right)^T\left(Q^{(I)}\right)^T\left(\Sigma_U^{(i)}\right)^{-1}\mu_U^{(i)}$$

$$= (I - B)^T R^{(I)}\left(\Sigma_\varepsilon^{(i)}\right)^{-1}R^{(I)}AX^{(i)} + \left(Q^{(I)}\right)^T\left(\Sigma_U^{(i)}\right)^{-1}\mu_U^{(i)} \tag{2.B.5}$$

Now we calculate $(\xi^{(i)})^T(K^{(i)})^{-1}\xi^{(i)}$. Using $\xi^{(i)} = K^{(i)}\mu_Y^{(i)}$ gives

$$\left(\xi^{(i)}\right)^T\left(K^{(i)}\right)^{-1}\xi^{(i)} = (\mu_Y^{(i)})^T\xi^{(i)}. \tag{2.B.6}$$

Recall that

$$\left(I - R^{(I)}B\right)^{-1} = I + R^{(I)}B + \dots + \left(R^{(I)}B\right)^{p-1}. \tag{2.B.7}$$

Substituting Equation 2.B.5 into Equation 2.B.6 and using Equation 2.B.7, we obtain

$$(\mu_Y^{(i)})^T\xi^{(i)} = (\mu_Y^{(i)})^T\left(I - R^{(I)}B\right)^T R^{(I)}\left(\Sigma_\varepsilon^{(i)}\right)^{-1}R^{(I)}AX^{(i)}$$

$$+ (\mu_Y^{(i)})^T\left(Q^{(I)}\right)^T\left(\Sigma_U^{(i)}\right)^{-1}\mu_U^{(i)}, \tag{2.B.8}$$

Using the expansion equation $Q^{(I)} R^{(I)} = 0$ and $R^{(I)}R^{(I)} = R^{(I)}$ yields

$$\left(\mu_Y^{(i)}\right)^T \left(I - R^{(I)}B\right)^T R^{(I)} \left(\Sigma_\varepsilon^{(i)}\right)^{-1} R^{(I)}AX^{(i)} = \left[\left(I - R^{(I)}B\right)^{-1}R^{(I)}AX^{(i)}\right.$$

$$\left. + (I - R^{(I)}B)^{-1}\left(Q^{(I)}\right)^T \mu_U^{(i)} \right]^T$$

$$* \left(I - R^{(I)}B\right)^T R^{(I)} \left(\Sigma_\varepsilon^{(i)}\right)^{-1} R^{(I)}AX^{(i)}$$

$$= \left(X^{(i)}\right)^T A^T R^{(I)} \left(I - R^{(I)}B\right)^{-T}$$

$$\left(I - R^{(I)}B\right)^T R^{(I)} \left(\Sigma_\varepsilon^{(i)}\right)^{-1} R^{(I)}AX^{(i)}$$

$$+ \left(\mu_U^{(i)}\right)^T Q^{(I)} \left(I - R^{(I)}B\right)^{-T}$$

$$\left(I - R^{(I)}B\right)^T R^{(I)} \left(\Sigma_\varepsilon^{(i)}\right)^{-1} R^{(I)}AX^{(i)}$$

$$= \left(X^{(i)}\right)^T A^T R^{(I)} \left(\Sigma_\varepsilon^{(i)}\right)^{-1} R^{(I)}AX^{(i)}.$$

$$(2.B.9)$$

Using Equation 2.121 gives

$$\left(\mu_Y^{(i)}\right)^T \left(Q^{(I)}\right)^T \left(\Sigma_U^{(i)}\right)^{-1} \mu_U^{(i)} = \left[\left(I - R^{(I)}B\right)^{-1}R^{(I)}AX^{(i)} + \left(I - R^{(I)}B\right)^{-1}\left(Q^{(I)}\right)^T \mu_U^{(i)}\right]^T$$

$$\left(Q^{(I)}\right)^T \left(\Sigma_U^{(i)}\right)^{-1} \mu_U^{(i)}$$

$$= \left(X^{(i)}\right)^T A^T R^{(I)} \left(I - R^{(I)}B\right)^{-T} \left(Q^{(I)}\right)^T \left(\Sigma_U^{(i)}\right)^{-1} \mu_U^{(i)}$$

$$+ \left(\mu_U^{(i)}\right)^T Q^{(I)} (I - R^{(I)}B)^{-T}$$

$$* \left(Q^{(I)}\right)^T \left(\Sigma_U^{(i)}\right)^{-1} \mu_U^{(i)}. \qquad (2.B.10)$$

Using Equation 2.B.7, $Q^{(I)}R^{(I)} = 0$ and $Q^{(I)}(Q^{(I)})^T = I$, we obtain

$$R^{(I)} \left(I - R^{(I)}B\right)^{-T} \left(Q^{(I)}\right)^T = 0$$

$$Q^{(I)} \left(I - R^{(I)}B\right)^{-T} \left(Q^{(I)}\right)^T = I.$$

$$(2.B.11)$$

Substituting Equation 2.B.11 into Equation 2.B.10 yields

$$\left(\mu_Y^{(i)}\right)^T \left(Q^{(l)}\right)^T \left(\Sigma_U^{(i)}\right)^{-1} \mu_U^{(i)} = \left(\mu_U^{(i)}\right)^T \left(\Sigma_U^{(i)}\right)^{-1} \mu_U^{(i)}. \qquad (2.B.12)$$

Combining Equations 2.B.6, 2.B.9, and 2.B.12, we have

$$\left(\xi^{(i)}\right)^T \left(K^{(i)}\right)^{-1} \xi^{(i)}$$

$$= \left(X^{(i)}\right)^T A^T R^{(l)} \left(\Sigma_\varepsilon^{(i)}\right)^{-1} R^{(l)} A X^{(i)} + \left(\mu_U^{(i)}\right)^T \left(\Sigma_U^{(i)}\right)^{-1} \mu_U^{(i)}. \qquad (2.B.13)$$

Now we calculate the determinant of $K^{(i)}$. Consider the example in Figure 2.8. The matrix $K^{(i)}$ is

$$K^{(i)} = \begin{bmatrix} \sigma_1^{-2} & 0 & 0 & 0 & 0 \\ 0 & \sigma_2^{-2} & 0 & 0 & 0 \\ -\dfrac{\beta_{31}}{\sigma_1^2} & -\dfrac{\beta_{32}}{\sigma_2^2} & \tau_1^{-2} & -\dfrac{\beta_{34}}{\sigma_4^2} & 0 \\ 0 & -\dfrac{\beta_{42}}{\sigma_2^2} & 0 & \sigma_4^{-2} & 0 \\ 0 & 0 & 0 & 0 & \tau_2^{-2} \end{bmatrix}.$$

Its determinant is $|K^{(i)}| = \sigma_1^{-2}\sigma_2^{-2}\sigma_4^{-2}\tau_1^{-2}\tau_2^{-2}$. In general, we can show that

$$|K^{(i)}| = \prod_{j \in I} \sigma_j^{-2} \left| \left(\Sigma_U^{(i)}\right)^{-1} \right|. \qquad (2.B.14)$$

Combining Equations 2.B.2, 2.B.5, 2.B.13, and 2.B.14, we obtain the negative likelihood that does not involve the model parameters:

$$-l_D(A, B, T, \Sigma_\varepsilon) \approx \frac{1}{2} \sum_{i=1}^n [\mathrm{Tr}\ (K^{(i)} Y^{(i)} \left(Y^{(i)}\right)^T) - \left(Y^{(i)}\right)^T (I - B)^T R^{(l)}$$

$$\left(\Sigma_\varepsilon^{(i)}\right)^{-1} R^{(l)} A X^{(i)}$$

$$+ \mathrm{Tr}\ (A^T R^{(l)} \left(\Sigma_\varepsilon^{(i)}\right)^{-1} R^{(l)} A X^{(i)} \left(X^{(i)}\right)^T - \sum_{j \notin I} \log \sigma_j^{-2}].$$

$$(2.B.15)$$

Removing the terms that do not contain the model parameters in the matrix $K^{(l)}$, Equation 2.125 can be further reduced to

$$-l_D(A,B,T,\Sigma_\varepsilon) \approx \frac{1}{2}\sum_{I\in\mathbf{T}}\left\{ n^{(I)}\text{Tr}\left( S^{(I)}(I-B)^T R^{(I)}\left(\Sigma_\varepsilon^{(i)}\right)^{-1} R^{(I)}(I-B)\right)\right.$$

$$-n^{(I)}\sum_{j\notin I}\log \sigma_j^{-2}$$

$$-2\sum_{i:T^{(i)}=I}\left[\left(Y^{(i)}\right)^T (I-B)^T R^{(I)}\left(\Sigma_\varepsilon^{(i)}\right)^{-1} R^{(I)}AX^{(i)}\right.$$

$$\left.\left. -\text{Tr}\left(A^T R^{(I)}\left(\Sigma_\varepsilon^{(i)}\right)^{-1}R^{(I)}AX^{(i)}\left(X^{(i)}\right)^T\right]\right\}.\qquad (2.B.16)$$

Define $n^{(-k)} = \sum_{I\in\mathbf{T}:k\notin I} n^{(I)}$ and $S^{(-k)} = \sum_{I\in\mathbf{T}:k\in I}\frac{n^{(I)}}{n^{(-k)}}S^{(I)}$, where $S^{(-k)} = 0$, if $n^{(-k)} = 0$. Now we decompose the terms in Equation 2.B.16. Using the above notations, we have

$$\sum_{I\in\mathbf{T}} n^{(I)}\text{Tr}\left( S^{(I)}(I-B)^T R^{(I)}\left(\Sigma_\varepsilon^{(i)}\right)^{-1} R^{(I)}(I-B)\right)$$

$$= \sum_{i=1}^{n}\text{Tr}\left[ R^{(T^{(i)})}\left(\Sigma_\varepsilon^{(i)}\right)^{-1} R^{(T^{(i)})}(I-B)Y^{(i)}\left(Y^{(i)}\right)^T (I-B)^T\right].\qquad (2.B.17)$$

It is clear that

$$\left( R^{(T^{(i)})}(I-B)\right)_{(k)} = \begin{cases} (I-B)_{k.} & k\notin T^{(i)} \\ 0 & k\in T^{(i)}, \; k=1,...,p, \end{cases}\qquad (2.B.18)$$

where $(I-B)_{k.}$ represents the *kth* row of the matrix *I–B*, which consists of only the entries of the node *k* and its parents. Substituting Equation 2.B.18 into Equation 2.B.17 gives

$$\sum_{i=1}^{n}\text{Tr}\left[ R^{(T^{(i)})}\left(\Sigma_\varepsilon^{(i)}\right)^{-1} R^{(T^{(i)})}(I-B)Y^{(i)}\left(Y^{(i)}\right)^T (I-B)^T\right]$$

$$= \sum_{i=1}^{n}\sum_{k\notin T^{(i)}}\sigma_k^{-2}(I-B)_{k.}Y^{(i)}\left(Y^{(i)}\right)^T ((I-B)_{k.})^T$$

$$= \sum_{k\notin T^{(i)}}\sigma_k^{-2}(I-B)_{k.}\left(\sum_{i=1}^{n}Y^{(i)}\left(Y^{(i)}\right)^T\right)((I-B)_{k.})^T$$

$$= \sum_{k=1}^{p} n^{(-k)}\sigma_k^{-2}(I-B)_{k.}S^{(-k)}((I-B)_{k.})^T.\qquad (2.B.19)$$

Next, we calculate the second term in Equation 2.B.16. By the similar arguments, we have

$$\sum_{l \in \mathbf{T}} n^{(l)} \sum_{k \notin I} \log \sigma_k^{-2} = \sum_{i=1}^{n} \sum_{k \notin T^{(i)}} \log \sigma_k^{-2}$$

$$= \sum_{j=1}^{p} \sum_{i'' k \in T^{(i)}} \log \sigma_k^{-2}$$

$$= \sum_{j=1}^{p} \left( \sum_{I, k \notin I} n^{(l)} \right) \log \sigma_k^{-2} \qquad \text{(2.B.20)}$$

$$= \sum_{k=1}^{p} n^{(-k)} \log \sigma_k^{-2}.$$

Now we calculate the third term in Equation 2.B.16. It follows from Equation 2.B.16 that

$$\sum_{i=1}^{n} \left[ \left( Y^{(i)} \right)^T (I - B)^T R^{(T^{(i)})} \left( \Sigma_{\varepsilon}^{(i)} \right)^{-1} R^{(T^{(i)})} A X^{(i)} \right.$$

$$\left. - \operatorname{Tr} \left( A^T R^{(T^{(i)})} \left( \Sigma_{\varepsilon}^{(i)} \right)^{-1} R^{(T^{(i)})} A X^{(i)} \left( X^{(i)} \right)^T \right) \right]$$

$$= \sum_{i=1}^{n} \operatorname{Tr} \left( \left( \Sigma_{\varepsilon}^{(i)} \right)^{-1} R^{(T^{(i)})} A X^{(i)} \left( Y^{(i)} \right)^T (I - B)^T \left( R^{(T^{(i)})} \right) \right.$$

$$\left. - \operatorname{Tr} \left( \left( \Sigma_{\varepsilon}^{(i)} \right)^{-1} R^{(l)} A X^{(i)} \left( X^{(i)} \right)^T A^T R^{(T^{(i)})} \right) \right]$$

$$= \sum_{i=1}^{n} \sum_{(l \notin T^{(i)}) \cap E_C} \sigma_l^{-2} \left[ A_l . X^{(i)} \left( Y^{(i)} \right)^T ((I - B)_l.)^T \qquad \text{(2.B.21)} \right.$$

$$\left. - A_l . X^{(i)} \left( X^{(i)} \right)^T (A_l.)^T \right]$$

$$= \sum_{(l \notin T^{(i)} \cap E_c)} \sigma_l^{-2} \left[ A_l . \sum_{i=1}^{n} X^{(i)} \left( Y^{(i)} \right)^T ((I - B)_l.)^T \right.$$

$$\left. - A_l . \sum_{i=1}^{n} X^{(i)} \left( X^{(i)} \right)^T (A_l.)^T \right]$$

$$= \sum_{l=1}^{p} n^{(-l)} \sigma_l^{-2} \left[ A_l . S_{XY}^{(-l)} ((I - B)_l.)^T - A_l . S_{XX}^{(-l)} (A_l.)^T \right],$$

where $E_c = \{i \mid \text{at least one } a_{ij} \neq 0\}$ is the set of nodes that are connected with the exogenous variables, $n^{(-l)} = \sum_{l \in \mathbf{T} : (l \in E_c) \cap (l \notin I)} n^{(l)}$, the number of samples in which the node $l$ connecting with exogenous variables is not intervened, $S_{XY}^{(l)} = \frac{1}{n^{(l)}} \sum_{i : T^{(i)} = l} X^{(i)} (Y^{(i)})^T$, $S_{XX}^{(l)} = \frac{1}{n^{(l)}} \sum_{i : T^{(i)} = l} X^{(i)} (X^{(i)})^T$, $S_{XY}^{(-l)} = \sum_{l \in \mathbf{T} : (l \in E_c) \cap (l \notin I), i : T^{(i)} = l} \frac{n^{(l)}}{n^{(-l)}} S_{XY}^{(l)}$, and $S_{XX}^{(-l)} = \sum_{l \in \mathbf{T} \cdot (l \in E_c) \cap (l \notin I), i : T^{(i)} = l} \frac{n^{(l)}}{n^{(-l)}} S_{XX}^{(l)}$.

In summary, combining Equations 2.B.19, 2.B.20, and 2.B.21, we obtain the negative likelihood decomposition:

$$-l_D(A, B, T, \Sigma_\varepsilon) = \sum_{k=1}^{p} l_k\left(A_{k.}, B_{k.}, \sigma_k^2, T, Y, X\right), \qquad (2.B.22)$$

where

when the node has connections with exogenous variables,

$$l_k\left(A_{k.}, B_{k.}, \sigma_k^2, T, Y, X\right) = \frac{1}{2} n^{(-k)} \left\{ \sigma_k^{-2}(I - B)_{k.} S^{(-k)}((I - B)_{k.})^T - \log \sigma_k^{-2} \right.$$

$$\left. - 2\sigma_k^{-2} \left[ A_{k.} S_{XY}^{(-k)}((I - B)_{k.})^T - A_{k.} S_{XX}^{(-k)}(A_{k.})^T \right] \right\}$$

$$(2.B.23)$$

When the node does not have connections with any exogenous variables,

$$l_k\left(B_{k.}, \sigma_k^2, T, Y, X\right) = \frac{1}{2} n^{(-k)} \left\{ \sigma_k^{-2}(I - B)_{k.} S^{(-k)}((I - B)_{k.})^T - \log \sigma_k^{-2} \right\} \qquad (2.B.24)$$

Since the negative log-likelihood function can be decomposed into a summation of the log-likelihood function of each node, the parameters $A$, $B$ and variance of the errors can be estimated separately for each node. We first estimate the parameters of the system without exogenous variables. Setting the partial derivative of log-likelihood in Equation 2.B.24 with respect to $B_{k.}^T$ equal to zero, we obtain

$$\frac{\partial l}{\partial B_{k.}^T} = S^{(-k)}((I - B)_{k.})^T = 0. \qquad (2.B.25)$$

Note that

$$((I - B)_{k.})^T = \begin{bmatrix} 0 \\ \vdots \\ 1 \\ \vdots \\ 0 \end{bmatrix} - \begin{bmatrix} \beta_{k1} \\ \vdots \\ \beta_{kk} \\ \vdots \\ \beta_{kp} \end{bmatrix}. \qquad (2.B.26)$$

Substituting Equation 2.B.26 into Equation 2.B.25 gives

$$S_{.k}^{(-k)} - S^{(-k)}(B_{k.})^T = 0.$$

Solving Equation 2.B.26 for the vector of parameters $B_{k.}^T$, we obtain

$$B_{k.}^T = \left(S^{(-k)}\right)^{-1} S_{.k}^{(-1)}. \qquad (2.B.27)$$

Since the matrix $B$ is sparse, we only need to consider the node and its parents. Let $B_{k,pa_D(k)}$ be a row vector consisting of only the entry of the node and its parents in the parameter matrix $B$. Equation 2.B.27 can be further reduced to

$$B_{k,pa_D(k)} = S_{k,pa_D(k)}^{(-k)} \left( S_{pa_D(k),pa_D(k)}^{(-k)} \right)^{-1}. \tag{2.B.28}$$

Similarly, we have

$$\frac{\partial l_k}{\partial \sigma_k^{-2}} = \frac{1}{2} n^{(-k)} \left\{ (I-B)_{k.} S^{(-k)} ((I-B)_{k.})^T - \frac{1}{\sigma_k^{-2}} \right\} = 0,$$

which implies that

$$\hat{\sigma}_k^2 = \left( I - \hat{B} \right)_{k.} S^{(-k)} \left( \left( I - \hat{B} \right)_{k.} \right)^T. \tag{2.B.29}$$

Substituting Equation 2.B.28 and 2.B.29 into Equation 2.B.24, we obtain the score for the node $k$:

$$\text{Score}_D(k) = \frac{1}{2} n^{(-k)} \left( 1 + \log \hat{\sigma}_k^2 \right). \tag{2.B.30}$$

Next, consider the log-likelihood defined in Equation 2.B.23. Similar to Equation 2.B.25, we have

$$\frac{\partial l_k}{\partial B_{k.}} = n^{(-k)} \sigma_k^{(-2)} \left\{ -(I-B)_{k.} S^{(-k)} + A_{k.} S_{XY}^{(-k)} \right\} = 0. \tag{2.B.31}$$

Taking partial derivative of negative log-likelihood with respect to $A_{k.}^T$ and setting it equal to zero gives

$$\frac{\partial l_k}{\partial A_{k.}^T} = -n^{(-k)} \sigma_k^{(-2)} \left[ S_{XY}^{(-k)} ((I-B)_{k.})^T - S_{XX}^{(-k)} (A_{k.})^T \right] = 0. \tag{2.B.32}$$

Rearranging Equations 2.B.31 and 2.B.32, we obtain

$$\begin{bmatrix} S^{(-k)} & S_{YX}^{(-k)} \\ S_{XY}^{(-k)} & S_{XX}^{(-k)} \end{bmatrix} \begin{bmatrix} (B_{k.})^T \\ (A_{k.})^T \end{bmatrix} = \begin{bmatrix} \left( S_{k.}^{(-k)} \right)^T \\ \left( S_{XY}^{(-k)} \right)_{.k} \end{bmatrix}. \tag{2.B.33}$$

Solving Equation 2.B.33 gives the solution:

$$\begin{bmatrix} (B_{k.})^T \\ (A_{k.})^T \end{bmatrix} = \begin{bmatrix} S^{(-k)} & S_{YX}^{(k)} \\ S_{XY}^{(-k)} & S_{XX}^{(-k)} \end{bmatrix}^{-1} \begin{bmatrix} \left( S_{k.}^{(-k)} \right)^T \\ \left( S_{XY}^{(-k)} \right)_{.k} \end{bmatrix}. \tag{2.B.34}$$

## Exercises

Exercise 1. Calculate the score function $Score_D(G)$ for the network in Example 2.1.

Exercise 2. For Example 2.1 and Figure 2.3 show

$$P(x_1 = 0, x_2 = 0, x_3 = 0) = \frac{1}{2}\frac{1}{2}\frac{3}{4} = \frac{3}{16}.$$

Exercise 3. Derive the mixed structural equation model for Figure 2.4.

Exercise 4. Derive the formula for the variance var $(\hat{\delta}_i)$ in Equation 2.31 for Figure 2.4.

Exercise 5. Express the endogenous variables $Y_1$, $Y_2$, $Y_3$, and $Y_4$ in the causal graph in Figure 2.4 in terms of the random effects $u_1$, $u_2$, $u_3$, and $u_4$, and errors $e_1$, $e_2$, $e_3$, and $e_4$.

Exercise 6. Consider two random variables $y_1$ and $y_2$, and two DAGs: $y_1 \rightarrow y_2$ and $y_2 \rightarrow y_1$. Write two nonlinear additive SEMs for two DAGs.

Exercise 7. Assume that $H_0$ is a finite dimensional space with orthonormal basis functions $\varphi_1(x), ..., \varphi_p(x)$. Show that that RK $R_0(x, z)$ is given by

$$R_0(x,z) = \sum_{j=1}^{p} \varphi_j(x)\varphi_j(z).$$

Exercise 8. Assume that the RKHS $H_0$ and RKHS $H_1$ are formed, respectively, by

$$H_0 = span\{1\},$$

$$H_1 = \left\{ f : f = 0, \int_a^b \left(f'\right)^2 dx < \infty \right\}$$

Find the corresponding RKs: $R_0(x, z)$ and $R_1(x, z)$.

Exercise 9. Consider a general multivariate nonlinear function model:

$$Y_d^i = L_i f_d + \varepsilon_i,$$

$$i = 1, ..., n,$$

where $Y_d^i$ is the observed $Y_d$ in the $i$th sample, $L_i$ are continuous functionals, and $\varepsilon_i$ are zero-mean independent random errors with common variance $\sigma_e^2$. Write a penalized likelihood function for the above nonlinear functional model to estimate the smoothing spline.

Exercise 10. Consider the smoothing spline regressions for the partially linear structural equation model:

$$y_j = \mu_j + \sum_{l=1}^{l_j} \alpha_{jl} w_l + \sum_{i=1}^{i_j} L_{ji} f_{ji} + \varepsilon_j,$$

where $L_{ji}$ is a continuous functional defined on RKHS $H^i$, $f_{ji}$ represents a general unknown function, and $\varepsilon_i$ are zero-mean independent random errors with a common variance $\sigma_j^2$. Write the objective function for the regression coefficient estimators and smoothing spline estimators of the function $f_{ji}$ in the above model.

Exercise 11. Consider the example in Figure 2.8. Write the negative log-likelihood for the DAG with both observational and interventional data.

Exercise 12. Write score function for the node 4 of the DAG in Figure 2.8.

Exercise 13. Show that

$$\left( R^{(T^{(i)})}(I - B) \right)_{(k)} = \begin{cases} (I - B)_k. & k \notin T^{(i)} \\ 0 & k \in T^{(i)}, k = 1, ..., p. \end{cases}$$

Exercise 14. Show that the partial derivative of log-likelihood in Equation 2.B.24 with respect to $B_k^T$ is given by

$$\frac{\partial l}{\partial B_k^T} = -n^{(-k)} \sigma_k^{-2} S^{(-k)} \left( (I - B)_k. \right)^T$$

and calculate $\dfrac{\partial l}{\partial B_k^T}$ for the DAG in Figure 2.8.

Exercise 15. Find feasible integral points for the following integer linear programming:

$$\text{Min} \qquad x + 2y$$
$$2x + y \le 6$$
$$-x + y \le 2$$
$$x \ge 0, y \ge 0$$

Exercise 16. Find all parent sets in the DAG showing in Figure 2.11.

Exercise 17. Calculate objective function for a DAG in Figure 2.11.

Exercise 18. Show that cluster constraints in Figure 2.14a are satisfied, but in Figure 2.14b are violated.

Exercise 19. Write down all convexity constraints and cluster constraints for the graph in Figure 2.14a.

# 3

# Wearable Computing and Genetic Analysis of Function-Valued Traits

Low-cost, non-invasive, wireless, wearable biosensors have rapidly developed. The wearable biosensors allow for the development of mobile health (mHealth) technologies that can continuously monitor patients, athletes, premature infants, children, psychiatric patients, people who need long-term care, elderly, and people in impassable regions far from health and medical services (Ajami and Teimouri 2015). Wearable biosensors can remotely collect the data for tracking physical activity, measure electrocardiograms (ECG), and record the electrical and photoplethsymogram (PPG), away from traditional hospital settings (Sun et al. 2016). Wearable biosensors can detect real time changes in health status of individuals, even deliver continuous measurement of the molecular physiology of an individual's body including metabolites and electrolytes and monitor what is happening inside the body (Gao et al. 2016). Wearable computing has the potential to revolutionize health care and disease management.

The ability to collect real time clinical and physiological phenotypes is an essential feature of the wearable biosensors. In other words, the data collected by wearable biosensors take curve or functional form. Although the traditional disease prediction and detection based on the classical biosensors and the genetic study of quantitative traits has seen wide application and extensive technical development, the disease prediction and detection based on time course data and quantitative genetic analysis of function-valued trait is comparatively less development (Hansen et al. 2011). There is increasing evidence that standard multivariate statistical analysis often fails with functional data. Functional and dynamic models will be used for disease classification and genetic studies of functional-valued traits. First, three methods: functional principal component analysis, ordinary differential equations, and deep learning will be discussed for classification of wearable biosensor data. Then, the function regression model with functional responses and functional predictors (FLMFF) for quantitative genetic analysis of function-valued traits with NGS data will be developed and analyzed. Finally, the gene–gene interaction models for function-valued traits will be discussed.

## 3.1 Classification of Wearable Biosensor Data

### 3.1.1 Introduction

Wearable biosensors allow continuous measurement of health-related physiology including electrocardiogram (ECG), seimocardiography (SCG), oxygen saturation levels, heart rate, skin temperature, blood pressure, and physical activities (Li et al. 2017). These real-time measurements are curves and functions. Figure 3.1 plots the ECG QRS complex. The ECG is a time-varying signal which reflects the ionic current flow which causes the cardiac fibers to contract and subsequently relax. QRS complex contains information about the electrical function of the heart by changing the shape of its component waves, P, QRS, and T waves. Three letters Q, R, and S stand for the three main phases of a cardiac cycle (Ravier et al. 2007). The QRS complex is considered as the most striking waveform within the ECG and the morphology, magnitudes, and duration time provide valuable information about the biological processes of the heartbeat.

When a patient's upper airway is blocked, his or her oxygen concentration will drop. The oxygen saturation is biologically related to the sleep apnea (Varvarigou et al. 2011). There are some sleep apnea studies which use oxygen saturation information. The oxygen saturation signals are usually measured by seconds. Figure 3.2 shows a typical oxygen concentration curve over a night where there are 35,280 oxygen measurements. These time course physiological phenotype data collected from the wearable biosensors provide important information on the health status of humans and can be used to monitor and diagnose diseases.

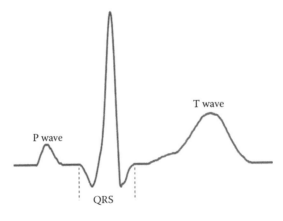

**FIGURE 3.1**
QRS complex example.

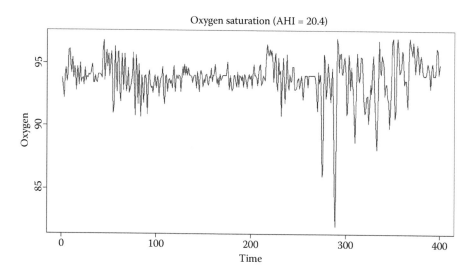

**FIGURE 3.2**
Concentration of oxygen in the blood.

### 3.1.2 Functional Data Analysis for Classification of Time Course Wearable Biosensor Data

A basic feature of the functional data is that the functional data have infinite dimensions and are highly correlated, which makes classification of time course data difficult. There is increasing evidence that standard multivariate statistical analysis often fails with functional data. The key for the success of classification of time course data is to reduce the dimension of the data. A common solution is to choose a finite dimensional basis and project the functional curve onto this basis. Then the resulting basis coefficients form a finite dimensional representation. Therefore, the widely used statistical methods for functional data classification consist of two steps.

At the first step the functional data are projected into a finite-dimensional space of basis functions or eigenfunction (Luo et al. 2013; Ramsay and Silverman 2005).

Let $\phi_j(t)$ be the set of orthogonal basis functions including Fourier functions, eigenfunctions, or functional principal component functions. Then each function can be written as a linear combination of the basis functions.

$$x_i(t) = \sum_{j=1}^{\infty} \xi_{ij} \phi_j(t), \tag{3.1}$$

the expansion coefficients $\xi_{ij}$ are estimated by

$$\xi_{ij} = \int_T x_i(t)\phi_j(t)dt. \tag{3.2}$$

At the second step, all the methods for feature selection, including sufficient dimension reduction, and classification of multivariate data, including linear discriminant analysis (LDA), logistic regression, support vector machine, can be used to classify the coefficients in the expansions.

To more efficiently use functional dependent information, highlight some features of the data and improve the classification accuracy. Functional data are often transformed in several ways. The data are, in general, centered and normalized. The most important data transformation is the derivation. A linear combination of the original functional curves and their different order of derivatives are then projected onto basis functions or eigenfunctions (Alonso et al. 2012). Incorporating dynamic information (derivative information) into feature will capture the dependence characteristics of the functional data and hence improve classification.

### 3.1.3 Differential Equations for Extracting Features of the Dynamic Process and for Classification of Time Course Data

Although incorporating a functional derivative into features increases the potential to improve classification accuracy, taking functional derivatives as features will substantially increase the number of features. We hope that we will not increase the number of features while employing dynamic information in classifying functional curves. Using the parameters in differential equations that model the dynamics of continuously changing the process such as QRS complex in the ECG data as features can serve this purpose. Differential equations are widely used powerful tools to describe dynamic systems in many physical, chemical, and biological processes. Parameters in differential equations capture fundamental behaviors of the real dynamic processes and are consistent with the available data. Only a few parameters are needed to capture essential dynamic features of the systems. Therefore, in this section we discuss how to use parameters in differential equations that model dynamic processes in time course data for classification.

#### 3.1.3.1 Differential Equations with Constant and Time-Varying Parameters for Modeling a Dynamic System

We assume that $x(t)$ is a state variable in a dynamic system which can be modeled by the following second order ordinary differential equation (ODE) with constant parameter:

$$L(x(t)) = \frac{d^2x(t)}{dt^2} + w_1\frac{dx(t)}{dt} + w_0x(t) = 0 \qquad \text{or} \tag{3.3}$$

the second-order differential equation with time-varying parameters:

$$L(x(t)) = \frac{d^2x(t)}{dt^2} + w_1(t)\frac{dx(t)}{dt} + w_0(t)x(t) = 0, \tag{3.4}$$

where $w_1$ ($w_1(t)$) and $w_0$ ($w_0(t)$) are weighting coefficients or parameters in the ODE. The state $x(t)$ is hidden. Its observations $y(t)$ often have measurement errors:

$$y(t) = x(t) + e(t), \tag{3.5}$$

where $e(t)$ is the measurement error at the time $t$.

### 3.1.3.2 Principal Differential Analysis for Estimation of Parameters in Differential Equations

The estimators of the parameters in the ODE can be obtained by principal differential analysis (Poyton et al. 2006). The purpose of parameter estimation is to attempt to determine the appropriate parameter values that make the errors between the predicted response values and the measured data as small as possible. The predicted response values can be obtained by solving ODE for modeling the dynamic system. One way to solve ODE is to first expand the function $x(t)$ in terms of basis functions. Let $x_i(t)$ be the state variable at time $t$ of the $i$-th sample satisfying ODE (3.3) or (3.4) and $y_i(t)$ be its observation ($i = 1,...,n$). Then, $x_i(t)$ can be expanded as

$$x_i(t) = \sum_{j=1}^{K} c_{ij}\phi_j(t) = C_i^T \phi(t), \tag{3.6}$$

where $C_i = [c_{i1}, ..., c_{iK}]^T$ and $\phi(t) = [\phi_1(t), ..., \phi_K(t)]^T$.
    Similarly, the parameters $w_1(t)$ and $w_0(t)$ can be expanded as

$$w_1(t) = \sum_{j=1}^{K} h_{1j}\phi_j(t) = h_1^T \phi(t) \quad \text{and}$$

$$w_0(t) = \sum_{j=1}^{K} h_{0j}\phi_j(t) = h_0^T \phi(t), \tag{3.7}$$

where $h_1 = \begin{bmatrix} h_{11} \\ \vdots \\ h_{1K} \end{bmatrix}$ and $h_0 = \begin{bmatrix} h_{01} \\ \vdots \\ h_{0K} \end{bmatrix}$.

Let

$$\psi(t) = \frac{d^2\phi}{dt^2} + \frac{d\phi}{dt}\phi^T(t)h_1 + \phi(t)\phi^T(t)h_0$$

$$= \frac{d^2\phi}{dt^2} + G(t)h,$$

and

$$J_{\phi h} = \int_T \psi(t)\psi^T(t)dt,$$

where $G(t) = \begin{bmatrix} \dfrac{d\phi}{dt}\phi^T(t) & \phi(t)\phi^T(t) \end{bmatrix}, h = \begin{bmatrix} h_1 \\ h_0 \end{bmatrix}$.

The differential operator is given by $L(x_i(t)) = C_i^T \psi(t)$. The penalty term is defined as

$$\lambda \int_T L(x_i(t))L^T(x_i(t))dt = \lambda C_i^T J_{\phi h} C_i. \tag{3.8}$$

We estimate the state function $x(t)$ from the observation data $y(t)$ by minimizing the following objective function which consists of the sum of the squared errors between the observations and the states and the penalty terms:

$$\sum_{i=1}^{n} \left\{ \sum_{j=1}^{T} [y_i(t_j) - x_i(t_j)]^2 + \lambda \int_T L(x_i(t))L^T(x_i(t))dt \right\}$$

$$= \sum_{i=1}^{n} \left\{ \sum_{j=1}^{T} [y_i(t_j) - \phi^T(t_j)C_i]^2 + \lambda C_i^T J_{\phi h} C_i \right\}. \tag{3.9}$$

Let

$$Y_i = \begin{bmatrix} y_i(t_1) \\ \vdots \\ y_i(t_T) \end{bmatrix}, \tilde{\phi} = \begin{bmatrix} \phi^T(t_1) \\ \vdots \\ \phi^T(t_T) \end{bmatrix}, Y = \begin{bmatrix} Y_1 \\ \vdots \\ Y_n \end{bmatrix}, \Phi = \begin{bmatrix} \tilde{\phi} & 0 & \cdots & 0 \\ \vdots & \vdots & \ddots & 0 \\ 0 & 0 & \cdots & \tilde{\phi} \end{bmatrix},$$

$$J = \begin{bmatrix} J_{\phi h} & 0 & \cdots & 0 \\ 0 & J_{\phi h} & \cdots & 0 \\ \vdots & \vdots & \ddots & \vdots \\ 0 & 0 & \cdots & J_{\phi h} \end{bmatrix}, C = \begin{bmatrix} C_1 \\ \vdots \\ C_n \end{bmatrix}$$

The problem (3.9) can then be reduced to

$$\min_{C} \quad (Y-\Phi C)^T(Y - \Phi C) + \lambda C^T J C. \tag{3.10}$$

The Least Square Estimators of the expansion coefficients are then given by

$$C = (\Phi^T \Phi + \lambda J)^{-1} \Phi^T Y. \tag{3.11}$$

Next, we estimate the parameters in the ODE. The parameters in the ODE can be estimated by minimizing the following least squares objective function:

$$\min_{h} \quad SSE_p = \int_T L^T(X(t))L(X(t))dt, \tag{3.12}$$

where $L(X(t)) = [L(x_1(t)), ..., L(x_n(t))]^T$. Since $L(x_i(t)) = C_i^T \psi(t)$, the $L(X(t))$ can be expressed in terms of the estimated expansion coefficients as

$$L(X(t)) = \hat{C}\Psi(t).$$

Therefore, problem (3.3) can be reduced as

$$\min_{h} \quad SSE_p = \int_T \psi^T(t)\hat{C}^T\hat{C}\psi(t)dt, \tag{3.13}$$

where the matrix $\hat{C}$ is estimated and hence fixed in the minimization problem (3.13). Setting the partial derivative of $SSE_p$ to be zero:

$$\frac{\partial SSE_p}{\partial h} = \int_T G^T(t)\hat{C}^T\hat{C}\left[\frac{d^2\phi(t)}{dt^2} + G(t)h\right]dt = 0. \tag{3.14}$$

Solving Equation 3.14 for $h$, we obtain

$$h = -\left[\int_T G^T(t)\hat{C}^T\hat{C}G(t)dt\right]^{-1}\int_T G^T(t)\hat{C}^T\hat{C}\frac{d^2\phi}{dt^2}dt. \tag{3.15}$$

In summary, we iteratively determine the expansion coefficients for fixed parameters in the ODE by Equation 3.11 and estimate the parameters in the ODE for fixed expansion coefficients by Equation 3.15 until convergence.

### 3.1.3.3 QRS Complex Example

#### 3.1.3.3.1 Accuracy of the ODE Fitting the Data

Automatic ECG analysis has wide applications in heartbeat classification. Automatic ECG analysis can monitor cardiac activity for timely detection of abnormal heart conditions and diagnosis of cardiac arrhythmias. The heartbeats are also often referred to QRS complex shapes. Three letters Q, R, and S stand for the three main phases of a cardiac cycle (Ravier et al. 2007). We apply the proposed methods to estimate the parameters in Equations 3.4 and 3.5 for modeling the QRS complex. We first evaluate how well the B-spline fits the signal $x(t)$ in the QRS complex. The data were from the MIT-BIH arrhythmia database, which contained 48 half-hour excerpts of two channel ambulatory ECG signals (lead MLII and lead V5). Only lead MLII was used from 47 subjects. The ECG signals were band-pass filtered at 0.1–100 Hz and sampled at 360 Hz per channel with 3-bit resolution over a 10 mV range (http://www.physionet.org/). Two or more cardiologists independently annotated each record; disagreements were resolved to obtain the computer-readable reference annotations for each beat included in the database. An annotation file was associated with each record to provide the reference annotations like QRS location and heartbeat categories which were used to extract QRS segments and true labels for each heartbeat, respectively.

ECG data were measured and recorded by skin electrodes. These data were easily contaminated by various types of noises due to power-line interference, muscle contraction and electrode movements, and electromyography. To remove unnecessary noise and improve the signal-to-noise ratio, the ECG waveform is band-pass filtered at 5–12 Hz (low-band and high-band pass filtered) to remove baseline wander and powerline interference. The filtered ECG signal is used for dynamical model fitting. Figures 3.3 and 3.4 plotted the observed signal, the fitted curve by the cubic B-spline with a uniform knot spacing of 0.012 second, and the trajectories of solutions to the second ODE with constant parameters and time-varying parameters for normal and abnormal QRS complex, respectively. The tuning parameter was chosen by cross-validation at each iteration of the smoothing. Both Figures 3.3 and 3.4 showed that the cubic B-spline and the second ODE with the time-varying

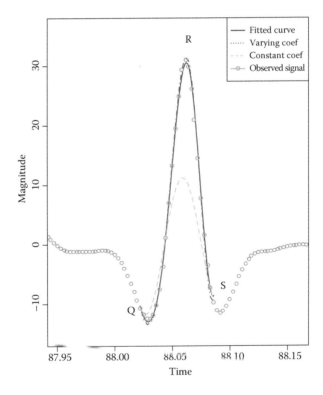

**FIGURE 3.3**
Fitted curve by the second ODE.

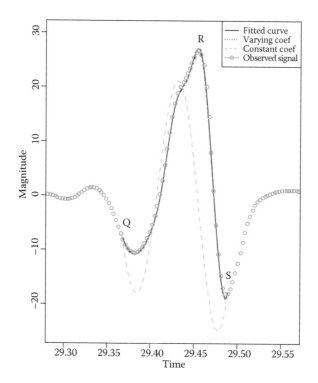

**FIGURE 3.4**
Fitted curve of the abnormal QRS complex by the second varying ODE.

parameters can approximate the observed signals very well in both normal and abnormal complexes. However, the trajectories of the solutions to the second ODE with constant parameters cannot approximate the observed signals very well and are only similar to the shapes of the curves of the observed signals.

### 3.1.3.3.2 Stability and Transient-Response Analysis

Although the second ODE with the constant parameters can only approximate the shape of the curves of the observed signals, it can characterize the dynamic behaviors of the QRS complex remarkably well. Dynamic properties include stability and transient-response, which determine how the systems maintain their functions and performance under a broad range of random internal and external perturbations and their responses to changes in environments.

The most important dynamic property of biological systems is concerned with stability. A dynamic system is called stable if their state variables return to, or toward their original states or equilibrium states following internal and external perturbations (Kremling and Saez-Rodriguez 2007). The stability of the system is a property of the system itself. One of the methods for assessing the stability of the dynamic system is to analyze eigenvalues of the eigen-equation of the high-order ODE which models the linear dynamic systems. If the real parts of all eigenvalues of the dynamic system are negative, then the system is stable and if the real parts of all eigenvalues of the dynamic system are positive, then the system is unstable. The coefficients of the second ODE for the normal QRS complex in Figure 3.3 are $w_1 = 2.598$ and $w_0 = 9394.2$ and its eigenvalues are $\lambda_1 = -1.30 + 96.9i$ and $\lambda_2 = -1.30 - 96.9i$. The coefficients of the second ODE for the abnormal QRS complex in Figure 3.4 are $w_1 = -6.97$ and $w_0 = 4535.9$ and its eigenvalues are $\lambda_1 = 3.48 + 67.26i$ and $\lambda_2 = 3.48 - 67.26i$. We clearly observed that the real parts of all eigenvalues of the second ODE for the normal QRS complex were negative, but the real parts of all eigenvalues for the abnormal QRS complex were positive. This showed that the normal QRS complex was stable, but the abnormal complexes were unstable. The dynamic system in the heart will remain at steady state until occurrence of external perturbation. Depending on dynamic behavior of the system after perturbation of environments, the steady-states of the system are either stable (the system returns to the initial state) or unstable (the system leaves the initial equilibrium state). For any practical purpose, the dynamic QRS complex must be stable. An unstable dynamic system underlying abnormal QRS complex will lead to the irregular activities in the heart or even to heart failure.

The dynamic behavior of the cardiac underlying QRS complex is encoded in the temporal evolution of its states. Response of a biological system to perturbation of internal and external stimuli has two parts: the transient and the steady state response. The process generated in going from the initial state to the final state in the response to the perturbation of the internal and external stimuli is called transient response. Steady-state response studies the system behavior at infinite time. Transient-response analysis of the cardiac system can be used to quantify their dynamics. It can reveal how fast the dynamic system in the heart responds to perturbation of environments and how accurately the system in the heart can finally achieve the desired steady-state values. It can also be used to study damped vibration behavior and stability of the cardiac dynamic system.

The transient response of a dynamic system depends on the input signals. Different signals will cause different responses. There are numerous types of signals in practice. For the convenience of comparison, we consider two types of signals: (1) unit-step signal and (2) unit-impulse signal.

The transfer function of the response of the cardiac dynamic system underlying the QRS complex to unit-step and unit-impulse input signals are given by $Y(s) - \dfrac{G(s)}{s}$ and $Y(s) = G(s)$, respectively, where $G(s)$ is the transfer function of the dynamic system. The transient-response analysis of the dynamic system can be performed by inverse Laplace transformation. We performed the transient-response analysis with MATLAB® (Ogata 1997). Figures 3.5 and 3.6 showed the unit-step response curves of the cardiac

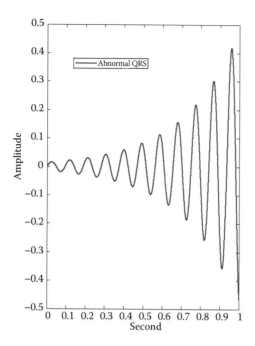

**FIGURE 3.5**
Unit-step response curves of cardiac system underlying normal QRS complexes.

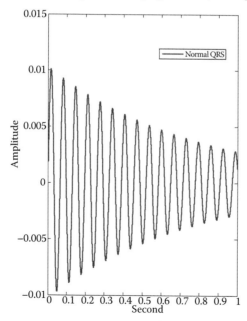

**FIGURE 3.6**
Unit-step response curves of cardiac system underlying abnormal QRS complexes.

system underlying normal and abnormal QRS complexes, respectively. They showed that the unit-step response cardiac system underlying normal and abnormal QRS complexes were substantially different. Although we observed signal oscillations in both normal and abnormal QRS complex, while the signals from the normal QRS complex after the unit-step perturbation quickly reach the steady states, the signals from the abnormal QRS complex to respond to the perturbation were highly oscillated and would never reach the steady-state values. This phenomenon suggested that dynamic responses of abnormal QRS complexes to environmental stimuli were irregular. Figures 3.7 and 3.8 showed the unit-impulse response curves of the cardiac system underlying normal and abnormal QRS complexes, respectively. We observed similar patterns to that of the unit-step response of QRS complexes.

### 3.1.3.3.3 Classification of Signals from Dynamic Systems

Previous studies strongly demonstrated that the second-order ODE can capture dynamic features of QRS complexes very effectively. Next, we evaluate the performance of using the parameters in the second-order ODE to classify QRS complexes. We used three measures to evaluate the classification performance: sensitivity, specificity, and accuracy. Sensitivity is defined as the

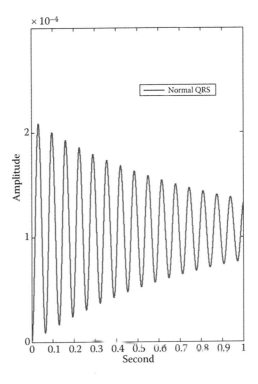

**FIGURE 3.7**
Unit-impulse response curves of the cardiac system underlying normal QRS complexes.

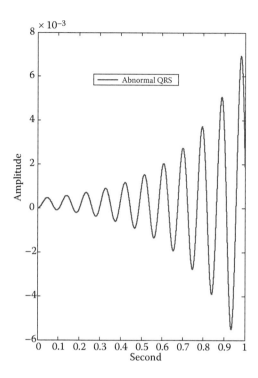

**FIGURE 3.8**
Unit-impulse response curves of the cardiac system underlying abnormal QRS complexes.

percentage of the true abnormal signals correctly classified as abnormal. Specificity is defined as the percentage of the true normal signals correctly classified as normal. The classification accuracy is defined as the percentage of the correctly classified normal and abnormal signals. We used 19 ECG records of the MIT/BIH database (Ravier et al. 2007). The parameters in the second-order ODE were used as features of the QRS complexes and input into the support vector machine (SVM) (Chang and Lin 2011) to classify QRS complexes. We also compare our results with the neural network classifier that was used in Ravier et al. (2007). For the time-varying parameters in the second-order ODE, we used functional principal component scores (Ramsay and Silverman 2005) of the time-varying parameter functions $w_0(t)$ and $w_1(t)$ in the second-order ODE as features of the SVM. Specifically, the QRS complex signals were expanded in terms of the Fourier series. The Fourier expansion coefficients of the QRS complexes were used as features for a multilayer neural network (NN) with 16 input nodes, 4 neurons in the hidden layer, and 1 output neuron. For both ODEs with constant and time-varying parameters, we also used the height of the R point and width of the QRS complex as their features.

The results of classification were summarized in Table 3.1 where ODE is referred to as the ODE with constant parameters, ODET, the ODE with

**TABLE 3.1**

Accuracy of ODE Constant and Time Varying Parameters, and Neural Networks for Classifying QRS Complex

| File # | # of Normal QRS | # of Abnormal QRS | Sensitivity | | | Specificity | | | Accuracy | | |
|---|---|---|---|---|---|---|---|---|---|---|---|
| | | | ODE | ODET | NN | ODE | ODET | NN | ODE | ODET | NN |
| 100 | 2233 | 34 | 0.059 | 0.118 | 0.235 | 0.999 | 0.999 | 0.995 | 0.985 | 0.985 | 0.983 |
| 101 | 1851 | 8 | 0.250 | 0.000 | 0.167 | 1.000 | 1.000 | 1.000 | 0.997 | 0.996 | 0.997 |
| 102 | 99 | 2082 | 1.000 | 1.000 | 0.998 | 0.980 | 0.909 | 0.788 | 0.999 | 0.996 | 0.988 |
| 104 | 163 | 2060 | 0.969 | 0.992 | 0.983 | 0.699 | 0.675 | 0.783 | 0.950 | 0.969 | 0.964 |
| 105 | 2484 | 110 | 0.745 | 0.273 | 0.622 | 0.988 | 1.000 | 0.996 | 0.978 | 0.969 | 0.990 |
| 106 | 1500 | 511 | 0.990 | 0.990 | 0.939 | 0.999 | 0.970 | 0.978 | 0.997 | 0.975 | 0.968 |
| 119 | 1539 | 444 | 1.000 | 1.000 | 0.995 | 1.000 | 0.995 | 0.992 | 1.000 | 0.996 | 0.993 |
| 200 | 1736 | 847 | 0.919 | 0.939 | 0.867 | 0.997 | 0.929 | 0.963 | 0.971 | 0.932 | 0.931 |
| 201 | 1604 | 272 | 0.853 | 0.673 | 0.747 | 0.999 | 0.993 | 0.991 | 0.978 | 0.946 | 0.956 |
| 202 | 2044 | 69 | 0.391 | 0.188 | 0.306 | 0.999 | 0.997 | 0.991 | 0.979 | 0.971 | 0.971 |
| 205 | 2564 | 83 | 0.867 | 0.084 | 0.790 | 1.000 | 1.000 | 0.995 | 0.995 | 0.971 | 0.988 |
| 208 | 1573 | 1356 | 0.973 | 0.982 | 0.945 | 0.987 | 0.941 | 0.960 | 0.980 | 0.960 | 0.953 |
| 209 | 2616 | 384 | 0.444 | 0.447 | 0.482 | 0.981 | 0.978 | 0.976 | 0.912 | 0.910 | 0.912 |
| 210 | 2410 | 212 | 0.882 | 0.269 | 0.770 | 0.995 | 0.999 | 0.954 | 0.986 | 0.940 | 0.940 |
| 212 | 919 | 1824 | 0.967 | 0.971 | 0.958 | 0.984 | 0.928 | 0.958 | 0.973 | 0.957 | 0.958 |
| 213 | 2634 | 609 | 0.849 | 0.875 | 0.863 | 0.995 | 0.983 | 0.987 | 0.968 | 0.963 | 0.964 |
| 215 | 2412 | 120 | 0.958 | 0.196 | 0.800 | 0.997 | 0.904 | 0.986 | 0.995 | 0.705 | 0.979 |
| 217 | 242 | 1959 | 0.995 | 0.993 | 0.987 | 0.946 | 0.946 | 0.964 | 0.990 | 0.988 | 0.984 |
| 219 | 1539 | 51 | 0.824 | 0.278 | 0.667 | 1.000 | 0.862 | 0.994 | 0.994 | 0.697 | 0.986 |

time-varying parameters and NN, neural networks. Several features emerge from the results in Table 3.1. First, in most cases, the ODE with constant parameters has the highest accuracy. Second, it is surprisingly observed that although the trajectory of the solution to ODE with time-varying parameters approximates the observed signal curve of the QRS complex, it is much better than that of the ODE with constant parameters, its performance for classifying QRS complex is not as good as that of the ODE with constant parameters in many cases. Third, when the numbers of the sampled normal and abnormal QRS complexes were highly imbalanced (records: 51, 100, 101, 202) either sensitivity or specificity of all three classifiers was clearly not accurate because the training samples did not provide enough information on distinguishing normal QRS complex from abnormal QRS complex. In general, when the number of sampled normal QRS complexes was substantially larger than that of abnormal QRS complexes, the specificity of classification was high. In contrast, if the number of abnormal complexes is larger than that of normal QRS complexes we can obtain high sensitivity. Fourth, when the numbers of both normal and abnormal QRS complexes were balanced, both sensitivity and specificity can reach high values.

### 3.1.4 Deep Learning for Physiological Time Series Data Analysis

Wearable biosensors continuously monitor individual's health status and generate real time course data. We can measure various kinds of physiological time-series data through wearable biosensors at any place (Wang et al. 2016) and are interested in prediction, segmentation, classification, and searching for abnormalities in blood pressure, heart rate, and sleep patterns from the observed time course data (Cristian and Gamboa 2017). Alternatives to traditional approaches including autoregression, linear dynamic systems, and Hidden Markov Model (some of which were discussed in Sections 3.1.2 and 3.1.3), deep neural networks can be applied to physiological time series data analysis. Feature extraction and selections are the key to the success of risk prediction and abnormality detection. Conventional techniques for time-series data analysis are either limited in processing natural data of the raw form or dependent on the manually extracted features. Features of the signals are the internal representation of the data that are needed for abnormality detection and classification.

Deep learning methods are modern machine learning techniques. They amplify the features that are important for discrimination and suppress irrelevant variations in the data via automatic representation learning with multiple levels of hierarchical representations (LeCun et al. 2015). In other words, deep learning is a paradigm of machine learning that uses multiple hierarchically organized processing layers to extract and select discriminating information from complicated time series and imaging data (Ravi et al. 2016). In recent years, the realization has been well recognized that deep learning has great potential in discovering intricate features in high-dimensional data.

Deep learning has found large applications in genomics, wearable computing, disease prevention, diagnosis, and management (Albarqouni et al. 2016; Wang et al. 2017; Cheng et al. 2015; Jin and Dong 2016; Marblestone et al. 2016; Sathyanarayana et al. 2016). Convolutional neural networks are designed to process the data in the form of signals and arrays, for example, images. In this section, we will focus on convolutional network-based time course data analysis.

### 3.1.4.1 Procedures of Convolutional Neural Networks for Time Course Data Analysis

General procedures for application of convolutional neural networks (CNNs) to time-series data analysis are shown in Figure 3.9 where RNN denotes recurrent neural networks. The physiological time course data are first normalized and preprocessed. Then, the CNNs will be used to extract discriminating features from the physiological data. The extracted features will be either sent to classification and abnormality detection or RNN for time series forecasting. Many machine learning techniques developed in Chapter 8 in the book *Big Data in Omics and Imaging: Association Analysis* can be directly used for classification.

### 3.1.4.2 Convolution is a Powerful Tool for Liner Filter and Signal Processing

Convolution is a linear filter and used to describe the response of a linear, time-invariant (LTI) system under an external signal perturbation. In general, an input signal can be expanded as a weighted sum of basis signals, each of these basis signals is passed through a linear system and lead to the response of the linear system. The basic signals include Fourier series and delta (impulse) function. In this section we will focus on delta function.

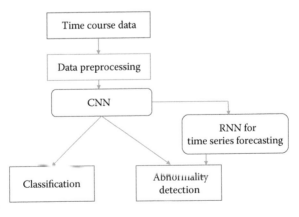

**FIGURE 3.9**
Procedures for application of CNN to time series data analysis.

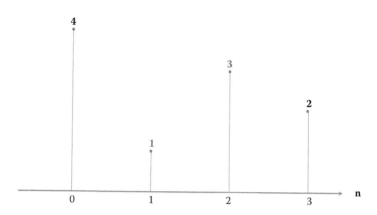

**FIGURE 3.10**
An example of signal decomposition.

A delta (impulse) function, denoted as $\delta(t)$, is defined as

$$\delta(t) = \begin{cases} 1 & 0 \\ 0 & t \neq 0. \end{cases} \quad (3.16)$$

Any signal can be expanded as a sum of delta functions $\delta(t)$. Consider an example in Figure 3.10. Let $x(0) = 4$, $x(1) = 1$, $x(2) = 3$, and $x(3) = 2$. It is clear that signals $x(0)$, $x(1)$, $x(3)$, and $x(4)$ can be expressed as

$$x(n) = x(0) = 4\delta(0) = x(0)\delta(n)$$
$$x(n) = x(1) = 1\delta(0) = x(1)\delta(n-1)$$
$$x(n) = x(2) = 3\delta(0) = x(2)\delta(n-2) \quad (3.17)$$
$$x(n) = x(3) = 2\delta(0) = x(3)\delta(n-3).$$

Expression 3.17 can be summarized as a weighted summation:

$$x(n) = x(0)\delta(n-0) + x(1)\delta(n-1) + x(2)\delta(n-2) + x(3)\delta(n-3). \quad (3.18)$$

Similar to Equation 3.18, in general, a signal can be decomposed as

$$x(n) = \sum_{j=-\infty}^{\infty} x(j)\delta(n-j). \quad (3.19)$$

Next we discuss a linear filter or the impulse response of linear systems to impulse signal inputs. As shown in Figure 3.11, the impulse response of a linear system is the output of the system under the delta function perturbation. We observed from Figure 3.11 that the output of a time-shifted delta function is also an impulse response function shifted at the same amount of time and that output of a scaled delta input signal is the scaled impulse response of the original input signal. The impulse response of the signals $4\delta(0) + \delta(n-1) + 3\delta(n-2) + 2\delta(n-3)$ in Figure 3.10 is $4h(0) + h(n-1) + 3h(n-2) + 2h(n-3)$.

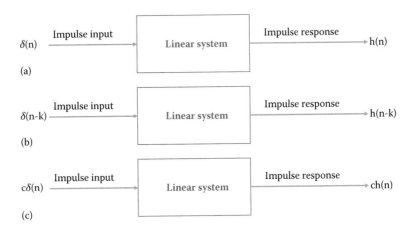

**FIGURE 3.11**
Impulse response.

Therefore, if the general signal $x(n)$ is expanded as the summation of delta functions by Equation 3.19 then the response of the system will be (Figure 3.12)

$$y(n) = \sum_{j=-\infty}^{\infty} x(j)h(n-j). \tag{3.20}$$

The output of the linear system can also be viewed as equal to the input signal convolved with the system's delta function. Now we formally define convolution.

### Definition 3.1: Discrete Convolution

The convolution of discrete functions $x(n)$ and $h(n)$ which is written as $x*h$, is defined as

$$(x*h)(n) = \sum_{j=-\infty}^{\infty} x(j)h(n-j)$$

$$= \sum_{j=-\infty}^{\infty} x(n-j)h(n). \tag{3.21}$$

$$x(n) = \sum_{j=-\infty}^{j=\infty} x(j)\delta(n-j) \longrightarrow \boxed{\begin{array}{c}\textbf{Linear system}\\\textbf{h(n)}\end{array}} \longrightarrow y(n) = \sum_{j=-\infty}^{\infty} x(j)h(n-j)$$

$$x(n) \longrightarrow \boxed{\begin{array}{c}\textbf{Linear system}\\\textbf{h(n)}\end{array}} \longrightarrow \begin{array}{c}y(n) = \sum_{j=-\infty}^{\infty} x(j)h(n-j)\\= x(n) * h(n)\end{array}$$

Convolution

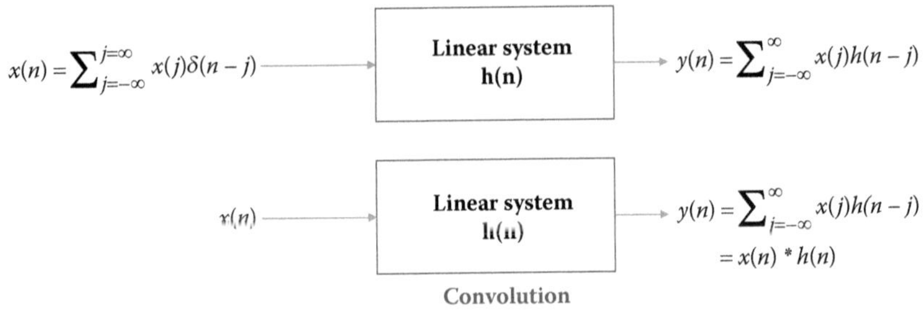

**FIGURE 3.12**
The impulse response of a linear system and convolution.

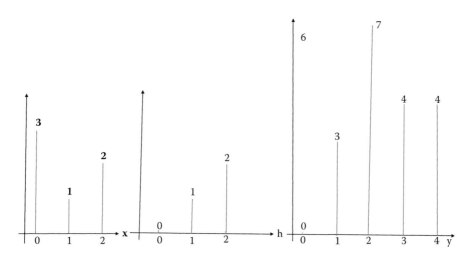

**FIGURE 3.13**
Example of convolution.

The system is completely determined by the impulse response. Different impulse responses determine the different behaviors of the linear system.

**Example 3.1**

Consider input signal $x(n)$ = [3,1,2] and impulse response vector $h(n)$ = [0,1,2] (Figure 3.13). Now we calculate the convolution of $x*h$ using Equation 3.21:

$$y(0) = \sum_{j=-\infty}^{\infty} x(j)h(0-j) = x(0)y(0) = 3^*0 = 0,$$

$$y(1) = \sum_{j=-\infty}^{\infty} x(j)h(1-j) = x(0)h(1) + x(1)h(0) = 3^*1 + 1^*0 = 3,$$

$$y(2) = \sum_{j=-\infty}^{\infty} x(j)h(2-j) = x(0)h(2) + x(1)h(1) + x(2)h(0) = 3^*2 + 1^*1 + 2^*0 = 7,$$

$$y(3) = \sum_{j=-\infty}^{\infty} x(j)h(3-j) = x(0)h(3) + x(1)h(2) + x(2)h(1) = 3^*0 + 1^*2 + 2^*1 = 4,$$

$$y(4) = \sum_{j=-\infty}^{\infty} x(j)h(4-j) = x(0)h(4) + x(1)h(3) + x(2)h(2) + x(3)h(1) + x(4)h(0)$$

$$= 2^*2 = 4.$$

The output is $y$ = [0 3 7 4 4].

### 3.1.4.3 Architecture of CNNs

CNNs consist of a sequence of layers: input layer, convolutional layers, fully connected layers, and classifier (Figure 3.14). Each layer of a CNN transforms a set of activations to another via distinctive functions. The convolutional

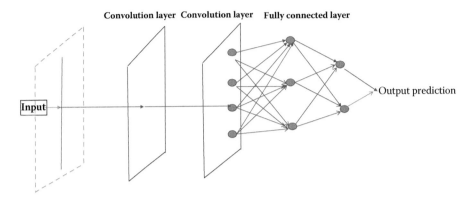

**FIGURE 3.14**
Architecture of convolutional neural networks.

layers in the CNNs receive the input signals in the form of multiple time series and extract features from the signals through convolutional layers. The CNNs have four key ideas: local connection, shared weights, pooling, and presence of multiple layers. For the convenience of discussion, convolution layers are further divided into convolution layers and sub-sampling layers. Each convolutional layer or sub-sampling includes filter operations and non-linear activation operations.

Input signals are time series. The parameters of typical time series data include their length and transition channels. Neurons in the convolutional layer are connected to local regions in the input. The designed filters that function as linear systems for filtering the input data are convolved with the input signals to extract features. The nonlinearity layer applies an elementary activation function. The sub-sampling (pooling) layer performs a nonlinear down-sampling operation to progressively reduce the spatial size of the feature representation and control overfitting. The fully connected layer computes the class scores that are sent to classifiers for classification and prediction. Convolutional layer, pooling layer, and fully connected layer are stacked to form a full CNN architecture.

In summary, a CNN consists of three main types of layers: (1) convolutional layers, (2) sub-sampling (pooling) layers, and (3) an output multiple perceptron (MLP) including a fully connected layer. One convolution layer and one sub-sampling layer are present in pair. Each convolution layer is followed by a sub-sampling layer, and the last convolutional layer is followed by the output fully connected layer. Each layer has many feature maps. The depth of CNN can be extended by adding an arbitrary number of the pairs of convolution layer and sub-sampling layer. We assume that there is a total of $L = 2a + 2$ layers, where $a$ is a positive integer number. All convolutional layers are indexed by $l = 1,3,\ldots,2a + 1$ and sub-sampling (pooling) layers are indexed by $l = 2,4,\ldots,2a$. The last layer is a fully connected MLP.

### 3.1.4.4 Convolutional Layer

Convolutional layer is the essential component of a CNN. It includes a filter bank layer, non-linearity layer, and feature pooling layer. The parameters of the filter bank layer consist of a set of learnable filters (or kernels), which are small spatially, but extend through all channels of the input time series. Figure 3.15 shows that a filter has size $5 \times 3$ ($m_1 \times m_2$) (i.e., 5 time points and 3 channels). During the forward pass, each filter is convolved across the length of the input time series, computing the dot product between the entries of the filter and the input at any position and producing a one-dimensional feature map of the filter with the length of $n$ which shows the responses of the filter at every position of the time point. We design various filters that give their response to capture the specific features of the signals when the filters detect the input signals. Stacking the feature maps for all filters across the sequence of time series forms the full output volume of the convolution layer.

#### 3.1.4.4.1 Filter

In the convolutional layer, each feature map is connected to one or more feature maps of the preceding layer. A connection is associated with a filter (kernel). The input of each filter in the $l^{th}$ layer is a 2D array with $n_1^l$ 1D feature map of size $n_2^l$. In the first input layer, the feature map is a univariate time series. The $j^{th}$ activation component in the $u^{th}$ feature map of the $l^{th}$ layer is denoted $a_{uj}^l$, $u = 1, 2, ..., n_1^l$, $j = 1, 2, ..., n_2^l$.

Next, we consider the size of the filter. In time series data analysis, a filter is a vector. The element of the filter is often called the neuron. In general, the dimensions of the input data and features in the feature maps are very high.

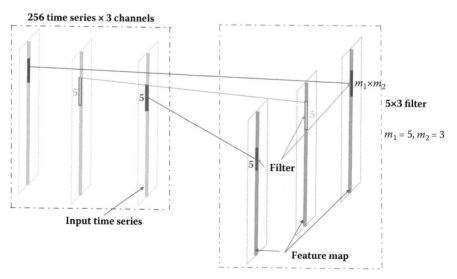

**FIGURE 3.15**
Convolution layer.

When dealing with high-dimensional input and feature data, it is infeasible in practice to connect neurons in the filter to all neurons in the input layer or the previous $(l-1)^{th}$ layer. To overcome these limitations, each neuron will be connected to only a small local region of the input or previous layer. The extent of this connectivity is a hyperparameter that is often called the receptive field of the neuron or simply called the filter size, denoted as $m^l$. The connections are local along the length of the series, but always full along the entire depth of the input layer.

Now we calculate the convolution. Let $w_{u,j}^l(i)$ be the filter from the feature map $u$ in the $(l-1)^{th}$ layer to the feature map $j$ in the $l^{th}$ layer where

$$w_{u,j}^l(i) = [w_{u,j}^l(i-k^l), \ldots, w_{u,j}^l(i), \ldots, w_{u,j}^l(i+k^l)]^T.$$

The output of the $i^{th}$ location in the $j^{th}$ feature map of the $l^{th}$ layer, denoted $z_i^l(j)$, is computed by convolution of $a_u^{l-1}$ with the filter $w_{u,j}^l$:

$$z_j^l(i) = \sum_{u=1}^{n_1^{l-1}} a_u^{l-1} * w_j^l + b_j^l$$

$$= \sum_{u=1}^{n_1^{l-1}} \sum_{v=-k^l}^{k^l} a_u^{l-1}(i+v) w_{u,j}^l(i+v) + b_j^l, \quad i = 1, 2, \ldots, n_2^l,$$

(3.22)

where the * denotes the convolution operator and $b_j^l$ is a trainable bias parameter. Each filter detects a particular feature and outputs its activation at every location on the input or feature maps. Convolution preserves the correlation between time points by learning features of time series using small segments of input data.

Now we calculate the size of the feature map. The size of the feature map is determined by three parameters: the depth, stride, and zero-padding. The depth of the feature map is defined as the number of filters we would like to use for extracting the various features. In Figure 3.15, we used three different filters, producing three feature maps. The depth of the feature map is three. The stride is defined by the number of time points by which we slide the filter over the input sequence. When the stride is 1, then we slide the filters one time point at a time. When the stride is 2, then the filters jump 2 time points at a time as we move them. In general, the larger the stride, the smaller the feature maps produced. Zero-padding is to pad the input time series with zeros around the end of interval. The size of zero-padding is defined as the number of zeros padded to the end of the time series.

Let $W$ be the size of the input data (the number of time points of input time series), $F$ be the size of the filter, $S$ be the stride, and $P$ be the size of zero-padding. Then, the size of the feature map is equal to $\dfrac{(W-F+2P)}{S} + 1$.

**Example 3.2**

Consider one time series $\{1,2,-1,0,2\}$ with the size $W = 5$. We pad zero at each end of the time series ($P = 1$). Consider three filters $\{1,0,-1\},\{0,1,-1\},$

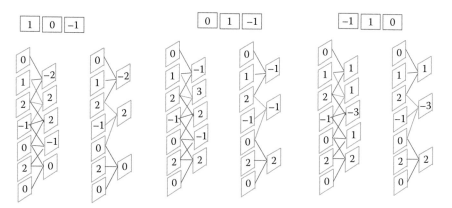

**FIGURE 3.16**
Illustration of Example 3.2.

and $\{-1,1,0\}$, each filter has size $F = 3$. The data are shown in Figure 3.16. Two types of strides $S = 1$ and $S = 2$ are used. Therefore, the sizes of two types of feature maps are $\dfrac{(W - F + 2P)}{S} + 1 = \dfrac{5 - 3 + 2}{1} + 1 = 5$ and $\dfrac{5 - 3 + 2}{2} + 1 = 3$, respectively. A filter slides over the input time series (convolution operation) to produce a feature map. For example, in the beginning, filter $\{1,0,-1\}$ convolves $\{0,1,2\}$, which outputs $-2$ by performing elementwise multiplication between two vectors. Then, the filter $\{1,0,-1\}$ slides over the input time series with $S = 1$, producing feature map $\{-2,2,2, -1,0\}$. If the stride is $S = 2$, then the filter produces a feature map $\{-2,2,0\}$ with small size.

Figure 3.16 also plots the feature maps produced by two other filters. We observe that different filters produce different feature maps. In general, the convolution of different filters, over the same time series data give different feature maps. It is important to recognize that the convolution operation captures the local correlations in the original time series.

Suppose that the size of input to the convolution layer is $W_1 \times D_1$, where $W_1$ is the length of input time series and $D_1$ is the number of channels (the number of time series). Assume four hyperparameters: the number of filters $K$, the size of filter $F$, the strides $S$, and the number of zero padding $P$. The convolution layer will produce $K$ feature maps with size $W_2$, where $W_2 = \dfrac{W_1 - F + 2p}{S} + 1$.

### 3.1.4.4.2 Non-Linearity Activation Layer

After every convolution operation, a non-linear activation layer and non-linear operation will be introduced. Non-linear functions include

1. Sigmoid: $sigmoid(x) = \dfrac{1}{1 + e^{-x}}$,

2. Tanh: $\tanh(x) = \dfrac{e^x - e^{-x}}{e^x + e^{-x}}$,

3. Rectified linear unit (ReLU): $ReLU(x) = max(0,x)$.

Since convolution is a linear operation – element-wise matrix multiplication and addition and many real data are nonlinear, we add a non-linearity activation layer to account for non-linearity in the data. ReLU is a novel and simple nonlinear activation function. Many learning practices show that ReLU can improve generalization, avoid vanishing gradient issues, and increase computational efficiency.

### 3.1.4.4.3 Pooling Layer

Pooling is also called subsampling or down-sampling. The function of a pooling layer is to reduce the dimensionality of each feature map, the number of parameters and computation, control overfitting, and making it robust to variations of previous learned features. The pooling operations treat each feature map separately. The pooling includes three major types of operations: max, average, and sum. The pooling operation is performed in a window (a neighborhood) with size 2 and stride 2. The window is often constructed by splitting the input feature maps into equal length subsequences.

The average pooling operation is to compute the average value in each window at different positions with stride. The average value is then taken as a new feature. Max pooling computes the maximum from the rectified feature map within that window. Max pooling selects superior invariant features and increases convergence rate, which leads to improving generalization. Sum pooling operation computes the sum of all elements in the window of the rectified feature map.

In summary, the pooling layer

1. Accepts an input of size $W_1 \times D_1$ with two parameters: the length of window $F$ and stride $S$,

2. Produces a feature map of size $W_2 \times D_2$ where $D_2 = D_1$ and

$$W_2 = \frac{(W_1 - F)}{S} + 1.$$

### 3.1.4.4.4 Fully Connected Layer

The output from the convolutional and pooling layers represents high-level features of the input time series. The purpose of the fully connected layer is to use these features for classifying the input time series into various classes based on the training dataset. The fully connected layer fully connects all neurons in the previous layer to all neurons in the fully connected layer. Their output or activations can be similarly computed with a matrix multiplication followed by a bias correction. Each neuron in the fully connected layer will output an $m$-dimensional vector where $m$ is the number of classes that the classifier must choose from. Each element in the vector represents the probability of a certain class (Figure 2.17). The fully connected layer extracts the features of the previous layer, for example, pooling layer, and determines which features most correlate to a particular class.

### 3.1.4.5 Parameter Estimation

#### 3.1.4.5.1 Lost Function

The last layer is the fully connected layer. Its output is the classes to which the input signals correspond (Figure 2.17). We begin with the fully connected layers (Bouvrie 2006; Zheng et al. 2015). These fully connected layers are one-dimensional layers. All features in the output map of the fully connected layer are organized into one vector.

Consider $c$ classes that correspond to the $c$ output units. Let $p_k^n$ be the probability of the $n^{th}$ time series (sample) being the $k^{th}$ class, $y_k^n$ be the predicted value of the $k^{th}$ output unit for the $n^{th}$ sample, and $z_k^n$ be the $k^{th}$ element of the target of the $n^{th}$ time series (Figure 2.17). $z_k^n$ forms a vector $Z^n = [z_1^n, ..., z_c^n]^T$. $z_k^n$ will be positive if the pattern $x^n$ of the $n^{th}$ time series belongs to class $k$ and the rest of the elements of the target vector $Z^n$ will be either zero or negative, depending on the output nonlinear activation function. The lost function can be defined either as

$$E = -\sum_{n=1}^{N}\sum_{k=1}^{c} y_k^n \log(p_k^n) \tag{3.23}$$

or

$$E = \frac{1}{2}\sum_{n=1}^{N}\sum_{k=1}^{c} ||z_k^n - y_k^n||^2. \tag{3.24}$$

The lost functions in both Equations 3.23 and 3.24 are just a sum over the individual terms on each sample. Therefore, in parameter estimation we will first focus on a single pattern. Let $Y^n = [y_1^n, ..., y_c^n]^T$. Then, the lost functions in Equation 3.23 and 3.24 will be reduced to

$$E_n(x_n, W) = -\sum_{k=1}^{c} y_k^n \log(p_k^n) \tag{3.25}$$

and

$$E_n(x_n, W) = \frac{1}{2}||Z^n - Y^n||_2^2, \tag{3.26}$$

respectively.

### 3.1.4.5.2 Multilayer Feedforward Pass

The fully connected layers in the last stage of the convolutional neural network (CNN) are a multilayer feedforward neural network. The final output layer of the fully connected neural network is denoted layer $L$. Let $l$ denote the current layer. Consider a network with $p$ input neurons and $q$ output neurons. The weight for the connection from the $k^{th}$ neuron in the input layer to the $j^{th}$ neuron in the first layer is denoted by $w_{jk}^{(1)}$. The bias of the $j^{th}$ neuron in the first layer is denoted by $b_j^{(1)}$. The total number of neurons in the first layer is $m_1$. The weighted input to the $j^{th}$ neuron and activation of the $j^{th}$ neuron in the first layer are denoted by $z_j^{(1)}$ and $a_j^{(1)}$, respectively. Given $x^{(1)} = [x_1^{(1)}, ..., x_p^{(1)}]^T$. Then, the weighted input $z_j^{(1)}$ from the data input layer to the first layer and activation $a_j^{(1)}$ of the $j^{th}$ neuron in the first layer are given by

$$z_j^{(1)} = \sum_{k=1}^{p} w_{jk}^{(1)} x_k^{(1)} + b_j^{(1)}, j = 1, ..., m_1, \quad \text{and} \quad a_j^{(1)} = \sigma\left(z_j^{(1)}\right).$$

Similarly, for the $l^{th}$ layer, we denote the weight for the connection from the $k^{th}$ neuron in the $(l-1)^{th}$ layer to the $j^{th}$ neuron in the $l^{th}$ layer by $w_{jk}^{(l)}$, the bias of the $j^{th}$ neuron in the $l^{th}$ layer by $b_j^{(l)}$. The weighted input to the $j^{th}$ neuron and activation of the $j^{th}$ neuron in the $l^{th}$ layer are denoted by $z_j^{(l)}$ and $a_j^{(l)}$, respectively. Then, we have

$$z_j^{(l)} = \sum_{k=1}^{m_{l-1}} w_{jk}^{(l)} a_k^{(l-1)} + b_j^{(l)}, j = 1, ..., m_l,$$

$$a_j^{(l)} = \sigma(z_j^{(l)}).$$

Finally, for the output layer, we have

$$z_j^{(L)} = \sum_{k=1}^{m_{L-1}} w_{jk}^{(L)} a_k^{L-1} + b_j^{(L)}, j = 1, ..., m_L,$$

$$a_j^{(L)} = \sigma(z_j^{(L)}).$$

### 3.1.4.5.3 Backpropagation Pass

A popular algorithm to minimize $E_n(x_n, W)$ in Equations 3.25 or 3.26 is a gradient descent. The idea is to update the weights along the direction of fastest descent of $E_n(x_n, W)$. We first work on a single training example and then work on the whole dataset by averaging over all training examples. The gradient descent algorithm for updating weights is

$$W^{t+1} - W^t - \eta_t \frac{\partial E_n(x^{(n)}, W))}{\partial W}, \tag{3.27}$$

where $\eta_t \in R_+$ is the learning rate.

The gradient $\dfrac{\partial E_n(x^{(n)}, W)}{\partial W}$ is computed by the back-propagation algorithm. Since $E_n(x^{(n)}, W)$ is a complicated composite function of weights $W$, a key for computing $\dfrac{\partial E_n(x^{(n)}, W)}{\partial W}$ is a chain rule (see Appendix 3.A).

We first examine how the changes in $z_j^L$ cause changes in the cost function $E_n(x^{(n)}, W)$. We define the error rate $\delta_j^L$ in the $j^{th}$ neuron in the final output layer of the fully connect layers as

$$\delta_j^L = \frac{\partial E_n(x^{(n)}, W)}{\partial z_j^L}.$$

In Appendix 3.A, we show that

$$\delta_j^L = (a_j^L - y_j^{(n)})\sigma'(z_j^L), \tag{3.28}$$

$$\delta_j^L = -\frac{y_j^{(n)}}{a_j^L}\sigma'(z_j^L), \tag{3.29}$$

and

$$\frac{\partial E_n(x_n, W)}{\partial b_j^{(L)}} = \delta_j^{(L)}. \tag{3.30}$$

We denote the rate of change of the cost with respect to any weight in the MLP by $\dfrac{\partial E_n}{\partial w_{jk}^{(L)}}$. It can be shown (Appendix 3.A) that

$$\frac{\partial E_n}{\partial w_{jk}^{(L)}} = \delta_j^{(L)} a_k^{(L-1)} \tag{3.31}$$

### 3.1.4.5.4 Convolutional Layer

*3.1.4.5.4.1 Last Convolution Layer: l = L − 1*   We first consider convolutional layer $l = L - 1$, which is connected to the fully connected layer $L$. The error sensitive derivative for the last convolutional layer is (Appendix 3.A)

$$\delta_j^{(L-1)} = \sigma'(z_j^{(L-1)})\sum_{m=1}^{m_L}\delta_m^{(L)}w_{mj}^{(L)}, \; j = 1, 2, \ldots, N_{L-1}. \tag{3.32}$$

The rates of change of the cost with respect to any weight $w_{mj}^{(L-1)}$ and bias are

$$\frac{\partial E_n(x_n, W)}{\partial w_{mj}^{(L-1)}(i)} = \delta_m^{(L-1)} a_j^{(L-2)} \tag{3.33}$$

and

$$\frac{\partial E_n(x_n, W)}{\partial b_j^{(L-1)}} = \sum_{i=1}^{n_2^{(L-1)}} \delta_j^{(L-1)}(i), \; j = 1, \ldots, n_1^L, \tag{3.34}$$

respectively.

*3.1.4.5.4.2 Last Sub-Sampling Layer*  Now we consider the last sub-sampling layer. Recall that in the last convolution layer, each feature map is connected to exactly the one preceding the feature map in the last sub-sampling layer. The sensitivity derivative in the last sub-sampling layer is

$$\delta_j^{(L-2)}(i) = \frac{\partial E_n(x_n, W)}{\partial z_j^{(L-2)}(i)}$$
$$= \delta_j^{(L-1)}(i) w_{jj}^{(L-1)} \sigma'(z_j^{(L-2)}). \tag{3.35}$$

The derivative of the cost function with respect to the weights and bias in the last sub-sampling layer are

$$\frac{\partial E_n(x_n, W)}{\partial w_{jj}^{(L-2)}} = \sum_{i'} \delta_j^{(L-2)}(i') a_j^{(L-3)}(i') \tag{3.36}$$

and

$$\frac{\partial E_n(x_n, W)}{\partial b_j^{(L-2)}} = \sum_{i'} \delta_j^{(L-2)}(i'), \tag{3.37}$$

respectively.

*3.1.4.5.4.3 Convolutional Layer 1*  Next, we consider a general convolutional layer $l$. Each feature map $j$ in the convolutional layer $l$ is connected to the feature map $j$ in the sub-sampling layer $l + 1$. The sensitive derivative of the cost function with respect to input in the convolutional layer $l$ is

$$\delta_j^{(l)}(i) = \delta_j^{(l+1)}(i') w_j^{(l+1)} \sigma'(Z_j^{(l)}(i)), \tag{3.38}$$

where $i' = \left\lfloor \dfrac{i}{2} \right\rfloor$.

Again, the derivative of the cost function with respect to the weights and bias are given by

$$\frac{\partial E_n(x_n, W)}{\partial w_{mj}^{(l)}(i)} = \delta_m^{(l)}(i) a_j^{(l-1)}(i). \tag{3.39}$$

and

$$\frac{\partial E_n(x_n, W)}{\partial b_j^{(l)}} = \sum_{i=1}^{n_2^{(l)}} \delta_j^{(l)}(i) \, , j = 1, \ldots, n_1^l, \tag{3.40}$$

respectively.

*3.1.4.5.4.4 Sub-Sampling Layer l* Finally, we calculate the sensitivity derivative for the general sub-sampling layer $l$ and gradient of the cost function with respect to the weights and bias. In Appendix 3.A, we show that

$$\delta_j^{(l)}(i) = \sigma'(z_j^l(i)) \sum_{m=1}^{m_{l+1}} \delta_m^{(l+1)}(i') \sum_{v=-k^l}^{k^l} w_{m,j}^{(l+1)}(i'+v), \tag{3.41}$$

$$\frac{\partial E_n(x_n, W)}{\partial w_{jj}^{(l)}} = \sum_{i'} \delta_j^{(l)}(i') a_j^{(l-1)}(i'), \tag{3.42}$$

and

$$\frac{\partial E_n(x_n, W)}{\partial b_j^{(l)}} = \sum_{i'} \delta_j^{(l)}(i'). \tag{3.43}$$

## 3.2 Association Studies of Function-Valued Traits

### 3.2.1 Introduction

Physiological traits such as electrocardiogram (ECG), phonocardiogram (PCG), seimocardiography (SCG), and oxygen saturation levels provide important information on the health status of humans and can be used to monitor, diagnose, and manage diseases. For example, ECG is a measurement of the electrical activity of the heart muscle obtained from the surface of the skin. It measures the rate and regularity of heartbeats. ECG is the most commonly performed cardiac test and is of great clinical value. It provides valuable information on the biological processes and current state of the heart, and can be used for diagnosis of arrhythmias, myocardial infarction, and other cardiovascular diseases. Oxygen saturation levels which are proportional to the reduction in airflow cause total or partial reduction in respiration of the Sleep Apnea-Hypopnea Syndrome (SAHS) during sleep. SAHS is a risk factor for cardiac and cerebral infarct, high arterial pressure, arrhythmias, and in general, several dysfunctions of the cardiorespiratory system. The rapidly developed NGS technologies have become the platform of choice for gene expression profiling (Sun and Zhu 2012). RNA-seq provides multiple layers of resolutions and transcriptome complexity: the expression at exon, SNP, and positional level, splicing, isoform, and allele-specific expression. RNA-seq profile across a gene is represented as many reads at each genomic position. These traits are observed either as continuous random functions of time or space, or on a dense grid are referred to as function-valued traits.

Function-valued traits are highly correlated data with their inherent order, spacing, smoothly varying structure, and functional nature which are ignored by traditional summary-based univariate and multivariate regression methods designed for quantitative genetic analysis of scalar trait and common variants. These methods use summary statistics to measure or represent physiological traits. For example, heart rate (HR), the P-R interval, QRS complex duration, QT, and QTc interval are often used as a trait in genetic analysis of the ECG. Physiological traits are time dependent and dynamic in nature. They are repeatedly measured at multiple time points and often described by functions or curves. The temporal pattern of genetic control for physiological traits should be compared across different stages of development. To capture the morphological shape and dynamic features of the physiological traits, methods that analyze all dynamic time points (traits) jointly and are often referred to as function valued QTL analysis have recently developed (Fusi and Listgarten 2017; Kwak et al. 2015; Hernandez 2015).

Parametric models and non-parametric models are two basic approaches to association analysis of function-valued traits. Parametric models include logistic growth models (Ma et al. 2002) and regression models (Kwak et al. 2014) and nonparametric models include expanding function-valued traits in terms of basis functions (Yang et al. 2009), wavelets (Shim and Stephens 2015), Legendre polynomial (Das et al. 2011), functional principal components (2015), and Gaussian process regression with a radial basis function (Fusi and Listgarten 2017).

In the past few years we have witnessed the rapid development of novel statistical methods for association studies using NGS data. However, these methods might not be appropriate for genetic analysis of function-valued traits. The quantitative genetic analysis of rare variants for function-valued traits remains challenging. In this section, we will introduce the function linear model with functional responses and functional predictors (FLMF) for quantitative genetic analysis of function-valued traits with NGS data. In the FLMF, the time varying values of physiological trait are taken as a functional response and the genotype profile across a genomic region or a gene can be modeled as a function of genomic location. The functional linear model with functional response and multivariate predictors which can be used for association analysis of function-valued traits with common variants will be developed as a spacial case of the FLMF. The FLMF has several remarkable features. First, the FLMF accounts for the continuous change in traits and preserves the intrinsic smooth structure and all the positional-level genetic information. Second, the FLMF simultaneously utilizes both correlation information among the trait at different times and among all variants in a genomic region. Third, the multicolinearity problems in the FLMF which may be presented in both trait and genetic variation is alleviated. Fourth, the FLMF expands both trait function and genotype function in terms of orthogonal eigenfunction, which leads to substantial dimension reduction.

### 3.2.2 Functional Linear Models with Both Functional Response and Predictors for Association Analysis of Function-Valued Traits

For the convenience of discussion, temporal or space trait is referred to as a function-valued trait. For simplicity, we consider a temporal trait $y_i(t)$, $t \in T = [0,T]$ of the $i$-th individual which varies in time and a genomic region (or gene) $[a,b]$. Let $s$ be a genomic position in the region $S = [a,b]$. Define a genotype profile $x_i(s)$ of the i-th individual as

$$
x_i(s) = \begin{cases} 2P_m(s), & \text{MM} \\ P_m(s)\text{-}P_M(s), & \text{Mm} \\ -2P_M(s), & \text{mm} \end{cases}
$$

where M and m are two alleles of the marker at the genomic position $s$, $P_M(s)$ and $P_m(s)$ are the frequencies of the alleles M and m, respectively. A functional linear model with both functional response and predictors is defined as

$$
y_i(t) = W_i^T \alpha(t) + \int_S x_i(s)\beta(s,t)ds + \varepsilon_i(t), \quad t = t_1, ..., t_T, i = 1, ..., n, \tag{3.44}
$$

where $W_i = [w_{i1}, ..., w_{id}]^T$ is a vector of covariates, $\alpha(t) = [\alpha_1(t), ..., \alpha_d(t)]^T$ is a vector of effects associated with the covariates, $\beta(s,t)$ is a genetic additive effect function in the genomic position $s$ and time $t$, and $\varepsilon_i(t)$ is the residual function of the noise and unexplained effect for the $i^{th}$ individual. Let $\eta(t) = [\eta_1(t), ..., \eta_k(t)]^T$ be a vector of basis functions. To transform the functional linear model (3.44) into the standard multivariate linear mode, we consider the functional expansions for the trait function $y_i(t)$, effect functions $\alpha(t), \beta(s,t)$, and genotype function $x_i(s)$. We assume that both phenotype and genotype profiles are centered. The trait function and genotype function are expanded in terms of the orthonormal basis function as

$$
y_i(t) = \sum_{l=1}^{k_y} y_{il}\eta_l(t) = y_i^T \eta(t), \tag{3.45}
$$

and

$$x_i(s) = \sum_{m=1}^{k_x} x_{im} \theta_m(s),$$  (3.46)

where the expansion coefficients $y_{il}$ and $x_{im}$ are estimated by

$$y_{il} = \int_0^T y_i(t)\eta_l(t)dt$$  (3.47)

and

$$x_{im} = \int_S x_i(s)\theta_m(s)ds.$$  (3.48)

The sets of eigenfunctions $\eta_l(t)$ and $\theta_m(s)$ are estimated from the phenotype functions $y_i(t)$ and $x_i(s)$, respectively. The covariate effect functions $\alpha_j(t)$ can be expanded in terms of eigenfunctions $\eta_l(t)$ as

$$\alpha_j(t) = \sum_{l=1}^{k_y} \alpha_{jl}\eta_l(t) = \alpha_j^T \eta(t),$$  (3.49)

where $\alpha_j = [\alpha_{j1}, ..., \alpha_{jk_y}]^T$.

Define a vector of covariate effect functions and a matrix of expansion coefficients as

$$\alpha(t) = \begin{bmatrix} \alpha_1(t) \\ \vdots \\ \alpha_d(t) \end{bmatrix} \text{ and } \alpha = \begin{bmatrix} \alpha_{11} & \cdots & \alpha_{1k_y} \\ \vdots & \vdots & \vdots \\ \alpha_{d1} & \cdots & \alpha_{dk_y} \end{bmatrix}, \text{ respectively.}$$

Then we have

$$\alpha(t) = \alpha\eta(t).$$  (3.50)

Similarly, the error functions $\varepsilon_i(t)$ can also be expanded as

$$\varepsilon_i(t) = \sum_{l=1}^{k_y} \varepsilon_{il}\eta_l(t) = \varepsilon_i^T \eta(t),$$  (3.51)

where $\varepsilon_i = [\varepsilon_{i1}, ..., \varepsilon_{ik_y}]^T$.

By the similar arguments, the genetic effect functions $\beta(s,t)$ can be expanded in terms of two sets of orthogonal eigenfunctions $\theta_k(s)$ and $\eta_l(t)$ as

$$\beta(s,t) = \sum_{k=1}^{k_x} \sum_{l=1}^{k_y} b_{kl}\theta_k(s)\eta_l(t) = \theta^T(s)B\eta(t),$$  (3.52)

where $\theta(s) = [\theta_1(s), ..., \theta_{k_b}(s)]^T$, $B = (b_{kl})_{k_x \times k_y}$ is a matrix of expansion coefficients of the genetic additive effect function. Thus, using Equation 3.46, the integral $\int_S x_i(s)\beta(s,t)ds$ can be expanded as (Exercise 10)

$$\int_S x_i(s)\beta(s,t)ds = \int_S x_i(s)\theta^T(s)dsB\eta(t) = x_i^T B\eta(t).$$

Substituting these expansions into Equation 3.44, we obtain

$$y_i^T \eta(T) = W_i^T \alpha\eta(t) + x_i^T B\eta(t) + \varepsilon_i^T \eta(t), i = 1, ..., n. \tag{3.53}$$

Since Equation 3.53 should hold for all $t$, we must have

$$y_i^T = W_i^T \alpha + x_i^T B + \varepsilon_i^T, i = 1, ..., n. \tag{3.54}$$

The model (3.54) is a standard linear model. Instead of using the observed data as the values of the response and predictor variables, we use their expansion coefficients as the values of the response and predictor variables in the linear model (3.54). Equation 3.54 can be further written in a matrix form:

$$Y = W\alpha + XB + \varepsilon, \tag{3.55}$$
$$= A\gamma + \varepsilon,$$

where $Y = [y_1, ..., y_n]^T$, $W = [W_1, ..., W_n]^T$, $X = [x_1, ..., x_n]^T$, $\varepsilon = [\varepsilon_1, ..., \varepsilon_n]^T$, $A = [W, X]$, and $\gamma = [\alpha^T, B^T]^T$.

To estimate the parameters in the matrix $\gamma$, we define the total squares of errors between the true $Y$ and predicted values $A\hat{\gamma}$ as the objective function to be minimized:

$$F = \text{Tr}((Y - A\gamma)^T(Y - A\gamma)) = \text{Tr}(\varepsilon^T \varepsilon). \tag{3.56}$$

Using trace derivative formula, we can obtain

$$\frac{\partial F}{\partial \gamma} = -A^T(Y - A\gamma).$$

Setting $\frac{\partial F}{\partial \gamma} = 0$, we obtain the least square estimates of the parameter vector $\gamma$:

$$\hat{\gamma} = (A^T A)^{-1} A^T Y. \tag{3.57}$$

The last $k_x$ rows of the estimated matrix $\hat{\gamma}$ form the estimator of the matrix $B$. Using Equations 3.52 and 3.57 we obtain the genetic additive effect function:

$$\hat{\beta}(s, t) = \theta^T(s)\hat{B}\eta(t). \tag{3.58}$$

### 3.2.3 Test Statistics

An essential problem in association analysis of the functional quantitative trait is to test the association of a genomic region with the functional quantitative trait. Formally, we investigate the problem of testing the following hypothesis:

$$H_0 : \beta(s, t) = 0, \ \forall s \in S, t \in T \tag{3.59}$$

against

$$H_a : \beta(s, t) \neq 0.$$

If the genetic effect function $\beta(s,t)$ is expanded in terms of the basic functions:

$$\beta(s, t) = \theta^T(s)B\eta(t),$$

then, testing the null hypothesis $H_0$ in Equation 3.59 is equivalent to testing the hypothesis:

$$H_0 : B = 0. \tag{3.60}$$

To derive the test statistic, we first calculate variance of the estimated expansion coefficient matrix of the genetic additive effect function. Let vec denote the vector operation. Then, from Equation 3.57, we have

$$
\begin{aligned}
vec(\hat{\gamma}) &= \left[ I_{k_y} \otimes \left( A^T A \right)^{-1} A^T \right] vec(y) \\
&= vec(\gamma) + \left[ I_{k_y} \otimes \left( A^T A \right)^{-1} A^T \right] vec(\varepsilon).
\end{aligned}
\tag{3.61}
$$

Note that

$$var(vec(\varepsilon)) = \Sigma_{k_y} \otimes I_n, \tag{3.62}$$

where

$$
\Sigma_{k_y} =
\begin{bmatrix}
\sigma_{11} & \cdots & \sigma_{1k_y} \\
\vdots & \vdots & \vdots \\
\sigma_{k_y 1} & \cdots & \sigma_{k_y k_y}
\end{bmatrix}.
\tag{3.63}
$$

From Equation 3.55, we obtain

$$y_{ij} = \sum_{k=1}^{d} W_{ik}\alpha_{kj} + \sum_{k=1}^{k_y} x_{ik}b_{kj} + \varepsilon_{ij}, i = 1, \ldots, n, j = 1, \ldots, k_y. \tag{3.64}$$

Variance $\sigma_{ul}$ can be estimated by

$$\sigma_{ul} = \frac{1}{nk_y - dk_y - k_yk_x} \sum_{i=1}^{n} \left( y_{iu} - \sum_{k=1}^{d} W_{ik}\hat{\alpha}_{ku} - \sum_{k=1}^{k_y} x_{ik}b_{ku} \right)$$

$$\left( y_{il} - \sum_{k=1}^{d} W_{ik}\hat{\alpha}_{kl} - \sum_{k=1}^{k_y} x_{ik}b_{kl} \right), u = 1, \ldots, k_y, l = 1, \ldots, k_y.$$

Then, $\text{var}(vec(\hat{\gamma}))$ is given by

$$
\begin{aligned}
\text{var}(vec(\hat{\gamma})) &= \left[ I_n \otimes (A^TA)^{-1}A^T \right] \text{cov}(vec(\varepsilon)) \left[ I_n \otimes A(A^TA)^{-1} \right] \\
&= \left[ I_n \otimes (A^TA)^{-1}A^T \right] \left[ \Sigma_{k_y} \otimes I_n \right] \left[ I_n \otimes A(A^TA)^{-1} \right] \qquad (3.65) \\
&= \Sigma_{k_y} \otimes (A^TA)^{-1}.
\end{aligned}
$$

Let

$$\hat{b} = vec(\hat{B})$$

and $\Lambda$ be the matrix that is obtained from the last $k_yk_x$ rows and $k_yk_x$ columns of the matrix $\text{var}(vec(\hat{\gamma}))$. We can define the following statistic for testing the association of a genomic region with the functional trait:

$$T_F = \hat{b}^T \Lambda^{-1} \hat{b}. \qquad (3.66)$$

Under the null hypothesis of no association, the statistic $T_F$ will be distributed as a central $\chi^2_{(k_yk_x)}$ distribution.

### 3.2.4 Null Distribution of Test Statistics

In the previous section, we have shown that the test statistics $T_F$ are asymptotically distributed as a central $\chi^2_{(k_yk_\beta)}$ distribution. To examine the validity of this statement, we performed a series of simulation studies to compare their empirical levels with the nominal ones (Lee 2015).

We calculated the type I error rates for rare alleles, and both rare and common alleles. We assumed the following model to generate a functional quantitative trait for type 1 error calculations:

$$y_i(t_j) = \mu + \varepsilon_i(t_j),$$

where $y_i(t_j)$ is the trait value of the $i^{th}$ individual at the time $t_j$, $\mu$ is a constant for all $i$ and $t_j$, $\varepsilon_i(t_j)$ is the error term of $i^{th}$ individual at the time $t_j$, and this error term is generated by independent standard Brownian motion.

We first considered both common and rare variants, that is, entire allelic spectrum of variants. We generated 1,000,000 chromosomes by resampling from 2225 individuals with variants in five genes (*CDC2L1, GBP3, IQGAP3, TNN, ACTN2*) selected from the NHLBI's Exome Sequencing Project (ESP). The five genes included 461 SNPs. The number of sampled individuals from populations of 1,000,000 chromosomes ranged from 1000 to 2000. The time points taking trait measurements for type 1 error calculations were 15, 20, 30, and 40. A total of 5000 simulations were repeated. Tables 3.2 and 3.3 summarized the average type I error rates of the test statistics for testing the association of rare variants (MAF < 0.05) and all common and rare variants over five genes, respectively, at the nominal levels $\alpha = 0.05$, $\alpha = 0.01$, and $\alpha = 0.001$. Tables 3.2 and 3.3 showed that, in general, the type I error rates of the test statistics in the functional quantitative trait analysis were not appreciably different from the nominal alpha levels.

**TABLE 3.2**

Average Type 1 Error Rates of the Statistics for Testing Association of a Gene That Consists of Rare Variants (MAF < 0.05) with a Function Quantitative Trait over 5 Genes

| Time | Sample Size | 0.001 | 0.01 | 0.05 |
|------|-------------|-------|------|------|
| 15 | 1000 | 0.00156 | 0.01232 | 0.05640 |
| | 1250 | 0.00088 | 0.01032 | 0.05320 |
| | 1500 | 0.00120 | 0.01264 | 0.05428 |
| | 1750 | 0.00152 | 0.01296 | 0.05484 |
| | 2000 | 0.00148 | 0.01144 | 0.05492 |
| 20 | 1000 | 0.00108 | 0.01164 | 0.05856 |
| | 1250 | 0.00088 | 0.01272 | 0.05876 |
| | 1500 | 0.00088 | 0.01152 | 0.05152 |
| | 1750 | 0.00144 | 0.01056 | 0.05508 |
| | 2000 | 0.00088 | 0.01068 | 0.05220 |
| 30 | 1000 | 0.00136 | 0.01200 | 0.05560 |
| | 1250 | 0.00112 | 0.01108 | 0.05232 |
| | 1500 | 0.00092 | 0.01032 | 0.05108 |
| | 1750 | 0.00136 | 0.01032 | 0.05204 |
| | 2000 | 0.00084 | 0.01020 | 0.05116 |
| 40 | 1000 | 0.00116 | 0.01184 | 0.05612 |
| | 1250 | 0.00128 | 0.01048 | 0.05344 |
| | 1500 | 0.00124 | 0.01144 | 0.05228 |
| | 1750 | 0.00096 | 0.01032 | 0.04876 |
| | 2000 | 0.00112 | 0.01100 | 0.05092 |

Time: The number of time points when taking trait measurement.

**TABLE 3.3**

Average Type 1 Error Rates of the Statistics for Testing Association of a Gene
That Consists of All Variants with a Function Quantitative Trait over 5 Genes

| Time | Sample Size | 0.001 | 0.01 | 0.05 |
|------|-------------|-------|------|------|
| 15 | 1000 | 0.0014 | 0.0130 | 0.0594 |
|    | 1500 | 0.0010 | 0.0102 | 0.0552 |
|    | 2000 | 0.0006 | 0.0116 | 0.0506 |
| 20 | 1000 | 0.0008 | 0.0100 | 0.0532 |
|    | 1500 | 0.0014 | 0.0078 | 0.0490 |
|    | 2000 | 0.0008 | 0.0090 | 0.0418 |
| 30 | 1250 | 0.0012 | 0.0138 | 0.0544 |
|    | 1500 | 0.0006 | 0.0092 | 0.0458 |
|    | 1750 | 0.0010 | 0.0076 | 0.0426 |
| 40 | 1250 | 0.0010 | 0.0126 | 0.0518 |
|    | 1500 | 0.0008 | 0.0094 | 0.0482 |
|    | 1750 | 0.0008 | 0.0086 | 0.0418 |

Time: The number of time points when taking trait measurement.

### 3.2.5 Power

To evaluate the performance of the functional linear models with both functional response and predictors for testing the association of a genomic region with a function-valued trait, we used simulated data to estimate their power to detect a true association. A true functional quantitative genetic model is given as follows. Consider $L$ trait loci that are located at the genomic positions $s_1,...,s_L$. Let $A_s$ be a risk allele at the $s^{th}$ trait locus. Let $t_j$ be the $j$-th time point when the trait measurement is taken. The following multiple linear regression is used as an additive genetic model for a quantitative trait:

$$y_i(t_j) = \mu + \sum_{s=1}^{L} x_{is} b_s(t_j) + \varepsilon_i(t_j),$$

where $y_i(t_j)$ is the trait value of $i^{th}$ individual measured in the time $t_j$, $\mu$ is an overall mean, $x_{is}$ is an indicator variable for the genotype of $i^{th}$ individual at the $s^{th}$ trait locus, $b_s(t_j)$ is the genetic additive effect of the SNP at the $s^{th}$ trait locus and the time $t_j$, the error term $\varepsilon_i(t_j)$ is generated by independent standard Brownian motion process. The genetic effect $b_s(t_j)$ is modeled as $b_s(t_j) = b_s b(t_j)$, where $b(t_j) = 10^{-6} e^t$. We considered two genetic models for $b_s$: recessive and multiplicative. The relative risks across all variant sites are assumed to be equal and the variants were assumed to influence the trait independently (i.e., no epistasis). Let $f_0 = 1$ be a baseline penetrance that is defined as the contribution of the wild genotype to the trait variation and $r$ be a risk parameter. The genetic additive effects for the two trait models are defined as follows:

recessive model: $b_s = P_s(r - 1)f_0$ and multiplicative model: $b_s = (rP_s + 1 - P_s)$ $(r - 1)f_0$, where $P_s$ is the frequency of the risk allele located at the genomic position $s$.

For power comparisons, we also consider cross-section trait models. The genetic effects for the cross-section trait models is defined as the average of the genetic effect function over the time where the phenotype values were measured at 20 time points: $\bar{b}_s = b_s b(t_{med})$, where $b(t_{med})$ is the median of the function of $b(t_j)$, $j = 1,...,20$. The trait value for the cross-sectional model is generated by

$$y_i = \mu + \sum_{s=1}^{L} x_{is}\bar{b}_s + \varepsilon_i.$$

We generate 100,000 individuals by resampling from 2225 individuals of European origin with variants in gene *TNN* (88 rare variants and 18 common variants) selected from the ESP dataset. We randomly selected 10% of the variants as risk variants. A total of 1000 individuals for the multiplicative trait models and 2000 individuals for the recessive trait model were sampled from the populations. A total of 1000 simulations were repeated for the power calculation. We compared the power of six methods. For the time course trait data, we considered the FLM with trait function and genotype function, the multivariate regression for multiple phenotypes and simple regression for multiple phenotypes. For the cross-sectional data, we considered the FLM with the scalar trait and genotype function, multivariate regression for single phenotype and simple regression for single phenotype.

We compare the power curves of FLM with cross-sectional models, multivariate regression model and simple linear regression model in this study. We repeat 1000 simulations for all the comparisons. Also, we assume that all variances are independently and equally influencing the trait. That is, we assume there are no interactions. Figures 3.17 and 3.18 plot the power curves of six statistic models: the functional linear model with both functional response and predictors for function-valued traits (FLMF), the multiple linear model for function-valued trait (MLMF), the simple regression model for function-valued traits (SRGF), the functional linear model with scalar response and functional predictors for cross-section marginal genetic model (FLMC), multiple linear model for cross-section marginal genetic model and

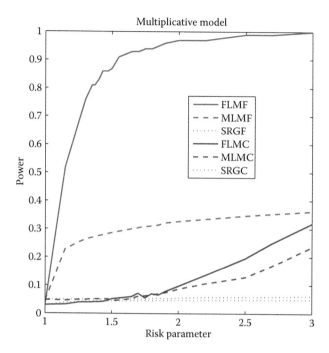

**FIGURE 3.17**
The power curve as a function of the risk parameter of six models under the multiplicative model.

simple regression for cross-section marginal genetic model (SRGC) for testing association of rare variants in the genomic region under multiplicative, and recessive models, respectively.

These power curves are a function of the risk parameter at the significance level $\alpha = 0.05$. Several features emerged from these figures. First, the power of the FLMF was the highest. Except for the recessive models, the FLMF could still detect association of a gene with the function-valued trait even using sample sizes of 1000. Second, power difference between the FLMF and other five models was substantial. Third, the power of simple regression for both function-valued traits and cross-section marginal models (SRGF and SRGC) was extremely low. In most scenarios, the simple regression does not have the power to detect association. Fourth, in general, the power of tests using a function-valued approach was higher than that using a traditional cross-section approach.

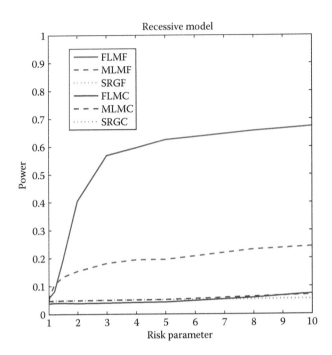

**FIGURE 3.18**
The power curve as a function of the risk parameter of six models under recessive model.

### 3.2.6 Real Data Analysis

To further evaluate its performance, the FLMF was applied to oxygen saturation studies in Starr County, Texas. The oxygen saturation signals were measured by seconds. A total of 35,280 measurements were taken over a night. Oxygen saturation provides important information on the sleep quality for those with obstructive sleep apnea (Lee 2015). A total of 406,299 SNPs in 22,670 genes were typed for 833 individuals of Mexican Americans origin from Starr County. Since the FLMF requires to expand genotype function in terms of eigenfunction, which need to have at least 3 SNPs in the gene, we exclude the gene with only one or two SNPs in it. The left total number of genes for analysis was 17,258. Therefore, the P-value for declaring significance after applying the Bonferroni correction for multiple tests was $2.90 \times 10^{-6}$.

To reduce the number of measurements included in the analysis, we used the mean of the oxygen saturation taken every 10 seconds as the trait values. SNP's in 5*kb* flanking region of the gene are assumed to belong to the gene. To ensure the numerical stability, we used single value decomposition to calculate the inverse of the matrix. We selected the number of single values such

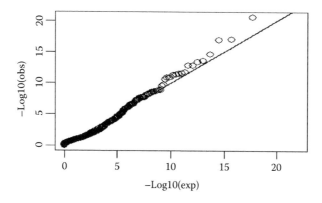

**FIGURE 3.19**
QQ plot.

that it can account for 99% of the total variation. To examine the behavior of the FLMF, we plotted QQ of the test (Figure 3.19) where P-values were calculated after adjusting for sex, age, and BMI in the model.

The QQ plots showed that the false positive rate of the FLMF for detection of association with the oxygen saturation trait is controlled. In total, we identified 65 genes that were significantly associated with oxygen saturation function-valued traits with P-values ranging from $2.4 \times 10^{-6}$ to $2.5 \times 10^{-21}$. To compare with other methods for association analysis of function-valued traits, we provided Table 3.4 in which we also listed minimum P-values of 65 significant genes over all observed time periods which were calculated using MLM and SRG for each time point.

Several remarkable features were observed from this real data analysis. First, the FLMF utilizes the merits of taking both phenotype and genotype as functions. It decomposes time varying phenotype function into orthogonal eigenfunctions of time and position varying genotype function into orthogonal eigenfunctions of genomic position. The FLMF reduces the dimensions due to both phenotype variation and genotype variation (only a few eigenfunctions are used to model variation), which in turn increases statistical power of the test. This real data example showed that the function-value (time course data) approach can achieve much stronger significance than the scalar value (cross-section study) approach. Second, to further illustrate that the function-valued statistical methods can be more powerful than the traditional quantitative genetic analysis, we presented Table 3.4 showing that the P-values of the FLMF were smaller than the minimum of P-values of the MLM and SRG over all observed time intervals at night. Third, genetic variants in a gene might make only mild contributions to the oxygen saturation variation at individual time points, these genetic variants may show significant association with the oxygen

**TABLE 3.4**

P-Values of 65 Significant Genes Calculated Using FLMF, MFMF, and SRGF

| Gene | P-value | | |
|------|---------|---|---|
|      | FMLF | MLM(min) | SRG(mim) |
| MAN1B1 | 2.53E-21 | 1.10E-05 | 6.57E-03 |
| TMEM57 | 8.90E-18 | 8.50E-02 | 3.97E-02 |
| OR5H15 | 1.28E-17 | 4.21E-03 | 3.29E-02 |
| PABPC4L | 2.66E-15 | 1.09E-01 | 7.95E-02 |
| ANKLE1 | 2.51E-14 | 6.52E-07 | 2.98E-03 |
| TTI2 | 4.64E-14 | 6.82E-02 | 4.57E-03 |
| KRTAP4-7 | 1.67E-13 | 1.12E-02 | 5.45E-03 |
| WDR90 | 1.73E-13 | 1.49E-10 | 4.56E-04 |
| ZER1 | 1.67E-12 | 2.28E-03 | 1.42E-02 |
| DPH2 | 3.05E-12 | 3.09E-03 | 2.12E-02 |
| B9D2 | 3.43E-12 | 1.07E-01 | 2.87E-02 |
| GGT1 | 4.29E-12 | 7.59E-23 | 8.89E-03 |
| SGSH | 4.90E-12 | 3.69E-04 | 4.42E-02 |
| FAM211B | 1.03E-11 | 1.97E-23 | 8.89E-03 |
| FBXO27 | 1.37E-11 | 9.53E-03 | 5.00E-02 |
| COA6 | 1.44E-11 | 1.59E-03 | 2.85E-03 |
| MAK16 | 2.66E-11 | 3.91E-03 | 9.30E-03 |
| CDKN2AIP | 1.69E-10 | 1.62E-01 | 8.54E-02 |
| RRM2 | 3.07E-10 | 6.14E-02 | 1.66E-02 |
| DTX3L | 1.24E-09 | 2.41E-02 | 4.96E-03 |
| C17orf75 | 1.41E-09 | 4.38E-02 | 9.86E-03 |
| TAS2R5 | 1.72E-09 | 2.75E-01 | 4.57E-02 |
| GIPC1 | 1.93E-09 | 2.29E-02 | 4.41E-02 |
| CDC14C | 2.09E-09 | 2.39E-04 | 2.40E-02 |
| MIR4520A | 2.49E-09 | 1.06E-03 | 1.44E-01 |
| MIR4520B | 2.49E-09 | 1.06E-03 | 1.44E-01 |
| PROX2 | 3.63E-09 | 5.24E-03 | 1.15E-02 |
| MAFF | 4.04E-09 | 2.17E-03 | 9.18E-04 |
| UQCRQ | 5.95E-09 | 1.55E-01 | 2.20E-02 |
| LYRM1 | 6.58E-09 | 1.25E-02 | 2.47E-03 |
| ZFPM1 | 7.81E-09 | 2.78E-03 | 9.92E-04 |
| TMEM50B | 9.59E-09 | 3.14E-23 | 1.27E-03 |
| KCNK15 | 1.78E-08 | 1.19E-01 | 7.12E-03 |
| EEF1B2 | 1.99E-08 | 4.60E-03 | 1.08E-03 |
| SNORA41 | 1.99E-08 | 4.60E-03 | 1.08E-03 |
| SNORD51 | 1.99E-08 | 4.60E-03 | 1.00E-03 |
| LDLRAP1 | 2.01E-08 | 3.94E-02 | 2.69E-02 |
| NEK4 | 2.17E-08 | 3.57E-05 | 5.34E-05 |

*(Continued)*

**TABLE 3.4 (CONTINUED)**

P-Values of 65 Significant Genes Calculated Using FLMF, MFMF, and SRGF

| | P-value | | |
|---|---|---|---|
| Gene | FLMF | MLM(min) | SRG(mim) |
| COMMD7 | 2.29E-08 | 1.41E-02 | 1.38E-02 |
| EEF1A1 | 3.39E-08 | 3.29E-07 | 1.14E-02 |
| MIR1-1 | 4.51E-08 | 7.91E-02 | 3.37E-02 |
| HMGN4 | 4.63E-08 | 2.05E-03 | 3.47E-03 |
| EVPLL | 5.48E-08 | 3.53E-03 | 1.13E-02 |
| C22orf26 | 5.60E-08 | 1.17E-02 | 1.52E-03 |
| CDC42EP5 | 7.88E-08 | 4.91E-02 | 1.81E-02 |
| MIR29C | 1.30E-07 | 2.86E-02 | 8.72E-03 |
| LHX2 | 1.95E-07 | 6.47E-03 | 2.77E-04 |
| ZNF284 | 2.09E-07 | 1.06E-01 | 7.87E-02 |
| RBAKDN | 2.34E-07 | 2.04E-03 | 3.55E-04 |
| BAIAP2L2 | 2.44E-07 | 1.91E-02 | 2.14E-03 |
| P2RX5 | 2.92E-07 | 6.68E-03 | 1.62E-02 |
| P2RX5-TAX | 2.92E-07 | 6.68E-03 | 1.62E-02 |
| RELB | 3.01E-07 | 1.10E-02 | 1.29E-02 |
| TREML3P | 5.42E-07 | 5.01E-02 | 4.94E-03 |
| TSPAN10 | 5.86E-07 | 5.43E-02 | 5.33E-02 |
| RPS16 | 6.17E-07 | 2.63E-03 | 6.00E-04 |
| GNLY | 8.09E-07 | 1.99E-02 | 1.10E-02 |
| LRRC48 | 8.17E-07 | 1.45E-01 | 1.35E-02 |
| WSB1 | 9.64E-07 | 7.57E-03 | 6.63E-03 |
| GFPT1 | 1.09E-06 | 4.92E-12 | 5.56E-03 |
| MIR3677 | 1.13E-06 | 8.67E-02 | 1.40E-02 |
| MIR940 | 1.13E-06 | 8.67E-02 | 1.40E-02 |
| TOE1 | 1.13E-06 | 6.48E-05 | 6.86E-02 |
| TMEM41A | 2.12E-06 | 5.98E-03 | 3.55E-02 |
| TPM4 | 2.40E-06 | 1.20E-01 | 8.28E-03 |

saturation curve as shown in Figure 3.20 where the P-value for testing the association of the gene *ANKLE1* with the oxygen saturation curve using the FLMF was $2.51 \times 10^{-14}$ and the P-values for the tests using the MLM at the individual time point ranges from $6.52 \times 10^{-7}$ to 0.9265. There was a total of 3528 time points. We observed a total of 188 time points with P-value < 0.05 when using the MLM to test association at the individual time points. None of the 3528 tests showed strong evidence of association, but indeed we observed strong association of the gene *ANKLE1* with the oxygen saturation curve due to using all information about correlation and continuity of underlying structure of phenotype function. Fourth, unlike traditional quantitative genetic analysis

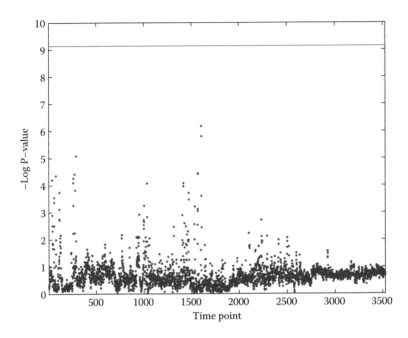

**FIGURE 3.20**
P-value for testing the association of gene ANKLE1 with the oxygen saturation curve at the individual time point.

where a single constant P-value for the test is calculated, in the association analysis of function-valued traits we can observe the time varying P-values. To illustrate this, we plotted Figure 3.21 showing the P-values of the MLMT for testing the association of all SNPs within the gene *TMEM50B* with the oxygen saturation at each time point over night as a function of time $t$. There was the rapid changes of P-value of the MLM test over time. We observed two peaks showing significant association with the oxygen saturation. At most times during the night, the genetic variation in the gene *TMEM50B* did not have a large impact on the variation of the oxygen saturation.

The genetic effect in the FLMF is characterized by its spatiotemporal pattern. The genetic effect is a function of both time $t$ and genomic position $s$. Similar to the concept of probability density function in the probability theory, the genetic effect function is viewed as the average genetic effect in a unit interval of time (or index value) and the genomic region. The genetic effect function is more interpretable than the scattered spatiotemporal genetic effect points of the SNPs within the gene. It often consists of several peaks and valleys where the values at the peak of the genetic effect function are the synthesized genetic effects of the individual SNPs in the region due to the correlation between the peak and nearby time and SNPs. To illustrate this,

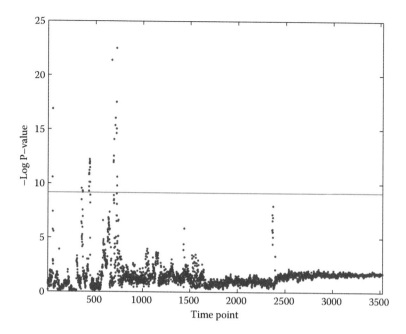

**FIGURE 3.21**
P-values of the MLMT for testing the association of all SNPs within the gene *TMEM50B* with the oxygen saturation at each time point overnight as a function of time *t*.

we plotted Figure 3.22 showing the genetic effect function $\beta(s,t)$ of the gene *KRTAP4-7* (P-value $<1.67 \times 10^{-13}$) as a function of time and the genomic position in the FLMF model. The genetic effect function surface $\beta(s,t)$ provides full and detailed spatiotemporal information on how and what genetic variants affect the development of the biological process, which will lead to new biological insights.

### 3.2.7 Association Analysis of Multiple Function-Valued Traits

Simple functional linear models can be easily extended to multiple functional linear models for association analysis of function-valued traits. Extension of the model (3.44) to multiple function-valued traits gives

$$y_{ik}(t) = W_i^T \alpha_k(t) + \int_S x_i(s)\beta_k(s,t)ds + \varepsilon_{ik}(t), t = t_1, ..., t_T, k = 1, ..., K, i$$

$$= 1, ..., n, \tag{3.67}$$

where $y_{ik}(t)$ is the $k^{th}$ function-valued trait of the $i^{th}$ individual, $\alpha_k(t) = [\alpha_{k1}(t), ..., \alpha_{kd}(t)]^T$ is a vector of effects associated with the covariates and $k^{th}$

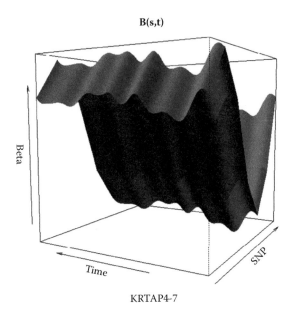

**FIGURE 3.22**
The genetic effect function of the gene *KPTAP4-7*.

function-valued trait, $\beta_k(s,t)$ is the $k^{th}$ genetic additive effect function in the genomic position $s$ and time $t$, and $\varepsilon_{ik}(t)$ is the residual function of the noise and unexplained effect for the $i^{th}$ individual.

The trait function and genotype function are expanded in terms of the orthonormal basis function as

$$y_{ik}(t) = \sum_{l=1}^{k_y} y_{ikl}\eta_l(t) = y_{ik}^T \eta(t), \tag{3.68}$$

where $y_{ikl} = \int_0^T y_{ik}(t)\eta_l(t)dt$ and $y_{ik} = [y_{ik1}, ..., y_{ikk_y}]^T$.

The covariate effect functions $\alpha_{kj}(t)$ can be expanded in terms of eigenfunctions $\eta_l(t)$ as

$$\alpha_{kj}(t) = \sum_{l=1}^{k_y} \alpha_{kjl}\eta_l(t) = \alpha_{kj}^T \eta(t), \tag{3.69}$$

where $\alpha_{kj} = [\alpha_{kj1}, ..., \alpha_{kjk_y}]^T$.

Define

$$\alpha_k = \left( \alpha_{kjl} \right)_{d \times k_y} \quad \text{and} \quad \alpha = [\alpha_1, .., \alpha_K].$$

Similarly, the error functions $\varepsilon_{ik}(t)$ can also be expanded as

$$\varepsilon_{ik}(t) = \sum_{l=1}^{k_y} \varepsilon_{ikl}\eta_l(t) = \varepsilon_{ik}^T \eta(t), \tag{3.70}$$

where $\varepsilon_{ik} = [\varepsilon_{ik1}, ..., \varepsilon_{iky}]^T$.

Define

$$\varepsilon_k = \begin{bmatrix} \varepsilon_{1k}^T \\ \vdots \\ \varepsilon_{nk}^T \end{bmatrix} = \begin{bmatrix} \varepsilon_{1k1} & \cdots & \varepsilon_{1kk_y} \\ \vdots & \vdots & \vdots \\ \varepsilon_{nk1} & \cdots & \varepsilon_{nkk_y} \end{bmatrix} \tag{3.71}$$

The genetic effect functions $\beta_k(s,t)$ can be expanded in terms of two sets of orthogonal eigenfunctions $\theta_k(s)$ and $\eta_l(t)$ as

$$\beta_k(s,t) = \sum_{j=1}^{k_{xx}} \sum_{l=1}^{k_y} b_{kjl}\theta_j(s)\eta_l(t) = \theta^T(s)B_k\eta(t), \tag{3.72}$$

where $B_k = (b_{kjl})_{k_x \times k_y}$.

Therefore, Equation 3.54 can be extended to

$$y_{ik}^T = W_i^T \alpha_k + x_i^T B_k + \varepsilon_{ik}^T, i = 1, ..., n. \tag{3.73}$$

Model (3.73) can be rewritten in a matrix form:

$$Y = W\alpha + XB + \varepsilon,$$
$$= A\gamma + \varepsilon, \tag{3.74}$$

where

$$Y = [Y_1 \cdots Y_K], \alpha = [\alpha_1 \cdots \alpha_K], B = [B_1 \cdots B_K], \varepsilon = [\varepsilon_1 \cdots \varepsilon_K], \gamma$$

$$= \begin{bmatrix} \alpha \\ B \end{bmatrix}, W, X$$

and $A$ are defined as before. Note that the parameters can be decomposed to

$$\gamma = [\gamma_1 \cdots \gamma_K] = \begin{bmatrix} \alpha_1 & \cdots & \alpha_K \\ B_1 & \cdots & B_K \end{bmatrix}.$$

Again, the least square estimates of the parameter vector $\gamma$ is (Exercise 3)

$$\hat{\gamma} = (A^T A)^{-1} A^T Y, \tag{3.75}$$

and

$$\hat{\gamma}_k = (A^T A)^{-1} A^T Y_k, \ k = 1, ..., K. \tag{3.76}$$

Define

$$vec(\varepsilon) = \begin{bmatrix} vec(\varepsilon_1) \\ \vdots \\ vec(\varepsilon_K) \end{bmatrix}, \Sigma = var(vec(\varepsilon)) = \begin{bmatrix} \Sigma_{11} & \cdots & \Sigma_{1K} \\ \vdots & \vdots & \vdots \\ \Sigma_{K1} & \cdots & \Sigma_{KK} \end{bmatrix},$$

where

$$\Sigma_{kv} = \Sigma^{kv} \otimes I_n, \Sigma^{kv} = \begin{bmatrix} \sigma_{11}^{(k,v)} & \cdots & \sigma_{1k_y}^{(k,v)} \\ \vdots & \vdots & \vdots \\ \sigma_{k_y 1}^{(k,v)} & \cdots & \sigma_{k_y k_y}^{(k,v)} \end{bmatrix}, \Sigma = \Sigma_* \otimes I_n$$

and $\Sigma_* = \begin{bmatrix} \Sigma^{11} & \cdots & \Sigma^{1K} \\ \vdots & \vdots & \vdots \\ \Sigma^{K1} & \cdots & \Sigma^{KK} \end{bmatrix}.$

Model (3.74) can also be written as

$$y_{ikj} = \sum_{l=1}^{d} w_{il} \alpha_{klj} + \sum_{l=1}^{k_y} x_{il} b_{klj} + \varepsilon_{ikj}, \ i = 1, ..., n, k = 1, ..., K, j = 1, ..., k_y. \tag{3.77}$$

Variance $\sigma_{ul}^{(k,v)}$ can be estimated by $\hat{\sigma}_{ul}^{(k,v)} = \dfrac{1}{nk_y - dk_y - k_y k_x} \sum_{i=1}^{n} \hat{\varepsilon}_{iku} \hat{\varepsilon}_{ivl}, \ u = 1, ..., k_y, l = 1, ..., k_y, k = 1, ..., K, v = 1, ..., K.$ Equation 3.75 can be written as

$$\hat{\gamma} = \begin{bmatrix} \hat{\alpha} \\ \hat{B} \end{bmatrix} = \begin{bmatrix} C_1 \\ C_2 \end{bmatrix} Y, \tag{3.78}$$

where $C_1$ and $C_2$ are generated by the first $d$ rows and last $k_x$ rows of the matrix $(A^T A)^{-1} A^T$, respectively.

Thus, Equation 3.78 gives

$$\hat{B} = C_2 Y. \tag{3.79}$$

Let

$$\hat{b} = vec(\hat{B}) = (I \otimes C_2)vec(Y).$$  (3.80)

$$\text{Note that } cov(vec(Y)) = cov(vec(\varepsilon)).$$  (3.81)

Using Equations 3.80 and 3.81 gives

$$\Lambda = var(vec(\ddot{b})) = [I \otimes C_2]cov(vec(Y))[I \otimes C_2^T]$$
$$= [I \otimes C_2][\Sigma_* \otimes I_n][I \otimes C_2^T]$$  (3.82)
$$= \Sigma_* \otimes (C_2 . C_2^T).$$

Now we investigate the problem of testing the following hypothesis:

$$H_0 : \beta_k(s, t) = 0, \forall s \in S, \ t \in T, \ k = 1, ..., K$$  (3.83)

against

$$H_a : \beta_k(s, t) \neq 0 \text{ at least one } k \in (1, ..., K).$$

Testing the null hypothesis $H_0$ in Equation 3.83 is equivalent to testing the hypothesis:

$$H_0 : B_k = 0 \ , \ k = 1, ..., K.$$  (3.84)

We can define the following statistic for testing the association of a genomic region with the $K$ functional traits:

$$T_F = \hat{b}^T \Lambda^{-1} \hat{b}.$$  (3.85)

Under the null hypothesis of no association, the statistic $T_F$ will be distributed as a central $\chi^2_{(k_y k_x K)}$ distribution.

---

## 3.3 Gene–Gene Interaction Analysis of Function-Valued Traits

### 3.3.1 Introduction

Although many physiological traits are measured as a function, the widely used methods for interaction analysis of function-valued traits in humans are the same as that for the traditional single-valued quantitative traits where a single number is taken as a quantitative trait. These methods use summary statistics to measure or represent function-valued traits. Therefore, all the methods developed for interaction analysis of scale quantitative traits in the past chapters can be applied for the interaction analysis of the function-valued traits. Quantitative genetic analysis of function-valued traits enhances our understanding of the genetic control of the whole dynamic process of the traits and improves the statistical power to detect epistasis. Although the genetic study of quantitative traits has seen wide application and extensive technical development, the quantitative genetic analysis of function-valued trait is comparatively less developed.

Fast and less expansive next generation sequencing (NGS) technologies will generate unprecedentedly massive and highly dimensional genetic variation data that allow nearly complete evaluation of genetic variation including both common and rare variants. This provides a powerful tool to comprehensively catalog human genetic variation. Despite their promise, critical barrier in interaction analysis with NGS data is that most traditional statistical methods use SNP-based tests for pair-wise epistasis are originally designed for testing the interaction for common variants and are difficult to apply to rare variants for their high type 1 error rates, prohibitive computational time, a major statistical challenge of severe multiple testing, and low power to detect interaction between rare variants.

To fill this gap, in this section we introduce novel approaches based on nonlinear functional regression models with both function response and function predictors for epistasis analysis in the function-valued trait, which allows simultaneous capture of all temporal and space information hidden in the traits and genomic region, but with substantially reduced dimensions. The functional valued-functional regression (F-FRG) models collectively test interaction between all possible pairs of SNPs within two genome regions (or genes).

As we discussed in Section 7.5, CCA and functional CCA can be used as a unified framework for testing gene–gene and gene–environment interaction. Finally, we will briefly discuss how the functional CCA will be used for interaction analysis of function-valued traits.

### 3.3.2 Functional Regression Models

Consider a temporal trait $y_i(\tau)$, $\tau \in T_\tau = [0,T_\tau]$ of the $i^{th}$ individual which varies in time and two genomic regions (or genes) $[a_1,b_1]$ and $[a_2,b_2]$. Let $x_i(t)$ and $x_i(s)$ be the genotypic functions of the $i^{th}$ individual defined in the regions $[a_1,b_1]$ and $[a_2,b_2]$, respectively. Let $t$ and $s$ be a genomic position in the first and second genomic regions, respectively. The genotype functions $x_i(t)$ and $x_i(s)$ are defined as

$$X_i(t) = \begin{cases} 2P_m(t), & MM \\ P_m(t)\text{-}P_M(t), & Mm, \\ -2P_M(t), & mm \end{cases} \quad X_i(s) = \begin{cases} 2P_m(s), & MM \\ P_m(s)\text{-}P_M(s), & Mm, \\ -2P_M(s), & mm \end{cases}$$

where M and m are two alleles of the marker at the genomic position $t$ or $s$, $P_M(t)$ and $P_m(t)$, and $P_M(s)$ and $P_m(s)$ are the frequencies of the alleles M and m at the genomic positions $t$ and $s$, respectively. Consider a functional regression model:

$$Y_i(\tau) = \mu(\tau) + W_i^T \omega(\tau) + \int_T x_i(t)\alpha(t,\tau)dt + \int_S z_i(s)\beta(s,\tau)ds$$

$$+ \int_T \int_S x_i(t)z_i(s)\gamma(t,s,\tau)dsdt + \varepsilon_i(\tau), \qquad (3.86)$$

where $\mu(\tau)$ is an overall mean function at time $\tau$, $W_i$ a vector of covariates for the $i^{th}$ individual, $\omega(\tau)$ is a vector of effects associated with the covariates, $\alpha(t,\tau)$ is a genetic additive effect function at genomic position $t$ of the first gene and time $\tau$, $\beta(s,t)$ is a genetic additive effect function at genomic positions s of the second gene and time $\tau$, $\gamma(t,s,\tau)$ is an interaction effect function between two putative quantitative trait loci (QTLs) located at the genomic positions $t$ and $s$ at the time $\tau$, and $\varepsilon_i(\tau)$ is a residual function of the unexplained effect for the $i^{th}$ individual at time $\tau$. The interaction function is measured by double integrals of the genotype function in two genes.

### 3.3.3 Estimation of Interaction Effect Function

We assume that both phenotype and genotype functions are centered. The genotype functions $x_i(t)$ and $x_i(s)$ are expanded in terms of the orthonormal basis function as:

$$x_i(t) = \sum_{j=1}^{\infty} \xi_{ij}\phi_j(t) \quad \text{and}$$

$$x_i(s) = \sum_{l=1}^{\infty} \eta_{il}\psi_l(s), \qquad (3.87)$$

where $\phi_j(t)$ and $\psi_l(s)$ are sequences of the orthonormal basis functions. The expansion coefficients $\xi_{ij}$ and $\eta_{il}$ are estimated by

$$\xi_{ij} = \int_T x_i(t)\phi_j(t)dt \quad \text{and}$$

$$\eta_{il} = \int_S x_i(s)\psi_l(s)ds. \qquad (3.88)$$

In practice, numerical methods for the integral will be used to calculate the expansion coefficients.

Substituting Equation 3.87 into Equation 3.86, we obtain

$$Y_i(\tau) = \mu(\tau) + W_i^T \omega(\tau) + \int_T \sum_{j=1}^J \xi_{ij} \phi_j(t) \alpha(t, \tau) dt + \int_S \sum_{l=1}^L \eta_{il} \psi_l(s) \beta(s, \tau) ds$$

$$+ \int_T \int_S \sum_{j=1}^J \xi_{ij} \phi_j(t) \sum_{l=1}^L \eta_{il} \psi_l(s) \gamma(t, s, \tau) ds dt + \varepsilon_i(\tau)$$

$$= \mu(\tau) + W_i^T \omega(\tau) + \sum_{j=1}^J \xi_{ij} \int_T \phi_j(t) \alpha(t, \tau) dt + \sum_{l=1}^L \eta_{il} \int_S \psi_l(s) \beta(s, \tau) ds \quad (3.89)$$

$$+ \sum_{j=1}^J \sum_{l=1}^L \xi_{ij} \eta_{il} \int_T \int_S \phi_j(t) \psi_l(s) \gamma(t, s, \tau) ds dt + \varepsilon_i(\tau)$$

$$= \mu(\tau) + W_i^T \omega(\tau) + \sum_{j=1}^J \xi_{ij} \alpha_j(\tau) + \sum_{l=1}^L \eta_{il} \beta_l(\tau)$$

$$+ \sum_{j=1}^J \sum_{l=1}^L \xi_{ij} \eta_{il} \gamma_{jl}(\tau) + \varepsilon_i(\tau),$$

where

$$\alpha_j(\tau) = \int_T \alpha(t, \tau) \phi_j(t) dt, \beta_l(\tau) = \int \beta(s, \tau) \psi_l(s) ds \text{ and } \gamma_{jl}(\tau)$$

$$= \int_T \int_S \gamma(t, s, \tau) \phi_j(t) \psi_l(s) dt ds.$$

The parameters $\alpha_j(\tau), \beta_l(\tau)$ and $\gamma_{jl}(\tau)$ are referred to as genetic additive and additive × additive effect score functions. These score functions can also be viewed as the expansion coefficients of the genetic effect functions with respect to orthonormal basis functions:

$$\alpha(t, \tau) = \sum_j \alpha_j(\tau) \phi_j(t), \beta(s, \tau) = \sum_l \beta_l(\tau) \psi_l(s) \text{ and } \gamma(s, t)$$

$$= \sum_j \sum_l \gamma_{jl}(\tau) \phi_j(s) \psi_l(t).$$

Let

$$Y(\tau) = \begin{bmatrix} Y_1(\tau) \\ \vdots \\ Y_n(\tau) \end{bmatrix}, e = \begin{bmatrix} 1 \\ \vdots \\ n \end{bmatrix}, W = \begin{bmatrix} W_{11} & \cdots & W_{1d} \\ \vdots & \ddots & \vdots \\ W_{n1} & \cdots & W_{nd} \end{bmatrix},$$

$$\omega(\tau) = \begin{bmatrix} \omega_1(\tau) \\ \vdots \\ \omega_d(\tau) \end{bmatrix}, \xi = \begin{bmatrix} \xi_{11} & \cdots & \xi_{1J} \\ \vdots & \ddots & \vdots \\ \xi_{n1} & \cdots & \xi_{nJ} \end{bmatrix},$$

$$\eta = \begin{bmatrix} \eta_{11} & \cdots & \eta_{1L} \\ \vdots & \ddots & \vdots \\ \eta_{n1} & \cdots & \eta_{nL} \end{bmatrix}, \xi_i = \begin{bmatrix} \xi_{i1} \\ \vdots \\ \xi_{iJ} \end{bmatrix}, \eta_i = \begin{bmatrix} \eta_{i1} \\ \vdots \\ \eta_{iL} \end{bmatrix}, \alpha(\tau) = \begin{bmatrix} \alpha_1(\tau) \\ \vdots \\ \alpha_J(\tau) \end{bmatrix},$$

$$\Gamma = \begin{bmatrix} \xi_1^T \otimes \eta_1^T \\ \vdots \\ \xi_n^T \otimes \eta_n^T \end{bmatrix} = \begin{bmatrix} \xi_{11}\eta_{11} & \cdots & \xi_{11}\eta_{1L} & \cdots & \xi_{1J}\eta_{11} & \cdots & \xi_{1J}\eta_{1L} \\ \cdots & \cdots & \cdots & \cdots & \cdots & \cdots & \cdots \\ \xi_{n1}\eta_{n1} & \cdots & \xi_{n1}\eta_{nL} & \cdots & \xi_{nJ}\eta_{n1} & \cdots & \xi_{nJ}\eta_{nL} \end{bmatrix}, \beta(\tau) = \begin{bmatrix} \beta_1(\tau) \\ \vdots \\ \beta_L(\tau) \end{bmatrix},$$

$$\varepsilon(\tau) = \begin{bmatrix} \varepsilon_1(\tau) \\ \vdots \\ \varepsilon_n(\tau) \end{bmatrix}.$$

Equation 3.89 can be written in a matrix form:

$$Y(\tau) = e\mu(\tau) + W\omega(\tau) + \xi\alpha(\tau) + \eta\beta(\tau) + \Gamma\gamma(\tau) + \varepsilon(\tau). \tag{3.90}$$

Expanding $Y(\tau)$, $\mu(\tau)$, $\omega(\tau)$, $\alpha(\tau)$, $\beta(\tau)$, $\gamma(\tau)$ and $\varepsilon(\tau)$ in terms of orthogonal basis functions and substituting their expansions into Equation 3.90 yields

$$Y_i(\tau) = \sum_{k=1}^{K} y_{ik}\theta_k(\tau), \mu(\tau) = \sum_{k=1}^{K}\mu_k\theta_k(\tau), \omega_j(\tau) = \sum_{k=1}^{K}\omega_{jk}\theta_k(\tau), \alpha_j(\tau)$$

$$= \sum_{k=1}^{K}\alpha_{jk}\theta_k(\tau),$$

$$\beta_j(\tau) = \sum_{k=1}^{K}\beta_{jk}\theta_k(\tau), \gamma_{jl}(\tau) = \sum_{k=1}^{K}\gamma_{jlk}\theta_k(\tau), \text{ and } \varepsilon_i(\tau) = \sum_{k=1}^{K}\varepsilon_{ik}\theta_k(\tau).$$

Define expansion coefficient vectors and matrices as follows.

$$Y = \begin{bmatrix} y_{11} & \cdots & y_{1K} \\ \vdots & \ddots & \vdots \\ y_{n1} & \cdots & y_{nK} \end{bmatrix}, \mu = \begin{bmatrix} \mu_1 \\ \vdots \\ \mu_K \end{bmatrix}, E = \begin{bmatrix} 1 & \cdots & 1 \\ \vdots & \ddots & \vdots \\ 1 & \cdots & 1 \end{bmatrix}, \omega = \begin{bmatrix} \omega_{11} & \cdots & \omega_{1K} \\ \vdots & \ddots & \vdots \\ \omega_{d1} & \cdots & \omega_{dK} \end{bmatrix},$$

$$\alpha = \begin{bmatrix} \alpha_{11} & \cdots & \alpha_{1K} \\ \vdots & \ddots & \vdots \\ \alpha_{J1} & \cdots & \alpha_{JK} \end{bmatrix}, \beta = \begin{bmatrix} \beta_{11} & \cdots & \beta_{1K} \\ \vdots & \ddots & \vdots \\ \beta_{L1} & \cdots & \beta_{LK} \end{bmatrix}, \gamma = \begin{bmatrix} \gamma_{111} & \cdots & \gamma_{11K} \\ \vdots & \ddots & \vdots \\ \gamma_{JL1} & \cdots & \gamma_{JLK} \end{bmatrix},$$

$$\text{and } \varepsilon = \begin{bmatrix} \varepsilon_{11} & \cdots & \varepsilon_{1K} \\ \vdots & \ddots & \vdots \\ \varepsilon_{n1} & \cdots & \varepsilon_{nK} \end{bmatrix}.$$

Thus, Equation 3.90 can be transformed into

$$Y\theta(\tau) = \mu\theta(\tau) + W\omega\theta(\tau) + \xi\alpha\theta(\tau) + \eta\beta\theta(\tau) + \Gamma\gamma\theta(\tau) + \varepsilon\theta(\tau). \qquad (3.91)$$

Since Equation 3.91 holds for every time point $\tau$, the coefficients on both sides of Equation 3.91 should be equal. Therefore, functional regression model 3.91 can be further transformed to the standard multivariate multiple regression:

$$Y = E\mu + W\omega + \xi\alpha + \eta\beta + \Gamma\gamma + \varepsilon. \qquad (3.92)$$

Let

$$A = [\, E \; W \; \xi \; \eta \; \Gamma \,] \text{ and } b = \begin{bmatrix} \mu \\ \omega \\ \alpha \\ \beta \\ \gamma \end{bmatrix}.$$

Equation 3.92 can be rewritten as

$$Y = Ab + \varepsilon. \qquad (3.93)$$

Similar to Equation 3.56, the total squares of errors between the true $Y$ and predicted values $A\hat{b}$ is defined as

$$F = \text{Tr}((Y - Ab)^T (Y - Ab)) = \text{Tr}(\varepsilon^T \varepsilon).$$

Setting $\dfrac{\partial F}{\partial b} = -A^T (Y - Ab) = 0$, we obtain the least square estimates of the parameter vector $b$:

$$\hat{b} = (A^T A)^{-1} A^T Y. \qquad (3.94)$$

The covariance matrix $\Sigma$ is estimated by

$$\hat{\Sigma} = \frac{(Y - A\hat{b})^T (Y - A\hat{b})}{(n - (J + L + JL))k}.$$

### 3.3.4 Test Statistics

An essential problem in genetic interaction studies of the quantitative traits is to test the interaction between two genomic regions (or genes). Formally, we investigate the problem of testing the following hypothesis:

$$\gamma(t, s, \tau) = 0, \; \forall\, t \in [a_1, b_1], s \in [a_2, b_2], \tau \ni [0, T_\tau],$$

which is equivalent to testing the hypothesis:

$$H_0 : \gamma = 0. \tag{3.95}$$

Let Vec denote the vector operation. To develop test statistics, we begin with calculating the covariance matrix of the $vec(\hat{b})$. We assume that

$$var(vec(\varepsilon)) = \Sigma \otimes I_n. \tag{3.96}$$

Recall that

$$vec(\hat{b}) = [I_n \otimes (A^T A)^{-1} A^T] vec(Y).$$

Therefore, we have

$$
\begin{aligned}
var(vec(\hat{b})) &= [I_K \otimes (A^T A)^{-1} A^T](\Sigma \otimes I_n)[I_K \otimes A(A^T A)^{-1}] \\
&= \Sigma \otimes (A^T A)^{-1}.
\end{aligned} \tag{3.97}
$$

Let $\Lambda$ be a matrix consisting of the last $JLK$ columns and $JLK$ rows of the covariance matrix $var(vec(\hat{b}))$ and $\hat{\gamma}$ be the estimators of interaction which can be obtained by extracting the last $JL$ rows of the estimators of the matrix $\hat{b}$. Define the test statistic for testing the interaction between two genomic regions $[a_1, b_1]$ and $[a_2, b_2]$ as

$$T_I = (vec(\hat{\gamma})^T \Lambda^{-1} vec(\hat{\gamma}). \tag{3.98}$$

Then, under the null hypothesis $H_0 : \gamma = 0$, $T_1$ is asymptotically distributed as a central $\chi^2_{(JLK)}$ distribution with degrees of freedom $JLK$

### 3.3.5 Simulations

#### 3.3.5.1 Type 1 Error Rates

To examine the null distribution of test statistics, a series of simulation studies to compare their empirical levels with the nominal ones was performed. We calculated the type I error under three models. We first assumed the model with no marginal effects:
**Model 1** (no marginal effect):

$$Y_i(\tau_l) = \mu + \varepsilon_i(\tau_l), l = 1, \ldots, L,$$

where error $\varepsilon_i(t_l)$ is generated by the standard Brownian motion, the points $t_l$ were equally distributed between interval $[0,1]$, and $L = 203,040$ were considered.

Then, we considered the model with marginal genetic effects at one gene:

**Model 2** (a marginal effect at the first gene):

$$Y_i(\tau_l) = \mu + \sum\nolimits_{j=1}^{J} x_{ij}\alpha_j(\tau_l) + \varepsilon_i(\tau_l),$$

where $x_{ij}$ is an indicator variable for the genotype of the $i^{th}$ individual in the $j^{th}$ SNP at the first gene, genetic additive effect function is assumed to be equal to $\alpha_j(\tau_l) = \alpha_j\alpha(\tau_l)$, $\alpha_j = (1 - P_j)(r_1 - 1)f_0$, $\alpha(\tau_l) = 0.05e^{0.05\tau_l}$, $P_j$ is the frequency of the minor allele in the $j^{th}$ genomic position at the first gene, $\gamma_1 = 1.01$ is a risk parameter for the first gene, $f_0 = 1$ is the baseline penetrance, again error $\varepsilon_i(t_l)$ is generated by the standard Brownian motion, the points $t_l$ were equally distributed between interval [0,1], and $L = 20,30,40$ were considered.

**Model 3** (marginal effects at both the first and the second genes):

$$Y_i(\tau_l) = \mu + \sum\nolimits_{j=1}^{J} x_{ij}\alpha_j(\tau_l) + \sum\nolimits_{k=1}^{K} z_{ik}\beta_k(\tau_l) + \varepsilon_i(\tau_l),$$

where $z_{ik}$ is an indicator variable for the genotype of the $i^{th}$ individual in the $k^{th}$ SNP at the second gene, genetic additive effect function $\beta_k(\tau_l)$ is assumed to be equal to $\beta_k(\tau_l) = \beta_k\beta(\tau_l)$, $\beta_k = (1 - P_k)(r_2 - 1)f_0$, $\beta(\tau_l) = 0.05e^{0.05\tau_l}$, $P_k$ is the frequency of the minor allele in the $k^{th}$ genomic position at the second gene, $r_2 = 1.01$ is a risk parameter in the second gene, and other parameters are defined as that in the model 2.

We generated 1,000,000 chromosomes by resampling from 2225 individuals with variants in five genes CDC2L1, GBP3, IQGAP3, TNN, and ACTN2 selected from the NHLBI's Exome Sequencing Project (ESP). We randomly selected 10% of the SNPs as causal variants. The number of sampled individuals from populations of 1,000,000 chromosomes ranged from 1000 to 3000. Three numbers of time points 20, 30, and 40 were considered. We calculated average type 1 error rates over 10 pairs of genes selected from the above five genes. A total of 5000 simulations were repeated.

Table 3.5 summarized the average type I error rates of the test statistics for testing the interaction between two genes under three models consisting of only rare variants with 30 time points, respectively, over 10 pairs of genes at the nominal levels $\alpha = 0.05$, $\alpha = 0.01$, and $\alpha = 0.001$. Table 3.6 summarized the average type I error rates of the test statistics for testing the interaction between two genes under three models consisting of both rare and common variants 30 time points, respectively, over 10 pairs of genes at the nominal levels $\alpha = 0.05$, $\alpha = 0.01$, and $\alpha = 0.001$. These results clearly showed that the type I error rates of the FRG-based test statistics for testing interaction between two genes with function valued traits with or without marginal effects were not appreciably different from the nominal levels.

### 3.3.5.2 *Power*

To test the power of FRG with both functional response and predictors for detecting the true interaction between two genetic regions or genes for a function valued trait, we used simulated data to estimate their power. A true

**TABLE 3.5**

Average Type I Error Rates for Testing the Interaction between First Genomic Region and the Second Genomic Region with Rare SNPs

|         | Sample Size | 0.05 | 0.01 | 0.001 |
|---------|-------------|------|------|-------|
| Model 1 | 1000 | 0.04206 | 0.00774 | 0.00048 |
|         | 1500 | 0.04564 | 0.00860 | 0.00098 |
|         | 2000 | 0.04796 | 0.00978 | 0.00104 |
|         | 2500 | 0.04788 | 0.01000 | 0.00110 |
|         | 3000 | 0.04728 | 0.00926 | 0.00086 |
| Model 2 | 1000 | 0.04208 | 0.00774 | 0.00048 |
|         | 1500 | 0.04564 | 0.00860 | 0.00098 |
|         | 2000 | 0.04796 | 0.00978 | 0.00104 |
|         | 2500 | 0.04788 | 0.01000 | 0.00110 |
|         | 3000 | 0.04722 | 0.00926 | 0.00086 |
| Model 3 | 1000 | 0.0421 | 0.00774 | 0.00048 |
|         | 1500 | 0.04564 | 0.00860 | 0.00098 |
|         | 2000 | 0.04796 | 0.00978 | 0.00104 |
|         | 2500 | 0.04784 | 0.01000 | 0.00110 |
|         | 3000 | 0.04722 | 0.00926 | 0.00086 |

The number of the measurements is 30.

**TABLE 3.6**

Average Type I Error Rates for Testing the Interaction between First Genomic Region and the Second Genomic Region with All SNPs

|         | Sample Size | 0.05 | 0.01 | 0.001 |
|---------|-------------|------|------|-------|
| Model 1 | 1000 | 0.04316 | 0.00864 | 0.00106 |
|         | 1500 | 0.04460 | 0.00884 | 0.00074 |
|         | 2000 | 0.04584 | 0.00826 | 0.00066 |
|         | 2500 | 0.04756 | 0.00964 | 0.00080 |
|         | 3000 | 0.04632 | 0.00904 | 0.00102 |
| Model 2 | 1000 | 0.04316 | 0.00864 | 0.00106 |
|         | 1500 | 0.04456 | 0.00884 | 0.00074 |
|         | 2000 | 0.04588 | 0.00828 | 0.00066 |
|         | 2500 | 0.04758 | 0.00966 | 0.00080 |
|         | 3000 | 0.04634 | 0.00904 | 0.00102 |
| Model 3 | 1000 | 0.04312 | 0.00862 | 0.00106 |
|         | 1500 | 0.04464 | 0.00884 | 0.00076 |
|         | 2000 | 0.04584 | 0.00832 | 0.00066 |
|         | 2500 | 0.04764 | 0.00968 | 0.00080 |
|         | 3000 | 0.04642 | 0.00900 | 0.00104 |

The number of the measurements is 30.

functional quantitative genetic model is given as follows. Consider two genes. Assume that the first gene had $k_1$ SNPs and the second gene had $k_2$ SNPs. There was a total of $k_1k_2$ SNPs from two genes. For the $h^{th}$ pair of SNPs, let $Q_{h_1}$ and $q_{h_1}$ be two alleles at the SNP in the first gene, and $Q_{h_2}$ and $q_{h_2}$ be two alleles at the SNP in the second gene. Let $u_{ijkl}^h$ denote his or her genotypes of the $h^{th}$ pair of SNPs, where $ij \in Q_{h_1}Q_{h_1}, Q_{h_1}q_{h_1}, q_{h_1}q_{h_1}$ and $kl \in Q_{h_2}Q_{h_2}, Q_{h_2}q_{h_2}, q_{h_2}q_{h_2}$. Let $g_{u_{ijkl}}^h(\tau)$ denote his or her genotypic value in the $h^{th}$ pair of SNPs at time $\tau$. Then we can use the following multiple regression model to generate the temporal quantitative trait of the $u^{th}$ individual of the $h^{th}$ pair of SNPs at time $\tau$.

$$Y_u(\tau) = \sum_{h=1}^{k_1k_2} g_{u_{ijkl}}^h(\tau) + \varepsilon_u(\tau), \tag{3.99}$$

where $u = 1,...,n$, $g_{u_{ijkl}}^h(\tau) = r*25e^{-0.25\tau}$, $r$ is a risk parameter which is determined by the gene interaction model (Table 3.7), $\varepsilon_u(\tau)$ is the error term of the $u^{th}$ individual at time $\tau$, and is generated by the standard Brownian motion process.

To compare the power with the cross-sectional approach, we take the average of the genetic interaction effect function over the period of times studied as a cross-sectional genetic interaction effect, $\bar{\gamma}$. In other words, the cross-sectional genetic interaction effect was calculated as follows.

$$\bar{\gamma} = \frac{1}{T}\int_T \gamma(\tau)d\tau = \frac{\sum_{j=1}^m \gamma(\tau_j)}{m},$$

where $T$ is the total time considered, $m$ is the number of measurements, and $\tau_j = j\frac{T}{m}$.

**TABLE 3.7**

The Table of the Risk Parameter r in the $h^{th}$ Pair of SNPs by Four Different Interaction Models

| First Locus | Second Locus | Dominant or Dominant | Dominant and Dominant | Recessive or Recessive | Threshold |
|---|---|---|---|---|---|
| $Q_{h_1}Q_{h_1}$ | $Q_{h_2}Q_{h_2}$ | r | r | r | r |
| $Q_{h_1}q_{h_1}$ | $Q_{h_2}Q_{h_2}$ | r | r | r | r |
| $q_{h_1}q_{h_1}$ | $Q_{h_2}Q_{h_2}$ | r | 0 | r | 0 |
| $Q_{h_1}Q_{h_1}$ | $Q_{h_2}q_{h_2}$ | r | r | r | r |
| $Q_{h_1}q_{h_1}$ | $Q_{h_2}q_{h_2}$ | r | r | 0 | 0 |
| $q_{h_1}q_{h_1}$ | $Q_{h_2}q_{h_2}$ | r | 0 | 0 | 0 |
| $Q_{h_1}Q_{h_1}$ | $q_{h_2}q_{h_2}$ | r | 0 | r | 0 |
| $Q_{h_1}q_{h_1}$ | $q_{h_2}q_{h_2}$ | r | 0 | 0 | 0 |
| $q_{h_1}q_{h_1}$ | $q_{h_2}q_{h_2}$ | 0 | 0 | 0 | 0 |

We generate 1,000,000 individuals by resampling from 2225 individuals of European origin with variants in two genes *KANK4* with 68 SNPs (57 rare and 3 common SNPs) and *GALNT2* with 57 SNPs (48 rare and 9 common SNPs) selected from the ESP dataset. We randomly selected 20% of the variants as causal variants. A total of 1000 individuals for the four interaction models were sampled from the populations. A total of 5000 simulations were repeated for the power calculation. We compare the power of FRG with both functional response and predictors (F-FRG) with five other statistical models: mean value cross-sectional functional regression model in which mean value of $Y(\tau)$ was used as the response variable (MC-FRG), point-wise cross-sectional functional regression model in which sample points of $Y(\tau)$ were used as the response variable (PC-FRG), regression on functional principle component score (FPC-FPC) in which a top FPC score of the value $Y(\tau)$ was taken as the response and FPC scores of the genotype functions were taken as predictors (interacting unit), point-wise cross-sectional and pair-wise regression interaction model in which the sample points of $Y(\tau)$ were taken as responses and pair-wise of SNPs were taken as predictors (interacting unit) (PC-pair wise), mean value cross-sectional and pair-wise regression interaction model in which the mean value of $Y(\tau)$ was taken as a response and pair-wise of SNPs were taken as predictors (interacting unit) (MC-pair-wise).

Figures 3.23 and 3.24 plotted the power curves of six statistics: F-FRG, MC-FRG, PC-FRG, FPC-FPC, PC-pair-wise, and MC-pair-wise under Dominant

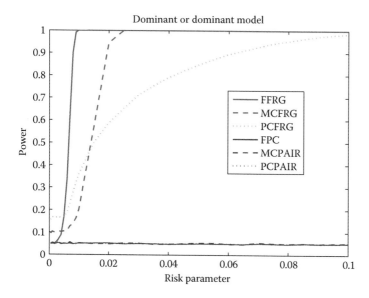

**FIGURE 3.23**
The power curves of six statistics: F-FRG, MC-FRG, PC-FRG, FPC-FPC, PC-pair-wise, and MC-pair-wise under Dominant or Dominant model.

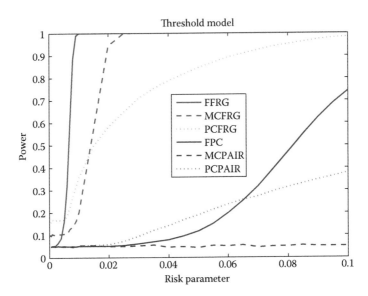

**FIGURE 3.24**
The power curves of six statistics: F-FRG, MC-FRG, PC-FRG, FPC-FPC, PC-pair-wise, and MC-pair-wise under the Threshold model.

or Dominant and Threshold models, respectively. Permutations were used to adjust for multiple testing for testing interactions between two genomic regions for two pair-wise test statistics. These power curves are a function of the risk parameter at the significance level $\alpha = 0.05$. From these figures we observed several remarkable features. First, under all four interaction models the test based on the FRG model had the highest power. The F-FRG utilizes the merits of taking both phenotype and genotype as functions. It decomposes time varying phenotype functions into orthogonal eigenfunctions of time and position, varying genotype functions into orthogonal eigenfunctions of genomic position. The F-FRG reduces the dimensions due to both phenotype variation and genotype variation (only a few eigenfunctions are used to model variation), which in turn increases statistical power of the test. Second, the power difference between the F-FRG and the other five models was substantial. Third, our data consisted of trait values and genotype values. Both trait and genotype values can take three types of values: function (curve) value, mean value, and point-wise value (time point or genomic position). We clearly observed that the power order of the tests was F-FRG, MCFRG, and PCFRG if we took a function as a genotypic value. However, if we considered the genotype value at each SNP, then the power of the PC-pair-wise was higher than that of the MC-pair-wise test. Fourth, the power of the traditional mean-valued cross-sectional pair-wise test (MCPAIR) was much lower than the F-FRG. In many cases, the MCPAIR did not have power to detect interaction between two genes.

### 3.3.6 Real Data Analysis

To further evaluate its performance, the F-FRG was applied to oxygen saturation studies in Starr County, Texas. The oxygen saturation signals were measured by seconds. A total of 35,280 measurements was taken over a night. Oxygen saturation provides important information on the sleep quality of the obstructive sleep apnea (Nieto et al. 2000). A total of 406,299 SNPs in 20,763 genes was typed for 833 individuals of Mexican American origin from Starr County, Texas. Since the F-FRG requires to expand genotype function in terms of eigenfunction, which need to have at least 3 SNPs in the gene, we exclude the genes with only one or two SNPs in them. The left total number of genes for analysis was 17,258, and there were 148,910,653 different pairs of genes. Therefore, the P-value for declaring significance after applying the Bonferroni correction for multiple tests was $3.36 \times 10^{-10}$. To reduce the number of measurements included in the analysis, we used the mean of the oxygen saturation in every 10 seconds as the trait values. SNPs in the *5kb* flanking region of the gene were assumed to belong to the gene. To ensure the numerical stability, we used single value decomposition to calculate the inverse of the matrix. We selected the number of single values such that they can account for 99% of the total variation.

In total, we identified 13 pairs of significantly interacted genes consisting of 23 genes, with P-values ranging from $2.27 \times 10^{-10}$ to $4.85 \times 10^{-43}$ (Table 3.8). Instead of using the whole oxygen saturation curve as a functional response,

**TABLE 3.8**

P-Values of 13 Pairs of Significantly Interacted Genes Identified by FRG

| | | | | | | P-values | |
| Chr | Gene1 | Chr | Gene2 | # of Pairs | F-FRG | Cross-Sectional FRM | SRG (minimum) |
|---|---|---|---|---|---|---|---|
| 9 | MIR4520A | 9 | MIR4520B | 9 | 4.85E-43 | 3.92E-36 | 0.00469 |
| 1 | DPH2 | 1 | ATP6V0B | 9 | 2.55E-37 | 2.86E-31 | 5.93E-05 |
| 22 | MAK16 | 22 | TTI2 | 25 | 7.59E-33 | 1.86E-21 | 0.01894 |
| 4 | TAS2R5 | 4 | TAS2R31 | 32 | 2.99E-27 | 2.01E-25 | 0.03168 |
| 16 | FAM211B | 16 | GGT1 | 12 | 4.42E-23 | 2.09E-11 | 0.01246 |
| 17 | OR5H15 | 17 | OR5H14 | 25 | 4.47E-17 | 3.92E-07 | 0.00042 |
| 22 | MIR378D2 | 22 | PDP1 | 9 | 3.10E-14 | 1.28E-07 | 0.00896 |
| 9 | MIR4520A | 9 | C17orf100 | 12 | 6.41E-14 | 9.95E-02 | 0.00020 |
| 9 | MIR4520B | 9 | C17orf100 | 12 | 6.41E-14 | 9.95E-02 | 0.00020 |
| 8 | WDR90 | 8 | RHOT2 | 9 | 2.43E-13 | 1.57E-04 | 0.00022 |
| 11 | B9D2 | 11 | TMEM91 | 18 | 1.15E-11 | 5.38E-08 | 7.76E-14 |
| 8 | KCTD1 | 8 | TAOK2 | 20 | 9.52E-11 | 6.74E-04 | 0.00550 |
| 21 | MIR489 | 21 | MIR653 | 9 | 2.27E-10 | 1.10E-02 | 0.00691 |

we also used their mean as a scalar response variable and applied the MC-FRG to test for interaction. In Table 3.8, we included the minimum P-values from using the instantaneous value of the trait as a phenotype and the simple regression interaction model for testing interaction between all possible pairs of SNPs across two genes where each SNP in the pair were located in different genes, over all observed time periods. We observed that the P-values of the F-FRG for testing interaction were much smaller than that of the point-pair wise test (MC-FRG). The F-FRG utilizes the merits of taking both phenotype and genotype as functions and decomposes time varying phenotype functions into orthogonal eigenfunctions of time and position varying genotype function into orthogonal eigenfunctions of genomic position. Only a few eigenfunctions that capture major information on the trait function and genotype function were used to model the trait variation and genetic variation. This substantially reduced the dimension in both phenotype and genotype variation of the data.

## Appendix 3.A    Gradient Methods for Parameter Estimation in the Convolutional Neural Networks

### 3.A.1    Multilayer Feedforward Pass

The fully connected layers in the last stage of the convolutional neural network (CNN) are a multilayer feedforward neural network. The final output layer of the fully connected neural network is denoted layer $L$. Let $l$ denote the current layer. The final stage of the CNN consists of one input layer, one output layer, and $L - 1$ hidden layers (Figure 3.A.1).

Consider a network with $p$ input neurons and $q$ output neurons. The weight for the connection from the $k^{th}$ neuron in the input layer to the $j^{th}$ neuron in the first layer is denoted by $w_{jk}^{(1)}$. The bias of the $j^{th}$ neuron in the first layer is denoted by $b_j^{(1)}$. The weighted input to the $j^{th}$ neuron and activation of the $j^{th}$ neuron in the first layer are denoted by $z_j^{(1)}$ and $a_j^{(1)}$, respectively. Given $x^{(1)} = [x_1^{(1)}, ..., x_p^{(1)}]^T$. Then, the weighted input $z_j^{(1)}$ from the data input layer to the first layer and activation $a_j^{(1)}$ of the $j^{th}$ neuron in the first layer are given by

$$z_j^{(1)} = \sum_{k=1}^{p} w_{jk}^{(1)} x_k^{(1)} + b_j^{(1)}, j = 1, ..., m_1, \qquad (3.A.1)$$

and

$$a_j^{(1)} = \sigma(z_j^{(1)}), \qquad (3.A.2)$$

respectively.

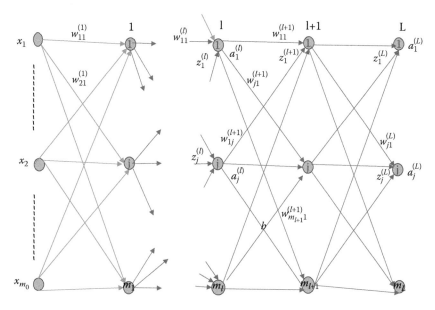

**FIGURE 3.A.1**
Multilayer feedforward pass.

Let

$$Z^{(1)} = \begin{bmatrix} z_1^{(1)} \\ \vdots \\ z_{m_1}^{(1)} \end{bmatrix}, \; W^{(1)} = \begin{bmatrix} w_{11}^{(1)} & \cdots & w_{1p}^{(1)} \\ \vdots & \vdots & \vdots \\ w_{m_1 1}^{(1)} & \cdots & w_{m_1 p}^{(1)} \end{bmatrix}, \; b^{(1)} = \begin{bmatrix} b_1^{(1)} \\ \vdots \\ b_{m_1}^{(1)} \end{bmatrix}, \; a^{(1)} = \begin{bmatrix} a_1^{(1)} \\ \vdots \\ a_{m_1}^{(1)} \end{bmatrix}.$$

Then, in a matrix form, Equations 3.A.1 and 3.A.2 can be written as

$$Z^{(1)} = W^{(1)} x^{(1)} + b^{(1)}, \tag{3.A.3}$$

and

$$a^{(1)} = \sigma(Z^{(1)}), \tag{3.A.4}$$

where

$$\sigma(Z^{(1)}) = \begin{bmatrix} \sigma(z_1^{(1)}) \\ \vdots \\ \sigma(z_{m_1}^{(1)}) \end{bmatrix}.$$

Similarly, for the $l^{th}$ layer, we denote the weight for the connection from the $k^{th}$ neuron in the $(l-1)^{th}$ layer to the $j^{th}$ neuron in the $l^{th}$ layer by $w_{jk}^{(l)}$, the bias of the $j^{th}$ neuron in the $l^{th}$ layer by $b_j^{(l)}$. The weighted input to the $j^{th}$ neuron and activation of the $j^{th}$ neuron in the $l^{th}$ layer are denoted by $z_j^{(l)}$ and $a_j^{(l)}$, respectively. Equations 3.A.1 and 3.A.2 are, in general, expressed as

$$z_j^{(l)} = \sum_{k=1}^{m_{l-1}} w_{jk}^{(l)} a_k^{(l-1)} + b_j^{(l)}, j = 1, \ldots, m_l, \tag{3.A.5}$$

$$a_j^{(l)} = \sigma(z_j^{(l)}). \tag{3.A.6}$$

Define

$$Z^{(l)} = \begin{bmatrix} z_1^{(l)} \\ \vdots \\ z_{m_l}^{(l)} \end{bmatrix}, W^{(l)} = \begin{bmatrix} w_{11}^{(l)} & \cdots & w_{1m_{l-1}}^{(l)} \\ \vdots & \vdots & \vdots \\ w_{m_l 1}^{(l)} & \cdots & w_{m_l m_{l-1}}^{(l)} \end{bmatrix}, b^{(l)} = \begin{bmatrix} b_1^{(l)} \\ \vdots \\ b_{m_l}^{(l)} \end{bmatrix}, a^{(l)} = \begin{bmatrix} a_1^{(l)} \\ \vdots \\ a_{m_l}^{(l)} \end{bmatrix}, \sigma(z^{(l)})$$

$$= \begin{bmatrix} \sigma(z_1^{(l)}) \\ \vdots \\ \sigma(z_{m_l}^{(l)}) \end{bmatrix}.$$

Similarly, in a matrix form, Equations 3.A.5 and 3.A.6 can be rewritten as

$$Z^{(l)} = W^{(l)} a^{(l-1)} + b^{(l)}, \tag{3.A.7}$$

and

$$a^{(l)} = \boldsymbol{\sigma}(Z^{(l)}). \tag{3.A.8}$$

Finally, for the output layer, we have

$$z_j^{(L)} = \sum_{k=1}^{m_{L-1}} w_{jk}^{(L)} a_k^{L-1} + b_j^{(L)}, j = 1, \ldots, m_l, \tag{3.A.9}$$

$$a_j^{(L)} = \sigma(z_j^{(L)}). \tag{3.A.10}$$

The activation of the $j^{th}$ neuron in the output layer is the composition of several activation function in the previous layers and is given by

$$a_j^L = \sigma(\sum_{k=1}^{m_{L-1}} w_{jk}^{(L)} \sigma(\sum_{u=1}^{m_{L-2}} w_{km}^{(L-1)} \sigma(\ldots) + b_k^{(L-1)}) + b_j^{(L)}). \tag{3.A.11}$$

Again, define the following vector and matrix notation:

$$W^{(L)} = \begin{bmatrix} w_{11}^{(L)} & \cdots & w_{1m_{L-1}}^{(L)} \\ \vdots & \vdots & \vdots \\ w_{c1}^{(L)} & \cdots & w_{cm_{L-1}}^{(L)} \end{bmatrix}, b^{(l)} = \begin{bmatrix} b_1^{(l)} \\ \vdots \\ b_c^{(l)} \end{bmatrix}, a^{(L)} = \begin{bmatrix} a_1^{(L)} \\ \vdots \\ a_c^{(L)} \end{bmatrix}, Z^{(L)} = \begin{bmatrix} z_1^{(L)} \\ \vdots \\ z_c^{(L)} \end{bmatrix} \text{ and }$$

$$\sigma(Z^L) = \begin{bmatrix} \sigma(z_1^{(L)}) \\ \vdots \\ \sigma(z_c^{(L)}) \end{bmatrix}.$$

Equations 3.A.9 and 3.A.10 can be rewritten in a matrix form:

$$Z^{(L)} = W^{(L)}a^{(L-1)} + b^{(L)}, \tag{3.A.12}$$

$$a^{(L)} = \sigma(Z^{(L)}). \tag{3.A.13}$$

### 3.A.2   Backpropagation Pass

Neural networks can be viewed as a general class of nonlinear functions from a vector $x$ of input variables to a vector $y$ of output variables where $y = [y_1, ..., y_c]^T$, $y_k$ will be 1 if the pattern $x$ of the input time series belongs to class $k$, otherwise $y_k$ is equal to zero. Our goal is to approximate output variables as accurately as possible using neural networks. Given a set of input variables $x^{(n)}$ and output variables $y^{(n)}$, two cost functions can be defined as

$$E(W) = \frac{1}{2} \sum_{n=1}^{N} \sum_{k=1}^{c} ||y_k^{(n)} - a_k^L(x^{(n)}, W)||^2, \tag{3.A.14}$$

and

$$E(W) = -\sum_{n=1}^{N} \sum_{k=1}^{c} y_k^{(n)} \log(a_k^L(x^{(n)}, W)), \tag{3.A.15}$$

where $N$ is the number of training samples, $W$ are weights in the network, and $a_k^n(x^{(n)}, W)$ is the vector of activations output from the network or can be interpreted as the probability of the $n^{th}$ time series (sample) being the $k^{th}$ class when input data $x^{(n)}$ is given.

The lost functions in both Equations 3.A.14 and 3.A.15 are just a sum over the individual terms on each sample. Therefore, in parameter estimations we will first focus on a single pattern. Then, the lost functions in Equations 3.A.14 and 3.A.15 can be reduced to

$$E_n(x^{(n)}, W) = \frac{1}{2} \sum_{k=1}^{c} ||y_k^{(n)} - a_k^L(x^{(n)}, W)||^2 \qquad (3.A.16)$$

and

$$E_n(x^{(n)}, W) = -\sum_{k=1}^{c} y_k^{(n)} \log(a_k^L(x^{(n)}, W)). \qquad (3.A.17)$$

The weights $W$ can be estimated by minimizing

$$\min_W \ E_n(x^{(n)}, W) = \frac{1}{2} \sum_{k=1}^{c} ||y_k^{(n)} - a_k^L(x^{(n)}, W)||^2 \qquad (3.A.18)$$

or

$$\min_W \ E_n(x^{(n)}, W) = -\sum_{k=1}^{c} y_k^{(n)} \log(a_k^L(x^{(n)}, W)). \qquad (3.A.19)$$

A popular algorithm to minimize $E(W)$ is gradient descent. The idea is to update the weights along the direction of the fastest descent of $E(W)$. We first work on a single training example and then work on the whole dataset by averaging over all training examples. The gradient descent algorithm for updating weights is

$$W^{t+1} = W^t - \eta_t \frac{\partial E_n(x^{(n)}, W))}{\partial W}, \qquad (3.A.20)$$

where $\eta_t \in R_+$ is the learning rate.

The gradient $\dfrac{\partial E_n(x^{(n)}, W)}{\partial W}$ is computed by the back-propagation algorithm. Since $E_n(x^{(n)}, W)$ is a complicated composite function of weights $W$, a key for computing $\dfrac{\partial E_n(x^{(n)}, W)}{\partial W}$ is a chain rule.

We first examine how the changes in $z_j^L$ causes changes in the cost function $E_n(x^{(n)}, W)$. We define the error rate $\delta_j^L$ in the $j^{th}$ neuron in the final output layer of the fully connected layers as

$$\delta_j^L = \frac{\partial E_n(x^{(n)}, W)}{\partial z_j^L}. \qquad (3.A.21)$$

Using Equations 3.A.12, 3.A.18, and 3.A.21, we obtain

$$\delta_j^L = (a_j^L - y_j^{(n)})\frac{\partial a_j^L}{\partial z_j^L}$$

$$= (a_j^L - y_j^{(n)})\sigma'(z_j^L).$$

(3.A.22)

Using Equations 3.A.12, 3.A.19, and 3.A.21, we obtain

$$\delta_j^L = -\frac{y_j^{(n)}}{a_j^L}\sigma'(z_j^L).$$

(3.A.23)

To write Equations 3.A.22 and 3.A.23 in a vector form, we introduce the Hadamard product that is defined as the element-wise product of the two vectors:

$$u \otimes v = \begin{bmatrix} u_1 \\ \vdots \\ u_m \end{bmatrix} \otimes \begin{bmatrix} v_1 \\ \vdots \\ v_m \end{bmatrix} = \begin{bmatrix} u_1 v_1 \\ \vdots \\ u_m v_m \end{bmatrix}.$$

(3.A.24)

Equation 3.A.22 can then be rewritten as

$$\delta^L = (a^L - y^{(n)}) \otimes \sigma'(z^L),$$

(3.A.25)

where

$$a^L = \begin{bmatrix} a_1^L \\ \vdots \\ a_c^L \end{bmatrix}, y^{(n)} = \begin{bmatrix} y_1^{(n)} \\ \vdots \\ y_c^{(n)} \end{bmatrix}, \text{ and } \sigma'(z^L) = \begin{bmatrix} \sigma'(z_1^L) \\ \vdots \\ \sigma'(z_c^L) \end{bmatrix}.$$

Now we calculate the rate of change $\dfrac{\partial E_n}{\partial b_j^{(L)}}$ of the cost with respect to any bias in the network. Since the cost is a function of $z_j^{(L)}$, by definition and chain rule, we have

$$\frac{\partial E_n}{\partial b_j^{(L)}} = \frac{\partial E_n}{\partial z_j^{(L)}}\frac{\partial z_j^{(L)}}{\partial b_j^{(L)}}$$

$$= \delta_j^{(L)}\frac{\partial z_j^{(L)}}{\partial b_j^{(L)}}.$$

(3.A.26)

Using Equation 3.A.9 gives

$$\frac{\partial z_j^{(l)}}{\partial b_j^{(l)}} = 1.$$

(3.A.27)

Substituting Equation 3.A.27 into Equation 3.A.26, we obtain

$$\frac{\partial E_n}{\partial b_j^{(L)}} = \delta_j^{(L)}. \tag{3.A.28}$$

Equation 3.A.28 can be rewritten in a matrix form as

$$\frac{\partial E_n}{\partial b} = \delta. \tag{3.A.29}$$

This shows that the error $\delta_j^{(L)}$ is exactly equal to the rate of change $\frac{\partial E_n}{\partial b_j^{(L)}}$ of the cost with respect to bias.

Finally, we calculate the rate of change of the cost with respect to any weight in the network $\frac{\partial E_n}{\partial w_{jk}^{(L)}}$.

It is clear that the cost involves the weight $w_{jk}^{(L)}$ via the following function:

$$E_n(z_j^{(L)}) = \sum_{k=1}^{m_{l-1}} w_{jk}^{(L)} a_k^{(L-1)} + b_j^{(L)}).$$

Therefore, using chain rule, we obtain

$$\frac{\partial E_n}{\partial w_{jk}^{(L)}} = \frac{\partial E_n}{\partial z_j^{(L)}} \frac{\partial z_j^{(L)}}{\partial w_{jk}^{(L)}} = \delta_j^{(L)} a_k^{(L-1)}. \tag{3.A.30}$$

### 3.A.3 Convolutional Layer

#### 3.A.3.1 *Last convolution Layer: l = L − 1*

We first consider convolutional layer $l = L - 1$, which is connected to the fully connected layer L. It is clear that $E_n$ is a function of the variables $Z_1^L, ..., Z_{m_L}^L$. The error sensitive derivative for the last convolutional layer is

$$\delta_j^{(L-1)} = \frac{\partial E_n(x_n, W)}{\partial z_j^{(L-1)}} = \sum_{m=1}^{m_L} \frac{\partial E_n(x_n, W)}{\partial z_m^{(L)}} \frac{\partial z_m^{(L)}}{\partial z_j^{(L-1)}}$$

$$= \sum_{m=1}^{m_L} \delta_m^{(L)} \frac{\partial z_m^{(L)}}{\partial z_j^{(L-1)}}. \tag{3.A.31}$$

It follows from Equations 3.A.6 and 3.A.9 that

$$\frac{\partial z_m^{(L)}}{\partial z_j^{(L-1)}} = w_{mj}^{(L)} \sigma'(z_j^{(L-1)}). \tag{3.A.32}$$

Substituting Equation 3.A.32 into Equation 3.A.31 gives

$$\delta_j^{(L-1)} = \sigma'(z_j^{(L-1)}) \sum_{m=1}^{m_L} \delta_m^{(L)} w_{mj}^{(L)} \, , j = 1, 2, ..., N_{L-1}. \tag{3.A.33}$$

Again, we calculate the rate of change of the cost with respect to any weight $w_{mj}^{(L-1)}$. By definition and chain rule, we have

$$
\begin{aligned}
\frac{\partial E_n(x_n, W)}{\partial w_{mj}^{(L-1)}(i)} &= \sum_{u=1}^{m_{L-1}} \frac{\partial E_n(x_n, W)}{\partial z_u^{(L-1)}(i)} \frac{\partial z_u^{(L-1)}(i)}{\partial w_{mj}^{(L-1)}(i)} \\
&= \sum_{u=1}^{m_{L-1}} \delta_u^{(L-1)}(i) \frac{\partial z_u^{(L-1)}(i)}{\partial w_{mj}^{(L-1)}(i)},
\end{aligned}
\tag{3.A.34}
$$

where $\delta_u^{(L-1)}(i) = \dfrac{\partial E_n(x_n, W)}{\partial z_u^{(L-1)}(i)}$.

Recall that

$$z_u^{(L-1)}(i) = \sum_{k=1}^{m_{L-1}} w_{uk}^{(L-1)}(i) a_k^{(L-2)} + b_u^{(L-1)}, \quad u = 1, ..., m_{L-1}. \tag{3.A.35}$$

It follows from Equation 3.A.41 that

$$\frac{\partial z_u^{(L-1)}}{\partial w_{mj}^{(L-1)}(i)} = \left\{ \begin{array}{ll} a_j^{(L-2)} & u = m \\ 0 & u \ne m \end{array} \right\}. \tag{3.A.36}$$

Substituting Equation 3.A.36 into Equation 3.A.34 gives

$$\frac{\partial E_n(x_n, W)}{\partial w_{mj}^{(L-1)}(i)} = \delta_m^{(L-1)} a_j^{(L-2)}. \tag{3.A.37}$$

Similarly, we can obtain

$$
\begin{aligned}
\frac{\partial E_n(x_n, W)}{\partial b_j^{(L-1)}} &= \sum_{i=1}^{n_2^{(L-1)}} \frac{\partial E_n(x_n, W)}{\partial z_j^{(L-1)}(i)} \frac{\partial z_j^{(L-1)}(i)}{\partial b_j^{(L-1)}} \\
&= \sum_{i=1}^{n_2^{(L-1)}} \delta_j^{(L-1)}(i), \, j = 1, ..., n_1^L.
\end{aligned}
\tag{3.A.38}
$$

### 3.A.3.2  Last Sub-Sampling Layer

Now we consider the last sub-sampling layer $l = L - 2$. Recall that in the last convolution layer, each feature map is connected to exactly one preceding feature map in the last sub-sampling layer. This implies that

$$z_j^{(L-1)}(i) = w_{jj}^{(L-1)} a_j^{(L-2)}(i) + b_j^{(L-1)}, \qquad (3.A.39)$$

and

$$a_j^{(L-2)}(i) = \sigma(z_j^{(L-2)}(i)). \qquad (3.A.40)$$

Therefore, the sensitivity derivative in the last sub-sampling layer is

$$
\begin{aligned}
\delta_j^{(L-2)}(i) &= \frac{\partial E_n(x_n, W)}{\partial z_j^{(L-2)}(i)} \\
&= \frac{\partial E_n(x_n, W)}{\partial z_j^{(L-1)}(i)} \frac{\partial z_j^{(L-1)}(i)}{\partial z_j^{(L-2)}(i)}.
\end{aligned}
\qquad (3.A.41)
$$

Using Equations 3.A.39 and 3.A.40, we obtain

$$\frac{\partial z_j^{(L-1)}(i)}{\partial z_j^{(L-2)}(i)} = w_{jj}^{(L-1)} \sigma'(z_j^{(L-2)}(i)). \qquad (3.A.42)$$

Substituting Equation 3.A.42 into Equation 3.A.41 gives

$$\delta_j^{(L-2)}(i) = \delta_j^{(L-1)}(i) w_{jj}^{(L-1)} \sigma'(z_j^{(L-2)}). \qquad (3.A.43)$$

Now we calculate the derivative of the cost function with respect to the weights and bias in the last sub-sampling layer. By definition and chain rule, we obtain

$$
\begin{aligned}
\frac{\partial E_n(x_n, W)}{\partial w_{jj}^{(L-2)}} &= \sum_{i'} \frac{\partial E_n(x_n, W)}{\partial z_j^{(L-2)}(i')} \frac{\partial z_j^{(L-2)}(i')}{\partial w_{jj}^{(L-2)}} \\
&= \sum_{i'} \delta_j^{(L-2)}(i') \frac{\partial z_j^{(L-2)}(i')}{\partial w_{jj}^{(L-2)}},
\end{aligned}
\qquad (3.A.44)
$$

where $i' = 1, ..., \left\lfloor \dfrac{i}{2} \right\rfloor$.

Note that

$$z_j^{(L-2)}(i') = w_{jj}^{(L-2)} a_j^{(L-3)}(i') + b_j^{(L-2)}, \qquad (3.A.45)$$

which implies

$$\frac{\partial z_j^{(L-2)}(i')}{\partial w_{jj}^{(L-2)}} = a_j^{(L-3)}(i'). \tag{3.A.46}$$

Substituting Equation 3.A.46 into Equation 3.A.44 gives

$$\frac{\partial E_n(x_n, W)}{\partial w_{jj}^{(L-2)}} = \sum_{i'} \delta_j^{(L-2)}(i') a_j^{(L-3)}(i'). \tag{3.A.47}$$

Similarly, we have

$$\frac{\partial E_n(x_n, W)}{\partial b_j^{(L-2)}} = \sum_{i'} \delta_j^{(L-2)}(i'). \tag{3.A.48}$$

### 3.A.3.3  Convolutional Layer I

Next, we consider a general convolutional layer $l$. Each feature map $j$ in the convolutional layer $l$ is connected to the feature map $j$ in the sub-sampling layer $l + 1$. This implies that

$$z_j^{(l+1)}(i') = z_j^{(l)}(i') w_j^{(l+1)} + b_j^{(l+1)}(i'), \tag{3.A.49}$$

where

$$\begin{aligned} z_j^{(l)}(i') &= a_j^{(l)}(2i'-1) + a_j^{(l)}(2i') \\ &= \sigma(z_j^{(l)}(2i'-1)) + \sigma(z_j^{(l)}(2i')) \end{aligned} \tag{3.A.50}$$

or

$$z_j^{(l)}(i') = \max\left(\sigma(z_j^{(l)}(2i'-1)), \sigma(z_j^{(l)}(2i'))\right). \tag{3.A.51}$$

The sensitive derivative of the cost function with respect to input in the convolutional layer $l$ is

$$\begin{aligned} \delta_j^{(l)}(i) &= \frac{\partial E_n(x_n, W)}{\partial z_j^{(l)}(i)} \\ &= \frac{\partial E_n(x_n, W)}{\partial z_j^{(l+1)}(i')} \frac{\partial z_j^{(l+1)}(i')}{\partial z_j^{(l)}(i)} \\ &= \delta_j^{(l+1)}(i') \frac{\partial z_j^{(l+1)}(i')}{\partial z_j^{(l)}(i)}. \end{aligned} \tag{3.A.52}$$

However, it follows from Equations 3.A.49, 3.A.50, and 3.A.51 that

$$\frac{\partial z_j^{(l+1)}(i')}{\partial z_j^{(l)}(i)} = w_j^{(l+1)}\frac{\partial z_j^{(l)}(i')}{\partial z_j^{(l)}(i)} \tag{3.A.53}$$

$$= w_j^{(l+1)}\sigma'(z_j^{(l)}(i)).$$

Substituting Equation 3.A.53 into Equation 3.A.52 gives

$$\delta_j^{(l)}(i) = \delta_j^{(l+1)}(i')w_j^{(l+1)}\sigma'(Z_j^{(l)}(i)), \tag{3.A.54}$$

where $i' = \left\lfloor \dfrac{i}{2} \right\rfloor$.

Next, we calculate the derivative of the cost function with respect to the weights and bias. Similar to the discussion in the last convolutional layer section, we can obtain

$$\frac{\partial E_n(x_n, W)}{\partial w_{mj}^{(l)}(i)} = \delta_m^{(l)}a_j^{(l-1)}. \tag{3.A.55}$$

and

$$\frac{\partial E_n(x_n, W)}{\partial b_j^{(l)}} = \sum_{i=1}^{n_2^{(l)}}\delta_j^{(l)}(i)\ , j = 1, ..., n_1^l. \tag{3.A.56}$$

### 3.A.3.4  Sub-Sampling Layer l

Finally, we calculate the sensitivity derivative for the general sub-sampling layer $l$ and gradient of the cost function with respect to the weights and bias.

By definition and chain rule, we have

$$\delta_j^{(l)}(i) = \frac{\partial E_n(x_n, W)}{\partial z_j^{(l)}(i)}$$

$$= \sum_{m=1}^{m_{l+1}}\frac{\partial E_n(x_n, W)}{\partial z_m^{(l+1)}(i')}\frac{\partial z_m^{(l+1)}(i')}{\partial z_j^{(l)}(i)} \tag{3.A.57}$$

$$= \sum_{m=1}^{m_{l+1}}\delta_m^{(l+1)}(i')\frac{\partial z_m^{(l+1)}(i')}{\partial z_j^{(l)}(i)}.$$

It follows from Equation 3.22 that

$$\frac{\partial z_m^{(l+1)}(i')}{\partial z_j^{(l)}(i)} = \sum_{v=-k^l}^{k^l} \sigma'(z_j^l(i)) w_{m,j}^{(l+1)}(i'+v). \tag{3.A.58}$$

Substituting Equation 3.A.57 into Equation 3.A.58 yields

$$\delta_j^{(l)}(i) = \sigma'(z_j^l(i)) \sum_{m=1}^{m_{l+1}} \delta_m^{(l+1)}(i') \sum_{v=-k^l}^{k^l} w_{m,j}^{(l+1)}(i'+v). \tag{3.A.59}$$

Again, we calculate the rate of change of the cost with respect to any weights in general sub-sampling layers. By definition and chain rule, we obtain

$$\frac{\partial E_n(x_n, W)}{\partial w_{jj}^{(l)}} = \sum_{i'} \frac{\partial E_n(x_n, W)}{\partial z_j^{(l)}(i')} \frac{\partial z_j^{(l)}(i')}{\partial w_{jj}^{(L-2)}}$$

$$= \sum_{i'} \delta_j^{(l)}(i') \frac{\partial z_j^{(l)}(i')}{\partial w_{jj}^{(l)}}, \tag{3.A.60}$$

where $i' = 1, ..., \left\lfloor \dfrac{i}{2} \right\rfloor$.

Note that

$$z_j^{(l)}(i') = w_{jj}^{(l)} a_j^{(l-1)}(i') + b_j^{(l)}, \tag{3.A.61}$$

which implies

$$\frac{\partial z_j^{(l)}(i')}{\partial w_{jj}^{(l)}} = a_j^{(l-1)}(i'). \tag{3.A.62}$$

Substituting Equation 3.A.62 into Equation 3.A.60 gives

$$\frac{\partial E_n(x_n, W)}{\partial w_{jj}^{(l)}} = \sum_{i'} \delta_j^{(l)}(i') a_j^{(l-1)}(i'). \tag{3.A.63}$$

Similarly, we have

$$\frac{\partial E_n(x_n, W)}{\partial b_j^{(l)}} = \sum_{i'} \delta_j^{(l)}(i'). \tag{3.A.64}$$

## Exercises

Exercise 1. Show that the expansion coefficients $\xi_{ij}$ can be estimated by

$$\xi_{ij} = \int_T x_i(t)\phi_j(t)dt.$$

Exercise 2. Show that the problem (3.9) can then be reduced to

$$\min_C \ (Y - \Phi C)^T(Y - \Phi C) + \lambda C^T JC.$$

Exercise 3. Show that the solution to Equation 3.14 is

$$h = -\left[\int_T G^T(t)\hat{C}^T\hat{C}G(t)dt\right]^{-1}\int_T G^T(t)\hat{C}^T\hat{C}\frac{d^2\phi}{dt^2}dt.$$

Exercise 4. Show that the eigenvalues for the normal QRS complex in Figure 3.3 are $\lambda_1 = -1.30 + 96.9i$ and $\lambda_2 = -1.30 - 96.9i$.

Exercise 5. Show that the eigenvalues for the abnormal QRS complex in Figure 3.4 are $\lambda_1 = 3.48 + 67.26i$ and $\lambda_2 = 3.48 - 67.26i$.

Exercise 6. Show that the transfer function of the response of the cardiac dynamic system underlying the QRS complex to unit-step and unit-impulse input signals are given by $Y(s) = \dfrac{G(s)}{S}$ and $Y(s) = G(s)$, respectively, where $G(s)$ is the transfer function of the dynamic system.

Exercise 7. Consider input signal $x(n) = [2,1,4]$ and impulse response vector $h(n) = [1,0,2]$. Calculate the convolution of $x*h$ using Equation 3.21.

Exercise 8. Consider one time series $\{1,0,-1,2\}$ with the size $W = 4$. We pad zero at each end of the time series ($P = 1$). Consider three filters $\{1,1,-1\}$. Calculate the feature map produced by convolution of the filter with the input time series.

Exercise 9. Show $\dfrac{\partial E_n(x_n, W)}{\partial b_j^{(L-2)}} = \sum_{i'} \delta_j^{(L-2)}(i').$

Exercise 10. Show

$$\int_S x_i(s)\beta(s,t)ds = \int_S x_i(s)\theta^T(s)dsB\eta(t) = x_i^T B\eta(t).$$

# 4

# RNA-Seq Data Analysis

## 4.1 Normalization Methods on RNA-Seq Data Analysis

### 4.1.1 Gene Expression

RNA is a dynamic and diverse biological molecules. Gene expression plays an essential role in various biological processes. The information in the gene is encoded in the sequence of nucleotides. Gene expression transfers the DNA information contained in genes into the production of protein, which is achieved via message RNA (mRNA). (https://www2.stat.duke.edu/courses/Spring04/sta278/refinfo/Gene_Expression.pdf). The mRNA is identical to the one of the DNA sense strands except for replacing T with U and is complimentary to the template strand (see Figure 4.1a).

The transcription process consists of three essential steps. At the first step, the transcription factors interact with the binding sites in the promoters and guide RNA polymerase (see Figure 4.1b). Additional transcription factors may bind to enhancers that are located further upstream or downstream from the gene (see Figure 4.1b). At the second stage, the polymerase II (basis transcription apparatus) must initiate the transcription process, starting at a specific site in the gene which is determined by both the RNA polymerases and transcription factors. At the third stage, the polymerase completes and terminates the transcription of the gene. Termination of transcription can take place heterogeneously over a broad region of the gene.

The regulation of transcription initiation is the key step of gene control. The regulation of transcription initiation mainly consists of the regulation of transcription factors that interact with the cis-acting sequences in the promoter and enhancer, determining the frequency of RNA polymerase binding to the gene transcription.

Following the initiation of the transcription, RNA processing is performed which involves splicing protein coding transcripts, joining exon sequences together, and removing intro sequences (Figure 4.2). Splicing reaction begins with recognizing specific sequences within the RNA at the splice site junctions. Then, cleave the RNA at the exon 3′ end/intron 5′ end border,

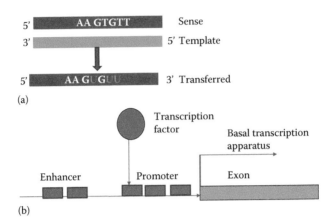

**FIGURE 4.1**
Transcription process.

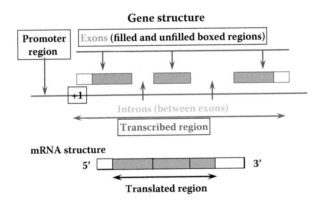

**FIGURE 4.2**
RNA processing.

subsequently, cleave the RNA at the 3′ end of the intron and allow the two exons to be ligated. Figure 4.3 shows that the two mRNA isoforms are generated via splicing. The mRNA 1 is formed by joining exon 1 and exon 3 and the mRNA 2 is formed by joining all three exons: exon 1, exon 2, and exon 3. The final step of the transcription is to modify the 3′ terminus of the RNA by polyadenylation with 200 adenosine residues. This poly A serves as a binding site for the protein taking part in protein synthesis and to protect the RNA from degradation. The transcript is produced in the nucleus. The final processed product (mRNA) is transported from the nucleus to the cytoplasm for its translation to protein with the aid of the ribosomes.

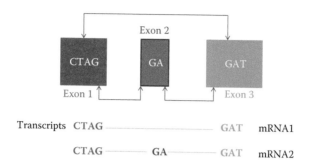

**FIGURE 4.3**
Splicing.

## 4.1.2 RNA Sequencing Expression Profiling

Despite the great progress made in the genetic studies of complex diseases, information on the function of the identified genetic variation in association studies has been limited. Gene expression variation will cause phenotype variation. Gene expression analyses are important sources to study the function of genetic variation and are increasingly acquiring an important role in unraveling the mechanism of complex traits. Next generation sequencing (NGS) technologies have revolutionized advances in the study of the transcriptome. The newly developed deep-sequencing technologies are becoming the platform of choice for gene expression profiling. By measuring messenger RNA levels for all genes in a sample, RNA-seq for expression profiling offers a comprehensive picture of the transcriptome and provides an attractive option to characterize the global changes in transcription (Hong et al. 2013).

RNA-seq technologies have made many significant qualitative and quantitative improvements on gene expression analysis over the microarray (Zyprych-Walczak et al. 2015). First, RNA-seq provides multiple layers of resolutions and transcriptome complexity: the expression at the exon, SNP, and positional level; splicing; post-transcriptional RNA editing across the entire gene; and isoform and allele-specific expression. Second, RNA-seq data have less background noise and a greater dynamic range for detection (Hrdlickova et al. 2017). Third, RNA-seq data allow detecting alternative splicing isoforms.

To generate RNA-seq data, the complete set of mRNA is first extracted from an RNA sample and then shattered and reverse transcribed into a library of cDNA fragments with adaptors attached (see Figure 4.4). These short pieces of cDNA are amplified by polymerase chain reaction and sequenced by machine, producing millions of short reads. These reads are then mapped to a reference genome or reference transcript. The number of reads within a region of interest is used as a measure of abundance. The reads can also be assembled de novo without the genomic sequence to create a transcription map (Li and Xiong 2015).

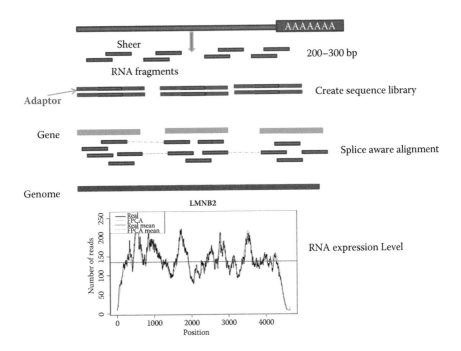

**FIGURE 4.4**
RNA-seq expression profiling.

### 4.1.3 Methods for Normalization

Due to their complexity, the estimation of mRNA abundance from RNA-seq data is not a simple task. The analysis methods are a key to the accurate data interpretation (Li et al. 2015). We often observe many artifacts and biases that affect quantification of expression from RNA-seq data. Therefore, normalization is a crucial step in downstream RNA-seq data analysis, for example, gene differential expression, eQTL analysis, and gene regulatory network analysis, just to name a few. The purpose of normalization is to identify and remove systematic technical differences between samples that take place in the data to ensure that most technical biases have been removed. Many normalization methods for RNA-seq data have been developed. They include raw count (RC), upper quartile (UQ), median (Med), trimmed mean of M-value (TMM) normalization (Robinson and Oshlack 2010), DESeq (Anders and Huber 2010), quantile (Q) (Bolstad et al. 2003), reads per kilobase per million mapped reads (RPKM) (Mortazavi et al. 2008), RNA-seq by expectation-maximization (RSEM) (Li et al. 2011), and Sailfish (Patro et al. 2014).

The major reason for normalization is the sample differences in the total number of aligned reads. We first consider global normalization methods where only a single factor $C_j$ is used to scale the counts of all the genes for each sample $j$. The purpose of the normalization is to make the read counts of different samples with difference sequence depths comparable. Define

$n_{g_j}$: observed count for gene $g$ in sample $j$,

$D_j = \sum_{g=1}^{G} n_{g_j}$: total number of reads in sample $j$, where $G$ is the number of genes and

$C_j$: normalization factor associated with sample $j$.

### 4.1.3.1 Total Read Count Normalization

We assume that read counts are proportional to expression level and sequencing depth. Normalization of the data is to make sure that the scaled total number of reads in each sample is equal. Therefore, the normalization factor should be chosen such that

$$C_1 D_1 = \ldots = C_N D_N = 10^6, \tag{4.1}$$

where $N$ is the total number of samples.

The normalization factor for sample $j$ is calculated by

$$C_j = \frac{10^6}{D_j}. \tag{4.2}$$

The normalized read count for gene $g$ is

$$n_{g_j}^* = C_j n_{g_j} = \frac{n_{g_j}}{D_j} 10^6. \tag{4.3}$$

In other words, the normalized read count is equal to dividing the transcript read count by the total number of reads and rescale the factors to counts per million.

#### Example 4.1

Consider a toy example 4.1 in Table 4.1. There were 3 samples and observed counts of 10 genes. Since the number of genes is small we used 100 to replace 1 million. The normalized read counts of each gene were calculated using Equation 4.3. The normalization factors and normalized data are summarized in Table 4.1.

### 4.1.3.2 Upper Quantile Normalization

The total read count normalization method uses the total read count for each sample to calculate the scalar factor for data normalization and does not consider the distribution of the read count for each sample. To match the between sample distribution of the read counts of the gene, the upper quantile

**TABLE 4.1**

The Raw and Normalized Read Count

|  | Raw Data | | | Normalized Data | | |
|---|---|---|---|---|---|---|
|  | NA19099 | NA18510 | NA18520 | NA19099 | NA18510 | NA18520 |
|  | 26 | 13 | 27 | 11.76 | 12.87 | 15.25 |
|  | 15 | 7 | 15 | 6.79 | 6.93 | 8.47 |
|  | 22 | 15 | 16 | 9.95 | 14.85 | 9.04 |
|  | 3 | 0 | 8 | 1.36 | 0 | 4.52 |
|  | 2 | 0 | 1 | 0.91 | 0 | 0.57 |
|  | 11 | 11 | 17 | 4.98 | 10.89 | 9.6 |
|  | 27 | 6 | 29 | 12.22 | 5.94 | 16.38 |
|  | 85 | 28 | 38 | 38.46 | 27.72 | 21.47 |
|  | 7 | 7 | 3 | 3.17 | 6.93 | 1.69 |
|  | 23 | 14 | 23 | 10.41 | 13.86 | 12.99 |
| Sum | 221 | 101 | 177 | 100 | 100 | 100 |
| Normalization factor | 0.4525 | 0.9901 | 0.565 | | | |

normalization method is developed (Bullard et al. 2010). Let $Q_j^{(p)}$ be the upper quantile ($p^{th}$ percentile) of sample $j$. Replacing $D_j$ by $D_j Q_j^{(p)}$ in Equation 4.1, we obtain

$$C_1 D_1 Q_1^{(p)} = \ldots = C_N D_N Q_N^{(p)} = C. \tag{4.4}$$

If we assume that the constant $C$ is equal to the geometric mean $\left(\prod_{l=1}^{N} D_l Q_l^{(p)}\right)^{\frac{1}{N}}$, then Equation 4.4 is reduced to

$$C_1 D_1 Q_1^{(p)} = \ldots = C_N D_N Q_N^{(p)} = \left(\prod_{l=1}^{N} D_l Q_l^{(p)}\right)^{\frac{1}{N}}$$

$$= \exp\left\{\frac{1}{N} \sum_{l=1}^{N} \log\left(D_l Q_l^{(p)}\right)\right\}. \tag{4.5}$$

Therefore, the normalization factor associated with sample $j$ is calculated by

$$C_j = \frac{\exp\left\{\frac{1}{N} \sum_{l=1}^{N} \log\left(D_l Q_l^{(p)}\right)\right\}}{D_j Q_j^{(p)}}. \tag{4.6}$$

We often take $p = 0.75$ for upper quantile normalization.

### 4.1.3.3 Relative Log Expression (RLE)

The normalization factor ratio $\frac{C_{j_1}}{C_{j_2}}$ can also be viewed as the size ratio. If gene $g$ is not differentially expressed or the samples $j_1$ and $j_2$ are replicates, then the ratio $\frac{n_{gj_1}}{n_{gj_2}}$ of the expected counts of the same gene $g$ in different samples $j_1$ and $j_2$ would be equal to the size ratio $\frac{C_{j_1}}{C_{j_2}}$. In the previous section, we used the total number of reads $D_j$ to estimate the size $C_j$. However, we often observed that a few highly differentially expressed genes have a large influence on the total read counts, which may bias the estimation of total read counts or the estimation of the ratio of expected counts.

If we take the geometric mean across samples $\left(\prod_{l=1}^{N} n_{gl}\right)^{\frac{1}{N}}$ as the gene expression level of a pseudo-reference sample and assume that the expression levels of all genes in the pseudo-reference sample are equal to the geometric mean across the sample, then we have

$$\frac{n_{gj_1}}{n_{gj_2}} = \frac{n_{gj_1}}{\left(\prod_{l=1}^{N} n_{gl}\right)^{\frac{1}{N}}} = \frac{C_{j_1}}{C_{j_2}}. \tag{4.7}$$

If we assume that the normalization factor $C_{j_2}$ of the pseudo-reference sample is equal 1, then Equation 4.7 is reduced to

$$C_{j_1} = \frac{n_{gj_1}}{\left(\prod_{l=1}^{N} n_{gl}\right)^{\frac{1}{N}}}. \tag{4.8}$$

Taking the median on both sides of Equation 4.8, we obtain

$$C_j = \underset{g}{\text{median}} \ \frac{n_{gj}}{\left(\prod_{l=1}^{N} n_{gl}\right)^{\frac{1}{N}}}. \tag{4.9}$$

The normalization factors should multiple to 1, therefore, finally, the normalization factors are defined as

$$C_j = \frac{\exp\left(\frac{1}{N} \sum_{l=1}^{N} \log(C_l)\right)}{C_j}. \tag{4.10}$$

We can easily check

$$\prod_{j=1}^{N} C_j = \frac{\exp\left(\sum_{l=1}^{N} \log(C_l)\right)}{\prod_{j=1}^{N} C_j} = \frac{\prod_{j=1}^{N} C_j}{\prod_{j=1}^{N} C_j} = 1.$$

**TABLE 4.2**

The Raw and Normalized Read Count Using RLE

|  | Raw Data | | | Normalized Data | | |
|---|---|---|---|---|---|---|
|  | **NA19099** | **NA18510** | **NA18520** | **NA19099** | **NA18510** | **NA18520** |
|  | 26 | 13 | 27 | 23.5 | 15.41 | 25.2 |
|  | 15 | 7 | 15 | 13.56 | 8.3 | 14 |
|  | 22 | 15 | 16 | 19.89 | 17.78 | 14.93 |
|  | 3 | 0 | 8 | 2.71 | 0 | 7.47 |
|  | 2 | 0 | 1 | 1.81 | 0 | 0.93 |
|  | 11 | 11 | 17 | 9.94 | 13.04 | 15.87 |
|  | 27 | 6 | 29 | 24.4 | 7.11 | 27.07 |
|  | 85 | 28 | 38 | 76.83 | 33.19 | 35.47 |
|  | 7 | 7 | 3 | 6.33 | 8.3 | 2.8 |
|  | 23 | 14 | 23 | 20.79 | 16.6 | 21.47 |
| Normalization factor | 0.4525 | 0.9901 | 0.565 | 0.9039 | 1.1854 | 0.9333 |

**Example 4.2**

Consider RNA-seq data of three samples in Example 4.1. The vector of normalization factors is $C = [0.9039, 1.1854, 0.9333]$. The normalized data were listed in Table 4.2. We observed some normalization improvement over Table 4.1.

### 4.1.3.4 Trimmed Mean of M-Values (TMM)

It is clear that the total read count heavily depends on a few highly expressed genes. In the previous section, we assume that most genes are not differentially expressed. We need to remove the upper and lower expressed genes before normalization of the data. The trimmed mean of M-value (TMM) procedure starts with defining two quantities: log-fold-changes and absolute intensity. Let $r$ index a reference sample. The log-fold-changes $M_g(j, r)$ (sample $j$ relative to reference sample $r$ for gene $g$) is defined as

$$M_g(j, r) = \log_2\left(\frac{n_{gj}}{D_j}\right) - \log_2\left(\frac{n_{gr}}{D_r}\right). \tag{4.11}$$

The absolute intensity $A_g(j, r)$ of gene $g$ is defined as

$$A_g(j, r) = \frac{1}{2}\left[\log_2\left(\frac{n_{gj}}{D_j}\right) + \log_2\left(\frac{n_{gr}}{D_r}\right)\right]. \tag{4.12}$$

Let $G^*$ be the set of genes with valid $M_g$ and $A_g$ values after trimming the 30% more extreme $M_g$ and 5% more extreme $A_g$. Define the normalization statistic $TMM(j, r)$ as

$$TMM(j,r) = \frac{\sum_{g \in G^*} W_g(j,r) M_g(j,r)}{\sum_{g \in G^*} W_g(j,r)}, \tag{4.13}$$

where $W_g(j, r)$ is the weight, which is defined as the inverse of variance of the log-fold-changes $M_g(j, r)$.

We can show that the variance of $M_g(j, r)$ can be approximated by (Exercise 4.2)

$$var\left( M_g(j,r) \right) = \frac{D_j - n_{gj}}{D_j n_{gj}} + \frac{D_r - n_{gr}}{D_r n_{gr}}.$$

Therefore, the weight $W_g(j, r)$ is given by

$$W_g(j,r) = \left( \frac{D_j - n_{gj}}{D_j n_{gj}} + \frac{D_r - n_{gr}}{D_r n_{gr}} \right)^{-1}. \tag{4.14}$$

The normalization factor for the sample $j$ is defined as

$$C_j = 2^{TMM(j,r)}. \tag{4.15}$$

Finally, to ensure that the normalization factors multiple to 1, the normalization factor $C_j$ is defined as

$$C_j = \frac{\exp\left( \frac{1}{N} \sum_{l=1}^{N} \log(C_l) \right)}{C_j}. \tag{4.16}$$

### 4.1.3.5 RPKM, FPKM, and TPM

The RPKM approach quantifies gene expression from RNA-seq data by normalizing for the total transcript length and the number of sequencing reads. Let $C$ be the number of mappable reads that fell onto the exons of the gene, $N$ be the total number of mappable reads in the experiment, and $L$ be the sum of the exons in base pairs. Then, RPKM (reads per kilobase transcript per million reads) for quantifying gene expression level is given by

$$RPKM = \frac{C}{\frac{N}{10^6} \frac{L}{10^3}} = \frac{10^9 C}{NL}. \tag{4.17}$$

**TABLE 4.3**

(A) The Raw Data and (B) the Normalized Data Using RPKM

|  | Counts (Rep1) | Counts (Rep2) | Counts (Rep3) |
|---|---|---|---|
| **A** |  |  |  |
| A(2kb) | 10 | 14 | 20 |
| B(6kb) | 30 | 40 | 60 |
| C(1kb) | 5 | 10 | 15 |
|  | 45 | 64 | 97 |
|  | 4.5 | 6.4 | 9.7 |
| **B** |  |  |  |
| A(2kb) | 1.111 | 1.094 | 1.031 |
| B(6kb) | 1.111 | 1.042 | 1.031 |
| C(1kb) | 1.111 | 1.563 | 1.546 |

**Example 4.3**

Consider three genes. The simplified data are summarized in Table 4.3A making it easy to read, instead of dividing $10^9$, we divide $NL$ by 10. Table 4.3B shows the normalized gene expression levels of three genes using RPKM.

RPKM is designed for single-end RNA-seq. FPKM (fragments per kilobase million) is designed for paired-end RNA. For paired-end RNA-seq, there are two scenarios: (1) two reads correspond to a single fragment and (2) one read corresponds to a single fragment, if one read in the pair did not map. It is essentially analogous to RPKM, but unlike RPKM where every read corresponds to a single fragment that is sequenced. Rather than using read counts to approximate the relative abundance of transcripts, we use the fragments. In other words, FPKM considers that two reads can map to one fragment (and so it doesn't count this fragment twice) by using Equation 4.17 to calculate the FPKM.

TPM (transcripts per million) is a technology-independent abundance measure. TPM counts the number of copies of the transcript that would exist in a collection of 1 million transcripts. Calculation of TPM is similar to the calculation of RPKM and FPKM. It consists of three steps:

Divide the read counts by the gene length in kilobases which is referred to as reads per kilobase (RPK); summarize all the RPK values in a sample and divide this number by 1 million which is referred to as "per million" scaling factor; and divide the RPK values by the "per million" scaling factor which leads to TPM.

**Example 4.4**

Consider data in Example 4.4 and use step 1 to calculate the RPK values which were listed in Table 4.4. Then, use step 2 to summarize all RPK values for each sample and step 3 to calculate the scaling factor (here divide by 10 instead of dividing by 1,000,000). The scaling factors are also

**TABLE 4.4**

RPK, Normalized by Gene Length

|  | RPK (Rep1) | RPK (Rep2) | RPK (Rep3) |
|---|---|---|---|
| A(2kb) | 5 | 7 | 10 |
| B(6kb) | 5 | 6.667 | 10 |
| C(1kb) | 5 | 10 | 15 |
| Scaling Factor | 1.5 | 2.367 | 3.5 |

**TABLE 4.5**

TPM, Finally Normalized Gene Expression Level

|  | TPM (Rep1) | TPM (Rep2) | TPM (Rep3) |
|---|---|---|---|
| A(2kb) | 3.333 | 2.958 | 2.857 |
| B(6kb) | 3.333 | 2.817 | 2.857 |
| C(1kb) | 3.333 | 4.225 | 4.286 |

listed in Table 4.4. Finally, all the RPK values were divided by the scaling factors. The normalized gene expression level or TPM values were summarized in Table 4.5.

### 4.1.3.6 Isoform Expression Quantification

#### 4.1.3.6.1 Generative Model for Reads with Ungapped Alignment of RNA-Seq Data

A major challenge in expression quantification is that RNA-seq reads often map to multiple genes or isoforms (Li and Dewey 2011). RNA-seq gene expression quantification is to use the RNA-seq data to measure the copy number of transcripts in a sample. To accurately quantify the expression using RNA-seq data, we need to consider read mapping uncertainty, sequencing non-uniformity, and multiple isoforms (Li et al. 2010). In the presence of alternative splicing, different exons may be shared by different numbers of isoforms. Many reads are unlikely to be uniquely aligned to the isoform from which they originate (Liu et al. 2015). Therefore, reads mapped to the shared exons should be deconvoluted. Popular methods for deconvolution of reads are latent generative models.

Figure 4.5 shows the gene structure and isoforms. This example gene contains three exons and three transcripts (isoforms). An additional transcript that includes all annotated exons and junctions is also presented. Assume that $N$ independent and identically distributed reads of length $L$ are sampled. The read sequences are observed data. The $n^{th}$ read sequences are denoted by the $R_n$ random variables. The directed graphical model or Bayesian network (Li et al. 2010) is used to model the read generating process and is shown in

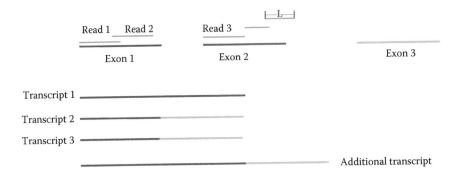

**FIGURE 4.5**
An example of gene structure and isoforms.

Figure 4.6. Consider $M$ isoforms and a vector of parameters $\theta = [\theta_0, \theta_1, ..., \theta_M]$ where $\theta_0$ is associated with a noise isoform and $\theta_i$, $1 \le i \le M$ is associated with the $i^{th}$ isoform. To generate the observed read sequence $R_n$, we assume three hidden variables: isoform variable $G_n$ representing the isoform choice of read $n$, start position variable $S_n$ representing the start position of read $n$, and orientation variable $O_n$ representing what strand is sequenced. Assuming Bayesian network as shown in Figure 4.6, the joint probability of the variables $G_n$, $S_n$, $O_n$, and $R_n$ is

$$P(g, s, o, r \mid \theta) = \prod_{n=1}^{N} P(g_n \mid \theta) P(s_n \mid g_n) P(o_n \mid g_n) P(r_n \mid g_n, s_n, o_n). \tag{4.18}$$

Now we calculate each term in Equation 4.18. The isoform random variable takes values in $[0, M]$. Assume that the proportion of the $i^{th}$ form in the population of isoforms is $\theta_i$. Thus, the probability of selecting the $i^{th}$ form is $\theta_i$, that is, $P(G_n = i \mid \theta) = \theta_i$ and $\sum_{i=0}^{M} \theta_i = 1$ (Figure 4.7a). Let $l_i$, $(i = 0, ..., M)$ be the length of the $i^{th}$ form. The range of the start position of the read is the interval $[1, \max_i l_i]$. Assume that the reads start uniformly across transcripts and that

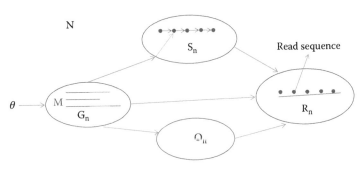

**FIGURE 4.6**
Graphic generative model for the RNA-seq data.

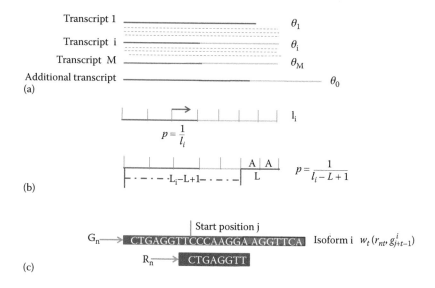

**FIGURE 4.7**
Illustration of a generative model. (a) Probability of selecting the $i^{th}$ form. (b) Probability that the read starts generation with $j$ along the $i^{th}$ transcript. (c) Probability of observing the read sequence, given the selected isoform and start position.

the reads can start at the last position of isoforms and extend into the poly(A) tails. In this scenario, the probability that the read starts generation with $j$ along the $i^{th}$ transcript is $P(S_n = j|G_n = i) = \dfrac{1}{l_i}$ (see Figure 4.7b). If we assume that no Poly(A) tails are presented at the end of mRNA, then we have the probability $P(S_n = j|G_n = i) = \dfrac{1}{l_i - L + 1}$.

There are two approaches to RNA-seq: the standard and non-stranded RNA-seq protocol and strand specific RNA-seq. For the standard RNA-seq protocol, the information on which strand the original mRNA template for sequencing is coming from is lost. For the stranded specific RNA-seq, the sequence reads are generated from the first strand and hence the strand information is retained throughout the sequencing process (Zhao et al. 2015). The orientation variable $O_n$ is a binary variable. Assume that $O_n = 0$ indicates that the orientation of the sequence of read $n$ is the same as that of its parental isoform and $O_n = 1$ indicates that the sequence of $n$ is reverse complemented. Therefore, for a strand-specific protocol, $O_n = 0$, the probability $P(O_n = 0 | G_n \neq 0) = 1$ is assumed. For the non-stranded RNA-seq, no strand information is available. The orientation variable $O_n$ takes either 0 or 1 with equal probability, that is, $P(O_n = 0 | G_n \neq 0) = P(O_n = 1 | G_n \neq 0) = 0.5$.

Finally, we calculate the probability $P(R_n = r_n | G_n = i, S_n = j, O_n = k)$ of observing the read sequence, given the selected isoform and start position.

The calculation of this probability is based on the comparison of the observed sequence and isoforms and their alignment score (see Figure 4.7c). Define the indicator random variable $Z_{nijk}$:

$$Z_{nijk} = \begin{cases} 1 & G_n = i, S_n = j, O_n = k \\ 0 & \text{Otherwise} \end{cases}. \tag{4.19}$$

Then, the probability $P(R_n = r_n \mid G_n = i, S_n = j, O_n = k)$ can be written as $P(R_n = r_n \mid Z_{nijk} = 1)$ and can be calculated by

$$P(R_n = r_n \mid Z_{nijk} = 1) = \begin{cases} \prod_{t=1}^{L} w_t\left(r_{nt}, g_{j+t-1}^i\right), k = 0 \\ \prod_{t=1}^{L} w_t\left(r_{nt}, \bar{g}_{j+t-1}^i\right), k = 1 \end{cases}, \tag{4.20}$$

where $w_t(a, b)$ is a position-specific substitution matrix, $r_{nt}$ is the $t^{th}$ base of the $n^{th}$ read, $g_{j+t-1}^i$ is the $(j + t - 1)^{th}$ base of the $i^{th}$ isoform, $\bar{g}_{j+t-1}^i \bar{g}_{j+t-1}^i$ is the $(j + t - 1)^{th}$ base of the reverse complement of the $i^{th}$ isoform. A substitution matrix assigns each pair of bases a score for match or mismatch (substitution). The value of $w_t(a, b)$ is the score of aligning $a$ and $b$ at the position $t$, where the score is determined by the position-specific substitution matrix. The position-specific substitution matrix or score not only considers the match/mismatch score, but also includes base-call errors, polymorphism, and reference sequence errors. All these factors will lead to the observed substitution between an isoform sequence and an observed read.

If the reads are derived from the noise isoform, then the reads are generated from a position-independent background distribution $\beta$. We set $G_n = 0$, the start position $j = 1$ and orientation variable $O_n = 0$. Equation 4.20 is then reduced to

$$P(R_n = r_n \mid Z_{n010} = 1) = \prod_{t=1}^{L} \beta(r_{nt}). \tag{4.21}$$

### 4.1.3.6.2 Generative Model for Reads with Gapped Alignment of RNA-Seq Data

The limitation of the previous generative model for RNA-seq data is that it cannot deal with the gapped alignment of reads. To overcome this limitation, we modify the previous generative model to allow the gapped alignment of the reads (Nariai et al. 2013). We extend the generative model from the un-gapped alignment of RNA-seq data in Figure 4.6 to the gapped alignment of RNA-seq data in Figure 4.8 where two nodes: $A_n$ node for dealing alignment and $Q_n$ node for incorporating sequence quality scores are included. Let $G_n$, $S_n$, $O_n$, $A_n$, $Q_n$, and $R_n$ be the transcript isoform selection variable, transcript start position variable, read orientation variable, alignment state variable, sequence quality score variable and read variable, respectively. The parameters $\theta$ are

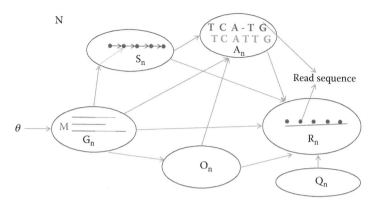

**FIGURE 4.8**
Graphic generative model for the RNA-seq data with gapped alignment.

defined in Section 4.1.3.6.1. The alignment state variable $A_n$ indicates the alignment state (match/mismatch and indel at each alignment position) between read $n$ and reference isoform $i$. The sequence quality score variable $Q_n$ takes Phred quality scores as its values (Nariai et al. 2013). The joint probability of the random variables $G_n$, $S_n$, $O_n$, $A_n$, $Q_n$ and $R_n$ for the directed graphical model in Figure 4.8 is given by

$$P(G_n, S_n, O_n, A_n, Q_n, R_n | \theta) = P(Q_n)P(G_n|\theta)P(S_n|G_n)P(O_n|G_n)P(A_n|G_n, S_n, O_n)$$
$$\times P(R_n|G_n, O_n, S_n, A_n, Q_n).$$

$$(4.22)$$

The probabilities $P(G_n | \theta), P(S_n | G_n)$ and $P(O_n | G_n)$ are defined in Section 4.1.3.6.1. Now we calculate the probability $P(A_n | G_n, S_n, O_n)$. The alignment state variable $A_n$ represents the alignment states at each alignment position between the read $n$ and reference isoform $i$. The alignment states consist of match (M), mismatch (N), insertion/deletion, or gap (G). The alignment between the read $R_n$ and isoform $G_n = i$ starting at the position $j$ is shown in Figure 4.9. The alignment states are denoted by variable $A_n$ (see Figure 4.9). Let $x$ denote the alignment position, $X$ be the total number of alignment positions, $a[x]$ denote the alignment state at the position $x$, and $trans(a[x], a[x + 1])$ denote the probability of transition from the alignment state at the position $x$ to the alignment state at the position $x + 1$. Assume that $trans(a[X], a[X + 1])$ is equal to the probability of alignment ending with $a[X]$. Denote the probability of starting with the state $a[1]$ by $start(a[1])$. The probability $P(A_n | G_n, S_n, O_n)$ can be calculated by (Nariai et al. 2013)

$$P(A_n|G_n, S_n, O_n) = start(a[1])\prod_{x=1}^{X} trans(a[x], a[x + 1]). \qquad (4.23)$$

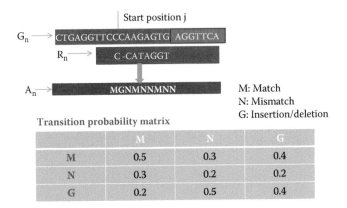

**FIGURE 4.9**
Illustration of alignment between read and isoform.

**Example 4.5**

The selected isoform $G_n$, starting position and read $n$ are shown in Figure 4.9. The alignment between the read $n$ and isoform $G_n$ is

$$C\ C\ A\ A\ G\ A\ G\ T\ G$$
$$C-\ \ C\ AT\ A\ G\ G\ T.$$

The alignment states $A_n$ is

$$MGNMNNMNN.$$

The toy transition probability matrix is shown in Figure 4.9. Assume *start* $(a[1]) = 0.5$. Then, the probability $P(A_n \mid G_n, S_n, O_n)$ is

$$P(A_n \mid G_n, S_n, O_n) = 0.5 \times 0.4 \times 0.5 \times 0.3 \times 0.3 \times 0.2 \times 0.3 \times 0.3 \times 0.2$$

$$= 3.24 \times 10^{-5}.$$

Finally, we will calculate the probability $P(R_n \mid G_n, O_n, S_n, A_n, Q_n)$ that represents the conditional probability of the read $n$, given the isoform choice $G_n = i$, orientation $O_n = k$, start position $S_n = j$, alignment state $A_n = a$, and score of the read. We start with defining the indicator variable $Z_{nikja}$ for the hidden variables:

$$Z_{nikja} = \begin{cases} 1, & G_n = i, O_n = k, S_n = j, A_n = a \\ 0 & \text{otherwise} \end{cases}. \qquad (4.24)$$

Then, we define

$$P(R_n|G_n, O_n, S_n, A_n, Q_n) = P(R_n|Z_{nikja}, Q_n)$$
$$= \prod_{x=1}^{X} emit(r[x], q[x], c[x], a[x]) \tag{4.25}$$

where $x$ denotes the alignment position, $X$ is the total number of alignment position, *emit* is the emission probability defined as

$$emit(r[x], q[x], c[x], a[x]) = \begin{cases} subst(r[x], q[x], c[x]) & a[x] = "M" \\ insert(r[x]) & a[x] = "G" \\ delet(c[x]) & a[x] = "D", \end{cases} \tag{4.26}$$

where "M" denotes the match/mismatch, "G" denotes the insertion, "D" denotes the deletion, *subst* is a substitution matrix taking the Phred quality score into account in defining alignment score where the score aligning the nucleotide $r[x]$ of the observed read over the nucleotide $c[x]$ of the reference sequence at the alignment position $x$ with the aid of the quality score $q[x]$ is assigned, *insert* represents a position independent insertion probability and defined for each nucleotide, and *delet* represents a position independent deletion probability that is defined for each nucleotide.

Again, if the reads are derived from the noise isoform, then the reads are generated from a position-independent background distribution $\beta$. We set $G_n = 0$, the start position $j = 1$ and orientation variable $O_n = 0$. Then, the conditional probability of reads generated from a noise isoform is

$$P(R_n = r_n|Z_{n010} = 1) = \prod_{t=1}^{L} \beta(read[x]), \tag{4.27}$$

where $read[x]$ represents a nucleotide at the position $x$ and $\beta$ is defined as before.

### 4.1.3.6.3 Variational Bayesian Methods for Parameter Estimation and RNA-Seq Data Normalization

RNA-seq data normalization is to estimate the proportion, that is, the parameter vector $\theta = [\theta_0, \theta_1, ..., \theta_M]^T$, of each isoform abundance among the total mRNA, given the observed reads. In other words, we maximize the posterior probability $P(\theta|R_n)$, which can be expressed as

$$P(\theta|R_n) = \frac{P(\theta)P(R_n|\theta)}{P(R_n)}. \tag{4.28}$$

Therefore, a posterior maximum likelihood estimator (MAP) is then given by

$$\theta_{MAP} = \arg\max_{\theta} P(\theta)P(R_n|\theta). \qquad (4.29)$$

However, MAP examines only probability density, rather than mass, which leads to overlooking potentially large contributions to the integral or cumulative distribution (www.cse.buffalo.edu/faculty/mbeal/thesis/beal03_2.pdf.).

The models in Figures 4.6 and 4.8 assume the hidden variables that are organized into the Bayesian network. To estimate the parameters, we also need to use structured hidden variable information. Variational approach for the Bayesian network will be used to estimate the parameters (Nariai et al. 2013). Based on the variation approach, we can introduce the expectation-maximization (EM) algorithm and variational Bayesian EM algorithms for estimation of the parameters in the model. In Appendix 4.A, we show that the EM algorithm for the un-gapped and gapped alignment are given as follows.

**Result 4.1:   EM Algorithm for the Un-Gapped Alignment of RNA-Seq Data**

E step:

$$q^{(u+1)}(Z_n) = \frac{\dfrac{\theta_i^{(u)}}{l_i} P(R_n = r_n | Z_{nijk} = 1, \theta^{(u)})}{\displaystyle\sum_{i'=1}^{M}\sum_{j'=1}^{[1,\max_{i'}l_{i'}]} \dfrac{\theta_{i'}^{(u)}}{l_{i'}} P(R_n = r_n | Z_{ni'j'k} = 1, \theta^{(u)})}, \qquad (4.30)$$

$$n = 1, 2, \ldots, N,$$

and
M step:

$$\theta_i^{(u+1)} = \frac{\displaystyle\sum_{n=1}^{N}\sum_j\sum_k P\left(Z_{nijk} | R_n, \theta^{(u)}\right)}{N}, \; i = 1, 2, \ldots, M \qquad (4.31)$$

**Result 4.2:   EM Algorithm for the Gapped Alignment of RNA-Seq Data**

E step is defined as
E step:

$$q^{(u+1)}(Z_n) = P\left(Z_{nijk} | R_n, Q_n, \theta^{(u)}\right)$$

$$= \frac{P\left(Z_{nijk} = 1, R_n = r_n, Q_n, \theta^{(u)}\right)}{\displaystyle\sum_{(i',k',j',a') \in \pi_n} P\left(Z'_{ni'j'k} = 1, R_n = r_n, Q_n, \theta^{(u)}\right)}, \; n = 1, 2, \ldots, N,$$

$$(4.32)$$

where $\pi_n$ is a set of $i, k, j, a$ for all possible alignments of read $n$ and $(i,k,j,a) \in \pi_n$, and M step is defined as

M step:

$$\theta_i^{(u+1)} = \frac{\sum_{n=1}^{N} \sum_{i,k,j,a} P\left(Z_{nikja} \mid R_n, Q_n, \theta^{(u)}\right)}{N}, \quad i = 1, 2, \ldots, M, \quad (4.33)$$

where $(i,k,j,a) \in \pi_n$, $\pi_n$ is a set of $i, k, j, a$ for all possible alignments of read $n$ and

$$P\left(Z_{nikja} \mid R_n, Q_n, \theta^{(u)}\right) = \frac{P\left(Z_{nikja} = 1, R_n = r_n, Q_n, \theta^{(u)}\right)}{\sum_{(i',k',j',a') \in \pi_n} P\left(Z'_{ni'k'j'a} = 1, R_n = r_n, Q_n, \theta^{(u)}\right)},$$

$$n = 1, 2, \ldots, N.$$

The variational methods for the EM algorithm can be extended to Bayesian learning. The variational Bayesian (VB) for the estimation of the transcript isoform abundance of the un-gapped and gapped alignment of RNA-seq are summarized in Results 4.3 and 4.4, respectively (Appendix 4.A; Nariai et al. 2013).

### Result 4.3: Variational Bayesian Algorithm for the Un-Gapped Alignment of RNA-Seq Data

Step 1. Initialization

For each transcript isoform, set initial value $\alpha_i^{(0)}, i = 1, \ldots, M$ of the parameters in the Dirichlet distribution.

Step 2. **VBE step**

Using the current estimate of $E_\theta[\theta^{(u)}]$, compute the density function

$$q_Z^{(u+1)}(Z) = \prod_n \prod_i \prod_j \prod_k \left(\eta_{nijk}\right)^{Z_{nijk}}, \quad (4.34)$$

where

$$\eta_{n,i,j,k} = \frac{\rho_{n,i,j,k}}{\sum_{(i',j',k') \in \pi_n} \rho_{ni',i',j',k'}} \quad \text{or} \quad (4.35)$$

$$E_Z\left[Z_{nijk}\right] = \eta_{nijk},$$

$$\log \rho_{nijk} = E_\theta\left[\log \theta_i\right] + \log P(S_n = j \mid G_n = i) + \log P(O_n = k \mid G_n = i)$$

$$+ \log P(R_n = r_n \mid Z_{nijk} = 1),$$

where

$$E_\theta\left[\log\theta_i\right] = \psi(\alpha_i) - \psi\left(\sum_{j=1}^M \alpha_j\right), \; \psi(\alpha) = \frac{\dfrac{d\Gamma(\alpha)}{d\alpha}}{\Gamma(\alpha)} \text{ is the digamma function .}$$

Step 3. **VBM step**

Using the current estimate $q_z^{(u+1)}(Z)$, calculate

$$E_\theta\left[\theta_i^{(u+1)}\right] = \frac{\alpha_i^{(u+1)}}{\sum_{i=0}^M \alpha_i^{(u+1)}}, \tag{4.36}$$

where

$$\begin{aligned}
\alpha_i^{(u+1)} &= \alpha_i^{(u)} + \sum_{(n,j,k)\in\pi_n} E_Z\left[Z_{n,i,j,k}\right] \\
&= \alpha_i^{(u)} + \sum_{(n,j,k)\in\pi_n} \eta_{nijk}\,.
\end{aligned} \tag{4.37}$$

Step 4. Stop criterion.

If $\| E_\theta[\theta^{(u+1)} - \theta^{(u)}] \|_2^2 < \varepsilon$, stop. Otherwise, return to Step 2.

**Result 4.4:   Variational Bayesian Algorithm for the Gapped Alignment of RNA-Seq Data**

Step 1. Initialization

For each transcript isoform, set initial value $\alpha_i^{(0)}, i = 1, \ldots, M$ of the parameters in the Dirichlet distribution.

Step 2. **VBE step**

Using the current estimate of $E_\theta[\theta^{(u)}]$, compute the density function

$$q_z^{(u+1)}(Z) = \prod_n\prod_i\prod_j\prod_k \left(\eta_{nijka}\right)^{Z_{nijka}}, \tag{4.38}$$

where

$$\eta_{n,i,j,k,a} = \frac{\rho_{n,i,j,k,a}}{\sum_{(i\prime,j\prime,k\prime)\in\pi_n}\rho_{n\prime,i\prime,j\prime,k\prime,a\prime}} \text{ or} \tag{4.39}$$

$$E_Z\left[Z_{n,i,j,k,a}\right] = \eta_{n,i,j,k,a},$$

$$\log\rho_{n,i,j,k,a} = E_\theta\left[\log\theta_i\right] + \log P(S_n = j|G_n = i) + \log P(O_n = k|G_n = i),$$

$$+\log P(A_n = a|G_n = i, S_n = j, O_n = k) + +\log P(R_n = r_n|Z_{nijk} = 1, Q_n)$$

where

$$E_\theta\left[\log\theta_i\right] = \psi(\alpha_i) - \psi\left(\sum_{j=1}^M \alpha_j\right), \; \psi(\alpha) = \frac{\dfrac{d\Gamma(\alpha)}{d\alpha}}{\Gamma(\alpha)} \text{ is the digamma function.}$$

Step 3. **VBM step**
Using the current estimate $q_z^{(u+1)}(Z)$, calculate

$$E_\theta\left[\theta_i^{(u+1)}\right] = \frac{\alpha_i^{(u+1)}}{\sum_{i=0}^{M}\alpha_i^{(u+1)}}, \tag{4.40}$$

where

$$\begin{aligned}
\alpha_i^{(u+1)} &= \alpha_i^{(u)} + \sum_{(n,j,k,a)\in\pi_n} E_Z\left[Z_{n,i,j,k},a\right] \\
&= \alpha_i^{(u)} + \sum_{(n,j,k,a)\in\pi_n} \eta_{n,i,j,k,a}.
\end{aligned} \tag{4.41}$$

Step 4. Stop criterion.
If $\| E_\theta[\theta^{(u+1)} - \theta^{(u)}]\|_2^2 < \varepsilon$, stop. Otherwise, return to Step 2.

### 4.1.3.7 Allele-Specific Expression Estimation from RNA-Seq Data with Diploid Genomes

#### 4.1.3.7.1 Generative Models

Differences in the expression of two alleles, that is, allele-specific expression (ASE) is often observed. RNA-seq techniques provide a powerful tool for identifying ASE. Figure 4.10 presents a Bayesian network for a read generative model with ASE (most materials in this section are from Nariai et al. 2016). Four variables: isoform choice variable $G_n$, haplotype choice variable $H_n$, start position variable $S_n$, and observed read variable $R_n$ are included in the model. The variable $G_n = i$ is defined as before. It indicates that the read $n$

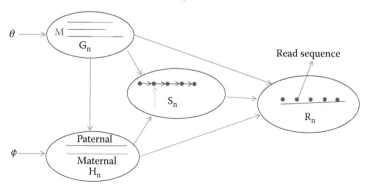

**FIGURE 4.10**
Graphic generative model for ASE.

is generated from the $i^{th}$ isoform. The variable $H_n$ represents what haplotype the read $n$ is generated from. We assume that $H_n = 0$ indicates that the read $n$ is generated from the paternal haplotype and $H_n = 1$ indicates that the read $n$ is generated from the maternal haplotype. The variable $S_n$ represents the start position of the read $n$. In the model we consider two vectors of parameters: $\theta = [\theta_0, \theta_1, ..., \theta_M]^T$ that represent the isoform abundance and $\phi = [\phi_0, \phi_1, ..., \phi_M]^T$ that represent the proportion of the paternal haplotype for each isoform. We assume that $\sum_{i=0}^{M} \theta_i = 1, 0 \le \theta_i \le 1$ and $0 \le \phi_i \le 1$.

Now we calculate the complete likelihood of the data. We assume that the variables in the nodes (Figure 4.10) form a Bayesian network, given the vectors of parameters $\theta$ and $\phi$. Following the rule for the Bayesian network, the likelihood of the data is decomposed into the product of factor probabilities:

$$L(\theta, \phi) = P(G_n, H_n, S_n, R_n | \theta, \phi)$$
$$= P(G_n|\theta)P(H_n|G_n, \phi)P(S_n|G_n, H_n)P(R_n|G_n, H_n, S_n). \tag{4.42}$$

To calculate the likelihood, we first calculate its components. As before, $P(G_n = i | \theta)$ is the probability that the read $n$ is generated from the $i^{th}$ isoform. Thus, $P(G_n = i | \theta) = \theta_i$ is defined.

The probability $P(H_n = 0 | G_n = i, \phi)$ is defined as $\phi_i$ and the probability $P(H_n = 1 | G_n = i, \phi)$ is defined as $1 - \phi_i$. Let $l_{ih}$ be the length of the isoform $i$ of haplotype $h$. The probability $P(S_n = j | G_n = i, H_n = h)$ is defined as the probability that the read $n$ is generated from position $j$, given it is from isoform $i$ and haplotype $h$, which is calculated as $P(S_n = j | G_n = i, H_n = h) = \dfrac{1}{l_{ih} - L + 1}$, where $L$ is the length of read $n$. For the convenience of discussion, we define an indicator variable $Z_{nihj}$ as

$$Z_{nihj} = \begin{cases} 1 & G_n = i, H_n = h, S_n = j \\ 0 & \text{otherwise.} \end{cases} \tag{4.43}$$

Let $\pi_n$ be the set of $(i,h,j)$ tuples for all possible alignments of read $n$. The probability $P(R_n | G_n, H_n, S_n)$ in Equation 4.42 can be expressed as

$$P(R_n|G_n, H_n, S_n) = P(R_n = r_n | Z_{nihj} = 1)$$
$$= \prod_{x=1}^{L} subst(r_n[x], q_n[x], c_{ih}[x]), \tag{4.44}$$

where $subst(r,q,c)$ is the substitution matrix involving quality score, $x$ is the position of the alignment, $r_n[x]$ is the nucleotide of the read $n$ at the position $x$, $q_n[x]$ is the quality score at the position $x$, and $c_{ih}[x]$ is the nucleotide of the cDNA reference sequence of isoform $i$ of haplotype $h$. The substitution matrix *subst(r,q,c)* can be estimated either using the Phred base quality score or using the best alignment of read $n$ over the reference cDNA sequence. It is clear that the likelihood $L(\theta,\phi)$ in Equation 4.42 of the model can fully generate read $n$.

### 4.1.3.7.2 Variational Bayesian Methods for ASE Estimation

Using variational Bayesian methods introduced in Section 4.1.3.6.3, we can estimate the model parameters and ASE. In this section we introduce the variational Bayesian methods for the model parameter and ASE estimation proposed in Nariai et al. (2016).

The variational Bayesian methods require the specification of the prior distribution. Again, we use the Dirichlet distribution

$$P(\theta) = \frac{\Gamma\left(\sum_{i=0}^{M}\alpha_i\right)}{\prod_{i=0}^{M}\Gamma(\alpha_i)}\theta_i^{\alpha_i-1} \tag{4.45}$$

for the prior distribution of $\theta$, where $\Gamma(.)$ is the gamma function. When $\alpha_i - 1 \geq 0$, $\alpha_i - 1$ represents the prior count of reads that are assigned to isoforms; while $\alpha_i - 1 < 0$, it indicates that the isoform abundance is equal to zero. It is easy to show that

$$E[\theta_i] = \frac{\alpha_i}{\sum_{j=0}^{M}\alpha_j}. \tag{4.46}$$

The prior distribution for the parameters $\phi$ is the Beta distribution:

$$P(\phi_i) = \frac{1}{B(\beta_{i1}, \beta_{i2})}\phi_i^{\beta_{i1}-1}(1 - \phi_i)^{\beta_{i2}-1}, \tag{4.47}$$

where $B(\beta_{i1}, \beta_{i2}) = \dfrac{\Gamma(\beta_{i1})\Gamma(\beta_{i2})}{\Gamma(\beta_{i1} + \beta_{i2})}$ is the Beta function, $\beta_{i1} > 0$ and $\beta_{i2} > 0$ are two shape parameters of the Beta distribution. The Beta distribution is often used to model the distribution of allele frequencies in population genetics and serve the conjugate prior probability distribution for the binomial distribution. The parameters $\beta_{i1}$ and $\beta_{i2}$ are used to indicate the prior counts of reads that are assigned to the paternal and maternal haplotypes. We can show

$$E[\phi] = \frac{\beta_{i1}}{\beta_{i1} + \beta_{i2}}. \tag{4.48}$$

Similar to Result 4.3, we can develop the variational Bayesian algorithm for the estimation of the parameters $\theta$ and $\phi$ as follows (Nariai et al. 2016).

**Result 4.5:   Variational Bayesian Algorithm for the Estimation of ASE**

Step 1. Initialization

For each transcript isoform, set initial value $\alpha_i^{(0)} = \alpha_0, i = 1, \ldots, M$ of the parameters in the Dirichlet distribution and initial values $\beta_{i1}^{(0)} = 1, \beta_{i2}^{(0)} = 1, i = 1, \ldots, M$ of the parameters in the Beta distribution.

Step 2. **VBE step**

Using the current estimates of $E_\theta[\theta^{(u)}]$ and $E_\phi[\phi^{(u)}]$, compute the density function

$$q_z^{(u+1)}(Z) = \prod_n \prod_i \prod_h \prod_k \left(\eta_{nihj}\right)^{Z_{nihj}}, \tag{4.49}$$

where

$$\eta_{n,i,h,j} = \frac{\rho_{n,i,h,j}}{\sum_{(i\prime,h\prime,j\prime)\in\pi_n} \rho_{n\prime,i\prime,h\prime,j\prime}} \text{ or}$$

$$E_Z\left[Z_{nihj}\right] = \eta_{nihj},$$

$$\log\rho_{nihj} = \begin{cases} E_\theta\left[\log\theta_i^{(u)}\right] + E_\phi\left[\log\phi_i^{(u)}\right] + \log P(S_n|G_n) & H_n = 0 \\ \quad +\log P(R_n|G_n, H_n, S_n) & \\ E_\theta\left[\log\theta_i^{(u)}\right] + E_\phi\left[\log\left(1 - \phi_i^{(u)}\right)\right] + \log P(S_n|G_n) & H_n = 1, \\ \quad +\log P(R_n|G_n, H_n, S_n) & \end{cases}$$

where

$$E_\theta\left[\log\theta_i^{(u)}\right] = \psi\left(\alpha_i^{(u)}\right) - \psi\left(\sum_{j=1}^M \alpha_j^{(u)}\right), \tag{4.50}$$

$$\psi(\alpha) = \frac{\dfrac{d\Gamma(\alpha)}{d\alpha}}{\Gamma(\alpha)} \text{ is the digamma function,}$$

$$E_\phi\left[\log\phi_i^{(u)}\right] = \psi\left(\beta_{i1}^{(u)}\right) - \psi\left(\beta_{i1}^{(u)} + \beta_{i2}^{(u)}\right), \text{ and} \tag{4.51}$$

$$E_\phi\left[\log\left(1 - \phi_i^{(u)}\right)\right] = \psi\left(\beta_{i1}^{(u)}\right) - \psi\left(\beta_{i1}^{(u)} + \beta_{i2}^{(u)}\right).$$

Step 3. **VBM step**

Using the current estimate $q_z^{(u+1)}(Z)$, calculate

$$E_\theta\left[\theta_i^{(u+1)}\right] = \frac{\alpha_i^{(u+1)}}{\sum_{i=0}^M \alpha_i^{(u+1)}}, \tag{4.52}$$

where

$$\alpha_i^{(u+1)} = \alpha_i^{(u)} + \sum_{(n\prime,i,h\prime,j\prime)\in\pi_n} E_Z\left[Z\prime_{n\prime,i,h\prime,j}\right]$$

$$= \alpha_i^{(u)} + \sum_{(n\prime,i,h\prime,j\prime)\in\pi_n} \eta_{n\prime ih\prime j\prime}. \tag{4.53}$$

Similarly, using the current estimation of $q^{(u+1)}(Z)$, we calculate

$$E_\phi[\phi_i] = \frac{\beta_{i1}^{(u+1)}}{\beta_{i1}^{(u+1)} + \beta_{i2}^{(u+1)}}, \tag{4.54}$$

where

$$\beta_{i1}^{(u+1)} = \beta_{i1}^{(u)} + \sum_{n\prime,t,h\prime=0,j\prime} E_Z\left[Z_{n\prime,t,h\prime,j\prime}\right], \text{ and}$$

$$\beta_{i2}^{(u+1)} = \beta_{i2}^{(u)} + \sum_{n\prime,t,h\prime=1,j\prime} E_Z\left[Z_{n\prime,t,h\prime,j\prime}\right].$$

Step 4. Stop criterion.

If $\| E_\theta[\theta^{(u+1)} - \theta^{(u)}] \|_2^2 < \varepsilon$, stop. Otherwise, return to Step 2.

## 4.2 Differential Expression Analysis for RNA-Seq Data

Identification of significant differential expression between groups is an essential initial step in the RNA-seq data analysis. Unlike microarray gene expression data where a gene expression level is quantified by a real number, a unique feature of biased discrete sequencing reads of RNA-seq data makes differential expression analysis nontrivial. Major challenges faced by differential expression analysis are due to limitations inherent by NGS technologies (Zhang et al. 2014). In general, NGS technologies will generate the bias and errors in the library preparation, in sequence quality and error rate and in abundance measures including the effects of nucleotide composition and the varying length of genes or transcripts. In addition, the combination of technical and biological variation will compromise the estimation of real biological differences between groups.

Two major approaches have been developed to address these challenges. One approach is to accurately model the generating processes and distribution of sequencing read counts across technical replicates and biological samples. Poisson, negative binomial, and beta binomial distributions are used to model the counts of sequence reads (Huang et al. 2015). The second approach is a nonparametric approach that can model the counts of sequence reads as a function of genomic position (Xiong et al. 2014). Both approaches have their merits and limitations and will be introduced in this section.

### 4.2.1 Distribution-Based Approach to Differential Expression Analysis

#### 4.2.1.1 Poisson Distribution

Consider gene $g$ (red color) and a number of other genes (green color) in the genome (see Figure 4.11). Assume that the reads that are assigned to the same gene take the same color in Figure 4.11. Suppose that the total number of reads sampled from the genome is $n$ and $n_g$ is the number of sequenced reads that are independently sampled from gene$g$. Let $p_g$ be the probability that a read is sampled from the gene $g$. Clearly, the number of reads $n_g$ sampled from the gene $g$ follows a binomial distribution:

$$P\left(N_g = n_g\right) = \binom{n}{n_g} p_g^{n_g}\left(1 - p_g\right)^{n-n_g}. \tag{4.55}$$

This will be denoted as $n_g \sim Bin(n,p_g)$.

When the total number of sequenced reads $n$ becomes large, the binomial distribution can be approximated by a Poisson distribution. In fact, the mean of binomial distribution is $\lambda = np_g$ or $p_g = \dfrac{\lambda}{n}$. It follows from Equation 4.55 that

$$\begin{aligned}
P\left(N_g = n_g\right) &= \binom{n}{n_g} p_g^{n_g}\left(1 - p_g\right)^{n-n_g} \\
&= \frac{n(n-1)..(n-n_g+1)}{n_g!}\left(\frac{\lambda}{n}\right)^{n_g}\left(1 - \frac{\lambda}{n}\right)^{n-n_g} \\
&= \frac{n(n-1)..(n-n_g+1)}{n_g!n^{n_g}}\lambda^{n_g}\left(1 - \frac{\lambda}{n}\right)^{\frac{n\lambda}{\lambda}}\left(1 - \frac{\lambda}{n}\right)^{-n_g}.
\end{aligned} \tag{4.56}$$

**FIGURE 4.11**
The reads are assigned to genes.

Note that for large $n$,

$$\frac{n(n-1)\ldots(n-n_g+1)}{n^{n_g}} \approx 1, \tag{4.57}$$

$$\left(1-\frac{\lambda}{n}\right)^{-n_g} \approx 1, \text{ and} \tag{4.58}$$

$$\left(1-\frac{\lambda}{n}\right)^{\frac{n}{\lambda}} \approx e^{-1}. \tag{4.59}$$

Substituting Equations 4.57, 4.58, and 4.59 into Equation 4.56 gives

$$P\left(N_g = n_g\right) \approx \frac{\lambda^{n_g}e^{-\lambda}}{n_g!}. \tag{4.60}$$

This shows that the number of reads follow a Poisson distribution. Therefore, we have

$$E\left[N_g\right] = \lambda \text{ and } Var\left(N_g\right) = \lambda.$$

The log linear model can be used to test the association of the gene with the phenotype or differential expression (Li et al. 2012).

First, we define notations. Consider $k$ genes and $n$ samples. Let $N_{ig}$ be the count of reads for gene $g$ and sample. Define $N_{i.} = \sum_{g=1}^{k} N_{ig}$, $N_{.g} = \sum_{i=1}^{n} N_{ig}$ and $N_{..} = \sum_{i=1}^{n}\sum_{g=1}^{k} N_{ig}$. Assume that $N_{ig}$ follows a Poisson distribution $N_{ig} \sim \text{Poisson}(\lambda_{ig})$. Consider two types of outcome variables: categorical variables with $M$ classes $\{C_1,\ldots,C_M\}$ and quantitative variables $y_i$. Assume that the variable $y_i$ is centered $\left(\sum_{i=1}^{n} y_i = 0\right)$. The log linear model is defined as

1. for the categorical variable

$$\log\lambda_{ig} = \log d_i + \log\beta_g + \sum_{m=1}^{M}\alpha_{gm}I_{(i\in C_m)}, \text{ and} \tag{4.61}$$

2. for the quantitative variable $y_i$

$$\log\lambda_{ig} = \log d_i + \log\beta_g + \alpha_g y_i, \tag{4.62}$$

where $d_i\left(\sum_{i=1}^{n} d_i = 1\right)$ represents the sequence depth for sample $i$, $\beta_g$ represents the expression level of gene $g$, $\alpha_g$ represents the association coefficient of the gene $g$ with the phenotype and $\alpha_{gm}$ represents the association coefficient of the gene $g$ with the class $m$.

Since $N_{ig}$ follows a Poisson distribution, the log-likelihood is given by

$$l(\lambda_{ig}) = \sum_{i=1}^{n}\sum_{g=1}^{k}\left[N_{ig}\log\lambda_{ig} - \lambda_{ig} - N_{ig}!\right]. \tag{4.63}$$

The maximum likelihood estimator of the parameter $\lambda_{ig}$ is

$$\hat{\lambda}_{ig} = \frac{N_{ig}}{N_{...}}. \tag{4.64}$$

We are unable to simultaneously estimate the parameters in the log linear model using likelihood function. Li et al. (2012) proposed two stage procedures for estimation of parameters in the log linear model.

**Stage 1.** We first fit the null model, if no gene is associated with the outcome:

$$\log\lambda_{ig} = \log d_i + \log\beta_g, i = 1,\ldots,n, g = 1,\ldots,k. \tag{4.65}$$

The log likelihood based on model (4.65) is

$$l\left(N_{ig}\mid d_i, \beta_g\right) = \sum_{i=1}^{n}\sum_{g=1}^{k}\left[N_{ig}\left(\log d_i + \log\beta_g\right) - d_i\beta_g - N_{ig}!\right]. \tag{4.66}$$

Setting $\dfrac{\partial l}{\partial d_i} = 0$ and $\dfrac{\partial l}{\partial \beta_g} = 0$ gives

$$\sum_{g=1}^{k}\left(N_{ig} - \beta_g d_i\right) = 0, \tag{4.67}$$

$$\sum_{i=1}^{n}N_{ig} - \left(\sum_{i=1}^{n}d_i\right)\beta_g = 0. \tag{4.68}$$

Recall that $\sum_{i=1}^{n}d_i = 1$, which implies

$$\hat{\beta}_g = N_{.g}. \tag{4.69}$$

Substituting Equation 4.69 into Equation 4.67 yields

$$\hat{d}_i = \frac{N_{i.}}{N_{..}}. \tag{4.70}$$

If we only consider the null model (4.65) with a set $S$ of genes that are not differentially expressed, then Equation 4.67 is changed to

$$\sum_{g\in S}\left(N_{ig} - \beta_g d_i\right) = 0, \text{ which implies}$$

$$\hat{d}_i = \frac{\sum_{g \in S} N_{ig}}{\sum_{g \in S} N_{\cdot g}}. \tag{4.71}$$

Next, we introduce using the goodness-of-fit statistic to determine the set $S$.

**Algorithm 4.1: Algorithm for $S$ Selection**

Step 1. Initialization.

Set initial value: $d^{(0)} = \dfrac{N_{i\cdot}}{N_{\cdot\cdot}}$.

Step 2. Compute the goodness-of-fit statistic for each gene.

$$\text{GOF}_g = \sum_{i=1}^{n} \frac{\left(N_{ig} - d_i^{(u)} N_{\cdot g}\right)^2}{d_i^{(u)} N_{\cdot g}}. \tag{4.72}$$

Step 3. Set $S$ selection.

Select genes whose $\text{GOF}_g$ values are in the $(\varepsilon, 1 - \varepsilon)$ quantile of all $\text{GOF}_g$ values, where $\varepsilon \in (0, 0.5)$ is a fixed constant.

Step 4. Update the estimate of sequence depth.

$$d_i^{(u+1)} = \frac{\sum_{g \in S} N_{ig}}{\sum_{g \in S} N_{\cdot g}}, \quad i = 1, \dots, n.$$

Step 5. Check for convergence.

If $\| d^{(u+1)} - d^{(u)} \|_2 < \delta$ then stop. Otherwise, go to step 2, where $d^{(u)} = [d_1^{(u)}, \dots, d_n^{(u)}]^T$ and $\delta$ is a pre-specified error.

**Stage 2.** At Stage 2, we include an additional term to model the association of genes with the outcome:

$$\log \lambda_{ig} = \log d_i + \log \beta_g + \alpha_g y_i. \tag{4.73}$$

In the case of a quantitative outcome

$$\log \lambda_{ig} = \log d_i + \log \beta_g + \sum_{m=1}^{M} \alpha_{gm} I_{(i \in C_m)} \tag{4.74}$$

In the case of categorical outcome, at Stage 1, we estimated $\hat{d}_i$ and $\hat{\beta}_g$. Define

$$N_{ig}^{(0)} = \log \hat{d}_i + \log \hat{\beta}_g. \tag{4.75}$$

The models (4.73) and (4.74) can be, respectively, reduced to

$$\log\lambda_{ig} = \log\hat{N}_{ig}^{(0)} + \alpha_g y_i, \tag{4.76}$$

and

$$\log\lambda_{ig} = \log\hat{N}_{ig}^{(0)} + \sum_{m=1}^{M} \alpha_{gm} I_{(i \in C_m)}. \tag{4.77}$$

The log-likelihood for the models (4.76) and (4.77) are, respectively, given by

$$l\left(N_{ig} \mid \alpha_g\right) = \sum_{i=1}^{n}\sum_{g=1}^{k}\left[N_{ig}\left(\log N_{ig}^{(0)} + \alpha_g y_i\right) - N_{ig}^{(0)} e^{\alpha_g y_i} - N_{ig}!\right], \tag{4.78}$$

and

$$l\left(N_{ig} \mid \alpha_{gm}\right) = \sum_{i=1}^{n}\sum_{g=1}^{k}\left[N_{ig}\left(\log N_{ig}^{(0)} + \sum_{m=1}^{M}\alpha_{gm} I_{(i \in C_M)}\right)\right.$$
$$\left. - N_{ig}^{(0)} \exp\left(\sum_{m=1}^{M}\alpha_{gm} I_{(i \in C_M)}\right) - N_{ig}!\right]. \tag{4.79}$$

We can obtain (Exercise 9) that

$$\frac{\partial l\left(N_{ig} \mid \alpha_g\right)}{\partial \alpha_g} = \sum_i\left[N_{ig} y_i - N_{ig}^{(0)} y_i \exp\left(y_i \alpha_g\right)\right] = 0$$

$$\frac{\partial l\left(N_{ig} \mid \alpha_{gm}\right)}{\partial \alpha_{gm}} = \sum_i\left[N_{ig} I_{(i \in C_m)} - N_{ig}^{(0)} I_{(i \in C_m)} \exp\left(\sum_m \alpha_{gm} I_{(i \in C_m)}\right)\right] = 0. \tag{4.80}$$

Equation 4.80 are nonlinear equations. There are no analytic expressions for the maximum likelihood of the association coefficients $\alpha_g$ and $\alpha_{gm}$, $m = 1, \ldots, M$. Li et al. (2012) proposed to use a score statistic that does not need estimation of the parameters to test the association or the differential expression. Let $\theta$ be the parameters. Recall that score function is defined as the derivative of the log-likelihood with respect to $\theta$:

$$U(\theta) = \frac{\partial l(\theta)}{\partial \theta}. \tag{4.81}$$

It can be shown that

$$E[U(\theta)] = 0 \tag{4.82}$$

and

$$\text{var}(U(\theta)) = -E\left[\frac{\partial^2 U}{\partial\,\theta\,\partial\,\theta^T}\right] = I(\theta), \tag{4.83}$$

where $I(\theta)$ is the Fisher information matrix.

To test $H_0\text{:}\theta = \theta_0$ against $H_a\text{:}\theta \neq 0$ for a $k$ dimensional restricted parameter $\theta_0$, we define the score test:

$$S(\theta) = U^T(\theta)I^{-1}(\theta)U(\theta). \tag{4.84}$$

Under the null hypothesis, the score statistic $S(\theta)$ is distributed as a central $\chi^2_{(k)}$ distribution. Now we apply the score test to differential expression analysis. First, we consider quantitative outcome.

It can be shown that

$$U(\alpha_g) = \left.\frac{\partial l(\alpha_g)}{\partial\,\alpha_g}\right|_{\alpha_g=0} = \sum_{i=1}^{n}\left(N_{ig}y_i - N_{ig}^{(0)}y_i\right), \tag{4.85}$$

and

$$I(\alpha_g) = -E\left[\frac{\partial^2 l(\alpha_g)}{\partial\,\alpha_g^2}\right]_{\alpha_g=0} = \sum_{i=1}^{n}y_i^2 N_{ig}^{(0)}. \tag{4.86}$$

The score test for association of gene expression with a quantitative trait is

$$S_{qg} = \frac{U^2(\alpha_g)}{I(\alpha_g)} = \frac{\sum_{i=1}^{n}\left(N_{ig}y_i - N_{ig}^{(0)}y_i\right)^2}{\sum_{i=1}^{n}y_i^2 N_{ig}^{(0)}}. \tag{4.87}$$

Next, we consider two or multiple-class outcome. Again, we can show its score function:

$$U(\alpha_{GM}) = \left.\frac{\partial l(\alpha_{GM})}{\partial\,\alpha_{GM}}\right|_{\alpha_{GM}=0} = \left.\begin{bmatrix}\dfrac{\partial l(\alpha_{GM})}{\partial\,\alpha_{g1}}\\ \vdots\\ \dfrac{\partial l(\alpha_{GM})}{\partial\,\alpha_{gm}}\end{bmatrix}\right|_{\alpha_{GM}=0}$$

$$= \begin{bmatrix}\sum_{i\in C_1}\left(N_{ig} - N_{ig}^{(0)}\right)\\ \vdots\\ \sum_{i\in C_M}\left(N_{ig} - N_{ig}^{(0)}\right)\end{bmatrix}. \tag{4.88}$$

The Fisher information matrix for two or multiple-class outcomes is a diagonal matrix

$$
I(\alpha_{GM}) = \begin{bmatrix} \sum_{i \in C_1} N_{ig}^{(0)} & \cdots & 0 \\ \vdots & \vdots & \vdots \\ 0 & \cdots & \sum_{i \in C_M} N_{ig}^{(0)} \end{bmatrix}.
$$

(4.89)

Thus, the score statistics for testing association of gene $g$ with two or multiple-class outcomes is (Exercise 4.11)

$$
\begin{aligned}
S_{cg} &= U^T(\alpha_{GM}) I^{-1}(\alpha_{GM}) U(\alpha_{GM}) \\
&= \sum_{m=1}^{M} \frac{\left[ \sum_{i \in C_m} \left( N_{ig} - N_{ig}^{(0)} \right) \right]^2}{\sum_{i \in C_m} N_{ig}^{(0)}}.
\end{aligned}
$$

(4.90)

In summary, the following score statistics can be used to test differential expressions.

### Result 4.6: Score Test for Differential Expressions with Poisson Distribution of Reads

The score test for association of gene expression with a quantitative trait is

$$
S_{qg} = \frac{U^2(\alpha_g)}{I(\alpha_g)} = \frac{\left[ \sum_{i=1}^{n} \left( N_{ig} y_i - N_{ig}^{(0)} y_i \right) \right]^2}{\sum_{i=1}^{n} y_i^2 N_{ig}^{(0)}}.
$$

(4.91)

Under the null hypothesis that no association of gene $g$ with the quantitative trait, the score statistic $S_{qg}$ is asymptotically distributed as a central $\chi^2_{(1)}$ distribution.

The score statistics for testing association of gene $g$ with two or multiple-class outcome is

$$
S_{cg} = \sum_{m=1}^{M} \frac{\left[ \sum_{i \in C_m} \left( N_{ig} - N_{ig}^{(0)} \right) \right]^2}{\sum_{i \in C_m} N_{ig}^{(0)}}.
$$

(4.92)

Under the null hypothesis with no differential expressions between conditions, the score test $S_{cg}$ is asymptotically distributed as a central $\chi^2_{(M-1)}$ distribution where $M$ is the number of classes.

### 4.2.1.2 Negative Binomial Distribution

#### 4.2.1.2.1 Negative Binomial Distribution for Modeling RNA-Seq Count Data

The Poisson distribution for modeling the count data assumes that the mean and variance are equal. However, in practice due to heterogeneity we often observe that the variance of the count data in the RNA-seq is larger than the mean. This indicates that we need to modify the Poisson distribution for modeling the sequence readers. We consider two variations in the count data: technology variation and biology variation and then develop a statistical model for the count data incorporating two variations (McCarthy 2012).

We first consider the technology variation. We assume that the same biology sample is repeatedly sequenced multiple of times. Let $\pi_{gi}$ be the fraction of all cDNA fragments in sample $i$, sampled from gene $g$. The fraction $\pi_{gi}$ varies from replicate to replicate. Let $\sqrt{\phi_g}$ be the coefficient of variation (CV) of $\pi_{gi}$ between the replicates, defined as the standard deviation of $\pi_{gi}$ divided by its mean. The total number of mapped reads in sample $i$ and the number of reads mapped to gene $g$ are denoted by $N_i$ and $y_{gi}$, respectively. It is clear that

$$E\left[y_{gi}\right] = \mu_{gi} = N_i\pi_{gi}. \tag{4.93}$$

We can show (Appendix 4.B) that the variance of the read counts $y_{gi}$ of gene $g$ is

$$\text{var}(y_{gi}) = \mu_{g_i} + \phi_g\mu_{g_i}^2, \tag{4.94}$$

where $\phi_g$ is often called the dispersion.

Equation 4.94 indicates that the variance of $y_{gi}$ is larger than its mean.

Now we consider a compound Poisson process with mean $Z$. The mean $Z$ itself is a random variable and follows a gamma distribution with shape $\alpha = \dfrac{1}{\phi_g}$ and rate $\beta = \dfrac{1}{\phi_g\mu_{gi}}$. In Appendix 4.B we show that the marginal probability $P(Y = y_{gi})$ for the compound Poisson process is

$$P\left(Y = y_{gi}\right) = \frac{\Gamma(y_{gi} + \alpha)}{y_{gi}!\,\Gamma(\alpha)} \left(\frac{1}{1+\beta}\right)^{y_{gi}} \left(\frac{\beta}{1+\beta}\right)^{\alpha} \tag{4.95}$$

or

$$P\left(Y = y_{gi}\right) = \frac{\Gamma(y_{gi} + \alpha)}{y_{gi}!\,\Gamma(\alpha)} p^{y_{gi}}(1 - p)^{\alpha}, \tag{4.96}$$

where $p = \dfrac{1}{1+\beta}$.

The negative binomial distribution is often written as $NB(\alpha,p)$. Its mean and variance are

$$E[y_{gi}] = \mu_{gi} \text{ and}$$

$$\text{var}(y_{gi}) = \mu_{gi}\left(1 + \phi_g \mu_{gi}\right)$$

### 4.2.1.2.2 Log-Linear Model

Assume the following log-linear model for testing the differential expression (McCarthy et al. 2012):

$$\log \mu_{gi} = x_i^T \gamma_g + \log N_i, \tag{4.97}$$

where $x_i$ is a vector of covariates that specifies the state of the RNA sample $i$ or the treatment condition applied to the RNA sample $i$ and $\gamma_g$ is a $q$ dimensional vector of regression coefficients associated with gene $g$. The log-likelihood is proportional to

$$l\left(\gamma_g, \phi_g\right) \approx \left(\sum_{i=1}^n y_{gi}\right) \log \phi_g$$

$$+ \sum_{i=1}^n \left[ y_{gi}\left(x_i^T \gamma_g + \log N_i\right) - \left(y_{gi} + \frac{1}{\phi_g}\right) \log\left(1 + \phi_g N_i e^{x_i^T \gamma_g}\right)\right]. \tag{4.98}$$

The Newton–Raphson iteration procedure for the maximum likelihood estimation of the parameters $\gamma_g$ (Appendix 4.B) is

$$\gamma_g^{new} = \gamma_g^{old} + \left(XWX^T\right)^{-1} X z_g, \tag{4.99}$$

where

$$X = [x_1, \ldots, x_n], \ z_{gi} = \frac{y_{gi} - \mu_{gi}}{1 + \phi_g \mu_{gi}}, \mu_{gi} = N_i e^{x_i^T \gamma_g}, \ i = 1, \ldots, n, \ z_g = \left[z_{g1}, \ldots, z_{gn}\right]^T \text{ and}$$

$$W = \begin{bmatrix} \dfrac{\mu_{g1} y_{g1} \phi_g}{\left(1 + \phi_g \mu_{g1}\right)^2} & \cdots & 0 \\ \vdots & \vdots & \vdots \\ 0 & \cdots & \dfrac{\mu_{gn} y_{gn} \phi_g}{\left(1 + \phi_g \mu_{gn}\right)^2} \end{bmatrix}.$$

Although in general, the iteration increases the likelihood function, in practice, the iteration might produce solutions that decrease the likelihood. To ensure that the iterations will always produce solutions increasing the likelihood function we can modify the iteration (4.99) via line search. Let $\delta = (XWX^T)^{-1}Xz_g$. Define the new iteration procedure:

$$\gamma_g^{new} = \gamma_g^{old} + \alpha\delta, \tag{4.100}$$

where $\alpha$ is a stepsize constant. Taking $\delta$ as a search direction and using a linear search, we can determine the stepsize $\alpha$ to ensure that iteration always increases the likelihood. To simplify computation of search direction $\delta$, iteration procedure (4.100) can be reduced to

$$\gamma_g^{new} = \gamma_g^{old} + \alpha_0 Xz_g, \tag{4.101}$$

where $\alpha_0$ is a constant to determine the stepsize in each iteration. Again, using a linear search, we can ensure that the likelihood increases as the iterations proceed.

### 4.2.1.2.3 Cox–Reid Adjusted Profile Likelihood

The most statistical inferences are provided by the observed likelihood function. In the presence of nuisance parameters, the statistical inferences are carried out via the adjusted profile likelihood function. In Section 4.2.1.2.2, we assume that the dispersion parameter is known. However, in practice, the dispersion parameters are often unknown and need to be estimated. In this section, we will introduce the Cox–Reid adjusted profile likelihood and present the estimation methods for dispersion parameter $\phi$.

The adjusted profile likelihood (APL) for $\phi_g$ is defined as the penalized log-likelihood

$$APL_g\left(\phi_g, \hat{\gamma}_g\right) = l\left(\phi_g; y_g, \hat{\gamma}_g\right) - \frac{1}{2}\log|I_g|,$$

where

$$I_g = XWX, \text{ and}$$

$$l\left(\phi_g; y_g, \hat{\gamma}_g\right) = \sum_{i=1}^{n}\left[\sum_{l=1}^{y_{gi}-1}\log\left(1+l\phi_g\right) - y_{gi}\log\phi_g - \log y_{gi}! + y_{gi}\log\left(\phi_g\mu_{gi}\right)\right.$$
$$\left. - \left(y_{gi}+\frac{1}{\phi_g}\right)\log\left(1+\phi_g\mu_{gi}\right)\right].$$

The dispersion parameter $\phi_g$ can be estimated by the Newton–Raphson iteration algorithm:

$$
\phi_g^{(new)} = \phi_g^{(old)} - \frac{\left.\dfrac{\partial\, APL(\phi_g)}{\partial\, \phi_g}\right|_{\phi_g^{(old)}}}{\left.\dfrac{\partial^2 APL(\phi_g)}{\partial\, \phi_2^2}\right|_{\phi_g^{old}}}
\tag{4.102}
$$

$$
= \phi_g^{(old)} + \delta,
$$

where $\delta$ is listed in Appendix 4.B.

The Newton–Raphson algorithm usually generates an increase in the likelihood function. However, it is possible that the iteration also generates a decrease in the likelihood function. Again, to ensure that the iterations always produce increases in the likelihood function, we introduce a slack constant. Equation 4.102 can be reduced to

$$
\phi_g^{(new)} = \phi_g^{(old)} + \alpha\delta,
\tag{4.103}
$$

where $\alpha$ is a slack constant. Using a line search we find $\alpha$ such that $APL_g$ $(\phi_g^{(new)}) \ge APL_g(\phi_g^{(old)})$. In both Equations 4.100 and 4.102, we assume that the estimator $\hat{\gamma}_g$ is available. However, to estimate $\gamma_g$, we also need a dispersion parameter $\phi_g$. Therefore, we need to iterate between $\gamma_g$ and $\phi_g$.

### 4.2.1.2.3 Test Statistics

We are interested in testing the significance of coefficients $\gamma_g$ in the log-linear model. The null hypothesis is

$$
H_0 : \gamma_g = 0.
$$

Define test statistics (Huang et al. 2015):

In the absence of overdispersion, the likelihood ratio statistics for testing the null hypothesis $H_0 : \gamma_g = 0$ is defined as

$$
LR_g = 2\left[ l\left(\hat{\gamma}_g, \hat{\phi}_g\right) - l\left(\gamma_g^0, \hat{\phi}_g\right)\right],
\tag{4.104}
$$

where

$$
l\left(\hat{\gamma}_g, \phi_g\right) \approx \left(\sum_{i=1}^{n} y_{gi}\right) \log \phi_g
$$
$$
+ \sum_{i=1}^{n}\left[ y_{gi}\left(\mathbf{x}_i^T \hat{\gamma}_g + \log N_i\right) - \left(y_{gi} + \frac{1}{\phi_g}\right)\log\left(1 + \phi_g N_i e^{\mathbf{x}_i^T \hat{\gamma}_g}\right)\right].
$$

In the presence of overdispersion, the likelihood ration statistics for testing the null hypothesis $H_0 : \gamma_g = 0$ is defined as

$$LR_{gd} = \frac{2\left[l\left(\hat{\gamma}_g, \hat{\phi}_g\right) - l\left(\gamma_g^0, \hat{\phi}_g\right)\right]}{\hat{\phi}_g}. \tag{4.105}$$

Under the null hypothesis $H_0 : \gamma_g = 0$, the test statistics $LR_g$ and $LR_{gd}$ are asymptotically distributed as a central $\chi_{(q)}^2$ and $F_{(q, n-q-1)}$ distribution, respectively.

A natural way to test the differential expression is to compare the difference in expressions between cases and controls. Consider two groups: group $A$ (cases) and group $B$ (controls). Let $n_{gA}$ and $n_{gB}$ be the number of reads for gene $g$ in group $A$ and group $B$, respectively. Let $n_A$ and $n_B$ be the total number of reads in group $A$ and group $B$, respectively, and $n_A - n_{gA}$ and $n_B - n_{gB}$ be the number of reads for the remaining genes in group $A$ and group $B$, respectively. Let $\mu_{gA}$ and $\mu_{gB}$ be the average number of reads for gene $g$ in group $A$ and group $B$, respectively. The null hypothesis of no difference in expression for gene $g$ between group $A$ and group $B$ is

$$H_0 : \mu_{gA} = \mu_{gB}.$$

To test the differential expression, we can use an exact test similar to Fisher's exact test for testing differential expression. Consider a $2 \times 2$ contingency Table 4.6 where $n_g = n_{gA} + n_{gB}$ is the marginal row total for gene $g$, and $n = n_A + n_B$ is the grand total. Assume that $m_A$ subjects are sampled from group $A$ and $m_B$ subjects are sampled from group $B$. Let $n_{gA} = a$ and $n_{gB} = b$. Denote the probability of observing the events $n_{gA} = a$ and $n_{gB} = b$ by $p(a, b)$ for any pair of numbers $a$ and $b$. The P-value, $P_g$ of a pair of observed number of reads $(n_{gA}^*, n_{gB}^*)$, is defined as the sum of all probabilities $p(a, b)$ less than or equal to $p(n_{gA}^*, n_{gB}^*)$ of observing the events $n_{gA}^*$ and $n_{gB}^*$, given that the overall sum is $n_g$ (Gonzalez 2014):

$$P_g = \sum_{\substack{a+b = n_g \\ p(a,b) \le p\left(n_{gA}^*, n_{gB}^*\right)}} p(a, b), \tag{4.106}$$

**TABLE 4.6**

$2 \times 2$ Contingency Table for Gene $g$

|  | Group $A$ | Group $B$ | Total |
|---|---|---|---|
| Gene $g$ | $n_{gA}$ | $n_{gB}$ | $n_g$ |
| Remaining genes | $n_A - n_{gA}$ | $n_B - n_{gB}$ | $n - n_g$ |
| Total | $n_A$ | $n_B$ | $n$ |

where $a,b = 0,1,\ldots,n_g$, and $p(a,b) = p(n_{gA} = a)p(n_{gB} = b)$. In Appendix 4.B, we show that the probabilities $P(n_{gA} = a)$ and $P(n_{gB} = b)$ can be calculated by

$$P\left(n_{gA} = a\right) = \frac{\Gamma\left(a + \dfrac{m_A}{\phi_{gA}}\right)}{a!\,\Gamma\left(\dfrac{m_A}{\phi_{gA}}\right)} \left(\phi_{gA}\mu_{gA}\right)^a \left(\frac{1}{1 + \phi_{gA}\mu_{gA}}\right)^{a + \frac{m_A}{\phi_{gA}}}, \qquad (4.107)$$

and

$$P\left(n_{gB} = b\right) = \frac{\Gamma\left(b + \dfrac{m_B}{\phi_{gB}}\right)}{b!\,\Gamma\left(\dfrac{m_B}{\phi_{gB}}\right)} \left(\phi_{gB}\mu_{gB}\right)^b \left(\frac{1}{1 + \phi_{gB}\mu_{gB}}\right)^{b + \frac{1}{\phi_{gB}}}. \qquad (4.108)$$

## 4.2.2 Functional Expansion Approach to Differential Expression Analysis of RNA-Seq Data

As we discussed in Section 4.2.1, a popular strategy for differential expression analysis with RNA-seq data consists of (1) developing statistical models distribution of read counts, (2) estimating an overall expression level of a gene based on distribution models of read counts, and (3) comparing differences in overall expressions between two conditions to identify differentially expressed genes. However, expression is inherently a stochastic process. Intrinsic and extrinsic classes of noise cause complicated cell-cell variation in gene expression. Varying the usage of splice sites, transcription start sites and polyadenylation sites is further confounded with biological variability. Furthermore, tissues samples contain many different types of cells. The transcription start points, transcription rates, and splicing sites may vary from cell to cell. The observed read counts are position dependent curves. The pattern of the number of reads across the gene is too complicated to accurately and comprehensively model.

Developing nonparametric statistical methods that take distribution of read counts, biological variation, and sequencing technology biases into account will improve differential expression analysis of RNA-seq data. In this section, we introduce the functional expansion approached to differential expression analysis of RNA-seq data in which we model base-level read counts as a random expression function of genomic position and expand random functions in terms of orthogonal functional principal components through Karhunen–Loeve decomposition (Xiong et al. 2014). A formal functional principal component analysis-based statistic for testing differential expressions between two conditions will be introduced.

### 4.2.2.1 Functional Principal Component Expansion of RNA-Seq Data

We first define the read counts in a gene as an expression function of genomic position in the gene. Let $t$ be the position of a nucleotide within a genomic region and $T$ be the length of the genomic region being considered. We consider two conditions: case and control. Assume that $n_A$ cases and $n_G$ controls are sampled and their mRNA are sequenced. We define an expression function $x_i(t)$ of the $i^{th}$ individual in cases as the number of reads of the $i^{th}$ individual which overlaps the nucleotide at the genomic position $t$. We can similarly define the expression function $y_i(t)$ for the $i^{th}$ individual in the controls.

Pooling the expression functions in cases and controls, we can construct orthogonal functional principal components (basis functions) $\{\beta_j(t)\}$. By the Karhunen–Loéve expansion, $X_i(t)$ and $Y_i(t)$ can be expressed as

$$X_i(t) = \sum_{j=1}^{k} \xi_{ij} \beta_j(t) \tag{4.109}$$

and

$$Y_i(t) = \sum_{j=1}^{k} \eta_{ij} \beta_j(t), \tag{4.110}$$

where $\xi_{ij} = \int_T X_i(t)\beta_j(t)dt$ and $\eta_{ij} = \int_T Y_i(t)\beta_j(t)dt$, $\xi_{ij}$ and $\eta_{ij}$ are uncorrelated random variables with zero mean and variances $\lambda_j$ with $\sum \lambda_j < \infty$. Define the averages $\bar{\xi}_j$ and $\bar{\eta}_j$ of the principal component scores $\xi_{ij}$ and $\eta_{ij}$ in the cases and controls. Then, the statistic for testing the differential expression of the gene between cases and controls is defined as

$$T_{FPC} = \frac{1}{\dfrac{1}{n_A} + \dfrac{1}{n_G}} \sum_{j=1}^{k} \frac{\left(\bar{\xi}_j - \bar{\eta}_j\right)^2}{S_j}, \tag{4.111}$$

where $S_j = \dfrac{1}{n_A + n_G - 2}\left[\sum_{i=1}^{n_A}(\xi_{ij} - \bar{\xi}_j)^2 + \sum_{i=1}^{n_G}(\eta_{ij} - \bar{\eta}_j)^2\right]$.

Under the null hypothesis of no differential expression of the gene between cases and controls, the test statistic $T_{FPC}$ is asymptotically distributed as a central $\chi^2_{(k)}$ distribution where $k$ is the number of functional principal components.

### Example 4.6

To assess the accuracy of prediction using the Poisson distribution, a negative binomial distribution and FPCA to fit the read count data of RNA-seq, read and count data of gene *LMNB2* with 465 samples were taken from the GEUVADIS project (http://www.ebi.ac.uk/array express/files/E-GEUV-3/). Poisson distribution, negative binomial distribution and FPCA were used to fit read count data of gene *LMNB2*. Figure 4.12 showed the observed mean count curve, fitted mean count curves over all 465 samples by these three methods: Poisson distribution, negative binomial distribution and FPCA. Figure 4.12 also presented the observed mean overall expression level, fitted mean overall expression level by three methods. We observed that although the observed and fitted mean overall expression level of gene *LMNB2* was very close, the prediction accuracies by three methods were quite different. The Poisson distribution and negative distribution fitted the count data poorly, but the FPCA fitted the data quite well.

**Example 4.7**

To evaluate their performance for testing differential expressions, the FPCA and summary statistic based on the overall expression level were applied to RNA-seq data from the TCGA-Ovarian Cancer Project where a total of 15,104 genes in 233 ovarian cancer tissue samples with 70 drug resistance and 163 drug response samples were sequenced. Figure 4.13 presented expression profiles of gene *CHST10* in the TCGA ovarian cancer dataset. The P-values for testing differential expression of gene *CHST10* between drug resistant and drug response using FPCA and overall expression level (RPKM) were 0.00003 and 0.1674, respectively. Mean difference in overall expression level between drug resistance and drug response was 67.32. From Figure 4.13 we did not observer a large difference in overall expression level of gene *CHST10*, but we observed its difference in expression profiles between drug resistance and drug response samples.

## 4.2.3 Differential Analysis of Allele Specific Expressions with RNA-Seq Data

Identifying differential allele-specific expression (ASE) is of considerable importance in integrating genome and transcriptome data to unravel mechanisms of disease. However, variable ASE has complicated expression patterns. Nucleotide sequence variation will differentially influence the changes in gene expressions at the gene, isoform, exon, and genomic position and allelic levels. In recent years, RNA-seq technology that is a high-throughput sequencing assay provides a powerful tool to measure values, characterize features and unravel mechanisms of gene expressions and hence to reveal complex patterns of gene expressions. Next-generation sequencing (NGS)-based expression profile methods can simultaneously identify genetic polymorphisms and assess quantities of allele specific expression (ASE) and capture a comprehensive picture of transcriptome and discover differentially expressed alleles. RNA-seq is emerging as a major method for ASE study.

Current RNA-seq differential analysis methods are to compare difference in gene expression values. They attempt to accurately estimate gene, isoform

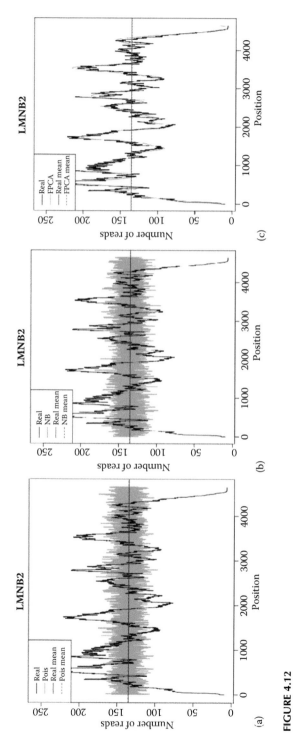

**FIGURE 4.12**

(a) Fitted Poisson distribution, (b) fitted negative binomial distribution, and (c) fitted expression curve using FPCA.

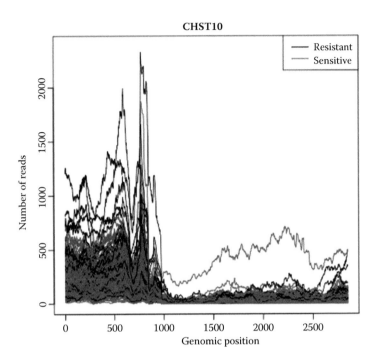

**FIGURE 4.13**
Expression profiles of gene CHST10 in the TCGA ovarian cancer dataset.

and allele specific expression values (Trapnell et al. 2013). Accurate estimation of expression values largely depends on the assumed models. Due to allele-specific alternative splicing, allele-specific transcription start sites, ending sites, allele-specific polyadenylation and natural selection or other unknown mechanisms (Skelly et al. 2011), we observe substantial expression variation across nucleotides and SNPs. Gene expression patterns at multiple layers of levels: gene, exon, isoform, nucleotides and SNPs are extremely complex. To fully model the gene expressions is difficult or impossible.

In this section, we mainly introduce bivariate functional principal component analysis (BFPCA) for testing significant difference in ASE between two conditions which allows levels of ASE to vary across SNPs and can consider complicated patterns of ASE. The existing methods for identifying ASE often require knowledge of the haplotypes. However, the number of haplotypes with NGS data is extremely large. The estimation errors of haplotypes which include rare variants are very high. To avoid haplotype inference, we introduce a vector of ASE functions which are defined as the number of reads corresponding to three genotypes at each SNP across the gene and hence are functions of genomic position. ASE functions are random functions. We extended a single variate FPCA to multi-variate FPCA. We expand random functions in terms of orthogonal functional principal components through Karhunen–Loeve decomposition and develop a novel BFCA-based statistic for testing significant difference in ASE between two conditions. Instead of comparing difference in the overall level of ASE or in the parameters of the models that are used to fit the measured ASE from RNA-seq data, we compare the difference in functional principal component scores of the ASE functions across the experimental conditions. Therefore, the BFPCA-based statistical methods allow levels of ASE to vary across SNPs.

### 4.2.3.1 Single-Variate FPCA for Testing ASE or Differential Expression

We briefly introduce FPCA-based statistics for testing ASE or differential expression. An expression function is defined in Section 4.2.2. We use the pooled expression functions $X_i(t)$ of condition 1 and $Y_i(t)$ of condition 2 to estimate the orthonormal principal component function $\phi_j(t)$ (15). By the Karhunen–Loéve expansion, $X_i(t)$ and $Y_i(t)$ can be expressed as

$$X_i(t) = \sum_{j=1}^{k} \xi_{ij} \beta_j(t) \text{ and}$$

$$Y_i(t) = \sum_{j=1}^{k} \eta_{ij} \beta_j(t),$$

where

$$\xi_{ij} = \int_T X_i(t) \beta_j(t) dt \text{ and}$$

$$\eta_{ij} = \int_T Y_i(t) \beta_j(t) dt,$$

$\xi_{ij}$ and $\eta_{ij}$ are uncorrelated random variables with zero mean and variances $\lambda_j$ with $\sum_j \lambda_j < \infty$. Define the averages $\bar{\xi}_j$ and $\bar{\eta}_j$ of the principal component scores $\xi_{ij}$ and $\eta_{ij}$ in the condition 1 and condition 2. Then, the statistic for testing the differential expression of the gene or ASE between two conditions is defined as

$$T_{FPC} = \frac{1}{\dfrac{1}{n_A} + \dfrac{1}{n_G}} \sum_{j=1}^{k} \frac{\left(\bar{\xi}_j - \bar{\eta}_j\right)^2}{S_j},$$

where $S_j = \dfrac{1}{n_A + n_G - 2} \left[ \sum_{i=1}^{n_A} (\xi_{ij} - \bar{\xi}_j)^2 + \sum_{i=1}^{n_G} (\eta_{ij} - \bar{\eta}_j)^2 \right].$

Under the null hypothesis of no differential expression of the gene or ASE between two conditions, the test statistic $T_{FPC}$ is asymptotically distributed as a central $\chi^2_{(k)}$ distribution.

### 4.2.3.2 Allele-Specific Differential Expression by Bivariate Functional Principal Component Analysis

#### 4.2.3.2.1 Definition of Vectors of Allelic Expression Function

Let $t$ be the genomic position of a SNP within a gene and $T$ be the length of the gene being tested. We consider two conditions: cases and controls. Suppose that the SNP located at genomic position $t$ has two alleles. For the $i^{th}$ individual in cases and the SNP located at the genomic position $t$ we define $x_{i1}(t)$ and $x_{i2}(t)$ be the counts of reads from the alleles $A$ and $a$ of the SNP located at $t$, respectively. We define $x_i(t) = [x_{i1}(t), x_{i2}(t)]^T$. Similarly, we can define $y_j(t)$ for the $j^{th}$ individual in controls.

#### 4.2.3.2.2 Formulation of ASE Analysis of RNA-Seq Data as Bivariate Functional Principal Component Analysis

Let $\beta(t) = [\beta_1(t), \beta_2(t)]^T$ be a vector of basis functions and define an inner product as the sum of the component-wise inner product:

$$f_i = \langle x_i, \beta \rangle = \langle x_{i1}, \beta_1 \rangle + \langle x_{i2}, \beta_2 \rangle$$

$$= \int_T x_{i1}(t)\beta_1(t)dt + \int_T x_{i2}(t)\beta_2(t)dt = \int_T = \beta^T(t)x_i(t)dt.$$

By the formula for the variance of stochastic integral, we can calculate the variance of $f_i$:

$$\text{var}(f_i) = \int_T \int_T \beta^T(s)R(s,t)\beta(t)dsdt,$$

where $R(s,t)$ is the matrix-valued covariance function. It can be written as $[R(s,t)]_{ij} = R_{ij}(s,t)$, that is, its $ij^{the}$ element is the covariance function of $x_i(t)$ and $x_j(t)$. Let $\|\beta\|_2 = \int_T \beta^T(t)\beta(t)dt$. The functional principal component is to seek maximizing the variance of $f_i$ subject to constraint $\|\beta\|_2 = 1$:

$$\max_{\beta} \quad \int_T \int_T \beta^T(s)R(s,t)\beta(t)dsdt$$

$$\text{s.t.} \quad \int_T \beta^T(t)\beta(t)dt = 1. \tag{4.112}$$

By the Lagrange multiplier, we reformulate the constrained optimization problem (4.112) into the following non-constrained optimization problem:

$$\max_{\beta} \quad F(\beta, \lambda) = \int_T \int_T \beta^T(s)R(s,t)\beta(t)dsdt + \lambda\left(1 - \int_T \beta^T(t)\beta(t)dt\right).$$

Setting the differential of $F(\beta,\lambda)$ equal to zero, we have

$$2\int_T\int_T\beta^T(s)R(s,t)\,\partial\beta(t)dsdt - 2\lambda\int_T\beta^T(t)\beta(t)dt = 0. \qquad (4.113)$$

Since Equation 4.113 must hold for all values of $\partial\beta(t)$, it follows that we must have the following integral eigenequation:

$$\int_T R(t,s)\beta(s)ds = \lambda\beta(t), \qquad (4.114)$$

for all values of $t$. This is the optimality condition.

Suppose there are $n$ amples observed across some interval, then we can write all observations at genomic position $t$ as a matrix

$$X(t) = [x_1(t), x_2(t)] = \begin{bmatrix} x_{11}(t) & x_{12}(t) \\ \vdots & \vdots \\ x_{n1}(t) & x_{n2}(t) \end{bmatrix}.$$

Writing each sample's scalar component function in basis expansion form,

$$X(t) = [Z_1\Phi(t), Z_2\Phi(t)] = [Z_1, Z_2]\begin{bmatrix} \Phi(t) & 0 \\ 0 & \Phi(t) \end{bmatrix} = Z(I_2 \otimes \Phi(t)),$$

where $Z_j \begin{bmatrix} Z_{1j}^1 & \cdots & Z_{1j}^k \\ \vdots & \ddots & \vdots \\ Z_{nj}^1 & \cdots & Z_{nj}^k \end{bmatrix}$ is a matrix of coefficients and $\Phi(t) = \begin{bmatrix} \phi_1(t) \\ \vdots \\ \phi_k(t) \end{bmatrix}$ is a vector

of basis functions, $I_2$ is a two-dimensional identify matrix and $\otimes$ denotes the Kronecker product of two matrices.

The variance estimation is then

$$R(s,t) \approx \frac{1}{n}X(s)^T X(t) = \frac{1}{n}(I_2 \otimes \Phi(s))^T Z^T Z(I_2 \otimes \Phi(t)).$$

Suppose we approximate eigenfunction $\beta(t)$ by the same basis functions $\Phi(t)$,

$$\beta(t) = [\beta_1(t), \beta_2(t)]^T = [B_1^T\Phi(t), B_2^T\Phi(t)]^T = (I_2 \otimes \Phi^T(t))B, \text{ where } B = [B_1^T, B_2^T]^T.$$

Plugging them into the optimality condition (4.114), we get

$$\frac{1}{n}(I_2 \otimes \Phi^T(t))Z^T Z B = \lambda(I_2 \otimes \Phi^T(t))B, \qquad (4.115)$$

If $\Phi(t)$ is orthonormal. Sine equation (4.115) holds for every $t$, Equation 4.114 implies that

$$\frac{1}{n} Z^T Z B = \lambda B,$$

or

$$\frac{1}{n} \begin{bmatrix} Z_1^T Z_1 & Z_1^T Z_2 \\ Z_2^T Z_1 & Z_2^T Z_2 \end{bmatrix} \begin{bmatrix} B_1 \\ B_2 \end{bmatrix} = \lambda \begin{bmatrix} B_1 \\ B_2 \end{bmatrix}. \tag{4.116}$$

Let $B^j \begin{bmatrix} B_1^j \\ B_2^j \end{bmatrix}$ be the $j^{th}$ eigenvector of eigenequation (4.116). Then, eigenfunction $\beta_j(t)$ is given by

$$\beta_j(t) \begin{bmatrix} \Phi^T(t) B_1^j \\ \Phi^T(t) B_2^j \end{bmatrix} = \begin{bmatrix} \beta_{j1}(t) \\ \beta_{j2}(t) \end{bmatrix}. \tag{4.117}$$

Therefore the coefficients of eigenfunction $\beta(t)$ can be computed from $Z^T Z / n$, if the basis functions $\Phi(t)$ are orthonormal.

Let $x_i(t)$ be expanded in terms of eigenfunctions:

$$x_i(t) = \sum_{j=1}^{J} \xi_{ij} \beta_j(t).$$

Then, the functional principal component score of the allelic expression function of the $i^{th}$ individual can be calculated by

$$\xi_{ij} = \langle x_i(t), \beta_j(t) \rangle = \sum_{k=1}^{2} \int x_{ik}(t) \beta_{jk}(t) dt = \xi_{ij}^{(1)} + \xi_{ij}^{(2)}, \tag{4.118}$$

where $\xi_{ij}^{(1)} = \int_T x_{i1}(t) \beta_{j1}(t) dt$ and $\xi_{ij}^{(2)} = \int_T x_{i2}(t) \beta_{j2}(t) dt$.

In summary, the algorithms for computing functional principal component scores are given as follows.

For each individual, we define two-dimensional allelic expression curves. For the major allele and minor alleles, we can define allelic expression curves separately. For the major allele, we have expansion: $X_1(t) = Z_1 \Phi(t)$ and for the minor allele, we have expansion: $X_2(t) = Z_2 \Phi(t)$.

Define the matrix $Z = [Z_1 \ Z_2]$. Solving the eigenequation (4.116), we obtain the eigenvectors. Using these eigenvectors, we obtain the following eigenfunctions:

$$\beta_j(t) = \begin{bmatrix} \Phi^T(t) B_1^j \\ \Phi^T(t) B_2^j \end{bmatrix} = \begin{bmatrix} \beta_{j1}(t) \\ \beta_{j2}(t) \end{bmatrix}.$$

Using Equation 4.118, we obtain

$$\xi_{ij} = \xi_{ij}^{(major)} + \xi_{ij}^{(minor)}.$$

### 4.2.3.2.3 Test Statistics

Similar to the previous section, we define the statistic for testing differential ASE by comparing functional principal component scores. Define the averages $\bar{\xi}_j$ and $\bar{\eta}_j$ of the principal component scores $\xi_{ij} = \xi_{ij}^{(major)} + \xi_{ij}^{(minor)}$ and $\eta_{ij} = \eta_{ij}^{(major)} + \eta_{ij}^{(minor)}$ which are calculated by Equation 4.118 in the condition 1 and condition 2. Then, the statistic for testing the differential ASE between two conditions is defined as

$$T_{BFPC} = \frac{1}{\frac{1}{n_A} + \frac{1}{n_G}} \sum_{j=1}^{k} \frac{\left(\bar{\xi}_j - \bar{\eta}_j\right)^2}{S_j},$$

where $S_j = \frac{1}{n_A + n_G - 2} \left[ \sum_{i=1}^{n_A}(\xi_{ij} - \bar{\xi}_j)^2 + \sum_{i=1}^{n_G}(\eta_{ij} - \bar{\eta}_j)^2 \right]$ and $k$ is the number of functional principal components.

Under the null hypothesis of no difference in ASE between two conditions, the test statistic $T_{BFPC}$ is asymptotically distributed as a central $\chi^2_{(k)}$ distribution.

### 4.2.3.3 Real Data Application

#### 4.2.3.3.1 Data Set

To illustrate its application, the proposed BFPCA was applied to a schizophrenia RNA-seq study that sequenced mRNA in 31 schizophrenia and 26 normal samples and RNA-seq data of ovarian cancer with 233 high-grade serious ovarian adenocarcinomas tumor samples (163 samples sensitive to treatments and 70 samples resistant to treatment) from TCGA. In the schizophrenia RNA-seq study, the RNA samples were from postmortem brain tissues; the brain region is the anterior cingulate cortex, also called Brodmann's Area 24. The RNA-seq data were produced on the IlluminaHiSeq platform. Datasets were preprocessed using the SeqWare Pipeline project. A total of 11,266 genes were analyzed after filtering out genes with number of SNPs less than 10. In the TCGA Ovarian Cancer Project, the RNA-seq data were produced on the IlluminaHiSeq platform where 233 ovarian cancer patients with raw Bam dataset are obtained to generate the expression curve profile, among which 163 patients are sensitive to chemotherapy, and 70 are chemo-resistant. Platinum status is defined as resistant if the patient recurred within six months or else sensitive if the platinum free interval is six months or greater, and there is no evidence of progression or recurrence, and the follow-up interval is at least six months from the date of last primary platinum treatment defined by the TCGA committee (2011). A total of 15,104 genes were analyzed after filtering out genes with number of SNPs less than 10.

### 4.2.3.3.2 Differential ASE Analysis

Read counts from two alleles across all SNPs in each gene form two expression profiles that are referred to as ASE curves or functions when SNPs are densely distributed across the gene. We define a two-dimensional vector of ASE functions for each gene which takes two expression curves as its two component functions. For schizophrenia and normal samples (or ovarian treatment sensitive and resistant tumor samples), we defined vectors of ASE functions separately. When differences in ASE between two conditions is significant, we can observe significant difference in ASE curves. Therefore, testing for differences in ASE between two conditions can be reformulated as testing difference in ASE curves between two conditions. Since dimension of the ASE curve is very high we use FPCA to reduce the dimension of ASE curve. By comparing differences in functional principal component scores of the ASE curves between two conditions, we test for significant difference in ASE between them. The total number of genes being tested in the schizophrenia study and ovarian cancer study are 11,226 and 1871, respectively. The thresholds to declare genome-wide significance after the Bonferroni correction for schizophrenia and ovarian cancer studies are $4.44 \times 10^{-6}$ and $2.67 \times 10^{-5}$, respectively. We identified a total of 16 genes showing significant differences in ASE between schizophrenia and normal samples (Table 4.7). In the ovarian cancer study, no differential ASE gene reached whole genome significance. We listed P-values of the top 10 differential ASE genes for testing differential ASE between treatment sensitive and resistant samples in Table 4.8.

### 4.2.3.3.3 Features of Differential ASE

Differential ASE consists of two parts. The first part is ASE. In other words, expression is allele dependent. The second part is presence of differences in ASE between two conditions. We observed differences in read counts between two alleles across SNPs in each gene within groups (schizophrenia or normal; treatment sensitive or resistant) and difference in read counts between two conditions for each allele across the SNPs in differential ASE. Since NGS data can generate many haplotypes, measuring expressions of parental haplotypes of individuals in the population is not easy or impossible when the number of individuals is large. We quantify ASE by counting the number of reads for each allele across SNPs in each gene.

Figures 4.14 and 4.15 showed the gene expression curves for major and minor alleles in gene *SLC13A3* with significant difference in ASE between schizophrenia and normal samples, respectively. Figures 4.16 and 4.17 plotted the RNA seq expression curves for major and minor alleles in gene *PARP14* with differences in ASE between treatment resistant and sensitive in

**TABLE 4.7**

P-Values of Top 16 Differential ASE Genes for Testing Differential
Expression between Schizophrenia and Normal Samples and ASE
in Schizophrenia or Normal Samples

| | P-value | | |
|---|---|---|---|
| **Gene** | **Differential ASE** | **Normal** | **Schizophrenia** |
| SLC13A3 | 1.43E-08 | 1.71E-02 | 1.16E-07 |
| TTF1 | 4.43E-08 | 2.86E-06 | 2.99E-13 |
| AC019205 | 1.15E-07 | 4.82E-07 | <1.0E-17 |
| PSMA5 | 2.14E-07 | 1.22E-03 | 1.17E-09 |
| ANXA4 | 3.02E-07 | 2.84E-03 | 9.66E-06 |
| LRRTM3 | 4.12E-07 | 3.46E-04 | 5.39E-10 |
| FAM123A | 5.82E-07 | 4.56E-06 | <1.0E-17 |
| MAGI1 | 6.96E-07 | 1.70E-08 | <1.0E-17 |
| XRN2 | 8.70E-07 | 6.22E-06 | 1.87E-14 |
| KIF2A | 1.49E-06 | 3.25E-08 | 4.22E-15 |
| NUDT15 | 1.85E-06 | 4.71E-03 | 1.09E-11 |
| ZNF815 | 1.87E-06 | 8.52E-05 | 6.81E-11 |
| KALRN | 2.09E-06 | 2.34E-07 | 1.47E-12 |
| RRP15 | 2.46E-06 | 2.98E-07 | 1.89E-15 |
| ST3GAL5 | 3.21E-06 | 5.25E-06 | 3.20E-13 |
| CTNNAL1 | 3.67E-06 | 3.50E-06 | 8.97E-13 |

**TABLE 4.8**

P-Values of Top 10 Differential ASE Genes for Testing Differential
ASE between Treatment Sensitive and Resistant Samples and ASE
in Treatment Resistant or Sensitive Samples in Ovarian Cancer

| | P-value | | |
|---|---|---|---|
| **Gene** | **Differential ASE** | **Resistant** | **Sensitive** |
| PARP14 | 1.55E-04 | <1.00E-17 | 1.51E-04 |
| DMTF1 | 4.13E-04 | <1.00E-17 | 2.44E-08 |
| DDX60L, SNORA51 | 7.17E-04 | <1.00E-17 | 4.81E-08 |
| SULT6B1, CEBPZ | 8.07E-04 | <1.00E-17 | 3.38E-07 |
| CEP89 | 1.13E-03 | <1.00E-17 | 1.58E-06 |
| PDE4DIP | 2.09E-03 | <1.00E-17 | 1.14E-05 |
| RNPEP,ELF3 | 2.44E-03 | <1.00E-17 | 1.78E-15 |
| RP4-758J18.9 | 2.67E-03 | 3.23E-02 | 2.59E-01 |
| RQCD1 | 2.68E-03 | 4.33E-15 | 1.74E-03 |
| COL18A1, SLC19A1 | 2.82E-03 | 3.00E-13 | 1.38E-05 |

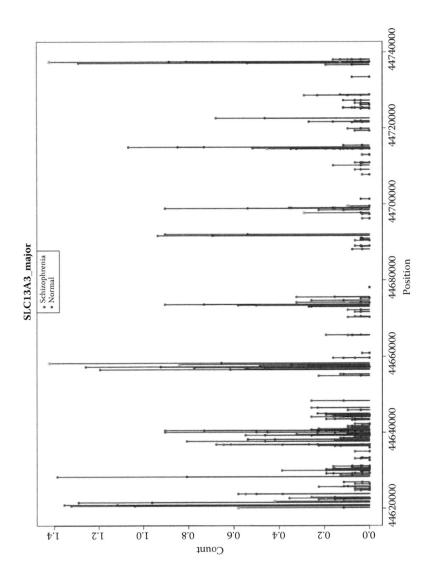

**FIGURE 4.14**
Differential expression of major allele in gene *SLC13A3*.

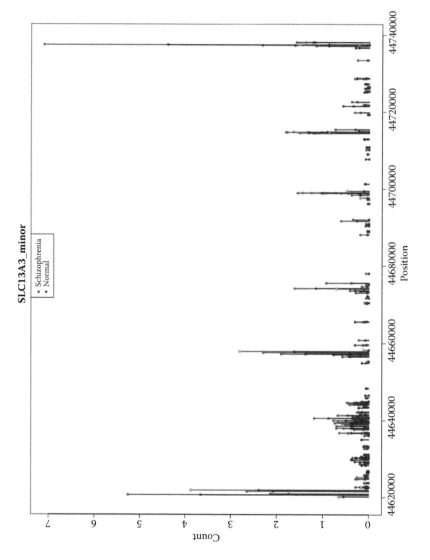

**FIGURE 4.15**
Differential expression of minor allele in gene *SLC13A3*.

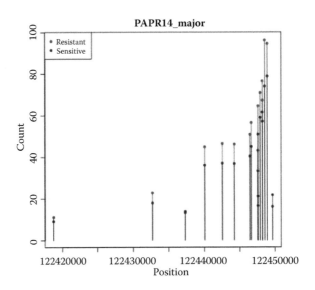

**FIGURE 4.16**
Differential expression of major allele in gene *PAPR14*.

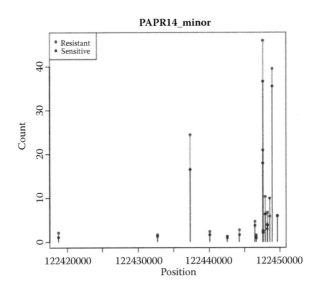

FIGURE 4.17
Differential expression of minor allele in gene *PAPR14*.

the ovarian cancer study, respectively. Several features emerge from these figures. First, levels of allelic expression dramatically vary across SNPs. We observed complicated patterns of ASE which may be due to alternative transcription start or end sites, alternative allele-specific splicing, multiple regulatory variants, biological variability, and sequencing technical variation. Directly summing read counts from the alleles across SNPs in each gene as overall expression values of ASE or estimating them from statistical models for erratic patterns of ASE is not trivial. Second, we observed significant differences in expression levels between major and minor alleles in many SNPs across the gene. Although the P-values for testing differential ASE in the ovarian cancer study were not small, we still observed substantial differences in expression levels between major and minor alleles in the treatment resistant and sensitive samples. Third, we observed difference both in the read counts of major alleles and in the read counts of minor alleles at many SNPs in *SLC13A3* between schizophrenia and normal samples and in *PARP14* between the treatment resistant and sensitive samples. We formally tested for difference in ASE across all SNPs in *SLC13A3* between schizophrenia and normal samples. P-values of the top 10 SNPs for testing differences in ASE of *SLC13A3* between schizophrenia and normal samples are listed in Table 4.9. We observed from Table 4.9 that an individual SNP may make mild

**TABLE 4.9**

P-Values of Top 10 SNPs in Gene *SLC13A3* for Testing Difference in ASE between Schizophrenia and Normal Samples

| SNP | | |
|---|---|---|
| **Genomic Position** | **RS #** | **P-value** |
| 44674193 | | 0.00265 |
| 44658276 | | 0.0072 |
| 44737451 | | 0.00911 |
| 44737452 | | 0.01741 |
| 44699440 | rs6124837 | 0.02602 |
| 44620155 | rs12617 | 0.02958 |
| 44622520 | rs910063 | 0.03634 |
| 44697695 | rs2425889 | 0.0424 |
| 44714575 | | 0.05169 |
| 44621840 | rs4809591 | 0.05195 |

contributions to differences in ASE, but they jointly cause significant differences in ASE between schizophrenia and normal samples for gene SLC13A3. This shows that using functional data analysis as a data reduction tool, even genotype-based tests maintain a high power to detect significant differences in ASE.

## 4.3 eQTL and eQTL Epistasis Analysis with RNA-Seq Data

RNA-seq provides multiple layers of resolutions and transcriptome complexity: the expression at exon, SNP, and positional level, splicing, isoform and allele-specific expression. The RNA-seq profile across a gene is represented as many reads at each genomic position. Consequently, the RNA-seq profile is a function of genomic position and can be taken as a function-valued trait. All methods for QTL and epistasis analysis of function-valued traits in Chapter 3 can be applied to eQTL and eQTL epistasis with RNA-seq data. In this section, we introduce an additional method that is called quadratically regularized functional canonical correlation analysis (QRFCCA) for eQTL and eQTL epistasis analysis with RNA-seq data. In Section 8.4, we introduced low rank approximation and generalized regulation. The principals discussed in Section 8.4 (*Big Data in Omics and Imaging: Association Analysis*, Xiong 2018) can be applied to QRFCCA.

The pipeline for QRFCCA is shown in Figure 4.18.

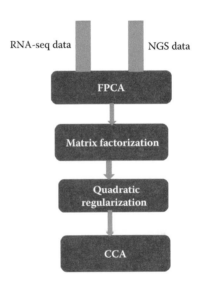

**FIGURE 4.18**
Pipe line for QRFCCA.

### 4.3.1 Matrix Factorization

Consider a data matrix $A \in R^{n \times q}$ consisting of $n$ samples with $q$ features (variables, FPCA scores). The data matrix $A$ represents the genotype data, the FPC scores or the phenotype data. The phenotype data include both continuous and discrete values. The $i^{th}$ row of $A$ is a vector of $q$ features for the $i^{th}$ sample, and the $j^{th}$ column of $A$ is a vector of the $j^{th}$ feature across the set of $n$ samples. Matrix factorization is used as a general framework to embed the genetic, RNA-seq and phenotype data into the low dimensional vector space to reduce the data dimension and remove anomalous or noise data points (Udell et al. 2016). To accomplish this, we first seek the best rank-$l$ approximation to the matrix $A$ by factorizing it into a product of two low rank matrices.

Let $G \in R^{n \times l}$ and $H \in R^{l \times q}$. Assume that the rank of $A$ is $r$. Therefore, $r \leq min(n,q)$. Matrix factorization attempts to minimize the approximation error:

$$\min_{G,H} \quad \| A - GH \|_F^2, \tag{4.119}$$

where $\| \cdot \|_F$ denotes the Frobenius norm of a matrix.

Let $g_i \in R^{1 \times l}$ be the $i^{th}$ row of $G$, $h_j \in R^l$ be the $j^{th}$ column of $H$ and $A_{ij}$ be the value of the $j^{th}$ feature in the $i^{th}$ sample. Define $g_i h_j = (GH)_{ij}$ as an inner product. The objective function in problem (4.118) can be rewritten as

$$\sum_{i=1}^{n} \sum_{j=1}^{q} \left( A_{ij} - g_i h_j \right)^2. \tag{4.120}$$

A solution to problem (4.119) can be found by truncating the singular value decomposition (SVD) of $A_{ij}$ (Udell et al. 2016). Let the SVD of $A$ be given by

$$A = U \Lambda V^T, \tag{4.121}$$

where $U = [u_1,...,u_r] \in R^{n \times r}$, $V = [v_1,...,v_r] \in R^{q \times r}$, $U^T U = I_{r \times r}$, $V^T V = I_{r \times r}$, and $\Lambda = diag(\lambda_1,...,\lambda_r) \in R^{r \times r}$ with $\lambda_1 \geq \lambda_2 \geq ... \geq \lambda_r > 0$. The columns of $U$ and $V$ are referred to as the left and right singular vectors of $A$, respectively, and $\lambda_1,...,\lambda_r$ are referred to as the singular values of $A$.

Let $\Lambda_l = diag(\lambda_1,...,\lambda_l)$, $U_l = [u_1,...,u_l]$ and $V_l = [v_1,...,v_l]$. Define $G = U_l \Lambda_l^{1/2}$ and $H = \Lambda_l^{1/2} V_l$. The best rank-$l$ approximation to the matrix $A$ or the $l$-rank matrix factorization of $A$ is then given by (Udell et al. 2016):

$$A \approx GH. \tag{4.122}$$

The matrix factorization compresses the $q$ features (variables) in the original data set to $l < q$ new features and hence reduces the data dimension.

### 4.3.2 Quadratically Regularized Matrix Factorization and Canonical Correlation Analysis

The power of the test statistics in CCA depends on the squared canonical correlations or eigenvalues of the matrix $R^2$. We wish to increase the power via changing the distribution of the canonical correlations and data reduction. In matrix factorization, for fixed rank $l$, we want to approximate the matrix $A$ by the product of two factor matrices $G$ and $H$ as accurately as possible. However, the Frobenius norm of the matrices $G$ and $H$ may be large. We need to balance the approximation accuracy and the Frobenius norm of the factor matrices. Specifically, we add the Frobenius norm of the factor matrices to the objective in Equation 4.119. The optimization problem (4.119) is now transformed to the quadratically regularized matrix factorization problem:

$$\min_{G,H} \quad F = \| A - GH \|_F^2 + \mu \| G \|_F^2 + \mu \| H \|_F^2. \tag{4.123}$$

From Equation 4.122, the matrices $G$ and $H$ have the forms:

$$G = U_l \Lambda_l^{1/2} \text{ and } H = \Lambda_l^{1/2} V_l. \tag{4.124}$$

Substituting Equation 4.124 into Equation 4.123 gives

$$\min_{\Lambda_l} \quad F = \| \Lambda - \Lambda_l \|_F^2 + 2\mu \text{Tr}(\Lambda_l), \tag{4.125}$$

where $\Lambda = diag(\lambda_1,...,\lambda_k)$ and $\Lambda_l = diag(\tau_1,...,\tau_l,0,...,0)$. Expanding the Frobenius norm of the matrices, problem (4.125) can be further reduced to

$$\min_{\tau} \quad F = \sum_{i=1}^{l}(\lambda_i - \tau_i)^2 + \sum_{i=l+1}^{r}\lambda_i^2 + 2\mu \sum_{i=1}^{l} \tau_i. \tag{4.126}$$

To minimize $F$, we set

$$\frac{\partial F}{\partial \tau_j} = -2\left(\lambda_j - \tau_j\right) + 2\mu = 0. \tag{4.127}$$

Solving Equation 4.127 for $\tau_j$ yields

$$\tau_j = \lambda_j - \mu. \tag{4.128}$$

To ensure that the factor matrices are non-negative, $\tau_j$ must be non-negative. Thus, we have the solution:

$$\tau_j = \left(\lambda_j - \mu\right)_+, \tag{4.129}$$

where $(a)_+ = \max(a,0)$.

In practice, a singular value is selected as a penalty parameter $\mu$ such that the sum from the selected singular value to the smallest singular value accounted for 20% of total singular values.

Define the matrix $\Lambda_l$ as

$$\Lambda_l = diag\left(\left(\lambda_1 - \mu\right)_+, \ldots, \left(\lambda_l - \mu\right)_+\right).$$

Then, factor matrices $G = U_l\Lambda_l^{1/2}$ and $H = \Lambda_l^{1/2}V_l$ are the solution to the minimization problem (4.42). We use truncation of the SVD to keep only the top $l$ singular values and soft-thresholding on the singular values to change distribution of the singular values. When $\mu$ increases beyond some singular values $\lambda_m$, $l - m + 1$ singular values of $GH$ will disappear.

### 4.3.3 QRFCCA for eQTL and eQTL Epistasis Analysis of RNA-Seq Data

#### 4.3.3.1 QRFCCA for eQTL Analysis

Both RNA-seq data and NGS genotype data can be taken as functional data. FPCA is first applied to both functional RNA-seq and genotype data to obtain FPC scores for both gene expression and genotypes. Quadratically regularized matrix factorization is applied to FPC scores of both functional RNA-seq and genotype data. The outcomes are $G$ for RNA-seq and $H$ for genotype data in Equation 4.123. Finally, matrices $G$ and $H$ will be used as two matrices $\widetilde{G} = H$ and $\widetilde{Y} = G$ in Equation 4.123.

#### 4.3.3.2 Data Structure for Interaction Analysis

Before we conduct eQTL epistasis of the RNA-seq using QRFCCA, we first define the data structure. Consider two genomic regions or two genes. There are $p$ FPC scores and $q$ FPC scores in the first and second genomic regions, respectively. Assume that $n$ individuals are sampled. For the $i^{th}$ individual, two vectors of the genetic variation data (FPC scores) for the first and second genomic regions are denoted by $x_i = [x_{i1},\ldots,x_{ip}]$ and $z_i = [z_{i1},\ldots,z_{iq}]$, respectively, where $x_{i1},\ldots,x_{ip}$ are FPC scores for the first genomic regions and $z_{i1},\ldots,$ $z_{iq}$ are the FPC scores for the second genomic region. Let $\xi_i = [x_{i1}z_{i1},\ldots x_{i1}z_{iq},\ldots,$ $x_{ip}z_{i1},\ldots,x_{ip}z_{iq}]$.

Define the matrix

$$\xi = \begin{bmatrix} \xi_{11} & \cdots & \xi_{1pq} \\ \vdots & \vdots & \vdots \\ \xi_{n1} & \cdots & \xi_{npq} \end{bmatrix}.$$

FPCA of the RNA-seq data is performed. Assume that there are $K$ FPC scores of the RNA-seq expression profiles and $y_{i1},\ldots,y_{iK}$ are their FPC scores.

Define $y_i = [y_{i1}, ...., y_{iK}]^T$ as a vector of FPC scores of the RNA-seq expression function.

### 4.3.3.3 Multivariate Regression

Multivariate regression of FPC scores $y_i$ on the FPC scores of two genes $x_i$ and $z_i$ will be used to remove the main effects. Define the multivariate regression model as

$$y_{ik} = \mu_k + \sum_{m=1}^{M} \tau_{im} \theta_{mk} + \sum_{j=1}^{p} x_{ij} \alpha_{kj} + \sum_{l=1}^{q} z_{il} \beta_{kl} + \varepsilon_{ik}, \quad k = 1, ..., K, \quad (4.130)$$

where $\mu_k$ is an overall mean of the $k^{th}$ FPC score, $\theta_{mk}$ is the regression coefficient associated with the covariate, $\tau_{im}, \alpha_{kj}$ is the main genetic additive effect of the $j^{th}$ FPC score of the first gene for the $k^{th}$ FPC score of the RNA-seq expression function, and $\beta_{kl}$ is the main genetic additive effect of the $l^{th}$ FPC score of the second gene for the $k^{th}$ FPC score of the RNA-seq expression function.

Let $\hat{y}_{ik}$ be the predicted value by the fitted model:

$$\hat{y}_{ik} = \hat{\mu}_k + \sum_{m=1}^{M} \tau_{im} \hat{\theta}_{mk} + \sum_{j=1}^{p} x_{ij} \hat{\alpha}_{kj} + \sum_{l=1}^{q} z_{il} \hat{\beta}_{kl}, \quad k = 1, ..., K. \quad (4.131)$$

The residual is defined as

$$\eta_{ik} = y_{ik} - \hat{y}_{ik}, \quad k = 1, ..., K.$$

Define residual matrix:

$$\eta = \begin{bmatrix} \eta_{11} & \cdots & \eta_{1K} \\ \vdots & \vdots \\ \eta_{n1} & \cdots & \eta_{nK} \end{bmatrix}.$$

Data matrix $\xi$ is defined as before.

### 4.3.3.4 CCA for Epistasis Analysis

For the convenience, we assume that $K \leq pq$. Define the covariance matrix between the vectors $\xi$ and $\eta$ for epistasis analysis:

$$\Sigma = \begin{bmatrix} \Sigma_{\xi\xi} & \Sigma_{\xi\eta} \\ \Sigma_{\eta\xi} & \Sigma_{\eta\eta} \end{bmatrix}.$$

The matrix $\Sigma$ will be estimated by

$$\hat{\Sigma} = \begin{bmatrix} \hat{\Sigma}_{\xi\xi} & \hat{\Sigma}_{\xi\eta} \\ \hat{\Sigma}_{\eta\xi} & \hat{\Sigma}_{\eta\eta} \end{bmatrix} = \frac{1}{n} \begin{bmatrix} \xi^T\xi & \xi^T\eta \\ \eta^T\xi & \eta^T\eta \end{bmatrix}. \tag{4.132}$$

Solution to CCA starts with defining the $R^2$ matrix Equation 1.220 in Xiong 2018:

$$R^2 = \hat{\Sigma}_{\eta\eta}^{-1/2}\hat{\Sigma}_{\eta\xi}\hat{\Sigma}_{\xi\xi}^{-1}\hat{\Sigma}_{\xi\eta}\hat{\Sigma}_{\eta\eta}^{-1/2}. \tag{4.133}$$

Let

$$W = \hat{\Sigma}_{\xi\xi}^{-1/2}\hat{\Sigma}_{\xi\eta}\hat{\Sigma}_{\eta\eta}^{-1/2}. \tag{4.134}$$

Suppose that the singular value decomposition (SVD) of the matrix is given by

$$W = U\Lambda V^T, \tag{4.135}$$

where $\Lambda = diag\ (\lambda_1,\dots,\lambda_d)$ and $d = min(K,pq)$. It is clear that

$$W^T W = R^2 = V\Lambda^2 V^T. \tag{4.136}$$

The matrices of canonical covariates are defined as

$$A = \Sigma_{\xi\xi}^{-1/2}U,$$
$$B = \Sigma_{\eta\eta}^{-1/2}V. \tag{4.137}$$

The vector of canonical correlations are

$$CC = [\lambda_1, \dots, \lambda_d]^T. \tag{4.138}$$

Canonical correlations between the interaction terms and FPC scores of the RNA-seq expression functions measure the strength of the interaction. The CCA produces multiple canonical correlations. However, we wish to use a single number to measure the interaction. We propose to use the summation of the square of the singular values as a measure to quantify the interaction:

$$r = \sum_{i=1}^{d}\lambda_i^2 = \mathrm{Tr}\left(\Lambda^2\right) = \mathrm{Tr}\left(R^2\right). \tag{4.139}$$

To test the interaction between two genes is equivalent to testing independence between $\xi$ and $\eta$ or to test the hypothesis that each variable in the set $\xi$ is uncorrelated with each variable in the set $\eta$. The null hypothesis of no interaction can be formulated as

$$H_0 : \Sigma_{\xi\eta} = 0.$$

The likelihood ratio for testing $H_0{:}\Sigma_{\xi\eta} = 0$ is

$$\Lambda_r = \frac{|\Sigma|}{|\Sigma_{\xi\xi}||\Sigma_{\eta\eta}|} = \prod_{i=1}^{d}(1 - \lambda_i^2), \tag{4.140}$$

which is equal to the Wilks' lambda $\Lambda$. This demonstrates that testing for interaction using multivariate linear regression can be treated as special case of CCA.

We usually define the likelihood ratio test statistic for testing the interaction as:

$$T_{CCA} = -N\sum_{i=1}^{d} \log(1 - \lambda_i^2). \tag{4.141}$$

For small $\lambda_i^2$, $T_{CCA}$ can be approximated by $N\sum_{i=1}^{d}\lambda_i^2 = Nr$, where $r$ is the measure of interaction between two genes. The stronger the interaction, the higher the power that the test statistic can test the interaction. Under the null hypothesis $H_0{:}\Sigma_{\xi\eta} = 0$, $T_{CC}$ is asymptotically distributed as a central $\chi^2_{(Kpq)}$. When sample size is large, Bartlett (1939) suggests using the following statistic for hypothesis testing:

$$T_{CCA} = -\left[N - \frac{(d+3)}{2}\right]\sum_{i=1}^{d} \log(1 - \lambda_i^2). \tag{4.142}$$

### 4.3.4 Real Data Analysis

In this section, to illustrate how to perform epistasis of RNA-seq data, we present epistasis analysis results of the 1000 Genome Project using the functional regression methods with both functional response and functional predictors studied in Section 3.3.

#### 4.3.4.1 RNA-Seq Data and NGS Data

The function regression model with both functional response and functional predictors (BFGM) was applied to the RNA-seq data in the GEUVADIS RNA Sequencing Project (Lappalainen et al. 2013; Xu et al. 2017) and the WGS data in the 1000 Genomes Project. A total of 350 samples with European origin was shared between the GEUVADIS RNA Sequencing Project and the 1000 Genomes Project. RNA-seq data of the 15,656 genes and 2,566,261 SNPs in 18,986 genes were included in the analysis. DESeq (Anders and Huber 2010) was used to normalize the RNA-seq data.

#### 4.3.4.2 Cis-Trans Interactions

The target gene selected from the 15,656 gene expressions was referred to as gene1. One of the remaining 18,985 genotyping genes was selected as gene2.

We used BFGM to test for the interactions between gene1 and gene2 influencing the expression of the target gene1. The total number of gene pairs tested for interactions which included both common and rare variants was 297,229,160. A P-value for declaring significant interaction after applying the Bonferroni correction for multiple tests was $1.68 \times 10^{-10}$.

For comparisons, the functional regression model with scalar response and functional predictors (SFGM) was also applied to the dataset. RPKM and DESeq were used to compute the overall expression value of genes from the RNA-seq data. All the expression values were processed by the rank-based inverse normal transformation. For both common and rare variants, in total, 162,361, 260 and 51 significant *cis*–trans interactions regulating the gene expressions were identified by the BFGM, SFGM with the RPKM and DESeq, respectively (Xu et al. 2017). We observed 9,846 genes whose expressions were influenced by 16,2361 *cis*–trans interactions. We found that the average number of epistasis influencing each gene was 16. A total of 3,505 gene expressions were influenced by one significant *cis*–trans gene–gene interactions, 169 gene expressions were influenced by more than 100 *cis*–trans gene–gene interactions.

To investigate the pattern of interactions, the P-values of the top 20 interactions between genes ranked by the BFGM method were summarized in Table 4.10 where P-values for testing interactions between genes by the SFGM (RPKM, DESeq and RNAmin) and min P-values were also listed. The SFGM (RPKM) or SFGM (DESeq) represented the functional regression model with RPKM or the summary statistic calculated by DESeq as the scalar response and genotype functions as the predictors for interaction analysis, the RNAmin denoted the minimum of P-values computed by the SFGM method with the number of reads at each genome position of the gene as the scalar response in the functional regression model. The min P-values denoted to take the minimum of all P-values for testing all possible pairs of SNPs between two genes using the functional regression model with functional response and scalar predictors.

Table 4.10 showed several remarkable features. First, the pair-wise interaction between rare and rare variants (34.38%), and rare and common variants (59.38%) was often observed. Less observed was the significant pair-wise interaction between common and common variants (6.25%). Second, significant interactions between two genes often indicated that at least one significant pair of SNPs in two genes could be observed (min P-values were small). However, it was observed that pairs of SNPs between two genes jointly had significant interaction effects, but individually each pair of SNPs mildly contributed to the interaction effects. Third, the BFGM often had a much smaller P-value to detect interaction than other tests. Fourth, it was observed that genes may not show even mild marginal association, but they did demonstrate significant evidence of interaction. If only the interactions between two marginally significant genes are tested, some significant

**TABLE 4.10**

P-Values of Top 20 Genes Ranked by the BFGM Methods

| | | | | | | | P-value (Interaction) | | | |
| | | | | | | | | SFGM | | |
| Gene Expression | Gene 1 | Marginal | Gene 2 | Marginal | BFGM | RPKM | DESeq | RNA-min | min P-value |
|---|---|---|---|---|---|---|---|---|---|
| ULK4 | ULK4 | 1.5E-01 | C19orf70 | 8.0E-06 | 0.0E+00 | 4.2E-01 | 5.0E-01 | 0.0E+00 | 0.0E+00 |
| ULK4 | ULK4 | 1.5E-01 | OR10A2 | 6.0E-15 | 4.1E-305 | 4.3E-05 | 5.0E-03 | 0.0E+00 | 0.0E+00 |
| CCDC13 | CCDC13 | 5.6E-01 | TMEM121 | 9.9E-24 | 2.9E-302 | 5.0E-03 | 1.7E-02 | 1.4E-304 | 2.1E-12 |
| ULK4 | ULK4 | 1.5E-01 | PSMC5 | 7.3E-23 | 5.7E-267 | 2.8E-05 | 1.5E-03 | 0.0E+00 | 0.0E+00 |
| ULK4 | ULK4 | 1.5E-01 | COX5B | 1.2E-03 | 2.5E-242 | 1.8E-03 | 3.1E-02 | 2.5E-259 | 4.94E-323 |
| NKX2-5 | NKX2-5 | 3.4E-01 | TP53TG3D | 5.8E-10 | 2.2E-226 | 6.7E-02 | 7.6E-02 | 1.1E-228 | 0.0E+00 |
| ASIC2 | ASIC2 | 2.2E-02 | RPS16P5 | 4.1E-05 | 3.4E-226 | 1.1E-01 | 7.8E-02 | 1.8E-158 | 8.0E-237 |
| TMEM132E | TMEM132E | 8.3E-02 | LOC100144602 | 9.5E-51 | 2.0E-213 | 1.3E-01 | 1.3E-01 | 2.3E-142 | 1.1E-215 |
| TMEM98 | TMEM98 | 6.7E-01 | LOC100144602 | 9.3E-51 | 4.9E-213 | 8.1E-02 | 1.1E-01 | 4.1E-144 | 6.0E-216 |
| SPACA3 | SPACA3 | 7.1E-02 | LOC100144602 | 1.4E-50 | 9.2E-211 | 9.7E-03 | 1.2E-02 | 8.8E-141 | 4.9E-214 |
| ASIC2 | ASIC2 | 2.2E-02 | OR5B12 | 3.1E-05 | 2.9E-205 | 2.0E-01 | 1.2E-01 | 3.6E-141 | 1.2E-259 |
| CCL1 | CCL1 | 6.5E-02 | TINF2 | 3.6E-22 | 3.9E-205 | 1.2E-01 | 1.3E-01 | 1.5E-157 | 3.3E-210 |
| SCN2A | SCN2A | 1.4E-01 | DEFB4B | 1.3E-32 | 5.5E-203 | 1.9E-02 | 4.9E-02 | 1.3E-104 | 2.7E-236 |
| ZNF254 | ZNF254 | 4.4E-01 | OR2V1 | 2.2E-13 | 1.2E-183 | 2.4E-02 | 4.9E-02 | 5.5E-225 | 2.9E-259 |
| KRT5 | KRT5 | 2.5E-02 | OR5K1 | 1.0E-28 | 3.3E-177 | 1.2E-02 | 2.3E-02 | 2.3E-201 | 5.8E-199 |
| FNDC8 | FNDC8 | 4.8E-01 | LOC100144602 | 8.1E-49 | 3.5E-172 | 4.4E-03 | 6.2E-02 | 1.2E-124 | 8.0E-175 |
| CCT6B | CCT6B | 3.3E-01 | LOC100144602 | 8.8E-49 | 4.2E-172 | 7.5E-03 | 3.6E-02 | 3.1E-125 | 1.7E-177 |
| TMEM163 | TMEM163 | 1.2E-01 | HIST1H4H | 1.8E-09 | 8.5E-170 | 6.7E-02 | 4.8E-02 | 3.4E-112 | 1.2E-25 |
| ASIC2 | ASIC2 | 2.2E-02 | LOC100144602 | 9.7E-51 | 6.1E-165 | 9.2E-01 | 8.7E-01 | 3.8E-104 | 9.5E-236 |
| KRT5 | KRT5 | 2.5E-02 | IFNA7 | 3.7E-16 | 2.4E-163 | 2.3E-02 | 1.8E-02 | 2.2E-178 | 2.1E-211 |

interactions may be missed. The fifth, the BFGM tremendously reduced the computation burden.

## 4.4 Gene Co-Expression Network and Gene Regulatory Networks

Regulation of gene expression is a complex biological process (Lelli et al. 2012). Large-scale regulatory networks involve thousands of interacting proteins and RNA components to regulate gene transcriptions and coordinate cell functions (Äijö and Bonneau 2017). Large-scale regulatory network inference provides a general framework for comprehensively learning regulatory interactions, understanding the biological activity, devising effective therapeutics, identifying drug targets of complex diseases and discovering the novel pathways. Constructing gene regulatory networks from experiments have serious limitations. It is time consuming, tedious, expensive and particularly lacks reproducibility (Liu et al. 2016). Uncovering and modeling gene regulatory networks are one of the long-standing challenges in genomics and computational biology. Various statistical methods and computational algorithms for network inference have been developed. Method diversity in regulatory network inferences arise from the diversity in the regulatory network modeling, genomic and epigenomic technologies producing the data and experimental designs. In this section, we focus on introducing regulatory network inference from expression data.

There are basic approaches to gene regulatory network inferences: unsupervised and supervised approaches. The unsupervised approach includes the co-expression network (Wang et al. 2017; Hong et al. 2013), ordinary differential equation (Noh and Gunawan 2016; Zhu et al. 2012), graphical Gaussian model (Menéndez et al. 2010; Krämer et al. 2009; Belilovsky et al. 2015; Kalyagin et al. 2017), and Bayesian networks (Liu et al. 2016; Akutekwe and Seker 2015).

### 4.4.1 Co-Expression Network Construction with RNA-Seq Data by CCA and FCCA

Gene co-expression networks are often used to extract important information about groups of co-regulated genes which play a central role in the regulatory processes. Co-expression networks can comprehensively capture the relationships of individual components of the transcriptome perturbed by environments and hence provide a powerful tool to gain new insights into the function of genes, biological processes, the global structure of the transcriptome, and mechanism of complex diseases.

Weighted correlation network analysis, mutual information relevance networks, covariance selection, sparse graphical models, and partial correlation methods for co-expression network construction are mainly designed for microarray expression data where all these methods use a single value of summarizing statistics to represent gene expression level and overlook all information on expression difference in exons, genomic position and alleles. To explore observed expression variation in exons or in genomic position across the genes, canonical correlation analysis (CCA) which quantify the correlation between a linear combination of the expressions at the exon levels or position levels in one gene and another such combination of expressions in another gene can be used to construct co-expression networks. To reduce data dimension, functional canonical correlation analysis (FCCA) can be used to construct co-expression networks. To model allele specific expression (ASE), bivariate CCA to construct co-expression networks with ASE data, allowing levels of ASE to vary across SNPs and to consider complicated patterns of ASE due to allele-specific splicing and alternative transcription start sites will also be introduced (Hong et al. 2013).

### 4.4.1.1 CCA Methods for Construction of Gene Co-Expression Networks

A gene co-expression network is considered as an undirected graph, where a gene is represented as a node and each edge connecting two nodes is regarded as the co-expression relationship of the two connected genes. Construction of co-expression networks is often carried out by detecting the pair-wise correlation of gene co-expression. The CCA is to seek maximization of the correlation between two linear combination of the variables in the datasets (Hong et al. 2013). Suppose that we have $p$ exons or positions in one gene and $q$ exons or positions in another gene. Let $x_j^{(1)}$ denote the expression of the $j$-th exon or the count of reads at the $j$-th genomic position within the first gene. We can similarly define $x_j^{(2)}$ for the second gene. Let $X^{(1)} = [x_1^{(1)}, ..., x_p^{(1)}]^T$ and $X^{(2)} = [x_1^{(2)}, ..., x_q^{(2)}]^T$. For the convenience of presentation, we assume that $p \leq q$. Let

$$X = \begin{bmatrix} X^{(1)} \\ X^{(2)} \end{bmatrix} \text{ and } \Sigma = \text{cov}(X, X) = \begin{bmatrix} \Sigma_{11} & \Sigma_{12} \\ \Sigma_{21} & \Sigma_{22} \end{bmatrix}.$$

Construction of co-expression networks is implemented by seeking maximization of correlation coefficients between linear combination $U = a^T X^{(1)}$ for the first gene and linear combination $V = b^T X^{(2)}$:

$$\max_{a,b} \quad corr(U, V) = \frac{a^T \Sigma_{12} b}{\sqrt{a^T \Sigma_{11} a} \sqrt{b^T \Sigma_{22} b}}. \tag{4.143}$$

Solutions to the optimization problem (4.143) are the eigenvalues $\lambda_1 \geq \lambda_2 \geq ... \geq \lambda_p$ and their corresponding eigenvectors of the Rayleigh quotient matrix:

$$R = \Sigma_{11}^{-1/2} \Sigma_{12} \Sigma_{22}^{-1} \Sigma_{21} \Sigma_{11}^{-1/2}.$$

Let $P_k$ be the P-value of the test statistic

$$T_k = -\left[n - \frac{1}{2}(p+q)\right]\sum_{i=k+1}^{p}\log\left(1 - \hat{\lambda}_i^2\right)$$

with distribution $\chi^2_{(p-k)(q-k)}$, where $n$ is sample size for testing the null hypothesis $H_0:\lambda_k = \ldots = \lambda_p = 0$. We assign a weight to the edge connecting two genes:

$$w = \frac{\sum_{i=1}^{p}\lambda_i I(\log P_i)}{\sum_{i=1}^{p}I(\log P_i)}, \tag{4.144}$$

where $I(\log P) = \begin{cases} 1 & P > 0.05 \\ -\log P & P \leq 0.05 \end{cases}$.

The rank procedure is used to construct the network. The weights are ranked. A threshold is specified. The edges with rank larger than threshold are retained in the network.

### 4.4.1.2 Bivariate CCA for Construction of Co-Expression Networks with ASE Data

The bivariate CCA can be used to construct co-expression networks with ASE data. Let $x_j^{(1)}$ and $x_j^{(2)}$ be the counts of reads of the major and minor allele at the $j$-the SNP in the gene, respectively. We can similarly define $y_j^{(1)}$ and $y_j^{(2)}$ for the other gene. Let $X = [x_1^{(1)}, x_1^{(2)}, \ldots, x_p^{(1)}, x_p^{(2)}]^T$ and $Y = [y_1^{(1)}, y_1^{(2)}, \ldots, y_q^{(1)}, y_q^{(2)}]^T$. Define linear combinations $U = a^T X$ and $V = b^T X$, where $a = [\alpha_1^{(1)}, \alpha_1^{(2)}, \ldots, \alpha_p^{(1)}, \alpha_p^{(2)}]^T$ and $b = [\beta_1^{(1)}, \beta_1^{(2)}, \ldots, \beta_q^{(1)}, \beta_q^{(2)}]^T$. These linear combinations can be rewritten as

$$U = \left[\alpha^{(1)}\right]^T X^{(1)} + \left[\alpha^{(2)}\right]^T X^{(2)} \text{ and } V = \left[\beta^{(1)}\right]^T Y^{(1)} + \left[\beta^{(2)}\right]^T Y^{(2)},$$

where $\alpha^{(1)} = [\alpha_1^{(1)}, \ldots, \alpha_p^{(1)}]^T$, $\alpha^{(2)} = [\alpha_1^{(2)}, \ldots, \alpha_p^{(2)}]^T$, $\beta^{(1)} = [\beta_1^{(1)}, \ldots, \beta_q^{(1)}]^T$, $\beta^{(2)} = [\beta_1^{(2)}, \ldots, \beta_q^{(2)}]^T$, $X^{(1)} = [x_1^{(1)}, \ldots, x_p^{(1)}]^T$, $X^{(2)} = [x_1^{(2)}, \ldots, x_p^{(2)}]^T$, $Y^{(1)} = [y_1^{(1)}, \ldots, y_q^{(1)}]^T$ and $Y^{(2)} = [y_1^{(2)}, \ldots, y_q^{(2)}]^T$.

Define the covariance matrices:

$$\Sigma = \begin{bmatrix} \Sigma_{xx} & \Sigma_{xy} \\ \Sigma_{yx} & \Sigma_{yy} \end{bmatrix}, \text{ where}$$

$$\Sigma_{xx} = \begin{bmatrix} \Sigma_{x^{(1)}x^{(1)}} & \Sigma_{x^{(1)}x^{(2)}} \\ \Sigma_{x^{(2)}x^{(1)}} & \Sigma_{x^{(2)}x^{(2)}} \end{bmatrix}, \; \Sigma_{xy} = \Sigma_{yx}^T = \begin{bmatrix} \Sigma_{x^{(1)}y^{(1)}} & \Sigma_{x^{(1)}y^{(2)}} \\ \Sigma_{x^{(2)}y^{(1)}} & \Sigma_{x^{(2)}y^{(2)}} \end{bmatrix} \text{ and } \Sigma_{yy}$$

$$= \begin{bmatrix} \Sigma_{y^{(1)}y^{(1)}} & \Sigma_{y^{(1)}y^{(2)}} \\ \Sigma_{y^{(2)}y^{(1)}} & \Sigma_{y^{(2)}y^{(2)}} \end{bmatrix}.$$

The CCA seeks to maximize

$$\max_{\alpha,\beta} \; corr(U,V) = \frac{\alpha^T \Sigma_{xy} \beta}{\sqrt{\alpha^T \Sigma_{xx} \alpha} \sqrt{\beta^T \Sigma_{yy} \beta}}, \qquad (4.145)$$

where $\alpha = \begin{bmatrix} \alpha^{(1)} \\ \alpha^{(2)} \end{bmatrix}$ and $\beta = \begin{bmatrix} \beta^{(1)} \\ \beta^{(2)} \end{bmatrix}$.

The solutions to the optimization problem (4.145) are the eigenvectors of the matrix with the eigenvalues $\lambda_1 \geq \lambda_2 \geq \dots \geq \lambda_{2p}$:

$$R = \Sigma_{xx}^{-1/2} \Sigma_{xy} \Sigma_{yy}^{-1} \Sigma_{yx} \Sigma_{xx}^{-1/2}.$$

Again, let $P_k$ be the P-value of the test statistic

$$T_k = -[n - (p+q)] \sum_{i=k+1}^{2p} \log\left(1 - \hat{\lambda}_i^2\right)$$

with distribution $\chi^2_{(2p-k)(2q-k)}$, where $n$ is sample size for testing the null hypothesis $H_0 : \lambda_k = \dots = \lambda_{2p} = 0$. We assign a weight to the edge connecting two genes:

$$w = \frac{\sum_{i=1}^{2p} \lambda_i I(\log P_i)}{\sum_{i=1}^{2p} I(\log P_i)}, \qquad (4.146)$$

where $I(\log P) = \begin{cases} 1 & P > 0.05 \\ -\log P & P \leq 0.05 \end{cases}$

Similar to single CCA for construction of co-expression networks, after we rank the weights we also use rank procedure to prune the networks.

### 4.4.2 Graphical Gaussian Models

In the previous section, we used correlation and canonical correlation which measures both direct and indirect interactions between pairs of genes to construct co-expression networks. Now we introduce graphical Gaussian models for modeling genetic networks where partial correlations are used to quantify the direct interaction only (Krämer et al. 2009). In Section 1.1 we introduced graphical Gaussian models for modeling genotype networks. The graphical Gaussian models can also be used to model gene regulatory networks. Recall that a graphical Gaussian model is an undirected graph where a node represents a variable and an edge connecting two nodes indicates that the connected two variables are conditionally dependent, given all other variables. In the context of gene regulatory networks, each gene is represented by a node or random variable in the graph and gene regulatory relationship is represented by an edge. An edge is missing in the network if and only if two gene expressions are independent given all other gene expressions in the network.

Three statistical tools can be used to estimate graphical Gaussian models: partial correlations, concentration matrix and sparse regressions which are briefly discussed here. Detailed descriptions were introduced in Section 1.1. We assume that the vector of $q$ gene expression variables follow a normal distribution $N(0, \Sigma)$.

Suppose that a graph $G = (V, E)$ consists of nodes $V = \{1,..., q\}$ and undirected edges. Define notation $-(i, j) \equiv \{k : 1 \le k \ne i, j \le q\}$. A partial correlation between the expressions of genes $i$ and $j$ is defined as the conditional correlation between the expressions of genes $i$ and $j$, given other gene expressions in the network:

$$\rho^{ij} = Corr\left(y_i, y_j \mid y_{-(i,j)}\right) = \frac{\mathrm{cov}\left(y_i, y_j \mid y_{-(i,j)}\right)}{\left[\mathrm{var}\left(y_i \mid y_{-(i,j)}\right)\mathrm{var}\left(y_j \mid y_{-(i,j)}\right)\right]^{\frac{1}{2}}}, \qquad (4.147)$$

for $1 \le i < j \le q$.

If we assume the normality of the variables (gene expressions), the presence of edge between the nodes $i$ and $j$ that indicates two gene expressions $y_i$ and $y_j$ are conditionally dependent, given all other gene expressions if and only if $\rho^{ij} \ne 0$. Therefore, the partial correlation can be used to estimate the structure of gene regulatory networks.

Let $\Sigma^{-1} = (\sigma^{ij})_{q \times q}$ be the concentration or precision matrix. In Section 1.1, we showed in Equation (9.A.4) that

$$\rho^{ij} = -\frac{\sigma^{ij}}{\sqrt{\sigma^{ii}\sigma^{jj}}}. \qquad (4.148)$$

Equation 4.148 indicates that the element $\sigma^{ij} = 0$ in the concentration matrix implies the absence of edge connecting the nodes $i$ and $j$ or the conditional independence between expressions of genes $i$ and $j$, given all other gene expressions.

When sample size is much larger than the number of variables, the concentration matrix can be directly estimated from the inverse of the sampling covariance matrix. However, when the number of variables in the network is larger than the sample size, the inverse of sampling covariance matrix may not exist. Graphical lasso that is an algorithm for learning the structure in an undirected Gaussian graphical model, using $l_1$ regularization has been developed for network modeling via estimating the sparse concentration matrix (Friedman et al. 2007).

The third approach to estimate graphic Gaussian models is sparse regression (Dobriban and Fan 2016). It is well known that the relationship between the partial correlation and regression exists. In other words, if $y_i$ is expressed as $y_i = \beta_{ij} y_j + \sum_{k \neq i,j} \beta_{ik} y_k + \varepsilon_i$ and $\varepsilon_i$ is independent of $y_{(-ij)}$, then $\beta_{ij} = \rho^{ij} \sqrt{\dfrac{\sigma^{jj}}{\sigma^{ii}}}$.

Therefore, search for nonzero partial correlations can be formulated as a variable selection problem in regression in which the $l_1$ penalty on a loss function is imposed. Specifically, the sparse regression for estimating the concentration matrix is formulated as

$$L(\beta, \lambda) = \frac{1}{2} \left( \sum_{i=1}^{q} \| Y_i - \sum_{j \neq i} \beta_{ij} Y_j \|^2 \right) + \lambda \sum_{1 \leq i < j \leq q} |\beta_{ij}|, \qquad (4.149)$$

where $Y_i = [y_{i1}, \ldots, y_{in}]^T$ is the sample of the $i^{th}$ variable and $\lambda$ is a penalty parameter which controls the size of the penalty. A larger value of $\lambda$ leads to a sparse regression that fits the data less well and smaller $\lambda$ leads to regression that fits the data well but is less sparse.

### 4.4.3 Real Data Applications

Inferring the gene co-expression network can be simply summarized to detect the pair-wise correlation of gene co-expression, three measurements of graph topology: centralization, assortativity and degree distribution were used to evaluate the performance of the methods for construction of the gene co-expression networks. Three measures are briefly introduced below.

*Network centralization* measures the level of a network as to how centralized it is around a particular nodes. It is based on the node centrality definition. Denote an undirected graph by $G = (V, E)$. The network centralization $C_G$ is defined as:

$$C_G = \sum_{v \in V} \left( \max_{v' \in V} (c(v')) - c(v) \right),$$

where $c(v)$ denotes the centrality of node $v$.

*Assortativity* is a very important index of assessing network topology, which is used to describe the degree-degree correlations for the networks. If the assortativity coefficient was positive, the network is said to be assortative. On the other hand, if the assortativity coefficient is negative, the network is recognized as disassortative. It had been widely recognized that the most social networks are likely to be assortative, whereas biological networks are usually disassortative. For random networks, the assortativity coefficient tended to be nearly zero. In an undirected graph, the assortativity coefficient can be calculated as follows:

$$r = \frac{\sum_{jk} jk\left(e_{jk} - q_j q_k\right)}{\sum_k k^2 q_k - \left(\sum_k k q_k\right)^2},$$

where $q_k = \sum_j e_{jk}$, $q_j = \sum_k e_{jk}$, $e_{jk}$ denotes the connect probability of two nodes that have degrees $j$ and $k$, separately.

*Degree distribution* For many real networks, the degree distribution follows a power-law distribution. Networks that have this property are also called scale-free networks in which some nodes have a much higher degree than others. The scale-free property implies that the degrees of the nodes in the network are proportional to exponential:

$$P(d = k) \infty A k^{-\alpha},$$

where $P(d = k)$ denotes the probability of a node with a $k$ degree, and $\alpha$ denotes the exponent parameter.

Four methods: CCA, weighted CCA (WCCA), Graphical Lasso (glasso), and Sparse Joint Regression (Sparse-regression) were applied to the LUSC (Lung Squamous Cell Cancer) dataset (242 samples, 20,532 genes and 239,886 exons) downloaded from TCGA. The lung cancer related KEGG pathway that is named hsa05223 consists of 54 genes. After discarding the isolated nodes in the pathway and matching the TCGA LUSC RNA-seq filtered data, a KEGG curated network consisting of 44 vertices and 14 edges was generated. The four methods were used to construct gene co-expression networks using the LUSC expression data of matched 44 genes and their related exons in the lung cancer pathway. The topology properties of the estimated gene co-expression networks using four methods were summarized in Table 4.11. The results

**TABLE 4.11**

Topology Property Comparison of the Estimated Gene Co-Expression Networks by Four Methods

| Method | Assortativity | Centralization | Power Law |
|---|---|---|---|
| CCA | −0.350819 | 0.735585 | 1.532607 |
| Sparse Regression | 0.03498 | 0.649426 | 1.7882 |
| Glasso | −0.004059 | 0.63419 | 1.792665 |
| WCCA | −0.394349 | 0.742659 | 1.529924 |
| KEGG | −0.230487 | 0.725692 | 1.826356 |

indicate that the topology of the gene co-expression networks using WCCA and CCA was closed to the topology of the pathway in the KEGG database. The co-expression network estimated using the WCCA and CCA, and pathway in the KEGG database had higher centrality than that estimated using the Glasso and sparse joint regression method. The power-law results showed that the co-expression networks estimated using all four methods and the pathway in the KEGG had a scale free property. The results of assortativity comparison showed that only the CCA and WCCA methods discovered the disassortative property of the biological network. On the other hand, networks inferred by the Glasso and sparse joint regression methods had a nearly zero (either positive or negative) assortativity coefficient, which usually appeared in random networks.

## 4.5 Directed Graph and Gene Regulatory Networks

In the previous sections, we studied the undirected graphical models for gene regulation. However, the undirected graphical models are lack of semantic interpretability. The directed edges in the graph bear a natural causal implication. Therefore, the directed graph is a powerful tool for modeling regulatory networks. In this section, we primarily introduce the score-based linear and non-linear causal inference for construction of gene regulatory networks with both microarray and RNA-seq measured expression data. These methods can provide a deeper understanding of the regulatory mechanism and infer the directions of regulations in the network.

### 4.5.1 General Procedures for Inferring Genome-Wide Regulatory Networks

The most current methods for causal inference can only estimate directed acyclic graphics up to dozens or hundreds of nodes and hence cannot be used to construct genome-wide regulatory networks. The incomplete maps of regulatory networks will limit our understanding of the molecular mechanisms underlying complex traits and human diseases (Barzel and Barabási 2013). With the rapid progress in genomic technologies, we are expected to provide complete genome-wide regulatory networks. Therefore, there is the urgent need to develop tools for inferring the genome-wide regulatory networks.

A general procedure for inferring genome-wide regulatory networks is summarized in Figure 4.19 (Liu et al. 2016). The procedure consists of four major components: (1) construct initial directed acyclic graphics (DAG) for regulatory networks, (2) select sets of nodes for constructions of DAGs, (3) construct causal subnetworks (DAGs), and (4) integrate sets of causal

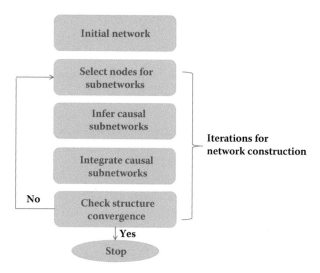

**FIGURE 4.19**
A general procedure for construction of large-scale regulatory networks.

subnetworks into a connected causal network. Steps (2), (3), and (4) are repeated until the structure of the inferred causal networks converges. The final causal network will be taken as a whole-genome regulatory network.

Now we briefly describe four major components as follows:

**Step 1: Construct the initial network.** The initial network can be either undirected network or directed network. The initial undirected network can be inferred by correlation and mutual information (MI) (Liu et al. 2016). The initial directed network can be constructed by the network score-based integer programming (Section 2.6) or functional models for bivariate causal discovery (Section 2.5.1.1).

**Step 2: Select nodes with degrees $d$ as central nodes of the subnetworks.** For each central node, select the nodes with the shorted path to the central node less than or equal to $k$. The central and selected nodes form a set of nodes for local subnetwork construction. Suppose that a total of $l$ sets of nodes are selected. We can infer $l$ local subnetworks.

**Step 3: Construct local subnetworks.** Algorithms for inferring DAGs can be used to construct local subnetworks for each set of the selected nodes. For example, we use the sparse structural equation models to calculate the score function for each node. Then, integer programming algorithms will be used to select the best local network with the optimal score. A total of $l$ local subnetworks will be constructed.

**Step 4: Integrate local causal subnetworks.** Take the union of all local causal subnetworks and integrate them into a connected global network. If directions from multiple local causal networks are different, functional models for bivariate causal discovery will be used to determine the direction. The output of step 4 is a tentative global regulatory network.

Repeat step 2 to step 4 until convergence in the topological structure of the network is achieved.

### 4.5.2 Hierarchical Bayesian Networks for Whole Genome Regulatory Networks

In general, many genes are measured in gene expression profiling. Current algorithms for construction of DAGs are unable to jointly construct a whole genome regulatory network. In this section, we introduce hierarchical Bayesian networks for modeling the whole genome regulatory network (Figure 4.20). The hierarchical Bayesian networks can also be used to construct an initial regulatory network that was discussed in Section 4.5.1. The procedure for inferring a hierarchical Bayesian network consists of three steps. The first step is determining by the sets of nodes (genes) for generating subnetworks. If the genes form a pathway in pathway database, then all genes in the pathway are taken as a group of genes for construction of local causal subnetworks. However, large number of genes is not in the pathway database. For those genes, cluster analysis methods can be used to group the genes into clusters. The genes in each cluster are taken as a group for inferring a local causal subnetwork. Summary statistics, for example, PCA and matrix completion, are then used to represent the expressions of all genes in the group. Summary statistics are used for each group and score-based integer programming to infer the DAG for modeling the causal relations among the groups of gene expressions. The second step is to construct local regulatory network using the score-based integer programming method for casual network construction. The third step is to integrate all regulatory subnetworks into a connected whole genome regulatory network.

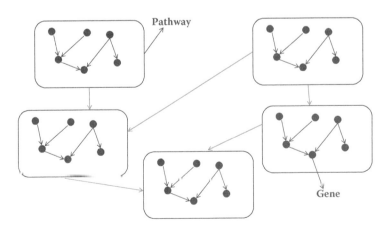

**FIGURE 4.20**
Schematic diagram of initial Bayesian networks.

### 4.5.2.1 Summary Statistics for Representation of Groups of Gene Expressions

Generalized low rank models including matrix completion (given the incomplete entries of a low-rank matrix, fill in the missing values), PCA, and robust PCA (given a matrix which has been corrupted, recover the original low rank matrix) (Liu et al. 2013; Udell et al. 2016; Cand`es et al. 2009) provide effective methods to reduce data dimensions, segment (cluster) data and summarize the gene expression data in pathways and clusters. In this section, we will mainly introduce low rank matrix representation models and take robust PCA as its special case.

Consider an observational gene expression matrix $X \in R^{d \times n}$ with $d$ rows and $n$ samples. For the microarray measured gene expression data, a row of the matrix $X$ represents expressions of a gene across $n$ samples. For RNA-seq expression data, if an overall expression level is used to summarize the expressions across the gene region, then a row of the matrix $X$ also represents expression of a gene across $n$ samples. RNA-seq data can also be expanded in terms of functional principal components. The expression of a gene will be represented by many functional principal component scores. The matrix $X$ contains the functional principal component scores. A row of the matrix $X$ consists of functional principal component score of the gene of $n$ samples. Suppose that the matrix $X$ corrupted with noises can be decomposed into the following equation (Figure 4.21):

$$X = AZ + E, \tag{4.150}$$

where $A \in R^{n \times r}$ is a "dictionary" matrix that linearly spans the data space, the matrix $Z \in R^{r \times n}$ is the lowest rank representation of the gene expression data matrix $X$ with respect to a dictionary matrix $A$ and $E \in R^{d \times n}$ is a noisy matrix. Our goal is to remove the noises $E$ and recover the original low rank matrix $X_0$ where

$$X = X_0 + E. \tag{4.151}$$

Equation 4.151 indicates that the original low rank matrix $X_0$ is corrupted by noises, which leads to the observed matrix $X$. The goal can be

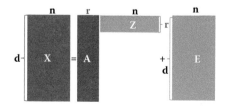

**FIGURE 4.21**
Low rank model.

accomplished by

$$\min_{Z,E} \quad \| Z \|_* + \lambda \| E \|_{2,1}$$

$$\text{s.t.} \quad X = AZ + E,$$

(4.152)

where $\| \cdot \|_*$ denotes the nuclear norm (the sum of singular values of a matrix) and $\| E \|_{2,1} = \sum_j \| [E]_{:,j} \|_2$, $\| \cdot \|_2$ denotes the $L_2$ norm, $[E]_{.j}$ is the $j^{th}$ column of the matrix $E$, and $\lambda$ is a penalty parameter. In practice, we often use the nuclear norm to replace the rank function. Here, the $L_{2,1}$ norm is used to characterize the error term $E$. The aim of penalty $\lambda \| E \|_{2,1}$ is to remove the outliers due to sample-specific corruptions. To remove the small Gaussian noise, $\| E \|_F^2$ will be used as a penalty. However, to remove the random corruptions we will use the $L_1$ norm $\| E \|_1$. After solving the optimization problem (4.152), we will use the solution $Z^*$ to represent the expressions of the genes in the group where the $j^{the}$ column of the matrix $Z^*$ with $r$ components is a vector of representation scores and represent the summarized expressions of all the genes in the group of the $j^{th}$ sample. The original low matrix $X_0$ can be recovered from $AZ^*$ or $Z - E^*$ where $E^*$ is a solution to the error matrix in the optimization problem (4.152).

The problem can be further be transformed to

$$\min_{Z,E,J} \quad \| J \|_* + \lambda \| E \|_{2,1}$$

$$\text{s.t.} \quad X = AZ + E, Z = J.$$

(4.153)

The problem (4.153) can be solved by the alternative direction of multiplier method (ADMM) (Boyd et al. 2011), which minimizes the following augmented Lagrangian function:

$$L(Z,E,J,Y_1,Y_2,\lambda,\mu) = \| J \|_* + \lambda \| E \|_{2,1} + \text{tr}\left(Y_1^T(X - AZ - E)\right) + \text{tr}\left(Y_2^T(Z - J)\right)$$

$$+ \frac{\mu}{2}\left(\| X - AZ - E \|_F^2 + \| Z - J \|_F^2\right).$$

(4.154)

The ADMM algorithm for solving the optimization problem (4.154) is summarized as follows (for detailed derivations, see Appendix 4.C).

**Algorithm 4.2: ADMM Algorithm**

**Step 1:** Initialize: $Z = J = 0$, $E = 0$, $Y_1 = 0$, $Y_2 = 0$, $\mu = 10^{-6}$, $\mu_{max} = 10^{-6}$, $\rho = 1.1$, $\varepsilon = 10^{-8}$.

**Step 2:** Repeat until convergence

a. Update the matrix $J$. Perform single value decomposition: $Z^{(k)} + \frac{1}{\mu}Y_2^{(k)} = U^{(k+1)}\Sigma^{(k+1)}\left(V^{(k+1)}\right)^T$ where $\Sigma^{(k+1)} = diag\left(\sigma_1^{(k+1)}, \ldots, \sigma_r^{(k+1)}\right)$.

Update: $J^{(k+1)} = U^{(k+1)} diag \left( \max \left( \sigma_1^{(k+1)} - \frac{1}{\mu}, 0 \right), \ldots, \max \left( \sigma_r^{(k+1)} - \frac{1}{\mu}, 0 \right) \right) \left( V^{(k+1)} \right)^T$.

b. Update $Z$: $Z^{(k+1)} = (I + A^T A)^{-1} \left[ A^T (X - E^{(k)}) + J^{(k+1)} + \frac{1}{\mu} (A^T Y_1^{(k)} - Y_2^{(k)}) \right]$.

c. Update $E$: Define $Q = X - AZ^{(k+1)} + \frac{1}{\mu} Y_1^{(k)}$

$$\left[ E^{(k+1)} \right]_{:,j} = \frac{\left( \| [Q]_{:,j} \|_2 - \frac{\lambda}{\mu} \right)_+}{\| [Q]_{:,j} \|_2} [Q]_{:,j}$$

d. Update the multipliers:

$$Y_1^{(k+1)} = Y_1^{(k)} + \mu \left( X - AZ^{(k+1)} - E^{(k+1)} \right)$$
$$Y_2^{(k+1)} = Y_2^{(k)} + \mu \left( Z^{(k+1)} - J^{(k+1)} \right)$$

e. Update the parameter $\mu = \min(\rho\mu, \mu_{max})$.

f. Check the convergence conditions:
   If $\| X - AZ^{(k+1)} - E^{(k+1)} \|_\infty < \varepsilon$, $\| Z^{(k+1)} - J^{(k+1)} \|_\infty < \varepsilon$, then stop; otherwise $k \leftarrow k + 1$, go to step 2(a).

End

Next, we study how to use low rank representation models to segment the data. By selecting $A = X$ in Equation 4.152, the optimization problem (4.152) is reduced to

$$\min_{Z,E} \quad \| Z \|_* + \lambda \| E \|_{2,1}$$

$$\text{s. t.} \quad X = XZ + E. \tag{4.155}$$

The ADMM algorithm can be similarly applied to solve the specific optimization problem (4.155). After obtaining the optimization solutions $(Z^*, E^*)$, the following algorithm can be used to segment (cluster) the data.

**Algorithm 4.3: Algorithm for Subspace Segmentation (Cluster Analysis)**

**Step 1:** Obtain the minimizer $Z^*$ and compute the skinny SVD of $Z^*$ as $U^* \Sigma^* (V^*)^T$.

**Step 2:** Construct an affinity matrix W:
Let $\tilde{U} = U^* (\Sigma^*)^{1/2}$. Then, W is given by $[W]_{ij} = ([\tilde{U}\tilde{U}^T]_{ij})^2$. The purpose of using $(.)^2$ is to ensure that the values of the affinity matrix $W$ are positive because the matrix $\tilde{U}\tilde{U}^T$ may have negative values.

**Step 3:** Use the affinity matrix W to perform Ncut for segmenting the data samples into $k$ clusters.

a. Let $d_i = \Sigma_j[W]_{ij}$. Define a diagonal matrix $D = diag(d_1,..., d_D)$. Solve the following eigenequation:

$$(D - W)e = \lambda D e \text{ or}$$

$$D^{-1/2}(D - W)D^{-1/2}g = \lambda g, \text{ where } g = D^{-1/2}e.$$

b. Use the eigenvector with the second smallest eigenvalue to bipartition the graph.

Specifically, compute the median $m$ of the components of the second smallest eigenvector $e$, and place all points whose component in $e$ is greater than the median $m$ in cluster $G_1$, and the rest in $G_2$. Repeatedly partitioning the subsets in this fashion, we can perform hierarchical clustering.

**Step 4:** Estimate the space number $k$ (number of clusters).

$\hat{k} = n - \text{int}\left( \sum_{i=1}^{n} f_\tau(\sigma_i) \right)$, where int(.) is the function that output the nearest integer of a real number, $\sigma_i$, $i = 1,..., n$ are the singular values of the Laplacian matrix $L = D - W$, $0 < \tau < 1$ is a parameter and

$$f_\tau(\sigma) = \begin{cases} 1 & \sigma > \tau \\ \log_2\left(1 + \dfrac{\sigma^2}{r^2}\right) & \text{otherwise.} \end{cases}$$

### 4.5.2.2 Low Rank Presentation Induced Causal Network

Suppose that there are $K$ pathways or clusters. For the $k^{th}$ pathway or cluster, consider $m_k$ genes. Define the data matrix $Y$:

$$Y = \begin{bmatrix} y_{11}^1 & \cdots & y_{1m_1}^1 & y_{11}^2 & \cdots & y_{1m_2}^2 & \cdots & y_{11}^K & \cdots & y_{1m_K}^K \\ y_{21}^1 & \cdots & y_{2m_1}^1 & y_{21}^2 & \cdots & y_{2m_2}^2 & \cdots & y_{21}^K & \cdots & y_{2m_K}^K \\ \vdots & \ddots & \ddots & \ddots & \ddots & \ddots & \ddots & \ddots & \ddots & \vdots \\ y_{n1}^1 & \cdots & y_{nm_1}^1 & \cdots & y_{n1}^2 & \cdots & y_{nm_2}^2 & y_{n1}^K & \cdots & y_{nm_K}^K \end{bmatrix}_{n \times M}, \tag{4.156}$$

where $M = \sum_{i=1}^{K} m_i$.

The coefficient matrix $\Gamma$ can be decomposed into $K \times K$ submatrices corresponding to $K$ pathways or clusters:

$$\Gamma = \begin{bmatrix} \Gamma^{11} & \Gamma^{12} & \cdots & \Gamma^{1K} \\ \Gamma^{21} & \Gamma^{22} & \cdots & \Gamma^{2K} \\ \vdots & \vdots & \vdots & \vdots \\ \Gamma^{K1} & \Gamma^{K2} & \cdots & \Gamma^{KK} \end{bmatrix}_{M \times M}, \tag{4.157}$$

where

$$\Gamma^{ij} = \begin{bmatrix} \gamma_{11}^{ij} & \gamma_{12}^{ij} & \cdots & \gamma_{1m_j}^{ij} \\ \gamma_{21}^{ij} & \gamma_{22}^{ij} & \cdots & \gamma_{2m_j}^{ij} \\ \vdots & \vdots & \cdots & \vdots \\ \gamma_{m_i1}^{ij} & \gamma_{m_i2}^{ij} & \cdots & \gamma_{m_im_j}^{ij} \end{bmatrix}, \ i,j = 1,2,\ldots,K. \tag{4.158}$$

Low rank representation of the expression data matrix $Y^T \in R^{M \times n}$ in Equation 4.150 can be written as

$$Y^T = AZ + E, \tag{4.159}$$

where

$$A = \begin{bmatrix} A^1 & 0 & \cdots & 0 \\ 0 & A^2 & \cdots & 0 \\ \vdots & \vdots & \vdots & \vdots \\ 0 & 0 & \cdots & A^K \end{bmatrix}_{M \times r}, \ A^k = \begin{bmatrix} a_{11}^k & a_{12}^k & \cdots & a_{1r_k}^k \\ a_{21}^k & a_{22}^k & \cdots & a_{2r_k}^k \\ \vdots & \vdots & \vdots & \vdots \\ a_{m_k1}^k & a_{m_k2}^k & \cdots & a_{m_kr_k}^k \end{bmatrix}_{m_k \times r_k}, \ k = 1,\ldots,K, \tag{4.160}$$

$$Z = \begin{bmatrix} Z^1 \\ Z^2 \\ \vdots \\ Z^K \end{bmatrix}_{r \times n}, \ Z^k = \begin{bmatrix} z_{11}^k & z_{12}^k & \cdots & z_{1n}^k \\ z_{21}^k & z_{22}^k & \cdots & z_{2n}^k \\ \vdots & \vdots & \vdots & \vdots \\ z_{r_k1}^k & z_{r_k2}^k & \cdots & z_{r_kn}^k \end{bmatrix}_{r_k \times rn}, \ r = \sum_{k=1}^{K} r_k. \tag{4.161}$$

Let

$$G = Z^T = \left[ \left(Z^1\right)^T \left(Z^2\right)^T \cdots \left(Z^K\right)^T \right]$$
$$= \left[ G^1 G^2 \cdots G^K \right]_{n \times r}, \tag{4.162}$$

where

$$G^k = \begin{bmatrix} g_{11}^k & g_{12}^k & \cdots & g_{1r_k}^k \\ g_{21}^k & g_{22}^k & \cdots & g_{2r_k}^k \\ \vdots & \vdots & \vdots & \vdots \\ g_{n1}^k & g_{n2}^k & \cdots & g_{nr_k}^k \end{bmatrix}_{n \times r_k} = \begin{bmatrix} z_{11}^k & z_{21}^k & \cdots & z_{r_k1}^k \\ z_{12}^k & z_{22}^k & \cdots & z_{r_k2}^k \\ \vdots & \vdots & \vdots & \vdots \\ z_{1n}^k & z_{2n}^k & \cdots & z_{r_kn}^k \end{bmatrix}.$$

Let

$$\alpha = A^T \Gamma. \tag{4.163}$$

Substituting Equations 4.156 and 4.159 into Equation 4.163 gives

$$\alpha = \begin{bmatrix} \left(A^1\right)^T \Gamma^{11} & \left(A^1\right)^T \Gamma^{12} & \cdots & \left(A^1\right)^T \Gamma^{1K} \\ \left(A^2\right)^T \Gamma^{21} & \left(A^2\right)^T \Gamma^{22} & \cdots & \left(A^2\right)^T \Gamma^{2K} \\ \vdots & \vdots & \vdots & \vdots \\ \left(A^K\right)^T \Gamma^{K1} & \left(A^K\right)^T \Gamma^{K2} & \cdots & \left(A^K\right)^T \Gamma^{KK} \end{bmatrix}_{r \times M} \tag{4.164}$$

$$= \begin{bmatrix} \alpha^{11} & \alpha^{12} & \cdots & \alpha^{1K} \\ \alpha^{21} & \alpha^{22} & \cdots & \alpha^{2K} \\ \vdots & \vdots & \vdots & \vdots \\ \alpha^{K1} & \alpha^{K2} & \cdots & \alpha^{KK} \end{bmatrix}_{r \times M} ,$$

where

$$\alpha^{kj} = \left(A^k\right)^T \Gamma^{kj} = \begin{bmatrix} \alpha_{11}^{kj} & \alpha_{12}^{kj} & \cdots & \alpha_{1m_j}^{kj} \\ \alpha_{21}^{kj} & \alpha_{22}^{kj} & \cdots & \alpha_{2m_j}^{kj} \\ \vdots & \vdots & \vdots & \vdots \\ \alpha_{r_k1}^{kj} & \alpha_{r_k2}^{kj} & \cdots & \alpha_{r_km_j}^{kj} \end{bmatrix}. \tag{4.165}$$

A low rank representation induced linear structural equation model can be formulated as

$$G\alpha + XB + E = 0, \tag{4.166}$$

where $G$ and $\alpha$ are defined as before,

$$X = \begin{bmatrix} X^1 & X^2 & \cdots & X^L \end{bmatrix}$$

$$= \begin{bmatrix} x_{11} & x_{12} & \cdots & x_{1L} \\ x_{21} & x_{22} & \cdots & x_{2L} \\ \vdots & \vdots & \ddots & \vdots \\ x_{n1} & x_{n2} & \cdots & x_{nL} \end{bmatrix},$$

$$B = \begin{bmatrix} B^1 & B^2 & \cdots & B^K \end{bmatrix},$$

$$B^i = \begin{bmatrix} \beta^i_{11} & \beta^i_{12} & \cdots & \beta^i_{1r_i} \\ \beta^i_{21} & \beta^i_{22} & \cdots & \beta^i_{2r_i} \\ \vdots & \vdots & \vdots & \vdots \\ \beta^i_{L1} & \beta^i_{L2} & \cdots & \beta^i_{Lr_i} \end{bmatrix},$$

and

$$E = \begin{bmatrix} E^1 & E^2 & \cdots & E^K \end{bmatrix},$$

$$E^i = \begin{bmatrix} e^i_{11} & e^i_{12} & \cdots & e^i_{1r_i} \\ e^i_{21} & e^i_{22} & \cdots & e^i_{2r_i} \\ \vdots & \vdots & \vdots & \vdots \\ e^i_{n1} & e^i_{n2} & \cdots & e^i_{nr_i} \end{bmatrix},$$

The difference between the low rank representation induced linear structural equation models and the classical scale linear structural equation models is that unlike the scale structural equation models where each node represents a random variable, each node in the low rank representation induced structural equation models represents a vector of random variables. Therefore, the matrix equation corresponding to the node $i$ are

$$G\alpha^i + XB^i + E^i = 0 \tag{4.167}$$

or

$$G^i = G^1\alpha^{1i} + \ldots + G^{i-1}\alpha^{i-1i} + G^{i+1}\alpha^{i+1i} + \ldots \; G^K\alpha^{Ki} + XB^i + E^i,$$

where $\alpha^{-i}$ is the matrix of the path coefficients of the all other nodes connecting the node $i$. The matrix $\alpha^{ji}$ are path coefficients that measure the strength of the causal relationships from the vector $G^j$ to $G^i$ or from the node $j$ to the node $i$, $\beta^i_l$ are path coefficients from the exogenous variable to the endogenous variable which measures the causal effect of the exogenous variable $x_l$ on the low rank representation score vector $G^i$. The coefficients $\| \alpha^{ji} \|_{2,1} = 0$ and $\| \beta^i_l \|_{2,1} = 0$ imply the zero direct influence of $G^j$ and $x_l$ on $G^i$, respectively, and are usually omitted from the equation.

Let $g^i = \text{vec}(G^i)$ be a $nr_i$ dimensional vector of low rank representation score of the gene expression of the node (pathway)$i$, $e_i = \text{vec}(E^i)$ be a $nr_i$ dimensional vector of random errors associated with the node (pathway)$i$, $\delta_{ji} = \text{vec}(\alpha^{ji})$, $j = 1,\ldots, i - 1, i + 1,\ldots, K$ be a $r_j r_i$ dimensional vector of path coefficients connecting the node $j$ to the node $i$,$\delta_i = [\, \delta^T_{1i} \; \cdots \; \delta^T_{i-1i} \; \delta^T_{i+1i} \; \cdots \; \delta^T_{Ki} \,]^T$ be a $J_i = (r_1 + \ldots + r_{i\,1} + r_{i+1} + \ldots + r_K)r_i$ dimensional vector path coefficients connecting all other nodes to the node $i$, and $b^i = vec(B^i) = [(b_1^i)^T \ (b_2^i)^T \ \cdots \ (b_{r_i}^i)^T]^T$ be a $r_iL$ dimensional vector of path coefficients connecting all the exogenous (genotype and environment nodes) to the node $i$. In Appendix 4.D, we show that the model can be written as

$$g^i = W_i \delta_i + H_i b^i + e_i, \tag{4.168}$$

where $e_i$ is a $r_i n$ dimensional vector of random variables with mean zeros and covariance matrix $\sigma_{ii} I_{r_i n}$.

Multiplying both sides of Equation 4.168 by $(I_{r_i} \otimes X)^T$, we obtain

$$(I_{r_i} \otimes X)^T g^i = (I_{r_i} \otimes X)^T \left( W_i \delta_i + H_i b^i \right) + (I_{r_i} \otimes X)^T e_i, \tag{4.169}$$

where

$$cov\left( (I_{r_i} \otimes X)^T e_i, (I_{r_i} \otimes X)^T e_i \right) = \sigma_{ii} (I_{r_i} \otimes X^T X).$$

Using weighted least square and $l_2$-norm penalization of Equation 4.169, we can form the following optimization problem:

$$\min_{\delta_i, b^i} \ f\left(\delta_i, b^i\right) + \lambda_1 \left( \sum_{j=1}^{i-1} \| \delta_{ji} \|_2 + \sum_{j=i+1}^{K} \| \delta_{ji} \|_2 \right) + \lambda_2 \sum_{l=1}^{r_i} \| b_l^i \|_2, \tag{4.170}$$

where

$$f\left(\delta_i, b^i\right) = \left[ (I_{r_i} \otimes X)^T g^i - (I_{r_i} \otimes X)^T \left( W_i \delta_i + H_i b^i \right) \right]^T (I_{r_i} \otimes X^T X)^{-1})$$
$$* \left[ (I_{r_i} \otimes X)^T g^i - (I_{r_i} \otimes X)^T \left( W_i \delta_i + H_i b^i \right) \right],$$

$$\| \alpha^{-i} \|_{2,1} = \sum_{j=1}^{i-1} \| \delta_{ji} \|_2 + \sum_{j=i+1}^{K} \| \delta_{ji} \|_2, \quad \| B^i \|_{2,1} = \sum_{i=1}^{r_i} \| b_l^i \|_2,$$

and the norm $\|M\|_{2,1}$ of the matrix $M$ is defined as

$$\| M \|_{2,1} = \sum_j \| [M]_{.j} \|_2,$$

$[M]_{.j}$ is the $j^{th}$ column vector of the matrix $M$.

To separate smooth optimization $\min f(\delta_i, b^i)$ from nonsmooth optimization $\| \alpha^{-i} \|_{2,1}$ and $\| B^i \|_{2,1}$, we can introduce constraints $\delta_{ji} - Z_j^1 = 0$ and $b_l^i - Z_l^2 = 0$, and transform the unconstrained optimization problem (4.170) into the constrained optimization problem:

$$\min_{\delta_i, b^i} \ f\left(\delta_i, b^i\right) + \lambda_1 \left( \sum_{j=1}^{i-1} \| Z_j^1 \|_2 + \sum_{j=i+1}^{K} \| Z_j^1 \|_2 \right) + \lambda_2 \sum_{l=1}^{r_i} \| Z_l^2 \|_2 \tag{4.171}$$

Subject to

$$\delta_{ji} - Z_j^1 = 0,\ j = 1, \ldots, i-1, i+1, \ldots, K,$$

$$b_l^i - Z_l^2 = 0,\ l = 1, \ldots, r_i.$$

In Appendix 4.B, we show that the optimization problem (4.171) can be solved by the alternating direction method of multipliers (ADMM) algorithm that is summarized as Result 4.7.

**Result 4.7: ADMM Algorithm for Solving Low Rank Representation Induced SEMs (LRRISEM) is Summarized as Follows**

For $i = 1, \ldots, K$

**Step 1:** Initialization

$$u^0 := 0,\ v^0 := 0,$$

$$\delta_i^{(0)} = \left\{ W_i^T \left[ I_{r_i} \otimes X (X^T X)^{-1} X^T \right] W_i + \rho_1 I \right\}^{-1} W_i^T \left[ I_{r_i} \otimes X (X^T X)^{-1} X^T \right] g^i,$$

$$\left( b^i \right)^{(0)} = \left\{ H_i^T \left[ I_{r_i} \otimes X (X^T X)^{-1} X^T \right] H_i + \rho_2 I \right\}^{-1} H_i^T \left[ I_{r_i} \otimes X (X^T X)^{-1} X^T \right] g^i,$$

$$\left( Z_j^1 \right)^{(0)} = \frac{\left( \| \delta_{ji}^{(0)} \|_2 - \dfrac{\lambda_1}{\rho_1} \right)_+}{\| \delta_{ji}^{(0)} \|_2} \delta_{ji}^{(0)},\ j = 1, \ldots, i-1, i+1, \ldots, K,$$

$$\left( Z_l^2 \right)^{(0)} = \frac{\left( \| (b_l^i)^{(0)} \|_2 - \dfrac{\lambda_2}{\rho_2} \right)_+}{\| (b_l^i)^{(0)} \|_2} \left( b_l^i \right)^{(0)},\ l = 1, \ldots, r_i.$$

Carry out steps 2, 3, and 4 until convergence

**Step 2:**

$$\delta_i^{(k+1)} = \left\{ W_i^T \left[ I_{r_i} \otimes X (X^T X)^{-1} X^T \right] W_i + \rho_1 I \right\}^{-1}$$

$$\left\{ W_i^T \left[ I_{r_i} \otimes X (X^T X)^{-1} X^T \right] g^i + \rho_1 ((Z^1)^{(k)} - u^{(k)}) \right\},$$

$$\left( b^i \right)^{k+1} = \left\{ H_i^T \left[ I_{r_i} \otimes X (X^T X)^{-1} X^T \right] H_i + \rho_2 I \right\}^{-1}$$

$$\left\{ H_i^T \left[ I_{r_i} \otimes X (X^T X)^{-1} X^T \right] g^i + \rho_2 ((Z^2)^{(k)} - v^{(k)}) \right\}.$$

**Step 3:**

$$\left( Z_j^1 \right)^{(k+1)} = \frac{\left( \| \delta_{ji}^{(k+1)} + u_j^{(k)} \|_2 - \dfrac{\lambda_1}{\rho_1} \right)_+}{\| \delta_{ji}^{(k+1)} + u_j^{(k)} \|_2} \left( \delta_{ji}^{(k+1)} + u_j^{(k)} \right), j$$

$$-1, \ldots, i-1, i+1, \ldots, K,$$

$$\left(Z_l^2\right)^{(k+1)} = \frac{\left(\|\left(b_l^i\right)^{(k+1)} + v_l^{(k)}\|_2 - \frac{\lambda_2}{\rho_2}\right)_+}{\|\left(b_l^i\right)^{(k+1)} + v_l^{(k)}\|_2}\left(\left(b_l^i\right)^{(k+1)} + v_l^{(k)}\right), l = 1, \dots, r_i.$$

**Step 4:**

$$u_j^{(k+1)} : == u_j^{(k+1)} + \left(\delta_{ji}^{(k+1)} - \left(Z_j^1\right)^{(k+1)}\right), j = 1, \dots, i-1, i+1, \dots, K,$$

$$v_l^{(k+1)} : == v_l^{(k)} + \left(\left(b_l^i\right)^{(k+1)} - \left(Z_l^2\right)^{(k+1)}\right), l = 1, \dots, r_i.$$

**Example 4.8**

The proposed Low Rank Representation Induced SEMs coupled with the integer programming (LRRISEMIG) algorithm was applied to RNA-seq data with 24 signal transduction pathways including 1728 genes which were measured in 447 samples. The estimated regulatory network was shown in Figure 4.22. The LRRISEMIG estimated 50 directed edges, 25 of them were correctly aligned with paths in the KEGG pathway database. There was a total of 69 paths in the KEGG pathway database. We compared the LRRISEMIG with a Bayesian network and random network. We performed 10,000 simulations. We average the total number of detected directed edges, correctly identified directed edges, and false detected directed edges over 10,000 simulations. The detection power was defined as

Detection Power

= the number of true paths detected/the number of edges estimated

and

False Positive Rate

= the number of estimated false edges/the number of labeled edges

We compare the LRRISEMIG with R package "bnlearn" that implements constraint-based (GS, IAMB, Inter-IAMB, Fast-IAMB, MMPC, Hiton-PC), pairwise (ARACNE and Chow-Liu), score-based (Hill-Climbing and Tabu Search) and hybrid (MMHC and RSMAX2) structure learning algorithms for discrete, Gaussian and conditional Gaussian networks, along with many score functions and conditional independence tests (Scutari 2017) and random networks. The results were summarized in Table 4.12. We observed that the LRRISEMIG performs much better than the randomly constructed network, which indicated that the inferred regulatory network by the LRRISEMIG algorithm made biological sense. We also observed that the LRRISEMIG algorithm had a high detection power and less false positive rates than the recently released package "bnlearn" for Bayesian network inference.

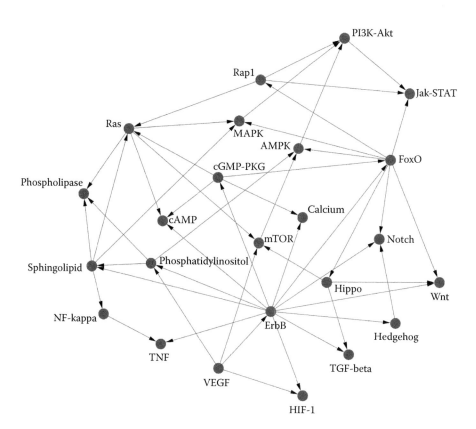

**FIGURE 4.22**

A regulatory network for 24 signal transduction pathways estimated by low rank representation induced structural equation models combined with integer programing.

**TABLE 4.12**

Detection Power and False Positive Rate

|  | Edges Detected | Edges Aligned with KEGG | False Positive | Detection Power | False Positive Rate |
|---|---|---|---|---|---|
| LRRISEMIG | 58 | 29 | 29 | 50.00% | 42.00% |
| bnlearn | 58 | 25 | 33 | 43.10% | 47.80% |
| Random network | 58 | 15 | 43 | 25.90% | 62.30% |

### 4.5.3 Linear Regulatory Networks

In Section 4.5.2, we investigated the low rank presentation induced causal network that can be used for assessing the regulatory relationships among the pathways or sets of genes. Next step is to estimate the regulatory network of the genes within one or several pathways. If the RNA-seq expression data are viewed as functional curves, causal network estimation with the nodes represented by curves will be investigated in Chapter 5 since both RNA-seq and methylation data can be treated as function curves. The statistical methods for causal inference of function-valued traits and their application to causal networks with RNA-seq and methylation data will be developed in Chapter 5. If the gene expression were measured by microarray or represented by summary statistics for RNA-seq data, their causal networks can be estimated by statistical methods introduced in Section 2.4. The computational algorithms that were developed in Section 2.4 can be used to construct gene regulatory networks for up to several hundreds of genes. These estimated gene regulatory networks that are subnetworks are then integrated as whole genome regulatory networks.

### 4.5.4 Nonlinear Regulatory Networks

In Section 2.5.1 we developed functional models for bivariate causal discovery and nonlinear structural equations for causal network discovery. Both of these can be used to estimate the nonlinear regulatory networks. Similar to correlation analysis where co-expressions of a pair of genes can be assessed by correlations, the causal relations between two genes can be investigated by functional models for bivariate causal discovery which discovers causal-effect structure. We can construct causal network by assessing pair-wise causal effect relationships.

**Example 4.9**

To evaluate its performance, we present Example 4.9 for testing the cause-effect relations between a pair of variables using an additive noise model. We first consider a nonlinear functional additive noise model with normal noise distributions. We simulated the data using the two-dimensional additive noise model (Nowzohour and Bühlmann 2016):

$$X = \varepsilon_1,$$
$$Y = X + b \log (X) + \varepsilon_2,$$

where the parameter $b$ was in the range $[-1, 1]$, which controlled the linearity of the model. The noise errors $\varepsilon_1$, $\varepsilon_2$ were distributed as normal: $\varepsilon_k \sim N(0,100)$.

Three possible causal relations between two variables generate the following models (Nowzohour and Bühlmann 2016):
Model 1: $X$ causes $Y$

$$X = \varepsilon_1$$
$$Y = f(x) + \varepsilon_2.$$

Model 2: $Y$ causes $X$

$$X = g(Y) + \varepsilon_1$$
$$Y = \varepsilon_2.$$

Model 3: $X$ and $Y$ are independent

$$X = \varepsilon_1$$
$$Y = \varepsilon_2.$$

The additive noise model for bivariate causal discovery in Section 2.5.1.1 is used to select one of three models. B-spline is used to fit the functional model. Figure 4.23 plots false decision rate where the errors are normally distributed. We observe from Figure 4.23 that when $b = 0$ the model is linear, the false decision rate is close to 0.5. This is equivalent to random guess. However, when $b$ is away from zero, the model becomes nonlinear. The false decision rates are close to zero. This indicated that when the models are nonlinear we have high probability to make correct causal inferences.

Next, we examine the linear model with non-normal error distributions. We simulated the data using the following linear model:

$$X = \varepsilon_1$$
$$Y = X + \varepsilon_2.$$

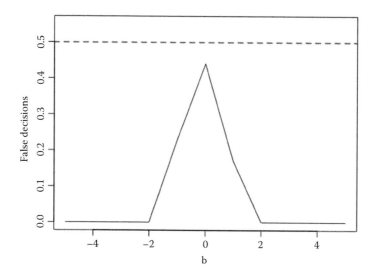

**FIGURE 4.23**
False decision rates for bivariate causal discovery assuming nonlinear model and normal error distributions.

The noise terms $\varepsilon_1$, $\varepsilon_2$ have the following distributions:

$$\varepsilon_k = \text{sgn}(v_k)|v_k|^q, \quad v_k \sim N(0, 25),$$

where the parameter $q$ is in the range $[0.5, 2]$ and thus Gaussianity is controlled (Nowzohour and Bühlmann 2016). Again the additive noise model with B-spline for fitting the nonlinear function is used to assess the causality. The false decision rates are shown in Figure 4.24. Figure 4.24 shows that when the model is linear and the error distribution is normal ($q = 1$) the probability of making wrong causal inference is very high. However, even when the model is linear if the error distribution is non-normal (when $q$ moves away from 1) we still have a high probability to make correct causal inferences.

**Example 4.10**

To illustrate their applications for estimation of gene regulatory networks, nonlinear additive noise models and glasso for construction of co-expression networks are applied to Wnt signaling pathway with RNA-Seq of 79 genes measured in 447 tissue samples. A total of 49 directions estimated by the functional model for bivariate causal discovery are shown in Figure 4.25 where directions in blue color indicate the shared edges with the same direction in the KEGG pathway database, the direction in the green color indicates that the estimated path and the path with direct effects in the KEGG are the same, the direction in the red color indicates that the estimated path and the path with indirect effects in the KEGG are the same, the direction in the brown color indicates that the estimated direction is the binding/association in the KEGG, the direction in the grey color indicates that the estimated direction is absent in the KEGG and the direction in the deep red color indicates that the estimated edge is the edge with opposite direction in the KEGG. The detection power of the functional model, glasso and random network is

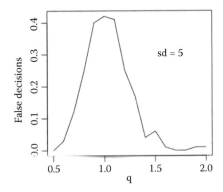

**FIGURE 4.24**
False decision rates for bivariate causal discovery assuming linear model and non-normal error distributions.

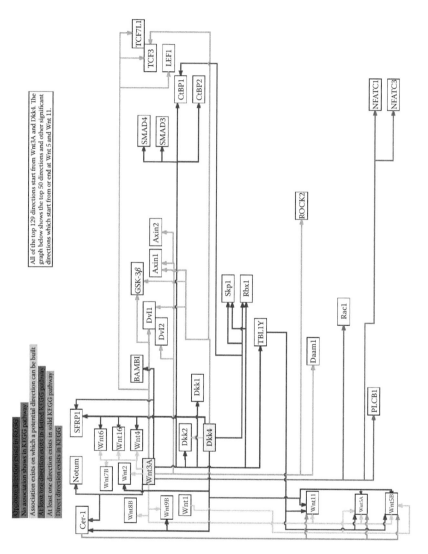

**FIGURE 4.25**

Inferred regulatory network for the Wnt signal pathway.

**TABLE 4.13**

Detection Power of Nonlinear Causal Inference

| Methods | Detection Power | |
|---|---|---|
| | **Directed Path** | **Paths without Considering Direction** |
| Functional model | 38% | 46% |
| Glasso | X | 24% |
| Random network | X | 32% |

summarized in Table 4.13. These results show that the performance of the functional nonlinear models for estimation of gene regulatory network is more precise than glasso.

## 4.6 Dynamic Bayesian Network and Longitudinal Expression Data Analysis

The dynamic Bayesian network is a probabilistic framework for modeling gene regulatory networks with time-course and longitudinal expression data (Nemati S and Adams 2014). We consider two types of dynamic Bayesian networks: stationary dynamic Bayesian network and non-stationary dynamic Bayesian networks. A stationary dynamic Bayesian network assumes that the structures and parameters of the dynamic Bayesian networks are fixed over time. A non-stationary dynamic Bayesian network assumes that the structures and parameters of the dynamic Bayesian networks vary over time. We first study the non-stationary dynamic Bayesian network and then the stationary Bayesian network.

In general, gene regulatory and signal transduction processes in cells change in response to environmental stimuli and growth, such as immune response, developmental processes and disease progression. The non-stationary dynamic Bayesian networks can be used to model non-stationary gene expression data. Consider $n$ samples. For each sample, we observe the expressions of the $M$ genes at $T$ time points. Let $y_{ij}^t$ be the expression value of the $j^{th}$ gene in the $i^{th}$ sample at the time point $t$. We also consider $K$ exogenous variables including genotypes of genes. Let $x_{ik}$ be the $k^{th}$ exogenous variable of the $i^{th}$ sample. The scheme of the dynamic Bayesian network for modeling the genotype–expression network is shown in Figure 4.26. We observed edges directed from expressed gene to expressed genes, and from genotypes to expressed genes. Directions in the networks consists of two parts. We observed directed connections between nodes for each fixed time point $t$ and directed connections from nodes at the preceding time $t-1$ to the nodes at the current time $t$.

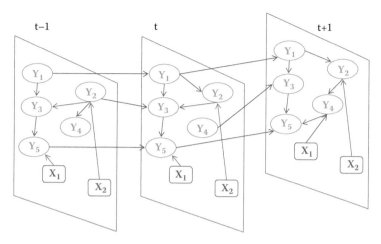

**FIGURE 4.26**
Scheme of the dynamic Bayesian network.

The introduced dynamic structural equation model coupled with integer programming algorithm for modeling dynamic Bayesian network consists of two steps. The dynamic structural equation model at the first step serves two purposes: generating initial dynamic directed graphs and calculate score function for each node. Integer linear programming at the second step is used to search the best dynamic Bayesian network with the optimal score.

### 4.6.1 Dynamic Structural Equation Models with Time-Varying Structures and Parameters

A dynamic structural equation model with the $M$ equations that represent the causal relationships among the jointly $M$ time course gene expressions as endogenous variables, the genotypes, drugs, environments and other pre-determined variables as the $K$ exogenous variables, and the random errors are given by

$$y_1^t \gamma_{11}^t + \ldots + y_M^t \gamma_{M1}^t + y_1^{t-1} \alpha_{11}^t + \ldots + y_M^{t-1} \alpha_{M1}^t + x_1 \beta_{11}^t + \ldots + x_M \beta_{K1}^t + e_1^t = 0$$

$$y_1^t \gamma_{12}^t + \ldots + y_M^t \gamma_{M2}^t + y_1^{t-1} \alpha_{12}^t + \ldots + y_M^{t-1} \alpha_{M2}^t + x_1 \beta_{12}^t + \ldots + x_M \beta_{K2}^t + e_2^t = 0$$

$$\vdots \qquad \qquad \vdots \qquad \qquad \vdots \qquad \qquad \vdots$$

$$y_1^t \gamma_{1M}^t + \ldots + y_M^t \gamma_{MM}^t + y_1^{t-1} \alpha_{1M}^t + \ldots + y_M^{t-1} \alpha_{MM}^t + x_1 \beta_{1M}^t + \ldots + x_M \beta_{KM}^t + e_M^t = 0$$

$$t = 1, 2, \ldots, T. \qquad (4.172)$$

In matrix notation, the dynamic structural equation model (4.172) can be written compactly as

$$\mathbf{Y^t \Gamma^t + Y^{t-1} A^t + X B^t + E^t = 0,} \tag{4.173}$$

where $\mathbf{0}$ is a $n \times M$ matrix of zeros,

$$\mathbf{Y^t} = \begin{bmatrix} y_{11}^t & y_{12}^t & \cdots & y_{1M}^t \\ y_{21}^t & y_{22}^t & \cdots & y_{2M}^t \\ \vdots & \vdots & \vdots \\ y_{n1}^t & y_{n2}^t & \cdots & y_{nM}^t \end{bmatrix} = \begin{bmatrix} \mathbf{y_1^t} & \mathbf{y_2^t} & \cdots & \mathbf{y_M^t} \end{bmatrix} \tag{4.174}$$

and

$$\mathbf{Y^{t-1}} = \begin{bmatrix} y_{11}^{t-1} & y_{12}^{t-1} & \cdots & y_{1M}^{t-1} \\ y_{21}^{t-1} & y_{22}^{t-1} & \cdots & y_{2M}^{t-1} \\ \vdots & \vdots & \vdots \\ y_{n1}^{t-1} & y_{n2}^{t-1} & \cdots & y_{nM}^{t-1} \end{bmatrix} = \begin{bmatrix} \mathbf{y_1^{t-1}} & \mathbf{y_2^{t-1}} & \cdots & \mathbf{y_M^{t-1}} \end{bmatrix} \tag{4.175}$$

are the $n \times M$ matrices of the sample values of the expression levels of the $M$ genes and $n$ samples at the time points $t$ and $t - 1$, respectively, matrix $\mathbf{X}$

$$\mathbf{X} = \begin{bmatrix} x_{11} & x_{12} & \cdots & x_{1K} \\ x_{21} & x_{22} & \cdots & x_{2K} \\ \vdots & \vdots & \vdots & \vdots \\ x_{n1} & x_{n2} & \cdots & x_{nK} \end{bmatrix} = \begin{bmatrix} \mathbf{x_1} & \mathbf{x_2} & \cdots & \mathbf{x_K} \end{bmatrix} \tag{4.176}$$

is the predetermined variables and

$$\mathbf{E^t} = \begin{bmatrix} e_{11}^t & e_{12}^t & \cdots & e_{1M}^t \\ e_{21}^t & e_{22}^t & \cdots & e_{2M}^t \\ \vdots & \vdots & \vdots & \vdots \\ e_{n1}^t & e_{n2}^t & \cdots & e_{nM}^t \end{bmatrix} = \begin{bmatrix} \mathbf{e_1^t} & \mathbf{e_2^t} & \cdots & \mathbf{e_M^t} \end{bmatrix} \tag{4.177}$$

is the matrix of unobserved values of the random variables at the time point $t$, the matrix

$$\mathbf{\Gamma^t} = \begin{bmatrix} \gamma_{11}^t & \gamma_{12}^t & \cdots & \gamma_{1M}^t \\ \gamma_{21}^t & \gamma_{22}^t & \cdots & \gamma_{2M}^t \\ \vdots & \vdots & \vdots & \vdots \\ \gamma_{M1}^t & \gamma_{M2}^t & \cdots & \gamma_{MM}^t \end{bmatrix} = \begin{bmatrix} \mathbf{\Gamma_1^t} & \mathbf{\Gamma_2^t} & \cdots & \mathbf{\Gamma_M^t} \end{bmatrix} \tag{4.178}$$

is the $M \times M$ matrix of gene regulatory coefficients at the time point $t$, the matrix

$$\mathbf{A}^t = \begin{bmatrix} \alpha_{11}^t & \alpha_{12}^t & \cdots & \alpha_{1M}^t \\ \alpha_{21}^t & \alpha_{22}^t & \cdots & \alpha_{2M}^t \\ \vdots & \vdots & \vdots & \vdots \\ \alpha_{M1}^t & \alpha_{M2}^t & \cdots & \alpha_{MM}^t \end{bmatrix} = \begin{bmatrix} \mathbf{A}_1^t & \mathbf{A}_2^t & \cdots & \mathbf{A}_M^t \end{bmatrix} \tag{4.179}$$

is the $M \times M$ matrix of coefficients of the regulation of the gene expression at the time point $t$ by the gene expression at the time point $t - 1$ and the matrix

$$\mathbf{B}^t = \begin{bmatrix} \beta_{11}^t & \beta_{12}^t & \cdots & \beta_{1M}^t \\ \beta_{21}^t & \beta_{22}^t & \cdots & \beta_{2M}^t \\ \vdots & \vdots & \vdots & \vdots \\ \beta_{K1}^t & \beta_{K2}^t & \cdots & \beta_{KM}^t \end{bmatrix} = \begin{bmatrix} \mathbf{B}_1^t & \mathbf{B}_2^t & \cdots & \mathbf{B}_M^t \end{bmatrix} \tag{4.180}$$

is a $K \times M$ matrix of coefficients of the contribution of the exogenous variables to the gene expressions at the time point $t$.

For the dynamic structural equation model we make the following assumptions. We assume that the random errors are generated by a non-stationary process with

$$E\left[e_i^t\right] = 0, i = 1, \ldots, M,$$

$$E\left[e_i^t \left(e_j^t\right)'\right] = \sigma_{ij}^t \mathbf{I}_n, \ i, j = 1, \ldots, M, \ t = 1, \ldots, T \tag{4.181}$$

and

$$E\left[e_i^{t_1} \left(e_j^{t_2}\right)'\right] = \sigma_{ij}^{t_1 t_2} \mathbf{I}_n, \ i, j = 1, \ldots, M, t_1, t_2 = 1, \ldots, T. \tag{4.182}$$

In matrix form, the above assumptions can be written as

$$E\left[vec(\mathbf{E}^{t_1}) \left(vec(E^{t_2})\right)'\right] = \begin{cases} \mathbf{\Sigma}_{tt} \otimes \mathbf{I}_n & t_1 = t_2 \\ \mathbf{\Sigma}_{t_1 t_2} \otimes \mathbf{I}_n & t_1 \neq t_2 \end{cases}, \tag{4.183}$$

where $vec(.)$ is a vector operator of the matrix, $\otimes$ denotes the Kronecker product, $\mathbf{I}_n$ is an $n$ dimensional identify matrix and

$$\mathbf{\Sigma}_{tt} = \begin{bmatrix} \sigma_{11}^t & \cdots & \sigma_{1M}^t \\ \vdots & \vdots & \vdots \\ \sigma_{M1}^t & \cdots & \sigma_{MM}^t \end{bmatrix} \text{ and }$$

$$\boldsymbol{\Sigma}_{t_1 t_2} = \begin{bmatrix} \sigma_{11}^{t_1 t_2} & \cdots & \sigma_{1M}^{t_1 t_2} \\ \vdots & \vdots & \vdots \\ \sigma_{M1}^{t_1 t_2} & \cdots & \sigma_{MM}^{t_1 t_2} \end{bmatrix}. \tag{4.184}$$

In addition, we also assume that the random errors and exogenous variables are uncorrelated.

In a matrix form, the $i^{th}$ equation in (4.171) can be written as

$$\mathbf{Y}^t \boldsymbol{\Gamma}_i^t + \mathbf{Y}^{t-1} \mathbf{A}_i^t + \mathbf{X} \mathbf{B}_i^t + \mathbf{E}_i^t = 0. \tag{4.185}$$

Traditionally, we often select one endogenous variable to appear on the left-hand side of the equation. Specifically, the i-th equation is

$$y_1^t \gamma_{1i}^t + \ldots + y_M^t \gamma_{Mi}^t + y_1^{t-1} \alpha_{1i}^t + \ldots + y_M^{t-1} \alpha_{Mi}^t + x_1 \beta_{1i}^t + \ldots + x_M \beta_{Ki}^t + e_i^t = 0. \tag{4.186}$$

Dividing both sides of the above equation by $-\gamma_{ii}^t$ and replacing $-\dfrac{\gamma_{ji}^t}{\gamma_{ii}^t}$, $\dfrac{\alpha_{li}^t}{\gamma_{ii}^t}$ and $-\dfrac{\beta_{ji}^t}{\gamma_{ii}^t}$ by $\gamma_{ji}^t$, $\alpha_{li}^t$ and $\beta_{ji}^t$, respectively, we obtain

$$y_i^t = y_1^t \gamma_{1i}^t + \ldots + y_{i-1,i}^t \gamma_{i-1i}^t + y_{i+1i}^t \gamma_{i+1}^t + \ldots + y_M^t \gamma_{Mi}^t + y_1^{t-1} \alpha_{1i}^t + \ldots + y_M^{t-1} \alpha_{Mi}^t$$
$$+ x_1 \beta_{1i}^t + \ldots + x_M \beta_{Ki}^t + e_i^t. \tag{4.187}$$

Let $(\mathbf{Y}_i^t)^*$, $(\mathbf{Y}_{li}^{t-1})^*$ and $(\mathbf{X}_i)^*$ be the matrices which contain those endogenous variables that are not present in the $i^{th}$ equation at the time points $t$ and $t-1$, and those exogenous variables that do not appear in the $i^{th}$ equation, respectively. Let $(\gamma_i^t)^*$, $(\alpha_i^t)^*$ and $(\beta_i^t)^*$ be their corresponding coefficients. Then, the vectors $\boldsymbol{\Gamma}_i^t$, $A_i^t$ and $B_i^t$ can be decomposed as

$$\boldsymbol{\Gamma}_i^t = \begin{bmatrix} -1 \\ \gamma_i^t \\ (\gamma_i^t)^* \end{bmatrix} = \begin{bmatrix} -1 \\ \gamma_i^t \\ 0 \end{bmatrix}, \ A_i^t = \begin{bmatrix} \alpha_i^t \\ (\alpha_i^t)^* \end{bmatrix} = \begin{bmatrix} \alpha_i^t \\ 0 \end{bmatrix} \text{ and } B_i^t = \begin{bmatrix} \beta_i^t \\ (\beta_i^t)^* \end{bmatrix} = \begin{bmatrix} \beta_i^t \\ 0 \end{bmatrix}.$$

Equation 4.187 can be reduced to

$$y_i^t = \mathbf{Y}_i^t \gamma_i^t + (\mathbf{Y}_i^t)^* (\gamma_i^t)^* + \mathbf{Y}_{li}^{t-1} \alpha_i^t + (\mathbf{Y}_{li}^{t-1})^* (\alpha_i^t)^* + \mathbf{X}_i \beta_i^t + \mathbf{X}_i^* (\beta_i^t)^* + e_i^t$$
$$= \mathbf{Y}_i^t \gamma_i^t + \mathbf{Y}_{li}^{t-1} \alpha_i^t + \mathbf{X}_i \beta_i^t + e_i^t$$
$$= \begin{bmatrix} \mathbf{Y}_i^t \mathbf{Y}_{li}^{t-1} \mathbf{X}_i \end{bmatrix} \begin{bmatrix} \gamma_i^t \\ \alpha_i^t \\ \beta_i^t \end{bmatrix} + e_i^t \tag{4.188}$$
$$= \mathbf{Z}_i^t \boldsymbol{\delta}_i^t + e_i^t, \ t = 1, \ldots, T,$$

where

$$Z_i^t = \begin{bmatrix} Y_i^t & Y_{li}^{t-1} & X_i \end{bmatrix},$$

$$\delta_i^t = \begin{bmatrix} \gamma_i^t \\ \alpha_i^t \\ \beta_i^t \end{bmatrix}.$$

Define

$$y_i = \begin{bmatrix} y_i^1 \\ y_i^2 \\ \vdots \\ y_i^T \end{bmatrix}, \ Z_i = \begin{bmatrix} Z_i^1 & 0 & \cdots & 0 \\ 0 & Z_i^2 & \cdots & 0 \\ \vdots & \vdots & \vdots & \vdots \\ 0 & 0 & \cdots & Z_i^T \end{bmatrix}, \ \delta_i = \begin{bmatrix} \delta_i^1 \\ \delta_i^2 \\ \vdots \\ \delta_i^T \end{bmatrix} \ \text{and} \ e_i = \begin{bmatrix} e_i^1 \\ e_i^2 \\ \vdots \\ e_i^T \end{bmatrix}.$$

Collectively, Equation 4.188 can be rewritten as

$$y_i = Z_i \delta_i + e_i. \tag{4.189}$$

From the previous assumption, we have

$$E[e_i] = 0 \text{ and}$$

$$V_i = \text{cov}(e_i, e_i) = \Sigma_{ii} \otimes I_n, \tag{4.190}$$

$$\Sigma_{ii} = \begin{bmatrix} \sigma_{ii}^{11} & \cdots & \sigma_{ii}^{1T} \\ \vdots & \vdots & \vdots \\ \sigma_{ii}^{T1} & \cdots & \sigma_{ii}^{TT} \end{bmatrix}.$$

We also assume that the covariance matrix $\Sigma_{ii}$ is non-singular. The weighted least square estimator of $\delta_i$ in (4.189) is

$$\hat{\delta}_i = \left(Z_i^T V_i^{-1} Z_i\right)^{-1} Z_i^T V_i^{-1} y_i. \tag{4.191}$$

To examine whether such estimation is unbiased we take its expectation:

$$\begin{aligned} E\left[\hat{\delta}_i\right] &= E\left[\left(Z_i^T V_i^{-1} Z_i\right)^{-1} Z_i^T V_i^{-1} y_i\right] \\ &= E\left[\left(Z_i^T V_i^{-1} Z_i\right)^{-1} Z_i^T V_i^{-1} (Z_i \delta_i + e_i)\right] \\ &= \delta_i + E\left[\left(Z_i^T V_i^{-1} Z_i\right)^{-1} Z_i^T V_i^{-1} e_i\right]. \end{aligned} \tag{4.192}$$

Since $e_i^t$ and $Y_i^t$ in matrix $Z_i^t$ is correlated the last term in Equation 4.192 will not be equal to zero. When $e_i^{t_1}$ and $e_i^{t_2}, t_1 \neq t_2$ are independent, Equation 4.191 will be reduced to

$$\hat{\delta}_i^t = \left( (Z_i^t)' Z_i^t \right)^{-1} (Z_i^t)' y_i^t, \ t = 1, ..., T, \tag{4.193}$$

where when $t = 1$ the terms involving $Y_i^{t-1}$ are not present in the matrix $Z_i^t$. Equation 4.193 indicates that when $e_i^{t_1}$ and $e_i^{t_2}, t_1 \neq t_2$ are independent, the parameters $\delta_i^t$ can be estimated separately.

Using Equation 4.171 and the assumption that matrix $\Gamma^t$ is nonsingular we can write gene expressions at the time point $t$ as a function of the gene expressions at the time point $t - 1$ and exogenous variables as follows.

$$Y^t = -Y_I^{t-1} A^t \left( \Gamma^t \right)^{-1} - X B^t \left( \Gamma^t \right)^{-1} - E^t \left( \Gamma^t \right)^{-1}$$

$$= \left[ Y_I^{t-1} X \right] \begin{bmatrix} -A^t \\ -B^t \end{bmatrix} \left( \Gamma^t \right)^{-1} - E^t \left( \Gamma^t \right)^{-1} \tag{4.194}$$

$$= W^t \Pi^t + \Psi^t,$$

where

$$W^t = \left[ Y_I^{t-1} \ X \right], \tag{4.195}$$

$$\Pi^t = -\begin{bmatrix} A^t \\ B^t \end{bmatrix} \left( \Gamma^t \right)^{-1} = \begin{bmatrix} \pi_{11}^t & \pi_{12}^t & \cdots & \pi_{1M}^t \\ \pi_{21}^t & \pi_{22}^t & \cdots & \pi_{2M}^t \\ \vdots & \vdots & \vdots & \vdots \\ \pi_{(K+M)1}^t & \pi_{(K+M)2}^t & \cdots & \pi_{(K+M)M}^t \end{bmatrix} = \begin{bmatrix} \pi_1^t & \pi_2^t & \cdots & \pi_M^t \end{bmatrix} \tag{4.196}$$

and

$$\Psi^t = -E^t \left( \Gamma^t \right)^{-1} = \begin{bmatrix} \psi_{11}^t & \psi_{12}^t & \cdots & \psi_{1M}^t \\ \psi_{21}^t & \psi_{22}^t & \cdots & \psi_{2M}^t \\ \vdots & \vdots & \vdots & \vdots \\ \psi_{n1}^t & \psi_{n2}^t & \cdots & \psi_{nM}^t \end{bmatrix} = \begin{bmatrix} \psi_1^t & \psi_2^t & \cdots & \psi_M^t \end{bmatrix}. \tag{4.197}$$

## 4.6.2 Estimation and Inference for Dynamic Structural Equation Models with Time-Varying Structures and Parameters

We extend the three popular methods: maximum likelihood method, two stage least squares (2SLS) method and three stage squares (3SLS) method for estimation of the parameters in the ordinary structural equation models to the parameter estimations for dynamic structural equation models (Judge et al. 1982).

### 4.6.2.1 Maximum Likelihood (ML) Estimation

The parameters in the dynamic structural equation models can be estimated by maximizing the likelihood function, assuming the structural errors are normally distributed. Let

$$
e_i^t = \begin{bmatrix} e_{i1}^t \\ \vdots \\ e_{iM}^t \end{bmatrix}, t = 1, \ldots, T \text{ and } e_i = \begin{bmatrix} e_i^1 \\ \vdots \\ e_i^T \end{bmatrix}, i = 1, \ldots, n,
$$

where

$$
e_i^t = -\left(\Gamma^t\right)' y_i^t - \left(A^t\right)' y_{1i}^{t-1} - \left(B^t\right)' x_i, \; i = 1, \ldots, n, \; t = 1, \ldots, T.
$$

Then, the vector of errors follow a multivariate normal distribution with mean zeros and covariance matrix $\Sigma$. The distribution of $e_i$ can be written as

$$
g(e_i) = (2\pi)^{-\frac{TM}{2}} |\Sigma|^{-\frac{1}{2}} \exp\left\{ -\frac{1}{2} e_i' \Sigma^{-1} e_i \right\},
$$

where

$$
\text{cov}(e_i, e_i) = \Sigma = \begin{bmatrix} \Sigma_{11} & \Sigma_{12} & \cdots & \Sigma_{1T} \\ \Sigma_{21} & \Sigma_{22} & \cdots & \Sigma_{2T} \\ \vdots & \vdots & \vdots & \vdots \\ \Sigma_{T1} & \Sigma_{T2} & \cdots & \Sigma_{TT} \end{bmatrix}, \; i = 1, \ldots, n, \text{ and}
$$

$$
\Sigma_{t\tau} = \text{cov}\left(e_i^t, e_i^\tau\right) = \begin{bmatrix} \sigma_{11}^{t\tau} & \cdots & \sigma_{1M}^{t\tau} \\ \vdots & \vdots & \vdots \\ \sigma_{M1}^{t\tau} & \cdots & \sigma_{MM}^{t\tau} \end{bmatrix}.
$$

The log-likelihood function is given in Result 4.8 (Appendix 4.E).

**Result 4.8**

The log likelihood function is defined as

$$
l(\Gamma, A, B, \Sigma | Y) = -\frac{nTM}{2} \log(2\pi) - \frac{n}{2} \log |\Sigma| + n \sum_{t=1}^{T} \log |\Gamma^t|
$$

$$
- \frac{1}{2} \sum_{i=1}^{n} e_i' \Sigma^{-1} e_i. \tag{4.198}
$$

If we assume that the errors $e_i^t$ and $e_i^\tau$ are uncorrelated, then, the likelihood function can be reduced to

$$l(\Gamma, A, B, \Sigma | Y) = -\frac{nTM}{2}\log(2\pi) - \frac{n}{2}\sum_{t=1}^{T}\log|\Sigma_{tt}|$$

$$+ n\sum_{t=1}^{T}\log|\Gamma^t| - \frac{1}{2}\sum_{i=1}^{n}\sum_{t=1}^{T}(e_i^t)'\Sigma_{tt}^{-1}e_i^t, \quad (4.199)$$

where

$$e_i^t = -(\Gamma^t)'y_i^t - (A^t)'y_{li}^{t-1} - (B^t)'x_i, \; i = 1, \ldots, n, \; t = 1, \ldots, T,$$

$$e_i^t = \begin{bmatrix} e_{i1}^t \\ \vdots \\ e_{iM}^t \end{bmatrix}, \; t = 1, \ldots, T \text{ and } e_i = \begin{bmatrix} e_i^1 \\ \vdots \\ e_i^T \end{bmatrix}, \; i = 1, \ldots, n.$$

The estimators $\hat{\Gamma}, \hat{A}, \hat{B}, \hat{\Sigma}$ of the parameters in the structural equations can be found by maximizing the log likelihood function in Equation 4.198 if the errors at different time points are correlated or in Equation 4.199 if the errors at different time points are uncorrelated.

### 4.6.2.2 Generalized Least Square Estimation

The previous maximum likelihood method needs to assume the normal distribution of the error stochastic processes. However, in practice normality assumptions on these variables will not hold. Therefore, we need to develop estimation methods that do not assume distributions of the error stochastic processes. Theoretic analysis indicates that the ordinary least square methods for estimation of the parameters in the structural equations, in general, will lead to inconsistent estimators. The statistical methods without assuming distribution are either single-equation methods, which can be applied to each equation of the structural equations, or complete systems methods, which are applied to all structural equations. Here, we extend the most important and widely used single equation method for the ordinary structural equation models to the dynamic structural equation models.

First, we introduce some notations for collection of the data at all time points. Let

$$Y = \begin{bmatrix} Y^1 & \cdots & Y^T \end{bmatrix}, \; W = \begin{bmatrix} W^1 & \cdots & W^T \end{bmatrix}, \; \Psi = \begin{bmatrix} \Psi^1 & \cdots & \Psi^T \end{bmatrix} \text{ and }$$

$$\Pi = \begin{bmatrix} \Pi^1 & \cdots & 0 \\ \vdots & \vdots & \vdots \\ 0 & \cdots & \Pi^T \end{bmatrix}.$$

From Equation 4.190 we obtain

$$\mathbf{Y} = \mathbf{W\Pi} + \mathbf{\Psi}, \qquad (4.200)$$

Equation 4.200 specifies the **reduced form** of the dynamic structural equation.

Now we calculate the covariance matrix of $\mathbf{\Psi}$. Recall that the matrix $\mathbf{\Psi}$ is

$$\mathbf{\Psi} = \begin{bmatrix} \mathbf{\Psi}^1 & \cdots & \mathbf{\Psi}^T \end{bmatrix}$$

$$= \begin{bmatrix} \psi_{11}^1 & \cdots & \psi_{1M}^1 & \cdots & \psi_{11}^T & \cdots & \psi_{1M}^T \\ \vdots & \ddots & & \ddots & & \vdots & \vdots \\ \psi_{j1}^1 & \cdots & \psi_{jM}^1 & \cdots & \psi_{j1}^T & \cdots & \psi_{jM}^T \\ \vdots & \ddots & & \ddots & & \vdots & \vdots \\ \psi_{n1}^1 & \cdots & \psi_{nM}^1 & \cdots & \psi_{n1}^T & \cdots & \psi_{nM}^T \end{bmatrix}. \qquad (4.201)$$

Let $\psi_j'$ be the $j^{th}$ row of $\mathbf{\Psi}$ and $e_j'$ be the $j^{th}$ row of $E$. It is to verify that

$$\left( \psi_j^t \right)' = -\left( e_j^t \right)' \left( \Gamma^t \right)^{-1} \qquad (4.202)$$

or

$$\psi_j^t = -\left( \left( \Gamma^t \right)^{-1} \right) e_t^j. \qquad (4.203)$$

Therefore, Equations 4.201 and 4.203 imply that

$$\psi_j = \begin{bmatrix} \psi_j^1 \\ \psi_j^2 \\ \vdots \\ \psi_j^T \end{bmatrix} = - \begin{bmatrix} \left( (\Gamma^1)^{-1} \right)' e_j^1 \\ \left( (\Gamma^2)^{-1} \right)' e_j^2 \\ \vdots \\ \left( (\Gamma^T)^{-1} \right)' e_j^T \end{bmatrix} \qquad (4.204)$$

$$= -\left( \Gamma^{-1} \right)' e_j,$$

where

$$\Gamma = \begin{bmatrix} \Gamma^1 & 0 & \cdots & 0 \\ 0 & \Gamma^2 & \cdots & 0 \\ \vdots & \vdots & \vdots & \vdots \\ 0 & 0 & \cdots & \Gamma^T \end{bmatrix} \text{ and } e_j = \begin{bmatrix} e_j^1 \\ e_j^2 \\ \vdots \\ e_j^T \end{bmatrix}.$$

The covariance matrix of $\psi_j$ is

$$\text{cov}\left(\psi_j, \psi_j\right) = \left(\Gamma^{-1}\right)'\Sigma\Gamma^{-1}, \tag{4.205}$$

where

$$\text{cov}(e_i, e_i) = \Sigma = \begin{bmatrix} \Sigma_{11} & \Sigma_{12} & \cdots & \Sigma_{1T} \\ \Sigma_{21} & \Sigma_{22} & \cdots & \Sigma_{2T} \\ \vdots & \vdots & \vdots & \vdots \\ \Sigma_{T1} & \Sigma_{T2} & \cdots & \Sigma_{TT} \end{bmatrix}, \quad i = 1, \ldots, n. \tag{4.206}$$

Since $\text{cov}(e_j, e_s) = 0$ we have

$$\text{cov}\left(\psi_j, \psi_s\right) = 0. \tag{4.207}$$

Combining Equations 4.205 and 4.207 gives

$$\text{cov}\left(vec(\Psi'), vec(\Psi')\right) = \left(\Gamma^{-1}\right)'\Sigma\Gamma^{-1} \otimes I_n. \tag{4.208}$$

From Equation 4.200 we obtain the least square estimator of the reduced form:

$$\Pi = \left(W'W\right)^{-1}W'Y. \tag{4.209}$$

In Appendix 4.H we derive the generalized least squares estimator of the parameter in the dynamic structural equation which is summarized in Result 4.9.

**Result 4.9**

Consider equation:

$$W'_*y_i = W'_*Z_i\delta_i + W'_*e_i, \tag{4.210}$$

where

$$W_* = \begin{bmatrix} W & \cdots & 0 \\ \vdots & \vdots & \vdots \\ U & \cdots & W \end{bmatrix}.$$

A generalized least square estimator of the parameters $\delta_i$ is

$$\hat{\delta}_i = \left[\left(W'_*Z_i\right)'\Lambda_i^{-1}W'_*Z_i\right]^{-1}\left(W'_*Z_i\right)'\Lambda_i^{-1}W'_*y_i, \tag{4.211}$$

where

$$\Lambda_i = \mathbf{W}'_* \mathbf{V}_i \mathbf{W}_*,$$

$$\mathbf{y}_i = \begin{bmatrix} \mathbf{y}_i^1 \\ \mathbf{y}_i^2 \\ \vdots \\ \mathbf{y}_i^T \end{bmatrix}, \mathbf{Z}_i = \begin{bmatrix} \mathbf{Z}_i^1 & 0 & \cdots & 0 \\ 0 & \mathbf{Z}_i^2 & \cdots & 0 \\ \vdots & \vdots & \vdots & \vdots \\ 0 & 0 & \cdots & \mathbf{Z}_i^T \end{bmatrix}, \boldsymbol{\delta}_i = \begin{bmatrix} \boldsymbol{\delta}_i^1 \\ \boldsymbol{\delta}_i^2 \\ \vdots \\ \boldsymbol{\delta}_i^T \end{bmatrix}, \boldsymbol{\Sigma} = \begin{bmatrix} \boldsymbol{\Sigma}_{11} & \boldsymbol{\Sigma}_{12} & \cdots & \boldsymbol{\Sigma}_{1T} \\ \boldsymbol{\Sigma}_{21} & \boldsymbol{\Sigma}_{22} & \cdots & \boldsymbol{\Sigma}_{2T} \\ \vdots & \vdots & \vdots & \vdots \\ \boldsymbol{\Sigma}_{T1} & \boldsymbol{\Sigma}_{T2} & \cdots & \boldsymbol{\Sigma}_{TT} \end{bmatrix},$$

$$\mathbf{Z}_i^t = \begin{bmatrix} \mathbf{Y}_i^t & \mathbf{Y}_{li}^{t-1} & \mathbf{X}_i \end{bmatrix}, \boldsymbol{\delta}_i^t = \begin{bmatrix} \boldsymbol{\gamma}_i^t \\ \boldsymbol{\alpha}_i^t \\ \boldsymbol{\beta}_i^t \end{bmatrix}, \mathbf{W}^t = \begin{bmatrix} \mathbf{Y}_i^{t-1} & \mathbf{X} \end{bmatrix}, \mathbf{W} = \begin{bmatrix} \mathbf{W}^1 & \cdots & \mathbf{W}^T \end{bmatrix},$$

$\mathbf{V}_i = \text{cov}(\mathbf{e}_i, \mathbf{e}_i) = \boldsymbol{\Sigma}_{ii} \otimes \mathbf{I}_n$ and $\boldsymbol{\Sigma}_{ii} = \begin{bmatrix} \sigma_{ii}^{11} & \cdots & \sigma_{ii}^{1T} \\ \vdots & \vdots & \vdots \\ \sigma_{ii}^{T1} & \cdots & \sigma_{ii}^{TT} \end{bmatrix}$. The covariance matrix
$\boldsymbol{\Sigma}_{ii}$ is estimated by

$$\hat{\boldsymbol{\Sigma}}_{ii} = \frac{1}{(nT - m_i + T - k_i - l_i)} (\mathbf{y}_i - \mathbf{Z}_i \boldsymbol{\delta}_i)(\mathbf{y}_i - \mathbf{Z}_i \boldsymbol{\delta}_i)'. \tag{4.212}$$

where $m_i = \sum_{t=1}^{T} m_i^t, l_i = \sum_{t=2}^{T} l_i^t, k_i = \sum_{t=1}^{T} k_i^t, m_i^t, l_i^t$ and $k_i^t$ be the number of endogenous at the time points $t$ and $t - 1$, and exogenous variables at the time point present $t$ in the $i^{th}$ equation, respectively.

The covariance matrix of the estimator $\hat{\boldsymbol{\delta}}_i$ is

$$\Sigma_\delta = \text{cov}\left(\hat{\boldsymbol{\delta}}_i, \hat{\boldsymbol{\delta}}_i\right) = \left[(\mathbf{W}'_* \mathbf{Z}_i)' \Lambda_i^{-1} \mathbf{W}'_* \mathbf{Z}_i\right]^{-1}. \tag{4.213}$$

### 4.6.3 Sparse Dynamic Structural Equation Models

In general, the dynamic structural Bayesian networks are sparse. Therefore, $\boldsymbol{\Gamma}^t, \mathbf{A}^t$ and $\mathbf{B}^t$ are sparse matrices. In order to obtain sparse estimates of $\boldsymbol{\Gamma}^t, \mathbf{A}^t$ and $\mathbf{B}^t$, the natural approach is the $l_1$-norm penalization.

#### 4.6.3.1 L₁-Penalized Maximum Likelihood Estimation

In order to obtain sparse estimates of $\boldsymbol{\Gamma}^t, \mathbf{A}^t$ and $\mathbf{B}^t, t = 1, 2, \ldots, T$, a natural approach is to maximize the log likelihood regularized by the $l_1$-norm terms $\|\boldsymbol{\Gamma}^t\|_1, \|\mathbf{A}^t\|_1$ and $\|\mathbf{B}^t\|_1$. The proposed $l_1$-penalized ML estimation approach is to maximize

$$\max_{\boldsymbol{\Gamma}^t, \mathbf{A}^t, \mathbf{B}^t} l(\boldsymbol{\Gamma}, \mathbf{A}, \mathbf{B}, \boldsymbol{\Sigma} | \mathbf{Y}) - \lambda_1 \sum_{t=1}^{T} \|\boldsymbol{\Gamma}^t\|_1 - \lambda_2 \sum_{t=1}^{T} \|\mathbf{A}^t\|_1$$

$$- \lambda_3 \sum_{t=1}^{T} \|\mathbf{B}^t\|_1, \tag{4.214}$$

where $\lambda_1, \lambda_2$ and $\lambda_3$ are penalty parameters,

$$l(\Gamma, A, B, \Sigma | Y) = -\frac{nTM}{2} \log(2\pi) - \frac{n}{2} \log|\Sigma| + n\sum_{t=1}^{T} \log|\Gamma^t| - \frac{1}{2}\sum_{i=1}^{n} e_i'\Sigma^{-1}e_i,$$

where

$$\Gamma = \begin{bmatrix} \Gamma^1 & 0 & \cdots & 0 \\ 0 & \Gamma^2 & \cdots & 0 \\ \vdots & \vdots & \vdots & \vdots \\ 0 & 0 & \cdots & \Gamma^T \end{bmatrix}, \ A = \begin{bmatrix} A^1 & 0 & \cdots & 0 \\ 0 & A^2 & \cdots & 0 \\ \vdots & \vdots & \vdots & \vdots \\ 0 & 0 & \cdots & A^T \end{bmatrix}, \ B = \begin{bmatrix} B^1 & 0 & \cdots & 0 \\ 0 & B^2 & \cdots & 0 \\ \vdots & \vdots & \vdots & \vdots \\ 0 & 0 & \cdots & B^T \end{bmatrix}, \ y_i = \begin{bmatrix} y_i^1 \\ y_i^2 \\ \vdots \\ y_i^T \end{bmatrix},$$

$$y_i^t = \begin{bmatrix} y_{i1}^t \\ y_{i2}^t \\ \vdots \\ y_{iM}^t \end{bmatrix}, \ y_{li} = \begin{bmatrix} y_{li}^0 \\ y_{li}^1 \\ \vdots \\ y_{li}^{T-1} \end{bmatrix}, \ y_{li}^{t-1} = \begin{bmatrix} y_{li1}^{t-1} \\ y_{li2}^{t-1} \\ \vdots \\ y_{liM}^{t-1} \end{bmatrix}, \ X_i = \begin{bmatrix} X_i \\ X_i \\ \vdots \\ X_i \end{bmatrix}, \ x_i = \begin{bmatrix} x_{i1} \\ x_{i2} \\ \vdots \\ x_{iK} \end{bmatrix},$$

$$e_i = \begin{bmatrix} e_i^1 \\ e_i^2 \\ \vdots \\ e_i^T \end{bmatrix}, \ e_i^t = \begin{bmatrix} e_{i1}^t \\ \vdots \\ e_{iM}^t \end{bmatrix},$$

$$e_i^t = -\left(\Gamma^t\right)'y_i^t - \left(A^t\right)'y_{li}^{t-1} - \left(B^t\right)'X_i, \ t = 1, \ldots, T, \ e_i = -\left(\Gamma'y_i + A'y_{li} + B'X_i\right) \text{ and}$$

$$e_i = \begin{bmatrix} e_i^1 \\ e_i^2 \\ \vdots \\ e_i^T \end{bmatrix}.$$

The objective function of the optimization problem (4.214) consists of two functions: convex differentiable functions and nonsmooth functions. The traditional Newton's method is an efficient tool for solving unconstrained smooth optimization problems, but is not suited for solving large nonsmooth convex problems. Proximal methods can be viewed as an extension of Newton's method from solving smooth optimization problems to nonsmooth optimization problems. In Appendix 4.G we derive primal gradient algorithm for $L_1$-penalized maximum likelihood estimation of parameters in the dynamic structural equation models. The algorithm is summarized as follows.

**Algorithm 4.4: Proximal Gradient Algorithm**

**Step 1.** Generate initial values using least square methods.
Let

$$\mathbf{W}^t = \begin{bmatrix} \mathbf{Y}_i^{t-1} \ \mathbf{X} \end{bmatrix}, \ \mathbf{W} = \begin{bmatrix} \mathbf{W}^1 & \cdots & \mathbf{W}^T \end{bmatrix}, \ \mathbf{Z}_i^t = \begin{bmatrix} \mathbf{Y}_i^t \ \mathbf{Y}_{li}^{t-1} \ \mathbf{X}_i \end{bmatrix},$$

$$t = 1, \ldots, T, \ \Lambda_i = \mathbf{W}'\mathbf{W}.$$

$$\mathbf{y}_i = \begin{bmatrix} \mathbf{y}_i^1 \\ \mathbf{y}_i^2 \\ \vdots \\ \mathbf{y}_i^T \end{bmatrix}, \ \mathbf{Z}_i = \begin{bmatrix} \mathbf{Z}_i^1 & 0 & \cdots & 0 \\ 0 & \mathbf{Z}_i^2 & \cdots & 0 \\ \vdots & \vdots & \vdots & \vdots \\ 0 & 0 & \cdots & \mathbf{Z}_i^T \end{bmatrix}, \ \boldsymbol{\delta}_i = \begin{bmatrix} \boldsymbol{\delta}_i^1 \\ \boldsymbol{\delta}_i^2 \\ \vdots \\ \boldsymbol{\delta}_i^T \end{bmatrix}, \ \boldsymbol{\delta}_i^t = \begin{bmatrix} \boldsymbol{\gamma}_i^t \\ \boldsymbol{\alpha}_i^t \\ \boldsymbol{\beta}_i^t \end{bmatrix}.$$

A least square estimator of the parameters $\boldsymbol{\delta}_i$ is

$$\hat{\boldsymbol{\delta}}_i = \left[ (\mathbf{W}'\mathbf{Z}_i)' \Lambda_i^{-1} \mathbf{W}'\mathbf{Z}_i \right]^{-1} (\mathbf{W}'\mathbf{Z}_i)' \Lambda_i^{-1} \mathbf{W}'\mathbf{y}_i, \ i = 1, \ldots, M. \tag{4.215}$$

From $\hat{\boldsymbol{\delta}}_i$ we can extract $\boldsymbol{\gamma}_i^t, \boldsymbol{\alpha}_i^t$ and $\boldsymbol{\beta}_i^t$. Assemble $\boldsymbol{\gamma}_i^t, \boldsymbol{\alpha}_i^t$ and $\boldsymbol{\beta}_i^t$ into matrices:

$$\boldsymbol{\Gamma}^t = \begin{bmatrix} \boldsymbol{\gamma}_1^t & \cdots & \boldsymbol{\gamma}_M^t \end{bmatrix}, \ \mathbf{A}^t = \begin{bmatrix} \boldsymbol{\alpha}_1^t & \cdots & \boldsymbol{\alpha}_M^t \end{bmatrix} \text{ and } \mathbf{B}^t = \begin{bmatrix} \boldsymbol{\beta}_1^t & \cdots & \boldsymbol{\beta}_M^t \end{bmatrix}.$$

Define

$$\boldsymbol{\Gamma}_0 = \begin{bmatrix} \boldsymbol{\Gamma}^1 & 0 & \cdots & 0 \\ 0 & \boldsymbol{\Gamma}^2 & \cdots & 0 \\ \vdots & \vdots & \vdots & \vdots \\ 0 & 0 & \cdots & \boldsymbol{\Gamma}^T \end{bmatrix}, \ \mathbf{A}_0 = \begin{bmatrix} \mathbf{A}^1 & 0 & \cdots & 0 \\ 0 & \mathbf{A}^2 & \cdots & 0 \\ \vdots & \vdots & \vdots & \vdots \\ 0 & 0 & \cdots & \mathbf{A}^T \end{bmatrix}, \ \mathbf{B}_0 = \begin{bmatrix} \mathbf{B}^1 & 0 & \cdots & 0 \\ 0 & \mathbf{B}^2 & \cdots & 0 \\ \vdots & \vdots & \vdots & \vdots \\ 0 & 0 & \cdots & \mathbf{B}^T \end{bmatrix}.$$

Set initial values of the matrix $W^0$ and vector $G^0$:

$$\mathbf{W}_\Gamma^0 = \boldsymbol{\Gamma}_0, \ \mathbf{W}_A^0 = \mathbf{A}_0, \ \mathbf{W}_B^0 = \mathbf{B}_0 \text{ and}$$

$$\mathbf{G}_\Gamma^0 = vec(\boldsymbol{\Gamma}_0), \ \mathbf{G}_A^0 = vec(\mathbf{A}_0) \text{ and } \mathbf{G}_B^0 = vec(\mathbf{B}_0).$$

Calculate the estimator of the covariance matrix $\Sigma$:

$$\hat{\Sigma}^1 = \frac{1}{n} \sum_{i=1}^n e_i e_i', \tag{4.216}$$

or

$$\hat{\Sigma}^1 = \frac{1}{(n - 2M - K)T} \sum_{i=1}^n e_i e_i', \tag{4.217}$$

where

$$e_i' = -\left( \mathbf{y}_i'\boldsymbol{\Gamma} + \mathbf{y}_{li}'\mathbf{A} + \mathbf{X}_i'\mathbf{B} \right)$$

and

$$\mathbf{e}_i = -(\mathbf{\Gamma}'\,\mathbf{y}_i + \mathbf{A}'\,\mathbf{y}_{li} + \mathbf{B}'\,\mathbf{X}_i)\,.$$

Set $\lambda_1^0 = 1, \lambda_2^0 = 1$ and $\lambda_3^0 = 1$, and parameter $\alpha_1, \alpha_2, \alpha_3 \in (0,1)$, for example $\alpha_1 = 0.5, \alpha_2 = 0.5, \alpha_3 = 0.5$. Set $k = 1, \lambda_1 = \lambda_1^{k-1}, \lambda_2 = \lambda_2^{k-1}$ and $\lambda_3 = \lambda_3^{k-1}$.

Set $l = 1$.

**Step 2:**

Repeat until convergence

1. Set

$$\mathbf{Z} = \begin{bmatrix} \mathbf{Z}_\Gamma \\ \mathbf{Z}_A \\ \mathbf{Z}_B \end{bmatrix} = \begin{bmatrix} \mathrm{Prox}_{\lambda_1 \Omega_\Gamma}\left(\mathbf{u}_\Gamma^k\right) \\ \mathrm{Prox}_{\lambda_2 \Omega_A}\left(\mathbf{u}_A^k\right) \\ \mathrm{Prox}_{\lambda_3 \Omega_B}\left(\mathbf{u}_B^k\right) \end{bmatrix},$$

where we calculate the proximal operator element-wise as follows:

$$\mathbf{u}_\Gamma^k = \mathbf{W}_\Gamma^k - \lambda_1 \frac{\partial f}{\partial \mathbf{W}_\Gamma}\bigg|_{W^k}, \quad \mathbf{u}_A^k = \mathbf{W}_A^k - \lambda_2 \frac{\partial f}{\partial \mathbf{W}_A}\bigg|_{W^k}, \quad \mathbf{u}_B^k = \mathbf{W}_B^k - \lambda_3 \frac{\partial f}{\partial \mathbf{W}_B}\bigg|_{W^k},$$

$$\frac{\partial f}{\partial \Gamma} = -n(\mathbf{\Gamma}')^{-1} - \sum_{i=1}^{n} y_i e_i'(\mathbf{\Sigma}^{-1}), \quad \frac{\partial f}{\partial \mathbf{A}} = -\sum_{i=1}^{n} y_{li} e_i' \mathbf{\Sigma}^{-1}, \quad \frac{\partial f}{\partial \mathbf{B}}$$

$$= -\sum_{i=1}^{n} \mathbf{X}_i e_i' \mathbf{\Sigma}^{-1},$$

$$\mathbf{e}_i' = -(\mathbf{y}_i' \mathbf{\Gamma} + \mathbf{y}_{li}' \mathbf{A} + \mathbf{X}_i' \mathbf{B}),$$

$$\frac{\partial f}{\partial \mathbf{W}_\Gamma}\bigg|_{W^k} = vec\left(\frac{\partial f}{\partial \mathbf{\Gamma}}\right), \quad \frac{\partial f}{\partial \mathbf{W}_A}\bigg|_{W^k} = vec\left(\frac{\partial f}{\partial \mathbf{A}}\right), \quad \frac{\partial f}{\partial \mathbf{W}_B}\bigg|_{W^k} = vec\left(\frac{\partial f}{\partial \mathbf{B}}\right),$$

all values of matrices and vectors are calculated at the $k^{th}$ iteration.

$$\mathbf{W}_\Gamma^k = \left(\gamma_{ij}^k\right)_{MT \times MT}, \quad \mathbf{W}_A^k = \left(\alpha_{ij}^k\right)_{MT \times MT}, \quad \mathbf{W}_B^k = \left(\beta_{ij}^k\right)_{KT \times MT},$$

$$\mathbf{u}_\Gamma^k = \left(u_{\Gamma ij}^k\right)_{MT \times MT}, \quad \mathbf{u}_A^k = \left(u_{Aij}^k\right)_{MT \times MT}, \quad \mathbf{u}_B^k = \left(u_{Bij}\right)_{KT \times MT},$$

$$\gamma_{ij}^z = \mathrm{sign}\left(u_{\Gamma ij}^k\right)\left(\left|u_{\Gamma ij}^k\right| - \lambda_1^k\right)_+, \, i = 1, \ldots, MT, j = 1, \ldots, MT, \quad (4.218)$$

$$\alpha_{ij}^z = \mathrm{sign}\left(u_{Aij}^k\right)\left(\left|u_{Aij}^k\right| - \lambda_2^k\right)_+, \, i = 1, \ldots, MT, j = 1, \ldots, MT, \quad (4.219)$$

$$\beta_{ij}^z = \mathrm{sign}\left(u_{Bij}^k\right)\left(\left|u_{Bij}^k\right| - \lambda_3^k\right)_+, \, i = 1, \ldots, KT, j = 1, \ldots, MT, \quad (4.220)$$

$$\mathbf{\Gamma}_z = \left(\gamma_{ij}^z\right)_{MT \times MT}, \ \mathbf{A}_z = \left(\alpha_{ij}^z\right)_{MT \times MT} \text{ and } \mathbf{B}_z = \left(\beta_{ij}\right)_{KT \times MT} \quad (4.221)$$

$$\mathbf{Z}_\Gamma = vec(\mathbf{\Gamma}_z), \ \mathbf{Z}_A = vec(\mathbf{A}_z) \text{ and } \mathbf{Z}_B = vec(\mathbf{B}_z).$$

2. Calculate the auxiliary function:

$$f_\lambda\left(\mathbf{Z}, \mathbf{G}^k\right) = f(\mathbf{Z}) + \left(\frac{\partial f}{\partial \mathbf{Z}}\right)^T \left(\mathbf{G}^k - \mathbf{Z}\right) + \frac{1}{2\lambda_1} \| \mathbf{G}_\Gamma^k - \mathbf{Z}_\Gamma \|_2^2$$

$$+ \frac{1}{2\lambda_2} \| \mathbf{G}_A^k - \mathbf{Z}_A \|_2^2 + \frac{1}{2\lambda_3} \| \mathbf{G}_B^k - \mathbf{Z}_B \|_2^2$$

3. Break and go to step 3 if $f(\mathbf{Z}) \leq \hat{f}_\lambda(\mathbf{Z}, \mathbf{G}^k)$.
4. Update $\lambda_1 = \alpha_1\lambda_1, \lambda_2 = \alpha_2\lambda_2$ and $\lambda_3 = \alpha_3\lambda_3$.

**Step 3:** Check convergence for the structural parameters.
If $\| \mathbf{\Gamma}_Z - \mathbf{W}_\Gamma^k \|_F > \varepsilon_1, \| \mathbf{A}_Z - \mathbf{W}_A^k \|_F > \varepsilon_2, \| B_Z - W_B^k \|_F > \varepsilon_3$ then
return $\lambda_1^k = \lambda_1, \lambda_2^k = \lambda_2, \lambda_3^k = \lambda_3, \mathbf{G}^{k+1} = \mathbf{Z}$ and

$$\mathbf{W}_\Gamma^{k+1} = \mathbf{\Gamma}_Z, \ \mathbf{W}_A^{k+1} = \mathbf{A}_Z, \ \mathbf{W}_B^{k+1} = \mathbf{B}_Z, \ k \leftarrow k + 1 \quad (4.222)$$

Go to step 2 (1).
Otherwise go to step 4
**Step 4:** Update the covariance matrix.

$$\hat{\Sigma}^{l+1} = \frac{1}{n} \sum_{i=1}^{n} e_i^l e_i^{l\prime}, \quad (4.223)$$

or

$$\hat{\Sigma}^{l+1} = \frac{1}{(n - 2M - K)T} \sum_{i=1}^{n} e_i^l e_i^{l\prime}, \quad (4.224)$$

where

$$e_i^{l\prime} = -\left(\mathbf{y}_i'\mathbf{\Gamma}_Z + \mathbf{y}_{li}'\mathbf{A}_Z + \mathbf{X}_i'\mathbf{B}_Z\right)$$

$$e_i^l = -\left(\mathbf{\Gamma}_Z'\mathbf{y}_i + \mathbf{A}_Z'\mathbf{y}_{li} + \mathbf{B}_Z'\mathbf{X}_i\right).$$

**Step 5:** Check for convergence of the covariance matrix.
If $\| \Sigma^{l+1} - \Sigma^l \|_F > \eta$ where $\eta$ is a pre-specified small tolerance error,
then $\hat{\Sigma}_l \leftarrow \hat{\Sigma}_{l+1}, l \leftarrow l + 1$, go to step 2.
Otherwise
Stop. $\mathbf{W}_\Gamma^{k+1} = \mathbf{\Gamma}_Z, \mathbf{W}_A^{k+1} = \mathbf{A}_Z, \mathbf{W}_B^{k+1} = \mathbf{B}_Z$ are estimated parameter matrices of $\mathbf{\Gamma}, \mathbf{A}$ and $\mathbf{B}$.

### 4.6.3.2 L₁ Penalized Generalized Least Square Estimator

Consider the $i^{th}$ equation:

$$\mathbf{y}_i = \mathbf{Z}_i\boldsymbol{\delta}_i + \mathbf{e}_i . \quad (4.225)$$

We can arrange the order of elements in the matrix $\mathbf{Z}_i$ and vector $\boldsymbol{\delta}_i$ to reduce Equation 4.225 to

$$\mathbf{y}_i = \mathbf{Z}_{\gamma i}\boldsymbol{\gamma}_i + \mathbf{Z}_{\alpha i}\boldsymbol{\alpha}_i + \mathbf{Z}_{\beta i}\boldsymbol{\beta}_i + \mathbf{e}_i, \tag{4.226}$$

where

$$\boldsymbol{\gamma}_i = \begin{bmatrix} \gamma_i^1 \\ \vdots \\ \gamma_i^T \end{bmatrix}, \boldsymbol{\alpha}_i = \begin{bmatrix} \alpha_i^1 \\ \vdots \\ \alpha_i^T \end{bmatrix}, \boldsymbol{\beta}_i = \begin{bmatrix} \beta_i^1 \\ \vdots \\ \beta_i^T \end{bmatrix}, \mathbf{Z}_{\gamma i} = \begin{bmatrix} \mathbf{Z}_{\gamma i}^1 & \cdots & 0 \\ \vdots & \vdots & \vdots \\ 0 & \cdots & \mathbf{Z}_{\gamma i}^T \end{bmatrix}, \mathbf{Z}_{\alpha i}$$

$$= \begin{bmatrix} \mathbf{Z}_{\alpha i}^1 & \cdots & 0 \\ \vdots & \vdots & \vdots \\ 0 & \cdots & \mathbf{Z}_{\alpha i}^T \end{bmatrix}, \mathbf{Z}_{\beta i} = \begin{bmatrix} \mathbf{Z}_{\beta i}^1 & \cdots & 0 \\ \vdots & \vdots & \vdots \\ 0 & \cdots & \mathbf{Z}_{\beta i}^T \end{bmatrix},$$

$\mathbf{Z}_{\gamma i}^t = \mathbf{Y}_i^t$, $\mathbf{Z}_{\alpha i}^t = \mathbf{Y}_{li}^{t-1}$, $\mathbf{Z}_{\beta i}^t = \mathbf{X}_i$ are corresponding endogenous variables (gene expressions) at the time points $t$ and $t - 1$, and exogenous variables which are not presented in the $i^{th}$ equation.

Using weighted least square and $l_1$-norm penalization of Equation 4.210, we can form the following optimization problem:

$$\min_{\delta_i} \quad f(\boldsymbol{\delta}_i) + \Omega(\boldsymbol{\delta}_i), \tag{4.227}$$

where

$$f(\boldsymbol{\delta}_i) = \left(\mathbf{W}' \mathbf{y_i} - \mathbf{W}' \mathbf{Z}_{\gamma i}\boldsymbol{\gamma}_i - \mathbf{W}' \mathbf{Z}_{\alpha i}\boldsymbol{\alpha}_i - \mathbf{W}' \mathbf{Z}_{\beta i}\boldsymbol{\beta}_i\right)' \boldsymbol{\Lambda}_i^{-1}$$

$$\left(\mathbf{W}' \mathbf{y_i} - \mathbf{W}' \mathbf{Z}_{\gamma i}\boldsymbol{\gamma}_i - \mathbf{W}' \mathbf{Z}_{\alpha i}\boldsymbol{\alpha}_i - \mathbf{W}' \mathbf{Z}_{\beta i}\boldsymbol{\beta}_i\right),$$

$$\boldsymbol{\Lambda_i} = \mathbf{W}' \mathbf{V}_i \mathbf{W}, \ \mathbf{W}^t = \begin{bmatrix} \mathbf{Y}_i^{t-1} & \mathbf{X} \end{bmatrix}, \ \mathbf{W}\begin{bmatrix} \mathbf{W}^1 & \cdots & \mathbf{W}^T \end{bmatrix},$$

$$\Omega(\boldsymbol{\delta}_i) = \lambda_1 \Omega_1(\boldsymbol{\gamma}_i) + \lambda_2 \Omega_2(\boldsymbol{\alpha}_i) + \lambda_3 \Omega_3(\boldsymbol{\beta}_i), \ \Omega_1(\boldsymbol{\gamma}_i) = \| \boldsymbol{\gamma}_i \|_1, \ \Omega_2(\boldsymbol{\alpha}_i)$$

$$= \| \boldsymbol{\alpha}_i \|_1, \ \Omega_3(\boldsymbol{\beta}_i) = \| \boldsymbol{\beta}_i \|_1.$$

Similar to Appendix 4.G, in Appendix 4.H, we derive the proximal gradient algorithm for solving the optimization problem (4.227). The algorithm is summarized as follows.

### Algorithm 4.5: Proximal Gradient Algorithm

**Step 1.** Generate initial values using least square methods.
A least square estimator of the parameters $\boldsymbol{\delta}_i$ is

$$\hat{\boldsymbol{\delta}}_i = \left[(\mathbf{W}' \mathbf{Z}_i)' \boldsymbol{\Lambda}_i^{-1} \mathbf{W}' \mathbf{Z}_i\right]^{-1} (\mathbf{W}' \mathbf{Z}_i)' \boldsymbol{\Lambda}_i^{-1} \mathbf{W}' \mathbf{y}_i, \tag{4.228}$$

where $\Lambda_i = \mathbf{W}'\mathbf{W}$, and $Z_i, y_i$ are defined in Equation 4.226.

From $\hat{\boldsymbol{\delta}}_i$ we can extract $\boldsymbol{\gamma}_i$, $\boldsymbol{\alpha}_i$ and $\boldsymbol{\beta}_i$. Define the set initial values $\boldsymbol{\gamma}_i^1 = \boldsymbol{\gamma}_i, \boldsymbol{\alpha}_i^1 = \boldsymbol{\alpha}_i$, $\boldsymbol{\beta}_i^1 = \boldsymbol{\beta}_i$. Let $\boldsymbol{\delta}_i^1 = \begin{bmatrix} \boldsymbol{\gamma}_i^1 \\ \boldsymbol{\alpha}_i^1 \\ \boldsymbol{\beta}_i^1 \end{bmatrix}$. Calculate the initial value of the covariance matrix $\boldsymbol{\Sigma}_{ii}$

$$\hat{\boldsymbol{\Sigma}}_{ii}^1 = \frac{1}{(n - m_i + 1 - l_i - k_i)T} \left(\mathbf{y}_i - \mathbf{Z}_i\boldsymbol{\delta}_i^1\right)\left(\mathbf{y}_i - \mathbf{Z}_i\boldsymbol{\delta}_i^1\right)', \tag{4.229}$$

$$\mathbf{V}_i^1 = \boldsymbol{\Sigma}_{ii}^1 \otimes \mathbf{I}_n.$$

Set $\lambda_1^1 = 1, \lambda_2^1 = 1$ and $\lambda_3^1 = 1$, and parameter $\alpha_1, \alpha_2, \alpha_3 \in (0,1)$, for example $\alpha_1 = 0.5$, $\alpha_2 = 0.5, \alpha_3 = 0.5$. Set $k = 1$, $\lambda_1 = \lambda_1^1$, $\lambda_2 = \lambda_2^1$ and $\lambda_3 = \lambda_3^1$. Set $k = 1$ and $l = 1$.

**Step 2:**

Repeat until convergence

1. Set

$$\boldsymbol{\delta}_i^z = \begin{bmatrix} \boldsymbol{\gamma}_i^z \\ \boldsymbol{\alpha}_i^z \\ \boldsymbol{\beta}_i^z \end{bmatrix} = \begin{bmatrix} \text{Prox}_{\lambda_1\Omega_1}\left(\mathbf{u}_{\gamma_i}^k\right) \\ \text{Prox}_{\lambda_2\Omega_2}\left(\mathbf{u}_{\alpha_i}^k\right) \\ \text{Prox}_{\lambda_3\Omega_3}\left(\mathbf{u}_{\beta_i}^k\right) \end{bmatrix},$$

where we calculate the proximal operator element-wise as follows:

$$\left.\frac{\partial f}{\partial \boldsymbol{\gamma}_i}\right|_{\boldsymbol{\delta}_i^k} = -2\mathbf{Z}'_{\gamma i}\mathbf{W}\Lambda_i^{-1}\left(\mathbf{W}'\mathbf{y_i} - \mathbf{W}'\mathbf{Z}_{\gamma i}\boldsymbol{\gamma}_i^k - \mathbf{W}'\mathbf{Z}_{\alpha i}\boldsymbol{\alpha}_i^k - \mathbf{W}'\mathbf{Z}_{\beta i}\boldsymbol{\beta}_i^k\right),$$

$$\frac{\partial f}{\partial \boldsymbol{\alpha}_i} = -2\mathbf{Z}'_{\alpha i}\mathbf{W}\Lambda_i^{-1}\left(\mathbf{W}'\mathbf{y_i} - \mathbf{W}'\mathbf{Z}_{\gamma i}\boldsymbol{\gamma}_i^k - \mathbf{W}'\mathbf{Z}_{\alpha i}\boldsymbol{\alpha}_i^k - \mathbf{W}'\mathbf{Z}_{\beta i}\boldsymbol{\beta}_i^k\right),$$

$$\left.\frac{\partial f}{\partial \boldsymbol{\beta}_i}\right|_{\boldsymbol{\delta}_i^k} = -2\mathbf{Z}'_{\beta i}\mathbf{W}\Lambda_i^{-1}\left(\mathbf{W}'\mathbf{y_i} - \mathbf{W}'\mathbf{Z}_{\gamma i}\boldsymbol{\gamma}_i^k - \mathbf{W}'\mathbf{Z}_{\alpha i}\boldsymbol{\alpha}_i^k - \mathbf{W}'\mathbf{Z}_{\beta i}\boldsymbol{\beta}_i^k\right),$$

$$\mathbf{u}_{\gamma_i}^k = \boldsymbol{\gamma}_i^k - \lambda_1^k\left.\frac{\partial f}{\partial \boldsymbol{\gamma}_i}\right|_{\boldsymbol{\delta}_i^k}, \quad \mathbf{u}_{\alpha_i}^k = \boldsymbol{\alpha}_i^k - \lambda_2^k\left.\frac{\partial f}{\partial \boldsymbol{\alpha}_i}\right|_{\boldsymbol{\delta}_i^k}, \quad \mathbf{u}_{\beta_i}^k = \boldsymbol{\beta}_i^k - \lambda_3^k\left.\frac{\partial f}{\partial \boldsymbol{\beta}_i}\right|_{\boldsymbol{\delta}_i^k},$$

$$\gamma_{ij}^z = \text{sign}\left(u_{\gamma_{ij}}^k\right)\left(\left|u_{\gamma_{ij}}^k\right| - \lambda_1^k\right)_+, j = 1, \ldots, m_i,$$

$$\alpha_{ij}^z = \text{sign}\left(u_{\alpha_{ij}}^k\right)\left(\left|u_{\alpha_{ij}}^k\right| - \lambda_2^k\right)_+, j = 1, \ldots, l_i,$$

$$\beta_{ij}^z = \text{sign}\left(u_{\beta_{ij}}^k\right)\left(\left|u_{\beta_{ij}}^k\right| - \lambda_3^k\right)_+, j = 1, \ldots, k_i.$$

Assemble $\gamma_{ij}^z$, $\alpha_{ij}^z$ and $\beta_{ij}^z$ into the vectors $\gamma_i^z$, $\alpha_i^z$ and $\beta_i^z$:

$$\gamma_i^z = \begin{bmatrix} \gamma_{i1}^z \\ \vdots \\ \gamma_{im_i}^z \end{bmatrix}, \ \alpha_i^z = \begin{bmatrix} \alpha_{i1}^z \\ \vdots \\ \alpha_{il_i}^z \end{bmatrix}, \ \beta_i^z = \begin{bmatrix} \beta_{i1}^z \\ \vdots \\ \beta_{ik_i}^z \end{bmatrix} \text{ and } \delta_i^z = \begin{bmatrix} \gamma_i^z \\ \alpha_i^z \\ \beta_i^z \end{bmatrix}.$$

2. Calculate the auxiliary function:

$$f_\lambda\left(\delta_i^z, \delta_i^k\right) = f(\delta_i^z) + \left(\frac{\partial f}{\partial \delta^i}\Big|_{\delta_i^k}\right)^T \left(\delta_i^k - \delta_i^z\right) + \frac{1}{2\lambda_1}\|\gamma_i^k - \gamma_i^z\|_2^2$$

$$+ \frac{1}{2\lambda_2}\|\alpha_i^k - \alpha_i^z\|_2^2 + \frac{1}{2\lambda_3}\|\beta_i^k - \beta_i^z\|_2^2.$$

3. Break and go to step 3 if $f(\delta_i^z) \leq \hat{f}_\lambda(\delta_i^z, \delta_i^k)$.
4. Update $\lambda_1 = \alpha_1\lambda_1, \lambda_2 = \alpha_2\lambda_2$ and $\lambda_3 = \alpha_3\lambda_3$.

**Step 3:** check convergence for the structural parameters.
If $\|\gamma_i^z - \gamma_i^k\|_2^2 > \varepsilon_1$, $\|\alpha_i^z - \alpha_i^k\|_2^2 > \varepsilon_2$, $\|\beta_i^z - \beta_i^k\|_2^2 > \varepsilon_3$ then
return $\lambda_1^k = \lambda_1, \lambda_2^k = \lambda_2, \lambda_3^k = \lambda_3$ and
$\gamma_i^{k+1} = \gamma_i^z, \alpha_i^{k+1} = \alpha_i^z, \beta_i^{k+1} = \beta_i^z, \delta_i^{k+1} = \delta_i^z$ and $k = k + 1$.
Go to **step 2** (1),
Otherwise go to **step 4.**
**Step 4:** update the covariance matrix.

$$\hat{\Sigma}_{ii}^{l+1} = \frac{1}{(n - m_i + 1 - l_i - k_i)T}(\mathbf{y}_i - \mathbf{Z}_i\delta_i^z)(\mathbf{y}_i - \mathbf{Z}_i\delta_i^z)',$$

$$\mathbf{V}_i^{l+1} = \Sigma_{ii}^{l+1} \otimes \mathbf{I}_n.$$

**Step 5:** check for convergence of covariance matrix.
If $\|\Sigma_{ii}^{l+1} - \Sigma_{ii}^l\|_F > \eta$ where $\eta$ is a pre-specified small tolerance error,
Then $\hat{\Sigma}_{ii}^l \leftarrow \Sigma_{ii}^{l+1}, V_i^l \leftarrow V_i^{l+1}, l \leftarrow l + 1$, go to **step 2.**
Otherwise stop. The final estimators of the parameter vectors of $\gamma_i, \alpha_i, \beta_i$
and $\delta_i$ are $\gamma_i^z, \alpha_i^z, \beta_i^z$ and $\delta_i^z$, respectively.

## 4.7 Single Cell RNA-Seq Data Analysis, Gene Expression Deconvolution, and Genetic Screening

Single-cell RNA-seq technologies allow sequencing the RNA of individual cells and investigating transcriptomic heterogeneity within cell populations. Some statistical methods and computational algorithms for analyzing RNA-seq data from bulk cell populations can be readily adapted to single cell RNA-seq data analysis (Stegle et al. 2015). However, single cell RNA-seq data analysis poses new computational and analytic challenges that require to develop new analytic strategies. In this section, we will focus on identification and characterization of cell types, discovery of cell types associated with complex phenotypes and diseases, estimation of cell type specific expressions, inference of gene regulatory networks across cell types and causal inference with interventional data generated by genetic screening.

### 4.7.1 Cell Type Identification

Classical gene expression analysis assumes that the gene expressions are measured from a homogeneous population. However, in biology, organisms are hierarchically organized. They are heterogeneous at multiple levels: organs, tissues and cell types (Pettit et al. 2014). A popular application of single cell RNA-seq analysis is to characterize and identify cell types using single cell transcriptomics. Unsupervised clustering methods are a major approach to cell type identification.

Wang et al. (2017) proposed to use multikernel-based single cell RNA-seq analysis method for reducing data dimensions and learning similarity. Then, spectral clustering uses the learning similarity to identify cell types.

Consider a $n \times k$ dimensional gene expression matrix $X$ with $n$ cells and $k$ genes. We start with introducing similarity graphs for spectral clustering. Consider a similarity graph $G = (V, E)$. Each node in the graph represents a data point (a cell) and each edge represents a similarity between two nodes connected by the edge, taking similarity measure as its associated weight. The problem of clustering cells can now be formulated using the similarity graph: to find a partition of the graph such that the edge between different groups have small weights and the edges within a group have large weights. Consider two cells (nodes): cell $i$ and cell $j$. Let $S_{ij}$ denotes the similarity between cell $i$ and cell $j$. Define the similarity matrix $S = (S_{ij})_{n \times n}$.

To find the best similarity graph that optimally clusters the cells we propose the following optimization problem:

$$\min_{S,L,W} \sum_{i=1}^{n}\sum_{j=1}^{n}D\left(x_i, x_j\right)S_{ij} + \beta \parallel S \parallel_F^2 + \gamma\mathrm{Tr}\left(L^T(I_n - S)L\right) + \rho\sum_l w_l \log w_l$$

$$\text{subject to} \quad D\left(x_i, x_j\right) = \sum_l w_l D_l\left(x_i, x_j\right), \sum_l w_l = 1, w_l \geq 0, L^T L = I_C, \sum_j S_{ij}$$

$$= 1, S_{ij} \geq 0,$$

$$(4.230)$$

where $x_i$ and $x_j$ are the gene expression vectors of the cells $i$ and $j$, i.e., the $i^{th}$ and $j^{th}$ row of the gene expression matrix $X$, $D_l(x_i, x_j)$ is the $l^{th}$ distance measure between cell $i$ and cell $j$ and $D(x_i, x_j)$ is the weighted summation of distances $D_l(x_i, x_j)$ between cell $i$ and cell $j$ with weights $w_l$, $I_n$ and $I_C$ are $n$ dimensional and $C$ dimensional identity matrices, respectively, $\parallel S \parallel_F$ is the Frobenius norm of the matrix $S$, and $\beta$ and $\gamma$ are non-negative penalty parameters. The objective function is minimized with respect to vector $w$ and matrices $S$ and $L$.

The first term in the objective function in Equation 4.7A5 indicates that if the distance between two cells is small, these two cells should be close and the similarity between them should be large. The second regularization term is to encourage the sparse similarity graph. The matrix $I_S - S$ in the third term is the graph Laplacian which encourages adjacent nodes induced by similarity measures to have close gene expression profiles. Thus, the third term coupled with the constraint on $L$ enforce the low rank cluster structure of $S$ with approximately $C$ connected components in a similarity graph in which the nodes represent the cells and edge weights represent the pair-wise similarity. The four terms impose a penalty to enforce selection of multiple kernels.

Substituting Equation 4.1.3 into Equation 4.1.6 gives the following optimization problem:

$$\min_{S,L,W} \quad -2\sum_{i=1}^{n}\sum_{j=1}^{n}w_l K_l\left(c_i, c_j\right)S_{ij} + \beta\,\|\,S\,\|_F^2 + \gamma \mathrm{Tr}\left(L^T(I_n - S)L\right)$$

$$+ \rho\sum_l w_l \log w_l$$

$$\text{subject to} \quad D\left(x_i, x_j\right) = \sum_l w_l D_l\left(x_i, x_j\right), \sum_l w_l = 1, w_l \geq 0, L^T L = I_C, \sum_j S_{ij}$$

$$= 1, S_{ij} \geq 0.$$

$$(4.231)$$

Gaussian kernels are often used in practice. Gaussian kernels are defined as

$$K\left(c_i, c_j\right) = \frac{1}{\sqrt{2\pi}\sigma_{ij}}\exp\left(-\frac{\|\,c_i, c_j\,\|_2^2}{2\sigma_{ij}^2}\right),\qquad (4.232)$$

where $c_i$ and $c_j$ are vectors of measured gene expressions from the cells $c_i$ and $c_j$, respectively, and $\|\,c_i - c_j\,\|_2$ is their Euclidean distance. The variance, $\sigma_{ij}$ are often calculated by

$$\sigma_{ij} = \frac{\sigma\left(\mu_i + \mu_j\right)}{2} \text{ and } \mu_i = \frac{\sum_{j \in KNN(c_i)}\|\,c_i - c_j\,\|_2}{k},\qquad (4.233)$$

where $KNN(c_i)$ denotes all cells that are top $k$ neighbors of the cell $c_i$, $k$ and $\sigma$ are parameters that generate the multiple kernels.

In Appendix 4.I, we present the similarity graph-based algorithm for solving the optimization problem (4.231). In summary, the algorithm is given below.

**Algorithm 4.6: Algorithm for Solving Optimization Problem (4.231)**

**Step 1:** Initialization of $S, w$ and $L$.
  The weight of multiple kernels is initiated as a uniform distribution vector: $w^{(0)} = \left[\dfrac{1}{G}, \dfrac{1}{G}, \dots, \dfrac{1}{G}\right]$, where $G$ is the number of kernels.

The initial similarity matrix is defined as $S_{ij} = \sum_{l=1}^{G} w_l K_l(c_i, c_j)$, $S = (S_{ij})_{C \times C}$. The top $C$ eigenvectors of the matrix $S$ is taken as an initial matrix $L$.

**Step 2:** Fixing $L$ and $w$ to update matrix $S$:

$$\min_{S} \quad -\sum_{i=1}^{n}\sum_{j=1}^{n}\left[\sum_{l=1}^{G} w_l K_l\left(c_i, c_j\right) + \gamma\left(LL^T\right)_{ij}\right] S_{ij} + \beta\| S \|_F^2$$

subject to $\sum_{j=1}^{n} S_{ij} = 1$ and $S_{ij} \geq 0$. $\qquad$ (4.234)

Let $s_i$ be the $i^{th}$ row of the matrix $S$ and

$$(v_i)_j = \frac{-1}{2\beta}\left[\gamma(LL^T)_{ij} - \sum_l w_l K_l\left(c_i, c_j\right)\right].$$

The solution to problem (4.231) is

$$(s_i)_j = \left((u_i)_j - \sigma^*\right)_+,$$

$$\sigma_j = \left(\sigma^* - (u_i)_j\right)_+, \text{ and}$$

$$f(\sigma^*) = \sigma^* - \frac{1}{n}\sum_{j=1}^{n}\left(\sigma^* - (u_i)_j\right)_+ = 0, \qquad (4.235)$$

where

Newton's method can be used to solve Equation 4.235 for $\sigma^*$.

**Step 3:** Update $L$ while fixing $S$ and $w$.

Fixing $S$ and $w$, and minimizing the objective function with respect to the latent matrix $L$ in (4.I.7) is equivalent to the following optimization problem:

$$\min_{L} \text{Tr}\left(L^T (S - I_n) L\right)$$

subject to $L^T L = I_C$. $\qquad$ (4.236)

The solution matrix $L^*$ are the eigenvectors corresponding to the $C$ largest eigenvalues of the matrix $S - I_n$:

$$(S - I_n)L = L\Lambda.$$

**Step 4:** Update $w$ while fixing $S$ and $L$.

To update $w$, we solve the following optimization problem with respect to $w$:

$$\min_{w} \quad -\sum_{l=1}^{G} w_l \sum_{i=1}^{n}\sum_{j=1}^{n} K_l\left(c_i, c_j\right) S_{ij} + \rho\sum_{l=1}^{G} w_l \log w_l$$

subject to $\sum_{l=1}^{G} w_l = 1, w_l \geq 0$. $\qquad$ (4.237)

The solution to the problem (4.237) is

$$w_l = \frac{\exp\left(\frac{1}{\rho}\sum_{i=1}^{n}\sum_{j=1}^{n} K_l\left(c_i, c_j\right) S_{ij}\right)}{\sum_{l=1}^{G} w_l \exp\left(\frac{1}{\rho}\sum_{i=1}^{n}\sum_{j=1}^{n} K_l\left(c_i, c_j\right) S_{ij}\right)}, \quad l = 1, \dots, G. \qquad (4.238)$$

**Step 5:** Similarity enhancement

View similarity between cells as information that can be spread over the similarity graph. Information diffusion theory can be used to model the similarity and hence to further enhance the similarity. Given the similarity matrix $S$, define a transition matrix: $P = (P_{ij})_{n \times n}$ where

$$P_{ij} = \frac{S_{ij} I_{\{j \in N_K(i)\}}}{\sum_l S_{il} I_{\{l \in N_K(i)\}}},$$

where $I_{\{.\}}$ represents the indicator function, and $N_K(i)$ represents the set of indices of cells that are $K$ top neighbors of cell $i$ measured by the learned distance metric. Using the transition matrix we can iteratively update the similarity matrix as follows:

$$H^{(t+1)} = \tau H^{(t)} P + (1 - \tau) I_n,$$

where $H^{(0)} = S$ and the matrix $H$ at the final iteration $T$ is used as the new similarity matrix $S$.

**Step 6:** Convergence checking

Let $\lambda_1 \le \lambda_2 \le \dots \le \lambda_n$ be the eigenvalues of the similarity matrix $S$ and $C$ be the number of clusters. Define

$$\text{eigengap}(C) = \lambda_{C+1} - \lambda_C.$$

When $\text{eigengap}(C)$ stops decreasing, in practice, the algorithm converges and iteration stops. Otherwise, go to Step 2. The algorithm repeats steps 2–6 until convergence.

The penalty parameters $\gamma$ and $\beta$ are determined from the data. Let $x_i$ denote the gene expression data of the $i^{th}$ cell, $x_i^j$ denote the top $j^{th}$ nearest neighbor of the $i^{th}$ cell and $m$ be a predefined parameter. In practice, for small datasets, we often set $m = 10$ and for large datasets, we set $m = 30$. The parameters $\gamma$ and $\beta$ are selected by

$$\gamma = \beta = \frac{1}{2n} \sum_{i=1}^{n} \sum_{j=1}^{m} \left( \| x_i - x_i^{m+1} \|_2^2 - \| x_i - x_i^j \|_2^2 \right).$$

The above algorithm reduces the data to a $C$ dimensional latent space. Next, we use the spectral clustering algorithms to assign each cell to the clusters.

**Algorithm 4.7: Spectral Clustering Algorithms**

**Step 1:** Compute a diagonal matrix $D = \text{diag}(d_1, d_2, \dots, d_n)$ where

$$d_i = \sum_{j=1}^{n} S_{ij}.$$

**Step 2:** Compute the Laplacian matrix

$$L = D - S.$$

**Step 3:** Let the eigenvalues of the Laplacian matrix $L$ be $0 = \lambda_1 \leq \lambda_2 \leq \ldots \leq \lambda_n$. Compute the first $C$ eigenvectors $u_1, \ldots, u_C$ of the Laplacian matrix $L$. Let $U \in R^{n \times C}$ be the matrix that consists of the vectors $u_1, \ldots, u_C$ as the column vectors.

**Step 4:** Let $y_i \in R^C$ be the vector of the $i^{th}$ row of the matrix $U$. Using k-means algorithm to cluster the points $\{y_1, \ldots, y_n\}$ into clusters $A_1, \ldots, A_C$.

## 4.7.2 Gene Expression Deconvolution and Cell Type-Specific Expression

The cellular composition of the biological samples is highly heterogeneous (Shai et al. 2013). Measuring cell type-specific gene expression is a key to understanding cell function, biological processes and mechanism of diseases (Nelms et al. 2016). However, obtaining cell specific gene expression profiles poses great technical challenges. One solution is computational deconvolution that infers cell type-specific expressions directly from heterogeneous samples. Cell-specific expressions control cell differentiation, define cell-specific phenotypes and provide the essential signature of cell identity. Deconvolution or source separation is a widely used method for estimating individual signal components from their mixtures (Mohammadi et al. 2017). The tasks of gene expression deconvolution are (1) to identify constituent cell types in a tissue, (2) discover relative proportions of the cell types in the tissue, and (3) estimate the precise gene expression levels for each cell type.

### 4.7.2.1 Gene Expression Deconvolution Formulation

The widely used gene expression deconvolution models assume that the expression of a gene in the mixture tissue is a sum of the expressions of its constitutive cell-types.

Let $A \in R^{n \times m}$ be a mixture matrix whose entry $a_{ij}$ is the raw expression of gene $i$ in the heterogeneous sample $j$. Each row represents expressions of a gene across the tissue samples and each column represents a vector of expressions of all genes being considered in a sample.

Let $X \in R^{n \times K}$ be a reference expression matrix where $K$ is the number of cell types considered, each column represents a vector of average gene expressions of the cell type and each row represents the expressions of the gene across the cell types.

Finally, we let $Y \in R^{K \times m}$ be a cell composition matrix where $y_{kj}$ is the proportion of the cell type $k$ in the sample $j$, rows correspond to cell types and columns represent samples in the mixture matrix. We aim to use the matrices $X$ and $Y$ for approximating the matrix $A$. Our goal is to find the reference expression matrix $X$ and cell composition matrix $Y$ such that their product is as close to the mixture matrix $A$ as possible. Thus, the expression deconvolution can be formulated as (Figure 4.27).

$$A = XY + E, \tag{4.239}$$

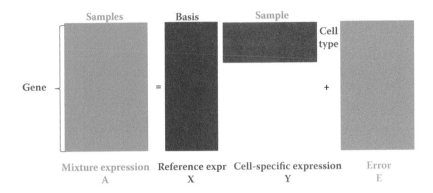

**FIGURE 4.27**
Scheme of the gene expression deconvolution.

where

$$A = \begin{bmatrix} a_{11} & \cdots & a_{1m} \\ \vdots & \vdots & \vdots \\ a_{n1} & \cdots & a_{am} \end{bmatrix}, X = \begin{bmatrix} x_{11} & \cdots & x_{1K} \\ \vdots & \vdots & \vdots \\ x_{n1} & \cdots & x_{iK} \end{bmatrix}, Y = \begin{bmatrix} y_{11} & \cdots & y_{1m} \\ \vdots & \vdots & \vdots \\ y_{K1} & \cdots & y_{Km} \end{bmatrix} E = \begin{bmatrix} e_{11} & \cdots & e_{1m} \\ \vdots & \vdots & \vdots \\ e_{n1} & \cdots & e_{nm} \end{bmatrix}.$$

We assume that $y_{kj}$ is non-negative and $\sum_{k=1}^{K} y_{kj} = 1$. Column vectors of the matrix $X$ can be viewed as a set of bases and columns of the matrix $Y$ and can be viewed as projection of the expressions in tissue sample on the bases or expansion of the expressions in the tissues in terms of bases. Therefore, deconvolution can be taken as dimension reduction problem.

When both $X$ and $Y$ are unknown, the deconvolution is called full deconvolution and when either $Y$ or $X$ is known the deconvolution is called partial deconvolution.

To estimate $X$ and $Y$ from the model (4.239) we need to define loss function and pose constraints. Since deconvolution can be viewed as dimension reduction, the deconvolution can be formally defined as the following generalized low rank model (Udell et al. 2016):

$$\text{minimize} \quad \sum_{(i,j)\in\Omega} L_{ij}\left(x_i y_j, a_{ij}\right) + \sum_{i=1}^{n} r_i(x_i) + \sum_{j=1}^{m} \tilde{r}_j(y_j), \quad (4.240)$$

where $x_i \in R^{1 \times K}$ is the $i^{th}$ row of $X$, $y_j \in R^K$ is the $j^{th}$ column of $Y$, $\Omega \subset \{1,\ldots,n\} \times \{1,\ldots,m\}$, $L_{ij}: R \times R \rightarrow R_+$ are given loss functions for $i = 1,\ldots,n$ and $j = 1,\ldots,m$, $\tilde{r}_i: R^K \rightarrow R \cup \{\infty\}$ and $\tilde{r}_j: R^K \rightarrow R \cup \{\infty\}$ for $i = 1,\ldots,n$ and $j = 1,\ldots,m$ are regularizers.

Regression-based methods and non-negative matrix factorization methods for deconvolution can be formulated as a generalized low rank model.

### 4.7.2.2 Loss Functions and Regularization

#### 4.7.2.2.1 Loss Functions

Loss functions depend on the data. Some loss functions are asymptotically optimal for a given noise distribution whereas others may perform very well when assumed noise distributions are violated. The widely used set of loss functions are summarized below (Mohammadi et al. 2017; Udell et al. 2016).

4.7.2.2.1.1 *Quadratic Loss Functions*  Assume that the distribution of noise is Gaussian and the classical squared loss function is often used. The squared loss function, denoted by $l_2$, is defined as

$$L_{ij}\left(x_i y_j, a_{ij}\right) = \left(a_{ij} - x_i y_j\right)^2. \tag{4.241}$$

4.7.2.2.1.2 *Absolute Deviation Loss Functions*  Since the absolute deviation loss function is less sensitive for extreme values, it is often used in the presence of outliers. The absolute deviation loss function, denoted by $l_1$, is formally defined as

$$L_{ij}\left(x_i y_j, a_{ij}\right) = \left|a_{ij} - x_i y_j\right|. \tag{4.242}$$

4.7.2.2.1.3 *Huber Loss Function*  The Huber Loss Function, denoted as $l_{Huber}$, is defined as

$$L_{ij}\left(x_i y_j, a_{ij}\right) = \begin{cases} \dfrac{1}{2}\left(a_{ij} - x_i y_j\right)^2 & \left|a_{ij} - x_i y_j\right| \leq 1 \\ \left|a_{ij} - x_i y_j\right| - \dfrac{1}{2}\left|a_{ij} - x_i y_j\right| & \left|a_{ij} - x_i y_j\right| > 1. \end{cases} \tag{4.243}$$

The Huber Loss Function is a combination of the $l_1$ and $l_2$ loss function. It is less sensitive to small errors than the $l_1$ loss function and less sensitive to outliers than the $l_2$ loss function.

4.7.2.2.1.4 *Poisson Loss Function*  Suppose that the expression data are read counts. The Poisson Loss Function is defined as

$$L_{ij}\left(x_i y_j, a_{ij}\right) = \exp\left(x_i y_j\right) - a_{ij} x_i y_j + a_{ij} \log a_{ij} - a_{ij}. \tag{4.244}$$

The Poisson Loss Function can be directly applied to the RNA-seq data deconvolution.

4.7.2.2.1.5 *Support Vector Machine Loss Function*  An $\varepsilon$- insensitive loss function can be used for support vector regression and is often called support vector machine (SVM) loss function. The SVM loss function penalizes errors when they are larger than a threshold. Formally, the SVM loss function is defined as

$$L_{ij}\left(x_i y_j, a_{ij}\right) = \max\left(0, \left|a_{ij} - x_i y_j\right| - \varepsilon\right), \tag{4.245}$$

where $\varepsilon$ is a pre-specified threshold.

*4.7.2.2.1.6 Hinge Loss Function*   The Hinge Loss Function is defined as

$$L_{ij}\left(x_iy_j, a_{ij}\right) = \max\left(0, 1 - a_{ij}x_iy_j\right) = \left|1 - a_{ij}x_iy_j\right|_+. \tag{4.246}$$

The empirical risk minimization of the Hinge Loss Function is equivalent to the classical formulation for SVMs.

### 4.7.2.2.2 Constraints and Regularization

Some cell types that appear in the reference profiles may not be present in the mixture samples. We also often observe that some constitutive cell-types are highly correlated. To identify the cell types in a particular sample and de-correlation between cell-types, in general, we regularize the factorization to obtain sparse solutions. Below we summarize the commonly used regularizers.

*4.7.2.2.2.1 $l_2$ Regularizer*   In the full deconvolution, $l_2$ or quadratic regularizer is defined as

$$\gamma\sum_{i=1}^{n}r_i(x_i) = \gamma\sum_{i=1}^{n}\| x_i \|_2^2 \quad \text{and} \quad \gamma\sum_{j=1}^{m}\tilde{r}_j\left(y_j\right) = \gamma\sum_{j=1}^{m}\| y_j \|_2^2. \tag{4.247}$$

In the partial deconvolution, the reference profiles for each cell type are known, we only need to regularize the cell proportion variables:

$$\gamma\sum_{j=1}^{m}\tilde{r}_j\left(y_j\right) = \gamma\sum_{j=1}^{m}\| y_j \|_2^2. \tag{4.248}$$

*4.7.2.2.2.2 $l_1$ Regularizer*   To identify the cell types presented in the tissue sample, we use $l_1$ to enforce sparsity on the cell type proportion vectors. Formally, $l_1$ regularizer is defined as

$$\gamma\| Y \|_1 = \gamma\sum_{j=1}^{m}\tilde{r}_j\left(y_j\right) = \gamma\sum_{j=1}^{m}\| y_j \|_1. \tag{4.249}$$

In general, $l_1$ regularizer shrinks some cell types to zero and leaves the cell types that are present in the tissue samples.

*4.7.2.2.2.3 Nonnegative Regularizer*   We can define the constraints $r = \mathbf{I}_+$ and $\tilde{r} = \mathbf{I}_+$ where $\mathbf{I}_+$ is the indicator function of the nonnegative values. The regularizers $r = \mathbf{I}_+$ and $\tilde{r} = \mathbf{I}_+$ are called nonnegative regularizers.
The generalized low rank model:

$$\min_{X,Y} \sum_{i=1}^{n}\sum_{j=1}^{m}\left(a_{ij} - x_iy_j\right)^2 + \lambda\sum_{i=1}^{n}r(x_i) + \lambda\sum_{j=1}^{m}\tilde{r}\left(y_j\right), \tag{4.250}$$

where $r(x_i) = \begin{cases} 0 & 0 \le x_{ik}, \sum_{k=1}^{K}x_{ik} = 1 \\ \infty & \text{otherwise} \end{cases}$ and $\tilde{r}(y_j) = \begin{cases} 0 & y_{kj} \ge 0, k = 1, \dots, K \\ \infty & \text{otherwise} \end{cases}$ is

referred to as the nonnegative matrix factorization (NNMF).

### 4.7.2.3 Algorithms for Fitting Generalized Low Rank Models

It is NP-hard to compute an exact solution to nonnegative matrix factorization (Vavasis 2009). In general, to find a global optimum of a generalized low rank model can be computationally intractable. In Section 8.4.3 (*Big Data in Omics and Imaging: Association Analysis*, Xiong 2018), we introduced a local optimization method based on alternating proximal gradient algorithm. Here we present a simple example.

**Example 4.11**

Consider the following nonnegative matrix factorization problem for gene expression deconvolution:

$$\min_{X,Y} \quad \| A - XY \|_F^2 + \lambda \| Y \|_1$$

$$\text{subject to} \quad x_{ij} \geq 0, \, y_{ij} \geq 0. \tag{4.251}$$

Nonnegative constraints can be formulated by indicator functions. Define indicator functions:

$$I_{R_{nK}^+}(X) = \begin{cases} 0 & \forall \, (i,j) \, x_{ik} \geq 0 \\ +\infty & \text{otherwise} \end{cases} \quad \text{and} \quad I_{R_{Km}^+}(Y) = \begin{cases} 0 & \forall \, (i,j) \, y_{kj} \geq 0 \\ +\infty & \text{otherwise} \end{cases}.$$

We can reformulate the constrained optimization problem (4.251 as the following unconstrained optimization problem (Rapin et al. 2012):

$$\min_{X,Y} \quad \| A - XY \|_F^2 + \lambda \| Y \|_1 + I_{R_{nK}^+}(X) + I_{R_{Km}^+}(Y). \tag{4.252}$$

Applying an alternating minimization algorithm to the optimization problem (4.251), we obtain

$$\min_Y \quad \| A - X^{(t)}Y \|_F^2 + \lambda \| Y \|_1 + I_{R_{Km}^+}(Y), \tag{4.253}$$

and

$$\min_X \quad \| A - XY^{(t+1)} \|_F^2 + I_{R_{nK}^+}(X). \tag{4.254}$$

Next we show that

$$\text{Prox}_{I_{R_+} + \lambda |y_{ij}|_1}\left( v_{ij} \right) = \left( v_{ij} - \lambda \right)_+. \tag{4.255}$$

Let $f(y_{ij}) = \lambda |y_{ij}|$ and $r(y_{ij}) = I_{R_+}(y_{ij})$. Then, by prox-prox method (Section 8.4.3.3 (*Big Data in Omics and Imaging: Association Analysis*, Xiong 2018)), we have

$$\text{Prox}_{I_{R_+} + \lambda |y_{ij}|_1}\left( v_{ij} \right) = \text{Prox}_r\left( \text{Prox}_f\left( v_{ij} \right) \right). \tag{4.256}$$

It follows from Equation 1.137 (Xiong 2018) that

$$\text{Prox}_{\lambda|y_{ij}|}\left(v_{ij}\right) = \begin{cases} v_{ij} - \lambda & v_{ij} > \lambda \\ 0 & |v_{ij}| \leq \lambda. \\ v_{ij} + \lambda & v_{ij} < -\lambda \end{cases} \qquad (4.257)$$

Recall that

$$\text{Prox}_{I_{R_+}}\left(w_{ij}\right) = \left(w_{ij}\right)_+. \qquad (4.258)$$

Combining Equations 4.256–4.258 gives

$$\text{Prox}_{I_{R_+} + \lambda|y_{ij}|_1}\left(v_{ij}\right) = \left(v_{ij} - \lambda\right)_+. \qquad (4.259)$$

Therefore, Equations 1.45 (Xiong 2018) and 4.259 lead to

$$\text{Prox}_{I_{R_{Km}^+} + \lambda\|Y\|_1}(W) = (W - \lambda\mathbf{1}_{Km})_+, \qquad (4.260)$$

where $\mathbf{1}_{Km}$ is the $K \times m$ dimensional matrix and full of ones. Now we solve the following optimization problem:

$$\min_{Y} \quad F = \|A - X^{(t)}Y\|_F^2. \qquad (4.261)$$

Setting $\dfrac{\partial F}{\partial Y} = 0$ and using matrix calculus yields

$$-\left(X^{(t)}\right)^T\left(A - X^{(t)}Y\right) = 0. \qquad (4.262)$$

Solving Equation 4.262 for the matrix $Y$ gives the solution:

$$Y^* = \left(\left(X^{(t)}\right)^T\left(X^{(t)}\right)\right)^{-1}\left(X^{(t)}\right)^T A. \qquad (4.263)$$

Using Equations 4.260 and 4.263, the update for solving Equation 4.253 is

$$Y^{(t+1)} = \left(\left(\left(X^{(t)}\right)^T\left(X^{(t)}\right)\right)^{-1}\left(X^{(t)}\right)^T A - \lambda\mathbf{1}_{Km}\right)_+. \qquad (4.264)$$

Similarly, we obtain the update for $X$:

$$X^{(t+1)} = \left(A\left(Y^{(t+1)}\right)^T\left(Y^{(t+1)}\left(Y^{(t+1)}\right)^T\right)^{-1}\right)_+. \qquad (4.265)$$

Algorithms are summarized as follows.

### Algorithm 4.8: Algorithms for Sparse Nonnegative Matric Factorization Used for Complete Expression Convolution

**Step 1:** Initialization

Assume that the number of cell types is $K$. Perform singular value decomposition (SVD) of the mixture matrix $A$:

$$A = U\Sigma V^T. \tag{4.266}$$

Define

$U_K = [u_1,...,u_K]$, $V_K = [v_1,...,v_K]$ and $\Sigma_K = \mathrm{diag}(\sigma_1,...,\sigma_K)$, where $u_k, v_k$ and $\sigma_k$ are the components of the matrices $U$, $V$ and $\Sigma$.

Define

$$X = U_K \Sigma_K^{1/2} \text{ and } Y = \Sigma_K^{1/2} V_K^T. \tag{4.267}$$

Select the initial penalty parameter $\lambda^{(0)}$.

**Step 2:** (Repeat until convergence)

a. Calculate

$$Y^{(t+1)} = \left( \left( \left( X^{(t)} \right)^T \left( X^{(t)} \right) \right)^{-1} \left( X^{(t)} \right)^T A - \lambda^{(t)} \mathbf{1}_{Km} \right)_+. \tag{4.268}$$

b. Calculate

$$X^{(t+1)} = \left( A \left( Y^{(t+1)} \right)^T \left( Y^{(t+1)} \left( Y^{(t+1)} \right)^T \right)^{-1} \right)_+. \tag{4.269}$$

**Step 3:** Convergence checking

If $\| Y^{(t+1)} - Y^{(t)} \|_F < \varepsilon$ and $\| X^{(t+1)} - X^{(t)} \|_F < \varepsilon$, where $\varepsilon$ is a pre-specified error, then

$Y: = Y^{(t+1)}$ and $X := X^{(t+1)}$; stop

otherwise

c. $t: = t + 1; Y^{(t)} := Y^{(t+1)}, X^{(t)} := X^{(t+1)}; \lambda^{(t)} = a\lambda^{(t)}$.

go to step 2

End

### Algorithm 4.9: Algorithms for Sparse Nonnegative Matric Factorization Used for Partial Expression Convolution

For $\lambda^1, \lambda^2,...,\lambda^T$ do

Calculate

$$Y^{(t)} = \left( X^T X \right)^{-1} \left( X^T A - \lambda^{(t)} \mathbf{1}_{Km} \right)_+. \tag{4.270}$$

End

## Software Package

Public sources for RNA-seq data include the Gene Expression Omnibus (GEO): http://www.ncbi.nlm.nih.gov/geo/ (both microarray and sequence data), Sequence Read Archive (SRA): http://www.ncbi.nlm.nih.gov/sra (all sequence data), and ArrayExpress: https://www.ebi.ac.uk/arrayexpress/ (European version of GEO).

Read mapping tool software "Bowtie", which is an ultrafast, memory-efficient short read aligner that can be downloaded from the website http://bowtie-bio.sourceforge.net/index.shtml. Software RSEM (RNA-seq by Expectation-Maximization), PennSeq and WemIQ for transcript and isoform quantification can be accessed from the website: https://deweylab.github.io/RSEM/, https://omictools.com/pennseq-tool, and https://omictools.com/weighted-log-likelihood-expectation-maximization-method-on-isoform-quantification-tool.

The edgeR package for differential expression analysis of RNA-seq data can be downloaded from http://bioconductor.org/packages/release/bioc/html/edgeR.html. R package DESeq2 for differential gene expression analysis based on the negative binomial distribution can be downloaded from http://bioconductor.org/packages/release/bioc/html/DESeq2.html. EBSeq, a R package for gene and isoform differential expression analysis of RNA-seq data with a Bayesian approach, can be downloaded from http://bioconductor.org/packages/release/bioc/html/EBSeq.html. NOISeq for Exploratory analysis and differential expression for the RNA-seq data using nonparametric approaches can be accessed from the website: https://bioconductor.org/packages/release/bioc/html/NOISeq.html.

CMGRN for constructing multilevel gene regulatory networks which uses the Bayesian network modeling to infer causal interrelationships can be downloaded from https://omictools.com/constructing-multilevel-gene-regulatory-networks-tool. ARACNE for the Reconstruction of Accurate Cellular Networks can be accessed from the website https://omictools.com/algorithm-for-the-reconstruction-of-accurate-cellular-networks-tool. GENIE3 for the inference of gene regulatory networks from expression data can be downloaded from https://omictools.com/genie3-tool. NARROMI, and a Matlab® code for inferring gene regulatory networks from gene expression data, can be downloaded from http://comp-sysbio.org/narromi.htm. The website https://omictools.com/gene-regulatory-networks-category collects gene regulatory network inference software tools.

Software "fastaqc", a quality control tool for high throughput sequence data, can be downloaded from the website: http://www.bioinformatics.babraham.ac.uk/projects/fastqc/. The scde package, a R package that implements a set of statistical methods for analyzing single-cell RNA-seq data, can be downloaded from http://pklab.med.harvard.edu/scde. Existing mapping tools GSNAP that can be used for both bulk RNA-seq and single cell RNA-seq can be downloaded from https://omictools.com/gsnap-tool and software for generating read counts, HTseq can be downloaded from http://www-huber .embl.de/HTSeq/doc/overview.html.

SIMLR (a tool for large-scale single-cell analysis by multi-kernel learning) is available on https://github.com/BatzoglouLabSU/SIMLR; https://bioconductor.org/packages/release/bioc/html/SIMLR.html).

PERT that uses non-negative least squares for gene expression deconvolution can be downloaded from the website http://qlab.faculty.ucdavis.edu/pert/. A R package 'DeconRNASeq' that employs a quadratic programming framework for gene expression deconvolution using $L_1$ loss function and Huber's Loss Function can be downloaded from https://www.bioconductor.org /packages/devel/bioc/manuals/DeconRNASeq/man/DeconRNASeq.pdf.

A R package "dcq: DCQ - Digital Cell Quantifier" that enforces sparsity by combining $L_1$ and $L_2$ regularization in gene expression deconvolution can be accessed in https://rdrr.io/cran/ComICS/man/dcq.html. A MATLAB implementation of fast optimization algorithms for computing nonnegative matrix and tensor factorizations can be downloaded from https://github .com/kimjingu/nonnegfac-matlab/blob/master/README.md. A package "CellMix" for gene expression deconvolution can be accessed online at: http:// web.cbio.uct.ac.za/~renaud/CRAN/web/CellMix. CIBERSORT that uses linear support vector regression (SVR), a machine learning approach highly robust to noise for expression deconvolution can be accessed from https:// github.com/zomithex/CIBERSORT. A R package for nonnegative matrix factorization with most of the built-in algorithms optimized in C++ can be accessed at https://cran.r-project.org/web/packages/NMF/index.html.

---

## Appendix 4.A  Variational Bayesian Theory for Parameter Estimation and RNA-Seq Normalization

### 4.A.1  Variational Methods for Expectation-Maximization (EM) Algorithm

Consider hidden variables $Z_n$, quality score $Q_n$ and observed variables $R_n$. The hidden variables are indicator variable $Z_{nijk}$ defined in Equation 4.19 or $Z_{nikja}$ defined in Equation 4.24. For simplicity, we use $Z_n$ to denote $Z_{nijk}$ or $Z_{nikja}$. Let $Z = [Z_1^T, ..., Z_N^T]^T$, $Q = [Q_1, ... Q_N]^T$ and $R = [R_1^T, ..., R_N^T]^T$. The likelihood function for generating the data $R_n$ and $Q_n$ is

$$P(R, Q|\theta) = \prod_{n=1}^{N} P(R_n, Q_n|\theta) = \prod_{n=1}^{N} \int P(Z_n, R_n, Q_n|\theta) dZ_n. \qquad (4.A.1)$$

Define the log-likelihood function:

$$L(\theta) = \log P(R, Q|\theta) = \sum_{n=1}^{N} \log P(R_n, Q_n|\theta)$$

$$= \sum_{n=1}^{N} \log \int P(Z_n, R_n, Q_n|\theta)dZ_n. \qquad (4.A.2)$$

Since the log-likelihood function involves a number of hidden variables, its evaluation is still difficult. To efficiently maximize the log-likelihood with respect to the parameters, we simplify the evaluation of log-likelihood by introducing the distribution function $q(Z)$ over the hidden variables. After introducing such auxiliary distribution over the hidden variables, the log-likelihood function in Equation 4.A.2 is reduced to

$$L(\theta) = \sum_{n=1}^{N} \log \int q(Z_n) \frac{P(Z_n, R_n, Q_n|\theta)}{q(Z_n)} dZ_n. \qquad (4.A.3)$$

Denote $E\left[\dfrac{P(Z_n, R_n, Q_n|\theta)}{q(Z_n)}\right] = \int q(Z_n) \dfrac{P(Z_n, R_n, Q_n|\theta)}{q(Z_n)} dZ_n$. Recall that Jensen's inequality implies that

$$\log E\left[\frac{P(Z_n, R_n, Q_n|\theta)}{q(Z_n)}\right] \geq E\left[\log \frac{P(Z_n, R_n, Q_n|\theta)}{q(Z_n)}\right]$$

$$= \int q(Z_n) \log \frac{P(Z_n, R_n, Q_n|\theta)}{q(Z_n)} dZ_n. \qquad (4.A.4)$$

Substituting Equation 4.A4 into Equation 4.A3 gives

$$L(\theta) \geq \sum_{n=1}^{N} \int q(Z_n) \log \frac{P(Z_n, R_n, Q_n|\theta)}{q(Z_n)} dZ_n$$

$$= \sum_{n=1}^{N} \left[\int q(Z_n) \log P(Z_n, R_n, Q_n|\theta)dZ_n - \int q(Z_n) \log q(Z_n)dZ_n\right]$$

$$= F(q(Z), \theta),$$

where

$$F(q(Z), \theta) = \sum_{n=1}^{N} \left[\int q(Z_n) \log P(Z_n, R_n, Q_n|\theta)dZ_n - \int q(Z_n) \log q(Z_n)dZ_n\right]$$

$$(4.A.5)$$

is a lower bound on the log-likelihood function $L(\theta)$. Our goal is to find a distribution $q^*(Z)$ such that $F(q^*(Z),\theta)$ is the largest lower bound. To seek such distributions, we solve the following optimization problem:

$$\max_{q(Z)} \quad F(q(Z), \theta)$$

$$\text{s.t.} \quad \int q(Z_n)dZ_n = 1, n = 1, \ldots, N. \tag{4.A.6}$$

Using the Lagrange multiplier methods, the optimization problem (4.A.6) can be reduced to

$$\bar{F}(q(Z), \theta, \lambda) = F(q(Z), \theta) + \sum_{n=1}^{N} \lambda_n \left[ 1 - \int q(Z_n)dZ_n \right], \tag{4.A.7}$$

where $\lambda_n$ is a Lagrange multiplier.

Similar to multivariate calculus where the necessary condition for maximum of a function $f(x)$ is $\dfrac{\partial f(x)}{\partial x} = 0$, by variation calculus, the necessary condition for a relative maximum of a functional is $\delta\bar{F}(h_n, \theta, \lambda) = 0$, which leads to

$$\delta\bar{F}(h_n, \theta, \lambda) = \frac{\partial \bar{F}(q(Z_n) + \varepsilon h_n, \theta, \lambda)}{\partial \varepsilon}$$

$$= \int [\log P(Z_n, R_n, Q_n|\theta) - \log q(Z_n) - 1 - \lambda_n] h_n dZ_n. \tag{4.A.8}$$

Substituting $h_n = [\log P(Z_n, R_n, Q_n | \theta) - \log q(Z_n) - 1 - \lambda_n]$ into Equation 4.A.8 gives

$$\delta\bar{F}(h_n, \theta, \lambda) = \int [\log P(Z_n, R_n, Q_n|\theta) - \log q(Z_n) - 1 - \lambda_n]^2 dZ_n. \tag{4.A.9}$$

Thus, using $\delta\bar{F}(h_n, \theta, \lambda) = 0$ and Equation 4.A.9, we obtain

$$\log P(Z_n, R_n, Q_n|\theta) - \log q(Z_n) - 1 - \lambda_n = 0. \tag{4.A.10}$$

Solving Equation 4.A.10 for $q(Z_n)$ gives

$$q(Z_n) = e^{-(\lambda_n+1)} P(Z_n, R_n, Q_n|\theta). \tag{4.A.11}$$

Now we solve Equation 4.A.10 for $\lambda_n$. Moving $-(1 + \lambda_n)$ from the left side of Equation 4.1A10 to the right side, we obtain

$$\lambda_n + 1 = \log P(Z_n, R_n, Q_n|\theta) - \log q(Z_n). \tag{4.A.12}$$

It follows from Equation 4.A.12 that

$$e^{\lambda_n+1} = \frac{P(Z_n, R_n, Q_n|\theta)}{q(Z_n)} \quad \text{or} \tag{4.A.13}$$

$$q(Z_n) = e^{-(\lambda_n+1)} P(Z_n, R_n, Q_n|\theta). \tag{4.A.14}$$

Substituting Equation 4.A.14 into $\int q(Z_n)dZ_n = 1, n = 1,...,N$ gives

$$e^{\lambda_n+1} = \int P(Z_n, R_n, Q_n|\theta)dZ_n = P(R_n, Q_n|\theta). \tag{4.A.15}$$

Substituting Equation 4.A.15 into Equation 4.1A.11, we obtain a final optimal solution

$$q(Z_n) = \frac{P(Z_n, R_n, Q_n|\theta)}{P(R_n, Q_n|\theta)} = P(Z_n|R_n, Q_n, \theta). \tag{4.A.16}$$

From Equation 4.A.5 we can rewrite the lower bound $F(q(Z),\theta)$ of the log-likelihood as

$$
\begin{aligned}
F(q(Z), \theta) &= \sum_{n=1}^{N} \int q(Z_n) \log \frac{P(Z_n, R_n, Q_n|\theta)}{q(Z_n)} dZ_n \\
&= \sum_{n=1}^{N} \int q(Z_n) \log \frac{P(R_n, Q_n|\theta)P(Z_n|R_n, Q_n, \theta)}{q(Z_n)} dZ_n \\
&= \sum_{n=1}^{N} \int q(Z_n) \log P(R_n, Q_n|\theta)dZ_n \\
&\quad + \int \sum_{n=1}^{N} q(Z_n) \log \frac{P(Z_n|R_n, Q_n, \theta)}{q(Z_n)} dZ_n \\
&= \sum_{n=1}^{N} \int q(Z_n) \log P(R_n, Q_n|\theta)dZ_n \\
&\quad - \sum_{n=1}^{N} KL[q(Z_n) \| P(Z_n|R_n, Q_n, \theta)] \\
&= \sum_{n=1}^{N} \log P(R_n, Q_n|\theta) - \sum_{n=1}^{N} KL[q(Z_n) \| P(Z_n|R_n, Q_n, \theta)],
\end{aligned}
\tag{4.A.17}
$$

where

$$KL[q(Z_n)\|P(Z_n|R_n, Q_n, \theta) = \int q(Z_n) \log \frac{q(Z_n)}{P(Z_n|R_n, Q_n, \theta)} dZ_n \tag{4.A.18}$$

is referred to as the Kullback–Leibler divergence between the variation distribution $q(Z_n)$ and the hidden variable posterior $P(Z_n | R_n, Q_n, \theta)$.

It follows from Equation 4.A.17 that

$$\sum_{n=1}^{N} \log P(R_n, Q_n \mid \theta) = F(q(Z), \theta) + \sum_{n=1}^{N} KL[q(Z_n)\|P(Z_n|R_n, Q_n, \theta), \quad (4.A.19)$$

where

$$F(q(Z), \theta) = \sum_{n=1}^{N} \int q(Z_n) \log \frac{P(Z_n, R_n, Q_n|\theta)}{q(Z_n)} dZ_n.$$

In summary, EM algorithm updates equations are given by

$$\mathbf{E \ step:} \ q^{(u=1)}(Z_n) = P\left(Z_n \mid R_n, Q_n, \theta^{(u)}\right), \ n = 1, 2, \dots, N, \quad (4.A.20)$$

$$\mathbf{M \ step:} \ \theta^{(u=1)} = \arg\max_{\theta} F\left(q^{(u+1)}(Z), \theta\right). \quad (4.A.21)$$

Now we consider un-gapped alignment and gapped alignment of the RNA-Seq data. For the un-gapped alignment, $Z_{nijk}$ is defined as

$$Z_{nijk} = \begin{cases} 1 & G_n = i, S_n = j, O_n = k \\ 0 & \text{Otherwise} \end{cases}.$$

The probability $P(Z_n|R_n, Q_n, \theta^{(u)})$ is reduced to $P(Z_{nijk}|R_n, \theta^{(u)})$. Using Equation 4.18, $P(Z_{nijk}|R_n, \theta^{(u)})$ is calculated as follows:

$$P\left(Z_{nijk}|R_n, \theta^{(u)}\right) = \frac{P\left(Z_{nijk}, R_n, \theta^{(u)}\right)}{P\left(R_n, \theta^{(u)}\right)}$$

$$= \frac{P\left(Z_{nijk} = 1, R_n = r_n, \theta^{(u)}\right)}{\sum_{i'=1}^{M} \sum_{j'=1}^{[1,\max_{i'} l_{i'}]} p\left(Z_{ni'j'k} = 1, R_n = r_n, \theta^{(u)}\right)}. \quad (4.A.22)$$

Recall that

$$P\left(Z_{nijk} = 1, \theta^{(u)}\right) = P(G_n = i|\theta^{(u)})P(S_n = j|G_n)P(O_n = k|G_n)$$

$$= \theta_i^{(u)} \frac{1}{l_i} P(O_n = k|G_n = i). \quad (4.A.23)$$

Substituting Equation 4.A.23 into Equation 4.A.22 gives

$$P\left(Z_{nijk}|R_n, \theta^{(u)}\right) = \frac{\dfrac{\theta_i^{(u)}}{l_i} P\left(R_n = r_n \mid Z_{nijk} = 1, \theta^{(u)}\right)}{\sum_{i'=1}^{M} \sum_{j'=1}^{[1,\max_{i'} l_{i'}]} \dfrac{\theta_{i'}^{(u)}}{l_{i'}} P\left(R_n = r_n \mid Z_{ni'j'k} = 1, \theta^{(u)}\right)}.$$

Therefore, the E step is
**E step:**

$$q^{(u+1)}(Z_n) = \frac{\dfrac{\theta_i^{(u)}}{l_i} P\left(R_n = r_n \mid Z_{nijk} = 1, \theta^{(u)}\right)}{\displaystyle\sum_{i'=1}^{M}\sum_{j'=1}^{[1,\max_{i'} l_{i'}]} \dfrac{\theta_{i'}^{(u)}}{l_{i'}} P\left(R_n = r_n \mid Z_{ni'j'k} = 1, \theta^{(u)}\right)}, \quad n$$

$$= 1, 2, \ldots, N. \tag{4.A.24}$$

Before we perform the M step, we need to evaluate $F(q^{(u+1)}(Z), \theta)$. Substituting Equation 4.A.24 into Equation 4.A.19, we obtain

$$F\left(q^{(u+1)}(Z), \theta\right) = \sum_{n=1}^{N}\sum_i\sum_j\sum_k P\left(Z_{nijk} \mid R_n, \theta^{(u)}\right) \log \theta_i + D, \tag{4.A.25}$$

where $D$ is not a function of $\theta$.

M step involves solving the following optimization problem:

$$\max_{\theta} \quad F\left(q^{(u+1)}(Z), \theta\right)$$

$$\text{s.t.} \quad \sum_i \theta_i = 1$$

or

$$\max_{\theta} \quad \sum_{n=1}^{N}\sum_i\sum_j\sum_k P\left(Z_{nijk} \mid R_n, \theta^{(u)}\right) \log \theta_i$$

$$\text{s.t.} \quad \sum_i \theta_i = 1 \tag{4.A.26}$$

Using Lagrange multiplier methods, the constrained optimization problem can be transformed into unconstrained optimization problem:

$$\max_{\theta, \lambda} \quad L = \sum_{n=1}^{N}\sum_i\sum_j\sum_k P\left(Z_{nijk} \mid R_n, \theta^{(u)}\right) \log \theta_i + \lambda\left(1 - \sum_i \theta_i\right), \tag{4.A.27}$$

where $\lambda$ is a Lagrange multiplier.

Set

$$\frac{\partial L}{\partial \theta_i} = \sum_{n=1}^{N}\sum_j\sum_k P\left(Z_{nijk} \mid R_n, \theta^{(u)}\right) \frac{1}{\theta_i} - \lambda = 0. \tag{4.A.28}$$

Multiplying both sides of Equation 4.A.28 gives

$$\sum_{n=1}^{N}\sum_j\sum_k P\left(Z_{nijk} \mid R_n, \theta^{(u)}\right) - \lambda\theta_i = 0. \tag{4.A.29}$$

Note that

$$\sum_i P\left(Z_{nijk} \mid R_n, \theta^{(u)}\right) = P\left(Z_{nijk} \mid R_n, \theta^{(u)}\right). \tag{4.A.30}$$

Summation Equation 4.A.29 over I and using Equation 4.A.30, we obtain

$$\lambda = \sum_{n=1}^{N} \sum_j \sum_k P\left(Z_{nijk} \mid R_n, \theta^{(u)}\right) = N. \tag{4.A.31}$$

Substituting Equation 4.A.31 into Equation 4.A.29 gives the solution

$$\theta_i^{(u+1)} = \frac{\sum_{n=1}^{N} \sum_j \sum_k P\left(Z_{nijk} \mid R_n, \theta^{(u)}\right)}{N}. \tag{4.A.32}$$

Therefore, M step is defined as
M step:

$$\theta_i^{(u+1)} = \frac{\sum_{n=1}^{N} \sum_j \sum_k P\left(Z_{nijk} \mid R_n, \theta^{(u)}\right)}{N}, \quad i = 1, 2, \ldots, M. \tag{4.A.33}$$

Next, we study the gapped alignment. For the gapped alignment, indicator variable $Z_{nikja}$ is defined as

$$Z_{nikja} = \begin{cases} 1, & G_n = i, O_n = k, S_n = j, A_n = a \\ 0 & \text{otherwise} \end{cases}. \tag{4.A.34}$$

The probability $P(Z_n \mid R_n, Q_n, \theta^{(u)})$ is reduced to $P(Z_{nikja} \mid R_n, Q_n, \theta^{(u)})$. Using Equation 4.22, $P(Z_{nikja} \mid R_n, Q_n, \theta^{(u)})$ is calculated as follows:

$$P\left(Z_{nikja} \mid R_n, Q_n, \theta^{(u)}\right) = \frac{P\left(Z_{nikja}, R_n, Q_n, \theta^{(u)}\right)}{P\left(R_n, Q_n, \theta^{(u)}\right)}$$

$$= \frac{P\left(Z_{nikja} = 1, R_n = r_n, Q_n, \theta^{(u)}\right)}{\sum_{(i',k',j',a') \in \pi_n} P\left(Z'_{ni'k'j'a} = 1, R_n = r_n, Q_n, \theta^{(u)}\right)}.$$

Thus, E step is defined as
E step:

$$q^{(u+1)}(Z_n) = P\left(Z_{nikja} \mid R_n, Q_n, \theta^{(u)}\right)$$

$$= \frac{P\left(Z_{nikja} = 1, R_n = r_n, Q_n, \theta^{(u)}\right)}{\sum_{(i',k',j',a') \in \pi_n} P\left(Z'_{ni'k'j'a} = 1, R_n = r_n, Q_n, \theta^{(u)}\right)}, \quad n = 1, 2, \ldots, N,$$

$$\tag{4.A.35}$$

where $\pi_n$ is a set of $i,k,j,a$ for all possible alignments of read $n$ and $(i,k,j,a) \in \pi_n$.

Now we investigate the M step. We first calculate $F(q^{(u+1)}(Z),\theta)$.

Before we perform the M step, we need to evaluate $F(q^{(u+1)}(Z),\theta)$. Substituting Equation 4.A.35 into Equation 4.A.19, we obtain

$$F\left(q^{(u+1)}(Z),\theta\right) = \sum_{n=1}^{N} \sum_{i,k,j,a} P\left(Z_{nikja} \mid R_n, Q_n, \theta^{(u)}\right) \log \frac{P\left(Z_{nikja}, R_n, Q_n \mid \theta\right)}{P\left(Z_{nikja} \mid R_n, Q_n, \theta^{(u)}\right)}$$

$$= \sum_{n=1}^{N} \sum_{i,k,j,a} P\left(Z_{nikja} \mid R_n, Q_n, \theta^{(u)}\right) \log P\left(Z_{nikja}, R_n, Q_n \mid \theta\right) -$$

$$- \sum_{n=1}^{N} \sum_{i,k,j,a} P\left(Z_{nikja} \mid R_n, Q_n, \theta^{(u)}\right) \log P\left(Z_{nikja} \mid R_n, Q_n, \theta^{(u)}\right)$$

$$= \sum_{n=1}^{N} \sum_{i,k,j,a} P\left(Z_{nikja} \mid R_n, Q_n, \theta^{(u)}\right) \log \theta_i + D.$$

$$(4.A.36)$$

By the same argument as that for the un-gapped alignment, we obtain the solution to maximize $F(q^{(u+1)}(Z),\theta)$:

$$\theta_i^{(u+1)} = \frac{\sum_{n=1}^{N} \sum_{i,k,j,a} P\left(Z_{nikja} \mid R_n, Q_n, \theta^{(u)}\right)}{N}, i = 1, 2, \ldots, M,$$

where $P(Z_{nikja} \mid R_n, Q_n, \theta^{(u)})$ is defined in Equation 4.A35.

Thus, the M step is defined as

M step:

$$\theta_i^{(u+1)} = \frac{\sum_{n=1}^{N} \sum_{i,k,j,a} P\left(Z_{nikja} \mid R_n, Q_n, \theta^{(u)}\right)}{N}, i = 1, 2, \ldots, M, \qquad (4.A.37)$$

where $(i,k,j,a) \in \pi_n$, $\pi_n$ is a set of $i,k,j,a$ for all possible alignments of read $n$ and

$$P\left(Z_{nikja} \mid R_n, Q_n, \theta^{(u)}\right) = \frac{P\left(Z_{nikja} = 1, R_n = r_n, Q_n, \theta^{(u)}\right)}{\sum_{(i\prime,k\prime,j\prime,a\prime) \in \pi_n} P\left(Z'_{nik\prime j\prime a\prime} = 1, R_n = r_n, Q_n, \theta^{(u)}\right)},$$

$n = 1, 2, \ldots, N.$

## 4.A.2   Variational Methods for Bayesian Learning

In this section, we extend the variational methods for the EM algorithm to Bayesian learning. We will introduce the variational Bayesian (VB) framework for estimation of parameters in the BN with both observational and hidden variables. The VB framework for parameter estimation is to approximate the distribution over both hidden variables and parameters with a simple distribution where we assume that the given data, the hidden states and parameters are independent (www.cse.buffalo.edu/faculty/mbeal /thesis/beal03_2.pdf.).

The VB methods for learning begin with approximating the marginal likelihood $P(R,Q)$. The lower bound of the marginal likelihood $P(R,Q)$ can be obtained by introducing any distributions over both hidden variables and parameters:

$$\log P(R,Q) = \log \int P(Z,R,Q,\theta) d\theta dZ$$

$$= \log \int q(Z,\theta) \frac{P(Z,R,Q,\theta)}{q(Z,\theta)} d\theta dZ \qquad (4.A.38)$$

$$\geq \int q(Z,\theta) \log \frac{P(Z,R,Q,\theta)}{q(Z,\theta)} d\theta dZ.$$

Assume that the distribution $q(Z,\theta)$ can be factorized to

$$q(Z,\theta) \approx q_z(Z) q_\theta(\theta). \qquad (4.A.39)$$

Substituting Equation 4.A.39 into Equation 4.A.38 gives

$$\log P(R,Q) \geq \int q_z(Z) q_\theta(\theta) \log \frac{P(Z,R,Q|\theta)P(\theta)}{q_z(Z) q_\theta(\theta)} d\theta dZ$$

$$= \int q_\theta(\theta) \left[ \int q_z(Z) \log \frac{P(Z,R,Q|\theta)}{q_z(Z)} dZ + \log \frac{P(\theta)}{q(\theta)} \right] \qquad (4.A.40)$$

$$= F(q_z(Z), q_\theta(\theta)),$$

where

$$F(q_z(Z), q_\theta(\theta)) = \int q_z(Z) q_\theta(\theta) \log \frac{P(Z,R,Q|\theta)P(\theta)}{q_z(Z) q_\theta(\theta)} d\theta dZ \qquad (4.A.41)$$

or

$$F(q_z(Z), q_\theta(\theta)) = \int q_\theta(\theta) \left[ \int q_z(Z) \log \frac{P(Z,R,Q|\theta)}{q_z(Z)} dZ + \log \frac{P(\theta)}{q(\theta)} \right]. \qquad (4.A.42)$$

Again, to derive the rule for the variational Bayesian EM (VBEM) we use variational calculus and Equation 4.A.42. It is clear that Equation 4.A.42 can be rewritten as

$$F(q_z(Z), q_\theta(\theta)) = \int d\theta q_\theta(\theta) \left[ \int (q_z(Z) \log P(Z,R,Q|\theta) \right.$$

$$\left. - q_z(Z) \log q_z(Z)) dZ + \log \frac{P(\theta)}{q(\theta)} \right]. \qquad (4.A.43)$$

Similar to optimization problem (4.A.6), we have

$$\max_{q(Z)} \quad F(q_z(Z), q_\theta(\theta))$$

$$\text{s.t.} \quad \int q(Z) dZ = 1 \tag{4.A.44}$$

By the same argument as before, we obtain the solution:

$$q_z^{(u+1)}(Z) = \frac{1}{C_z} \exp\left[\int q_\theta^{(u)}(\theta) \log P\left(Z, R, Q \mid \theta^{(u)}\right) d\theta\right], \tag{4.A.45}$$

where $C_z = \int \exp\left[\int q_\theta^{(u)}(\theta) \log P(Z, R, Q \mid \theta^{(u)}) d\theta\right] dZ$ is a normalization constant.

In general we assume that $Z_n$, $R_n$ and $Q_n$ are independent, the optimal $q_z^{(u+1)}(Z)$ can be factorized to

$$q_z^{(u+1)}(Z) = \prod_{n=1}^{N} q_{z_n}^{(u+1)}(Z_n), \tag{4.A.46}$$

where

$$q_{z_n}^{(u+1)}(Z) = \frac{1}{C_{z_n}} \exp\left[\int q_\theta^{(u)}(\theta) \log P\left(Z_n, R_n, Q_n \mid \theta^{(u)}\right) d\theta\right], \tag{4.A.47}$$

$$C_{Z_n} = \int \exp\left[\int q_\theta^{(u)}(\theta) \log P\left(Z_n, R_n, Q_n \mid \theta^{(u)}\right) d\theta\right] dZ_n, \tag{4.A.48}$$

$$C_Z = \prod_{N=1}^{n} C_{Z_n}. \tag{4.A.49}$$

Next we study how to update $q_\theta(\theta)$. Our goal is to find $q_\theta(\theta)$ that optimizes

$$\max_{q_\theta(\theta)} \quad F\left(q_z^{(u+1)}(Z), q_\theta(\theta)\right)$$

$$\text{s.t.} \quad \int q_\theta(\theta) d\theta = 1. \tag{4.A.50}$$

Using the same arguments (Lagrange multiplier methods) as before, we obtain the solution to optimization problem (4.A.50):

$$q_\theta^{(u+1)}(\theta) = \frac{1}{C_\theta} P\left(\theta^{(u)}\right) \exp\left[\int q_z\left(Z^{(u+1)}\right) \log P\left(Z, R, Q \mid \theta^{(u)}\right) dZ\right], \tag{4.A.51}$$

where $C_\theta = \int P(\theta^{(u)}) \exp\left[\int q_z(Z^{(u+1)}) \log P(Z, R, Q \mid \theta^{(u)}) dZ\right]$.

In summary, a general VBEM algorithm is given by
VBE step:

$$q_z^{(u+1)}(Z) = \prod_{n=1}^{N} q_{z_n}^{(u+1)}(\mathbf{Z}_n),$$
(4.A.52)

where

$$q_{z_n}^{(u+1)}(Z) = \frac{1}{C_{z_n}} \exp\left[\int q_\theta^{(u)}(\theta) \log P(\mathbf{Z}_n, \mathbf{R}_n, \mathbf{Q}_n \mid \theta) \, d\theta\right],$$
(4.A.53)

$$C_{Z_n} = \int \exp\left[\int q_\theta^{(u)}(\theta) \log P(\mathbf{Z}_n, \mathbf{R}_n, \mathbf{Q}_n \mid \theta) d\theta\right] dZ_n,$$
(4.A.54)

$$C_Z = \prod_{n=1}^{N} C_{Z_n}.$$
(4.A.55)

VBM step:

$$q_\theta^{(u+1)}(\theta) = \frac{1}{C_\theta} P\left(\theta^{(u)}\right) \exp\left[\int q_z\left(Z^{(u+1)}\right) \log P\left(Z, R, Q \mid \theta^{(u)}\right) dZ\right],$$
(4.A.56)

where $C_\theta = \int P(\theta^{(u)}) \exp\left[\int q_z(Z^{(u+1)}) \log P(Z, R, Q \mid \theta^{(u)}) dZ\right]$.

Again, we first consider un-gapped alignment and gapped alignment of the RNA-seq data (Nariai et al. 2013). For the un-gapped alignment, $Z_{nijk}$ is defined as

$$Z_{nijk} = \begin{cases} 1 & G_n = i, S_n = j, O_n = k \\ 0 & \text{Otherwise} \end{cases}.$$

The probability $P(Z_n, R_n, Q_n \mid \theta)$ is reduced to $P(Z_{nijk}, R_n \mid \theta)$. Define

$$\log \rho_{nijk} = E_\theta\left[\log \theta_i + \log P(S_n = j \mid G_n = i) + \log P(O_n = k \mid G_n = i)\right.$$
$$\left. + \log(R_n = r_n \mid Z_{nijk} = 1)\right]$$
$$= E_\theta\left[\log \theta_i\right] + \log P(S_n = j \mid G_n = i) + \log P(O_n = k \mid G_n = i)$$
$$+ \log(R_n = r_n \mid Z_{nijk} = 1),$$

where

$$E_\theta\left[\log \theta_i\right] = \psi(\alpha_i) - \psi\left(\sum_{j=1}^{M} \alpha_j\right), \ \psi(\alpha) = \frac{\dfrac{d\Gamma(\alpha)}{d\alpha}}{\Gamma(\alpha)} \text{ is the digamma function.}$$

Thus,

$$\int q_\theta \log P(\mathbf{Z}_n, \mathbf{R}_n \mid \theta) d\theta = \sum_{(i,j,k)\in\pi_n} \log \rho_{n,i,j,k}. \tag{4.A.57}$$

It follows from Equation 4.A.54 that

$$C_{Z_n} = \sum_{(i\prime,j\prime,k\prime)\in\pi_n} \rho_{n\prime,i\prime,j\prime,k\prime}. \tag{4.A.58}$$

Using Equation 4.A.53 gives

$$\begin{aligned}
q_{Z_n}^{(u+1)}(Z) &= \frac{1}{C_{Z_n}} \exp\left[\int q_\theta^{(u)}(\theta) \log P(\mathbf{Z}_n, \mathbf{R}_n, \mathbf{Q}_n \mid \theta)\, d\theta\right] \\
&= \frac{\exp\left[\sum_{(i,j,k)\in\pi_n} \log \rho_{n,i,j,k}\right]}{\sum_{(i\prime,j\prime,k\prime)\in\pi_n} \rho_{n\prime,i\prime,j\prime,k\prime}} \\
&= \prod_i \prod_j \prod_k \left(\frac{\rho_{n,i,j,k}}{\sum_{(i\prime,j\prime,k\prime)\in\pi_n} \rho_{n\prime,i\prime,j\prime,k\prime}}\right)^{Z_{nijk}} \\
&= \prod_i \prod_j \prod_k \left(\eta_{nijk}\right)^{Z_{nijk}},
\end{aligned} \tag{4.A.59}$$

where

$$\eta_{n,i,j,k} = \frac{\rho_{n,i,j,k}}{\sum_{(i\prime,j\prime,k\prime)\in\pi_n} \rho_{n\prime,i\prime,j\prime,k\prime}} \text{ and }$$

$$E_Z\left[Z_{nijk}\right] = \eta_{nijk}.$$

Therefore,

$$\begin{aligned}
q_Z^{(u+1)}(Z) &= \prod_{n=1}^{N} q_{Z_n}^{(u+1)}(\mathbf{Z}_n) \\
&= \prod_n \prod_i \prod_j \prod_k \left(\eta_{nijk}\right)^{Z_{nijk}}.
\end{aligned} \tag{4.A.60}$$

Now we study VBM step. Using Equation 4.A.56 gives

$$\begin{aligned}
\log q_\theta^{(u+1)}(\theta) &= \log P\left(\theta^{(u)}\right) + E_Z\left[\log P(\mathbf{Z}, \mathbf{R}\mid\theta^{(u)})\right] + D_0 \\
&= (\alpha_0 - 1)\sum_{i=0}^{M} \log \theta_i + \sum_{n,i,j,k} E_Z\left[Z_{n,i,j,k}\right] \log \theta_i + D \tag{4.A.61} \\
&= \sum_{i=0}^{M}\left(\alpha_0 - 1 + \sum_{n,j,k} E_Z\left[Z_{n,i,j,k}\right]\right)\log \theta_i + D.
\end{aligned}$$

Recall that the probability density function of the Dirichlet distribution is given by

$$P(\theta) = \frac{\prod_{i=0}^{M} \Gamma(\alpha_i)}{\Gamma\left(\sum_{i=0}^{M} \alpha_i\right)} \prod_{i=0}^{M} \theta_i^{\alpha_i - 1}. \tag{4.A.62}$$

Then, the logarithm of the Dirichlet distribution is

$$\log(P(\theta)) = \sum_{i=0}^{M} (\alpha_i - 1) \log \theta_i + \sum_{i=0}^{M} \log \Gamma(\alpha_i) - \log \Gamma\left(\sum_{i=0}^{M} \alpha_i\right). \tag{4.A.63}$$

Comparing Equation 4.A.61 with Equation 4.A.63, we infer that $q_\theta^{(u+1)}$ is the Dirichlet distribution with

$$\begin{aligned}
\alpha_i^{(u+1)} &= \alpha_i^{(u)} + \sum_{(n,j,k) \in \pi_n} E_Z\left[Z_{n,i,j,k}\right] \\
&= \alpha_i^{(u)} + \sum_{(n,j,k) \in \pi_n} \eta_{nijk}.
\end{aligned} \tag{4.A.64}$$

The mean of Dirichlet distribution is

$$E_\theta\left[\theta_i^{(u+1)}\right] = \frac{\alpha_i^{(u+1)}}{\sum_{i=0}^{M} \alpha_i^{(u+1)}}. \tag{4.A.65}$$

In summary, the variational Bayesian algorithm for the normalization of the RNA-seq with the ungapped alignment (Nariai et al. 2013) is

Step 1 Initialization
For each transcript isoform, set the initial value $\alpha_i^{(0)}$, $i = 1, \ldots, M$ of the parameters in the Dirichlet distribution.
Step 2 VBE step
Using the current estimate of $E_\theta[\theta^{(u)}]$, compute the density function

$$q_Z^{(u+1)}(Z) = \prod_n \prod_i \prod_j \prod_k \left(\eta_{nijk}\right)^{Z_{nijk}},$$

where

$$\eta_{n,i,j,k} = \frac{\rho_{n,i,j,k}}{\sum_{(i',j',k') \in \pi_n} \rho_{n',i',j',k'}} \quad \text{or}$$

$$E_Z\left[Z_{nijk}\right] = \eta_{nijk},$$

$$\log \rho_{nijk} = E_\theta\left[\log \theta_i\right] + \log P(S_n = j | G_n = i) + \log P(O_n = k | G_n = i)$$

$$+ \log(R_n = r_n | Z_{nijk} = 1),$$

where

$$E_\theta\big[\log\theta_i\big] = \psi(\alpha_i) - \psi\Big(\sum\nolimits_{j=1}^{M}\alpha_j\Big), \ \psi(\alpha) = \frac{\frac{d\Gamma(\alpha)}{d\alpha}}{\Gamma(\alpha)} \ \text{is the digamma function.}$$

Step 3 VBM step
    Using the current estimate $q_z^{(u+1)}(Z)$, calculate

$$E_\theta\Big[\theta_i^{(u+1)}\Big] = \frac{\alpha_i^{(u+1)}}{\sum_{i=0}^{M}\alpha_i^{(u+1)}},$$

where

$$\alpha_i^{(u+1)} = \alpha_i^{(u)} + \sum\nolimits_{(n,j,k)\in\pi_n} E_Z\Big[Z_{n,i,j,k}\Big]$$

$$= \alpha_i^{(u)}\sum\nolimits_{(n,j,k)\in\pi_n}\eta_{nijk}.$$

Step 4. Stop criterion
    If $\| E_\theta[\theta^{(u+1)} - \theta^{(u)}] \|_2^2 < \varepsilon$, stop. Otherwise, return to Step 2.
    The VB algorithms for the estimation of the transcript of the isoform abundance for the gapped alignment of the RNA-seq can be similarly derived.

---

## Appendix 4.B   Log-linear Model for Differential Expression Analysis of the RNA-Seq Data with Negative Binomial Distribution

### 4.B.1  Negative Binomial Distribution for Modeling the RNA-Seq Count Data

The Poisson distribution for modeling the count data assumes that the mean and variance are equal. However, in practice due to heterogeneity we often observe that the variance of the count data in RNA-seq is larger than the mean. This indicates that we need to modify the Poisson distribution for modeling the read counts. We consider two variations in the count data: technology variation and biology variation and develop statistical models for the count data incorporating two variations (McCarthy 2012).
    We first consider technology variation. We assume that the same biology samples are repeatedly sequenced multiple of times. Let $\pi_{gi}$ be the fraction of all cDNA fragments in sample $i$, sampled from gene $g$. The fraction $\pi_{gi}$ varies from replicate to replicate. Let $\sqrt{\phi_g}$ be the coefficient of variation (CV) of $\pi_{gi}$ between the replicates, defined as the standard deviation of $\pi_{gi}$ divided by its

mean. The total number of mapped reads in sample $i$ and the number of reads mapped to gene $g$ are denoted by $N_i$ and $y_{gi}$, respectively. It is clear that

$$E\left[y_{gi}\right] = \mu_{gi} = N_i \pi_{gi}. \tag{4.B.1}$$

Assume that $\pi_{gi}$ is a random variable. The variance of $y_{gi}$ comes from two parts. For fixed $\pi_{gi}$, we assume that $y_{gi}$ follows a Poisson process, and due to sequence technology variation, we can observe the conditional variance $\mathrm{var}(y|\pi)$. On the other hand, we will observe the variance from the variation of $\pi_{gi}$. Then, the total variance of $y_{gi}$ is

$$\begin{aligned}
\mathrm{var}\,(y_{gi}) &= E_\pi[\mathrm{var}(y|\pi)] + \mathrm{var}_\pi(E[y|\pi]) \\
&= E_\pi\left[\mu_{gi}\right] + \mathrm{var}_\pi(N_i \pi_{gi}) \\
&= \mu_{gi} + N_i^2 \mathrm{var}\,(\pi_{gi}).
\end{aligned} \tag{4.B.2}$$

By the definition of CV, we have

$$\mathrm{var}\,(\pi_{gi}) = \pi_{gi}^2 \phi_g. \tag{4.B.3}$$

Substituting Equation 4.B.3 into Equation 4.B.2 and using Equation 4.B.1, we obtain

$$\mathrm{var}\,(y_{gi}) = \mu_{g_i} + \phi_g \mu_{g_i}^2. \tag{4.B.4}$$

Now we consider a compound Poisson process with mean Z. The mean Z itself is a random variable and follows a gamma distribution with shape $\alpha = \dfrac{1}{\phi_g}$ and rate $\beta = \dfrac{1}{\phi_g \mu_{gi}}$ where its density function is $f(z) = \dfrac{\beta^\alpha z^{\alpha-1} e^{-\beta z}}{\Gamma(\alpha)}$. Then, the marginal probability $P(Y = y_{gi})$ is

$$\begin{aligned}
P\,(Y = y_{gi}) &= \int_0^\infty \frac{1}{y_{gi}!} e^{-z} z^{y_{gi}} \frac{1}{\Gamma(\alpha)} \beta^\alpha z^{\alpha-1} e^{-\beta z} dz \\
&= \frac{\beta^\alpha}{y_{gi}!\,\Gamma(\alpha)} \int_0^\infty e^{-(\beta+1)z} z^{y_{gi}+\alpha-1} dz.
\end{aligned} \tag{4.B.5}$$

Let $x = (\beta + 1)z$. Making change of variables in Equation 4.B.5 gives

$$\begin{aligned}
P\,(Y = y_{gi}) &= \frac{\beta^\alpha}{y_{gi}!\,\Gamma(\alpha)(\beta+1)^{y_{gi}+\alpha-1}} \int_0^\infty e^{-x} x^{y_{gi}+\alpha-1} dx \\
&= \frac{\beta^\alpha \Gamma(y_{gi}+\alpha)}{y_{gi}!\,\Gamma(\alpha)(\beta+1)^{y_{gi}+\alpha}} \\
&= \frac{\Gamma(y_{gi}+\alpha)}{y_{gi}!\,\Gamma(\alpha)} \left(\frac{1}{1+\beta}\right)^{y_{gi}} \left(\frac{\beta}{1+\beta}\right)^\alpha.
\end{aligned} \tag{4.B.6}$$

Define $p = \dfrac{1}{1+\beta}$. Then, Equation 4.B.6 can be reduced to

$$P(Y = y_{gi}) = \frac{\Gamma(y_{gi} + \alpha)}{y_{gi}!\,\Gamma(\alpha)} p^{y_{gi}}(1-p)^{\alpha}, \qquad (4.B.7)$$

which shows that the distribution of the read counts is a negative binomial distribution, parameterized by the probability parameter $p = \dfrac{1}{1+\beta}$ and dispersion parameter $\alpha$. We often write $y_{gi} \sim NB(\alpha, p)$. Its mean and variance are given by

$$E\left[y_{gi}\right] = \alpha \frac{p}{1-p} = \frac{\alpha}{\beta} = \frac{\frac{1}{\phi_g}}{\frac{1}{\phi_{gi}\mu_{gi}}} = \mu_{gi} \qquad (4.B.8)$$

and

$$\mathrm{var}\left(y_{gi}\right) = \alpha \frac{p}{(1-p)^2} = \alpha\left(1 + \frac{1}{\beta}\right) = \mu_{gi}\left(1 + \phi_g \mu_{gi}\right), \qquad (4.B.9)$$

respectively.

## 4.B.2  Log-Linear Model

Assume the following log-linear model for testing differential expression (McCarthy et al. 2012):

$$\log \mu_{gi} = x_i^T \gamma_g + \log N_i, \qquad (4.B.10)$$

where $x_i$ is a vector of covariates that specify the state of the RNA sample $i$ or the treatment condition applied to the RNA sample $i$ and $\gamma_g$ is a $q$ dimensional vector of regression coefficients associated with gene $g$. The log-likelihood is proportional to

$$
\begin{aligned}
l\left(\gamma_g, \phi_g\right) &\approx y_{gi}\log\phi_g + y_{gi}\log\mu_{gi} - \left(y_{gi} + \frac{1}{\phi_g}\right)\log\left(1 + \phi_g\mu_{gi}\right) \\
&= y_{gi}\log\phi_g + y_{gi}\left(x_i^T\gamma_g + \log N_i\right) - \left(y_{gi} + \frac{1}{\phi_g}\right)\log\left(1 + \phi_g N_i e^{x_i^T\gamma_g}\right),
\end{aligned}
\qquad (4.B.11)
$$

or

$$
\begin{aligned}
l\left(\gamma_g, \phi_g\right) &\approx \left(\sum_{i=1}^{n} y_{gi}\right)\log\phi_g + \sum_{i=1}^{n}\left[y_{gi}\left(x_i^T\gamma_g + \log N_i\right) - \left(y_{gi} + \frac{1}{\phi_g}\right)\right. \\
&\quad \left. \log\left(1 + \phi_g N_i e^{x_i^T\gamma_g}\right)\right].
\end{aligned}
\qquad (4.B.12)
$$

Next, we study the estimation of the parameters $\gamma_g$, assuming that $\phi_g$ is known. The necessary optimal condition of the maximum likelihood estimate of vector $\gamma_g$ is

$$\frac{\partial l}{\partial \gamma_g} = \sum_{i=1}^{n} \left[ y_{gi} \mathbf{x}_i - \left( y_{gi} + \frac{1}{\phi_g} \right) \frac{\phi_g \mathbf{x}_i N_i e^{\mathbf{x}_i^T \gamma_g}}{1 + \phi_g N_i e^{\mathbf{x}_i^T \gamma_g}} \right] = 0. \tag{4.B.13}$$

Equation 4.B.10 gives

$$\mu_{gi} = N_i e^{\mathbf{x}_i^T \gamma_g}. \tag{4.B.14}$$

Substituting Equation 4.B.14 into Equation 4.B.13 yields the derivative of log-likelihood with respect to the parameters $\gamma_g$ as a function of mean $\mu_{gi}$:

$$\begin{aligned}
\frac{\partial l}{\partial \gamma_g} &= \sum_{i=1}^{n} \left[ y_{gi} \mathbf{x}_i - \left( y_{gi} + \frac{1}{\phi_g} \right) \frac{\phi_g \mathbf{x}_i \mu_{gi}}{1 + \phi_g \mu_{gi}} \right] \\
&= \sum_{i=1}^{n} \frac{y_{gi} - \mu_{gi}}{1 + \phi_g \mu_{gi}} \mathbf{x}_i \\
&= \sum_{i=1}^{n} z_{gi} \mathbf{x}_i,
\end{aligned} \tag{4.B.15}$$

where

$$z_{gi} = \frac{y_{gi} - \mu_{gi}}{1 + \phi_g \mu_{gi}}. \tag{4.B.16}$$

Let the design matrix $X = [\mathbf{x}_1, \dots, \mathbf{x}_n]$ and vector $\mathbf{z}_g = [z_{g1}, \dots, z_{gn}]^T$. Then, Equation 4.B.15 can be written in a matrix form:

$$\frac{\partial l}{\partial \gamma_g} = X \mathbf{z}_g. \tag{4.B.17}$$

Therefore, the necessary optimal condition is

$$X \mathbf{z}_g = 0, \tag{4.B.18}$$

which is a nonlinear equation of $\gamma_g$. The Newton iterative method can be used to solve this nonlinear equation. First, we calculate the Fisher information matrix. Equation 4.B.14 gives

$$\frac{\partial \mu_{gi}}{\partial \gamma_g^T} = N_i \mathbf{x}_i^T e^{\mathbf{x}_i^T \gamma_g} = \mu_{gi} \mathbf{x}_i^T. \tag{4.B.19}$$

It follows from Equation 4.B.16 that

$$
\begin{aligned}
\frac{\partial z_{gi}}{\partial \gamma_g^T} &= \frac{-\dfrac{\partial \mu_{gi}}{\partial \gamma_g^T}\left[1 + \phi_g\mu_{gi} + \left(y_{gi} - \mu_{gi}\right)\phi_g\right]}{\left(1 + \phi_g\mu_{gi}\right)^2} \\[2em]
&= -\frac{\mu_{gi}\left[1 + \phi_g\mu_{gi} + \left(y_{gi} - \mu_{gi}\right)\phi_g\right]x_i^T}{\left(1 + \phi_g\mu_{gi}\right)^2} \\[2em]
&= -\frac{\mu_{gi}y_{gi}\phi_g x_i^T}{\left(1 + \phi_g\mu_{gi}\right)^2}.
\end{aligned}
\tag{4.B.20}
$$

Using Equation 4.B.17 and Equation 4.B.20, we obtain the Hessian matrix of the log-likelihood function:

$$
\begin{aligned}
\frac{\partial^2 l}{\partial \gamma_g \partial \gamma_g^T} &= X \frac{\partial \mathbf{z}_g}{\partial \gamma_g^T} \\[1em]
&= -XWX^T,
\end{aligned}
\tag{4.B.21}
$$

where

$$
W = \begin{bmatrix} \dfrac{\mu_{g1}y_{g1}\phi_g}{\left(1 + \phi_g\mu_{g1}\right)^2} & \cdots & 0 \\ \vdots & \vdots & \vdots \\ 0 & \cdots & \dfrac{\mu_{gn}y_{gn}\phi_g}{\left(1 + \phi_g\mu_{gn}\right)^2} \end{bmatrix}.
$$

The Hessian matrix is

$$
\begin{aligned}
I_g &= -\left[\frac{\partial^2 l}{\partial \gamma_g \partial \gamma_g^T}\right] \\[1em]
&= XWX^T.
\end{aligned}
\tag{4.B.22}
$$

By Taylor expansion, we have

$$
\frac{\partial l}{\partial \gamma_g} = \frac{\partial l}{\partial \gamma_g}\bigg|_{\gamma_g^{old}} + \frac{\partial^2 l}{\partial \gamma \partial \gamma^T}\left(\gamma_g - \gamma_g^{old}\right).
\tag{4.B.23}
$$

Optimal condition (4.B.11) implies that

$$\gamma_g - \gamma_g^{old} = -\left(\frac{\partial^2 l}{\partial\gamma\,\partial\gamma^T}\right)^{-1}\frac{\partial l}{\partial\gamma_g}\Bigg|_{\gamma_g^{old}}.$$

Using Equations 4.B.17 and 4.B.23, we obtain the following iteration procedure for the estimation of the parameters $\gamma_g$:

$$\gamma_g^{new} = \gamma_g^{old} + \left(XWX^T\right)^{-1}Xz_g. \tag{4.B.24}$$

Although in general, the iteration increases the likelihood function, the iteration might produce a solution that decreases the likelihood. To ensure that the iterations will always produce solutions increasing the likelihood function we can modify the iteration (4.B.24) via a line search. Let $\delta = (XWX^T)^{-1}Xz_g$. Define the new iteration procedure:

$$\gamma_g^{new} = \gamma_g^{old} + \alpha\delta, \tag{4.B.25}$$

where $\alpha$ is a stepsize constant. Taking $\delta$ as a search direction and using a linear search, we can determine the stepsize $\alpha$ to ensure that iteration always increases the likelihood. To simplify computation of the search direction $\delta$, we assume that $y_{gi} \approx \mu_{gi} + \dfrac{1}{\phi_g}$, $\mu_{g1} = \ldots = \mu_{gn} = \mu_g$ and $XX^T = I$. Then, $XWX^T \approx \dfrac{\mu_g}{1+\phi_g\mu_g}I$ and $\delta = \dfrac{\mu_g}{1+\phi_g\mu_g}Xz_g$. Let $\alpha_0 = \alpha\dfrac{\mu_g}{1+\phi_g\mu_g}$. Then, iteration procedure (4.A90) can be reduced to

$$\gamma_g^{new} = \gamma_g^{old} + \alpha_0 Xz_g. \tag{4.B.26}$$

Again, using a linear search, we can ensure that the likelihood increases as the iterations proceed.

### 4.B.3  Cox–Reid Adjusted Profile Likelihood

The widely used statistical inferences are provided by the observed likelihood function. In the presence of nuisance parameters, the statistical inferences are carried out via the adjusted profile likelihood function. In Section 4.2A2, we assume that the dispersion parameter is known. However, in practice, the dispersion parameters are often unknown and needs to be estimated. In this section, we will introduce the Cox–Reid adjusted profile likelihood and present the estimation methods for dispersion parameter $\phi$. For the easy calculation of the log-likelihood with respect to $\phi$, we change the form of the log-likelihood for a negative binomial.

Recall from Equation 4.B.5 that

$$P\left(Y = y_{gi}\right) = \frac{\Gamma\left(y_{gi} + \alpha\right)}{y_{gi}!\,\Gamma(\alpha)}\left(\frac{1}{1+\beta}\right)^{y_{gi}}\left(\frac{\beta}{1+\beta}\right)^{\alpha}$$

$$= \frac{\Gamma\left(y_{gi} + \dfrac{1}{\phi_g}\right)}{y_{gi}!\,\Gamma\left(\dfrac{1}{\phi_g}\right)}\left(\phi_g\mu_{gi}\right)^{y_{gi}}\left(\frac{1}{1+\phi_g\mu_{gi}}\right)^{y_{gi}+\frac{1}{\phi_g}}. \tag{4.B.27}$$

Note that

$$\frac{\Gamma\left(y_{gi} + \dfrac{1}{\phi_g}\right)}{y_{gi}!\,\Gamma\left(\dfrac{1}{\phi_g}\right)} = \frac{\left(y_{gi} + \dfrac{1}{\phi_g} - 1\right)\left(y_{gi} + \dfrac{1}{\phi_g} - 2\right)\cdots\left(\dfrac{1}{\phi_g} + 1\right)\dfrac{1}{\phi_g}\left(\dfrac{1}{\phi_g} - 1\right)\left(\dfrac{1}{\phi_g} - 2\right)\cdots}{y_{gi}!\left(\dfrac{1}{\phi_g} - 1\right)\left(\dfrac{1}{\phi_g} - 2\right)\cdots}$$

$$= \frac{\left(y_{gi} + \dfrac{1}{\phi_g} - 1\right)\left(y_{gi} + \dfrac{1}{\phi_g} - 2\right)\cdots\left(\dfrac{1}{\phi_g} + 1\right)\dfrac{1}{\phi_g}}{y_{gi}!}$$

$$= \frac{\left[\left(y_{gi} - 1\right)\phi_g + 1\right]\left[\left(y_{gi} - 2\right)\phi_g + 1\right]\cdots\left[\phi_g + 1\right]}{\phi_g^{y_{gi}}\,y_{gi}!}. \tag{4.B.28}$$

Substituting Equation 4.B.28 into Equation 4.B.27 gives

$$P\left(Y = y_{gi}\right) = \frac{\left[\left(y_{gi} - 1\right)\phi_g + 1\right]\left[\left(y_{gi} - 2\right)\phi_g + 1\right]\cdots\left[\phi_g + 1\right]}{\phi_g^{y_{gi}}\,y_{gi}!}$$

$$\left(\phi_g\mu_{gi}\right)^{y_{gi}}\left(\frac{1}{1+\phi_g\mu_{gi}}\right)^{y_{gi}+\frac{1}{\phi_g}}. \tag{4.B.29}$$

Using Equation 4.B.29 we obtain the log-likelihood

$$l\left(\phi_g; y_g, \hat{\gamma}_g\right) = \sum_{i=1}^{n}\left[\sum_{l=1}^{y_{gi}-1} \log\left(1 + l\phi_g\right) - y_{gi}\log\phi_g - \log y_{gi}!\right.$$

$$\left. + y_{gi}\log\left(\phi_g\mu_{gi}\right) - \left(y_{gi} + \frac{1}{\phi_g}\right)\log\left(1 + \phi_g\mu_{gi}\right)\right]. \tag{4.B.30}$$

The adjusted profile likelihood (APL) for $\phi_g$ is defined as the penalized log-likelihood

$$APL_g\left(\phi_g, \hat{\gamma}_g\right) = l\left(\phi_g; y_g, \hat{\gamma}_g\right) - \frac{1}{2}\log\left|I_g\right|. \qquad (4.B.31)$$

To calculate the first optimal condition for maximizing $APL_g(\phi_g)$, we begin by computing

$$\frac{\partial l\left(\phi_g; y_g, \hat{\gamma}_g\right)}{\partial \phi_g}.$$

It follows from Equation 4.B.30 that

$$\frac{\partial l\left(\phi_g; y_g, \hat{\gamma}_g\right)}{\partial \phi_g} = \sum_{i=1}^{n}\left[\sum_{l=1}^{y_{gi}-1}\frac{l}{1+l\phi_g} - \frac{y_{gi}}{\phi_g} + \frac{y_{gi}}{\phi_g} + \frac{\log\left(1+\phi_g\mu_g\right)}{\phi_g^2}\right.$$
$$\left. -\left(y_{gi}+\frac{1}{\phi_g}\right)\frac{\mu_{gi}}{1+\phi_g\mu_{gi}}\right]$$

$$= \sum_{i=1}^{n}\left[\sum_{l=1}^{y_{gi}-1}\frac{l}{1+l\phi_g} + \frac{\log\left(1+\phi_g\mu_g\right)}{\phi_g^2} - \left(y_{gi}+\frac{1}{\phi_g}\right)\frac{\mu_{gi}}{1+\phi_g\mu_{gi}}\right]. $$
$$(4.B.32)$$

Next we consider a second term in Equation 4.B.31. Using assumption that $y_{gi} \approx \mu_{gi} + \frac{1}{\phi_g}$, we obtain

$$W_1 = \frac{\partial W}{\partial \phi_g} = -\begin{bmatrix} \dfrac{\mu_{g1}^2}{\left(1+\phi_g\mu_{g1}\right)^2} & \cdots & 0 \\ \vdots & \vdots & \vdots \\ 0 & \cdots & \dfrac{\mu_{gn}^2}{\left(1+\phi_g\mu_{gn}\right)^2} \end{bmatrix}. \qquad (4.B.33)$$

Using Equation 1.160 in Xiong 2018 and chain rule, we obtain

$$\frac{\partial \log\left|I_g\right|}{\partial \phi_g} = -I_g^{-1}X\frac{\partial W}{\partial \phi_g}X^T = -I_g^{-1}XW_1X^T. \qquad (4.B.34)$$

Therefore, the derivative of APL with respect to $\phi_g$ is

$$\frac{\partial APL\left(\phi_g\right)}{\partial \phi_g} = \frac{\partial l\left(\phi_g; y_g, \hat{\beta}_g\right)}{\partial \phi_g} + \frac{1}{2}I_g^{-1}XW_1X^T. \qquad (4.B.35)$$

Next, we calculate the second derivative of APL with respect to $\phi_g$. Taking derivatives on both sides of Equation 4.B.32, we obtain

$$\frac{\partial^2 l\left(\phi_g; y_g, \hat{\beta}_g\right)}{\partial \phi_g^2} = \sum_{i=1}^{n}\left[-\sum_{l=1}^{y_{gi}-1}\frac{l^2}{\left(1 + l\phi_g\right)^2} + \frac{\mu_g\phi_g - 2\left(1 + \phi_g\mu_g\right)\log\left(1 + \phi_g\mu_g\right)}{\left(1 + \phi_g\mu_g\right)\phi_g^3}\right.$$
$$\left. + \frac{\mu_{gi}}{\left(1 + \phi_g\mu_g\right)\phi_{gi}^2} + \left(y_{gi} + \phi_g^{-1}\right)\frac{\mu_{gi}^2}{\left(1 + \phi_g\mu_g\right)^2}.\right.$$

(4.B.36)

Using Equation 4.B.33, we obtain

$$W_2 = \frac{\partial W_1}{\partial \phi_g} = \begin{bmatrix} \dfrac{2\mu_{g1}^3}{\left(1 + \phi_g\mu_{g1}\right)^3} & \cdots & 0 \\ \vdots & \vdots & \vdots \\ 0 & \cdots & \dfrac{\mu_{gn}^3}{\left(1 + \phi_g\mu_{gn}\right)^3} \end{bmatrix}.$$

(4.B.37)

Therefore, it follows from Equations 4.B.34 and 4.B.35 that

$$\frac{\partial^2 APL\left(\phi_g\right)}{\partial \phi_2^2} = \frac{\partial^2 l\left(\phi_g; y_g, \hat{\beta}_g\right)}{\partial \phi_g^2} - \frac{1}{2}\frac{\partial I_g^{-1}}{\partial \phi_g}XW_1X^T + \frac{1}{2}I_g^{-1}X\frac{\partial W_1}{\partial \phi_g}X^T$$
$$= \frac{\partial^2 l\left(\phi_g; y_g, \hat{\beta}_g\right)}{\partial \phi_g^2} - \frac{1}{2}\left(I_g^{-1}XW_1X^T\right)^2 + \frac{1}{2}I_g^{-1}XW_2X^T.$$

The dispersion parameter $\phi_g$ can be estimated by the Newton–Raphson iteration algorithm:

$$\phi_g^{(new)} = \phi_g^{(old)} - \frac{\left.\dfrac{\partial APL\left(\phi_g\right)}{\partial \phi_g}\right|_{\phi_g^{(old)}}}{\left.\dfrac{\partial^2 APL\left(\phi_g\right)}{\partial \phi_2^2}\right|_{\phi_g^{(old)}}}$$

(4.B.38)

$$= \phi_g^{(old)} + \delta,$$

where

$$\delta = -\frac{\dfrac{\partial l\left(\phi_g; y_g\hat{\beta}_g\right)}{\partial \phi_g} + \dfrac{1}{2}I_g^{-1}XW_1X^T}{\dfrac{\partial^2 l\left(\phi_g; y_g\hat{\beta}_g\right)}{\partial \phi_g^2} - \dfrac{1}{2}\left(I_g^{-1}XW_1X^T\right)^2 + \dfrac{1}{2}I_g^{-1}XW_2X^T},$$

$$\frac{\partial l\left(\phi_g; y_g, \hat{\beta}_g\right)}{\partial \phi_g} = \sum_{i=1}^{n}\left[\sum_{l=1}^{y_{gi}-1}\frac{l}{1+l\phi_g} + \frac{\log\left(1+\phi_g\mu_g\right)}{\phi_g^2} - \left(y_{gi}+\frac{1}{\phi_g}\right)\frac{\mu_{gi}}{1+\phi_g\mu_{gi}}\right],$$

$$I_g = XWX^T,$$

$$W = \begin{bmatrix} \dfrac{\mu_{g1}y_{g1}\phi_g}{\left(1+\phi_g\mu_{g1}\right)^2} & \cdots & 0 \\ \vdots & \vdots & \vdots \\ 0 & \cdots & \dfrac{\mu_{gn}y_{gn}\phi_g}{\left(1+\phi_g\mu_{gn}\right)^2} \end{bmatrix},$$

$$W_1 = \frac{\partial W}{\partial \phi_g} = -\begin{bmatrix} \dfrac{\mu_{g1}^2}{\left(1+\phi_g\mu_{g1}\right)^2} & \cdots & 0 \\ \vdots & \vdots & \vdots \\ 0 & \cdots & \dfrac{\mu_{gn}^2}{\left(1+\phi_g\mu_{gn}\right)^2} \end{bmatrix},$$

$$\frac{\partial^2 l\left(\phi_g; y_g, \hat{\beta}_g\right)}{\partial \phi_g^2} = \sum_{i=1}^{n}\left[-\sum_{l=1}^{y_{gi}-1}\frac{l^2}{\left(1+l\phi_g\right)^2} + \frac{\mu_g\phi_g - 2\left(1+\phi_g\mu_g\right)\log\left(1+\phi_g\mu_g\right)}{\left(1+\phi_g\mu_g\right)\phi_g^3}\right]$$

$$+ \frac{\mu_{gi}}{\left(1+\phi_g\mu_g\right)\phi_{gi}^2} + \left(y_{gi}+\phi_g^{-1}\right)\frac{\mu_{gi}^2}{\left(1+\phi_g\mu_g\right)^2},$$

$$W_2 = \frac{\partial W_1}{\partial \phi_g} = \begin{bmatrix} \dfrac{2\mu_{g1}^3}{\left(1+\phi_g\mu_{g1}\right)^3} & \cdots & 0 \\ \vdots & \vdots & \vdots \\ 0 & \cdots & \dfrac{2\mu_{gn}^3}{\left(1+\phi_g\mu_{gn}\right)^3} \end{bmatrix},$$

and

$$\mu_{gi} = N_i e^{x_i^T \hat{\gamma}_g}.$$

The Newton–Raphson algorithm usually generates an increase in the like-lihood function. However, it is possible that the iteration also generates a decrease in the likelihood function. To ensure that the iterations always produce increases in the likelihood function, we introduce a slack constant. Equation 4.B.38 can be reduced to

$$\phi_g^{(new)} = \phi_g^{(old)} + \alpha\delta, \tag{4.B.39}$$

where $\alpha$ is a slack constant. Using line search we find $\alpha$ such that $APL_g(\phi_g^{(new)}) \geq APL_g(\phi_g^{(old)})$. In both Equations 4.B.38 and 4.B.39, we assume that the estimator $\hat{\gamma}_g$ is available. However, to estimate $\gamma_g$, we also need dispersion parameter $\phi_g$. Therefore, we need to iterate between $\gamma_g$ and $\phi_g$.

### 4.B.4 Test Statistics

We are interested in testing the significance of coefficients $\gamma_g$ in the log-linear model. The null hypothesis is

$$H_0 : \gamma_g = 0.$$

Define test statistics (Huang et al. 2015):

In the absence of overdispersion, the likelihood ratio statistics for testing the null hypothesis $H_0 : \gamma_g = 0$. is defined as

$$LR_g = 2\left[l\left(\hat{\gamma}_g, \hat{\phi}_g\right) - l\left(\gamma_g^0, \hat{\phi}_g\right)\right], \tag{4.B.40}$$

where

$$l\left(\hat{\gamma}_g, \phi_g\right) = \left(\sum_{i=1}^n y_{gi}\right) \log \phi_g$$

$$+ \sum_{i=1}^n \left[y_{gi}\left(\mathbf{x}_i^T \hat{\gamma}_g + \log N_i\right) - \left(y_{gi} + \frac{1}{\phi_g}\right) \log\left(1 + \phi_g N_i e^{\mathbf{x}_i^T \hat{\gamma}_g}\right)\right].$$

In the presence of overdispersion, the likelihood ratio statistics for testing the null hypothesis $H_0 : \gamma_g = 0$. is defined as

$$LR_{gd} = \frac{2\left[l\left(\hat{\gamma}_g, \hat{\phi}_g\right) - l\left(\gamma_g^0, \hat{\phi}_g\right)\right]}{\hat{\phi}_g}. \tag{4.B.41}$$

Under the null hypothesis $H_0 : \gamma_g = 0$, the test statistics $LR_g$ and $LR_{gd}$ are asymptotically distributed as a central $\chi^2_{(q)}$ and $F_{(q, n-q-1)}$ distribution, respectively.

A natural way to test differential expression is to compare differences in expressions between cases and controls. Consider two groups: group $A$ (cases) and group $B$ (controls). Let $n_{gA}$ and $n_{gB}$ be the number of reads for gene $g$ in group $A$ and group $B$, respectively. Let $n_A$ and $n_B$ be the total number of reads in group $A$ and group $B$, respectively, and $n_A - n_{gA}$ and $n_B - n_{gB}$ be the number of reads for the remaining genes in group $A$ and group $B$, respectively. Let $\mu_{gA}$ and $\mu_{gB}$ be the average number of reads for gene $g$ in group $A$

and group $B$, respectively. The null hypothesis of no differences in expression for gene $g$ between group $A$ and group $B$ is

$$H_0 : \mu_{gA} = \mu_{gB} .$$

To test differential expression, we can use the exact test similar to Fisher's exact test for testing differential expression. Consider a $2 \times 2$ contingency Table 4.6 where $n_g = n_{gA} + n_{gB}$ is the marginal row total for gene $g$, and $n = n_A + n_B$ is the grand total. Assume that $m_A$ subjects are sampled from group $A$ and $m_B$ subjects are sampled from group $B$.

Define

$$n_{gA} = \sum_{i \in A} y_{gi} = \sum_{i=1}^{m_A} y_{gi} \tag{4.B.42}$$

and

$$n_{gB} = \sum_{j \in B} y_{gj} = \sum_{J=1}^{m_B} y_{gj} . \tag{4.B.43}$$

We can develop an exact test similar to the Fisher's exact test for contingency Table 4.6, but we need to replace the hypergeometric probabilities with negative binomial distribution. Let $n_{gA} = a$ and $n_{gB} = b$. Denote the probability of observing the events $n_{gA} = a$ and $n_{gB} = b$ by $p(a,b)$ for any pair of numbers $a$ and $b$. The P-value, $P_g$ of a pair of observed number of reads $(n_{gA}^*, n_{gB}^*)$, is defined as the sum of all probabilities $p(a,b)$ less than or equal to $p(n_{gA}^*, n_{gB}^*)$ of observing the events $n_{gA}^*$ and $n_{gB}^*$, given that the overall sum is $n_g$ (Gonzalez 2014):

$$P_g = \sum_{\substack{a+b = n_g \\ p(a,b) \leq p\left(n_{gA}^*, n_{gB}^*\right)}} p(a,b), \tag{4.B.44}$$

where $a,b = 0,1,\ldots,n_g$.

Now we study how to calculate $p(a,b)$. Under the null hypothesis of no difference in expression, counts (expression level) from different samples are assumed independent. Then, we obtain

$$p(a, b) = p\left(n_{gA} = a\right) p\left(n_{gB} = b\right).$$

We can show that if $y_{gi}$ is distributed as a negative binomial distribution $NB(r_i, p)$ then $n_{gA}$ is also distributed as a negative binomial distribution $NB\left(\sum_{i=1}^{m_A} r_i, p\right)$ (Exercise 4). Recall that $y_{gi}$ is distributed as a negative binomial distribution $NB\left(\phi_{gA}^{-1}, \dfrac{\phi_{gA}\mu_{gA}}{1 + \phi_{gA}\mu_{gA}}\right)$. Using Equation 4.B.42 and the above statements, we conclude that $n_{gA}$ is distributed as the negative bino-

mial distribution $NB\left(m_A \phi_{gA}^{-1}, \dfrac{\phi_{gA}\mu_{gA}}{1 + \phi_{gA}\mu_{gA}}\right)$. Similarly, $n_{gB}$ is distributed as the

negative binomial distribution $NB\left(m_B \phi_{gA}^{-1}, \dfrac{\phi_{gA}\mu_{gA}}{1 + \phi_{gA}\mu_{gA}}\right)$.

Using Equation 4.B.27, we calculate

$$P\left(n_{gA} = a\right) = \frac{\Gamma\left(a + \dfrac{m_A}{\phi_{gA}}\right)}{a!\,\Gamma\left(\dfrac{m_A}{\phi_{gA}}\right)} \left(\phi_{gA}\mu_{gA}\right)^a \left(\frac{1}{1 + \phi_{gA}\mu_{gA}}\right)^{a + \frac{m_A}{\phi_{gA}}}, \tag{4.B.45}$$

and

$$P\left(n_{gB} = b\right) = \frac{\Gamma\left(b + \dfrac{m_B}{\phi_{gB}}\right)}{b!\,\Gamma\left(\dfrac{m_B}{\phi_{gB}}\right)} \left(\phi_{gB}\mu_{gB}\right)^b \left(\frac{1}{1 + \phi_{gB}\mu_{gB}}\right)^{b + \frac{1}{\phi_{gB}}}. \tag{4.B.46}$$

## Appendix 4.C  Derivation of ADMM Algorithm

The first step of the ADMM algorithm (Boyd et al. 2011) to solve the optimization problem (4.154) is to update the matrix $J$ by

$$J = \arg\min_{J} \| J \|_* + \mathrm{Tr}\left(Y_2^T(Z - J) + \frac{\mu}{2}\| Z - J \|_F^2\right). \tag{4.C.1}$$

Note that

$$\mathrm{Tr}\left(Y_2^T(Z - J) + \frac{\mu}{2}\| Z - J \|_F^2\right) = \frac{\mu}{2}\mathrm{Tr}\left[(Z - J)^T(Z - J) + 2\left(\frac{Y_2}{\mu}\right)^T(Z - J)\right]$$

$$= \frac{\mu}{2}\mathrm{Tr}\left[\left(Z - J + \frac{Y_2}{\mu}\right)^T\left(Z - J + \frac{Y_2}{\mu}\right) - \frac{Y_2^T Y_2}{\mu^2}\right].$$

$$= \frac{\mu}{2}\mathrm{Tr}\left(Z - J + \frac{Y_2}{\mu}\right)^T\left(Z - J + \frac{Y_2}{\mu}\right) - \frac{Y_2^T Y_2}{2\mu}$$

$$\tag{4.C.2}$$

Substituting Equation 4.B.46 into Equation 4.C.1 gives

$$J = \arg\min_{J} \frac{1}{\mu} \| J \|_* + \frac{1}{2} \| J - \left( Z + \frac{Y_2}{\mu} \right) \|_F^2. \tag{4.C.3}$$

Now we solve the optimization problem (4.C.3). Suppose that the singular value decomposition of the matrix $Z + \frac{Y_2}{\mu}$ is

$$Z + \frac{Y_2}{\mu} = U\Sigma V^T. \tag{4.C.4}$$

Define the matrix

$$J = U\Lambda V^T. \tag{4.C.5}$$

Using Equations 4.C.4 and 4.C.5, we obtain

$$
\begin{aligned}
\frac{1}{\mu} \| J \|_* + \frac{1}{2} \| J - \left( Z + \frac{Y_2}{\mu} \right) \|_F^2 &= \frac{1}{\mu} \mathrm{Tr}(\Lambda) + \frac{1}{2} \| U\Lambda V^T - U\Sigma V^T \|_F^2 \\
&= \frac{1}{\mu} \mathrm{Tr}(\Lambda) + \frac{1}{2} \| U(\Lambda - \Sigma) V^T \|_F^2 \\
&= \frac{1}{\mu} \mathrm{Tr}(\Lambda) + \frac{1}{2} \| \Lambda - \Sigma \|_F^2 \\
&= \mathrm{Tr}\left[ \frac{1}{\mu} \Lambda + \frac{1}{2} (\Lambda - \Sigma)^T (\Lambda - \Sigma) \right] \\
&= \sum_{j=1}^{r} \left[ \frac{1}{\mu} \lambda_j + \frac{1}{2} \left( \lambda_j - \sigma_j \right)^2 \right],
\end{aligned}
\tag{4.C.6}
$$

where $\Lambda = \mathrm{diag}(\lambda_1, \ldots, \lambda_r)$ and $\Sigma = \mathrm{diag}(\sigma_1, \ldots, \sigma_r)$.

Therefore, the optimization problem (4.C.3) is reduced to

$$\min_{\lambda_j} \quad F = \sum_{j=1}^{r} \left[ \frac{1}{\mu} \lambda_j + \frac{1}{2} \left( \lambda_j - \sigma_j \right)^2 \right]. \tag{4.C.7}$$

Taking the partial derivative of the function $F$ with respect to the singular values $\lambda_j$ and setting them to zero give the necessary conditions for the optimal solutions to the problem (4.A118):

$$\frac{\partial F}{\partial \lambda_j} = \frac{1}{\mu} + \lambda_j - \sigma_j = 0, \ j = 1, \ldots, r. \tag{4.C.8}$$

Solving equation 4.C.8 for $\lambda_j (\lambda_j \geq 0)$ gives

$$\lambda_j = \max\left( \sigma_j - \frac{1}{\mu}, 0 \right).$$

Therefore, we obtain the solution to the optimization problem (4.C.1):

$$J^* = U \operatorname{diag}\left(\max\left(\sigma_1 - \frac{1}{\mu}, 0\right), \ldots, \max\left(\sigma_r^{(k+1)} - \frac{1}{\mu}, 0\right)\right) V^T \text{ or} \qquad (4.C.9)$$

$$J^{(k+1)} = U^{(k+1)} \operatorname{diag}\left(\max\left(\sigma_1^{(k+1)} - \frac{1}{\mu}, 0\right), \ldots, \max\left(\sigma_r^{(k+1)} - \frac{1}{\mu}, 0\right)\right)\left(V^{(k+1)}\right)^T.$$

Next we want to solve the following optimization problem and update the solution

$$Z : \min_Z F = \operatorname{tr}\left(Y_1^T(X - AZ - E)\right) + \operatorname{tr}\left(Y_2^T(Z - J)\right) + + \frac{\mu}{2}$$

$$\times \left(\| X - AZ - E \|_F^2 + \| Z - J \|_F^2\right). \qquad (4.C.10)$$

Using matrix calculus in Chapter 1, taking the partial derivative of function $F$ with respect to the matrix $Z$ and setting it to the zero yields

$$\frac{\partial F}{\partial Z} = -A^T Y_1 + Y_2 - \mu A^T(X - AZ - E) + \mu(Z - J) = 0. \qquad (4.C.11)$$

Solving Equation 4.C.9 for the matrix $Z$, we obtain

$$Z = \left(I + A^T A\right)^{-1}\left[A^T(X - E) + \frac{1}{\mu}\left(A^T Y_1 - Y_2\right) + J\right], \qquad (4.C.12)$$

or update formula for the matrix $Z$:

$$Z^{(k+1)} = \left(I + A^T A\right)^{-1}\left[A^T\left(X - E^{(k)}\right) + J^{(K+1)} + \frac{1}{\mu}\left(A^T Y_1^{(k)} - Y_2^{(k)}\right)\right].$$

Next update the matrix $E$ via solving the following minimization problem:

$$\min_E \quad F = \lambda\| E \|_{2,1} + \operatorname{Tr}\left(Y_1^T(X - AZ - E)\right) + \frac{\mu}{2}\| X - AZ - E \|_F^2. \qquad (4.C.13)$$

The function $F$ in Equation 4.C.13 can be reduced to

$$F = \mu\left\{\frac{\lambda}{\mu}\| E \|_{2,1} + \frac{1}{2}\left[\frac{2}{\mu}\operatorname{Tr}\left(Y_1^T(X - AZ - E)\right) + \| X - AZ - E \|_F^2\right]\right\}$$

$$\approx \mu\left[\frac{\lambda}{\mu}\| E \|_{2,1} + \frac{1}{2}\| E - \left(X - AZ + \frac{1}{\mu}Y\right)_1\|_F^2\right]. \qquad (4.C.14)$$

Let

$$Q = X - AZ + \frac{1}{\mu} Y. \tag{4.C.15}$$

Then, the optimization problem (4.A1244) can be rewritten as

$$\min_{E} \quad F = \alpha \| E \|_{2,1} + \frac{1}{2} \| E - Q \|_F^2, \tag{4.C.16}$$

where

$$\alpha = \frac{\lambda}{\mu}. \tag{4.C.17}$$

Taking the partial derivative of the function $F$ with respect to the column vector $E_{.j}$ gives

$$\frac{\partial F}{\partial E_{.j}} = \alpha \frac{E_{.j}}{\| E_{.j} \|_2} + E_{.j} - Q_{.j} = 0, \tag{4.C.18}$$

which can be reduced to

$$\left( \frac{\alpha}{\| E_{.j} \|_2} + 1 \right) E_{.j} = Q_{.j}. \tag{4.C.19}$$

Taking $\|.\|_2$ norm on both sides of Equation 4.C.19 yields

$$\left( \frac{\alpha}{\| E_{.j} \|_2} + 1 \right) \| E_{.j} \|_2 = \| Q_{.j} \|_2. \tag{4.C.20}$$

Solving Equation 4.C.20 for $\| E_{.j} \|_2$, we obtain

$$\| E_{.j} \|_2 = \| Q_{.j} \|_2 - \alpha, \tag{4.C.21}$$

which implies $\| Q_{.j} \|_2 \geq \alpha$.

Substituting Equation 4.C.21 into Equation 4.C.19, we obtain

$$E_{.j} = \frac{\| Q_{.j} \|_2 - \alpha}{\| Q_{.j} \|_2} Q_{.j} \text{ when } \| Q_{.j} \|_2 \geq \alpha. \tag{4.C.22}$$

Substituting Equation 4.C.17 into Equation 4.C.22 gives the solution:

$$E_{\cdot j} = \frac{\left( \| Q_{\cdot j} \|_2 - \dfrac{\lambda}{\mu} \right)_+}{\| Q_{\cdot j} \|_2} Q_{\cdot j}. \tag{4.C.23}$$

The update for $E_{\cdot j}^{(k)}$ is

$$E_{\cdot j}^{(k+1)} = \frac{\left( \| Q_{\cdot j} \|_2 - \dfrac{\lambda}{\mu} \right)_+}{\| Q_{\cdot j} \|_2} Q_{\cdot j}. \tag{4.C.24}$$

The ADMM (Boyd et al. 2011) gives the following update for the multipliers:

$$\begin{aligned}
Y_1^{(k+1)} &= Y_1^{(k)} + \mu\left( X - AZ^{(k+1)} - E^{(k+1)} \right) \\
Y_2^{(k+1)} &= Y_2^{(k)} + \mu\left( Z^{(k+1)} - J^{(k+1)} \right).
\end{aligned} \tag{4.C.25}$$

## Appendix 4.D   Low Rank Representation Induced Sparse Structural Equation Models

A low rank representation induced linear structural equation model can be formulated as

$$G\alpha + XB + E = 0, \tag{4.D.1}$$

where

$$G = \begin{bmatrix} G^1 & G^2 & \cdots & G^K \end{bmatrix}_{n \times r}$$

$$= \begin{bmatrix}
g_{11}^1 & \cdots & g_{1r_1}^1 & g_{11}^2 & \cdots & g_{1r_2}^2 & \cdots & g_{11}^K & \cdots & g_{1r_K}^K \\
g_{21}^1 & \cdots & g_{2r_1}^1 & g_{21}^2 & \cdots & g_{2r_2}^2 & \cdots & g_{21}^K & \cdots & g_{2r_k}^K \\
\vdots & \vdots & \vdots & \ddots & \cdots & \ddots & \cdots & \ddots & \cdots & \vdots \\
g_{n1}^1 & \cdots & g_{nr_1}^1 & g_{n1}^2 & \cdots & g_{nr_2}^2 & \cdots & g_{n1}^K & \cdots & g_{nr_K}^K
\end{bmatrix}_{n \times r} ,$$

$$\alpha = \begin{bmatrix} \alpha^{11} & \alpha^{12} & \cdots & \alpha^{1K} \\ \alpha^{21} & \alpha^{22} & \cdots & \alpha^{2K} \\ \vdots & \vdots & \vdots & \vdots \\ \alpha^{K1} & \alpha^{K2} & \cdots & \alpha^{KK} \end{bmatrix}_{r \times r}$$

$$= \begin{bmatrix} \alpha11 & \cdots & \alpha_{1r_1}^{11} & \alpha_{11}^{12} & \cdots & \alpha_{1r_2}^{12} & \cdots & \alpha_{11}^{1K} & \cdots & \alpha_{1r_K}^{1K} \\ \vdots & \ddots & \ddots & \ddots & \ddots & \ddots & \ddots & \ddots & \ddots & \vdots \\ \alpha_{r_11}^{11} & \cdots & \alpha_{r_1r_1}^{11} & \alpha_{r_11}^{12} & \cdots & \alpha_{r_1r_2}^{12} & \cdots & \alpha_{r_11}^{1K} & \cdots & \alpha_{r_1r_K}^{1K} \\ \alpha_{11}^{21} & \cdots & \alpha_{1r_1}^{21} & \alpha_{11}^{22} & \cdots & \alpha_{1r_2}^{22} & \cdots & \alpha_{11}^{2K} & \cdots & \alpha_{1r_K}^{2K} \\ \vdots & \ddots & \ddots & \ddots & \ddots & \ddots & \ddots & \ddots & \ddots & \vdots \\ \alpha_{r_21}^{21} & \cdots & \alpha_{r_2r_1}^{21} & \alpha_{r_21}^{22} & \cdots & \alpha_{r_2r_2}^{22} & \cdots & \alpha_{r_21}^{2K} & \cdots & \alpha_{r_2r_K}^{2K} \\ \vdots & \ddots & \ddots & \ddots & \ddots & \ddots & \ddots & \ddots & \ddots & \vdots \\ \alpha_{11}^{K1} & \cdots & \alpha_{1r_1}^{K1} & \alpha_{11}^{K2} & \cdots & \alpha_{1r_2}^{K2} & \cdots & \alpha_{11}^{KK} & \cdots & \alpha_{1r_K}^{KK} \\ \vdots & \ddots & \ddots & \ddots & \ddots & \ddots & \ddots & \ddots & \ddots & \vdots \\ \alpha_{r_K1}^{K1} & \cdots & \alpha_{r_Kr_1}^{K1} & \alpha_{r_K1}^{K2} & \cdots & \alpha_{r_Kr_2}^{K2} & \cdots & \alpha_{r_K1}^{KK} & \cdots & \alpha_{r_Kr_K}^{KK} \end{bmatrix},$$

$$= \begin{bmatrix} \alpha^1 & \alpha^2 & \cdots & \alpha^K \end{bmatrix}$$

$$\alpha^i = \begin{bmatrix} \alpha^{1i} \\ \alpha^{2i} \\ \vdots \\ \alpha^{Ki} \end{bmatrix}, \quad \alpha^{ii} = \begin{bmatrix} \alpha_{11}^{ii} & \cdots & \alpha_{1r_i}^{ii} \\ \vdots & \vdots & \vdots \\ \alpha_{r_i1}^{ii} & \cdots & \alpha_{r_ir_i}^{ii} \end{bmatrix},$$

$$X = \begin{bmatrix} X^1 & X^2 & \cdots & X^L \end{bmatrix}$$

$$= \begin{bmatrix} x_{11} & x_{12} & \cdots & x_{1L} \\ x_{21} & x_{22} & \cdots & x_{2L} \\ \vdots & \vdots & \ddots & \vdots \\ x_{n1} & x_{n2} & \cdots & x_{nL} \end{bmatrix},$$

$$B = \begin{bmatrix} \beta_{11}^1 & \cdots & \beta_{1r_1}^1 & \beta_{11}^2 & \cdots & \beta_{1r_2}^2 & \cdots & \beta_{11}^K & \cdots & \beta_{1r_K}^K \\ \beta_{21}^1 & \cdots & \beta_{2r_1}^1 & \beta_{21}^2 & \cdots & \beta_{2r_2}^2 & \cdots & \beta_{21}^K & \cdots & \beta_{2r_K}^K \\ \vdots & \ddots & \ddots & \ddots & \ddots & \ddots & \ddots & \ddots & \ddots & \vdots \\ \beta_{L1}^1 & \cdots & \beta_{Lr_1}^1 & \beta_{L1}^2 & \cdots & \beta_{Lr_2}^2 & \cdots & \beta_{L1}^K & \cdots & \beta_{Lr_K}^K \end{bmatrix},$$

$$= \begin{bmatrix} B^1 & B^2 & \cdots & B^K \end{bmatrix}$$

$$B^i = \begin{bmatrix} \beta_{11}^i & \beta_{12}^i & \cdots & \beta_{1r_i}^i \\ \beta_{21}^i & \beta_{22}^i & \cdots & \beta_{2r_i}^i \\ \vdots & \vdots & \vdots & \vdots \\ \beta_{L1}^i & \beta_{L2}^i & \cdots & \beta_{Lr_i}^i \end{bmatrix}$$

$$= \begin{bmatrix} \beta_1^i \\ \beta_2^i \\ \vdots \\ \beta_L^i \end{bmatrix}, \quad \beta_l^i = \begin{bmatrix} \beta_{l1}^i & \beta_{l2}^i & \cdots & \beta_{lr_i}^i \end{bmatrix},$$

or

$$B^i = \begin{bmatrix} b_1^i & b_2^i & \cdots & b_{r_i}^i \end{bmatrix}, \quad b_j^i = \begin{bmatrix} \beta_{1j}^i \\ \beta_{2j}^i \\ \vdots \\ \beta_{Lj}^i \end{bmatrix},$$

and

$$E = \begin{bmatrix} e_{11}^1 & \cdots & e_{1r_1}^1 & e_{11}^2 & \cdots & e_{1r_2}^2 & \cdots & e_{11}^K & \cdots & e_{1r_K}^K \\ e_{21}^1 & \cdots & e_{2r_1}^1 & e_{21}^2 & \cdots & e_{2r_2}^2 & \cdots & e_{21}^K & \cdots & e_{2r_K}^K \\ \vdots & \ddots & \ddots & \ddots & \ddots & \ddots & \ddots & \ddots & \ddots & \vdots \\ e_{n1}^1 & \cdots & e_{nr_1}^1 & e_{n1}^2 & \cdots & e_{nr_2}^2 & \cdots & e_{n1}^K & \cdots & e_{nr_K}^K \end{bmatrix},$$

$$= \begin{bmatrix} E^1 & E^2 & \cdots & E^K \end{bmatrix}$$

$$E^i = \begin{bmatrix} e_{11}^i & e_{12}^i & \cdots & e_{1r_i}^i \\ e_{21}^i & e_{22}^i & \cdots & e_{2r_i}^i \\ \vdots & \vdots & \vdots & \vdots \\ e_{n1}^i & e_{n2}^i & \cdots & e_{nr_i}^i \end{bmatrix}.$$

Difference between the low rank representation induced linear structural equation models and the classical scale linear structural equation models is that unlike the scale structural equation models where each node represents a random variable, each node in the low rank representation induced structural equation models represents a vector of random variables. Therefore, the matrix equation corresponding to the node $i$ are

$$G\alpha^i + XB^i + E^i = 0 \qquad (4.D.2)$$

or

$$G^1\alpha^{1i} + \ldots + G^{i-1}\alpha^{i-1i} + G^i\alpha^{ii} + G^{i+1}\alpha^{i+1i} + \ldots G^K\alpha^{Ki} + XB^i + E^i = 0. \qquad (4.D.3)$$

Similar to Equation 1.52 in Chapter 1, Equation 4.D.3 can be rewritten as

$$G^i = G^1\alpha^{1i} + \ldots + G^{i-1}\alpha^{i-1i} + G^{i+1}\alpha^{i+1i} + \ldots G^K\alpha^{Ki} + XB^i + E^i, \qquad (4.D.4)$$

where $\alpha^{-i}$ is the matrix of the path coefficients of the all other nodes connecting to node $i$. The matrix $\alpha^{ji}$ are path coefficients that measure the strength of the causal relationships from the vector $G^j$ to $G^i$ or from the node $j$ to the node $i$, $\beta^i_l$ are path coefficients from the exogenous variable to the endogenous variable which measures the causal effect of the exogenous variable $x_l$ on the low rank representation score vector $G^i$. The coefficients $\| \alpha^{ji} \|_{2,1} = 0$ and $\| \beta^i_l \|_{2,1} = 0$ imply the zero direct influence of $G^i$ and $x_l$ on $G^i$, respectively and are usually omitted from the equation.

Taking vector operations on both sides of Equation 4.B4, we obtain

$$
\begin{aligned}
\mathrm{vec}\left(G^i\right) &= \left(I_{r_i} \otimes G^1\right)\mathrm{vec}\left(\alpha^{1i}\right) + \ldots + \left(I_{r_i} \otimes G^{i-1}\right)\mathrm{vec}\left(\alpha^{i-1i}\right) \\
&\quad + \left(I_{r_i} \otimes G^{i+1}\right)\mathrm{vec}\left(\alpha^{i+1i}\right) + \ldots + \left(I_{r_i} \otimes G^K\right)\mathrm{vec}\left(\alpha^{Ki}\right) \\
&\quad + \left(I_{r_i} \otimes X\right)\mathrm{vec}\left(B^i\right) + \mathrm{vec}\left(E^i\right) \\
&= W_i\Delta_i + \mathrm{vec}\left(E^i\right),
\end{aligned}
\qquad (4.D.5)
$$

where

$$g^i = \mathrm{vec}\left(G^i\right) = \left[ \left(g^i_1\right)^T \left(g^i_2\right)^T \cdots \left(g^i_{r_i}\right)^T \right]^T, g^i_j = \left[ g^i_{1j}\ g^i_{2j} \cdots g^i_{nj} \right]^T,$$

$$W_i = \left[ \left(I_{r_i} \otimes G^1\right) \cdots \left(I_{r_i} \otimes G^{i-1}\right) \left(I_{r_i} \otimes G^{i+1}\right) \cdots \left(I_{r_i} \otimes G^K\right) \right],$$

$$H_i = \left(I_{r_i} \otimes X\right),$$

$$\delta_i = \left[ \left( vec\left( \alpha^{1i} \right) \right)^T \cdots \left( vec\left( \alpha^{i-1i} \right) \right)^T \left( vec\left( \alpha^{i+1i} \right) \right)^T \cdots \left( vec\left( \alpha^{Ki} \right) \right)^T \right]^T,$$

$$b^i = vec\left( B^i \right).$$

Let $g^i = vec(G^i)$ be a $nr_i$ dimensional vector of low rank representation score of the gene expression of the node (pathway)$i$, $e_i = vec(E^i)$ be a $nr_i$ dimensional vector of random errors associated with the node (pathway)$i$, $\delta_{ji} = vec(\alpha^{ji})$, $j = 1,\ldots,i-1, i+1,\ldots,K$ be a $r_j r_i$ dimensional vector of path coefficients connecting the node $j$ to the node $i$, $\delta_i = [ \delta_{1i}^T \cdots \delta_{i-1i}^T \ \delta_{i+1i}^T \cdots \delta_{Ki}^T ]^T$ be a $J_i = (r_1 + \ldots + r_{i-1} + r_{i+1} + \ldots + r_K)r_i$ dimensional vector path coefficients connecting all other nodes to the node $i$, and $b^i = vec(B^i) = [ (b_1^i)^T (b_2^i)^T \cdots (b_{r_i}^i)^T ]^T$ be a $r_i L$ dimensional vector of path coefficients connecting all the exogenous (genotype and environment nodes) to the node $i$ and $\Delta_i = \begin{bmatrix} \delta_i \\ b_i \end{bmatrix}$.

Equation 4.D.5 can then be written as

$$g^i = W_i \delta_i + H_i b^i + e_i, \tag{4.D.6}$$

where $e_i$ is a $r_i n$ dimensional vector of random variables with mean zeros and covariance matrix $\sigma_{ii} I_{r_i n}$.

Multiplying both sides of Equation 4.D.6 by $(I_{r_i} \otimes X)^T$, we obtain

$$\left( I_{r_i} \otimes X \right)^T g^i = \left( I_{r_i} \otimes X \right)^T \left( W_i \delta_i + H_i b^i \right) + \left( I_{r_i} \otimes X \right)^T e_i, \tag{4.D.7}$$

where

$$\mathrm{cov}\left( \left( I_{r_i} \otimes X \right)^T e_i, \left( I_{r_i} \otimes X \right)^T e_i \right) = \sigma_{ii} \left( I_{r_i} \otimes X^T X \right).$$

Using weighted least square and $l_2$-norm penalization of Equation 4.D.7, we can form the following optimization problem:

$$\min_{\delta_i, b^i} \quad f\left( \delta_i, b^i \right) + \lambda_1 \left( \sum_{j=1}^{i-1} \| \delta_{ji} \|_2 + \sum_{j=i+1}^{K} \| \delta_{ji} \|_2 \right) + \lambda_2 \sum_{l=1}^{r_i} \| b_l^i \|_2, \tag{4.D.8}$$

where
$$f(\delta_i, b^i) = [(I_{r_i} \otimes X)^T g^i - (I_{r_i} \otimes X)^T (W_i \delta_i + H_i b^i)]^T (I_{r_i} \otimes X^T X)^{-1} [(I_{r_i} \otimes X)^T g^i - (I_{r_i} \otimes X)^T (W_i \delta_i + H_i b^i)], \quad \| \alpha^{-i} \|_{2,1} = \sum_{j=1}^{i-1} \| \delta_{ji} \|_2 + \sum_{j=i+1}^{K} \| \delta_{ji} \|_2, \quad \| B^i \|_{2,1} = \sum_{l=1}^{r_i} \| b_l^i \|_2,$$ and the norm $\|M\|_{2,1}$ of the matrix $M$ is defined as $\| M \|_{2,1} = \sum_j \| [M]_{.j} \|_2$, $[M]_{.j}$ is the $j^{th}$ column vector of the matrix $M$.

To separate smooth optimization $\min f(\delta_i, b^i)$ from nonsmooth optimization $\| \alpha^{-i} \|_{2,1}$ and $\| B^i \|_{2,1}$, we can introduce constraints $\delta_{ji} - Z_j^1 = 0$ and $b_l^i - Z_l^2 = 0$, and transform the unconstrained optimization problem (4.D.8) into the constrained optimization problem:

$$\min_{\delta_i, b^i} \quad f\left(\delta_i, b^i\right) + \lambda_1 \left(\sum_{j=1}^{i-1} \| Z_j^1 \|_2 + \sum_{j=i+1}^{K} \| Z_j^1 \|_2\right) + \lambda_2 \sum_{l=1}^{r_i} \| Z_l^2 \|_2 \quad (4.D.9)$$

subject to

$$\delta_{ji} - Z_j^1 = 0, \, j = 1, \ldots, i-1, i+1, \ldots, K,$$

$$b_l^i - Z_l^2 = 0, \, l = 1, \ldots, r_i.$$

To solve the optimization problem (4.D.9), we form the augmented Lagrangian

$$
\begin{aligned}
L\left(\delta_i, b^i, Z\right) = {} & \frac{1}{2} f\left(\delta_i, b^i\right) + \lambda_1 \left(\sum_{j=1}^{i-1} \| Z_j^1 \|_2 + \sum_{j=i+1}^{K} \| Z_j^1 \|_2\right) \\
& + \lambda_2 \sum_{l=1}^{r_i} \| Z_l^2 \|_2 + \sum_{j=1}^{i-1} \mu_j^T \left(\delta_{ji} - Z_j^1\right) \\
& + \sum_{j=i+1}^{K} \mu_j^T \left(\delta_{ji} - Z_j^1\right) \\
& + \frac{\rho_1}{2} \left(\sum_{j=1}^{i-1} \| \delta_{ji} - Z_j^1 \|_2^2 + \sum_{j=i+1}^{K} \| \delta_{ji} - Z_j^1 \|_2^2\right) \\
& + \sum_{l=1}^{r_i} \pi_l^T \left(b_l^i - Z_l^2\right) + \frac{\rho_2}{2} \sum_{l=1}^{r_i} \| b_l^i - Z_l^2 \|_2^2.
\end{aligned} \quad (4.D.10)
$$

After some algebra, the optimization problem can be transformed to

$$
\begin{aligned}
L\left(\delta_i, b^i, Z\right) = {} & \frac{1}{2} f\left(\delta_i, b^i\right) + \lambda_1 \left(\sum_{j=1}^{i-1} \| Z_j^1 \|_2 + \sum_{j=i+1}^{K} \| Z_j^1 \|_2\right) \\
& + \lambda_2 \sum_{l=1}^{r_i} \| Z_l^2 \|_2 \\
& + \frac{\rho_1}{2} \left(\sum_{j=1}^{i-1} \| \delta_{ji} - Z_j^1 + u_j \|_2^2 + \sum_{j=i+1}^{K} \| \delta_{ji} - Z_j^1 + u_j \|_2^2\right) \\
& + \frac{\rho_2}{2} \sum_{l=1}^{r_i} \| b_l^i - Z_l^2 + v_l \|_2^2.
\end{aligned}
$$

$$(4.D.11)$$

The alternating direction method of multipliers (ADMM) for the optimization problem (4.D.11) consists of the iterations:

$$\delta_{ji}^{(k+1)} := \arg\min_{\delta_{ji}} \frac{1}{2} f\left(\delta_i, b^i\right) + \frac{\rho_1}{2}$$

$$\times \left(\sum_{j=1}^{i-1} \| \delta_{ji} - \left(Z_j^1\right)^k + u_j^{(k)} \|_2^2 + \sum_{j=i+1}^{K} \| \delta_{ji} - \left(Z_j^1\right)^{(k)} + u_j^{(k)} \|_2^2\right)$$

$$(j = 1, \ldots, i-1, i+1, \ldots, K), \tag{4.D.12}$$

$$\left(b_l^i\right)^{(k+1)} := \arg\min_{b_l^i} \ \frac{1}{2}f\left(\delta_i, b^i\right)$$

$$+ \frac{\rho_2}{2} \sum_{l=1}^{r_i} \| b_l^i - \left(Z_l^2\right)^{(k)} + v_l^{(k)} \|_2^2, \ (l = 1, \ldots, r_i), \tag{4.D.13}$$

$$\left(Z_j^1\right)^{(k+1)} := \arg\min_{Z_j^1} \ \lambda_1 \left( \sum_{j=1}^{i-1} \| Z_j^1 \|_2 + \sum_{j=i+1}^{K} \| Z_j^1 \|_2 \right)$$

$$+ \frac{\rho_1}{2} \left( \sum_{j=1}^{i-1} \| \delta_{ji}^{(k+1)} - Z_j^1 + u_j^{(k)} \|_2^2 + \sum_{j=i+1}^{K} \| \delta_{ji}^{(k+1)} - Z_j^1 + u_j^{(k)} \|_2^2 \right),$$

$$j = 1, \ldots i-1, i+1, \ldots, K. \tag{4.D.14}$$

$$\left(Z_l^2\right)^{(k+1)} := \arg\min_{Z_l^2} \ \lambda_2 \sum_{l=1}^{r_i} \| Z_l^2 \|_2 + \frac{\rho_2}{2} \sum_{l=1}^{r_i} \| b_l^i - Z_l^2 + v_l \|_2^2, l$$

$$= 1, \ldots, r_i. \tag{4.D.15}$$

$$u_j^{(k+1)} := = u_j^{(k+1)} + \left( \delta_{ji}^{(k+1)} - \left(Z_j^1\right)^{(k+1)} \right). \tag{4.D.16}$$

$$v_l^{(k+1)} := = v_l^{(k)} + \left( \left(b_l^i\right)^{(k+1)} - \left(Z_l^2\right)^{(k+1)} \right). \tag{4.D.17}$$

For some special forms of function $f(\Delta_i)$, the problem (9.90) can be solved in a closed form. To solve the minimization problem (4.D.4), we first need to calculate the following derivatives:

$$\frac{\partial f}{\partial \delta_i} = -2W_i^T \left(I_{r_i} \otimes X\right) \left(I_{r_i} \otimes X^T X\right)^{-1} \left[ \left(I_{r_i} \otimes X\right)^T g^i - \left(I_{r_i} \otimes X\right)^T \left(W_i \delta_i + H_i b^i\right) \right] \text{ and}$$

$$\frac{\partial \| \delta_i - Z^1 + u \|_2^2}{\partial \delta_i} = 2\left(\delta_i - Z^1 + u\right), \tag{4.D.18}$$

where

$$Z^1 = \begin{bmatrix} Z_1^1 \\ \vdots \\ Z_{i-1}^1 \\ Z_{i+1}^1 \\ \vdots \\ Z_K^1 \end{bmatrix}, u = \begin{bmatrix} u_1 \\ \vdots \\ u_{i-1} \\ u_{i+1} \\ \vdots \\ u_K \end{bmatrix}.$$

Setting $\dfrac{\partial l}{\partial \delta_i} = 0$ and Equation 4.D.18 we obtain

$$-W_i^T \left(I_{r_i} \otimes X\right)\left(I_{r_i} \otimes X^T X\right)^{-1}\left[\left(I_{r_i} \otimes X\right)^T g^i - \left(I_{r_i} \otimes X\right)^T \left(W_i \delta_i + H_i b^i\right)\right]$$
$$+ \rho_1 \left(\delta_i - Z^1 + u\right) = 0,$$

which implies

$$- W_i^T \left[I_{r_i} \otimes X\left(X^T X\right)^{-1} X^T\right]\left[g^i - \left(W_i \delta_i + H_i \left(b^i\right)^{(k)}\right)\right]$$
$$+ \rho_1 \left(\delta_i - \left(Z^1\right)^{(k)} + u^{(k)}\right) = 0. \tag{4.D.19}$$

Solving Equation 4.D.19 gives

$$\delta_i^{(k+1)} = \left\{ W_i^T \left[I_{r_i} \otimes X\left(X^T X\right)^{-1} X^T\right]W_i + \rho_1 I\right\}^{-1}$$
$$\left\{W_i^T \left[I_{r_i} \otimes X\left(X^T X\right)^{-1} X^T\right]g^i + \rho_1 \left(\left(Z^1\right)^{(k)} - u^{(k)}\right)\right\}. \tag{4.D.20}$$

Similarly, we have

$$\dfrac{\partial f}{\partial b^i} = -2 H_i^T \left(I_{r_i} \otimes X\right)\left(I_{r_i} \otimes X^T X\right)^{-1}\left[\left(I_{r_i} \otimes X\right)^T g^i - \left(I_{r_i} \otimes X\right)^T \left(W_i \delta_i + H_i b^i\right)\right]$$

and

$$\dfrac{\partial \| b^i - Z^2 + v \|_2^2}{\partial b^i} = 2\left(b^i - Z^2 + v\right).$$

Setting $\dfrac{\partial l}{\partial b^i} = 0$, we obtain

$$- H_i^T \left[I_{r_i} \otimes X\left(X^T X\right)^{-1} X^T\right]\left[g^i - \left(W_i \delta_i^{(k+1)} + H_i b^i\right)\right]$$
$$+ \rho_2 \left(\delta_i^{(k+1)} - \left(Z^2\right)^{(k)} + v^{(k)}\right) = 0. \tag{4.D.21}$$

Solving Equation 4.D.21 for $b_i$ gives

$$\left(b^i\right)^{k+1} = \left\{ H_i^T \left[I_{r_i} \otimes X\left(X^T X\right)^{-1} X^T\right]H_i + \rho_2 I\right\}^{-1}$$
$$\left\{H_i^T \left[I_{r_i} \otimes X\left(X^T X\right)^{-1} X^T\right]g^i + \rho_2 \left(\left(Z^2\right)^{(k)} - v^{(k)}\right)\right\}, \tag{4.D.22}$$

where

$$
b^i = \begin{bmatrix} b_1^i \\ b_2^i \\ \vdots \\ b_{r_i}^i \end{bmatrix}, \quad Z^2 = \begin{bmatrix} Z_1^2 \\ Z_2^2 \\ \vdots \\ Z_{r_i}^2 \end{bmatrix}, \quad \text{and } v = \begin{bmatrix} v_1 \\ v_2 \\ \vdots \\ v_{r_i} \end{bmatrix}.
$$

The optimization problem (4.D.14) is non-differentiable. Although the first two terms in (4.D.14) is not differentiable, we still can obtain a simple closed-form solution to the problem (4.D.14) using subdiffenrential calculus in Chapter 1.

When $\| Z_j^1 \|_2 \neq 0$, taking derivative of the objective function with respect to the vector $Z_j^1$ in the minimization problem (4.D.14) we obtain

$$
\lambda_1 \frac{Z_j^1}{\| Z_j^1 \|_2} - \rho_1 \left( \delta_{ji}^{(k+1)} - Z_j^1 + u_j^{(k)} \right) = 0, \text{ which implies that}
$$

$$
\left( \frac{\lambda_1}{\rho_1 \| Z_j^1 \|_2} + 1 \right) Z_j^1 = \delta_{ji}^{(k+1)} + u_j^{(k)}. \tag{4.D.23}
$$

Taking $l_2$ norm on both sides of Equation 4.D.23, we obtain

$$
\left( \frac{\lambda_1}{\rho_1 \| Z_j^1 \|_2} + 1 \right) \| Z_j^1 \|_2 = \| \delta_{ji}^{(k+1)} + u_j^{(k)} \|_2. \tag{4.D.24}
$$

Solving Equation 4.D.24 for $\| Z_j^1 \|_2$ yields

$$
\| Z_j^1 \|_2 = \left( \| \delta_{ji}^{(k+1)} + u_j^{(k)} \|_2 - \frac{\lambda_1}{\rho_1} \right)_+, \quad j = 1, \ldots, i-1, i+1, \ldots, K, \tag{4.D.25}
$$

where $(a)_+ = \max(a, 0)$.

Substituting Equation 4.D.25 into Equation 4.D.23 gives the solution:

$$
\left( Z_j^1 \right)^{(k+1)} = \frac{\left( \| \delta_{ji}^{(k+1)} + u_j^{(k)} \|_2 - \frac{\lambda_1}{\rho_1} \right)_+}{\| \delta_{ji}^{(k+1)} + u_j^{(k)} \|_2} \left( \delta_{ji}^{(k+1)} + u_j^{(k)} \right), \tag{4.D.26}
$$

$$
j = 1, \ldots, i-1, i+1, \ldots, K.
$$

Similarly, we can obtain the solution to the problem (4.D.15):

$$
\left( Z_l^2 \right)^{(k+1)} = \frac{\left( \| (b_l^i)^{(k+1)} + v_l^{(k)} \|_2 - \frac{\lambda_2}{\rho_2} \right)_+}{\| (b_l^i)^{(k+1)} + v_l^{(k)} \|_2} \left( (b_l^i)^{(k+1)} + v_l^{(k)} \right), \quad l = 1, \ldots, r_i. \tag{4.D.27}
$$

In summary, the algorithms are given below.
ADMM Algorithm for Solving Low Rank Representation Induced SEMs:
For $i = 1,...,K$
Step 1: Initialization

$$u^0 := 0, v^0 := 0,$$

$$\delta_i^{(0)} = \left\{ W_i^T \left[ I_{r_i} \otimes X (X^T X)^{-1} X^T \right] W_i + \rho_1 I \right\}^{-1} W_i^T \left[ I_{r_i} \otimes X (X^T X)^{-1} X^T \right] g^i,$$

$$\left( b^i \right)^{(0)} = \left\{ H_i^T \left[ I_{r_i} \otimes X (X^T X)^{-1} X^T \right] H_i + \rho_2 I \right\}^{-1} H_i^T \left[ I_{r_i} \otimes X (X^T X)^{-1} X^T \right] g^i,$$

$$\left( Z_j^1 \right)^{(0)} = \frac{\left( \| \delta_{ji}^{(0)} \|_2 - \frac{\lambda_1}{\rho_1} \right)_+}{\| \delta_{ji}^{(0)} \|_2} \delta_{ji}^{(0)}, \, j = 1,...,i-1,i+1,...,K,$$

$$\left( Z_l^2 \right)^{(0)} = \frac{\left( \| (b_l^i)^{(0)} \|_2 - \frac{\lambda_2}{\rho_2} \right)_+}{\| (b_l^i)^{(0)} \|_2} \left( b_l^i \right)^{(0)}, \, l = 1,...,r_i.$$

Carry out steps 2, 3 and 4 until convergence
Step 2:

$$\delta_i^{(k+1)} = \left\{ W_i^T \left[ I_{r_i} \otimes X (X^T X)^{-1} X^T \right] W_i + \rho_1 I \right\}^{-1}$$
$$\left\{ W_i^T \left[ I_{r_i} \otimes X (X^T X)^{-1} X^T \right] g^i + \rho_1 \left( (Z^1)^{(k)} - u^{(k)} \right) \right\},$$

$$\left( b^i \right)^{k+1} = \left\{ H_i^T \left[ I_{r_i} \otimes X (X^T X)^{-1} X^T \right] H_i + \rho_2 I \right\}^{-1}$$
$$\left\{ H_i^T \left[ I_{r_i} \otimes X (X^T X)^{-1} X^T \right] g^i + \rho_2 \left( (Z^2)^{(k)} - v^{(k)} \right) \right\}.$$

Step 3:

$$\left( Z_j^1 \right)^{(k+1)} = \frac{\left( \| \delta_{ji}^{(k+1)} + u_j^{(k)} \|_2 - \frac{\lambda_1}{\rho_1} \right)_+}{\| \delta_{ji}^{(k+1)} + u_j^{(k)} \|_2} \left( \delta_{ji}^{(k+1)} + u_j^{(k)} \right), \, j = 1,...,i-1,i+1,...,K,$$

$$\left( Z_l^2 \right)^{(k+1)} = \frac{\left( \| (b_l^i)^{(k+1)} + v_l^{(k)} \|_2 - \frac{\lambda_2}{\rho_2} \right)_+}{\| (b_l^i)^{(k+1)} + v_l^{(k)} \|_2} \left( (b_l^i)^{(k+1)} + v_l^{(k)} \right), \, l = 1,...,r_i.$$

Step 4:

$$u_j^{(k+1)} := == u_j^{(k+1)} + \left( \delta_{ji}^{(k+1)} - \left( Z_j^1 \right)^{(k+1)} \right), \, j = 1,...,i-1,i+1,...,K,$$

$$v_l^{(k+1)} : == v_l^{(k)} + \left( \left( b_l^i \right)^{(k+1)} - \left( Z_l^2 \right)^{(k+1)} \right), \, l = 1, \ldots, r_i .$$

## Appendix 4.E　Maximum Likelihood (ML) Estimation of Parameters for Dynamic Structural Equation Models

Let $(y_i^t)'$, $(y_i^{(t-1)}(x_i))'$ and $(e_i^t)'$ be the $i^{th}$ row of matrices $Y^t, Y^{t-1}, X$ and $E^{(t)}$, respectively. In other words, we define

$$\begin{aligned}
(y_i^t)' &= \left[ y_{i1}^t \, y_{i2}^t \, \cdots \, y_{iM}^t \right], \\
(y_{li}^{t-1})' &= \left[ y_{li1}^{t-1} \, y_{li2}^{t-1} \, \cdots \, y_{liM}^{t-1} \right], \\
x_i' &= \left[ x_{i1} \, x_{i2} \, \cdots \, x_{iK} \right], \\
(e_i^t)' &= \left[ e_{i1}^t \, e_{i2}^t \, \cdots \, e_{iM}^t \right], \, i = 1, \ldots, n.
\end{aligned} \tag{4.E.1}$$

From Equation 4.173 we obtain

$$(y_i^t)' \Gamma^t + (y_{li}^{t-1})' A^t + x_i' B^t + (e_i^t)' = 0. \tag{4.E.2}$$

Let

$$e_i^t = \begin{bmatrix} e_{i1}^t \\ \vdots \\ e_{iM}^t \end{bmatrix}, \, t = 1, \ldots, T \text{ and } e_i = \begin{bmatrix} e_i^1 \\ \vdots \\ e_i^T \end{bmatrix}, \, i = 1, \ldots, n.$$

The covariance between error variables is given by

$$\text{cov}(e_i, e_i) = \Sigma = \begin{bmatrix} \Sigma_{11} & \Sigma_{12} & \cdots & \Sigma_{1T} \\ \Sigma_{21} & \Sigma_{22} & \cdots & \Sigma_{2T} \\ \vdots & \vdots & \vdots & \vdots \\ \Sigma_{T1} & \Sigma_{T12} & \cdots & \Sigma_{TT} \end{bmatrix}, \, i = 1, \ldots, n, \tag{4.E.3}$$

where

$$\Sigma_{t\tau} = \text{cov}(e_i^t, e_i^\tau) = \begin{bmatrix} \sigma_{11}^{t\tau} & \cdots & \sigma_{1M}^{t\tau} \\ \vdots & \vdots & \vdots \\ \sigma_{M1}^{t\tau} & \cdots & \sigma_{MM}^{t\tau} \end{bmatrix}. \tag{4.E.4}$$

Assume that the matrix $\Gamma^t$ is nonsingular. Using Equation 4.E.2, the vector $(y_i^t)'$ can be written as a function of the errors $(e_i^t)'$:

$$(y_i^t)' = -(e_i^t)'(\Gamma^t)^{-1} - (y_{li}^{t-1})'A^t(\Gamma^t)^{-1} - x_i'B^t(\Gamma^t)^{-1}, i = 1, \ldots, n, t = 1, \ldots, T \quad (4.E.5)$$

or

$$y_i^t = -\left((\Gamma^t)'\right)^{-1}e_i^t - \left((\Gamma^t)'\right)^{-1}(A^t)'y_{li}^{t-1} - \left((\Gamma^t)'\right)^{-1}(B^t)'x_i, i = 1, \ldots, n, t$$

$$= 1, \ldots, T. \quad (4.E.6)$$

Let

$$y_i = \begin{bmatrix} y_i^1 \\ y_i^2 \\ \vdots \\ y_i^T \end{bmatrix}, y_{li} = \begin{bmatrix} 0 \\ y_{li}^1 \\ \vdots \\ y_{li}^{T-1} \end{bmatrix}, e_i = \begin{bmatrix} e_i^1 \\ e_i^2 \\ \vdots \\ e_i^T \end{bmatrix}, i = 1, \ldots, n. \quad (4.E.7)$$

Assume that the vector of errors follows a multivariate normal distribution with mean zeros and covariance matrix $\Sigma$. Then, the distribution of $e_i$ can be written as

$$g(e_i) = (2\pi)^{-\frac{TM}{2}}|\Sigma|^{-\frac{1}{2}}\exp\left\{-\frac{1}{2}e_i'\Sigma^{-1}e_i\right\}. \quad (4.E.8)$$

Since the vector of the gene expression $y_i^t$ is a function of the errors (Equation 4.E.6), its Jacobean matrix is

$$J = \frac{\partial y_i}{\partial e_i'} = \begin{bmatrix} \dfrac{\partial y_i^1}{\partial e_i'} \\ \dfrac{\partial y_i^2}{\partial e_i'} \\ \vdots \\ \dfrac{\partial y_i^T}{\partial e_i'} \end{bmatrix} = \begin{bmatrix} -\left((\Gamma^1)'\right)^{-1} & 0 & \cdots & 0 \\ 0 & -\left((\Gamma^2)'\right)^{-1} & \cdots & 0 \\ \vdots & \vdots & \vdots & \vdots \\ 0 & 0 & \cdots & -\left((\Gamma^T)'\right)^{-1} \end{bmatrix}, i$$

$$= 1, \ldots, n. \quad (4.E.9)$$

The density function of $y_i$ is then given by

$$f(y_i) = |J|^{-1}g(e_i)$$

$$= \prod_{t=1}^{T}|\Gamma^t|g(e_i) \quad (4.E.10)$$

$$= (2\pi)^{-\frac{TM}{2}}|\Sigma|^{-\frac{1}{2}}\prod_{t=1}^{T}|\Gamma^t|\exp\left\{-\frac{1}{2}e_i'\Sigma^{-1}e_i\right\},$$

where

$$e_i^t = -\left(\Gamma^t\right)' y_i^t - \left(A^t\right)' y_{li}^{t-1} - \left(B^t\right)' X_i, t = 1, \ldots, T \text{ and}$$

$$e_i = \begin{bmatrix} e_i^1 \\ e_i^2 \\ \vdots \\ e_i^T \end{bmatrix}.$$

Therefore, the joint probability of the gene expressions is

$$f(y_1, \ldots, y_n) = (2\pi)^{-\frac{nTM}{2}} |\Sigma|^{-\frac{n}{2}} \prod_{t=1}^T |\Gamma^t|^n \exp\left\{ -\frac{1}{2} \sum_{i=1}^n e_i' \Sigma^{-1} e_i \right\}. \qquad (4.E.11)$$

The log likelihood function is defined as

$$l(\Gamma, A, B, \Sigma | Y) = -\frac{nTM}{2} \log(2\pi) - \frac{n}{2} \log|\Sigma| + n \sum_{t=1}^T \log|\Gamma^t|$$

$$- \frac{1}{2} \sum_{i=1}^n e_i' \Sigma^{-1} e_i. \qquad (4.E.12)$$

If we assume that the errors $e_i^t$ and $e_i^\tau$ are uncorrelated, then the matrix $\Sigma$ becomes a diagonal block:

$$\Sigma = \begin{bmatrix} \Sigma_{11} & 0 & \cdots & 0 \\ 0 & \Sigma_{22} & \cdots & 0 \\ \vdots & \vdots & \vdots & \vdots \\ 0 & 0 & \cdots & \Sigma_{TT} \end{bmatrix}.$$

Then, the likelihood function can be reduced to

$$l(\Gamma, A, B, \Sigma | Y) = -\frac{nTM}{2} \log(2\pi) - \frac{n}{2} \sum_{t=1}^T \log|\Sigma_{tt}| + n \sum_{t=1}^T \log|\Gamma^t|$$

$$- \frac{1}{2} \sum_{i=1}^n \sum_{t=1}^T \left(e_i^t\right)' \Sigma_{tt}^{-1} e_i^t. \qquad (4.E.13)$$

The estimators $\hat{\Gamma}, \hat{A}, \hat{B}, \hat{\Sigma}$ of the parameters in the structural equations can be found by maximizing the log likelihood function in Equation 4.E.12 if the errors at different time points are correlated or in Equation 4.E.13 if the errors at different time points are uncorrelated.

## Appendix 4.F  Generalized Least Squares Estimator of the Parameters in Dynamic Structural Equation Models

Define the matrices

$$\Gamma = \begin{bmatrix} \Gamma^1 & 0 & \cdots & 0 \\ 0 & \Gamma^2 & \cdots & 0 \\ \vdots & \vdots & \vdots & \vdots \\ 0 & 0 & \cdots & \Gamma^T \end{bmatrix}, \Gamma_i = \begin{bmatrix} \Gamma_i^1 & 0 & \cdots & 0 \\ 0 & \Gamma_i^2 & \cdots & 0 \\ \vdots & \vdots & \vdots & \vdots \\ 0 & 0 & \cdots & \Gamma_i^T \end{bmatrix}, H = \begin{bmatrix} H^1 & 0 & \cdots & 0 \\ 0 & H^2 & \cdots & 0 \\ \vdots & \vdots & \vdots & \vdots \\ 0 & 0 & \cdots & H^T \end{bmatrix} \text{ and}$$

$$H_i = \begin{bmatrix} H_i^1 & 0 & \cdots & 0 \\ 0 & H_i^2 & \cdots & 0 \\ \vdots & \vdots & \vdots & \vdots \\ 0 & 0 & \cdots & H_i^T \end{bmatrix},$$

where $i$ indexes the $i^{th}$ equation, matrices $\Gamma^t$ and $\Gamma_i^t$ are defined in Equation 4.178, $H^t = \begin{bmatrix} A^t \\ B^t \end{bmatrix}$, $A^t$ is defined in Equation 4.179, and $B^t$ is defined in Equation 4.180, $A_i^t$, $B_i^t$ and $H_i^t$ denote matrices whose elements appear in the $i^{th}$ equation. Equation 4.196 defines

$$\Pi^t = -H^t \left( \Gamma^t \right)^{-1}, \tag{4.F.1}$$

and matrix $\Pi$ is defined in Equation 4.200. Post-multiplying both sides of Equation 4.F.1 by $\Gamma^t$ gives

$$\Pi^t \Gamma^t = -H^t. \tag{4.F.2}$$

Note hat

$$\Pi\Gamma = \begin{bmatrix} \Pi^1 & \cdots & 0 \\ \vdots & \vdots & \vdots \\ 0 & \cdots & \Pi^T \end{bmatrix} \begin{bmatrix} \Gamma^1 & 0 & \cdots & 0 \\ 0 & \Gamma^2 & \cdots & 0 \\ \vdots & \vdots & \vdots & \vdots \\ 0 & 0 & \cdots & \Gamma^T \end{bmatrix}$$

$$= \begin{bmatrix} \Pi^1 \Gamma^1 & 0 & \cdots & 0 \\ 0 & \Pi^2 \Gamma^2 & \cdots & 0 \\ \vdots & \vdots & \vdots & \vdots \\ 0 & 0 & \cdots & \Pi^T \Gamma^T \end{bmatrix}. \tag{4.F.3}$$

Substituting Equation 4.F.2 into Equation 4.F.3 gives

$$\Pi\Gamma = -\begin{bmatrix} H^1 & 0 & \cdots & 0 \\ 0 & H^2 & \cdots & 0 \\ \vdots & \vdots & \vdots & \vdots \\ 0 & 0 & \cdots & H^T \end{bmatrix} = -H. \tag{4.F.4}$$

Similarly, we have

$$\Pi\Gamma_i = \begin{bmatrix} \Pi^1\Gamma_i^1 & 0 & \cdots & 0 \\ 0 & \Pi^2\Gamma_i^2 & \cdots & 0 \\ \vdots & \vdots & \vdots & \vdots \\ 0 & 0 & \cdots & \Pi^T\Gamma_i^T \end{bmatrix} = -\begin{bmatrix} H_i^1 & 0 & \cdots & 0 \\ 0 & H_i^2 & \cdots & 0 \\ \vdots & \vdots & \vdots & \vdots \\ 0 & 0 & \cdots & H_i^T \end{bmatrix} = -H_i. \tag{4.F.5}$$

Substituting Equation 4.209 into Equation 4.F.5, we obtain

$$\Pi\Gamma_i = (W'W)^{-1}W'Y\Gamma_i$$

$$= (W'W)^{-1}W'[Y^1 \cdots Y^T]\begin{bmatrix} \Gamma_i^1 & 0 & \cdots & 0 \\ 0 & \Gamma_i^2 & \cdots & 0 \\ \vdots & \vdots & \vdots & \vdots \\ 0 & 0 & \cdots & \Gamma_i^T \end{bmatrix} \tag{4.F.6}$$

$$= (W'W)^{-1}W'[Y^1\Gamma_i^1 \ Y^2\Gamma_i^2 \ \cdots \ Y^T\Gamma_i^T] = -H_i$$

Multiplying both sides of Equation 4.F.7 by $W'W$ gives

$$W'[Y^1\Gamma_i^1 \ Y^2\Gamma_i^2 \ \cdots \ Y^T\Gamma_i^T] = -W'WH_i. \tag{4.F.7}$$

Note that

$$Y^t\Gamma_i^t = Y^t\begin{bmatrix} -1 \\ \gamma_i^t \\ 0 \end{bmatrix} = -y_i^t + Y_i^t\gamma_i^t. \tag{4.F.8}$$

Next, we calculate $WH_i$. By definition of $W$ and $H_i$, we have

$$WH_i = [W^1 \cdots W^T]\begin{bmatrix} H_i^1 & 0 & \cdots & 0 \\ 0 & H_i^2 & \cdots & 0 \\ \vdots & \vdots & \vdots & \vdots \\ 0 & 0 & \cdots & H_i^T \end{bmatrix} \tag{4.F.9}$$

$$= [W^1H_i^1 \ W^2H_i^2 \ \cdots \ W^TH_i^T].$$

Define

$$\mathbf{W}_i^t = \begin{bmatrix} \mathbf{Y}_{li}^{t-1} & \mathbf{X}_i \end{bmatrix}. \tag{4.F.10}$$

Then,

$$\mathbf{W}^t = \begin{bmatrix} \mathbf{Y}_i^{t-1} & \mathbf{X} \end{bmatrix} = \begin{bmatrix} \mathbf{Y}_{li}^{t-1} & (\mathbf{Y}_{li}^{t-1})^* & \mathbf{X}_i & (\mathbf{X}_i)^* \end{bmatrix}. \tag{4.F.11}$$

Recall that

$$\mathbf{H}_i^t = \begin{bmatrix} \boldsymbol{\alpha}_i^t \\ 0 \\ \boldsymbol{\beta}_i^t \\ 0 \end{bmatrix}. \tag{4.F.12}$$

Then, Equations 4.188, 4.194, 4.195, 4.F.10, 4.F.11, and 4.F.12 imply that

$$\mathbf{W}^t \mathbf{H}_i^t = \begin{bmatrix} \mathbf{Y}_{li}^{t-1} & (\mathbf{Y}_{li}^{t-1})^* & \mathbf{X}_i & (\mathbf{X}_i)^* \end{bmatrix} \begin{bmatrix} \boldsymbol{\alpha}_i^t \\ 0 \\ \boldsymbol{\beta}_i^t \\ 0 \end{bmatrix}$$

$$= \mathbf{Y}_{li}^{t-1} \boldsymbol{\alpha}_i^t + \mathbf{X}_i \boldsymbol{\beta}_i^t \tag{4.F.13}$$

$$= \mathbf{W}_i^t \begin{bmatrix} \boldsymbol{\alpha}_i^t \\ \boldsymbol{\beta}_i^t \end{bmatrix}, t = 1, .., \mathrm{T}.$$

It follows from Equation 4.F.7 that

$$\mathbf{W}'\mathbf{Y}^t \boldsymbol{\Gamma}_i^t = -\mathbf{W}'\mathbf{W}\mathbf{H}_i^t. \tag{4.F.14}$$

Using Equations 4.F.8, we obtain

$$\mathbf{W}'\mathbf{Y}^t \boldsymbol{\Gamma}_i^t = -\mathbf{W}'y_i^t + \mathbf{W}'\mathbf{Y}_i^t \boldsymbol{\gamma}_i^t. \tag{4.F.15}$$

Combining Equations 4.F.13, 4.F.14, and 4.F.15 gives

$$-\mathbf{W}'y_i^t + \mathbf{W}'\mathbf{Y}_i^t \boldsymbol{\gamma}_i^t = -\mathbf{W}'\mathbf{W}_i^t \begin{bmatrix} \boldsymbol{\alpha}_i^t \\ \boldsymbol{\beta}_i^t \end{bmatrix},$$

which implies

$$\mathbf{W}'y_i^t = \mathbf{W}'\mathbf{Y}_i^t\boldsymbol{\gamma}_i^t + \mathbf{W}'\mathbf{W}_i^t \begin{bmatrix} \alpha_i^t \\ \beta_i^t \end{bmatrix}$$

$$= \mathbf{W}'\left(\mathbf{Y}_i^t\boldsymbol{\gamma}_i^t + \mathbf{W}_i^t \begin{bmatrix} \alpha_i^t \\ \beta_i^t \end{bmatrix}\right)$$

$$= \mathbf{W}'\begin{bmatrix} \mathbf{Y}_i^t & \mathbf{W}_i^t \end{bmatrix} \begin{bmatrix} \boldsymbol{\gamma}_i^t \\ \alpha_i^t \\ \beta_i^t \end{bmatrix} \qquad (4.F.16)$$

$$= \mathbf{W}'\mathbf{Z}_i^t\boldsymbol{\delta}_i^t, \ t = 1, \dots, T.$$

Recall that

$$\mathbf{y}_i = \begin{bmatrix} \mathbf{y}_i^1 \\ \mathbf{y}_i^2 \\ \vdots \\ \mathbf{y}_i^T \end{bmatrix}, \ \mathbf{Z}_i = \begin{bmatrix} \mathbf{Z}_i^1 & 0 & \cdots & 0 \\ 0 & \mathbf{Z}_i^2 & \cdots & 0 \\ \vdots & \vdots & \vdots & \vdots \\ 0 & 0 & \cdots & \mathbf{Z}_i^T \end{bmatrix}, \ \boldsymbol{\delta}_i = \begin{bmatrix} \boldsymbol{\delta}_i^1 \\ \boldsymbol{\delta}_i^2 \\ \vdots \\ \boldsymbol{\delta}_i^T \end{bmatrix}. \qquad (4.F.17)$$

Combining Equations 4.F.16 and 4.F.17, we obtain

$$\mathbf{W}'_*\mathbf{y}_i = \mathbf{W}'_*\mathbf{Z}_i\boldsymbol{\delta}_i. \qquad (4.F.18)$$

Multiplying both sides of Equation 4.189 by $\mathbf{W}'_*$ gives

$$\mathbf{W}'_*\mathbf{y}_i = \mathbf{W}'_*\mathbf{Z}_i\boldsymbol{\delta}_i + \mathbf{W}'_*\mathbf{e}_i. \qquad (4.F.19)$$

It follows from Equation 4.190 that the covariance matrix of $\mathbf{W}'_*\mathbf{e}_i$ is

$$\Lambda_i = \text{cov}\left(\mathbf{W}'_*\mathbf{e}_i, \mathbf{W}'_*\mathbf{e}_i\right) = \mathbf{W}'_*\mathbf{V}_i\mathbf{W}_*, \qquad (4.F.20)$$

where

$$\mathbf{V}_i = \text{cov}(\mathbf{e}_i, \mathbf{e}_i) = \Sigma_{ii} \otimes \mathbf{I}_n, \qquad (4.F.21)$$

and

$$\Sigma_{ii} = \begin{bmatrix} \sigma_{ii}^{11} & \cdots & \sigma_{ii}^{1T} \\ \vdots & \vdots & \vdots \\ \sigma_{ii}^{T1} & \cdots & \sigma_{ii}^{TT} \end{bmatrix}. \qquad (4.F.22)$$

Applying generalized least squares to the model (4.F.19) gives

$$\hat{\boldsymbol{\delta}}_i = \left[(\mathbf{W}'_*\mathbf{Z}_i)'\Lambda_i^{-1}\mathbf{W}'_*\mathbf{Z}_i\right]^{-1}(\mathbf{W}'_*\mathbf{Z}_i)'\Lambda_i^{-1}\mathbf{W}'_*\mathbf{y}_i. \qquad (4.F.23)$$

The covariance matrix $\mathbf{\Sigma}_{ii}$ is estimated by

$$\hat{\mathbf{\Sigma}}_{ii} = \frac{1}{(n - m_1 + 1 - l_i - k_i)T} (\mathbf{y}_i - \mathbf{Z}_i \mathbf{\delta}_i)(\mathbf{y}_i - \mathbf{Z}_i \mathbf{\delta}_i)'. \qquad (4.F.24)$$

The covariance matrix of the estimator $\hat{\mathbf{\delta}}_i$ is

$$\begin{aligned}
\mathbf{\Sigma}_{\delta} = \mathrm{cov}\left(\hat{\mathbf{\delta}}_i, \hat{\mathbf{\delta}}_i\right) &= \left[(\mathbf{W}'_* \mathbf{Z}_i)' \mathbf{\Lambda}_i^{-1} \mathbf{W}'_* \mathbf{Z}_i\right]^{-1} (\mathbf{W}'_* \mathbf{Z}_i)' \mathbf{\Lambda}_i^{-1} \mathbf{W}'_* \mathrm{cov}(\mathbf{y}_i, \mathbf{y}_i) \\
&\quad * \mathbf{W}'_* \mathbf{\Lambda}_i^{-1} (\mathbf{W}'_* \mathbf{Z}_i) \left[(\mathbf{W}'_* \mathbf{Z}_i)' \mathbf{\Lambda}_i^{-1} \mathbf{W}'_* \mathbf{Z}_i\right]^{-1} \qquad (4.F.25) \\
&= \left[(\mathbf{W}'_* \mathbf{Z}_i)' \mathbf{\Lambda}_i^{-1} \mathbf{W}'_* \mathbf{Z}_i\right]^{-1}.
\end{aligned}$$

---

## Appendix 4.G   Proximal Algorithm for L$_1$-Penalized Maximum Likelihood Estimation of Dynamic Structural Equation Model

Let

$$f(\mathbf{\Gamma}, \mathbf{A}, \mathbf{B}) = -l(\mathbf{\Gamma}, \mathbf{A}, \mathbf{B}, \mathbf{\Sigma}|\mathbf{Y})$$

$$= \frac{nTm}{2} \log(2\pi) + \frac{n}{2} \log|\mathbf{\Sigma}| - n \log|\mathbf{\Gamma}| + \frac{1}{2} \sum_{i=1}^{n} \mathbf{e}'_i \mathbf{\Sigma}^{-1} \mathbf{e}_i, \qquad (4.G.1)$$

where

$$\mathbf{\Gamma} = \begin{bmatrix} \mathbf{\Gamma}^1 & 0 & \cdots & 0 \\ 0 & \mathbf{\Gamma}^2 & \cdots & 0 \\ \vdots & \vdots & \vdots & \vdots \\ 0 & 0 & \cdots & \mathbf{\Gamma}^T \end{bmatrix}, \quad \mathbf{A} = \begin{bmatrix} \mathbf{A}^1 & 0 & \cdots & 0 \\ 0 & \mathbf{A}^2 & \cdots & 0 \\ \vdots & \vdots & \vdots & \vdots \\ 0 & 0 & \cdots & \mathbf{A}^T \end{bmatrix}, \quad \mathbf{B} = \begin{bmatrix} \mathbf{B}^1 & 0 & \cdots & 0 \\ 0 & \mathbf{B}^2 & \cdots & 0 \\ \vdots & \vdots & \vdots & \vdots \\ 0 & 0 & \cdots & \mathbf{B}^T \end{bmatrix}, \quad \mathbf{y}_i = \begin{bmatrix} \mathbf{y}_i^1 \\ \mathbf{y}_i^2 \\ \vdots \\ \mathbf{y}_i^T \end{bmatrix},$$

$$\mathbf{y}_i^t = \begin{bmatrix} y_{i1}^t \\ y_{i2}^t \\ \vdots \\ y_{iM}^t \end{bmatrix}, \quad \mathbf{y}_{li} = \begin{bmatrix} \mathbf{y}_{li}^0 \\ \mathbf{y}_{li}^1 \\ \vdots \\ \mathbf{y}_{li}^{T-1} \end{bmatrix}, \quad \mathbf{y}_{li}^{t-1} = \begin{bmatrix} y_{li1}^{t-1} \\ y_{li2}^{t-1} \\ \vdots \\ y_{liM}^{t-1} \end{bmatrix}, \quad \mathbf{X}_i = \begin{bmatrix} \mathbf{X}_i \\ \mathbf{X}_i \\ \vdots \\ \mathbf{X}_i \end{bmatrix}, \quad \mathbf{x}_i = \begin{bmatrix} x_{i1} \\ x_{i2} \\ \vdots \\ x_{iK} \end{bmatrix},$$

$$\mathbf{e}_i = \begin{bmatrix} \mathbf{e}_i^1 \\ \mathbf{e}_i^2 \\ \vdots \\ \mathbf{e}_i^T \end{bmatrix}, \quad \mathbf{e}_i^t = \begin{bmatrix} e_{i1}^t \\ \vdots \\ e_{iM}^t \end{bmatrix},$$

$$e_i = -\left(\boldsymbol{\Gamma}'\mathbf{y}_i + \mathbf{A}'\mathbf{y}_{li} + \mathbf{B}'\mathbf{X}_i\right).$$

Our goal is to minimize the following function:

$$\min_{\boldsymbol{\Gamma}^t, A^t, B^t} \quad f(\boldsymbol{\Gamma}, \mathbf{A}, \mathbf{B}) + \lambda_1 \sum_{t=1}^{T} \|\boldsymbol{\Gamma}^t\|_1 + \lambda_2 \sum_{t=1}^{T} \|\mathbf{A}^t\|_1 + \lambda_3 \sum_{t=1}^{T} \|\mathbf{B}^t\|_1. \quad (4.G.2)$$

The objective function of the optimization problem (4.G.2) consists of two functions: convex differentiable functions and nonsmooth functions. The traditional Newton's method is an efficient tool for solving unconstrained smooth optimization problems but is not suited for solving large nonsmooth convex problems. Proximal methods can be viewed as an extension of Newton's method from solving smooth optimization problems to nonsmooth optimization problems.

The first step for applying the proximal methods to the optimization problem (4.G.2) is to take a partial derivative of the function $f(\boldsymbol{\Gamma}, \mathbf{A}, \mathbf{B})$ with respect to $\boldsymbol{\Gamma}$, $\mathbf{A}$ and $\mathbf{B}$. Applying matrix calculus (Section 1.4), we obtain

$$\frac{\partial f}{\partial \boldsymbol{\Gamma}'} = -n\boldsymbol{\Gamma}^{-1} - \sum_{i=1}^{n} \left(\boldsymbol{\Sigma}^{-1}\right)e_i\mathbf{y}_i'. \quad (4.G.3)$$

Taking transpose on both sides of Equation 4.G.3 gives

$$\frac{\partial f}{\partial \boldsymbol{\Gamma}} = -n\left(\boldsymbol{\Gamma}'\right)^{-1} - \sum_{i=1}^{n} \mathbf{y}_i e_i'\left(\boldsymbol{\Sigma}^{-1}\right). \quad (4.G.4)$$

Similarly, we have

$$\frac{\partial f}{\partial \mathbf{A}} = -\sum_{i=1}^{n} \mathbf{y}_{li} e_i' \boldsymbol{\Sigma}^{-1} \quad (4.G.5)$$

and

$$\frac{\partial f}{\partial \mathbf{B}} = -\sum_{i=1}^{n} \mathbf{X}_i e_i' \boldsymbol{\Sigma}^{-1}, \quad (4.G.6)$$

where

$$e_i' = -\left(\mathbf{y}_i'\boldsymbol{\Gamma} + \mathbf{y}_{li}'\mathbf{A} + \mathbf{X}_i'\mathbf{B}\right). \quad (4.G.7)$$

If we assume that the errors at the different time points are independent, the matrix $\boldsymbol{\Sigma}$ can be written as a diagonal block matrix $\boldsymbol{\Sigma} = \text{diag}(\boldsymbol{\Sigma}_{11}, \boldsymbol{\Sigma}_{22}, \ldots, \boldsymbol{\Sigma}_{TT})$. Equations 4.G.4–4.G.7 will be rewritten as

$$\frac{\partial f}{\partial \boldsymbol{\Gamma}^t} = -n\left(\left(\boldsymbol{\Gamma}^t\right)'\right)^{-1} - \sum_{i=1}^{n} \mathbf{y}_i^t e_i^{t'} \boldsymbol{\Sigma}_{tt}^{-1}, \quad (4.G.8)$$

$$\frac{\partial f}{\partial \mathbf{A}^t} = -\sum_{i=1}^{n} \mathbf{y}_{li}^t \mathbf{e}_i^{t\prime} \boldsymbol{\Sigma}_{tt}^{-1},$$ (4.G.9)

$$\frac{\partial f}{\partial \mathbf{B}^t} = -\sum_{i=1}^{n} \mathbf{x}_i \mathbf{e}_i^{t\prime} \boldsymbol{\Sigma}_{tt}^{-1}, \text{ and}$$ (4.G.10)

$$\mathbf{e}_i^{t\prime} = -\left(\mathbf{y}_i^{t\prime} \boldsymbol{\Gamma}^t + \mathbf{y}_{li}^{t\prime} \mathbf{A}^t + \mathbf{x}_i' \mathbf{B}^t\right).$$ (4.G.11)

Let

$$\mathbf{W}_\Gamma = vec(\boldsymbol{\Gamma}), \ \mathbf{W}_A = vec(\mathbf{A}), \ \mathbf{W}_B = vec(\mathbf{B}),$$

$$\frac{\partial f}{\partial \mathbf{W}_\Gamma} = vec\left(\frac{\partial f}{\partial \boldsymbol{\Gamma}}\right), \ \frac{\partial f}{\partial \mathbf{W}_A} = vec\left(\frac{\partial f}{\partial \mathbf{A}}\right), \ \frac{\partial f}{\partial \mathbf{W}_B} = vec\left(\frac{\partial f}{\partial \mathbf{B}}\right),$$

$$\mathbf{W} = \begin{bmatrix} \mathbf{W}_\Gamma \\ \mathbf{W}_A \\ \mathbf{W}_B \end{bmatrix} \text{ and } \frac{\partial f}{\partial \mathbf{W}} = \begin{bmatrix} \dfrac{\partial f}{\partial \mathbf{W}_\Gamma} \\[2mm] \dfrac{\partial f}{\partial \mathbf{W}_A} \\[2mm] \dfrac{\partial f}{\partial \mathbf{W}_B} \end{bmatrix}.$$ (4.G.12)

Let

$$f(\mathbf{W}) = f(\boldsymbol{\Gamma}, \mathbf{A}, \mathbf{B}), \ \Omega_\Gamma(\mathbf{W}_\Gamma) = \|\boldsymbol{\Gamma}\|_1, \ \Omega_A = \|\mathbf{A}\|_1, \ \Omega_B = \|\mathbf{B}\| \text{ and}$$

$$\Omega(\mathbf{W}) = \lambda_1 \Omega_\Gamma(\mathbf{W}_\Gamma) + \lambda_2 \Omega_A(\mathbf{W}_A) + \lambda_3 \Omega_B(\mathbf{W}_B).$$

Optimization problem (4.G.2) can be written as

$$\min_{\mathbf{W}} \ f(\mathbf{W}) + \Omega(\mathbf{W}).$$ (4.G.13)

Recall that the proximal operator $\text{Prox}_\Omega(u)$ of the function $\Omega(w)$ (1.42) is defined by

$$\text{Prox}_\Omega(\mathbf{u}) = \underset{w \in R^p}{\arg\min} \left(\Omega(\mathbf{W}) + \frac{1}{2} \|\mathbf{W} - \mathbf{u}\|_2^2\right).$$ (4.G.14)

Using the property of the proximal operator, we obtain

$$\text{Prox}_\Omega(\mathbf{u}) = \begin{bmatrix} \text{Prox}_{\lambda_1 \Omega_\Gamma}(\mathbf{u}_\Gamma) \\ \text{Prox}_{\lambda_2 \Omega_A}(\mathbf{u}_A) \\ \text{Prox}_{\lambda_3 \Omega_B}(\mathbf{u}_B) \end{bmatrix},$$ (4.G.15)

where

$$\mathbf{u}_\Gamma = \mathbf{W}_\Gamma - \frac{1}{L}\frac{\partial f}{\partial \mathbf{W}_\Gamma}, \ \mathbf{u}_A = \mathbf{W}_A - \frac{1}{L}\frac{\partial f}{\partial \mathbf{W}_A}, \ \mathbf{u}_B = \mathbf{W}_B - \frac{1}{L}\frac{\partial f}{\partial \mathbf{W}_B}. \qquad (4.G.16)$$

Let $u_{\Gamma_{ij}}$ be the $(i,j)^{th}$ element of the matrix $\mathbf{u}_\Gamma$, $u_{Aij}$ be the $(i,j)^{th}$ element of the matrix $\mathbf{u}_A$ and $u_{Bij}$ be the $(i,j)^{th}$ element of the matrix $\mathbf{u}_B$. Define

$$\mathbf{Z} = \begin{bmatrix} \mathbf{Z}_\Gamma \\ \mathbf{Z}_A \\ \mathbf{Z}_B \end{bmatrix} = \begin{bmatrix} \text{Prox}_{\lambda_1\Omega_\Gamma}(\mathbf{u}_\Gamma) \\ \text{Prox}_{\lambda_2\Omega_A}(\mathbf{u}_A) \\ \text{Prox}_{\lambda_3\Omega_B}(\mathbf{u}_B) \end{bmatrix}. \qquad (4.G.17)$$

Let

$$u^k_{\Gamma_{ij}} = \gamma^k_{ij} - \lambda^k_1\left(\frac{\partial f}{\partial \Gamma}\right)_{ij}, \ u^k_{Aij} = \alpha^k_{ij} - \lambda^k_2\left(\frac{\partial f}{\partial A}\right)_{ij} \text{ and } u^k_{Bij} = \beta^k_{ij} - \lambda^k_3\left(\frac{\partial f}{\partial B}\right)_{ij}.$$

Recall that the nonsmooth functions $\Omega_\Gamma$, $\Omega_A$ and $\Omega_B$ are $L_1$ norm. Using Equation 1.137 (Xiong 2018), we obtain the elements of $\text{Prox}_{\lambda_1\Omega_\Gamma}(\mathbf{u}_\Gamma)$, $\text{Prox}_{\lambda_2\Omega_A}(\mathbf{u}_A)$, and $\text{Prox}_{\lambda_3\Omega_B}(\mathbf{u}_B)$:

$$\gamma^z_{ij} = \text{sign}\left(u^k_{\Gamma_{ij}}\right)\left(\left|u^k_{\Gamma_{ij}}\right| - \lambda^k_1\right)_+, \ i = 1,\dots,MT, j = 1,\dots,MT, \qquad (4.G.18)$$

$$\alpha^z_{ij} = \text{sign}\left(u^k_{Aij}\right)\left(\left|u^k_{Aij}\right| - \lambda^k_2\right)_+, \ i = 1,\dots,MT, j = 1,\dots,MT, \qquad (4.G.19)$$

$$\beta^z_{ij} = \text{sign}\left(u^k_{Bij}\right)\left(\left|u^k_{Bij}\right| - \lambda^k_3\right)_+, \ i = 1,\dots,KT, j = 1,\dots,MT, \qquad (4.G.20)$$

where

$$\Gamma_z = \left(\gamma^z_{ij}\right)_{MT\times MT}, \ \mathbf{A}_z = \left(\alpha^z_{ij}\right)_{MT\times MT} \text{ and } \mathbf{B}_z = \left(\beta_{ij}\right)_{KT\times MT}.$$

Combining Equations 4.G.17–4.G.20, we have

$$\mathbf{Z}_\Gamma = vec(\Gamma_z), \ \mathbf{Z}_A = vec(\mathbf{A}_z) \text{ and } \mathbf{Z}_B = vec(\mathbf{B}_z). \qquad (4.G.21)$$

Let

$$\mathbf{G}^k_\Gamma = vec\left(\Gamma^k\right), \ \mathbf{G}^k_A = vec\left(\mathbf{A}^k\right) \text{ and } \mathbf{G}^k_B = vec\left(\mathbf{B}^k\right),$$

$$\mathbf{G}^k = \begin{bmatrix} \mathbf{G}^k_\Gamma \\ \mathbf{G}^k_A \\ \mathbf{G}^k_B \end{bmatrix},$$

where

$$\mathbf{\Gamma}^k = \left(\gamma_{ij}^k\right)_{MT \times MT}, \ \mathbf{A}^k = \left(\alpha_{ij}^k\right)_{MT \times MT}, \ \mathbf{B}^k = \left(\beta_{ij}^k\right)_{KT \times KT}.$$

Define an auxiliary function:

$$f_\lambda\left(\mathbf{Z}, \mathbf{G}^k\right) = f(\mathbf{Z}) + \left(\frac{\partial f}{\partial \mathbf{Z}}\right)^T \left(\mathbf{G}^k - \mathbf{Z}\right) + \frac{1}{2\lambda_1} \| \mathbf{G}_\Gamma^k - \mathbf{Z}_\Gamma \|_2^2$$

$$+ \frac{1}{2\lambda_2} \| \mathbf{G}_A^k - \mathbf{Z}_A \|_2^2 + \frac{1}{2\lambda_3} \| \mathbf{G}_B^k - \mathbf{Z}_B \|_2^2. \qquad (4.G.22)$$

### Algorithm 4.G.1: Proximal Gradient Algorithm

Step 1: Given $\mathbf{W}_\Gamma^k, \mathbf{W}_A^k, \mathbf{W}_B^k, \mathbf{G}^k, \mathbf{G}_\Gamma^k, \mathbf{G}_A^k, \mathbf{G}_B^k, \lambda_1^{k-1}, \lambda_2^{k-1}, \lambda_3^{k-1}$, and parameter $\alpha_1, \alpha_2, \alpha_3 \in (0,1)$. Set $\lambda_1 = \lambda_1^{k-1}$, $\lambda_2 = \lambda_2^{k-1}$ and $\lambda_3 = \lambda_3^{k-1}$.

Step 2:
 Repeat

1. Set

$$\mathbf{Z} = \begin{bmatrix} \mathbf{Z}_\Gamma \\ \mathbf{Z}_A \\ \mathbf{Z}_B \end{bmatrix} = \begin{bmatrix} \text{Prox}_{\lambda_1 \Omega_\Gamma}\left(\mathbf{u}_\Gamma^k\right) \\ \text{Prox}_{\lambda_2 \Omega_A}\left(\mathbf{u}_A^k\right) \\ \text{Prox}_{\lambda_3 \Omega_B}\left(\mathbf{u}_B^k\right) \end{bmatrix},$$

where we calculate the proximal operator element-wise as follows:

$$\mathbf{u}_\Gamma^k = \mathbf{W}_\Gamma^k - \lambda_1 \frac{\partial f}{\partial \mathbf{W}_\Gamma}\Big|_{W^k}, \ \mathbf{u}_A^k = \mathbf{W}_A^k - \lambda_2 \frac{\partial f}{\partial \mathbf{W}_A}\Big|_{W^k}, \ \mathbf{u}_B^k = \mathbf{W}_B^k - \lambda_3 \frac{\partial f}{\partial \mathbf{W}_B}\Big|_{W^k},$$

$$\mathbf{W}_\Gamma^k = \left(\gamma_{ij}^k\right)_{MT \times MT}, \ \mathbf{W}_A^k = \left(\alpha_{ij}^k\right)_{MT \times MT}, \ \mathbf{W}_B^k = \left(\beta_{ij}^k\right)_{KT \times MT},$$

$$\mathbf{u}_\Gamma^k = \left(u_{\Gamma ij}^k\right)_{MT \times MT}, \ \mathbf{u}_A^k = \left(u_{Aij}^k\right)_{MT \times MT}, \ \mathbf{u}_B^k = \left(u_{Bij}^k\right)_{KT \times MT},$$

$$\gamma_{ij}^z = \text{sign}\left(u_{\Gamma_{ij}}^k\right)\left(\left|u_{\Gamma_{ij}}^k\right| - \lambda_1^k\right)_+, \ i = 1, \dots, MT, j = 1, \dots, MT, \quad (4.G.23)$$

$$\alpha_{ij}^z = \text{sign}\left(u_{Aij}^k\right)\left(\left|u_{Aij}^k\right| - \lambda_2^k\right)_+, \ i = 1, \dots, MT, j = 1, \dots, MT, \quad (4.G.24)$$

$$\beta_{ij}^z = \text{sign}\left(u_{Bij}^k\right)\left(\left|u_{Bij}^k\right| - \lambda_3^k\right)_+, \ i = 1, \dots, KT, j = 1, \dots, MT \quad (4.G.25)$$

$$\mathbf{\Gamma}_z = \left(\gamma_{ij}^z\right)_{MT \times MT}, \ \mathbf{A}_z = \left(\alpha_{ij}^z\right)_{MT \times MT} \text{ and } \mathbf{B}_z = \left(\beta_{ij}\right)_{KT \times MT}. \quad (4.G.26)$$

$$\mathbf{Z}_\Gamma = vec(\mathbf{\Gamma}_z), \mathbf{Z}_A = vec(\mathbf{A}_z) \text{ and } \mathbf{Z}_B = vec(\mathbf{B}_z).$$

2. Calculate the auxiliary function:

$$f_\lambda\left(\mathbf{Z}, \mathbf{G}^k\right) = f(\mathbf{Z}) + \left(\frac{\partial f}{\partial \mathbf{Z}}\right)^T \left(\mathbf{G}^k - \mathbf{Z}\right) + \frac{1}{2\lambda_1} \| \mathbf{G}_\Gamma^k - \mathbf{Z}_\Gamma \|_2^2$$

$$+ \frac{1}{2\lambda_2} \| \mathbf{G}_A^k - \mathbf{Z}_A \|_2^2 + \frac{1}{2\lambda_3} \| \mathbf{G}_B^k - \mathbf{Z}_B \|_2^2$$

3. Break and go to step 3 if $f(\mathbf{Z}) \le \hat{f}_\lambda(\mathbf{Z}, \mathbf{G}^k)$.
4. Update $\lambda_1 = \alpha_1 \lambda_1, \lambda_2 = \alpha_2 \lambda_2$ and $\lambda_3 = \alpha_3 \lambda_3$.

Step 3: return $\lambda_1^k = \lambda_1, \lambda_2^k = \lambda_2, \lambda_3^k = \lambda_3$ and

$$\mathbf{G}^{k+1} = \mathbf{Z} \text{ and}$$

$$\mathbf{W}_\Gamma^{k+1} = \mathbf{\Gamma}_Z, \mathbf{W}_A^{k+1} = \mathbf{A}_Z, \mathbf{W}_B^{k+1} = \mathbf{B}_Z, k = k + 1. \qquad (4.G.27)$$

Finally, initial values of the matrices can be obtained by least square estimation. Let

$$\mathbf{W}^t = \left[ \mathbf{Y}_l^{t-1} \ \mathbf{X} \right], \ \mathbf{W} = \left[ \mathbf{W}^1 \ \cdots \ \mathbf{W}^T \right], \ \mathbf{Z}_i^t = \left[ \mathbf{Y}_i^t \ \mathbf{Y}_{li}^{t-1} \ \mathbf{X}_i \right], \ \mathbf{\Lambda}_i = \mathbf{W}'\mathbf{W}.$$

$$\mathbf{y}_i = \begin{bmatrix} \mathbf{y}_i^1 \\ \mathbf{y}_i^2 \\ \vdots \\ \mathbf{y}_i^T \end{bmatrix}, \ \mathbf{Z}_i = \begin{bmatrix} \mathbf{Z}_i^1 & 0 & \cdots & 0 \\ 0 & \mathbf{Z}_i^2 & \cdots & 0 \\ \vdots & \vdots & \vdots & \vdots \\ 0 & 0 & \cdots & \mathbf{Z}_i^T \end{bmatrix}, \ \delta_i = \begin{bmatrix} \delta_i^1 \\ \delta_i^2 \\ \vdots \\ \delta_i^T \end{bmatrix}, \ \delta_i^t = \begin{bmatrix} \gamma_i^t \\ \alpha_i^t \\ \beta_i^t \end{bmatrix}.$$

A least square estimator of the parameters $\delta_i$ is

$$\hat{\delta}_i = \left[ (\mathbf{W}'\mathbf{Z}_i)' \mathbf{\Lambda}_i^{-1} \mathbf{W}'\mathbf{Z}_i \right]^{-1} (\mathbf{W}'\mathbf{Z}_i)' \mathbf{\Lambda}_i^{-1} \mathbf{W}'\mathbf{y}_i, \ i = 1, \ldots, M. \qquad (4.G.28)$$

From $\hat{\delta}_i$ we can extract $\gamma_i^t, \alpha_i^t$ and $\beta_i^t$. Assemble $\gamma_i^t, \alpha_i^t$ and $\beta_i^t$ into matrices:

$$\mathbf{\Gamma}^t = \left[ \gamma_1^t \ \cdots \ \gamma_M^t \right], \ \mathbf{A}^t = \left[ \alpha_1^t \ \cdots \ \alpha_M^t \right] \text{ and } \mathbf{B}^t = \left[ \beta_1^t \ \cdots \ \beta_M^t \right].$$

Define

$$\mathbf{\Gamma}_0 = \begin{bmatrix} \mathbf{\Gamma}^1 & 0 & \cdots & 0 \\ 0 & \mathbf{\Gamma}^2 & \cdots & 0 \\ \vdots & \vdots & \vdots & \vdots \\ 0 & 0 & \cdots & \mathbf{\Gamma}^T \end{bmatrix}, \ \mathbf{A}_0 = \begin{bmatrix} \mathbf{A}^1 & 0 & \cdots & 0 \\ 0 & \mathbf{A}^2 & \cdots & 0 \\ \vdots & \vdots & \vdots & \vdots \\ 0 & 0 & \cdots & \mathbf{A}^T \end{bmatrix}, \ \mathbf{B}_0 = \begin{bmatrix} \mathbf{B}^1 & 0 & \cdots & 0 \\ 0 & \mathbf{B}^2 & \cdots & 0 \\ \vdots & \vdots & \vdots & \vdots \\ 0 & 0 & \cdots & \mathbf{B}^T \end{bmatrix}.$$

Set initial values of the matrix $W^0$ and vector $G^0$:

$$\mathbf{W}_\Gamma^0 = \mathbf{\Gamma}_0, \ \mathbf{W}_A^{01} = \mathbf{A}_0, \ \mathbf{W}_B^0 = \mathbf{B}_0 \text{ and}$$

$$\mathbf{G}_\Gamma^0 = vec(\mathbf{\Gamma}_0), \ \mathbf{G}_A^0 = vec(\mathbf{A}_0) \text{ and } \mathbf{G}_B^0 = vec(\mathbf{B}_0),$$

$$\mathbf{G}^0 \begin{bmatrix} \mathbf{G}_\Gamma^0 \\ \mathbf{G}_A^0 \\ \mathbf{G}_B^0 \end{bmatrix}.$$

## Appendix 4.H  Proximal Algorithm for $L_1$-Penalized Generalized Least Square Estimation of Parameters in the Dynamic Structural Equation Models

Our goal is to solve the following optimization problem:

$$\min_{\delta_i} \ f(\boldsymbol{\delta}_i) + \Omega(\boldsymbol{\delta}_i), \qquad (4.\text{H}.1)$$

where

$$f(\boldsymbol{\delta}_i) = \left(\mathbf{W}'\mathbf{y_i} - \mathbf{W}'\mathbf{Z}_{\gamma i}\boldsymbol{\gamma}_i - \mathbf{W}'\mathbf{Z}_{\alpha i}\boldsymbol{\alpha}_i - \mathbf{W}'\mathbf{Z}_{\beta i}\boldsymbol{\beta}_i\right)'\mathbf{\Lambda}_i^{-1}$$

$$\left(\mathbf{W}'y_i - \mathbf{W}'\mathbf{Z}_{\gamma i}\boldsymbol{\gamma}_i - \mathbf{W}'\mathbf{Z}_{\alpha i}\boldsymbol{\alpha}_i - \mathbf{W}'\mathbf{Z}_{\beta i}\boldsymbol{\beta}_i\right),$$

$$\mathbf{\Lambda_i} = \mathbf{W}'\mathbf{V_i}\mathbf{W},$$

$$\Omega(\boldsymbol{\delta}_i) = \lambda_1\Omega_1(\boldsymbol{\gamma}_i) + \lambda_2\Omega_2(\boldsymbol{\alpha}_i) + \lambda_3\Omega_3(\boldsymbol{\beta}_i), \ \Omega_1(\boldsymbol{\gamma}_i) = \|\boldsymbol{\gamma}_i\|_1, \ \Omega_2(\boldsymbol{\alpha}_i)$$

$$= \|\boldsymbol{\alpha}_i\|_1, \ \Omega_3(\boldsymbol{\beta}_i) = \|\boldsymbol{\beta}_i\|_1.$$

Similar to Appendix 4.G, the first step for applying the proximal methods to the optimization problem (4.H.1) is to take the partial derivative of the function $f(\boldsymbol{\delta}_i)$ with respect to $\boldsymbol{\delta}_i$. Again, applying matrix calculus, we obtain

$$\frac{\partial f}{\partial \boldsymbol{\gamma}_i} = -2\mathbf{Z}'_{\gamma i}\mathbf{W}\mathbf{\Lambda}_i^{-1}\left(\mathbf{W}'\mathbf{y_i} - \mathbf{W}'\mathbf{Z}_{\gamma i}\boldsymbol{\gamma}_i - \mathbf{W}'\mathbf{Z}_{\alpha i}\boldsymbol{\alpha}_i - \mathbf{W}'\mathbf{Z}_{\beta i}\boldsymbol{\beta}_i\right), \qquad (4.\text{H}.2)$$

$$\frac{\partial f}{\partial \boldsymbol{\alpha}_i} = -2\mathbf{Z}'_{\alpha i}\mathbf{W}\mathbf{\Lambda}_i^{-1}\left(\mathbf{W}'\mathbf{y_i} - \mathbf{W}'\mathbf{Z}_{\gamma i}\boldsymbol{\gamma}_i - \mathbf{W}'\mathbf{Z}_{\alpha i}\boldsymbol{\alpha}_i - \mathbf{W}'\mathbf{Z}_{\beta i}\boldsymbol{\beta}_i\right), \qquad (4.\text{H}.3)$$

$$\frac{\partial f}{\partial \boldsymbol{\beta}_i} = -2\mathbf{Z}'_{\beta i}\mathbf{W}\mathbf{\Lambda}_i^{-1}\left(\mathbf{W}'\mathbf{y_i} - \mathbf{W}'\mathbf{Z}_{\gamma i}\boldsymbol{\gamma}_i - \mathbf{W}'\mathbf{Z}_{\alpha i}\boldsymbol{\alpha}_i - \mathbf{W}'\mathbf{Z}_{\beta i}\boldsymbol{\beta}_i\right), \qquad (4.\text{H}.4)$$

$$\frac{\partial f}{\partial \boldsymbol{\delta}_i} = \begin{bmatrix} \dfrac{\partial f}{\partial \boldsymbol{\gamma}_i} \\[2ex] \dfrac{\partial f}{\partial \boldsymbol{\alpha}_i} \\[2ex] \dfrac{\partial f}{\partial \boldsymbol{\beta}_i} \end{bmatrix}. \tag{4.H.5}$$

Let

$$\mathbf{u}_{\gamma_i}^k = \boldsymbol{\gamma}_i^k - \lambda_1^k \frac{\partial f}{\partial \boldsymbol{\gamma}_i} \Big|_{\boldsymbol{\delta}_i^k}, \ \mathbf{u}_{\alpha_i}^k = \boldsymbol{\alpha}_i^k - \lambda_2^k \frac{\partial f}{\partial \boldsymbol{\alpha}_i} \Big|_{\boldsymbol{\delta}_i^k}, \ \mathbf{u}_{\beta_i}^k = \boldsymbol{\beta}_i^k - \lambda_3^k \frac{\partial f}{\partial \boldsymbol{\beta}_i} \Big|_{\boldsymbol{\delta}_i^k},$$

$$\boldsymbol{\delta}_i^k = \begin{bmatrix} \boldsymbol{\gamma}_i^k \\ \boldsymbol{\alpha}_i^k \\ \boldsymbol{\beta}_i^k \end{bmatrix}, \ \boldsymbol{\gamma}_i^k, \ \boldsymbol{\alpha}_i^k \text{ and } \boldsymbol{\beta}_i^k$$

be the values of vectors of the structural parameters for the endogenous variables at the time points $t$ and $t-1$, and the exogenous variables in the $k^{th}$ iteration, respectively.

Again, using Equation 1.137 (Xiong 2018), we obtain the elements of $\text{Prox}_{\lambda_1 \Omega_1}(\mathbf{u}_{\gamma i}^k)$, $\text{Prox}_{\lambda_2 \Omega_2}(\mathbf{u}_{\alpha i}^k)$ and $\text{Prox}_{\lambda_3 \Omega_3}(\mathbf{u}_{\beta i}^k)$:

$$\gamma_{ij}^z = \text{sign}\left(u_{\gamma_{ij}}^k\right)\left(\left|u_{\gamma_{ij}}^k\right| - \lambda_1^k\right)_+, \ j = 1, \dots, m_i, \tag{4.H.6}$$

$$\alpha_{ij}^z = \text{sign}\left(u_{\alpha_{ij}}^k\right)\left(\left|u_{\alpha_{ij}}^k\right| - \lambda_2^k\right)_+, \ j = 1, \dots, l_i, \tag{4.H.7}$$

$$\beta_{ij}^z = \text{sign}\left(u_{\beta_{ij}}^k\right)\left(\left|u_{\beta_{ij}}^k\right| - \lambda_3^k\right)_+, \ j = 1, \dots, k_i, \tag{4.H.8}$$

where $m_i = \sum_{t=1}^{T} m_i^t$, $l_i = \sum_{t=2}^{T} l_i^t$, $k_i = \sum_{t=1}^{T} k_i^t$, and $m_i^t$, $l_i^t$ and $k_i^t$ are the number of endogenous at the time points $t$ and $t-1$, and exogenous variables at the time point present $t$ in the $i^{th}$ equation, respectively.

Let

$$\boldsymbol{\gamma}_i^z = \begin{bmatrix} \gamma_{i1}^z \\ \vdots \\ \gamma_{im_i}^z \end{bmatrix}, \ \boldsymbol{\alpha}_i^z = \begin{bmatrix} \alpha_{i1}^z \\ \vdots \\ \alpha_{il_i}^z \end{bmatrix}, \ \boldsymbol{\beta}_i^z = \begin{bmatrix} \beta_{i1}^z \\ \vdots \\ \beta_{ik_i}^z \end{bmatrix} \text{ and } \boldsymbol{\delta}_i^z = \begin{bmatrix} \boldsymbol{\gamma}_i^z \\ \boldsymbol{\alpha}_i^z \\ \boldsymbol{\beta}_i^z \end{bmatrix}.$$

Now we define an auxiliary function (please see Equation 1.76, *Big Data in Omics and Imaging: Association Analysis*, Xiong (2018)) which implements majorization-minimization in the proximal gradient algorithm:

$$f_\lambda\left(\boldsymbol{\delta}_i^z, \boldsymbol{\delta}_i^k\right) = f(\boldsymbol{\delta}_i^z) + \left(\frac{\partial f}{\partial \boldsymbol{\delta}^i}\Big|_{\boldsymbol{\delta}_i^k}\right)^T \left(\boldsymbol{\delta}_i^k - \boldsymbol{\delta}_i^z\right) + \frac{1}{2\lambda_1}\parallel \boldsymbol{\gamma}_i^k - \boldsymbol{\gamma}_i^z \parallel_2^2$$

$$+ \frac{1}{2\lambda_2}\parallel \boldsymbol{\alpha}_i^k - \boldsymbol{\alpha}_i^z \parallel_2^2 + \frac{1}{2\lambda_3}\parallel \boldsymbol{\beta}_i^k - \boldsymbol{\beta}_i^z \parallel_2^2 . \qquad (4.H.9)$$

Now we apply the general proximal gradient algorithm to the $L_1$-penalized generalized least square estimation problem. The proximal gradient algorithm is given as follows.

### Algorithm 4.H.1: Proximal Gradient Algorithm

Step 1: Given $\boldsymbol{\delta}_i^z, \boldsymbol{\delta}_i^k, \lambda_1^{k-1}, \lambda_2^{k-1}, \lambda_3^{k-1}$, and parameter $\alpha_1, \alpha_2, \alpha_3 \in (0,1)$. Set $\lambda_1 = \lambda_1^{k-1}, \lambda_2 = \lambda_2^{k-1}$ and $\lambda_3 = \lambda_3^{k-1}$.

Step 2:
  Repeat

  1. Set

$$\boldsymbol{\delta}_i^z = \begin{bmatrix} \boldsymbol{\gamma}_i^z \\ \boldsymbol{\alpha}_i^z \\ \boldsymbol{\beta}_i^z \end{bmatrix} = \begin{bmatrix} \text{Prox}_{\lambda_1 \Omega_1}\left(\mathbf{u}_{\gamma_i}^k\right) \\ \text{Prox}_{\lambda_2 \Omega_2}\left(\mathbf{u}_{\alpha_i}^k\right) \\ \text{Prox}_{\lambda_3 \Omega_3}\left(\mathbf{u}_{\beta_i}^k\right) \end{bmatrix},$$

where we calculate the proximal operator element-wise as follows:

$$\mathbf{u}_{\gamma_i}^k = \boldsymbol{\gamma}_i^k - \lambda_1^k \frac{\partial f}{\partial \boldsymbol{\gamma}_i}\Big|_{\boldsymbol{\delta}_i^k}, \quad \mathbf{u}_{\alpha_i}^k = \boldsymbol{\alpha}_i^k - \lambda_2^k \frac{\partial f}{\partial \boldsymbol{\alpha}_i}\Big|_{\boldsymbol{\delta}_i^k}, \quad \mathbf{u}_{\beta_i}^k = \boldsymbol{\beta}_i^k - \lambda_3^k \frac{\partial f}{\partial \boldsymbol{\beta}_i}\Big|_{\boldsymbol{\delta}_i^k},$$

$$\gamma_{ij}^z = \text{sign}\left(u_{\gamma_{ij}}^k\right)\left(\left|u_{\gamma_{ij}}^k\right| - \lambda_1^k\right)_+, \ j = 1, \ldots, m_i,$$

$$\alpha_{ij}^z = \text{sign}\left(u_{\alpha_{ij}}^k\right)\left(\left|u_{\alpha_{ij}}^k\right| - \lambda_2^k\right)_+, \ j = 1, \ldots, l_i,$$

$$\beta_{ij}^z = \text{sign}\left(u_{\beta_{ij}}^k\right)\left(\left|u_{\beta_{ij}}^k\right| - \lambda_3^k\right)_+, \ j = 1, \ldots, k_i .$$

Assemble $\gamma_{ij}^z, \alpha_{ij}^z$ and $\beta_{ij}^z$ into the vectors $\boldsymbol{\gamma}_i^z, \boldsymbol{\alpha}_i^z$ and $\boldsymbol{\beta}_i^z$:

$$\boldsymbol{\gamma}_i^z = \begin{bmatrix} \gamma_{i1}^z \\ \vdots \\ \gamma_{im_i}^z \end{bmatrix}, \ \boldsymbol{\alpha}_i^z = \begin{bmatrix} \alpha_{i1}^z \\ \vdots \\ \alpha_{il_i}^z \end{bmatrix}, \ \boldsymbol{\beta}_i^z = \begin{bmatrix} \beta_{i1}^z \\ \vdots \\ \beta_{ik_i}^z \end{bmatrix} \text{ and } \boldsymbol{\delta}_i^z = \begin{bmatrix} \boldsymbol{\gamma}_i^z \\ \boldsymbol{\alpha}_i^z \\ \boldsymbol{\beta}_i^z \end{bmatrix}.$$

2. Calculate the auxiliary function:

$$\hat{f}_\lambda\left(\delta_i^z, \delta_i^k\right) = f(\delta_i^z) + \left(\frac{\partial f}{\partial \delta^i}\Big|_{\delta_i^k}\right)^T \left(\delta_i^k - \delta_i^z\right) + \frac{1}{2\lambda_1} \| \gamma_i^k - \gamma_i^z \|_2^2$$

$$+ \frac{1}{2\lambda_2} \| \alpha_i^k - \alpha_i^z \|_2^2 + \frac{1}{2\lambda_3} \| \beta_i^k - \beta_i^z \|_2^2 .$$

3. Break and go to step 3 if $f(\delta_i^z) \le \hat{f}_\lambda(\delta_i^z, \delta_i^k)$.
4. Update $\lambda_1 = \alpha_1\lambda_1, \lambda_2 = \alpha_2\lambda_2$ and $\lambda_3 = \alpha_3\lambda_3$.

Step 3: return $\lambda_1^k = \lambda_1, \lambda_2^k = \lambda_2, \lambda_3^k = \lambda_3$ and

$$\delta_i^{k+1} = \delta_i^z \text{ and}$$

$$\gamma_i^{k+1} = \gamma_i^z, \alpha_i^{k+1} = \alpha_i^z, \beta_i^{k+1} = \beta_i^z, k = k+1.$$

## Appendix 4.I  Multikernel Learning and Spectral Clustering for Cell Type Identification

Wang et al. (2017) proposed to use a multikernel-based single cell RNA-seq analysis method for reducing data dimensions and learning similarity. Then, spectral clustering uses the learning similarity to identify cell types. This appendix is to adapt the materials from Wang et al. (2017) to a self-contained introduction to multikernel learning.

Consider a $n \times k$ dimensional gene expression matrix $X$ with $n$ cells and $k$ genes. Define a distance measure between the $i^{th}$ cell and $j^{th}$ cell as

$$D\left(c_i, c_j\right) = \sum_l w_l \| \phi_l(c_i) - \phi_l\left(c_j\right) \|_2^2, \tag{4.I.1}$$

where $\phi_l(c_i)$ is the $l^{th}$ map of the cell $c_i$ from gene expression space to a higher dimensional feature space. The distance between the points in the feature space can be calculated by the kernel as follows:

$$\| \phi_l(c_i) - \phi_l\left(c_j\right) \|_2^2 = \phi_l(c_i)^T\phi_l(c_i) + \phi_l\left(c_j\right)^T\phi_l\left(c_j\right) - 2\phi_l(c_i)^T\phi_l\left(c_j\right)$$

$$= K_l(c_i, c_i) + K_l\left(c_j, c_j\right) - 2K_l\left(c_i, c_j\right) \tag{4.I.2}$$

$$= 2 - 2K_l\left(c_i, c_j\right),$$

where by convention we assume

$$K_l(c_i, c_i) = K_l\left(c_j, c_j\right) = 1.$$

Therefore, the distance between cell $i$ and cell $j$ is defined as the following kernel function:

$$D\left(c_i, c_j\right) = 2 - 2K_l\left(c_i, c_j\right). \tag{4.1.3}$$

Gaussian kernels are often used in practice. Gaussian kernels are defined as

$$K\left(c_i, c_j\right) = \frac{1}{\sqrt{2\pi}\sigma_{ij}} \exp\left(-\frac{\| c_i - c_j \|_2^2}{2\sigma_{ij}^2}\right), \tag{4.1.4}$$

where $c_i$ and $c_j$ are vectors of measured gene expressions from the cells $c_i$ and $c_j$, respectively, $\| c_i - c_j \|_2$ is their Euclidean distance. The variance, $\sigma_{ij}$ is often calculated by

$$\sigma_{ij} = \frac{\sigma\left(\mu_i + \mu_j\right)}{2} \text{ and } \mu_i = \frac{\sum_{j \in KNN(c_i)} \| c_i - c_j \|_2}{k}, \tag{4.1.5}$$

where $KNN(c_i)$ denotes all cells that are top $k$ neighbors of the cell $c_i$, $k$ and $\sigma$ are parameters that generate the multiple kernels.

We start by introducing similarity graphs for spectral clustering. Consider a similarity graph $G = (V, E)$. Each node in the graph represents a data point (a cell) and each edge represents a similarity between two nodes connected by the edge, taking similarity measure as its associated weight. The problem of clustering cells can now be formulated using the similarity graph: to find a partition of the graph such that the edge between different groups have small weights and the edges within a group have large weights. Consider two cells (nodes): cell $i$ and cell $j$. Let $S_{ij}$ denotes the similarity between cell $i$ and cell $j$. Define the similarity matrix $S = (S_{ij})_{n \times n}$.

To find the best similarity graph that optimally clusters the cells we propose the following optimization problem:

$$\min_{S,L,W} \sum_{i=1}^{n} \sum_{j=1}^{n} D\left(x_i, x_j\right) S_{ij} + \beta\| S \|_F^2 + \gamma \mathrm{Tr}\left(L^T(I_n - S)L\right) + \rho \sum_l w_l \log w_l$$

$$\text{subject to} \quad D\left(x_i, x_j\right) = \sum_l w_l D_l\left(x_i, x_j\right), \sum_l w_l = 1, w_l \geq 0, L^T L = I_C,$$

$$\sum_j S_{ij} = 1, S_{ij} \geq 0,$$

$$\tag{4.1.6}$$

where $x_i$ and $x_j$ are the gene expression vectors of the cells $i$ and $j$, that is, the $i^{th}$ and $j^{th}$ row of the gene expression matrix $X$, $D_l(x_i,x_j)$ is the $l^{th}$ distance measure between cell $i$ and cell $j$ and $D(x_i,x_j)$ is the weighted summation of distances $D_l$ $(x_i,x_j)$ between cell $i$ and cell $j$ with weights $w_l$, $I_n$ and $I_C$ are $n$ dimensional and $C$ dimensional identity matrices, respectively, $\|S\|_F$ is the Frobenius norm of the matrix $S$, and $\beta$ and $\gamma$ are non-negative penalty parameters. The objective function is minimized with respect to vector $w$ and matrices $S$ and $L$.

The first term in the objective function in (4.I.5) indicates that if the distance between two cells is small, these two cells should be close and similarity between them should be large. The second regularization term is to encourage the sparse similarity graph. The matrix $I_S - S$ in the third term is the graph Laplacian which encourages adjacent nodes induced by similarity measures to have close gene expression profiles. Thus, the third term coupled with the constraint on $L$ enforce the low rank cluster structure of $S$ with approximately $C$ connected components in a similarity graph in which the nodes represent the cells and edge weights represent the pair-wise similarity. The four terms impose penalty to enforce selection of multiple kernels.

Substituting Equation 4.I.3 into Equation 4.I.6 gives the following optimization problem:

$$\min_{S,L,W} \quad -2\sum_{i=1}^{n}\sum_{j=1}^{n} w_l K_l\left(c_i, c_j\right) S_{ij} + \beta\| S \|_F^2 + \gamma \mathrm{Tr}\left(L^T (I_n - S)L\right) + \rho\sum_l w_l \log w_l$$

$$\text{subject to} \quad D\left(x_i, x_j\right) = \sum_l w_l D_l\left(x_i, x_j\right), \ \sum_l w_l = 1, \ w_l \geq 0, \ L^T L = I_C,$$

$$\sum_j S_{ij} = 1, \ S_{ij} \geq 0.$$

$$(4.I.7)$$

Next we study how to solve the optimization problem (4.I.7).

**Algorithm 4.I.1: Algorithm for Solving Optimization Problem (4.I.7)**

**Step 1:** Initialization of $S$, $w$ and $L$.

The weight of multiple kernels is initiated as an uniform distribution vector: $w^{(0)} = \left[\dfrac{1}{G}, \dfrac{1}{G}, \ldots, \dfrac{1}{G}\right]$, where $G$ is the number of kernels.

The initial similarity matrix is defined as $S_{ij} = \sum_{l=1}^{G} w_l K_l(c_i, c_j)$, $S = (S_{ij})_{C\times C}$.

The top $C$ eigenvectors of the matrix $S$ is taken as an initial matrix $L$.

The optimization problem (4.A7) with all variables is a non-convex problem. However, iteratively applying an alternative convex optimization problem method to each sub-optimization problem in which one of the $S$, $w$ and $L$ is optimized while the other variables are being fixed, is an efficient algorithm. We first update $S$ while fixing $L$ and $w$.

Note that

$$-\gamma \ \mathrm{tr}\left(L^T S L\right) = -\gamma \ \mathrm{Tr}\left(SLL^T\right)$$

$$= -\gamma \sum_{i=1}^{C}\sum_{j=1}^{C} S_{ij}\left(LL^T\right)_{ij}. \tag{4.I.8}$$

Substituting Equation 4.5A8 into Equation 4.I.7 and absorbing 2 to $w_l$, we can obtain

**Step 2:** Fixing $L$ and $w$ to update matrix $S$:

$$\min_{S} \quad -\sum_{i=1}^{n}\sum_{j=1}^{n}\left[\sum_{l=1}^{G} w_l K_l\left(c_i, c_j\right) + \gamma\left(LL^T\right)_{ij}\right] S_{ij} + \beta\| S \|_F^2$$

$$\text{subject to} \quad \sum_{j=1}^{n} S_{ij} = 1 \text{ and } S_{ij} \geq 0.$$ (4.I.9)

Let $s_i$ be the $i^{th}$ row of the matrix $S$ and

$$(v_i)_j = \frac{-1}{2\beta}\left[\gamma\left(LL^T\right)_{ij} - \sum_l w_l K_l\left(c_i, c_j\right)\right].$$

Since each row of the matrix $S$ is independent, the optimization problem (4.I.9) can be decomposed to solving $n$ independent optimization problems:

$$\min_{S_i} \quad \frac{1}{2}\| s_i - v_i \|_2^2$$

$$\text{subject to} \quad s_i^T \mathbf{1} = 1,\ (s_i)_j \geq 0,\ j = 1, \ldots, n.$$ (4.I.10)

Using the Lagrangian multipliers, the optimization problem (4.I.10) can be transformed to the following unconstrained optimization problem:

$$L(s_i, \delta, \sigma) = \frac{1}{2}\| s_i - v_i \|_2^2 + \delta\left(s_i^T \mathbf{1} - 1\right) - \sigma^T s_i,$$ (4.I.11)

where $\delta$ and $\sigma$ are a scalar and a vector of Lagrangian multipliers. Using the Karush–Kuhn–Tucker (KKT) conditions (Chapter 1 in *Big Data in Omics and Imaging: Association Analysis*, (Xiong 2018)), we obtain

$$\frac{\partial L}{\partial s_i} = s_i - v_i + \delta\mathbf{1} - \sigma = 0,$$

$$s_i \geq 0,$$

$$\sigma \geq 0,$$ (4.I.12)

$$(s_i)_j \sigma_j = 0,$$

$$s_i^T \mathbf{1} - 1 = 0.$$

From Equation 4.I.12 we obtain

$$s_i = v_i + \sigma - \delta\mathbf{1}.$$ (4.I.13)

Taking inner product with $\mathbf{1}$ on both sides of Equation 4.I.13 and using the equality $s_i^T \mathbf{1} = 1$, we obtain

$$\mathbf{1}^T v_i + \mathbf{1}^T \sigma - \delta\mathbf{1}^T \mathbf{1} = 1.$$ (4.I.14)

Solving Equation 4.I.14 for $\delta$ gives

$$\delta = \frac{1}{n}\left(\mathbf{1}^T v_i + \mathbf{1}^T \sigma - 1\right).$$ (4.I.15)

Substituting Equation 4.I.15 into Equation 4.I.14, we obtain

$$
\begin{aligned}
s_i &= v_i + \sigma - \frac{1}{n}\mathbf{1}\left(\mathbf{1}^T v_i + \mathbf{1}^T \sigma - 1\right) \\
&= \left(I_n - \frac{1}{n}\mathbf{1}\mathbf{1}^T\right)v_i + \frac{1}{n}\mathbf{1} + \sigma - \frac{1}{n}\mathbf{1}\mathbf{1}^T\sigma \\
&= u_i + \sigma - \sigma^*\mathbf{1},
\end{aligned}
\tag{4.I.16}
$$

where

$$
u_i = \left(I_n - \frac{1}{n}\mathbf{1}\mathbf{1}^T\right)v_i + \frac{1}{n}\mathbf{1} \text{ and } \sigma^* = \frac{1}{n}\mathbf{1}^T\sigma.
$$

If $(s_i)_j > 0$ then using the KKT condition (4.5A4), we obtain

$$
\sigma_j = 0. \tag{4.I.17}
$$

Substituting Equation 4.I.17 into Equation 4.I.16 and using condition $(s_i)_j > 0$ yields

$$
(s_i)_j = \left((u_i)_j - \sigma^*\right)_+. \tag{4.I.18}
$$

Now suppose that $\sigma_j > 0$. Then, the KKT condition implies that

$$
(s_i)_j = 0. \tag{4.I.19}
$$

Substituting Equation 4.I.19 into Equation 4.I.16 gives

$$
\sigma_j = \sigma^* - (u_i)_j. \tag{4.I.20}
$$

Combining $\sigma_j > 0$ and Equation 4.I.20 leads to

$$
\sigma_j = \left(\sigma^* - (u_i)_j\right)_+. \tag{4.I.21}
$$

From $\sigma^* = \frac{1}{n}\mathbf{1}^T\sigma$ and Equation 4.I.21 we obtain the equation:

$$
f(\sigma^*) = \sigma^* - \frac{1}{n}\sum_{j=1}^{n}\left(\sigma^* - (u_i)_j\right)_+ = 0. \tag{4.I.22}
$$

Newton's method can be used to solve Equation 4.I.22 for $\sigma^*$. Next, we go to Step 3 to update the matrix $L$.

**Step 3:** Update $L$ while fixing $S$ and $w$.

Fixing $S$ and $w$, and minimizing the objective function with respect to the latent matrix $L$ in (4.I.7) is equivalent to the following optimization problem:

$$
\min_{L} \quad \text{Tr}\left(L^T(S - I_n)L\right)
$$
$$
\text{subject to} \quad L^T L = I_C. \tag{4.I.23}
$$

Using Lagrange multiplier method to solve the optimization problem (4.I.23), we obtain

$$\min_{L} \quad F = \text{Tr}\left(L^T(S - I_n)L\right) + \text{Tr}\left(\Lambda^T(I_C - L^TL)\right). \tag{4.I.24}$$

Using matrix calculus and setting the partial derivative of the function $F$ with respect to the matrix $L$ to zero gives

$$(S - I_n)L - L\Lambda = 0, \text{ which implies}$$

$$(S - I_n)L = L\Lambda. \tag{4.I.25}$$

Equation 4.I.25 is an eigenequation. It indicates that the solution matrix $L^*$ are the eigenvectors corresponding to the $C$ largest eigenvalues of the matrix $S - I_n$.

**Step 4:** Update $w$ while fixing $S$ and $L$.

To update $w$, we solve the following optimization problem with respect to $w$:

$$\min_{w} \quad -\sum_{l=1}^{G} w_l \sum_{i=1}^{n} \sum_{j=1}^{n} K_l\left(c_i, c_j\right) S_{ij} + \rho \sum_{l=1}^{G} w_l \log w_l$$

$$\text{subject to} \quad \sum_{l=1}^{G} w_l = 1, w_l \geq 0. \tag{4.I.26}$$

Using the Lagrange multiplier methods, the constrained optimization problem (4.I.13) can be transformed into the following unconstrained optimization problem:

$$\min_{w} \quad -\sum_{l=1}^{G} w_l \sum_{i=1}^{n} \sum_{j=1}^{n} K_l\left(c_i, c_j\right) S_{ij} + \rho \sum_{l=1}^{G} w_l \log w_l$$

$$+ \lambda\left(1 - \sum_{l=1}^{G} w_l\right) - \sum_{l=1}^{G} \mu_l w_l. \tag{4.I.27}$$

The necessary condition for the optimal solution to the optimization problem (4.I.27) is

$$-\sum_{i=1}^{n} \sum_{j=1}^{n} K_l\left(c_i, c_j\right) S_{ij} + \rho\left(1 + \log w_l\right) - \lambda - \mu_l = 0. \tag{4.I.28}$$

By KKT conditions, we have

$$w_l \mu_l = 0. \tag{4.I.29}$$

Thus, when $w_l \neq 0$ we have $\mu_l = 0$. Equation 4.I.28 is then reduced to

$$-\sum_{i=1}^{n} \sum_{j=1}^{n} K_l\left(c_i, c_j\right) S_{ij} + \rho\left(1 + \log w_l\right) - \lambda = 0. \tag{4.I.30}$$

Solving Equation 4.I.30 for $w_l$, we obtain

$$w_l = \exp\left(-\left(1 - \frac{\lambda}{\rho}\right)\right) \exp\left(\frac{1}{\rho} \sum_{i=1}^{n} \sum_{j=1}^{n} K_l\left(c_i, c_j\right) S_{ij}\right). \tag{4.I.31}$$

Summarizing over $l$ on both sides of Equation 4.I.31 gives

$$\exp\left(\left(1-\frac{\lambda}{\rho}\right)\right) = \sum_{l=1}^{G} w_l \exp\left(\frac{1}{\rho}\sum_{i=1}^{n}\sum_{j=1}^{n}K_l\left(c_i,c_j\right)S_{ij}\right). \quad (4.I.32)$$

Substituting Equation 4.I.32 into Equation 4.I.31 gives

$$w_l = \frac{\exp\left(\frac{1}{\rho}\sum_{i=1}^{n}\sum_{j=1}^{n}K_l\left(c_i,c_j\right)S_{ij}\right)}{\sum_{l=1}^{G}w_l \exp\left(\frac{1}{\rho}\sum_{i=1}^{n}\sum_{j=1}^{n}K_l\left(c_i,c_j\right)S_{ij}\right)}.$$

**Step 5:** Similarity enhancement

View the similarity between cells as information that can be spread over the similarity graph. Information diffusion theory can be used to model the similarity and hence to further enhance the similarity. Given the similarity matrix $S$, define a transition matrix: $P = (P_{ij})_{n \times n}$ where

$$P_{ij} = \frac{S_{ij}I_{\{j \in N_K(i)\}}}{\sum_l S_{il}I_{\{l \in N_K(i)\}}}, \quad (4.I.33)$$

where $I_{\{.\}}$ represents the indicator function, and $N_K(i)$ represents the set of indices of cells that are $K$ top neighbors of cell $i$ measured by the learned distance metric. Using transition matrix we can iteratively update the similarity matrix as follows:

$$H^{(t+1)} = \tau H^{(t)} P + (1 - \tau)I_n, \quad (4.I.34)$$

where $H^{(0)} = S$ and the matrix $H$ at the final iteration $T$ is used as the new similarity matrix $S$.

**Step 6:** Convergence checking

Let $\lambda_1 \leq \lambda_2 \leq \ldots \leq \lambda_n$ be the eigenvalues of the similarity matrix $S$ and $C$ be the number of clusters. Define

$$\text{eigengap}(C) = \lambda_{C+1} - \lambda_C.$$

When eigengap($C$) stops decreasing, in practice, the algorithm converges and iteration stops. Otherwise, go to Step 2. The algorithm repeats steps 2–6 until convergence.

The penalty parameters $\gamma$ and $\beta$ are determined from the data. Let $x_i$ denote the gene expression data of the $i^{th}$ cell, $x_i^j$ denote the top $j^{th}$ nearest neighbor of the $l^{th}$ cell and $m$ be a predefined parameter. In practice, for small datasets, we often set $m = 10$ and for large datasets, we set $m = 30$. The parameters $\gamma$ and $\beta$ are selected by

$$\gamma = \beta = \frac{1}{2n}\sum_{i=1}^{n}\sum_{j=1}^{m}\left(\|x_i - x_i^{m+1}\|_2^2 - \|x_i - x_i^j\|_2^2\right). \quad (4.I.35)$$

## Exercises

**Exercise 1.** Consider the raw count data in Example 4.1. Use upper quantile normalization to calculate the normalization factors and the normalized count, assuming $P = 0.75$.

**Exercise 2.** Show

$$\text{var}\left(M_g(j, r)\right) = \frac{D_j - n_{gj}}{D_j n_{gj}} + \frac{D_r - n_{gr}}{D_r n_{gr}}.$$

**Exercise 3.** Consider the raw count data in Example 4.1. Use the trimmed mean of M-values to calculate the normalization factors and the normalized count.

**Exercise 4.** Show that the solution to the optimization problem (4.1A49)

$$\max_{q_\theta(\theta)} \quad F\left(q_z^{(u+1)}(Z), q_\theta(\theta)\right)$$

$$\text{s.t.} \quad \int q_\theta(\theta) d\theta = 1,$$

is

$$q_\theta^{(u+1)}(\theta) = \frac{1}{C_\theta} P\left(\theta^{(u)}\right) \exp\left[\int q_z\left(Z^{(u+1)}\right) \log P(Z, R, Q | \theta^{(u)}) dZ\right]$$

where $C_\theta = \int P(\theta^{(u)}) exp[\int q_z(Z^{(u+1)}) log P(Z,R,Q | \theta^{(u)}) dZ]$.

**Exercise 5.** Show $E_\theta[\log \theta_i] = \psi(\alpha_i) - \psi(\sum_{j=1}^{M} \alpha_j)$, $\psi(\alpha) = \dfrac{\dfrac{d\Gamma(\alpha)}{d\alpha}}{\Gamma(\alpha)}$, where $\theta$ is distributed as Dirichlet distribution.

**Exercise 6.** Show the mean of the Dirichlet distribution (Equation 4.A64) is

$$E_\theta\left[\theta_i^{(u+1)}\right] = \frac{\alpha_i^{(u+1)}}{\sum_{i=0}^{M} \alpha_i^{(u+1)}}.$$

**Exercise 7.** Show Result 4.4 (variational Bayesian algorithm for the gapped alignment of RNA-seq data).

**Exercise 8.** Show that the mean of the Beta distribution is

$$E[\phi] = \frac{\beta_{i1}}{\beta_{i1} + \beta_{i2}}.$$

Exercise 9. Show the necessary conditions of the maximum of the log-likelihood for the log-linear model with the Poisson distribution are

$$\frac{\partial l\left(N_{ig} \mid \alpha_g\right)}{\partial \alpha_g} = \sum_i \left[N_{ig}y_i - N_{ig}^{(0)}y_i \exp\left(y_i\alpha_g\right)\right] = 0$$

$$\frac{\partial l\left(N_{ig} \mid \alpha_{gm}\right)}{\partial \alpha_{gm}} = \sum_i \left[N_{ig}I_{(i \in C_m)} - N_{ig}^{(0)}I_{(i \in C_m)} \exp\left(\sum_m \alpha_{gm}I_{(i \in C_m)}\right)\right]$$

$$= 0.$$

Exercise 10. Show

$$E[U(\theta)] = 0.$$

and

$$\text{var}(U(\theta)) = -E\left[\frac{\partial^2 U}{\partial \theta \partial \theta^T}\right] = I(\theta).$$

Exercise 11. Show that the score statistics for testing association of gene $g$ with two or multiple-class outcomes is

$$S_{cg} = \sum_{m=1}^M \frac{\left[\sum_{i \in C_m}\left(N_{ig} - N_{ig}^{(0)}\right)\right]^2}{\sum_{i \in C_m} N_{ig}^{(0)}}.$$

Exercise 12. Show that if $y_{gi}$ is distributed as a negative binomial distribution $NB(r_i, p)$ then $n_{gA}$ is also distributed as a negative binomial distribution $NB(\sum_{i=1}^{m_A} r_i, p)$.

Exercise 13. Show that

$$\| U\Lambda V^T - U\Sigma V^T \|_F^2 = \| \Lambda - \Sigma \|_F^2.$$

Exercise 14. Show that two optimization problems:

$$L\left(\delta_i, b^i, Z\right) = \frac{1}{2}f\left(\delta_i, b^i\right) + \lambda_1\left(\sum_{j=1}^{i-1}\|Z_j^1\|_2 + \sum_{j=i+1}^{K}\|Z_j^1\|_2\right)$$
$$+ \lambda_2\sum_{l=1}^{r_i}\|Z_l^2\|_2 + \sum_{j=1}^{i-1}\mu_j^T\left(\delta_{ji} - Z_j^1\right)$$
$$+ \sum_{j=i+1}^{K}\mu_j^T\left(\delta_{ji} - Z_j^1\right) + \frac{\rho_1}{2}\left(\sum_{j=1}^{i-1}\|\delta_{ji} - Z_j^1\|_2^2\right.$$
$$\left. + \sum_{j=i+1}^{K}\|\delta_{ji} - Z_j^1\|_2^2\right)$$
$$+ \sum_{l=1}^{r_i}\pi_l^T\left(b^i - Z_l^2\right) + \frac{\rho_2}{2}\sum_{l=1}^{r_i}\|b^i - Z_l^2\|_2^2$$

and

$$L\left(\delta_i, b^i, Z\right) = \frac{1}{2}f\left(\delta_i, b^i\right) + \lambda_1\left(\sum_{j=1}^{i-1}\|Z_j^1\|_2 + \sum_{j=i+1}^{K}\|Z_j^1\|_2\right)$$
$$+ \lambda_2\sum_{l=1}^{r_i}\|Z_l^2\|_2$$
$$+ \frac{\rho_1}{2}\left(\sum_{j=1}^{i-1}\|\delta_{ji} - Z_j^1 + u_j\|_2^2 + \sum_{j=i+1}^{K}\|\delta_{ji} - Z_j^1 + u_j\|_2^2\right)$$
$$+ \frac{\rho_2}{2}\sum_{l=1}^{r_i}\|b^i - Z_l^2 + v_l\|_2^2.$$

are equivalent.

# 5

## Methylation Data Analysis

DNA (CpG) methylation, an epigenetic mechanism, is an important regulator of many biological processes in humans, and at multiple biological levels including gene regulation, cellular differentiation, and organismal development (Vincent et al. 2017). DNA methylation data analysis discovers how environmental perturbations trigger cellular reprogramming, which in turn, affects cellular function (Lappalainen and Greally 2017) and provides epigenetic information that regulates gene expression (Li et al. 2015).

The widely used DNA methylation data analysis includes data preprocessing, normalization, differential methylation analysis (Fortin et al. 2017; Kurdyukov and Bullock 2016), and epigenome-wide association studies (EWAS) (Lappalainen and Greally 2017; Laird 2010). DNA methylation data preprocessing, normalization, and differential methylation analysis are, in principle, similar to gene expression data analysis. To save space, these will not be discussed in detail. The focus of this chapter is causal analysis of DNA methylation data.

## 5.1 DNA Methylation Analysis

DNA methylation is an epigenetic mechanism that regulates gene expression and modifies the function of the genes. DNA methylation adds methyl (CH3) groups to the DNA molecule, often to the fifth carbon atom of a cytosine ring, which leads to 5-methylcytosine (Figure 5.1). It is well known that each nucleotide is composed of one of four nucleobases—cytosine (C), guanine (G), adenine (A), or thymine (T). Two of them, cytosine and adenine, can be methylated. The methyl groups will inhibit transcription by preventing the binding of factors to the DNA that promote transcriptional activity. The addition of methyl groups is carried out by a family of DNA methyltransferases (DNMTs): DNMT1, DNMT2, DNMT3A, DNMT3B, and DNMT3L (Jin et al. 2011).

Although in mammals, DNA methylation can occur at cytosines in any context of the genome, more than 98% of DNA methylation takes place in a

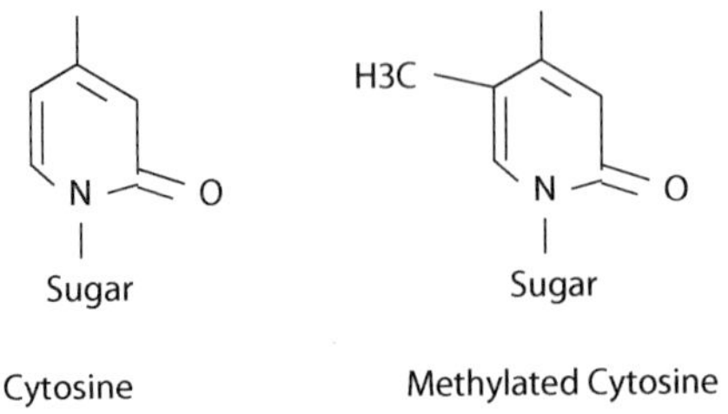

**FIGURE 5.1**
DNA methylation.

CpG dinucleotide region in somatic cells (Jin et al. 2011). DNA methylation patterns remain stable through somatic cell division and will be inherited from generation to generation (Krueger et al. 2012). To maintain methylation status at CpGs during DNA replication, CpG methylation usually takes place on both DNA strands, while non-CpG methylation must be remethylated de novo after each cell division. Methylated cytosines are, in general, under-represented in the genome and are often grouped in dense regions termed CpG islands. There is no formal definition of CpG islands. CpG islands are usually defined as a region with at least 200 bp with a GC percentage greater than 50% (https://en.wikipedia.org/wiki/CpG_site).

Methylation can be quantified as global methylcytosine content of DNA samples. Many methods have been developed to measure methylation levels. They include methylated DNA immunoprecipitation or methyl binding protein enrichment of methylated fragments, digestion with methylation-sensitive restriction enzymes and bisulfite modification of DNA (Krueger et al. 2012) and can be categorized into three groups: enzyme digestion, affinity enrichment, and bisulfite conversion (Yong et al. 2016). Bisulfite conversion is often used to measure methylation. The treatment of DNA with sodium bisulfite will convert cytosine (C) into uracil (U), while methylated C residues remain unchanged. A subsequent polymerase chain reaction (PCR) then converts U to thymine (T) (Figure 5.2). Comparing the modified DNA with the original sequence, we can detect and measure the methylation.

Illumina's Infinium Human Methylation450 BeadChip (HM450K) is the bisulfite conversion-based method. Each HM450K BeadChip can interrogate more than 450,000 methylation sites. However, the coverage of distal regulatory elements by HM450K is small. To overcome this limitation, Illumina developed the MethylationEPIC (EPIC) BeadChip that covers over 850,000 CpG sites with single-nucleotide resolution, including >90% of the CpGs from the HM450 and an additional 413,743 CpGs (Pidsley et al. 2016). Whole genome bisulfite sequencing (WGBS) is one of the current major bisulfite conversion-based methods (Krueger et al. 2012). The DNA fragments are treated by sodium bisulfite and then amplified by PCR. The resulting library is sequenced, which leads to detection and measurement of methylation.

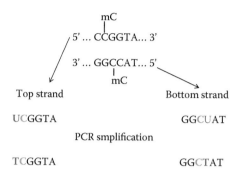

**FIGURE 5.2**
Bisulfite treatment.

WGBS can assess the methylation state of nearly every CpG site and discover all methylation information. WGBS is unbiased and does not require a digested template.

Let $m$ be the number of reads of methylated cytosines and $u$ be the number of reads of unmethylated cytosines. The methylation level of the locus is defined as $\dfrac{m}{m+u}$.

Table 5.1 lists the methylation level of four CpG sites for five individuals.

Pipelines of DNA methylation analysis include (1) filtering low-quality probes, (2) batch effect correction, (3) normalization, functional normalization, and beta-mixture quantile normalization, (4) principle component analysis, (5) singular value decomposition, (6) unsupervised learning and cluster analysis, (7) differential methylation analysis, (8) differentially methylated block (DMB) identification where DMBs are large-scale genomic

**TABLE 5.1**

Data Example of Methylation

| Gene | Methylation Level | | | |
| | AP1S1 | RPN1 | GNAS | C11orf24 |
| Subject | cg27665659 | cg27662379 | cg27661264 | cg27655905 |
| --- | --- | --- | --- | --- |
| 1 | 0.03107011 | 0.01847027 | 0.3440003 | 0.06395125 |
| 2 | 0.05291088 | 0.0406794 | 0.3748224 | 0.08925108 |
| 3 | 0.0495046 | 0.04009077 | 0.2824244 | 0.1097811 |
| 4 | 0.03407283 | 0.0218635 | 0.3223184 | 0.0663179 |
| 5 | 0.03349082 | 0.01669993 | 0.315316 | 0.0691947 |

regions (10 kb–Mb) containing hundreds of inter-genic CpG sites, and (9) detection of differentially methylated hotspots in user-defined gene networks (Tian et al. 2017).

## 5.2 Epigenome-Wide Association Studies (EWAS)

The current epigenome-wide association studies (EWAS) mainly identify DNA methylation signals associated with environments and diseases (Hachiya et al. 2017). EWAS tests the association of DNA methylation at individual or groups of adjacent cytosines in the genome with the phenotype of interest and diseases and discovers locus-specific DNA methylation (DNAm) in blood cells associated with various exposures, intermediate phenotypes, and diseases. DNA methylation variation can either cause disease or can be directly or indirectly a consequence of disease (Rakyan et al. 2011).

### 5.2.1 Single-Locus Test

Logistic regression is a widely used statistic for testing the association of DNA methylation with disease. Assume that $n$ individuals are sampled. Let $y_i = 1$ if the $i^{th}$ individual is affected, otherwise, $y_i = 0$. Consider $m$ covariate variables $x_{i1},...,x_{im}$. The methylation level at the CpG site being tested is denoted by $z_i$. Then, the logistic regression model for EWAS is given by

$$\log \frac{P(y_i = 1)}{1 - P(y_i = 1)} = \alpha_0 + \alpha^T X_i + \beta z_i, \tag{5.1}$$

where $\alpha = [\alpha_1, ..., \alpha_m]^T$ is the vector of regression coefficients for the covariates, $X_i = [x_{i1}, ..., x_{im}]^T$ is a vector of covariates, and $\beta$ is a regression coefficient for the methylation level at the CpG site. The logistic model (1) is a single-locus test. However, when the correlation among multiple CpG sites exist, we need to develop set (or gene)-based tests to utilize the correlation structure of the multiple CpG sites for power improvement.

### 5.2.2 Set-Based Methods

#### 5.2.2.1 Logistic Regression Model

The logistic regression model (1) can be extended to include multiple CpG sites:

$$\log \frac{P(y_i = 1)}{1 - P(y_i = 1)} = \alpha_0 + \alpha^T X_i + \beta^T Z_i, \tag{5.2}$$

where $\beta$ is a vector of coefficients for the methylation effects, $Z_i = [z_{i1}, ..., z_{ip}]^T$ is a vector of methylation levels of $p$CpG sites where variables and parameters are defined as before.

### 5.2.2.2 Generalized $T^2$ Test Statistic

Consider $k$ CpG sites. Let $x_{ij}$ and $y_{ij}$ be the methylation level of the $i^{th}$ individual at the $j^{th}$ CpG site, in cases and controls, respectively. Define two vectors:

$$X_i = (x_{i1}, .., x_{ik})^T, Y_i = (y_{i1}, .., y_{ik})^T.$$

For each CpG site, in cases and controls, we calculate the mean values $\bar{X}_j$ and $\bar{Y}_j$ of the methylation levels at the $j^{th}$ CpG site in cases and controls, respectively:

$$\bar{X}_j = \frac{1}{n_A} \sum_{i=1}^{n_A} X_{ij}, \bar{Y}_j = \frac{1}{n_G} \sum_{i=1}^{n_G} Y_{ij},$$

where $n_A$ and $n_G$ are the numbers of the sampled individuals in cases and controls, respectively.

Assembling all the mean values $\bar{X}_j$ and $\bar{Y}_j, j = 1, ..., k$ into vectors $\bar{X}$ and $\bar{Y}$: $\bar{X} = (\bar{X}_1, ..., \bar{X}_k)^T$, $\bar{Y} = (\bar{Y}_1, ..., \bar{Y}_k)^T$. We assume that the covariance matrices of the methylation profiles in cases and controls are equal. The pooled-sample variance-covariance matrix of the methylation variables is defined as

$$S = \frac{1}{n_A + n_G - 2} \left[ \sum_{i=1}^{n_A} (X_i - \bar{X})(X_i - \bar{X})^T + \sum_{i=1}^{n_G} (Y_i - \bar{Y})(Y_i - \bar{Y})^T \right].$$

Hotelling's (1931) $T^2$ statistic is then defined as

$$T^2 = \frac{1}{\frac{1}{n_A} + \frac{1}{n_G}} (\bar{X} - \bar{Y})^T S^{-1} (\bar{X} - \bar{Y}). \tag{5.3}$$

Under the null hypothesis of no association of methylation at the $k$CpG sites with the disease, the statistic $T^2$ is asymptotically distributed as a central $\chi^2_{(k)}$ distribution.

### 5.2.2.3 PCA

Principal component analysis (PCA) can be used to reduce the dimensions of the methylation level data. When the $p$ CpG sites are correlated, it is possible to use a few $k$ top principal components (PCs) to replace the original methylation variables at the $p$CpG sites in the logistic regression model (2) for EWAS. The PCA-based logistic model for EWAS is given by

$$\log \frac{P(y_i = 1)}{1 - P(y_i = 1)} = \alpha_0 + \alpha^T X_i + \beta^T PC_i, \tag{5.4}$$

where $PC_i = [PC_i^1, ..., PC_i^k]^T$ is a vector of PC scores.

### 5.2.2.4 Sequencing Kernel Association Test (SKAT)

Sequencing kernel association test (SKAT) (Wu et al. 2011) is a based logistic mixed effects model that can be used for rare variant association analysis. The SKAT for EWAS assumes the following logistic model

$$\log \frac{P(y_i = 1)}{1 - P(y_i = 1)} = \alpha_0 + \alpha^T X_i + \beta^T Z_i,$$

where variables are defined as before. One way to test the association of CpG sites within a region with disease is to test the null hypothesis:

$$H_0 : \beta = 0.$$

Since the number of CpG sites in a gene is often large and each CpG site makes small risk to the disease, the power of the standard likelihood ratio test is often low. To increase the power of the test, the SKAT aggregates the CpG sites across the genome region and tests the variance component under the null hypothesis:

$$H_0 : \tau = 0.$$

The SKAT statistic (Equation 3 in Wu et al. 2011)

$$Q = (y - \hat{\mu})K(y - \hat{\mu}), \tag{5.5}$$

where $K = ZWZ^T$, $Z$ is a matrix of methylation level at $p$CpG sites, and $W = \text{diag}(w_1, ..., w_p)$ with each weight $w_i$ prespecified.

Wu et al. (2011) showed that under the null hypothesis,

$$Q \sim \sum_{i=1}^{n} \lambda_i \chi_{1,i}^2, \tag{5.6}$$

where $\lambda_i$ are the eigenvalues of the matrix $P_0^{1/2} K P_0^{1/2}$, $P_0 = V - V\tilde{X}(\tilde{X}^T V \tilde{X})^{-1} \tilde{X}^T V$, $\tilde{X} = [1, X]$, $V = \text{diag}(\hat{\mu}_1(1 - \hat{\mu}_1), ..., \hat{\mu}_n(1 - \hat{\mu}_n))$ and $\chi_{1,i}^2$ represents independent $\chi_{(1)}^2$ random variables.

### 5.2.2.5 Canonical Correlation Analysis

Similar to GWAS, canonical correlation analysis (CCA) provides another statistical framework for testing the association of methylation with disease.

The goal of CCA is to seek optimal correlation between a disease and a linear combination of CpG sites within a genome region. The CCA measures the strength of association between the multiple CpG sites and the disease.

Consider a binary trait $y$ and $L$CpG sites with methylation variables $z_1,...,z_L$. Define the variance and covariance matrices:

$$\Sigma_{yy} = \sigma_y^2, \Sigma_{yz} = [\text{cov}(y, z_1) \cdots \text{cov}(y, z_L)] = \Sigma_{zy}^T, \text{ and}$$

$$\Sigma_{zz} = \begin{bmatrix} \sigma_{z_1 z_1} & \cdots & \sigma_{z_1 z_L} \\ \vdots & \ddots & \vdots \\ \sigma_{z_L z_1} & \cdots & \sigma_{z_L z_L} \end{bmatrix}.$$

Define

$$R = \Sigma_{yy}^{-1/2} \Sigma_{yz} \Sigma_{zz}^{-1} \Sigma_{zy} \Sigma_{yy}^{-1/2}.$$

Since $\Sigma_{yy} = \sigma_y^2$ is a number, the matrix $R$ is reduced to

$$R = \frac{\Sigma_{yz} \Sigma_{zz}^{-1} \Sigma_{zy}}{\sigma_y^2}.$$

The eigenvalue is

$$\lambda^2 = \frac{\Sigma_{yz} \Sigma_{zz}^{-1} \Sigma_{zy}}{\sigma_y^2}.$$

Let $\hat{\sigma}_y^2, S_{yg}, S_{gg},$ and $S_{gy}$ be sampling versions of $\sigma_y^2, \Sigma_{yz}, \Sigma_{zz},$ and $\Sigma_{zy}$. Then,

$$\hat{\lambda}^2 = \frac{S_{yz} S_{zz}^{-1} S_{zy}}{\hat{\sigma}_y^2}.$$

The statistic for testing association of the methylations in the genomic region with the disease is defined as

$$T_{CCA} = -N \log (1 - \hat{\lambda}^2) \tag{5.7}$$

Under the null hypothesis of no association of the methylations in the genomic region with the disease, $T_{CCA}$ is a central $\chi_{(L)}^2$ distribution.

---

## 5.3 Epigenome-Wide Causal Studies

### 5.3.1 Introduction

Despite significant progress in dissecting the genetic and epigenetic architecture of complex diseases by GWAS and EWAS, understanding the etiology and mechanism of complex diseases remains elusive. The current paradigm of genomic and epigenomic analysis is association and correlation analysis. Our experiences in association analysis strongly demonstrate that association analysis lacks power to discover the mechanisms of the diseases. The observed association may be in part due to chance, bias, and confounding. The recent study found that 'association signals tend to be spread across most of the genome,' which again shows that association signals provide limited information on causes of disease, which called the future of the GWAS into question (Boyle et al. 2017; Callaway 2017). An observed association may not lead to inferring a causal relationship and the lack of association may not be necessary to imply the absence of a causal relationship. The dominant use of association analysis for genetic and epigenetic studies of complex diseases is a key issue that hampers the theoretical development of genomic and epigenomic science and its application in practice. Causal analysis is more powerful than association analysis and allows estimation of the effect of intervention or distribution changes (Peters et al. 2017). Causal models can be used to predict the results of intervention, however, association usually cannot. Similar to EWAS and QTL (mQTL) analysis, we plan to develop novel statistical methods for epigenome-wide causal studies (EWCS) of both qualitative and quantitative traits to make paradigm changes of epigenetic studies of complex diseases from EWAS to EWCS. Causal inference is an essential component for discovery of disease mechanisms.

### 5.3.2 Additive Functional Model for EWCS

#### 5.3.2.1 Mathematic Formulation of EACS

In EWCA, we consider a binary trait that presents the disease status and a continuous variable that represents the methylation level at a CpG site. We investigate the causal direction from a continuous variable (methylation level) to a binary variable (disease variable). Let $Y$ denote a binary trait and $X$ denote a continuous methylation level variable. The binary trait $Y$ can be modeled as

$$Y = f(X, \varepsilon), \tag{5.8}$$

where $f$ is a nonlinear function and $\varepsilon$ is an error term and is independent from the hypothesized cause $X$. A special case of the nonlinear model (5.8) is an additive noise model (ANM):

$$Y = f(X) + \varepsilon, \tag{5.9}$$

where again, $f$ is a nonlinear function and $\varepsilon$ is an error term and is independent from the hypothesized cause $X$.

Additive noise models with binary effect and continuous cause have difficulty in estimation of errors. The effective methods for assessing causation between a continuous variable and a binary trait use discrete additive noise models to approximate the continuous additive noise models. Resulting discrete additive noise models can be used to assess causal relationships between $x$ and $y$.

Next, we investigate why simply applying continuous-discrete additive noise models for causal inference may meet some difficulties. Since a logistic sigmoid function is often used as a nonlinear function, the logistic sigmoid function will be used as a nonlinear function $f$ in Equation 5.9. Therefore, the ANM Equation 5.9 can be written as

$$Y = \sigma(\alpha X + b) + \varepsilon, \tag{5.10}$$

where $\alpha$ is a coefficient, $b$ is a bias, and $\sigma$ is a logistic sigmoid function and defined as

$$\sigma(t) = \frac{1}{1 + e^{-t}}. \tag{5.11}$$

The logistic sigmoid function consists of two components. The first component is a linear layer that computes $t = \alpha x + b$. The second component is to convert $t$ into a probability using the sigmoid function.

It is easy to see that

$$\frac{d}{dt}\sigma(t) = \left(\frac{1}{1 + e^{-t}}\right)^2 e^{-t} = \frac{1}{1 + e^{-t}}\frac{e^{-t}}{1 + e^{-t}} = \sigma(t)(1 - \sigma(t)). \tag{5.12}$$

### 5.3.2.2 Parameter Estimation

Let $\theta = [\alpha, b]^T$. The least square estimate of the parameters $\theta$ is to find the set of parameters $\theta$ that minimizes the sum of squared residuals:

$$F(\theta) = \sum_{i=1}^{n}[Y_i - \sigma(\alpha X_i + b)]^2 \tag{5.13}$$

Define

$$\frac{\partial F(\theta^{(k)})}{\partial \alpha} = -\sum_{i=1}^{n}[Y_i - \sigma(\alpha^{(k)}X_i + b^{(k)})]X_i$$

$$\frac{\partial F(\theta^{(k)})}{\partial b} = -\sum_{i=1}^{n}\sigma(\alpha^{(k)}X_i + b^{(k)})(1 - \sigma(\alpha^{(k)}X_i + b^{(k)}))[Y_i - \sigma(\alpha^{(k)}X_i + b^{(k)})]$$

$$\frac{\partial^2 F(\theta^{(k)})}{\partial \alpha^2} = \sum_{i=1}^{n}\sigma(\alpha^{(k)}X_i + b^{(k)})(1 - \sigma(\alpha^{(k)}X_i + b^{(k)})^2 X_i^2$$

$$\frac{\partial^2 F(\theta^{(k)})}{\partial \alpha \partial b} = \sum_{i=1}^{n}\sigma(\alpha^{(k)}X_i + b^{(k)})(1 - \sigma(\alpha^{(k)}X_i + b^{(k)})^2 X_i \tag{5.14}$$

$$\frac{\partial^2 F\left(\theta^{(k)}\right)}{\partial b \partial \alpha} = \sum_{i=1}^{n} \sigma(\alpha^{(k)} X_i + b^{(k)})^2 (1 - \sigma(\alpha^{(k)} X_i + b^{(k)})^2 X_i$$

$$\frac{\partial^2 F\left(\theta^{(k)}\right)}{\partial b^2} = \sum_{i=1}^{n} \sigma(\alpha^{(k)} X_i + b^{(k)})^2 (1 - \sigma(\alpha^{(k)} X_i + b^{(k)})^2$$

Let

$$H\left(\theta^{(k)}\right) = \begin{bmatrix} \dfrac{\partial^2 F\left(\theta^{(k)}\right)}{\partial \alpha^2} & \dfrac{\partial^2 F\left(\theta^{(k)}\right)}{\partial \alpha \partial b} \\[2ex] \dfrac{\partial^2 F\left(\theta^{(k)}\right)}{\partial b \partial \alpha} & \dfrac{\partial^2 F\left(\theta^{(k)}\right)}{\partial b^2} \end{bmatrix} \tag{5.15}$$

be a Hessian matrix of the sum of squared residuals $F$.

The necessary condition for minimizing the sum of squared residuals $F$ is

$$\frac{\partial F}{\partial \theta} = 0. \tag{5.16}$$

Equation 5.16 is a system of nonlinear equations. The Newton–Raphson method can be used to solve Equation 5.16. Specifically, the gradient of the sum of squared residuals $F$ can be approximated by

$$\frac{\partial F(\theta)}{\partial \theta} \approx \frac{\partial F\left(\theta^{(k)}\right)}{\partial \theta} + H\left(\theta^{(k)}\right)\left(\theta - \theta^{(k)}\right).$$

Setting $\dfrac{\partial F(\theta)}{\partial \theta} = 0$, we obtain

$$\frac{\partial F\left(\theta^{(k)}\right)}{\partial \theta} + H\left(\theta^{(k)}\right)\left(\theta - \theta^{(k)}\right) = 0. \tag{5.17}$$

Solving Equation 5.17 for the vector of parameters $\theta$ gives an approximate solution to Equation 5.17:

$$\theta^{(k+1)} = \theta^{(k)} - H^{-1}\left(\theta^{(k)}\right)\frac{\partial F\left(\theta^{(k)}\right)}{\partial \theta}. \tag{5.18}$$

Let

$$Y = \begin{bmatrix} Y_1 \\ \vdots \\ Y_n \end{bmatrix}, \quad H = \begin{bmatrix} X_1 & 1 \\ \vdots & \vdots \\ X_n & 1 \end{bmatrix}.$$

The Newton–Raphson algorithm for estimation of parameters $\theta$ is summarized in Result 5.1.

**Result 5.1: Newton–Raphson Algorithm**

Step 1: Initialization. Compute the initial value:

$$\theta^{(0)} = (H^t H)^{-1} H^T Y.$$

Step 2: Compute the gradient of the sum of squared residuals $F$:

$$\frac{\partial F(\theta^{(k)})}{\partial \theta} = \begin{bmatrix} -\sum_{i=1}^n \left[ Y_i - \sigma\left(\alpha^{(k)} X_i + b^{(k)}\right) \right] X_i \\ -\sum_{i=1}^n \sigma(\alpha^{(k)} X_i + b^{(k)})(1 - \sigma(\alpha^{(k)} X_i + b^{(k)}) \left[ Y_i - \sigma(\alpha^{(k)} X_i + b^{(k)}) \right] \end{bmatrix}.$$

Step 3: Compute the Hessian matrix of the sum of squared residuals $F$:

$$H\left(\theta^{(k)}\right) = \begin{bmatrix} \dfrac{\partial^2 F(\theta^{(k)})}{\partial \alpha^2} & \dfrac{\partial^2 F(\theta^{(k)})}{\partial \alpha \partial b} \\ \dfrac{\partial^2 F(\theta^{(k)})}{\partial b \partial \alpha} & \dfrac{\partial^2 F(\theta^{(k)})}{\partial b^2} \end{bmatrix},$$

which is defined in Equations 5.14 and 5.15.
Step 4: Update of the parameters $\theta$:

$$\theta^{(k+1)} = \theta^{(k)} - H^{-1}\left(\theta^{(k)}\right) \frac{\partial F(\theta^{(k)})}{\partial \theta}.$$

Step 5: Check convergence:
If $\| \theta^{(k+1)} - \theta^{(k)} \|_2 \le e$, stop; otherwise, $\theta^{(k)} \leftarrow \theta^{(k+1)}$, go to Step 2 where $e$ is a prespecified error.

### 5.3.2.3 Test for Independence

To test whether the methylation data satisfy the ANMs, we need to split the data into the training dataset and test dataset. By regressing the binary trait $Y$ on the methylation level $X$ using the training data $D_N$, we estimate $\hat{f}_Y$ for the regression function. Then, we use the test dataset $D_t$ to estimate the residual $\varepsilon = Y - \hat{f}_Y(X)$. Finally, we test the null hypothesis of independence of the residual $\varepsilon$ from the potential causal $X$. If independence is not rejected, then the data satisfy the ANM $X \rightarrow Y$.

Methods for testing independence between two random variables include kernel-based non-parameter independence tests (Zhang et al. 2017), entropy score-based methods (Kpotufe et al. 2014), Bayes score-based methods (Friedman and Nachman 2000), and Minimum Message Length score tests (Mooij and Janzing 2010). Widely used methods for tests of independence are kernel-based nonparametric methods which use representation of probability measure in reproducing kernel Hilbert space (RKHS) which provides a general mathematical framework for nonparametric tests of independence. An essential measurement is the Hilbert–Schmidt Independence Criterion (HSIC) that uses the distance between the RKHS embeddings of probability distributions (Gretton et al. 2005, 2015). We begin with a brief introduction of the key concepts in the RKHS embeddings.

### 5.3.2.3.1 RKHS Embeddings and HSIC

**Definition 5.1: Reproducing Kernel Hilbert Space (RKHS)**

Let $\mathcal{X}$ be a non-empty set. A Hilbert space $\mathcal{H}$ of functions $f: \mathcal{X} \to R$ defined on $\mathcal{X}$ is called a RKHS if evaluation functionals $\delta_x: f \mapsto f(x)$ are continuous for all $x \in \mathcal{X}$.

In general, high-dimensional feature spaces have much more complex structure. We often map the data in a low-dimensional space to a high-dimensional feature space from which the pattern of the data in the low-dimensional space can be unraveled. Consider a feature map that maps the data in two-dimensional space to the three-dimensional feature space:

$$\phi(x) = \begin{bmatrix} x_1 \\ x_2 \\ x_1 x_2 \end{bmatrix},$$

and two groups of the data:

$$A = \begin{bmatrix} -1 & -2 & 1 & 2 \\ -1 & -2 & 1 & 2 \end{bmatrix} \text{ and } B = \begin{bmatrix} -1 & -2 & 1 & 2 \\ 1 & 2 & -1 & -2 \end{bmatrix}.$$

We map the points in groups $A$ and $B$ to

$$\phi(A) = \begin{bmatrix} -1 & -2 & 1 & 2 \\ -1 & -2 & 1 & 2 \\ 1 & 4 & 1 & 4 \end{bmatrix} \text{ and } \phi(B) = \begin{bmatrix} -1 & -2 & 1 & 2 \\ 1 & 2 & -1 & -2 \\ -1 & -4 & -1 & -4 \end{bmatrix}.$$

There is no linear classifier that will separate the patterns in the group $A$ (in color blue) from the group $B$ (in color red) (Figure 5.3). However, in the

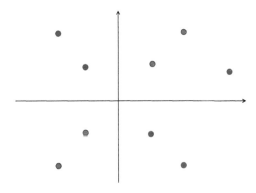

**FIGURE 5.3**
The points between the red and blue points can be separated by the linear classifiers.

three-dimensional feature space, we can easily separate them. Now we define a kernel (Gretton 2015).

**Definition 5.2: Kernel**

Let $\mathcal{X}$ be a non-empty set. A function $k : \mathcal{X} \times \mathcal{X} \to R$ is referred to as a kernel if there exists a Hilbert space and a feature map $\phi : \mathcal{X} \to \mathcal{H}$ such that for all $x, x' \in \mathcal{X}$, we have

$$k(x, y) = \langle \phi(x), \phi(y) \rangle_H. \tag{5.19}$$

Simple examples for kernel are

$$k(., x) = \phi(x) = \begin{bmatrix} x \\ x^2 \end{bmatrix}, k(x, y) = xy + x^2 y^2,$$

$$\text{and } k(., x) = \phi(x) = \begin{bmatrix} x_1 \\ x_2 \\ x_1 x_2 \end{bmatrix}, k(x, y) = x_1 y_1 + x_2 y_2 + x_1 x_2 y_1 y_2.$$

Three widely used kernels are

1. Linear kernel: $k(x, y) = x^T y$,
2. Polynomial kernel: $k(x, y) = (x^T y + 1)^d$, and
3. Gaussian kernel with bandwidth $\sigma > 0$: $k(x, y) = e^{\frac{\|x - y\|^2}{2\sigma^2}}$.

**Definition 5.3: RKHS**

Let $H$ be a Hilbert space of real-valued function defined on $\chi$. If a function $k: \chi \times \chi \to R$ satisfies the following two conditions:

1. $\forall\, x \in \chi, k(, x) \in H$  $\tag{5.20}$

2. $\forall x \in \chi, \forall f \in H, \langle f, k(., x) \rangle_H = f(x)$,

then $k$ is called a reproducing kernel of $H$ and $H$ is called a RKHS.
For example, consider a function $f(x) = ax + bx^2 + cx^3$. Define

$$k(., x) = \begin{bmatrix} x \\ x^2 \\ x^3 \end{bmatrix} \text{ and } f(.) = \begin{bmatrix} a \\ b \\ c \end{bmatrix}.$$

Then, we have

$$f(x) = \langle f(.), k(.x) \rangle = \begin{bmatrix} a \\ b \\ c \end{bmatrix}^T \begin{bmatrix} x \\ x^2 \\ x^3 \end{bmatrix}.$$

Consider a function $k(x,y)$. Let $f(.) = k(.,y)$ and $k(.,x)$ a reproducing kernel. Then, it follows from Equation 5.20 that

$$\langle k(.,y), k(.,x) \rangle_H = k(x,y). \tag{5.21}$$

It is well known that an inner product is a measure of similarity. Equation 5.21 gives a measure of similarity between points through the kernel. A traditional approach to representing points in a topological space is to embed them in a feature space $R^m$ via a feature map $\Phi(x)$:

$$\Phi(x) = [\,\phi_1(x) \,\cdots\, \phi_m(x)\,]^T.$$

Kernel functions are extensions of the feature map from finite dimensional space to infinite dimensional space. In particular, we consider a set of distributions. We can embed each distribution as a point to represent the distribution in the RKHS. Recall that from Equation 5.20, we have

$$E_p[f(x)] = E_P[\langle f(.), k(.,x) \rangle_H] = \left\langle f(.), E_P[k(.,x)] \right\rangle_H = \left\langle f(.), \mu_p \right\rangle_{H'} \tag{5.22}$$

which leads to the following definition of mean embedding (Gretton 2015).

## Definition 5.4: Mean Embedding

Let $P$ be a distribution and $k$ be a kernel in the RKHS $H$. The mean embedding of the probability distribution $P$ is defined as $E_P[k(.,x)] \in H$ such that

$$E_p[f(x)] = \left\langle f(.), E_p[k(.,x)] \right\rangle_H \tag{5.23}$$

Mean embedding provides a tool to measure distance between two probability distributions. Using mean embedding we can define maximum mean discrepancy (Zhang et al. 2017).

## Definition 5.5: Maximum Mean Discrepancy (MMD)

Suppose that $k$ is a kernel defined on $\chi$, $P$ and $Q$ are two probability distributions. Maximum mean discrepancy (MMD) between distributions $P$ and $Q$ with respect to $k$ is defined as the squared distance between the mean embedding of two distributions in the RKHS:

$$MMD_k(P,Q) = \|\, E_P[k(.,x)] - E_Q[k(.,x)] \,\|_{H_k}^2. \tag{5.24}$$

The estimators of MMD can be used to develop nonparametric two-sample test statistics (Zhang et al. 2017; Gretton et al. 2012, 2015). Before the mean embedding is extended to joint embedding, we first briefly introduce a concept of tensor product. Tensor product is also referred to as Kronecker product or direct product. Tensor product attempts to construct a large vector space out of two smaller vector spaces. First, we introduce the tensor product of two vectors. Consider an $n$ dimensional vector $X = [x_1, ..., x_n]^T$ and an $m$ dimensional vector $Y = [y_1, ..., y_m]^T$. The tensor product of $X$ and $Y$ is defined as

$$X \otimes Y = XY^T = \begin{bmatrix} x_1y_1 & \cdots & x_1y_m \\ \vdots & \vdots & \vdots \\ x_ny_1 & \cdots & x_ny_m \end{bmatrix}.$$

Next, we consider the tensor product of two matrices $A$ and $B$. Let

$$A = \begin{bmatrix} a_{11} & \cdots & a_{1m} \\ \vdots & \vdots & \vdots \\ a_{n1} & \cdots & a_{nm} \end{bmatrix} \text{ and } B = \begin{bmatrix} b_{11} & \cdots & b_{1l} \\ \vdots & \vdots & \vdots \\ b_{k1} & \cdots & b_{kl} \end{bmatrix}.$$

The tensor product of the matrix $A$ and matrix $B$ is defined as

$$A \otimes B = \begin{bmatrix} a_{11}B & \cdots & a_{1m}B \\ \vdots & \vdots & \vdots \\ a_{n1}B & \cdots & a_{nm}B \end{bmatrix}$$

$$= \begin{bmatrix} a_{11}b_{11} & \cdots & a_{11}b_{1l} & \cdots & \cdots & \cdots & a_{1m}b_{11} & \cdots & a_{1m}b_{1l} \\ \vdots & \vdots & \vdots & \vdots & \vdots & \vdots & \vdots & \vdots & \vdots \\ a_{11}b_{k1} & \cdots & a_{11}b_{kl} & \cdots & \cdots & \cdots & a_{1m}b_{k1} & \cdots & a_{1m}b_{kl} \\ \vdots & & & & \vdots & & & & \vdots \\ a_{n1}b_{11} & \cdots & a_{n1}b_{1l} & \cdots & \cdots & \cdots & a_{nm}b_{11} & \cdots & a_{nm}b_{1l} \\ \vdots & \vdots & \vdots & \vdots & \vdots & \vdots & \vdots & \vdots & \vdots \\ a_{n1} & \cdots & a_{n1}b_{kl} & \cdots & \cdots & \cdots & a_{nm}b_{k1} & \cdots & a_{nm}b_{kl} \end{bmatrix}.$$

Next, we define a rank-one operator (Gretton 2015). Let $a \in G$ and $b \in F$. A rank-one operator from $G$ to $F$ is defined as

$$(b \otimes a)f = \langle a, f \rangle_F b, \tag{5.25}$$

where $f \in F$.

If $a, b,$ and $f$ are vectors, then we have

$$(b \otimes a)f = ba^T f = \langle a, f \rangle b.$$

Now we define the Hilbert–Schmidt norm of the operator. Assume that $F$ and $H$ are two separable Hilbert spaces. Let $e_i, i \in I$ be an orthonormal basis for $F$ and $v_j, j \in J$ be an orthonormal basis for $H$. The index sets $I, J$ can be either finite or countable infinite. Let $A$ be an operator on a Hilbert space $H$. It is clear that $Av_j$ is in the Hilbert space $F$ and has a norm $\| Av_j \|_F$.

The element $Av_j$ in the space $F$ can also be expanded in terms of ortho-normal basis $e_i$:

$$Av_j = \sum_{i \in I} \left\langle Av_j, e_i \right\rangle e_i. \tag{5.26}$$

Now we define the Hilbert–Schmidt norm of the operator (Gretton et al. 2015).

**Definition 5.6: Hilbert–Schmidt Norm of the Operator and the Hilbert–Schmidt Inner Product of Two Operators**

The Hilbert–Schmidt norm of the operator $A$ is defined as

$$\begin{aligned} \| A \|_{HS}^2 &= \sum_{j \in I} \| Av_j \|_F^2 \\ &= \sum_{i \in I} \sum_{j \in J} \left| \left\langle Av_j, e_i \right\rangle_F \right|^2. \end{aligned} \tag{5.27}$$

If the Hilbert–Schmidt norm of the operator $A$ is finite, the operator $A$ is called a Hilbert–Schmidt operator. The Hilbert–Schmidt inner product of two Hilbert–Schmidt operators $A$ and $B$ is defined as

$$\langle A, B \rangle_{HS} = \sum_{j \in J} \left\langle Av_j, Bv_j \right\rangle_F. \tag{5.28}$$

We can show that (Exercise 3)

$$\langle A, B \rangle_{HS} = \sum_{i \in I} \sum_{j \in J} \left\langle Av_j, e_i \right\rangle_F \left\langle Bv_j, e_i \right\rangle_F. \tag{5.29}$$

Next, we calculate the Hilbert–Schmidt norm of the rank-one operator or tensor product.

Using Equations 5.25 and 5.27, we obtain

$$\begin{aligned} \| a \otimes b \|_{HS}^2 &= \sum_j \| (a \otimes b) v_j \|_F^2 \\ &= \sum_j \| \left\langle b, v_j \right\rangle_F a \|_F^2 \\ &= \sum_j \left( \left\langle b, v_j \right\rangle_F \right)^2 \| a \|_F^2 \\ &= \| a \|_F^2 \sum_j \left( \left\langle b, v_j \right\rangle_F \right)^2 \\ &= \| a \|_F^2 \| b \|_F^2. \end{aligned} \tag{5.30}$$

Therefore, we have Result 5.2 (Gretton 2015).

**Result 5.2: The Hilbert–Schmidt Norm of Rank-One Operator**

The rank-one operator is Hilbert–Schmidt. Its Hilbert–Schmidt norm is

$$\| a \otimes b \|_{HS}^2 = \| a \|_F^2 \| b \|_F^2. \tag{5.31}$$

Next, we introduce the Hilbert–Schmidt inner product of two rank-one operators.

Using Equation 5.28, we obtain

$$
\begin{aligned}
\langle L, a \otimes b \rangle_{HS} &= \sum_j \left\langle Lv_j, (a \otimes b)v_j \right\rangle_F \\
&= \sum_j \left\langle Lv, \left\langle b, v_j \right\rangle_F a \right\rangle_F \\
&= \sum_j \left\langle L(\left\langle b, v_j \right\rangle_F v_j, a \right\rangle_F \\
&= \left\langle L \sum_j \left\langle b, v_j \right\rangle_F v_j, a \right\rangle_F \\
&= \langle Lb, a \rangle_F.
\end{aligned}
\tag{5.32}
$$

Substituting $L = c \otimes d$ into Equation 5.32, we obtain

$$
\begin{aligned}
\langle c \otimes d, a \otimes b \rangle_{HS} &= \langle (c \otimes d)b, a \rangle_F \\
&= \langle \langle d, b \rangle_F c, a \rangle_F \\
&= \langle c, a \rangle_F \langle d, b \rangle_F.
\end{aligned}
\tag{5.33}
$$

Thus, we prove Result 5.3 (Gretton 2015).

**Result 5.3: Hilbert–Schmidt Inner Product of Rank-One Operators**

Let $L \in HS(,G,F)$ be a second Hilbert–Schmidt operator. Then, the Hilbert–Schmidt inner product of $L$ and $a \otimes b$ is given by

$$
\langle L, a \otimes b \rangle_{HS} = \langle Lb, a \rangle_F
\tag{5.34}
$$

and

$$
\langle c \otimes d, a \otimes b \rangle_{HS} = \langle c, a \rangle_F \langle d, b \rangle_F.
\tag{5.35}
$$

Now we introduce embedding of the joint distribution. Consider two random variables $X$ and $Y$ with joint distribution $P(x,y)$. Let $X \times Y$ be the product domain of $X$ and $Y$. Define a feature map:

$$
\Phi(x \otimes y) = \Phi(x) \otimes \Phi(y).
$$

Let $\Phi(x) = k(.,x)$ and $\Phi(y) = k(.,y)$. Then, we have

$$
\Phi(x \otimes y) = k(.,x) \otimes k(.,y).
\tag{5.36}
$$

Using Equation 5.35, we obtain

$$
\begin{aligned}
k &= k((x,y),(x',y')) \\
&= \| k(.,x) \otimes k(.,y) \|_{HS}^2 \\
&= \langle k(.,x) \otimes k(.,y), k(.,x) \otimes k(.,y) \rangle_{HS} \\
&= k(x,x')k(y,y').
\end{aligned}
\tag{5.37}
$$

Therefore, $k = k(.,x) \otimes k(.,y)$ is a valid kernel on the product domain $X \times Y$.

### Definition 5.7: Embedding of Joint Probability Distribution

Embedding of the joint distribution $P$ is defined as

$$P \mapsto C_{XY}(P) := E_P[k(.,x) \otimes k(.,y)]. \tag{5.38}$$

Similar to the mean embedding that represents the expectation operator, the joint embedding $C_{XY}$ can be viewed as the uncentered cross-covariance operator for any two functions $f, g \in H$ (Gretton, 2015). In fact, the expectation of the product of the nonlinear functions $f(X)g(Y)$ can be reduced to

$$E_P[f(X)g(Y)] = E_P[< f(.), k(,x) >_H < g(.), k(.,y) >_H]. \tag{5.39}$$

Using Equation 5.33, we obtain

$$E_p[\langle f(), k(,x) \rangle_H \langle g(), k(,y) \rangle_H] = E_P[\langle f \otimes g, k(.,x) \otimes k(.,y) \rangle_{HS}]$$

$$= \langle f \otimes g, E_P[k(.,x) \otimes k(.,y)] \rangle_{HS} \tag{5.40}$$

$$= \langle f \otimes g, C_{XY} \rangle_{HS}.$$

Then, it follows from Equation 5.32 that

$$\langle f \otimes g, C_{XY} \rangle_{HS} = \langle f, C_{XY}g \rangle_H. \tag{5.41}$$

Substituting Equations 5.40 and 5.41 into Equation 5.39, we obtain

$$E_P[f(X)g(Y)] = \langle f, C_{XY}g \rangle_H. \tag{5.42}$$

Similarly, we can define the auto-covariance operator (Lienart 2015).

### Definition 5.8: Auto-Covariance Operator

Auto-variance operators $C_{XX}$ and $C_{YY}$ are defined as

$$C_{XX} = E_P[k(.,x) \otimes k(.,x)] \tag{5.43}$$

and

$$C_{YY} = E_P[k(.,y) \otimes k(.,y)], \text{ respectively.} \tag{5.44}$$

We have defined a non-centered covariance operator that is an extension of the non-centered covariance. Similar to the covariance, we can define a general covariance operator, that is, a centered covariance operator.

**Definition 5.9: Centered Covariance Operator**

A general covariance operator or a centered covariance operator is defined as

$$\tilde{C}_{XY} = C_{XY} - \mu_X \otimes \mu_Y, \tag{5.45}$$

where $\mu_X = E_X[k(.,X)]$ and $\mu_Y = E_X[k(.,Y)]$.

In the previous discussion, we assume that both variables share the same kernel function. However, this restriction can be released. The cross-covariance operator can be defined on two different kernel functions (Lienart 2015).

Next, we introduce the Hilbert–Schmidt independence criterion (HSIC) for measuring dependence between two variables. If we assume that the RKHS embedding is injective (i.e., different points in the original space will map to the two different points in the feature space RKHS), then the distance in the RKHS can be used as a proxy for similarity in the distribution space (Lienart 2015; Zhang et al. 2017). In other words, the distance between the embedding $C_{XY}$ of the joint distribution $P(x,y)$ and the tensor product of the mean embeddings of two marginal distributions $P(x)$ and $P(y)$ can be used to measure dependence between two random variables $X$ and $Y$.

In classical statistics, if we assume that both variables $X$ and $Y$ follow a normal distribution, then $cov(x,y) = 0$ if and only if $X$ and $Y$ are independent. If $X$ and $Y$ are not normal variables, this statement will not hold. However, it is shown that for the general distributions of $X$ and $Y$, $\tilde{C}_{XY} = 0$ if and only if $X$ and $Y$ are independent (Mooij et al. 2016).

Now we mathematically define the HSIC.

**Definition 5.10: Hilbert–Schmidt Independence Criterion (HSIC)**

Let $P(x,y)$ be a joint distribution of the random variables $X$ and $Y$, $P(x)$ and $P(y)$ are a marginal distribution of $X$ and $Y$, respectively. Let $C_{XY}$ and $\tilde{C}_{XY}$ be non-centered and centered covariance operator of the variables $X$ and $Y$, respectively. Let $k(.,x)$ and $k(.,y)$ be kernels. The Hilbert–Schmidt independence criterion (HSIC) of $X$ and $Y$ is defined as

$$HSIC(X,Y) = \| C_{XY} - \mu_X \otimes \mu_Y \|^2_{H_{k_x} \times H_{k_y}}$$

$$= \| E_{P(x,y)}[k(.,X) \otimes k(.,Y)] - E_{P(x)}[k(.,X)] \otimes E_{P(y)}[k(.,Y)] \|^2_{H_{k_x} \times H_{k_y}}. \tag{5.46}$$

Result 5.4 for testing independence is adopted from Mooij et al. (2016) (Lemma 12, page 47).

**Result 5.4: HSIC and Independence Test**

Assume that the product kernel $k = k(.,x) \otimes k(.,y)$ is a characteristic kernel (Sriperumbudur et al. 2010). $HSIC(X,Y) = 0$ if and only if $X$ and $Y$ are independent.

*5.3.2.3.2 Estimation of HSIC*

Suppose that $n$ pairs of data $(x_i, y_i), i = 1,...,n$ are sampled from the joint distribution $P(x,y)$. A biased estimator of HSIC is given by (Zhang et al. 2017; Appendix 5.A):

$$HSIC_u(X, Y) = \frac{1}{n(n-3)}\left[ \text{tr}\left(\hat{K}_x\hat{K}_y\right) + \frac{\mathbf{1}^T\hat{K}_x\mathbf{1}\mathbf{1}^T\hat{K}_y\mathbf{1}}{(n-1)(n-2)} - \frac{2}{n-2}\mathbf{1}^T\hat{K}_x\hat{K}_y\mathbf{1} \right], \quad (5.47)$$

where $K_x = (k_x(x_i, x_j))_{n \times n}$, $K_y = (k_y(y_i, y_j))_{n \times n}$, $\hat{K}_x = K_x - diag(K_x)$, $\hat{K}_y = K_y - diag(K_y)$, $\mathbf{1}$ is a vector of 1s. A biased estimator of HSIC has the following simplified form (Zhang et al. 2017; Appendix 5.A):

$$HSIC_b(X, Y) = \frac{1}{n^2} \text{tr}\left(K_x H K_y H\right), \quad (5.48)$$

where $H = I_n - \frac{1}{n}\mathbf{1}\mathbf{1}^T$ is a centering matrix that centers the rows or columns.

*5.3.2.3.3 Asymptotic Distribution of the Estimators of HSIC*

The HSIC can be used to test for independence between two random variables. The null hypothesis being test is

$H_0$: Two random variables are independent.

The alternative hypothesis is

$H_a$: Two random variables are independent.

The asymptotic null distribution of the biased HSIC is given in Result 5.5 (Zhang et al. 2017).

**Result 5.5: Asymptotic Distribution of the Biased HSIC (Theorem 1 in Zhang et al. 2017)**

Under the null hypothesis of independence between two random variables, the asymptotic distribution of the biased HSIC is

$$nHSIC_b(X, Y) \xrightarrow{D} \sum_{i=1}^{\infty}\sum_{j=1}^{\infty} \lambda_i \eta_j N_{i,j}^2, \quad (5.49)$$

where $N_{i,j} \sim N(0,1)$ are i.i.d, and $\lambda_i, \eta_j, i = 1,2,....,j = 1,2,...$ are eigenvalues of the integral kernel equations

$$\int \bar{k}_x(x, w)g(w)dP_x(w) = \lambda g(x), \quad \bar{k}_x(x, x')$$

$$= \langle k_x(x, ) - E_W[k_x(W, )], k_x(x', .) - E_W[k_x(W, .)] \rangle, W \sim P_x$$

and

$$\int \bar{k}_y(y, w)g(w)dP_y(w) = \eta g(y), \bar{k}_y(y, y')$$

$$= \left\langle k_y(y, .) - E_W\left[k_y(W, .)\right], k_y(y', ), -E_W\left[k_y(W, .)\right]\right\rangle, W \sim P_y,$$

respectively.

Similarly, the asymptotic null distribution of the unbiased HSIC is given in Result 5.6 (Zhang et al. 2017).

**Result 5.6: Asymptotic Distribution of the Unbiased HSIC**

With the same notations as in Result 5.5, the asymptotic distribution of the unbiased HSIC under the null hypothesis of independence is

$$nHSIC_u(X, Y) \xrightarrow{D} \sum_{i=1}^{\infty}\sum_{j=1}^{\infty} \lambda_i \eta_j \left(N_{i,j}^2 - 1\right), \tag{5.50}$$

where $N_{i,j}$ are independently and identically distributed as $N(0,1)$.

### 5.3.2.3.4 The Block HSIC Statistic and its Asymptotic Null Distribution

We do not have closed analytical forms for the asymptotic null distribution of the biased and unbiased HSIC and hence it is difficult to calculate the P-values of the independence tests. To overcome these limitations, Zhang et al. (2017) proposed a block-based estimator of HSIC and derived its asymptotic distribution that has an analytical form.

Assume that the sample is split into $n/S(S << n)$ blocks, each block with size $S$. For each block, the dataset is given by $(x_i^{(b)}, y_i^{(b)}), i = 1, ..., S, b = 1, ..., n/S$. For each block, using Equation 5.47 we calculate the unbiased estimator of the HSIC:

$$\hat{c}_b = \frac{1}{S(S-3)}\left[\text{tr}\left(\hat{K}_x^{(b)}\hat{K}_y^{(b)}\right) + \frac{\mathbf{1}^T\hat{K}_x^{(b)}\mathbf{11}^T\hat{K}_y^{(b)}\mathbf{1}}{(S-1)(S-2)} - \frac{2}{S-2}\mathbf{1}^T\hat{K}_x^{(b)}\hat{K}_y^{(b)}\mathbf{1}\right]. \tag{5.51}$$

The block-based unbiased estimator of HSIC is the average of $\hat{c}_b$:

$$BHSIC = \frac{S}{n}\sum_{b=1}^{n/S}\hat{c}_b. \tag{5.52}$$

Since $\hat{c}_b$ is an unbiased estimator of the HSIC, under the null hypothesis of independence, we have $E[\hat{c}_b] = 0$. Using the central limit theorem, we can show the asymptotic Result 5.7 of the block-based unbiased estimator of HSIC (Zhang et al. 2017; Appendix 5.B).

**Result 5.7: Asymptotic Null Distribution of Block-Based HSIC**

When

$$n \to \infty, S \to \infty, \frac{n}{S} \to \infty, \sqrt{nS}BHSIC \xrightarrow{D} N(0, Var(W)) \quad \text{and} \qquad (5.53)$$

$$T_{BI} = \sqrt{ns}\frac{BHSIC}{\sqrt{Var(W)}} \qquad (5.54)$$

is asymptotically distributed as $N(0,1($ distribution, where $Var(W) = S^2 var(\hat{c}_b)$.

The variance of $W$ is estimated by

$$\hat{var}(W) = S^2 \frac{S}{n} \sum_{b=1}^{n/s} (\hat{c}_b - BHSIC)^2. \qquad (5.55)$$

### 5.3.2.4 Test Statistics for Epigenome-Wise Causal Studies

Now we are ready to give a general procedure for epigenome-wide causal studies (EWCS). The procedure is adopted from the ANMs for two continuous variables (Hoyer et al. 2009; Mooij et al. 2016). Assume that the data are split into training and test data sets. In the training dataset, we first regress $Y$ on $X$ (methylation level for a CpG Ysite) using the model (5.9) and calculate the predicted residuals. Then, in the training dataset, we test whether the methylation levels$X$ are independent of the predicted residuals using the HSIC or bloc HSIC-based statistics. The procedure is summarized in Algorithm 5.1.

**Algorithm 5.1: General Procedure for EWCS**

1. Split the data into training dataset $D_{TR} = (x_i, y_i), i = 1, ..., n_1$ and test dataset. $D_{TE} = (x'_i, y'_i), i = 1, ..., n_2$.
2. Use the models (5.9) or (5.10), parameter methods in Section 5.3.2.2, and training dataset. $D_{TR}$ to obtain estimates:

   $\hat{f}_Y$ of the nonlinear regression function $x \mapsto E[Y|X = x]$.

3. Calculate residuals:
   Residual calculation is a key issue to assess the causal relationships between a causal continuous variable and a discrete variable. Let $\bar{f}_Y$ be the average value of $\hat{f}_Y(x'_i)$. The residual errors are calculated by

   $$\hat{e}_i - \bar{f}_i - \hat{f}_Y(x'_i), i = 1, .., n_2.$$

4. Test causation $X \to Y$ in dataset $D_{TE}$.

   (a) The null hypothesis is
   $H_0$: no causation $X \to Y$ ($X$ and $e_Y$ are dependent).

(b) Test statistics

(i) Select the kernels $k_x(x_i,x_j)$ and $k_e(e_i,ej)$, and define the kernel matrices $K_x$ and $K_e$:

$$K_x = \begin{bmatrix} k_x(x_1,x_1) & \cdots & k_x(x_1,x_{n_2}) \\ \vdots & \vdots & \vdots \\ k_x(x_{n_2},x_1) & \cdots & k_x(x_{n_2},x_{n_2}) \end{bmatrix} \text{ and}$$

$$K_e = \begin{bmatrix} k_e(e_1,e_1) & \cdots & k_e(e_1,e_{n_2}) \\ \vdots & \vdots & \vdots \\ k_e(e_{n_2},e_1) & \cdots & k_e(e_{n_2},e_{n_2}) \end{bmatrix}.$$

(ii) Calculate the biased, unbiased estimators of the HSIC and the block HSIC using Equations 5.47, 5.48, and 5.52:

$$HSIC_u(X,e) = \frac{1}{n(n-3)} \left[ \mathrm{tr}\left(\hat{K}_x \hat{K}_e\right) + \frac{\mathbf{1}^T \hat{K}_x \mathbf{1} \mathbf{1}^T \hat{K}_e \mathbf{1}}{(n-1)(n-2)} - \frac{2}{n-2} \mathbf{1}^T \hat{K}_x \hat{K}_e \mathbf{1} \right],$$

where $\hat{K}_x = K_x - diag(K_x)$, $\hat{K}_e = K_e - diag(K_e)$, $\mathbf{1}$ is a vector of 1s;

$$HSIC_b(X,Y) = \frac{1}{n^2} \mathrm{tr}\left(K_x H K_y H\right),$$

where $H = I_n - \frac{1}{n}\mathbf{1}\mathbf{1}^T$ is a centering matrix that centers the rows or columns;

$$\hat{c}_b = \frac{1}{S(S-3)} \left[ \mathrm{tr}\left(\hat{K}_x^{(b)} \hat{K}_e^{(b)}\right) + \frac{\mathbf{1}^T \hat{K}_x^{(b)} \mathbf{1} \mathbf{1}^T \hat{K}_e^{(b)} \mathbf{1}}{(S-1)(S-2)} - \frac{2}{S-2} \mathbf{1}^T \hat{K}_x^{(b)} \hat{K}_e^{(b)} \mathbf{1} \right], b$$

$$= 1,2,...,n/s$$

$$BHSIC = \frac{S}{n} \sum_{b=1}^{n/S} \hat{c}_b.$$

(iii) Define three statistics:

$$T_{uHSIC} = -\log HSIC_u(X,e),$$
$$T_{bHSIC} = -\log HSIC_b(X,e),$$
$$T_{BHSIC} = -\log BHSIC.$$

(c) Calculate the P-values of the three tests.

The permutation/bootstrap approach can be used to calculate the P-values of three causal test statistics. Assume that three statistics $\hat{T}_{uHSIC}$,

$\hat{T}_{bHSIC}$, and $\hat{T}_{BHSIC}$ are calculated from the real test dataset $D_{TE}$. Further assume that the total number of permutations is $n_p$. For each permutation, we fix $x_i, i = 1,...,n_2$ and randomly permutate $y_i, i = 1,...,n_2$. Then, fit the ANMs (5.9) or (5.10). Finally, we calculate the residuals $e_i, i = 1,2,...,n_2$ and three statistics $\tilde{T}_{uHSIC}, \tilde{T}_{bHSIC}$, and $\tilde{T}_{BHSIC}$. Repeat $n_p$ times. The P-values are defined as the proportions of the three statistics $\tilde{T}_{uHSIC}, \tilde{T}_{bHSIC}$, and $\tilde{T}_{BHSIC}$ computed on the permuted data that are greater than or equal to $\hat{T}_{uHSIC}$, $\hat{T}_{bHSIC}$, and $\hat{T}_{BHSIC}$ computed on the original data $D_{TE}$.

## 5.4 Genome-Wide DNA Methylation Quantitative Trait Locus (mQTL) Analysis

DNA methylation is under the influence of genetic control. The genetic locus that is associated with DNA methylation is referred to as a methylation quantitative trait locus (mQTL). To identify the genetic variants that influence the methylation level is similar to the QTL and eQTL analysis (Hoffmann et al. 2016). Similar to eQTL analysis, the mQTL is also classified into *cis* mQTL and *trans* mQTL (McRae et al. 2017). All genetic variants within a genomic region spanning 2Mbp either side of the target CpG site are associated with the target CpG site is called *cis* mQTL. Other mQTLs are called *trans* mQTLs.

Both regression and canonical correlation analysis can be used to identify mQTL. Below we briefly introduce several models.

### 5.4.1 Simple Regression Model

The simple regression model is used to estimate the contribution of single SNP to the methylation variation of a single CpG site. Let $y_i$ be a methylation level of a CpG cite being tested from the $i^{th}$ individual and $x_i$ be an indicator variable for the genotype at a SNP of the $i^{th}$ individual. A simple regression for modeling mQTL is given by

$$y_i = \mu + \beta x_i + \varepsilon_i, i = 1, 2, .., n,$$

where $\beta$ is the genetic effect of the SNP on the methylation variation.

### 5.4.2 Multiple Regression Model

Consider $p$SNPs in a genomic region of a gene. A multiple regression that can be used to investigate the genetic contributions of multiple SNP's to the methylation level variation of a CpG site is given by

$$y_i = \mu + x_i^T \beta + \varepsilon_i,$$

where $x_i = [x_{i1}, ..., x_{ip}]^T$ is a vector of indicator variables for the genotype of $p$SNPs of the $i^{th}$ individual and $\beta$ is a vector of genetic effects on the methylation variation of the CpG site.

### 5.4.3 Multivariate Regression Model

Now consider $q$CpG sites. Let $Y_i = [y_{i1}, ..., y_{iq}]^T$ be a vector of methylation levels of the $q$CpG sites in a genomic region or a gene and $x_i$ be an indicator variable for the genotype at a SNP of the $i^{th}$ individual. A multivariate regression that can model the contribution of a single marker to the methylation levels of the multiple CpG sites in the genomic region or the gene is given by

$$Y_i = \mu + x_i\beta + \varepsilon_i,$$

where $\mu = [\mu_1, ..., \mu_q]^T$ is a vector of means, $\beta = [\beta_1, ..., \beta_q]^T$ is a vector of genetic effects of a single SNP to the $q$CpG sites, and $\varepsilon_i = [\varepsilon_{i1}, ..., \varepsilon_{iq}]^T$ is a vector of residuals.

### 5.4.4 Multivariate Multiple Regression Model

Now we extend multivariate regression models to the multiple SNPs. A multivariate multiple regression that can model the contributions of $p$SNPs to the $q$CpG sites is given by

$$Y_i^T = \mu^T + x_i^T B + \varepsilon_i^T,$$

where $B = \begin{bmatrix} \beta_{11} & \cdots & \beta_{1q} \\ \vdots & \vdots & \vdots \\ \beta_{p1} & \cdots & \beta_{pq} \end{bmatrix}$ is a matrix of the genetic effects of $p$SNPs to the $q$CpG sites and other parameters are defined as before.

### 5.4.5 Functional Linear Models for mQTL Analysis with Whole Genome Sequencing (WGS) Data

Functional linear models (FLMs) with a scalar response and functional predictors that were introduced in Chapter 3 can be used to investigate the contribution of a genomic region or a gene to the methylation level of the single CpG site with next-generation sequencing data.

Let $s$ be a genomic position in the region $S = [a,b]$. Define a genotype profile $x_i(s)$ of the i-th individual as

$$x_i(s) = \begin{cases} 2P_m(s), & MM \\ P_m(s)-P_M(s), & Mm \\ -2P_M(s), & mm \end{cases}$$

where M and m are two alleles of the marker at the genomic position $s$, $P_M(s)$, and $P_m(s)$ are the frequencies of the alleles M and m, respectively. A functional linear model (FLM) with scalar response and functional predictors is defined as

$$y_i = W_i^T \alpha + \int_S x_i(s)\beta(s)ds + \varepsilon_i, \quad i = 1,..,n,$$

where
$W_i = [w_{i1}, ..., w_{id}]^T$ is a vector of covariates, $\alpha = [\alpha_1, ..., \alpha_d]^T$ is a vector of effects associated with the covariates, $\beta(s)$ is a genetic additive effect function in the genomic position $s$, and $\varepsilon_i$ is an error variable.

### 5.4.6 Functional Linear Models with Both Functional Response and Predictors for mQTL Analysis with Both WGBS and WGS Data

With both WGBS and WGS data, CpG sites and SNPs are densely distributed across the genome. Both methylation profiles and genotype profiles can be modeled as a function of genomic position. We consider a methylation function $y_i(t)$, $t \in T = [0,T]$ of the methylation level at the $t$ genomic position in a genomic region (or gene)$[a,b]$ of the $i^{th}$ individual and $x_i(s)$ be a genotype function of the genomic position $s$ which is defined as before. A functional linear model with both functional response and predictors is defined as

$$y_i(t) = W_i^T \alpha(t) + \int_S x_i(s)\beta(s,t)ds + \varepsilon_i(t), \quad t = t_1,..,t_T, i = 1,..,n,$$

where $W_i = [w_{i1}, ..., w_{id}]^T$ is a vector of covariates, $\alpha(t) = [\alpha_1(t), ..., \alpha_d(t)]^T$ is a vector of effects associated with the covariates, $\beta(s,t)$ is a genetic additive effect function in the genomic position $s$ and $t$, and $\varepsilon_i(t)$ are the residual functions of the noise and unexplained effect for the $i^{th}$ individual. Methods for parameter estimation and test statistics were discussed in Chapter 3.

## 5.5 Causal Networks for Genetic-Methylation Analysis

Linear models for mQTL analysis ignore the dependence structures among methylations of multiple CpG sites. To incorporate the dependence structures into the genetic-methylation analysis, in this section we introduce causal network models. Networks are a powerful paradigm for representing relationships among variables and have been an essential tool for unravelling the structures of the complex biological systems (Ideker and Nussinov, 2017; Wang et al. 2017). We consider three linear causal networks: (1) networks with scalar variables, (2) networks with scalar endogenous variables and functional exogenous variables, and (3) networks with both functional endogenous variables exogenous variables. The causal networks with scalar variables were discussed in Chapters 9 and 10. In this section, we study two remaining linear causal networks for genetic-methylation or mQTL analysis which can also be applied to eQTL analysis.

### 5.5.1 Structural Equation Models with Scalar Endogenous Variables and Functional Exogenous Variables

#### 5.5.1.1 Models

Consider scalar methylation levels of $M$ CpG sites and next-generation sequencing data. Let $y_i$ be the methylation level of the $i^{th}$ CpG site and $x_k(t)$ be the genotype function of the SNP at the $t$ genomic position of the $k^{th}$ gene. The structural equations for the $M$ CpG sites and $K$ genes are given by

$$y_1 r_{11} + y_2 r_{21} + \cdots + y_M r_{M1} + \int_T x_1(t)\beta_{11}(t)dt + \cdots + \int_T x_K(t)\beta_{K1}(t)dt + e_1 = 0$$

$$\vdots \qquad \vdots \qquad \qquad \vdots \qquad \qquad \vdots \qquad \qquad \vdots$$

$$y_1 r_{1M} + y_2 r_{2M} + \cdots + y_M r_{MM} + \int_T x_1(t)\beta_{1M}(t)dt + \cdots + \int_T x_K(t)\beta_{KM}(t)dt + e_M = 0,$$

$$(5.56)$$

where $r$'s and $\beta(t)$ are the structure parameters of the mQTL analysis system that are unknown and will be estimated from the data. To transform functional structural equation models (FSEM) (5.56) to the standard multivariate SEMs, we expand $x_k(t)$ in terms of orthogonal eigenfunction $\psi_l(t)$, we obtain

$$x_k(t) = \sum_{l=1}^{L} \xi_{kl}\psi_l(t), \qquad (5.57)$$

where $\psi_l(t)$ is an eigenfunction and $\xi_{kl}$ is the expansion coefficients.

Substituting expansion (5.57) into the FSEMs (5.56) yields

$$y_1 \gamma_{11} + \cdots y_M \gamma_{M1} + \sum_{l=1}^{L} \xi_{1l} b_{l1}^{(1)} + \cdots + \sum_{l=1}^{L} \xi_{Kl} b_{lK}^{(1)} + e_1 = 0$$

$$\vdots \qquad\qquad\qquad\qquad\qquad\qquad\qquad\qquad (5.58)$$

$$y_1 \gamma_{1M} + \cdots y_M \gamma_{MM} + \sum_{l=1}^{L} \xi_{1l} b_{l1}^{(M)} + \cdots + \sum_{l=1}^{L} \xi_{Kl} b_{lK}^{(M)} + e_M = 0$$

where

$$b_{lk}^{(m)}(s) = \int_T \beta_{km}(t)\psi_l(t)dt, k = 1, .., K, m = 1, .., M.$$

Assume that $N$ individuals are sampled. Let $x_{nk}(t)$ be the genotype function of the $k^{th}$ gene from the $i^{th}$ individual. The eigenfunction expansion of the genotype function $x_{nk}(t)$ is

$$x_{nk}(t) = \sum_{l=1}^{L} \xi_{nkl}\psi_l(t), n = 1, .., N, \tag{5.59}$$

where $\psi_l(t)$ is eigenfunction and $\xi_{nkl}$ is expansion coefficients of the genotype function in the $k^{th}$ gene of the $n^{th}$ individual.

Define a methylation matrix $Y$, expansion coefficient matrix $X$ of the genotype functions, the coefficient matrix $\Gamma$ in the SEMs which connect the CpG sites, the coefficient matrix $B$ that measures the genetic effects of the genotype on the methylation levels and the error matrix $E$ as follows:

$$Y = \begin{bmatrix} y_{11} & \cdots & y_{1M} \\ \vdots & \vdots & \vdots \\ y_{T1} & \cdots & y_{TM} \end{bmatrix} = [y_1, y_2, \ldots, y_M]$$

$$X = \begin{bmatrix} \xi_{111} & \cdots & \xi_{11L} & \cdots & \xi_{1K1} & \cdots & \xi_{1KL} \\ \vdots & \vdots & \ddots & \vdots & \ddots & \vdots & \vdots \\ \xi_{T11} & \cdots & \xi_{T1L} & \cdots & \xi_{TK1} & \cdots & \xi_{TKL} \end{bmatrix} = [\xi_1, \ldots, \xi_K], \xi_K = \begin{bmatrix} \xi_{1k1} & \cdots & \xi_{1kL} \\ \vdots & \vdots & \vdots \\ \xi_{TK1} & \cdots & \xi_{TkL} \end{bmatrix}$$

$$\Gamma = \begin{bmatrix} r_{11} & r_{12} & \cdots & r_{1M} \\ \vdots & \vdots & & \vdots \\ r_{M1} & r_{M2} & \cdots & r_{MM} \end{bmatrix} = [\Gamma_1, \Gamma_2, \cdots, \Gamma_M]$$

$$B = \begin{bmatrix} b_{11}^{(1)} & \cdots & b_{11}^{(M)} \\ \vdots & \cdots & \vdots \\ b_{L1}^{(1)} & \cdots & b_{L1}^{(M)} \\ \vdots & \cdots & \vdots \\ b_{1K}^{(1)} & \cdots & b_{1K}^{(M)} \\ \vdots & \cdots & \vdots \\ b_{LK}^{(1)} & \cdots & b_{LK}^{(M)} \end{bmatrix} = [B_1, \ldots, B_M]$$

and

$$E = \begin{bmatrix} e_{11} & \cdots & e_{1M} \\ \vdots & \vdots & \vdots \\ e_{T1} & \cdots & e_{TM} \end{bmatrix} = [e_1, e_2, \ldots, e_M].$$

In matrix notation, the transformed SEMs (5.58) can be written as

$$Y\Gamma + XB + E = 0, \tag{5.60}$$

or

$$Y\Gamma_i + XB_i + e_i = 0, \ i = 1, \ldots, M. \tag{5.61}$$

### 5.5.1.2 The Two-Stage Least Squares Estimator

The parameters in the SEMs (5.60) can be estimated by the two-stage least squares method. After some algebra, the $i^{th}$ equation ($i = 1,\ldots,M$) in (5.61) can be written as

$$\begin{aligned} y_i &= Y_{-i}\gamma_i + \sum_{l=1}^{L} \xi_{1L} b_{L1}^{(i)} + \ldots + \sum_{l=1}^{L} \xi_{Kl} b_{LK}^{(i)} + e_i \\ &= Y_{-i}\gamma_i + XB_i + e_i \\ &= W_i \Delta_i + e_i, \end{aligned} \tag{5.62}$$

where $Y_{-i}$ is a vector of the endogenous (methylation) variables after removing variable $y_i$, $\gamma_i$ is a vector of the path coefficients associated with $Y_{-i}$, $W_i = [Y_{-i} \ \ X], \Delta_i = [\gamma_i^T \ \ B_i^T]^T$. Multiplying by the matrix $X^T$ on both sides of Equation 5.62, we obtain

$$X^T y_i = X^T Y_{-i}\gamma_i + X^T \left( \sum_{l=1}^{L} \xi_{1L} b_{L1}^{(i)} + \ldots + \sum_{l=1}^{L} \xi_{Kl} b_{LK}^{(i)} \right) + X^T e_i$$

or

$$X^T y_i = X^T Y_{-1}\gamma_i + X^T XB_i + X^T e_i, \tag{5.63}$$

where $\gamma_i = [\gamma_{1i}, \ldots, \gamma_{Mi}]^T$.

It is known that

$$\mathrm{cov}(X^T e_i, X^T e_i) = X^T X \sigma_{ii}.$$

Define objective function:

$$\begin{aligned} f(\Delta_i) &= \frac{1}{2} \left( X^T y_i - X^T Y_{-i}\gamma_i - X^T XB_i \right)^T \left[ \sigma_{ii} X^T X \right]^{-1} \left( X^T y_i - X^T Y_{-i}\gamma_i - X^T XB_i \right) \\ &= \frac{1}{2} \left( X^T y_i - X^T W_i \Delta_i \right)^T \left[ \sigma_{ii} X^T X \right]^{-1} \left( X^T y_i - X^T W_i \Delta_i \right). \end{aligned} \tag{5.64}$$

The generalized least square estimate $\hat{\Delta}_i$ is given by

$$
\begin{aligned}
\hat{\Delta}_i &= [W_i^T X (\sigma_i^2 X^T X)^{-1} X^T W_i]^{-1} W_i^T X (\sigma_i^2 X^T X)^{-1} X^T y_i \\
&= [W_i^T X (X^T X)^{-1} X^T W_i]^{-1} W_i^T X (X^T X)^{-1} X^T y_i.
\end{aligned}
\tag{5.65}
$$

The generalized least square estimate $\hat{\Delta}_i$ can be interpreted as a two-stage least square estimate (Judge et al. 1982; Section 1.2.3). The variance-covariance matrix of the estimate $\hat{\Delta}_i$ is given by

$$
\Lambda_i = \text{var}(\hat{\Delta}_i) = N \hat{\sigma}_i^2 [W_i^T X (X^T X)^{-1} X^T W_i]^{-1},
\tag{5.66}
$$

where $\hat{\sigma}_i^2$ is estimated by

$$
\hat{\sigma}_i^2 = \frac{1}{N - m_i + 1 - k_i} (y_i - W_i \hat{\Delta}_i)^T (y_i - W_i \hat{\Delta}_i),
\tag{5.67}
$$

with $m_i$ and $k_i$ being the number of endogenous and exogenous variables present in the $i^{th}$ equation, respectively.

Under fairly general conditions, $\sqrt{N}(\hat{\Delta}_i - \Delta_i)$ is asymptotically distributed as a normal distribution $N(0, \Lambda_i)$.

The presence of a directed edge can be tested. Suppose that we want to test whether the methylation level $y_j$ of the $j^{th}$ CpG site causes the changes of the methylation level $y_i$ of the $i^{th}$ CpG site. In other words, we want to test whether the parameter $\gamma_{ji}$ is equal to zero. Let $\Lambda_{ji}$ be the element of the variance-covariance matrix $\Lambda_i$ which corresponds to the variance of the estimator $\hat{\gamma}_{ji}$. If $j < i$, then $\Lambda_{ji}$ is the $j$th diagonal element of the matrix $\Lambda_i$ and if $j > i$, then $\Lambda_{ji}$ is the $j - 1$th diagonal element of the matrix $\Lambda_i$. Define the test statistics as

$$
T_\gamma = n \frac{\hat{\gamma}_{ji}^2}{\Lambda_{ji}}.
\tag{5.68}
$$

Under the null hypothesis of no directed connection from $y_j$ to $y_i$, the statistic $T_y$ is asymptotically distributed as a central $\chi_{(1)}^2$ distribution.

### 5.5.1.3 Sparse FSEMs

In general, the genotype-methylation networks are sparse. Therefore, $\Gamma$ and $B$ are sparse matrices. We use $l_1$ norm to penalize the parameters $r$'s connecting CpG sites and group lasso to penalize the parameters $b$'s that connect the genes to the CpG sites using a gene as a group. Therefore, the sparse functional SEMs can be formulated as solving the following optimization problem:

$$
\min \quad f(\Delta_i) + \lambda \left[ \| Z_{\gamma i} \|_1 + \sqrt{L} \sum_{k=1}^{K} \| Z_{ki} \|_2 \right]
\tag{5.69}
$$

subject to $\Delta_i - Z_i = 0$,

where

$$
B_i = \begin{bmatrix} b_{11}^{(i)} \\ \vdots \\ b_{L1}^{(i)} \\ \vdots \\ b_{1K}^{(i)} \\ \vdots \\ b_{LK}^{(i)} \end{bmatrix}, \Delta_i = \begin{bmatrix} \gamma_i \\ B_i \end{bmatrix} = \begin{bmatrix} \gamma_i \\ \Delta_{1i} \\ \vdots \\ \Delta_{Ki} \end{bmatrix}, \Delta_{ki} = \begin{bmatrix} \Delta_{1k}^{(i)} \\ \vdots \\ \Delta_{Lk}^{(i)} \end{bmatrix}, \Delta_{lk}^{(i)} = b_{lk}^{(i)},
$$

$$
Z_i = \begin{bmatrix} Z_{\gamma i} \\ Z_{1i} \\ \vdots \\ Z_{Ki} \end{bmatrix} = \Delta_i, Z_{\gamma_i} = \gamma_i, Z_{ki} = \Delta_{ki}, k = 1, \ldots, K, Z_{ki} = \begin{bmatrix} Z_{1k}^{(i)} \\ \vdots \\ Z_{LK}^{(i)} \end{bmatrix}.
$$

To solve the optimization problem (5.69), we form the augmented Lagrangian

$$
\begin{aligned}
L_\rho(\Delta_i, Z_i, \mu) &= f(\Delta_i) + \lambda \| Z_{\gamma i} \|_1 + \lambda \sum_{k=1}^{K} \sqrt{L} \| Z_{ki} \|_2 \\
&\quad + \mu^T (\Delta_i - Z_i) + \rho/2 \| \Delta_i - Z_i \|_2^2 \\
&= f(\Delta_i) + \lambda \| Z_{\gamma i} \|_1 + \lambda \sum_{k=1}^{K} \sqrt{L} \| Z_{ki} \|_2 \\
&\quad + \rho/2 \| \Delta_i - Z_i + \mu \|_2^2 - \rho/2 \| \mu \|_2^2
\end{aligned}
\tag{5.70}
$$

The alternating direction method of multipliers (ADMM) (Boyd et al. 2011) consists of the iterations:

$$
\Delta_i^{(l+1)} := \arg \min_{\Delta_i} L_\rho\left(\Delta_i, Z_i^{(l)}, \mu^{(l)}\right)
\tag{5.71}
$$

$$
Z_i^{(l+1)} := \arg \min_{Z_i} L_\rho\left(\Delta_i^{(l+1)}, Z_i, \mu^{(l)}\right)
\tag{5.72}
$$

$$
\mu^{(l+1)} := \mu^{(l)} + \Delta_i^{(l+1)} - Z_i^{(l+1)},
\tag{5.73}
$$

or

$$
\Delta_i^{(l+1)} := \arg \min_{\Delta_i} \left( f(\Delta_i) + \frac{\rho}{2} \| \Delta_i - Z_i^{(l)} + u^{(l)} \|_2^2 \right)
\tag{5.74}
$$

$$Z_i^{(l+1)} := \arg \min_{Z_i} \left( \lambda \| Z_{\gamma i} \|_1 + \lambda \sum_{k=1}^{K} \sqrt{L} \| Z_{ki} \|_2 + \rho/2 \| \Delta_i^{(l+1)} - Z_i + \mu^{(l)} \|_2^2 \right)$$

(5.75)

$$\mu^{(l+1)} := \mu^{(l)} + \Delta_i^{(l+1)} - Z_i^{(l+1)}.$$

(5.76)

Now we solve the minimization problem (5.71).

Let

$$\mu = \begin{bmatrix} \mu_\gamma \\ \mu_{1i} \\ \vdots \\ \mu_{Ki} \end{bmatrix}, \quad \mu_{ki} \begin{bmatrix} \mu_{1ki} \\ \vdots \\ \mu_{Lki} \end{bmatrix}$$

Set the partial derivative of $L_\rho$ to be equal to zero:

$$\frac{\partial L_\rho}{\partial r_i} = -W_i^T X [\sigma_{ii} X^T X]^{-1} (X^T y_i - X^T W \Delta_i) + \rho(\Delta_i - Z_i^{(l)} + \mu^{(l)})$$

$$= 0, \text{ which implies that } \Delta_i^{(l+1)}$$

$$= \left[ W_i^T X (X^T X)^{-1} X^T W_i + \rho I \right]^{-1} \left[ W_i^T X (X^T X)^{-1} X^T y_i + \rho \left( Z_i^{(l)} - \mu^{(l)} \right) \right]$$

(5.77)

The optimization problem (5.75) is non-differentiable. Although the first two terms in (5.75) are not differentiable, we still can obtain simple closed-form solutions to the problem (5.75) using subdifferential calculus. We first consider the generalized gradient of $\| Z_{\gamma i} \|_1$. Let $\Phi_i^{(m)}$ be a generalized derivative of the $m^{th}$ component of the vector $Z_{\gamma i}$:

$$\Phi_i^{(m)} = \begin{cases} 1 & Z_{\gamma i}^{(m)} = \gamma_i^{(m)} > 0 \\ [1,1] & Z_{\gamma i}^{(m)} = \gamma_i^{(m)} = 0 \\ -1 & Z_{\gamma i}^{(m)} = \gamma_i^{(m)} < 0 \end{cases}$$

Let $\Phi_i = \begin{bmatrix} \Phi_i^{(1)} \\ \vdots \\ \Phi_i^{(M)} \end{bmatrix}$.

Then, we have

$$\frac{\lambda}{\rho} \Phi_i + Z_{\gamma i} = \Delta_{\gamma i}^{(l+1)} + \mu_\gamma^{(l)},$$

which implies that

$$Z_{\gamma i}^{(l+1)} = \text{sgn}\left(\Delta_{\gamma i}^{(l+1)} + \mu_\gamma^{(l)}\right)\left(\left|\Delta_{\gamma i}^{(l+1)} + \mu_\gamma^{(l)}\right| - \frac{\lambda}{\rho}\right)_+, \qquad (5.78)$$

where

$$|x|_+ = \begin{cases} x & x \geq 0 \\ 0 & x < 0. \end{cases}$$

Next, we consider group lasso. The generalized gradient $\dfrac{\partial L_\rho}{\partial Z_{ki}}$ is given by

$$\frac{\partial L_\rho}{\partial Z_{ki}} = \lambda\sqrt{L}s + \rho\left(Z_{ki} - \left(\Delta_{ki}^{(l+1)} - \mu_{ki}^{(l)}\right)\right) = 0, \qquad (5.79)$$

where

$$s = \begin{cases} 0 & \|\Delta_{ki}^{(l+1)} - \mu_{ki}^{(l)}\|_2 < \dfrac{\lambda\sqrt{l}}{\rho} \\[3mm] \dfrac{Z_{ki}}{\|Z_{ki}\|_2} & \|\Delta_{ki}^{(l+1)} - \mu_{ki}^{(l)}\|_2 \geq \dfrac{\lambda\sqrt{l}}{\rho} \end{cases}$$

Equation (5.79) implies

$$\left(1 + \frac{\lambda\sqrt{L}}{\rho\|Z_{ki}\|_2}\right)Z_{ki} = \Delta_{ki}^{(l+1)} - \mu_{ki}^{(l)} \qquad (5.80)$$

or

$$\left(1 + \frac{\lambda\sqrt{L}}{\rho\|Z_{ki}\|_2}\right)\|Z_{ki}\|_2 = \|\Delta_{ki}^{(l+1)} - \mu_{ki}^{(l)}\|_2 \qquad (5.81)$$

Solving Equation 5.81, we obtain

$$\|Z_{ki}\|_2 = \|\Delta_{ki}^{(l+1)} - \mu_{ki}^{(l)}\|_2 - \frac{\lambda\sqrt{L}}{\rho}.$$

Thus, we have

$$\left(1 + \frac{\lambda\sqrt{L}}{\rho\|Z_{ki}\|_2}\right) = \frac{1}{1 - \dfrac{\lambda\sqrt{L}}{\rho\|\Delta_{ki}^{(l+1)} - \mu_{ki}^{(l)}\|_2}}. \qquad (5.82)$$

Combining Equations 5.80–5.82, we obtain

$$Z_{ki}^{(l+1)} = \left(1 - \frac{\lambda\sqrt{L}}{\rho\| \Delta_{ki}^{(l+1)} - \mu_{ki}^{(l)} \|_2}\right)_+ \left(\Delta_{ki}^{(l+1)} - \mu_{ki}^{(l)}\right), k = 1, \ldots, K. \tag{5.83}$$

Now the algorithm for parameter estimation in the sparse FSEMs can be summarized as Result 5.8.

**Result 5.8: Algorithm Construction of the Genotype-Methylation Networks Using FSEM**

For i =1,...,M
Step 1: Initialization

$$\mu^0 := 0$$
$$\Delta_i^0 := [W_i^T X(X^T X)^{-1} X^T W_i + \rho I]^{-1} W_i^T X(X^T X)^{-1} X^T y_i$$
$$Z_i^0 := \Delta_i^0,$$

where $\rho$ is a pre-specified parameter.
Carry out Steps 2, 3, and 4 until convergence.
Step 2:

$$\Delta_i^{(l+1)} = \left[W_i^T X\left(X^T X\right)^{-1} X^T W_i + \rho I\right]^{-1} \left[W_i^T X\left(X^T X\right)^{-1} X^T y_i + \rho\left(Z_i^{(l)} - \mu^{(l)}\right)\right]$$

Step 3:

$$Z_{\gamma i}^{(l+1)} = \text{sgn}\left(\Delta_{\gamma i}^{(l+1)} + \mu_\gamma^{(l)}\right)\left(\left|\Delta_{\gamma i}^{(l+1)} + \mu_\gamma^{(l)}\right| - \frac{\lambda}{\rho}\right)_+,$$

where

$$|x|_+ = \begin{cases} x & x \geq 0 \\ 0 & x < 0. \end{cases}$$

$$Z_{ki}^{(l+1)} = \left(-\frac{\lambda\sqrt{L}}{\rho\| \Delta_{ki}^{(l+1)} - \mu_{ki}^{(l)} \|_2}\right)_+ \left(\Delta_{ki}^{(l+1)} - \mu_{ki}^{(l)}\right), k = 1, \ldots, K.$$

Step 4:

$$\mu^{(l+1)} = \mu_i^{(l)} + \left(\Delta_i^{(l+1)} - Z_i^{(l+1)}\right).$$

## 5.5.2 Functional Structural Equation Models with Functional Endogenous Variables and Scalar Exogenous Variables (FSEMs)

### 5.5.2.1 Models

All RNA-seq, methylated DNA-seq (meth-seq), or simple densely distributed CpG site data can be taken as a function-valued trait (Chapter 3).

The read counts and methylated levels are treated as a function of genomic position. In this section, we propose to use functional structural equation models with functional endogenous variables and scalar exogenous variables (common SNPs) for construction of genotype-RNA-seq or genotype-meth-seq causal networks. For simplicity, we focus on the genotype-meth-seq causal network. However, the discussed models can be directly applied to the genotype-RNA-seq causal networks.

Suppose that there are $M$ methylated genes and $T$ individuals are sampled. Let $y_m(s_m), m = 1,...,M$ be the variable that represents the methylation level of the CpG site located at the genomic position $s_m$ and $y_{im}(s_m)$ be the value of $y_m(s_m)$ which the $i^{th}$ individual takes. We define $y_m(s_m)$ as a methylation function. Let $x_k$ be the indicator variable for the genotype of the $k$ SNP. Assume that all $M$ methylation functions are mapped into [0,1] interval.

Consider the FSEMs:

$$y_1(s)\gamma_{11} + ... + y_M(s)\gamma_{M1} + x_1\beta_{11}(s) + ... + x_K\beta_{K1}(s) + e_1(s) = 0$$

$$\vdots \qquad \vdots \qquad \vdots \qquad \vdots \qquad \vdots \qquad (5.84)$$

$$y_1(s)\gamma_{1M} + ... + y_M(s)\gamma_{MM} + x_1\beta_{1M}(s) + ... + x_K\beta_{KM}(s) + e_M(s) = 0,$$

where $s \in [0,1]$ is a mapped genomic position, $\beta_{km}(s)$ is the contribution of the $k$ SNP to the methylation level variation of the CpG site located at $s$ position of the $m^{th}$ gene, and $e_m(s)$ are the residuals.

We perform functional principle component analysis (FPCA) on all pooled methylation functions that are mapped into [0,1] interval. After FPCA, we obtain a set of eigenfunctions $\{\varphi_j(s)\}_{j=1}^{J}$. We expand $y_m(s)$, $\beta_{km}(s)$ and $e_m(s)$ in terms of eigenfunctions:

$$y_m(s) = \sum_{j=1}^{J} y_{mj}\varphi_j(s)$$

$$\beta_{km}(s) = \sum_{j=1}^{J} \eta_{kmj}\varphi_j(s) \qquad (5.85)$$

$$e_m(s) = \sum_{j=1}^{J} \varepsilon_{mj}\varphi_j(s),$$

where $y_{mj}, \eta_{kmj}, \varepsilon_{mj}$ are expansion coefficients of the methylation functions, the genetic effect functions, and residual functions.

Substituting expansions (5.85) into Equation 5.84 yields

$$\sum_{j=1}^{J} y_{1j}\varphi_j(s)r_{11} +...+ \sum_{j=1}^{J} y_{Mj}\varphi_j(s)r_{M1} + x_1\sum_{j=1}^{J}\eta_{11j}\varphi_j(s) + ... + x_K\sum_{j=1}^{J}\eta_{K1j}\varphi_j(s) + \sum_{j=1}^{J}\varepsilon_{1j}\varphi_j(s)=0$$

$$\vdots \qquad \vdots \qquad \vdots \qquad \vdots \qquad \vdots \qquad \vdots \qquad \vdots$$

$$\sum_{j=1}^{J} y_{1j}\varphi_j(s)r_{1M} +...+ \sum_{j=1}^{J} y_{Mj}\varphi_j(s)r_{MM} + x_1\sum_{j=1}^{J}\eta_{1Mj}\varphi_j(s) + ... + x_K\sum_{j=1}^{J}\eta_{KMj}\varphi_j(s) + \sum_{j=1}^{J}\varepsilon_{Mj}\varphi_j(s)=0$$

$$(5.86)$$

which implies that

$$y_{1j}r_{11} + \cdots + y_{Mj}r_{M1} + x_1\eta_{11j} + \cdots + x_K\eta_{K1j} + \varepsilon_{1j} = 0$$

$$\vdots \qquad\qquad \vdots \qquad\qquad \vdots \qquad\qquad \vdots \qquad\qquad (5.87)$$

$$y_{1j}r_{1M} + \cdots + y_{Mj}r_{MM} + x_1\eta_{1Mj} + \cdots + x_K\eta_{KMj} + \varepsilon_{Mj} = 0, \ j = 1,...,J.$$

Let

$$Y^{(j)} = \begin{bmatrix} y_{11j} & y_{12j} & \cdots & y_{1Mj} \\ \vdots & \vdots & & \vdots \\ y_{T1j} & y_{T2j} & \cdots & y_{TMj} \end{bmatrix} = \left[ y_1^{(j)}, y_2^{(j)}, \cdots, y_M^{(j)} \right]$$

$$X = \begin{bmatrix} X_{11} & X_{12} & \cdots & X_{1K} \\ \vdots & \vdots & & \vdots \\ X_{T1} & X_{T2} & \cdots & X_{TK} \end{bmatrix} = [x_1, x_2, \cdots, x_K]$$

$$E^{(j)} = \begin{bmatrix} \varepsilon_{11j} & \varepsilon_{12j} & \cdots & \varepsilon_{1Mj} \\ \vdots & \vdots & & \vdots \\ \varepsilon_{T1j} & \varepsilon_{T2j} & \cdots & \varepsilon_{TMj} \end{bmatrix} = \left[ \varepsilon_1^{(j)}, \varepsilon_2^{(j)}, \cdots, \varepsilon_M^{(j)} \right]$$

$$\Gamma = \begin{bmatrix} r_{11} & r_{12} & \cdots & r_{1M} \\ \vdots & \vdots & & \vdots \\ r_{M1} & r_{M2} & \cdots & r_{MM} \end{bmatrix} = [\Gamma_1, \Gamma_2, \cdots, \Gamma_M]$$

$$B^{(j)} = \begin{bmatrix} \eta_{11j} & \eta_{12j} & \cdots & \eta_{1Mj} \\ \eta_{21j} & l_{22j} & \cdots & \eta_{2Mj} \\ \vdots & \vdots & \ddots & \vdots \\ \eta_{K1j} & \eta_{K2j} & \cdots & \eta_{KMj} \end{bmatrix}.$$

After these definitions, Equation 5.87 can be rewritten in a matrix form:

$$Y^{(j)}\Gamma + XB^{(j)} + E^{(j)} = 0, \ j = 1, \cdots, J. \qquad (5.88)$$

### 5.5.2.2 The Two-Stage Least Squares Estimator

Similar to Section 5.5.1.2, the two-stage least squares method can be used to estimate the parameters in the FSEMs. The $j^{th}$ component and $i^{th}$ equation in 5.88 can be written as

$$y_i^{(j)} = Y_{-i}^{(j)}\gamma_i + \sum_{k=1}^{K} x_k \eta_{kij} + e_i^{(j)}$$

$$= Y_{-i}^{(j)}\gamma_i + X\eta_i^{(j)} + e_i^{(j)}, \ j = 1,...,J, . \qquad (5.89)$$

where $Y_{-i}^{(j)}$ is a vector of the $j^{th}$ component of the endogenous(methylation) variables after removing variable $y_i^{(j)}$ and $\gamma_i = [\gamma_{1i}, ..., \gamma_{Mi}]^T$.

Define

$$y_i = \begin{bmatrix} y_i^{(1)} \\ \vdots \\ y_i^{(J)} \end{bmatrix}, \ Y_{-i} = \begin{bmatrix} Y_{-i}^{(1)} \\ \vdots \\ Y_{-i}^{(J)} \end{bmatrix}, \ \eta_i = \begin{bmatrix} \eta_i^{(1)} \\ \vdots \\ \eta_i^{(J)} \end{bmatrix}, \ e_i = \begin{bmatrix} e_i^{(1)} \\ \vdots \\ e_i^{(J)} \end{bmatrix}, \ X_0$$

$$= \begin{bmatrix} X & \cdots & 0 \\ \vdots & \vdots & \vdots \\ 0 & \cdots & X \end{bmatrix}, \ W_i = [Y_{-i} \ X_0], \ \Delta_i = \begin{bmatrix} r \\ \eta_i \end{bmatrix}. \tag{5.90}$$

Then, Equation 5.89 can be written as

$$y_i = W_i \Delta_i + e_i. \tag{5.91}$$

Using the similar arguments as in Section 5.5.1.2, we obtain the following estimator:

$$\begin{aligned}
\hat{\Delta}_i &= \left[ W_i^T X_0 \left( \sigma_i^2 X_0^T X_0 \right)^{-1} X_0^T W_i \right]^{-1} W_i^T X_0 \left( \sigma_i^2 X_0^T X_0 \right)^{-1} X_0^T y_i \\
&= \left[ W_i^T X_0 \left( X_0^T X_0 \right)^{-1} X_0^T W_i \right]^{-1} W_i^T X_0 \left( X_0^T X_0 \right)^{-1} X_0^T y_i,
\end{aligned} \tag{5.92}$$

and

$$\Lambda_i = \mathrm{var}\left( \hat{\Delta}_i \right) = N \hat{\sigma}_i^2 \left[ W_i^T X_0 \left( X_0^T X_0 \right)^{-1} X_0^T W_i \right]^{-1}, \tag{5.93}$$

where $\hat{\sigma}_i^2$ is estimated by

$$\hat{\sigma}_i^2 = \frac{1}{NJ - m_i J + 1 - k_i} \left( y_i - W_i \hat{\Delta}_i \right)^T \left( y_i - W_i \hat{\Delta}_i \right), \tag{5.94}$$

with $m_i$ and $k_i$ being the number of endogenous and exogenous variables present in the $i^{th}$ equation, respectively.

Under fairly general conditions, $\sqrt{N}(\hat{\Delta}_i - \Delta_i)$ is asymptotically distributed as a normal distribution $N(0, \Lambda_i)$.

### 5.5.2.3 Sparse FSEMs

We use $l_1$ norm to penalize the parameters $r$'s connecting CpG sites and group lasso to penalize the parameters $\eta$'s that connect the SNPs to the methylated genes using a methylated gene as a group. Therefore, the sparse FSEMs can be formulated by solving the following optimization problem:

$$\min \quad f(\Delta_i) + \lambda\left[\|\,Z_{\gamma i}\,\|_1 + \sqrt{J}\sum_{k=1}^{K}\|\,Z_{ki}\,\|_2\right] \tag{5.95}$$

subject to $\Delta_i - Z_i = 0$,
where

$$f(\Delta_i) = \frac{1}{2}\sum_{j=1}^{J}\left(X^T y_i^{(j)} - X^T Y_{-i}^{(j)}\gamma_i - X^T X \eta_i^{(j)}\right)^T\left[\sigma_{ii}^{(j)} X^T X\right]^{-1}$$

$$\left(X^T y_i^{(j)} - X^T Y_{-i}^{(j)}\gamma_i - X^T X \eta_i^{(j)}\right)$$

$$\tilde{\Delta}_i = \begin{bmatrix}\tilde{\Delta}_{\gamma i}\\ \tilde{\Delta}_{1i}\\ \vdots\\ \tilde{\Delta}_{Ki}\end{bmatrix},\;\tilde{\Delta}_{\gamma i}=\gamma_i,\tilde{\Delta}_{ki}=\begin{bmatrix}\tilde{\Delta}_{ki}^{(1)}\\ \vdots\\ \tilde{\Delta}_{ki}^{(J)}\end{bmatrix},\;\tilde{\Delta}_{ki}^{(j)}=\eta_{ki}^{(j)},$$

$$Z_{ki}^{(j)} = \eta_{ki}^{(j)}, Z_i^{(j)} = \begin{bmatrix}Z_{1i}^{(j)}\\ \vdots\\ Z_{Ki}^{(j)}\end{bmatrix},\;Z_{ki}=\begin{bmatrix}Z_{ki}^{(1)}\\ \vdots\\ Z_{ki}^{(J)}\end{bmatrix},\;Z_i=\begin{bmatrix}Z_{\gamma i}\\ Z_{1i}\\ \vdots\\ Z_{Ki}\end{bmatrix}\text{ and }Z_{\gamma i}=\gamma_i$$

To solve the optimization problem (5.95), we form the augmented Lagrangian

$$\begin{aligned}L_\rho(\Delta_i, Z_i, \mu) &= f(\Delta_i) + \lambda\|\,Z_{\gamma i}\,\|_1 + \lambda\sum_{k=1}^{K}\sqrt{J}\|\,Z_{ki}\,\|_2\\ &\quad + \mu^T(\Delta_i - Z_i) + \rho/2\|\,\Delta_i - Z_i\,\|_2^2\\ &= f(\Delta_i) + \lambda\|\,Z_{\gamma i}\,\|_1 + \lambda\sum_{k=1}^{K}\sqrt{J}\|\,Z_{ki}\,\|_2\\ &\quad + \rho/2\|\,\Delta_i - Z_i + \mu\,\|_2^2 - \rho/2\|\,\mu\,\|_2^2.\end{aligned} \tag{5.96}$$

The alternating direction method of multipliers (ADMM) consists of the iterations:

$$\Delta_i^{(l+1)} := \arg\min_{\Delta_i} L_\rho\left(\Delta_i, Z_i^{(l)}, \mu^{(l)}\right) \tag{5.97}$$

$$Z_i^{(l+1)} := \arg\min_{Z_i} L_\rho\left(\Delta_i^{(l+1)}, Z_i, \mu^{(l)}\right) \tag{5.98}$$

$$\mu^{(l+1)} := \mu^{(l)} + \Delta_i^{(l+1)} - Z_i^{(l+1)}, \tag{5,99}$$

or

$$\Delta_i^{(l+1)} := \arg\min_{\Delta_i}\left(f(\Delta_i) + \frac{\rho}{2}\| \Delta_i - Z_i^{(l)} + \mu^{(l)} \|_2^2\right) \tag{5.100}$$

$$Z_i^{(l+1)} := \arg\min_{Z_i}\left(\lambda\| Z_{\gamma i} \|_1 + \lambda\sum_{k=1}^{K}\sqrt{J}\| Z_{ki} \|_2 + \rho/2\| \Delta_i^{(l+1)} - Z_i + \mu^{(l)} \|_2^2\right) \tag{5.101}$$

$$\mu^{(l+1)} := \mu^{(l)} + \Delta_i^{(l+1)} - Z_i^{(l+1)} \tag{5.102}$$

Let

$$\mu = \begin{bmatrix} \mu_\gamma \\ \mu_1 \\ \vdots \\ \mu_J \end{bmatrix} \text{ or } \mu = \begin{bmatrix} \mu_\gamma \\ \mu_{1i} \\ \vdots \\ \mu_{Ki} \end{bmatrix}.$$

We can show that the optimal solution to the problem (5.100) is (Exercise 7)

$$\hat{\gamma}_i = \left[A_1 - \sum_{j=1}^{J}B_1(j)(B_2(j) + \rho I)^{-1}A_2(j)\right]^{-1}D,$$

$$D = D_1 - \sum_{j=1}^{J}B_1(j)(B_2(j) + \rho I)^{-1}D_2(j), \tag{5.103}$$

$$\eta_i^{(j)} = [B_2(j) + \rho I]^{-1}[D_2(j) - A_2(j)\hat{\gamma}_i],$$

where

$$A_1 = \sum_{j=1}^{J}\left(Y_{-i}^{(j)}\right)^T X\left(\sigma_{ii}^{(j)}X^T X\right)^{-1}X^T Y_{-i}^{(j)},$$

$$A_2(j) = X^T X\left(\sigma_{ii}^{(j)}X^T X\right)^{-1}X^T Y_{-i}^{(j)},$$

$$B_1(j) = \left(Y_{-i}^{(j)}\right)^T X\left(\sigma_{ii}^{(j)}X^T X\right)^{-1}X^T X,$$

$$B_2(j) = X^T X\left(\sigma_{ii}^{(j)}X^T X\right)^{-1}X^T X,$$

$$D_1 = \sum_{j=1}^{J}\left(Y_{-i}^{(j)}\right)^T X\left(\sigma_{ii}^{(j)}X^T X\right)^{-1}X^T y_i^{(j)} + \rho\left(Z_{i\gamma}^{(l)} - \mu_\gamma^{(l)}\right),$$

$$D_2(j) = X^T X\left(\sigma_{ii}^{(j)}X^T X\right)^{-1}X^T y_i^{(j)} + \rho\left(Z_i^{(j)(l)} - \mu_j^{(l)}\right).$$

The optimization problem (5.101) is non-differentiable. Although the first two terms in (5.101) are not differentiable, we still can obtain simple closed-form solutions to the problem (5.101) using subdifferential calculus. We first consider the generalized gradient of $\| Z_{\gamma i} \|_1$. Let $\Phi_i^{(m)}$ be a generalized derivative of the $m$-th component of the vector $Z_{\gamma i}$:

$$\Phi_i^{(m)} = \begin{cases} 1 & Z_{\gamma i}^{(m)} = \gamma_i^{(m)} > 0 \\ [-1,1] & Z_{\gamma i}^{(m)} = \gamma_i^{(m)} = 0 \\ -1 & Z_{\gamma i}^{(m)} = \gamma_i^{(m)} < 0 \end{cases}$$

let $\Phi_i = \begin{bmatrix} \Phi_i^{(1)} \\ \vdots \\ \Phi_i^{(M)} \end{bmatrix}$.

Then, we can show that the optimal solution to the problem (5.101) is

$$Z_{\lambda i}^{(l+1)} = \text{sgn}(\Delta_{\gamma i}^{(l+1)} + \mu_\gamma^{(l)})(|\Delta_{\gamma i}^{(l+1)} + \mu_\gamma^{(l)}| - \frac{\lambda}{\rho})_+, \tag{5.104}$$

where

$$|x|_+ = \begin{cases} x & x \geq 0 \\ 0 & x < 0 \end{cases}.$$

Next, we consider group lasso. The generalized gradient $\dfrac{\partial L_\rho}{\partial Z_{ki}}$ is given by

$$\frac{\partial L_\rho}{\partial Z_{ki}} = \lambda \sqrt{J} s + \rho \left( Z_{ki} - \left( \Delta_{ki}^{(l+1)} - \mu_{ki}^{(l)} \right) \right) = 0, \tag{5.105}$$

where

$$s = \begin{cases} 0 & \| \Delta_{ki}^{(l+1)} - \mu_{ki}^{(l)} \|_2 < \dfrac{\lambda \sqrt{J}}{\rho} \\[2ex] \dfrac{Z_{ki}}{\| Z_{ki} \|_2} & \| \Delta_{ki}^{(l+1)} - \mu_{ki}^{(l)} \|_2 \geq \dfrac{\lambda \sqrt{J}}{\rho} \end{cases}.$$

We can show Equation 5.105 implies (Exercise 9)

$$\left( 1 + \frac{\lambda \sqrt{J}}{\rho \| Z_{ki} \|_2} \right) Z_{ki} = \Delta_{ki}^{(l+1)} - \mu_{ki}^{(l)}. \tag{5.106}$$

or

$$\left( 1 + \frac{\lambda \sqrt{J}}{\rho \| Z_{ki} \|_2} \right) \| Z_{ki} \|_2 = \| \Delta_{ki}^{(l+1)} - \mu_{ki}^{(l)} \|_2 \tag{5.107}$$

Solving Equation 5.107, we obtain

$$\| Z_{ki} \|_2 = \| \Delta_{ki}^{(l+1)} - \mu_{ki}^{(l)} \|_2 - \frac{\lambda \sqrt{J}}{\rho}.$$

Thus, we have

$$\left(1 + \frac{\lambda \sqrt{J}}{\rho \| Z_{ki} \|_2}\right) = \frac{1}{1 - \dfrac{\lambda \sqrt{J}}{\rho \| \Delta_{ki}^{(l+1)} - \mu_{ki}^{(l)} \|_2}} \tag{5.108}$$

Combining Equations 5.106–5.108, we obtain

$$Z_{ki}^{(l+1)} = \left(-\frac{\lambda \sqrt{J}}{\rho \| \Delta_{ki}^{(l+1)} - \mu_{ki}^{(l)} \|_2}\right)_+ \left(\Delta_{ki}^{(l+1)} - \mu_{ki}^{(l)}\right), k = 1, \ldots, K. \tag{5.109}$$

Summarizing the above derivation, we obtain Result 5.9.

**Result 5.9: Algorithm for Construction of Genotype-Methylation Causal Networks Where Methylation Level is Taken as a Function-Valued Trait**

For $i = 1, \ldots, M$,
Step 1: Initialization

$$B_1(j) = \left(Y_{-i}^{(j)}\right)^T X \left(\sigma_{ii}^{(j)} X^T X\right)^{-1} X^T X,$$

$$B_2(j) = X^T X \left(\sigma_{ii}^{(j)} X^T X\right)^{-1} X^T X,$$

$$\tilde{D}_1 = \sum_{j=1}^J \left(Y_{-i}^{(j)}\right)^T X \left(\sigma_{ii}^{(j)} X^T X\right)^{-1} X^T y_i^{(j)},$$

$$\tilde{D}_2 = X^T X \left(\sigma_{ii}^{(j)} X^T X\right)^{-1} X^T y_i^{(j)}.$$

$$\mu^0 := 0$$

$$\gamma_i^{(0)} = \left[A_1 - \sum_{j=1}^J B_1(j)(B_2(j) + \rho I)^{-1} A_2(j)\right]^{-1} \tilde{D},$$

$$\tilde{D} = \tilde{D}_1 - \sum_{j=1}^J B_1(j)(B_2(j) + \rho I)^{-1} \tilde{D}_2(j),$$

$$\eta_i^{(0)(j)} = [B_2(j) + \rho I]^{-1} \left[D_2(j) - A_2(j)\hat{\gamma}_i^{(0)}\right], j = 1, \ldots, J,$$

$$\Delta_i^{(0)} = \begin{bmatrix} \gamma_i^{(0)} \\ \eta_i^{(0)(1)} \\ \vdots \\ \eta_i^{(0)(J)} \end{bmatrix}$$

$$Z_i^{(0)} := \Delta_i^{(0)}$$

$$\tilde{\Delta}_i^{(0)} = \begin{bmatrix} \tilde{\Delta}_{\gamma i}^{(0)} \\ \tilde{\Delta}_{1i}^{(0)} \\ \vdots \\ \tilde{\Delta}_{Ki}^{(0)} \end{bmatrix}, \tilde{\Lambda}_{\gamma i}^{(0)} = \hat{\gamma}_i^{(0)}, \tilde{\Delta}_{ki}^{(0)} = \begin{bmatrix} \tilde{\Delta}_{ki}^{(0)(1)} \\ \vdots \\ \tilde{\Delta}_{ki}^{(0)(J)} \end{bmatrix}, \tilde{\Delta}_{ki}^{(0)(j)} = \eta_{ki}^{(0)(j)},$$

$$\mu^{(0)} = \begin{bmatrix} \mu_{\gamma}^{(0)} \\ \mu_1^{(0)} \\ \vdots \\ \mu_J^{(0)} \end{bmatrix}.$$

Carry out steps 2, 3, and 4 until convergence
Step 2:

$$A_1 = \sum_{j=1}^J \left(Y_{-i}^{(j)}\right)^T X \left(\sigma_{ii}^{(j)} X^T X\right)^{-1} X^T Y_{-i}^{(j)},$$
$$A_2(j) = X^T X \left(\sigma_{ii}^{(j)} X^T X\right)^{-1} X^T Y_{-i}^{(j)},$$

$$B_1(j) = \left(Y_{-i}^{(j)}\right)^T X \left(\sigma_{ii}^{(j)} X^T X\right)^{-1} X^T X,$$

$$B_2(j) = X^T X \left(\sigma_{ii}^{(j)} X^T X\right)^{-1} X^T X,$$

$$D_1 = \sum_{j=1}^J \left(Y_{-i}^{(j)}\right)^T X \left(\sigma_{ii}^{(j)} X^T X\right)^{-1} X^T y_i^{(j)} + \rho\left(Z_i^{(j)(l)} - \mu_{\gamma}^{(l)}\right),$$

$$D_2(j) = X^T X \left(\sigma_{ii}^{(j)} X^T\right)^{-1} X^T y_i^{(j)} + \rho\left(Z_i^{(j)(l)} - \mu_j^{(l)}\right)$$

$$\hat{\gamma}_i^{(l+1)} = \left[A_1 - \sum_{j=1}^J B_1(j)(B_2(j) + \rho I)^{-1} A_2(j)\right]^{-1} D,$$

$$D = D_1 - \sum_{j=1}^J B_1(j)(B_2(j) + \rho I)^{-1} D_2(j),$$

$$\eta_i^{(j)(l+1)} = [B_2(j) + \rho I]^{-1} \left[D_2(j) - A_2(j)\hat{\gamma}_i^{(l+1)}\right], j = 1, \dots, J,$$

$$\Delta_i^{(l+1)} = \begin{bmatrix} \hat{\gamma}_i^{(l+1)} \\ \eta_i^{(1)(l+1)} \\ \vdots \\ \eta_i^{(J)(l+1)} \end{bmatrix}$$

$$\tilde{\Delta}_i^{(l+1)} = \begin{bmatrix} \tilde{\Delta}_{\gamma i}^{(l+1)} \\ \tilde{\Delta}_{1i}^{(l+1)} \\ \vdots \\ \tilde{\Delta}_{Ki}^{(l+1)} \end{bmatrix}, \tilde{\Delta}_{\gamma i}^{(l+1)} = \hat{\gamma}_i^{(l+1)}, \tilde{\Delta}_{ki}^{(l+1)} = \begin{bmatrix} \tilde{\Delta}_{ki}^{(l+1)(1)} \\ \vdots \\ \tilde{\Delta}_{ki}^{(l+1)(J)} \end{bmatrix}, \tilde{\Delta}_{ki}^{(l+1)(j)} = \eta_{ki}^{(j)(l+1)},$$

Step 3:

$$\tilde{Z}_{\gamma i}^{(l+1)} = \text{sgn}\left(\tilde{\Delta}_{\gamma i}^{(l+1)} + \mu_\gamma^{(l)}\right)\left(\left|\tilde{\Delta}_{\gamma i}^{(l+1)} + \mu_\gamma^{(l)}\right| - \frac{\lambda}{\rho}\right)_+,$$

where

$$|x|_+ = \begin{cases} x & x \geq 0 \\ 0 & x < 0. \end{cases}$$

$$\tilde{Z}_{ki}^{(l+1)} = \left(-\frac{\lambda\sqrt{J}}{\rho\|\tilde{\Delta}_{ki}^{(l+1)} - \mu_{ki}^{(l)}\|_2}\right)_+ \left(\tilde{\Delta}_{ki}^{(l+1)} - \mu_{ki}^{(l)}\right), k = 1, ..., K.$$

$$\tilde{Z}_{ki}^{(l+1)} = \begin{bmatrix} \tilde{Z}_{\gamma i}^{(l+1)} \\ \tilde{Z}_{1i}^{(l+1)} \\ \vdots \\ \tilde{Z}_{Ki}^{(l+1)} \end{bmatrix}, \tilde{Z}_{ki}^{(l+1)} = \begin{bmatrix} \tilde{Z}_{ki}^{(1)(l+1)} \\ \vdots \\ \tilde{Z}_{ki}^{(J)(l+1)} \end{bmatrix}.$$

Let

$$Z_{ki}^{(j)(l+1)} = \tilde{Z}_{ki}^{(j)(l+1)}, Z_{\gamma i}^{(l+1)} = \tilde{Z}_{\gamma i}^{(l+1)}, Z_i^{(j)(l+1)} = \begin{bmatrix} Z_{1i}^{(j)(l+1)} \\ \vdots \\ Z_{Ki}^{(j)(l+1)} \end{bmatrix}, Z_i^{(l+1)} = \begin{bmatrix} Z_{\gamma i}^{(l+1)} \\ Z_i^{(1)(l+1)} \\ \vdots \\ Z_i^{(J)(l+1)} \end{bmatrix}.$$

Step 4:

$$\tilde{\mu}_i^{(l+1)} = \tilde{\mu}_i^{(l)} + \left(\tilde{\Delta}_i^{(l+1)} - \tilde{Z}_i^{(l+1)}\right),$$

where

$$\tilde{u}^{(l+1)} = \begin{bmatrix} \tilde{\mu}_\gamma^{(l+1)} \\ \tilde{\mu}_1^{(l+1)} \\ \vdots \\ \tilde{\mu}_K^{(l+1)} \end{bmatrix}, \tilde{\mu}_k^{(l+1)} = \begin{bmatrix} \tilde{\mu}_k^{(1)(l+1)} \\ \vdots \\ \tilde{\mu}_k^{(J)(l+1)} \end{bmatrix}, k = 1, ..., K.$$

Let

$$\mu^{(l+1)} = \begin{bmatrix} \mu_\gamma^{(l+1)} \\ \mu_1^{(l+1)} \\ \vdots \\ \mu_J^{(l+1)} \end{bmatrix}, \mu_\gamma^{(l+1)} = \tilde{\mu}_\gamma^{(l+1)}, \mu_j^{(l+1)} = \begin{bmatrix} \mu_1^{(j)(l+1)} \\ \vdots \\ \mu_K^{(j)(l+1)} \end{bmatrix}, \mu_k^{(j)(l+1)} = \tilde{\mu}_k^{(j)(l+1)}.$$

$l := l + 1$; go to step 2.

$$\mu^{(l+1)} = \mu_i^{(l)} + \left( \Delta_i^{(l+1)} - Z_i^{(l+1)} \right).$$

## 5.5.3 Functional Structural Equation Models with Both Functional Endogenous Variables and Exogenous Variables (FSEMF)

### 5.5.3.1 Model

Now we consider meth-seq and NGS genotype data. The genotype-methylation networks connect methylated genes to methylated genes, and genes that contain densely distributed SNPs to the methylated genes. Suppose that there are $M$ methylated genes, $K$ sequenced genes including rare variants or both common and rare variants, and $T$ individuals are sampled. Let $y_m(s_m)$, $m = 1,...,M$ be the variable that represents the methylation level of the CpG site located at the genomic position $s_m$ and $y_{im}(s_m)$ be the value of $y_m(s_m)$ which the $i^{th}$ individual takes. We define $y_m(s_m)$ as a methylation function. Let $x_k(t)$ be the indicator variable for the genotype of the SNP located at genome position $t$. Assume that all $M$ methylation functions and $K$ genotype functions are mapped into $[0,1]$ interval.

Define a FSEMF:

$$y_1(s)r_{11} + \cdots + y_M(s)r_{M1} + \underset{T}{\int} x_1(t)\beta_{11}(s, t)dt + \cdots + \underset{T}{\int} x_K(t)\beta_{K1}(s, t)dt + e_1(s) = 0$$

$$\vdots \qquad \vdots \qquad \vdots \qquad \vdots \qquad \vdots \qquad (5.110)$$

$$y_1(s)r_{1M} + \cdots + y_M(s)r_{MM} + \underset{T}{\int} x_1(t)\beta_{1M}(s, t)dt + \cdots + \underset{T}{\int} x_K(t)\beta_{KM}(s, t)dt + e_M(s) = 0,$$

where $s \in [0,1]$ is a mapped genomic position, $\beta_{km}(s,t)$ is the contribution of the SNP located at $t$ genomic position of the $k^{th}$ gene to the methylation level variation of the CpG site located at $s$ position of the $m^{th}$ gene.

To transform the FSEMF to the classical SEM, we first perform functional principle component analysis (FPCA) on all pooled genotype functions that are mapped into $[0,1]$ interval. After FPCA, we expand $x_k(t)$ in terms of orthonormal eigenfunction $\psi_l(t)$, we obtain

$$x_k(t) = \sum_{l=1}^{L} \xi_{kl} \psi_l(t). \qquad (5.111)$$

Substituting expansion (5.111) into model (5.110) yields

$$y_1(s)\gamma_{11} + ... y_M(s)\gamma_{M1} + \sum_{l=1}^{L} \xi_{1l} b_{l1}^{(1)}(s) + ... + \sum_{l=1}^{L} \xi_{Kl} b_{lK}^{(1)}(s) + e_1(s) = 0$$

$$\vdots \qquad \vdots \qquad \vdots \qquad \vdots \qquad \vdots \qquad (5.112)$$

$$y_1(s)\gamma_{1M} + ... y_M(s)\gamma_{MM} + \sum_{l=1}^{L} \xi_{1l} b_{l1}^{(M)}(s) + ... + \sum_{l=1}^{L} \xi_{Kl} b_{lK}^{(M)}(s) + e_M(s) = 0,$$

where

$$b_{lk}^{(m)}(s) = \int_T \beta_{km}(s,t)\psi_l(t)dt, \ m = 1,..,M, l = 1,..,L, k = 1,..,K. \tag{5.113}$$

Next, we perform FPCA on all pooled methylation functions, which leads to a set of eigenfunctions $\phi_j(s)$, $j = 1,...,J$. Then, we expand $y_i(s)$, $b_{lk}^{(m)}(s)$ and $e_i(s)$ in terms of eigenfunctions or other basis functions $\phi_j(s)$:

$$y_i(s) = \sum_{j=1}^{J} y_{ij}\phi_j(s), \ b_{lk}^{(m)}(s) = \sum_{j=1}^{J} \eta_{lkm}^{(j)}\phi_j(s) \text{ and } e_i(s) = \sum_{j=1}^{J} \varepsilon_{ij}\phi_j(s). \tag{5.114}$$

Substituting expansions (5.114) into Equation 5.112, we obtain

$$y_{1j}\gamma_{11} + \cdots + y_{Mj}\gamma_{M1} + \sum_{l=1}^{L}\xi_{1l}\eta_{l11}^{(j)} + \cdots + \sum_{l=1}^{L}\xi_{Kl}\eta_{lK1}^{(j)} + \varepsilon_{1j} = 0$$

$$\vdots \qquad\qquad \vdots \qquad\qquad \vdots \qquad\qquad \vdots \tag{5.115}$$

$$y_{1j}\gamma_{1M} + \cdots + y_{Mj}\gamma_{MM} + \sum_{l=1}^{L}\xi_{1l}\eta_{l1M}^{(j)} + \cdots + \sum_{l=1}^{L}\xi_{Kl}\eta_{lKM}^{(j)} + \varepsilon_{Mj} = 0, j = 1,..,J$$

For the $i^{th}$ individual, we expand $y_{ni}(s)$ and $x_{nk}(t)$, $n = 1,...,T$ in terms of eigenfunctions $\phi_j(s)$ and $\psi_l(t)$:

$$y_{ni}(s) = \sum_{j=1}^{L} y_{nij}\phi_j(s) \text{ and } x_{nk}(t) = \sum_{l=1}^{L}\xi_{nkl}\psi_l(t), n = 1,...,T.$$

Define the $j^{th}$ component expansion coefficient matrix $Y^{(j)}$ of the methylation functions, the expansion coefficient matrix $X$ of the genotype functions, structural parameter matrix $\Gamma$ connecting methylated genes, structural parameter matrix $B^{(j)}$ connecting genes to the methylated genes, and error matrix $E^{(j)}$ as follows.

$$Y^{(j)} = \begin{bmatrix} y_{11j} & y_{12j} & \cdots & y_{1Mj} \\ \vdots & \vdots & & \vdots \\ y_{T1j} & y_{T2j} & \cdots & y_{TMj} \end{bmatrix} = \begin{bmatrix} y_1^{(j)}, y_2^{(j)}, \cdots, y_M^{(j)} \end{bmatrix}$$

$$X = \begin{bmatrix} \xi_{111} & \cdots & \xi_{11L} & \cdots & \xi_{1K1} & \cdots & \xi_{1KL} \\ \vdots & \vdots & \ddots & \vdots & & \vdots & \vdots \\ \xi_{T11} & \cdots & \xi_{T1L} & \cdots & \xi_{TK1} & \cdots & \xi_{TKL} \end{bmatrix} = [\xi_1,...,\xi_K], \xi_k = \begin{bmatrix} \xi_{1k1} & \cdots & \xi_{1kL} \\ \vdots & \vdots & \vdots \\ \xi_{Tk1} & \cdots & \xi_{TkL} \end{bmatrix}$$

$$E^{(j)} - \begin{bmatrix} \varepsilon_{11j} & \varepsilon_{12j} & \cdots & \varepsilon_{1Mj} \\ \vdots & \vdots & & \vdots \\ \varepsilon_{T1j} & \varepsilon_{T2j} & \cdots & \varepsilon_{TMj} \end{bmatrix} = \begin{bmatrix} \varepsilon_1^{(j)}, \varepsilon_2^{(j)}, \cdots, \varepsilon_M^{(j)} \end{bmatrix}$$

$$\Gamma = \begin{bmatrix} r_{11} & r_{12} & \cdots & r_{1M} \\ \vdots & \vdots & & \vdots \\ r_{M1} & r_{M2} & \cdots & r_{MM} \end{bmatrix} = [\Gamma_1, \Gamma_2, \cdots, \Gamma_M]$$

$$B^{(j)} = \begin{bmatrix} \eta_{111}^{(j)} & \cdots & \eta_{11M}^{(j)} \\ \vdots & \cdots & \vdots \\ \eta_{L11}^{(j)} & \cdots & \eta_{L1M}^{(j)} \\ \vdots & \cdots & \vdots \\ \eta_{1K1}^{(j)} & \cdots & \eta_{1KM}^{(j)} \\ \vdots & \cdots & \vdots \\ \eta_{LK1}^{(j)} & \cdots & \eta_{LKM}^{(j)} \end{bmatrix} = \left[ \eta_1^{(j)}, \ldots, \eta_M^{(j)} \right].$$

The structural equation in (5.115) can be written in a matrix form:

$$Y^{(j)}\Gamma + XB^{(j)} + E^{(j)} = 0, \ j = 1, \cdots, J. \tag{5.116}$$

The $i^{th}$ equation consisting of $j^{th}$ component in Equation 5.116 is given by

$$y_i^{(j)} = Y_{-i}^{(j)}\gamma_i + \sum_{l=1}^{L} \xi_{1l}\eta_{l1i}^{(j)} + \ldots + \sum_{l=1}^{L} \xi_{Kl}\eta_{lKi}^{(j)} + e_i^{(j)}. \tag{5.117}$$

Let

$$\eta_i^{(j)} = \begin{bmatrix} \eta_{1i}^{(j)} \\ \vdots \\ \eta_{Ki}^{(j)} \end{bmatrix}, \eta_{ki}^{(j)} = \begin{bmatrix} \eta_{1ki}^{(j)} \\ \vdots \\ \eta_{Lki}^{(ij)} \end{bmatrix} \Delta_i = \begin{bmatrix} \gamma_i \\ \eta_i^{(1)} \\ \vdots \\ \eta_i^{(J)} \end{bmatrix}.$$

Then, Equation 5.117 can be reduced to

$$y_i^{(j)} = Y_{-i}^{(j)}\gamma_i + X\eta_i^{(j)} + e_i^{(j)}, \ j = 1, .., J, \tag{5.118}$$

where $\gamma_i = [\gamma_{1i}, \ldots, \gamma_{Mi}]^T$.

Multiplying by $X^T$ on both sides of Equation 5.118, we obtain

$$X^T y_i^{(j)} = X^T Y_{-i}^{(j)}\gamma_i + X^l X\eta_i^{(j)} + X^T e_i^{(j)}, j = 1, \ldots, J. \tag{5.119}$$

### 5.5.3.2 Sparse FSEMF for the Estimation of Genotype-Methylation Networks with Sequencing Data

Similar to Section 5.5.2.3, to formulate the sparse FSEMF we first define the objective function for fitting Equation 5.119:

$$f(\Delta_i) = \frac{1}{2} \sum_{j=1}^{J} \left( X^T y_i^{(j)} - X^T Y_{-i}^{(j)} \gamma_i - X^T X \eta_i^{(j)} \right)^T \left[ \sigma_{ii}^{(j)} X^T X \right]^{-1}$$

$$\left( X^T y_i^{(j)} - X^T Y_{-i}^{(j)} \gamma_i - X^T X \eta_i^{(j)} \right). \tag{5.120}$$

We use $l_1$ to penalize $\gamma_i$ that connect the methylated genes and group lasso to penalize $\eta_i^{(j)}$ that connect the genes to the methylated genes. The sparse FSEMF to accomplish this goal is given by

$$\min \quad f(\Delta_i) + \lambda \left[ \| Z_{\gamma i} \|_1 + \sqrt{JL} \sum_{k=1}^{K} \| Z_{ki} \|_2 \right] \tag{5.121}$$

subject to $\Delta_i - Z_i = 0$,
where

$$\tilde{\Delta}_i = \begin{bmatrix} \tilde{\Delta}_{\gamma_i} \\ \tilde{\Delta}_{1i} \\ \vdots \\ \tilde{\Delta}_{Ki} \end{bmatrix}, \tilde{\Delta}_{\gamma_i} = \gamma_i, \tilde{\Delta}_{ki} = \begin{bmatrix} \tilde{\Delta}_{1ki} \\ \vdots \\ \tilde{\Delta}_{Lki} \end{bmatrix}, \tilde{\Delta}_{lki} = \begin{bmatrix} \tilde{\Delta}_{lki}^{(1)} \\ \vdots \\ \tilde{\Delta}_{lki}^{(J)} \end{bmatrix}, \tilde{\Delta}_{lki}^{(j)} = \eta_{lki}^{(j)},$$

$$Z_{lki}^{(j)} = \eta_{lki}^{(j)}, Z_i^{(j)} = \begin{bmatrix} Z_{1i}^{(j)} \\ \vdots \\ Z_{Ki}^{(j)} \end{bmatrix}, Z_{ki} = \begin{bmatrix} Z_{1ki} \\ \vdots \\ Z_{Lki} \end{bmatrix}, Z_{lki} = \begin{bmatrix} Z_{lki}^{(1)} \\ \vdots \\ Z_{lki}^{(J)} \end{bmatrix}, Z_i = \begin{bmatrix} Z_{\gamma i} \\ Z_{1i} \\ \vdots \\ Z_{Ki} \end{bmatrix} \text{ and}$$

$$Z_{\gamma i} = \gamma_i$$

To solve the optimization problem (5.121), we form the augmented Lagrangian

$$L_\rho(\Delta_i, Z_i, \mu) = f(\Delta_i) + \lambda \| Z_{\gamma i} \|_1 + \lambda \sum_{k=1}^{K} \sqrt{J} \| Z_{ki} \|_2$$

$$+ \rho/2 \| \Delta_i - Z_i + \mu \|_2^2 - \rho/2 \| \mu \|_2^2 \tag{5.122}$$

The alternating direction method of multipliers (ADMM) for solving the optimization problem consists of the iterations:

$$\Delta_i^{(l+1)} := \arg\min_{\Delta_i} L_\rho\left(\Delta_i, Z_i^{(l)}, \mu^{(l)}\right) \tag{5.123}$$

$$Z_i^{(l+1)} := \arg\min_{Z_i} L_\rho\left(\Delta_i^{(l+1)}, Z_i, \mu^{(l)}\right) \tag{5.124}$$

$$\mu^{(l+1)} := \mu^{(l)} + \Delta_i^{(l+1)} - Z_i^{(l+1)}. \tag{5.125}$$

or

$$\Delta_i^{(l+1)} := \arg\min_{\Delta_i}\left(f(\Delta_i) + \frac{\rho}{2}\| \Delta_i - Z_i^{(l)} + u^{(l)} \|_2^2\right) \tag{5.126}$$

$$Z_i^{(l+1)} := \arg\min_{Z_i}\left(\lambda\| Z_{\gamma i} \|_1 + \lambda + \sum_{k=1}^{K} \sqrt{J}\| Z_{ki} \|_2 + \rho/2\| \Delta_i^{(l+1)} - Z_i + \mu^{(l)} \|_2^2\right) \tag{5.127}$$

$$\mu^{(l+1)} := \mu^{(l)} + \Delta_i^{(l+1)} - Z_i^{(l+1)}. \tag{5.128}$$

Now we solve minimization problem (5.126). The optimization problem (5.126) involves only differentiable functions. Simply using calculus, we can obtain the solutions.

Let

$$\mu = \begin{bmatrix} \mu_\gamma \\ \mu_1 \\ \vdots \\ \mu_J \end{bmatrix} \text{ or } \mu = \begin{bmatrix} \mu_\gamma \\ \mu_{1i} \\ \vdots \\ \mu_{Ki} \end{bmatrix}.$$

Setting the partial derivative of $L_\rho$ to be zero, we obtain

$$\frac{\partial L_\rho}{\partial r_i} = -\sum_{j=1}^{J}(Y_{-i}^{(j)})^T X(\sigma_{ii}^{(j)} X^T X)^{-1}(X^T y_i^{(j)} - X^T Y_{-i}^{(j)}\gamma_i - X^T X\eta_i^{(j)})$$

$$+ \rho(\gamma_i - Z_{\gamma i}^{(l)} + \mu_\gamma^{(l)}) = 0$$

$$\frac{\partial L_\rho}{\partial \eta_i^{(j)}} = -X^T X(\sigma_{ii}^{(j)} X^T X)^{-1}(X^T y_i^{(j)} - X^T Y_{-i}^{(j)}\gamma_i - X^T X\eta_i^{(j)}) \tag{5.129}$$

$$+ \rho(\eta_i^{(j)} - Z_i^{(j)(l)} + \mu_j^{(l)}) = 0, j = 1, .., J.$$

Equation 5.129 can be transformed to

$$(A_1 + \rho I)\gamma_i + \sum_{j=1}^{J} B_1(j)\eta_i^{(j)} = D_1$$

$$A_2(j)\gamma_i + [B_2(j) + \rho I]\eta_i^{(j)} = D_2(j), j = 1, ..., J, \tag{5.130}$$

where

$$A_1 = \sum_{j=1}^{J} \left( Y_{-i}^{(j)} \right)^T X \left( \sigma_{ii}^{(j)} X^T X \right)^{-1} X^T Y_{-i}^{(j)},$$

$$A_2(j) = X^T X \left( \sigma_{ii}^{(j)} X^T X \right)^{-1} X^T Y_{-i}^{(j)},$$

$$B_1(j) = \left( Y_{-i}^{(j)} \right)^T X \left( \sigma_{ii}^{(j)} X^T X \right)^{-1} X^T X,$$

$$B_2(j) = X^T X \left( \sigma_{ii}^{(j)} X^T X \right)^{-1} X^T X,$$

$$D_1 = \sum_{j=1}^{J} \left( Y_{-i}^{(j)} \right)^T X \left( \sigma_{ii}^{(j)} X^T X \right)^{-1} X^T y_i^{(j)} + \rho \left( Z_{i\gamma}^{(l)} - \mu_\gamma^{(l)} \right),$$

$$D_2(j) = X^T X \left( \sigma_{ii}^{(j)} X^T X \right)^{-1} X^T y_i^{(j)} + \rho \left( Z_i^{(j)(l)} - \mu_j^{(l)} \right)$$

The solutions to Equation 5.130 are given by

$$\hat{\gamma}_i = \left[ A_1 - \sum_{j=1}^{J} B_1(j)(B_2(j) + \rho I)^{-1} A_2(j) \right]^{-1} D,$$

$$D = D_1 - \sum_{j=1}^{J} B_1(j)(B_2(j) + \rho I)^{-1} D_2(j), \qquad (5.131)$$

$$\eta_i^{(j)} = [B_2(j) + \rho I]^{-1} [D_2(j) - A_2(j) \hat{\gamma}_i], \; j = 1, ..., J.$$

The optimization problem (5.127) is non-differentiable. Although the first two terms in (5.127) are not differentiable, we still can obtain simple closed-form solutions to the problem (5.127) using subdifferential calculus. We first consider the generalized gradient of $\| Z_{\gamma i} \|_1$. Let $\Phi_i^{(m)}$ be a generalized derivative of the $m^{th}$ component of the vector $Z_{\gamma i}$:

$$\Phi_i^{(m)} = \begin{cases} 1 & Z_{\gamma i}^{(m)} = \gamma_i^{(m)} > 0 \\ [-1, 1] & Z_{\gamma i}^{(m)} = \gamma_i^{(m)} = 0 \\ -1 & Z_{\gamma i}^{(m)} = \gamma_i^{(m)} < 0. \end{cases}$$

Let $\Phi_i = \begin{bmatrix} \Phi_i^{(1)} \\ \vdots \\ \Phi_i^{(M)} \end{bmatrix}$. Then, we have

$$\frac{\lambda}{\rho} \Phi_i + Z_{\gamma i} = \Delta_{\gamma i}^{(l+1)} + \mu_\gamma^{(l)},$$

which implies that

$$Z_{\lambda i}^{(l+1)} = \text{sgn} \left( \Delta_{\gamma i}^{(l+1)} + \mu_\gamma^{(l)} \right) \left( \left| \Delta_{\gamma i}^{(l+1)} + \mu_\gamma^{(l)} \right| - \frac{\lambda}{\rho} \right)_+, \qquad (5.132)$$

where

$$|x|_+ = \begin{cases} x & x \geq 0 \\ 0 & x < 0. \end{cases}$$

Next, we consider group lasso. The generalized gradient $\dfrac{\partial L_\rho}{\partial Z_{ki}}$ is given by

$$\frac{\partial L_\rho}{\partial Z_{ki}} = \lambda \sqrt{J} s + \rho \left( Z_{ki} - \left( \Delta_{ki}^{(l+1)} - \mu_{ki}^{(l)} \right) \right) = 0, \tag{5.133}$$

where

$$s = \begin{cases} 0 & \| \Delta_{ki}^{(l+1)} - \mu_{ki}^{(l)} \|_2 < \dfrac{\lambda \sqrt{J}}{\rho} \\[3mm] \dfrac{Z}{\| Z_{ki} \|_2} & \| \Delta_{ki}^{(l+1)} - \mu_{ki}^{(l)} \|_2 \geq \dfrac{\lambda \sqrt{J}}{\rho} \end{cases}$$

Equation 5.133 can be reduced to

$$\left( 1 + \frac{\lambda \sqrt{J}}{\rho \| Z_{ki} \|_2} \right) Z_{ki} = \Delta_{ki}^{(l+1)} - \mu_{ki}^{(l)} \tag{5.134}$$

or

$$\left( 1 + \frac{\lambda \sqrt{J}}{\rho \| Z_{ki} \|_2} \right) \| Z_{ki} \|_2 = \| \Delta_{ki}^{(l+1)} - \mu_{ki}^{(l)} \|_2 \tag{5.135}$$

Solving Equation 5.135, we obtain

$$\| Z_{ki} \|_2 = \| \Delta_{ki}^{(l+1)} - \mu_{ki}^{(l)} \|_2 - \frac{\lambda \sqrt{J}}{\rho}. \tag{5.136}$$

Substituting Equation 5.136 into Equation 5.135 leads to

$$\left( 1 + \frac{\lambda \sqrt{J}}{\rho \| Z_{ki} \|_2} \right) = \frac{1}{1 - \dfrac{\lambda \sqrt{J}}{\rho \| \Delta_{ki}^{(l+1)} - \mu_{ki}^{(l)} \|_2}}. \tag{5.137}$$

Combining Equations 5.134, 5.135, and 5.137, we obtain

$$Z_{ki}^{(l+1)} = \left( -\frac{\lambda \sqrt{J}}{\rho \| \Delta_{ki}^{(l+1)} - \mu_{ki}^{(l)} \|_2} \right)_+ \left( \Delta_{ki}^{(l+1)} - \mu_{ki}^{(l)} \right), k = 1, \ldots, K. \tag{5.138}$$

Finally, summarizing the above equations, we obtain Result 5.10.

**Result 5.10: Algorithm for Construction of Genotype-Methylation Causal Networks Where Both Methylation Levels and Genotype Functions are Taken as a Function-Valued Trait**

For $i = 1,...,M$

Step 1: Initialization

$$B_1(j) = \left(Y_{-i}^{(j)}\right)^T X \left(\sigma_{ii}^{(j)} X^T X\right)^{-1} X^T X,$$

$$B_2(j) = X^T X \left(\sigma_{ii}^{(j)} X^T X\right)^{-1} X^T X,$$

$$\tilde{D}_1 = \sum_{j=1}^{J} \left(Y_{-i}^{(j)}\right)^T X \left(\sigma_{ii}^{(j)} X^T X\right)^{-1} X^T y_i^{(j)},$$

$$\tilde{D}_2(j) = X^T X \left(\sigma_{ii}^{(j)} X^T X\right)^{-1} X^T y_i^{(j)}.$$

$$\mu^0 := 0$$

$$\gamma_i^{(0)} = \left[A_1 - \sum_{j=1}^{J} B_1(j)(B_2(j) + \rho I)^{-1} A_2(j)\right]^{-1} \tilde{D},$$

$$\tilde{D} = \tilde{D}_1 - \sum_{j=1}^{J} B_1(j)(B_2(j) + \rho I)^{-1} \tilde{D}_2(j),$$

$$\eta_i^{(0)(j)} = [B_2(j) + \rho I]^{-1} \left[D_2(j) - A_2(j)\hat{\gamma}_i^{(0)}\right], j = 1, ..., J,$$

$$\Delta_i^{(0)} = \begin{bmatrix} \gamma_i^{(0)} \\ \eta_i^{(0)(1)} \\ \vdots \\ \eta_i^{(0)(J)} \end{bmatrix}$$

$$Z_i^{(0)} :== \Delta_i^{(0)}$$

$$\tilde{\Delta}_i^{(0)} = \begin{bmatrix} \tilde{\Delta}_{\gamma i}^{(0)} \\ \tilde{\Delta}_{1i}^{(0)} \\ \vdots \\ \tilde{\Delta}_{Ki}^{(0)} \end{bmatrix}, \tilde{\Delta}_{\gamma i}^{(0)} = \hat{\gamma}_i^{(0)}, \tilde{\Delta}_{ki}^{(0)} = \begin{bmatrix} \tilde{\Delta}_{ki}^{(0)(1)} \\ \vdots \\ \tilde{\Delta}_{ki}^{(0)(J)} \end{bmatrix}, \tilde{\Delta}_{ki}^{(0)(j)} = \eta_{ki}^{(0)(j)},$$

$$\mu^{(0)} = \begin{bmatrix} \mu_\gamma^{(0)} \\ \mu_1^{(0)} \\ \vdots \\ \mu_J^{(0)} \end{bmatrix}.$$

Carry out steps 2, 3, and 4 until convergence.

Step 2:

$$A_1 = \sum_{j=1}^{J} \left(Y_{-i}^{(j)}\right)^T X \left(\sigma_{ii}^{(j)} X^T X\right)^{-1} X^T Y_{-i}^{(j)},$$

$$A_2(j) = X^T X \left(\sigma_{ii}^{(j)} X^T X\right)^{-1} X^T Y_{-i}^{(j)},$$

$$B_1(j) = \left(Y_{-i}^{(j)}\right)^T X \left(\sigma_{ii}^{(j)} X^T X\right)^{-1} X^T X,$$

$$B_2(j) = X^T X \left(\sigma_{ii}^{(j)} X^T X\right)^{-1} X^T X,$$

$$D_1 = \sum_{j=1}^{J} \left(Y_{-1}^{(j)}\right)^T X \left(\sigma_{ii}^{(j)} X^T X\right)^{-1} X^T y_i^{(j)} + \rho\left(Z_{\gamma i}^{(l)} - \mu_\gamma^{(l)}\right),$$

$$D_2(j) = X^T X \left(\sigma_{ii}^{(j)} X^T X\right)^{-1} X^T y_i^{(j)} + \rho\left(Z_i^{(j)(l)} - \mu_j^{(l)}\right).$$

$$\hat{\gamma}_i^{(l+1)} = \left[A_1 - \sum_{j=1}^{J} B_1(j)(B_2(j) + \rho I)^{-1} A_2(j)\right]^{-1} D,$$

$$D = D_1 - \sum_{j=1}^{J} B_1(j)(B_2(j) + \rho I)^{-1} D_2(j),$$

$$\eta_i^{(j)(l+1)} = [B_2(j) + \rho I]^{-1} \left[D_2(j) - A_2(j)\hat{\gamma}_i^{(l+1)}\right], j = 1, \ldots J,$$

$$\Delta_i^{(l+1)} = \begin{bmatrix} \hat{\gamma}_i^{(l+1)} \\ \eta_i^{(1)(l+1)} \\ \vdots \\ \eta_i^{(J)(l+1)} \end{bmatrix}$$

$$\tilde{\Delta}_i^{(l+1)} = \begin{bmatrix} \tilde{\Delta}_{\gamma i}^{(l+1)} \\ \tilde{\Delta}_{1i}^{(l+1)} \\ \vdots \\ \tilde{\Delta}_{Ki}^{(l+1)} \end{bmatrix}, \tilde{\Delta}_{\gamma i}^{(l+1)} = \hat{\gamma}_i^{(l+1)}, \tilde{\Delta}_{ki}^{(l+1)} = \begin{bmatrix} \tilde{\Delta}_{ki}^{(l+1)(1)} \\ \vdots \\ \tilde{\Delta}_{ki}^{(l+1)(J)} \end{bmatrix}, \tilde{\Delta}_{ki}^{(l+1)(j)} = \eta_{ki}^{(j)(l+1)},$$

Step 3:

$$\tilde{Z}_{\gamma i}^{(l+1)} = \text{sgn}\left(\tilde{\Delta}_{\gamma i}^{(l+1)} + \mu_\gamma^{(l)}\right)\left(\left|\tilde{\Delta}_{\gamma i}^{(l+1)} + \mu_\gamma^{(l)}\right| - \frac{\lambda}{\rho}\right)_+,$$

where

$$|x|_+ = \begin{cases} x & x \geq 0 \\ 0 & x < 0. \end{cases}$$

$$\tilde{Z}_{ki}^{(l+1)} = \left( -\frac{\lambda\sqrt{J}}{\rho\|\tilde{\Delta}_{ki}^{(l+1)} - \mu_{ki}^{(l)}\|_2} \right)_+ \left( \tilde{\Delta}_{ki}^{(l+1)} - \mu_{ki}^{(l)} \right), k = 1, ..., K.$$

$$\tilde{Z}_i^{(l+1)} = \begin{bmatrix} \tilde{Z}_{\gamma i}^{(l+1)} \\ \tilde{Z}_{1i}^{(l+1)} \\ \vdots \\ \tilde{Z}_{Ki}^{(l+1)} \end{bmatrix}, \tilde{Z}_{ki}^{(l+1)} = \begin{bmatrix} \tilde{Z}_{ki}^{(1)(l+1)} \\ \vdots \\ \tilde{Z}_{ki}^{(J)(l+1)} \end{bmatrix}.$$

Update $Z^{(l+1)}$:

$$Z_{ki}^{(j)(l+1)} = \tilde{Z}_{ki}^{(j)(l+1)}, Z_{\gamma i}^{(l+1)} = \tilde{Z}_{\gamma i}^{(l+1)}, Z_i^{(j)(l+1)} = \begin{bmatrix} Z_{1i}^{(j)(l+1)} \\ \vdots \\ Z_{Ki}^{(j)(l+1)} \end{bmatrix}, Z_i^{(l+1)} = \begin{bmatrix} Z_{\gamma i}^{(l+1)} \\ Z_i^{(1)(l+1)} \\ \vdots \\ Z_i^{(J)(l+1)} \end{bmatrix}.$$

Step 4:

$$\tilde{\mu}^{(l+1)} = \tilde{\mu}_i^{(l)} + \left( \tilde{\Delta}_i^{(l+1)} - \tilde{Z}_i^{(l+1)} \right),$$

where

$$\tilde{\mu}^{(l+1)} = \begin{bmatrix} \tilde{\mu}_\gamma^{(l+1)} \\ \tilde{\mu}_1^{(l+1)} \\ \vdots \\ \tilde{\mu}_K^{(l+1)} \end{bmatrix}, \tilde{\mu}_k^{(l+1)} = \begin{bmatrix} \tilde{\mu}_k^{(1)(l+1)} \\ \vdots \\ \tilde{\mu}_k^{(J)(l+1)} \end{bmatrix}, k = 1, ..., K.$$

Update $\mu^{(l+1)}$:

$$\mu^{(l+1)} = \begin{bmatrix} \mu_\gamma^{(l+1)} \\ \mu_1^{(l+1)} \\ \vdots \\ \mu_J^{(l+1)} \end{bmatrix}, \mu_\gamma^{(l+1)} = \tilde{\mu}_\gamma^{(l+1)}, \mu_j^{(l+1)} = \begin{bmatrix} \mu_1^{(j)(l+1)} \\ \vdots \\ \mu_K^{(j)(l+1)} \end{bmatrix}, \mu_k^{(j)(l+1)} = \tilde{\mu}_k^{(j)(l+1)}.$$

Go to step 2.

## Software Package

Package 'minfi' that analyzes Illumina DNA methylation array data from the Human Methylation450 ('450k') and EPIC platforms can be downloaded from https://github.com/kasperdanielhansen/minfi. ChAMP is an integrated analysis pipeline published in 2014, modified in 2017, and can be downloaded from https://bioconductor.org/packages/release/bioc/vignettes/ChAMP /inst/doc/ChAMP.html.

Illumina provides data analysis software that supports data integration, such as genotyping with gene expression for eQTL analysis and methylation for mQTL analysis (TECHNICAL NOTE: Illumina® SYSTEMS AND SOFT-WARE: QTL Analysis Software Tools for Illumina Data) 'meQTL mapping analysis cookbook' also provides software for meQTL analysis.

## Appendix 5.A    Biased and Unbiased Estimators of the HSIC

For completeness, we adopt the approach of Gretton (2015) for the estimation of the HSIC. It follows from Equation 5.46 that

$$
\begin{aligned}
HSIC(X, Y) &= \left\langle C_{XY} - \mu_x \otimes \mu_y, C_{XY} - \mu_x \otimes \mu_y \right\rangle \\
&= \left\langle C_{XY}, C_{XY} \right\rangle - 2\left\langle C_{XY}, \mu_x \otimes \mu_y \right\rangle + \left\langle \mu_x \otimes \mu_y, \mu_x \otimes \mu_y \right\rangle.
\end{aligned}
\tag{5.A.1}
$$

By definition of non-centered covariance operator, we obtain

$$
\begin{aligned}
\left\langle C_{XY}, C_{XY} \right\rangle &= \left\langle E_{p(x,y)}\left[ k_x(\cdot, X) \otimes k_y(\cdot, Y) \right], E_{p(x',y')}\left[ k_x(\cdot, X') \otimes k_y(\cdot, Y') \right] \right\rangle \\
&= E_{p(x,y)} E_{p(x',y')}\left[ \left\langle k_x(\cdot, X) \otimes k_y(\cdot, Y), k_x(\cdot, X') \otimes k_y(\cdot, Y') \right\rangle \right] \\
&= E_{p(x,y)} E_{p(x',y')}\left[ \left\langle k_x(\cdot, X), k_x(\cdot, X') \right\rangle \left\langle k_y(\cdot, Y), k_y(\cdot, Y') \right\rangle \right] \\
&= E_{p(x,y)} E_{p(x',y')}\left[ k_x(X, X') k_y(Y, Y') \right].
\end{aligned}
\tag{5.A.2}
$$

Using sampling formula for expectation, Equation 5.A.2 can be reduced to

$$
\begin{aligned}
\left\langle C_{XY}, C_{XY} \right\rangle &= \frac{1}{n^2} \sum_{i=1}^{n} \sum_{j=1}^{n} k_x\left( x_i, x_j \right) k_y\left( y_i, y_j \right) \\
&= \frac{1}{n^2} \sum_{i=1}^{n} \sum_{j=1}^{n} k_x\left( x_i, x_j \right) k_y\left( y_j, y_i \right) \\
&= \frac{1}{n^2} \operatorname{tr}\left( K_x K_y \right),
\end{aligned}
\tag{5.A.3}
$$

where $K_x = (k_x(x_i, x_j))_{n \times n}$ and $K_y = (k_y(y_i, y_j))_{n \times n}$.

Similarly, we have

$$
\begin{aligned}
\left\langle C_{XY}, \mu_x \otimes \mu_y \right\rangle &= \left\langle E_{p(x,y)} \left[ k_x(., X) \otimes k_y(., Y), \left( E_{p(x)} [k_x(., X')] \right) \right. \right. \\
&\quad \otimes \left. \left. \left( E_{p(y)} [k_y(., Y')] \right) \right] \right\rangle \\
&= E_{p(x,y)} \left[ E_{p(x)} [k_x(X, X')] E_{p(y)} [k_y(Y, Y')] \right].
\end{aligned}
\tag{5.A.4}
$$

Its sampling estimator is

$$
\begin{aligned}
\left\langle C_{XY}, \mu_x \otimes \mu_y \right\rangle &= \frac{1}{n^2} E_{p(x,y)} \left[ \sum_{j=1}^n k_x(X, x_j) \sum_{l=1}^n k_y(Y, y_l) \right] \\
&= \frac{1}{n^3} \sum_{i=1}^n \sum_{j=1}^n \sum_{l=1}^n k_x\left(x_i, x_j\right) k_y(y_i, y_l) \\
&= \frac{1}{n^3} \sum_{j=1}^n \sum_{i=1}^n \sum_{l=1}^n k_x\left(x_i, x_j\right) k_y(y_i, y_l) \\
&= \frac{1}{n^3} \mathbf{1}_n^T K_x K_y \mathbf{1}_n \\
&= \frac{1}{n^3} \operatorname{tr}\left( \mathbf{1}_n \mathbf{1}_n^T K_x K_y \right).
\end{aligned}
\tag{5.A.5}
$$

Finally, we can show

$$
\begin{aligned}
\left\langle \mu_x \otimes \mu_y, \mu_x \otimes \mu_y \right\rangle &= \langle \mu_x, \mu_x \rangle < \mu_y, \mu_y > \\
&= \left\langle E_x[k_x(., X)], E_{x'}[k_x(., X')] \right\rangle \left\langle E_y[k_y(., Y)], E_{y'}[k_y(., Y')] \right\rangle \\
&= E_x \left[ E_{x'}[k_x(X, X')] \right] E_y \left[ E_{y'}[k_y(Y, Y')] \right].
\end{aligned}
\tag{5.A.6}
$$

Again, its sampling estimator can be calculated as

$$
\begin{aligned}
\left\langle \mu_x \otimes \mu_y, \mu_x \otimes \mu_y \right\rangle &= \frac{1}{n^2} \sum_{i=1}^n \sum_{j=1}^n k_x\left(x_i, x_j\right) \frac{1}{n^2} \sum_{q=1}^n \sum_{l=1}^n k_y\left(y_q, y_l\right) \\
&= \frac{1}{n^4} \mathbf{1}_n^T K_x \mathbf{1}_n \mathbf{1}_n^T K_y \mathbf{1}_n \\
&= \frac{1}{n^4} \operatorname{tr}\left( \mathbf{1}_n \mathbf{1}_n^T K_x \mathbf{1}_n \mathbf{1}_n^T K_y \right).
\end{aligned}
\tag{5.A.7}
$$

Substituting Equations 5.A.3, 5.A.5, and 5.A.7 gives

$$
\begin{aligned}
HSIC(X, Y) &= \frac{1}{n^2} \left[ \operatorname{tr}\left( K_x K_y \right) - \frac{2}{n} \operatorname{tr}\left( \mathbf{1}_n \mathbf{1}_n^T K_x K_y \right) + \frac{1}{n^2} \operatorname{tr}\left( \mathbf{1}_n \mathbf{1}_n^T K_x \mathbf{1}_n \mathbf{1}_n^T K_y \right) \right] \\
&= \frac{1}{n^2} \left( K_x H K_y H \right),
\end{aligned}
\tag{5.A.8}
$$

where $H = I_n - \frac{1}{n}\mathbf{1}_n\mathbf{1}_n^T$.

Now we derive the unbiased estimator of the HSIC. In the previous calculations, we do not pay attention to the true samplings of $(x,y)$ and $(x',y')$. We completely treat samplings of $(x,y)$ and $(x',y')$ as independent. However, the true sampling of $(x,y)$ and $(x',y')$ should be sampling without replacement (Gretton 2015). Therefore, the unbiased sampling estimator of $\| C_{XY} \|_{HS}^2$ is

$$\langle C_{XY}, C_{XY} \rangle = E_{p(x,y)}E_{p(x',y')}\left[ k_x(X,X')k_y(Y,Y') \right]$$

$$= \frac{1}{n(n-1)}\sum_{i=1}^{n}\sum_{j \neq i}^{n} k_x\left( x_i, x_j \right)k_y\left( y_i, y_j \right) \tag{5.A.9}$$

To simplify notations, we denote $\mathbf{i}_q^n$ to be the set of all $q$-tuples drawn from $\{1,\ldots,n\}$ and $(n)_q = n(n-1)\ldots(n-p+1)$.

Under these notations, Equation 5.A.9 can be reduced to

$$\langle C_{XY}, C_{XY} \rangle = \frac{1}{(n)_2}\sum_{(i,j) \in \mathbf{i}_2^n} k_x\left( x_i, x_j \right)k_y\left( y_i, y_j \right) \tag{5.A.10}$$

Similarly, we obtain

$$\left\langle C_{XY}, \mu_x \otimes \mu_y \right\rangle = E_{p(x,y)}[E_{p(x)}[k_x(X,X')]E_{p(y)}[k_y(Y,Y')]]$$

$$= \frac{1}{n(n-1)}\sum_{j=1}^{n}\sum_{l \neq j}^{n} E_{p(x,y)}k_x(X,x_j)k_y(Y,y_l)$$

$$= \frac{1}{n(n-1)(n-2)}\sum_{i \neq j,l}^{n}\sum_{l \neq j}^{n}\sum_{j=1}^{n} k_x(x_i,x_j)k_y(y_i,y_l) \tag{5.A.11}$$

$$= \frac{1}{(n)_3}\sum_{(i,j,l) \in \mathbf{i}_3^n} k_x(x_i,x_j)k_y(y_i,y_l),$$

and

$$\left\langle \mu_x \otimes \mu_y, \mu_x \otimes \mu_y \right\rangle = E_x[E_{x'}[k_x(X,X')]]E_y[E_{y'}[k_y(Y,Y')]]$$

$$= \frac{1}{n(n-1)}\sum_{i=1}^{n}\sum_{j \neq i}^{n} E_x[k_x(X,x_i)]E_y[k_y(Y,y_j)]$$

$$= \frac{1}{n(n-1)(n-2)(n-3)} \tag{5.A.12}$$

$$\sum_{i=1}^{n}\sum_{i \neq j}^{n}\sum_{q \neq i,j}^{n}\sum_{l \neq i,j,q}^{n} k_x\left( x_q, x_i \right)k_y\left( y_l, y_j \right)$$

$$= \frac{1}{(n)_4}\sum_{(i,j,q,l) \in \mathbf{i}_4^n} k_x\left( x_i, x_q \right)k_y\left( y_j, y_l \right).$$

Although Equations 5.A.10–5.A.12 are unbiased estimators, their computational times are much more expensive than the biased estimators. Next, we reduce the computational time of the biased estimators. Let $\hat{K}_x$ and $\hat{K}_y$ denote the matrices of $K_x$ and $K_y$ with their diagonal terms replaced by zero. Then, Equation 5.A.9 can be rewritten as

$$
\begin{aligned}
\langle C_{XY}, C_{XY} \rangle &= \frac{1}{(n)_2} \sum_{i=1}^{n} \sum_{j \neq i}^{n} k_x\left(x_i, x_j\right) k_y\left(y_i, y_j\right) \\
&= \frac{1}{(n)_2} \sum_{i=1}^{n} \sum_{j=1}^{n} \hat{k}_x\left(x_i, x_j\right) \hat{k}_y\left(y_j, y_i\right) \\
&= \frac{1}{(n)_2} \operatorname{tr}\left(\hat{K}_x \hat{K}_y\right).
\end{aligned}
\tag{5.A.13}
$$

Similarly using Equation 5.A.11, we obtain

$$
\begin{aligned}
\left\langle C_{XY}, \mu_x \otimes \mu_y \right\rangle &= \frac{1}{n(n-1)(n-2)} \sum_{i=1}^{n} \sum_{j \neq i}^{n} \sum_{l \neq j,i}^{n} k_x(x_i, x_j) k_y(y_i, y_l) \\
&= \frac{1}{(n)_3} \sum_{j=1}^{n} \sum_{l \neq j}^{n} \sum_{i \neq j,l}^{n} k_x(x_j, x_i) k_y(y_i, y_l) \\
&= \frac{1}{(n)_3} \sum_{j=1}^{n} \sum_{i \neq j}^{n} \sum_{l \neq j,i}^{n} k_x(x_j, x_i) k_y(y_i, y_l) \\
&= \frac{1}{(n)_3} \sum_{j=1}^{n} \sum_{l \neq j}^{n} \sum_{i \neq j,l}^{n} k_x(x_j, x_i) k_y(y_i, y_l) \\
&= \frac{1}{(n)_3} \Bigg[ \sum_{j=1}^{n} \sum_{l=1}^{n} \sum_{i \neq j,l}^{n} k_x(x_j, x_i) k_y(y_i, y_l) \\
&\quad - \sum_{j=1}^{n} \sum_{i \neq j}^{n} k_x(x_j, x_i) k_y(y_i, y_j) \Bigg].
\end{aligned}
\tag{5.A.14}
$$

We can show (Exercise 4)

$$
\sum_{j=1}^{n} \sum_{l=1}^{n} \sum_{i \neq (j,l)}^{n} k_x(x_j, x_i) k_y(y_i, y_l) = \mathbf{1}_n^T \hat{K}_x \hat{K}_y \mathbf{1}_n,
\tag{5.A.15}
$$

and

$$
\sum_{j=1}^{n} \sum_{i \neq j}^{n} k_x\left(x_j, x_i\right) k_y\left(y_i, y_j\right) = \operatorname{tr}\left(\hat{K}_x \hat{K}_y\right).
\tag{5.A.16}
$$

Substituting Equations 5.A.15 and 5.A.16 gives

$$
\left\langle C_{XY}, \mu_x \otimes \mu_y \right\rangle = \frac{1}{(n)_3} \left[ \mathbf{1}_n^T \hat{K}_x \hat{K}_y \mathbf{1}_n - \operatorname{tr}\left(\hat{K}_x \hat{K}_y\right) \right].
\tag{5.A.17}
$$

Again, using Equation 5.A.12 we obtain

$$\left\langle \mu_x \otimes \mu_y, \mu_x \otimes \mu_y \right\rangle = \frac{1}{(n)_4} \sum_{(i,j,q,l)\in i_4^n} k_x\left(x_i, x_q\right) k_y\left(y_j, y_l\right). \tag{5.A.18}$$

Note that the terms in $\sum_{(i,q)\in i_2^n} k_x(x_i, x_q) \sum_{(j,l)\in i_2^n} k_y(y_j, y_l)$ can be expanded into (Gretton 2015, exercise 5):

$$\begin{aligned}
\sum_{(i,q)\in i_2^n} k_x\left(x_i, x_q\right) \sum_{(j,l)\in i_2^n} k_y\left(y_j, y_l\right) &= \sum_{(i,q,j)\in i_3^n} k_x(x_i, x_q) k_y(y_j, y_i) \\
&+ \sum_{(i,q,l)\in i_3^n} k_x(x_i, x_q) k_y(y_q, y_l) \\
&+ \sum_{(i,j,l)\in i_3^n} k_x(x_i, x_j) k_y(y_j, y_l) \\
&+ \sum_{(q,j,l)\in i_3^n} k_x(x_l, x_q) k_y(y_j, y_l) \quad (5.A.19) \\
&+ \sum_{(i,j,q,l)\in i_4^n} k_x(x_i, x_q) k_y(y_j, y_l) \\
&+ \sum_{(i,q)\in i_2^n} k_x(x_i, x_q) k_y(y_i, y_q) \\
&+ \sum_{(i,q)\in i_2^n} k_x(x_i, x_q) k_y(y_i, y_q).
\end{aligned}$$

It is clear that

$$\sum_{(i,q)\in i_2^n} k_x\left(x_i, x_q\right) = 1_n^T \hat{K}_x 1_n \quad \text{and} \quad \sum_{(j,l)\in i_2^n} k_y\left(y_j, y_l\right) = 1_n^T \hat{K}_y 1_n. \tag{5.A.20}$$

After applying simple algebra, we obtain

$$\begin{aligned}
1_n^T \hat{K}_x \hat{K}_y 1_n &= \sum_{i=1}^n \sum_{j\neq i}^n \sum_{l\neq j}^n k_x\left(x_i, x_j\right) k_y\left(y_j, y_l\right) \\
&= \sum_{i=1}^n \sum_{j\neq i}^n \sum_{l\neq(i,j)}^n k_x\left(x_i, x_j\right) k_y\left(y_j, y_l\right) \\
&\quad + \sum_{i=1}^n \sum_{j\neq i}^n k_x\left(x_i, x_j\right) k_y\left(y_j, y_i\right) \\
&= \sum_{(i,j,l)\in i_3^n} k_x\left(x_i, x_j\right) k_y\left(y_j, y_l\right) + \text{tr}\left(\hat{K}_x \hat{K}_y\right).
\end{aligned} \tag{5.A.21}$$

By similar arguments, we obtain

$$\begin{aligned}
1_n^T \hat{K}_x \hat{K}_y 1_n &= \sum_{(i,q,j)\in i_3^n} k_x\left(x_i, x_q\right) k_y\left(y_j, y_i\right) + \text{tr}\left(\hat{K}_x \hat{K}_y\right) \\
1_n^T \hat{K}_x \hat{K}_y 1_n &= \sum_{(i,q,l)\in i_3^n} k_x\left(x_i, x_q\right) k_y\left(y_q, y_l\right) + \text{tr}\left(\hat{K}_x \hat{K}_y\right) \\
1_n^T \hat{K}_x \hat{K}_y 1_n &= \sum_{(q,j,l)\in i_3^n} k_x\left(x_l, x_q\right) k_y\left(y_j, y_l\right) + \text{tr}\left(\hat{K}_x \hat{K}_y\right).
\end{aligned} \tag{5.A.22}$$

Substituting Equations 5.A.20–5.A.22 into Equation 5.A.19, we obtain

$$\left\langle \mu_x \otimes \mu_y, \mu_x \otimes \mu_y \right\rangle$$

$$= \frac{1}{(n)_4} \left[ \left( \mathbf{1}_n^T \hat{K}_x \mathbf{1}_n \right) \left( \mathbf{1}_n^T \hat{K}_y \mathbf{1}_n \right) - 4 \mathbf{1}_n^T \hat{K}_x \hat{K}_y \mathbf{1}_n + 2\mathrm{tr}\left( \hat{K}_x \hat{K}_y \right) \right] \qquad (5.A.23)$$

Finally, substituting Equations 5.A.13, 5.A.17, and 5.A.23 into equation 5.A.1, we obtain the unbiased estimator of HSIC:

$$HSIC_u(X, Y) = \frac{1}{(n)_2} \mathrm{tr}\left( \hat{K}_x \hat{K}_y \right) - \frac{2}{(n)_3} \left[ \mathbf{1}_n^T \hat{K}_x \hat{K}_y \mathbf{1}_n - \mathrm{tr}\left( \hat{K}_x \hat{K}_y \right) \right]$$

$$+ \frac{1}{(n)_4} \left[ \left( \mathbf{1}_n^T \hat{K}_y \mathbf{1}_n \right) \left( \mathbf{1}_n^T \hat{K}_y \mathbf{1}_n \right) - 4 \mathbf{1}_n^T \hat{K}_x \hat{K}_y \mathbf{1}_n + 2\mathrm{tr}\left( \hat{K}_x \hat{K}_y \right) \right]$$

$$= \left( \frac{1}{(n)_2} + \frac{2}{(n)_3} + \frac{1}{(n)_4} \right) \mathrm{tr}\left( \hat{K}_x \hat{K}_y \right) - \left( \frac{2}{(n)_3} + \frac{1}{(n)_4} \right) \mathbf{1}_n^T \hat{K}_x \hat{K}_y \mathbf{1}$$

$$+ \frac{1}{(n)_4} \left( \mathbf{1}_n^T \hat{K}_x \mathbf{1}_n \right) \left( \mathbf{1}_n^T \hat{K}_y \mathbf{1}_n \right)$$

$$(5.A.24)$$

$$= \frac{1}{n(n-3)} \mathrm{tr}\left( \hat{K}_x \hat{K}_y \right) - \frac{2}{n(n-2)(n-3)} \mathbf{1}_n^T \hat{K}_x \hat{K}_y \mathbf{1}$$

$$+ \frac{1}{n(n-1)(n-2)(n-3)} \left( \mathbf{1}_n^T \hat{K}_x \mathbf{1}_n \right) \left( \mathbf{1}_n^T \hat{K}_y \mathbf{1}_n \right)$$

$$= \frac{1}{n(n-3)} \left[ \mathrm{tr}\left( \hat{K}_x \hat{K}_y \right) - \frac{2}{n-2} \mathbf{1}_n^T \hat{K}_x \hat{K}_y \mathbf{1} \right.$$

$$\left. + \frac{1}{(n-1)(n-2)} \left( \mathbf{1}_n^T \hat{K}_x \mathbf{1}_n \right) \left( \mathbf{1}_n^T \hat{K}_y \mathbf{1}_n \right) \right].$$

---

## Appendix 5.B    Asymptotic Null Distribution of Block-Based HSIC

Note that

$$Var(BHSIC) = \frac{S}{n} Var(\hat{c}_b). \qquad (5.B.1)$$

Since $var(S\hat{c}_b) \approx var(W)$ and $W$ is distributed as $\sum_{i=1}^{\infty}\sum_{j=1}^{\infty} \lambda_i \eta_j (N_{i,j}^2 - 1)$, we obtain

$$Var(\hat{c}_b) \approx \frac{1}{S^2} var(W). \qquad (5.B.2)$$

It is well known that the variance of $\chi^2_{(1)}$ is 2. Since $N_{i,j}$ are independently and identically distributed as $N(0,1)$, we can obtain

$$Var(W) = 2\sum_{i=1}^{\infty}\sum_{j=1}^{\infty}\lambda_i^2\eta_j^2. \tag{5.B.3}$$

Let $\hat{K}_x(X, X')$ and $\hat{K}_y(Y, Y')$ be centered kernel matrices. Then, we obtain

$$\mathrm{tr}\left(\hat{K}_x^2\right) \rightarrow \sum_{i=1}^{\infty}\lambda_i^2 \quad \text{and} \quad \mathrm{tr}\left(\hat{K}_y^2\right) \rightarrow \sum_{j=1}^{\infty}\eta_j^2. \tag{5.B.4}$$

Recall that

$$\mathrm{tr}\left(\hat{K}_x^2\right) = \sum_{i=1}^{S}\sum_{j=1}^{S}k_x^2\left(x_i, x_j\right) = S^2\frac{1}{S^2}\sum_{i=1}^{S}\sum_{j=1}^{S}k_x^2\left(x_i, x_j\right)$$
$$\rightarrow S^2 E_{XX'}\left[\hat{k}_x^2(X, X')\right]$$

and

$$\mathrm{tr}\left(\hat{K}_y^2\right) \rightarrow S^2 E_{YY'}\left[\hat{k}_y^2(Y, Y')\right]. \tag{5.B.5}$$

Substituting Equations 5.B.4 and 5.B.5 into Equation 5.B.3, we obtain

$$var(W) = 2S^4 E_{XX'}\left[\hat{k}_x^2(X, X')\right]E_{YY'}\left[\hat{k}_y^2(Y, Y')\right]. \tag{5.B.6}$$

Combining Equations 5.B.1 and 5.B.2 gives

$$Var(BHSIC) = \frac{1}{ns}Var(W). \tag{5.B.7}$$

Therefore, define

$$T_{BI} = \sqrt{ns}\frac{BHSIC}{Var(W)}. \tag{5.B.8}$$

Under the null hypothesis of independence, $T_{BI}$ is asymptotically distributed as $N(0,1)$ distribution.

## Exercises

**Exercise 1.** Suppose that $k$ is a linear kernel, $P$ is a normal distribution N(0,1), and $Q$ is a binomial distribution. Calculate maximum mean discrepancy $MMD_k(P,Q)$.

**Exercise 2.** Let $A$ be a $n \times n$ dimensional matrix and $B$ be a $m \times m$ dimensional matrix. Show that

$$|A \otimes B| = |A|^n |B|^m \quad \text{and} \quad Tr(A \otimes B) = Tr(A)Tr(B),$$

where $Tr$ denotes the trace of the matrix.

**Exercise 3.** Show

$$\langle A, B \rangle_{HS} = \sum_{i \in I} \sum_{j \in J} \langle Av_j, e_i \rangle_F \langle Bv_j, e_i \rangle_F$$

**Exercise 4.** Show

$$\sum_{j=1}^{n} \sum_{l=1}^{n} \sum_{i \neq (j,l)}^{n} k_x(x_j, x_i) k_y(y_i, y_l)$$
$$= 1_n^T \hat{K}_x \hat{K}_y 1_n \text{ and } \sum_{j=1}^{n} \sum_{i \neq j} k_x(x_j, x_i) k_y(y_i, y_j) = \text{tr}(\hat{K}_x \hat{K}_y).$$

**Exercise 5.** Show

$$\sum_{(i,q) \in i_2^n} k_x(x_i, x_q) \sum_{(j,l) \in i_2^n} k_y(y_j, y_l) = \sum_{(i,q,j) \in i_3^n} k_x(x_i, x_q) k_y(y_j, y_i)$$
$$+ \sum_{(i,q,l) \in i_3^n} k_x(x_i, x_q) k_y(y_q, y_l)$$
$$+ \sum_{(i,j,l) \in i_3^n} k_x(x_i, x_j) k_y(y_j, y_l)$$
$$+ \sum_{(q,j,l) \in i_3^n} k_x(x_l, x_q) k_y(y_j, y_l)$$
$$+ \sum_{(i,j,q,l) \in i_4^n} k_x(x_i, x_q) k_y(y_j, y_l)$$
$$+ \sum_{(i,q) \in i_2^n} k_x(x_i, x_q) k_y(y_i, y_q)$$
$$+ \sum_{(i,q) \in i_2^n} k_x(x_i, x_q) k_y(y_i, y_q).$$

Exercise 6. Show Equations 5.92–5.94 in the text:

$$\hat{\Delta}_i = \left[ W_i^T X_0 (\sigma_i^2 X_0^T X_0)^{-1} X_0^T W_i \right]^{-1} W_i^T X_0 (\sigma_i^2 X_0^T X_0)^{-1} X_0^T y_i$$

$$= \left[ W_i^T X_0 (X_0^T X_0)^{-1} X_0^T W_i \right]^{-1} W_i^T X_0 (X_0^T X_0)^{-1} X_0^T y_i, \tag{5.92}$$

and

$$\Lambda_i = \mathrm{var}\left( \hat{\Delta}_i \right) = N\hat{\sigma}_i^2 \left[ W_i^T X_0 (X_0^T X_0)^{-1} X_0^T W_i \right]^{-1}, \tag{5.93}$$

where $\hat{\sigma}_i^2$ is estimated by

$$\hat{\sigma}_i^2 = \frac{1}{NJ - m_i J + 1 - k_i} \left( y_i - W_i \hat{\Delta}_i \right)^T \left( y_i - W_i \hat{\Delta}_i \right), \tag{5.94}$$

with $m_i$ and $k_i$ be the number of endogenous and exogenous variables present in the $i^{th}$ equation, respectively.

Exercise 7. Show that the optimal solution to the problem (5.99) in text is

$$\hat{\gamma}_i = \left[ A_1 - \sum_{j=1}^{J} B_1(j)(B_2(j) + \rho I)^{-1} A_2(j) \right]^{-1} D,$$

$$D = D_1 - \sum_{j=1}^{J} B_1(j)(B_2(j) + \rho I)^{-1} D_2(j),$$

$$\eta_i^{(j)} = [B_2(j) + \rho I]^{-1} [D_2(j) - A_2(j)\hat{\gamma}_i],$$

where

$$A_1 = \sum_{j=1}^{J} \left( Y_{-i}^{(j)} \right)^T X \left( \sigma_{ii}^{(j)} X^T X \right)^{-1} X^T Y_{-i}^{(j)},$$

$$A_2(j) = X^T X \left( \sigma_{ii}^{(j)} X^T X \right)^{-1} X^T Y_{-i}^{(j)},$$

$$B_1(j) = \left( Y_{-i}^{(j)} \right)^T X \left( \sigma_{ii}^{(j)} X^T X \right)^{-1} X^T X,$$

$$B_2(j) = X^T X \left( \sigma_{ii}^{(j)} X^T X \right)^{-1} X^T X,$$

$$D_1 = \sum_{j=1}^{J} \left( Y_{-i}^{(j)} \right)^T X \left( \sigma_{ii}^{(j)} X^T X \right)^{-1} X^T y_i^{(j)} + \rho \left( Z_{i\gamma}^{(l)} - \mu_\gamma^{(l)} \right),$$

$$D_2(j) = X^T X \left( \sigma_{ii}^{(j)} X^T X \right)^{-1} X^T y_i^{(j)} + \rho \left( Z_i^{(j)(l)} - \mu_j^{(l)} \right)$$

Exercise 8. Show that the optimal solution to the problem (5.101) is

$$Z_{\lambda i}^{(l+1)} = \text{sgn}\left(\Delta_{\gamma i}^{(l+1)} + \mu_\gamma^{(l)}\right)\left(\left|\Delta_{\gamma i}^{(l+1)} + \mu_\gamma^{(l)}\right| - \frac{\lambda}{\rho}\right)_+,$$

where

$$|x|_+ = \begin{cases} x & x \geq 0 \\ 0 & x < 0. \end{cases}$$

Exercise 9. Show Equation 5.105 implies (Exercise 8)

$$\left(1 + \frac{\lambda\sqrt{J}}{\rho\|Z_{ki}\|_2}\right)\|Z_{ki}\|_2 = \|\Delta_{ki}^{(l+1)} - \mu_{ki}^{(l)}\|_2.$$

# 6

## Imaging and Genomics

## 6.1 Introduction

Medical imaging is a visual representation of the interior of the body for clinical analysis and medical intervention. Currently popular medical imaging techniques include abdominal ultrasound (US) (Kulig et al. 2014), contrast-enhanced computer tomography (CT) (Cascio et al. 2012), magnetic resonance imaging (MRI) (Javery et al. 2013), pathology image, diffusion tensor imaging (DTI) (Garin-Muga and Borro 2014), photon emission tomography (PET) (Bailey et al. 2005), and functional magnetic resonance imaging (fMRI) (Glover 2011). These medical images study anatomic structures and function of tissue types/organs to identify their changes in tissue and organs. The imaging signals can be used for disease diagnosis, investigation of biological processes, and uncovering mechanisms of diseases.

Medical imaging data analysis includes image acquisition, storage, registration, retrieval, feature extraction, image segmentation, cluster analysis, classification, diagnosis, and surgery guidance. Machine learning and other statistical methods have been widely used in medical image analysis and emerges as a method of choice for image segmentation and diagnosis (Moeskops et al. 2017; Litjens et al. 2017). In particular, deep learning has high probability to automatically exploit hierarchical feature representations from data and hence can help to discover, classify, and measure physiological and clinical patterns in medical imaging (Shen et al. 2017).

Although biomedical imaging is playing an ever more important role in diagnosis of cancer, it does not consider the key biological processes that are involved in cancer development (Ahmed et al. 2014). There is increasing recognition that the ability to detect disease before symptoms arise by integrating imaging data and molecular profiles (ECR 2014 Press Release). Data sets from biomedical imaging, genomics and epigenomics are extremely high dimensional and high heterogeneous and are from multiple sources and multiple scales (Phan et al. 2012). Integrative analysis of high-dimensional imaging, miRNA-seq, and methylation-seq datasets poses great challenges. A key issue for integration of imaging, RNA-seq, miRNA-seq, and methylation-seq data is how to develop a unified representation of these multiple types of data, which makes intensity of imaging, the number of methylated reads, and number of expression reads comparable, and to reduce the dimensions of the data. In addition, Bayesian network and causal graphs will be used to develop novel and highly discriminating algorithms for combining mRNA, miRNA, methylation, and imaging classifiers.

To achieve these goal, the focus of this chapter is to introduce deep learning, in particular, convolutional neural networks for unsupervised and supervised image sematic segmentation. In principle, the introduced deep learning can be similarly applied to image classification and diagnosis. Then, we will develop statistical methods for imaging-genomic data analysis, in particular, using image segmentation as a framework. Most importantly, we will develop causal inference as a general framework and powerful tool for imaging-genomic data analysis. Finally, causal machine learning will be briefly introduced to combine causal inference and machine learning to improve accuracy of image classification.

## 6.2 Image Segmentation

Image segmentation involves clustering pixels or voxels into a set of image regions. Image segmentation plays an important role in prediction, diagnosis, treatment, and imaging-genomics data analysis (Gibson et al. 2017; Garcia-Garcia et al. 2017). Segmentation serves two purposes: (1) decomposition of the image into subregions for further analysis and (2) performing changes of representation. Various methods for image segmentation have been developed. In general, these methods can be classified into unsupervised learning methods and supervised learning methods. The traditional approach to image segmentation can be found in the book by Dhawan (2011). In this section, we will mainly introduce deep learning techniques for image segmentation (Garcia-Garcia et al. 2017; Hosseini-Asl 2016).

### 6.2.1 Unsupervised Learning Methods for Image Segmentation

#### 6.2.1.1 Nonnegative Matrix Factorization

*6.2.1.1.1 Data Matrix and Matrix Decomposition*

The task of image segmentation is to decompose the image into uniform and homogeneous components. Many tumors attempt to invade the nearby healthy tissues (Sauwen et al. 2017). As consequences, tumor boundaries may extend beyond what we can observe based on conventional MRI (cMRI). To overcome this limitation, other imaging modalities such as perfusion-weighted imaging (PWI) and diffusion-weighted imaging (DWI) will be included in the studies. The measurements of all imaging modalities need to be included into the data. Each pixel or voxel can be taken as a data point.

Consider a data matrix $X$:

$$X = [x_1 \quad \cdots \quad x_n] = \begin{bmatrix} x_{11} & \cdots & x_{1n} \\ \vdots & \vdots & \vdots \\ x_{m1} & \cdots & x_{mn} \end{bmatrix},$$

where $n$ is the number of data points and $m$ is the number of features for each data point. Each column of $X$ represents a set of features of one pixel or voxel, or a data point.

Similar to gene expression and methylation deconvolution, the image data matrix can also be decomposed to the product of two matrices: $G \in R^{m \times r}$ and $H \in R^{r \times n}$ (Figure 6.1). The matrix $G$ is a tissue type or subregion matrix. Each column of $W$ is a vector of tissue (or subregion) specific feature and defines a tissue type (or subregion). The matrix $H$ is a proportion matrix. Its column represents the proportions of tissue types (or subregions) which one pixel or voxel contains. Assume that the rank of $X$ is $k$. Therefore, $k \leq \min(m,n)$. Matrix factorization attempts to minimize the approximation error:

$$\min_{W,H} \quad \| X - GH \|_F^2 \tag{6.1}$$

where $\|.\|_F$ denotes the Frobenius norm of a matrix.

Let $g_i \in R^{1 \times r}$ be the $i^{th}$ row of $G$, $h_j \in R^r$ be the $j^{th}$ column of $H$ and $x_{ij}$ be the value of the $j^{th}$ feature in the $i^{th}$ sample. Define $g_i h_j = (GH)_{ij}$ as an inner product. The objective function in problem (6.1) can be rewritten as

$$\sum_{i=1}^m \sum_{j=1}^n \left( x_{ij} - g_i h_j \right)^2 \tag{6.2}$$

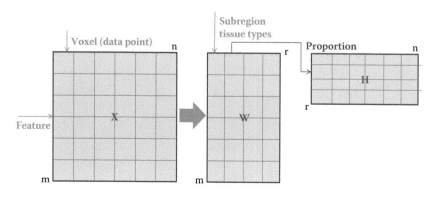

**FIGURE 6.1**
Matrix decomposition.

A solution to problem (6.1) can be found by truncating the singular value decomposition (SVD) of $X_{ij}$ (Udell et al. 2016). Let the SVD of $X$ be given by

$$X = U \Lambda V^T, \tag{6.3}$$

where $U = [u_1,...,u_r] \in R^{m \times r}$, $V = [v_1,...,v_r] \in R^{q \times n}$, $U^T U = I_{r \times r}$, $V^T V = I_{r \times r}$ and $\Lambda = dig(\lambda_1,...,\lambda_r) \in R^{r \times r}$ with $\lambda_1 \geq \lambda_2 \geq ... \geq \lambda_r > 0$. The columns of $U$ and $V$ are referred to as the left and right singular vectors of $X$, respectively, and $\lambda_1,...,\lambda_r$ are referred to as the singular values of $X$.

Substituting Equation 6.3 into Equation 6.1 gives

$$\min_{G,H} \quad \| U \Lambda V^T - GH \|_F^2,$$

which can be reduced to

$$\min_{G,H} \quad \| \Lambda - U^T GHV \|_F^2 \tag{6.4}$$

Let $Z = U^T GHV$ and the rank of $Z$ be $l$. Then, we have

$$\| \Lambda - Z \|_F^2 = \sum_{i=1}^{n} \sum_{j=1, j \neq i}^{r} Z_{ij}^2 + \sum_{i}^{r} (Z_{ii} - \lambda_i)^2 \tag{6.5}$$

To minimize $\| \Lambda - Z \|_F^2$, it must be $Z_{ij} = 0, i \neq j, Z_{ii} = 0, \forall i > l$ and $Z_{ii} = \lambda_i, i = 1,...,l$. In other words, the matrix which minimizes Equation 6.5 should be

$$Z = \begin{bmatrix} \Lambda_l & 0 \\ 0 & 0 \end{bmatrix},$$

where $\Lambda_l = diag(\lambda_1,...,\lambda_l)$. The error of $l$-rank matrix approximation to $X$ is

$$\| \Lambda - Z \|_F^2 = \sum_{i=l+1}^{r} \lambda_i^2 \tag{6.6}$$

Consequently,

$$U^T GHV = \begin{bmatrix} \Lambda_l & 0 \\ 0 & 0 \end{bmatrix} \quad \text{or}$$

$$GH = U \begin{bmatrix} \Lambda_l & 0 \\ 0 & 0 \end{bmatrix} V^T = U_l \Lambda_l V_l^T, \tag{6.7}$$

where $U_l = [u_1,...,u_l]$ and $V_l = [v_1,...,v_l]$.

Define $G = U_l \Lambda_l^{1/2}$ and $H = \Lambda_l^{1/2}$. The matrix factorization of $X$ is then given by

$$X \approx GH \tag{6.8}$$

### 6.2.1.1.2 *Sparse Coding and Nonnegative Matrix Factorization (NMF)*

In the matrix decomposition model (6.1), the tissue type (subregion) matrix $G$ and the proportion matrix $H$ are full matrices. However, each pixel or voxel will not have all tissue types or all subregions. Some elements in the proportion matrix $H$ should be zero. The matrix $H$ should be sparse. To achieve this, we can extend the matrix decomposition to allowing both small reconstruction error and sparseness.

Learning the parts of objects or performing the semantic segmentation requires imposing the non-negative constraints (Lee and Seung 1999). The non-negative constraints allow adding segments but prohibit segment subtraction. These non-negative constraints assume that all the elements of imaging data matrix $X$, the tissue or subregion matrix $G$, and proportion matrix $H$ are non-negative. The NMF problem can be formulated as

$$\min_{G,H} \frac{1}{2} \| X - GH \|_F^2 \tag{6.9}$$
$$\text{s.t. } G \geq 0, \ H \geq 0$$

The multiplicative gradient descent approach can be used to solve the optimization problem (6.9) (Lee and Seung 2001). By definition of the Frobenius norm of the matrix, $\frac{1}{2} \| X - GH \|_F^2$ can be expressed as

$$F = \frac{1}{2} \| X - GH \|_F^2 = \frac{1}{2} \operatorname{tr}\left((X - GH)^T (X - GH)\right)$$

To solve the problem (6.9), we can first set the partial derivatives of the function $F$ with respect to the matrix $H$ to zero:

$$\frac{\partial F}{\partial H} = -G^T (X - GH) = 0, \tag{6.10}$$

which implies

$$G^T GH = G^T X \tag{6.11}$$

To retain the negativity of the entry of the matrix, we use only element-wise multiplication and division. Element-wise multiplying by $H$ on both sites of Equation 6.11 obtains the following iterations:

$$\left(H^{k+1}\right) .* \left(G^T GH\right) = \left(H^k\right) .* \left(G^T X\right), \quad \text{or}$$
$$H^{k+1} = \left(H^k\right) .* \left(G^T X\right) . / \left(G^T GH\right) \tag{6.12}$$

Similarly, we obtain

$$G^{k+1} = G^k .* \left(XH^T\right) . / \left(GHH^T\right) \tag{6.13}$$

To enforce sparsity, we add a sparsity measure $f(H)$ to the objective function in the optimization problem (6.9), leading to

$$\min_{G,H} \frac{1}{2} \| X - GH \|_F^2 + \lambda f(H) \tag{6.14}$$

$$\text{s.t. } G \geq 0, \ H \geq 0,$$

where $\lambda \in R^+$ is a penalty parameter, which balances the reconstruction error and the sparsity measure.

In this section, we consider three sparsity measures: $L_1$, $L_{1/2}$, and $L_2$, which are, respectively, defined as

$$f_1(H) = (H)_{=1}\| H \|_1 = \sum_{i=1}^{r}\sum_{j=1}^{n}|h_{ij}|,$$

$$f_2(H) = \| H \|_{1/2} = \sum_{i=1}^{r}\sum_{j=1}^{n}\left(h_{ij}\right)^{1/2},$$

$$f_3(H) = \| H \|_2 = \sum_{i=1}^{r}\sum_{j=1}^{n}h_{ij}^2$$

The derivatives of the three norms of the matrix with the element $h_{ij}$ are given, respectively, by

$$\frac{\partial\| H \|_1}{\partial h_{ij}} = 1, \ h_{ij} > 0,$$

$$\frac{\partial\| H \|_{1/2}}{\partial h_{ij}} = \frac{1}{2\sqrt{h_{ij}}}, \tag{6.15}$$

$$\frac{\partial\| H \|_2}{\partial h_{ij}} = 2h_{ij}$$

Define objective functions:

$$F_i(H) = \frac{1}{2} \| X - GH \|_F^2 + \lambda f_i(H), \ i = 1,2,3 \tag{6.16}$$

Using Equation 6.15, we obtain

$$\frac{\partial F_1(H)}{\partial H} = -G^T(X - GH) + \lambda \mathbf{1}\mathbf{1}^T = 0,$$

$$\frac{\partial F_2(H)}{\partial H} = -G^T(X - GH) + \frac{\lambda}{2}H^{-1/2} = 0, \tag{6.17}$$

$$\frac{\partial F_3(H)}{\partial H} = -G^T(X - GH) + 2\lambda H = 0,$$

where $\mathbf{1} = [ 1 \ \cdots \ 1 ]^T$ and $H^{-1/2}$ denotes the reciprocal element-wise square root for each element in $H$.

Equation 6.16 can be further simplified to

$$1 = \left(G^T G\right). \big/ \left(G^T G H + \lambda \mathbf{11}^T\right),$$

$$1 = \left(G^T G\right). \big/ \left(G^T G H + \frac{\lambda}{2} H^{-1/2}\right), \qquad (6.18)$$

$$1 = \left(G^T G\right). \big/ \left(G^T G H + 2\lambda H\right).$$

Consider three penalty functions:

$$f(H) = \| H \|_1, f(H) = \| H \|_{1/2} \quad \text{and} \quad f(H) = \| H \|_2.$$

Using Equation 6.18 and similar techniques for deriving multiplicative iterative algorithms (6.12) and (6.13), we can obtain the algorithms for solving sparse NMF problems which are summarized in Result 6.1.

### Result 6.1 Algorithms for Sparse NMF

Step 1. Initialization. Use the matrix decomposition algorithm discussed in Section 6.2.1.1.1 to obtain the matrices $G^0$ and $H^0$ where all negative elements in the matrices are set to zero. Or, we can initialize W and H to random positive matrices.
Step 2. Update rules:
  $L_1$ NMF:

$$G^{k+1} = G^k_{\cdot} * \left(X\left(H^k\right)^T\right) \big/ \left(G^k H^k \left(H^k\right)^T\right), \qquad (6.19)$$

$$H^{k+1} = H^K_{\cdot} * \left(\left(G^{k+1}\right)^T G^{k+1}\right). \big/ \left(\left(G^{k+1}\right)^T G^{k+1} H^{k+1} + \lambda \mathbf{11}^T\right) \qquad (6.20)$$

  $L_{1/2}$ NMF:

$$G^{k+1} = G^k_{\cdot} * \left(X\left(H^k\right)^T\right). \big/ \left(G^k H^k \left(H^k\right)^T\right), \qquad (6.21)$$

$$H^{k+1} = H^K_{\cdot} * \left(\left(G^{k+1}\right)^T G^{k+1}\right). \big/ \left(\left(G^{k+1}\right)^T G^{k+1} H^{k+1} + \frac{\lambda}{2}\left(H^k\right)^{-1/2}\right) \qquad (6.22)$$

  $L_2$ NMF:

$$G^{k+1} = G^k_{\cdot} * \left(X\left(H^k\right)^T\right). \big/ \left(G^k H^k \left(H^k\right)^T\right), \qquad (6.23)$$

$$H^{k+1} = H^K_{\cdot} * \left(\left(G^{k+1}\right)^T G^{k+1}\right). \big/ \left(\left(G^{k+1}\right)^T G^{k+1} H^{k+1} + 2\lambda H^k\right) \qquad (6.24)$$

Step 3: Check for convergence
  Let $\varepsilon$ be a pre-specified error. If
  $\|G - G\|_F < \varepsilon$ and $\|H - H\|_F < \varepsilon$ then stop;
otherwise
  $k \leftarrow k + 1$, go to step 2.
  The parameter $\lambda$ can be set either by experience or by

$$\lambda = \frac{1}{\sqrt{m}} \sum_{k=1}^{n} \frac{\sqrt{n} - \dfrac{\|x_k\|_1}{\|x_k\|_2}}{\sqrt{n} - 1},$$

where $x_k$ is the $k^{th}$ column vector of the data matrix X.

### 6.2.1.2 Autoencoders

*6.2.1.2.1 Simple Autoencoders*

An autoencoder network compresses high-dimensional input data into low-dimensional data and discovers the hidden structure of high-dimensional input data. The autoencoder is an encoding function:

$$y = f(x, W, B) \approx x, \tag{6.25}$$

where $x \in R^n$ is an input vector, $W$ is a weight matrix, $B$ is a bias matrix, and $y$ is a vector of output. The architecture of the autoencoder is shown in Figure 6.2. The autoencoder consists of three layers: input layer with $n$ input neurons, hidden layer with $n_1$ neurons, and output layer with $n$ output neurons.

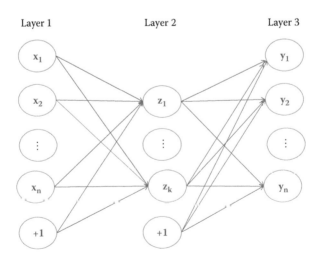

**FIGURE 6.2**
Autoencoder.

The weight for the connection from the $k^{th}$ neuron in the input layer to the $j^{th}$ neuron in the hidden layer is denoted by $w_{jk}^{(1)}$. The bias of the $j^{th}$ neuron in the hidden layer is denoted by $b_j^{(1)}$. The weighted input to the $j^{th}$ neuron and activation of the $j^{th}$ neuron in the hidden layer are denoted by $z_j^{(1)}$ and $a_j^{(1)}$, respectively. Then, the weighted input $z_j^{(1)}$ from the data input layer to the hidden layer and activation $a_j^{(1)}$ of the $j^{th}$ neuron in the hidden layer are given by

$$z_j^{(1)} = \sum_{k=1}^{n} w_{jk}^{(1)} x_k + b_j^{(1)}, j = 1, \dots n_1, \qquad (6.26)$$

and

$$a_j^{(1)} = \sigma\left(z_j^{(1)}\right), \qquad (6.27)$$

respectively. Logistic sigmoid is usually taken as $\sigma(z) = \dfrac{1}{1 + \exp(-z)}$. Let $W^{(1)} = (w_{jk}^{(1)})_{n \times n_1}$ be a weight matrix connecting the input data to the hidden neurons and $b^{(1)} = [b_1^{(1)} \quad \cdots \quad b_{n_1}^{(1)}]^T$ be a vector of biases in the hidden layer. In a matrix form, Equations 6.26 and 6.27 can be rewritten as

$$z^{(1)} = W^{(1)}x + b^{(1)}, \qquad (6.28)$$

and

$$a^{(1)} = \sigma\left(z^{(1)}\right), \qquad (6.29)$$

where $z^{(1)} = [z_1^{(1)} \quad \cdots \quad z_{n_1}^{(1)}]^T$, $a^{(1)} = [a_1^{(1)} \quad \cdots \quad a_{n_1}^{(1)}]^T$ and $\sigma(z)$ denotes an element-wise operation of a nonlinear function including the logistic sigmoid function. Similarly, for the output layer, we denote the weight for the connection from the $k^{th}$ neuron in the hidden layer to the $j^{th}$ neuron in the output layer by $w_{jk}^{(2)}$, the bias of the $j^{th}$ neuron in the output layer by $b_j^{(2)}$. The weighted input to the $j^{th}$ neuron and activation of the $j^{th}$ neuron in the output layer are denoted by $z_j^{(2)}$ and $a_j^{(2)}$, respectively. Similar to Equations 6.26 and 6.27, we obtain

$$z_j^{(2)} = \sum_{k=1}^{n_1} w_{jk}^{(2)} a_k^1 + b_j^{(2)}, j = 1, \dots, n_1, \qquad (6.30)$$

$$a_j^{(2)} = \sigma\left(z_j^{(2)}\right) \qquad (6.31)$$

Again, in a matrix form, Equations 6.30 and 6.31 can be written as

$$z^{(2)} = W^{(2)}x + b^{(2)}, \qquad (6.32)$$

and

$$a^{(2)} = \sigma\left(z^{(2)}\right), \tag{6.33}$$

where $z^{(2)} = [z_1^{(2)} \quad \cdots \quad z_n^{(2)}]^T$, $a^{(2)} = [a_1^{(2)} \quad \cdots \quad a_n^{(2)}]^T$, $W^{(2)} = (w_{jk}^{(2)})_{n_1 \times n}$ and $b^{(2)} = [b_1^{(2)} \quad \cdots \quad b_n^{(2)}]^T$.

The activations of the neurons in the output layer are the composition of activation functions in the hidden and input layers and are given by

$$y = \sigma\left(W^{(2)}\sigma\left(W^{(1)}x + b^{(1)}\right) + b^{(2)}\right) \tag{6.34}$$

Neural networks can be viewed as a general class of nonlinear functions from a vector $x$ of input variables to a vector $y$ of output variables. Our goal is to approximate output variables as accurately as possible using neural networks. Given a set of input variables $x^{(u)}$ and output variables $y^{(u)}$, $u = 1,\ldots,m$, a cost function for measuring the approximation error is defined as

$$C_E(W, b) = \frac{1}{2m}\sum_{u=1}^{m}\| y^{(u)} - x^{(u)} \|^2 \tag{6.35}$$

To estimate the parameters in the model (6.25), we minimize the cost function $C(W,b)$. Sparse representation serves (1) to reduce the dimension of the data and (2) to discover the structure hidden in the data. To achieve these, we often limit the activation of hidden neurons using the KL distance (Ehsan Hosseini 2016). The average activation of the hidden neuron $j$ is defined as

$$\hat{p}_j = \frac{1}{m}\sum_{u=1}^{m}a_j^{(1)}\left(x^{(u)}\right), \tag{6.36}$$

where

$$a_j^{(1)} = \sigma\left(\sum_{k=1}^{n}w_{jk}^{(1)}x_k^{(u)} + b_j^{(1)}\right)$$

Assume that $p$ is a sparsity parameter that is often selected as a small positive number near zero. To enforce sparsity, we set $\hat{p}_j = p$ and require that the average activation $\hat{p}_j$ of the hidden neuron should be close to the sparsity parameter $p$ as close as possible (Ehsan Hosseini 2016). Define the KL distance between $\hat{p}_j$ and $p$ as

$$C_{KL}(p\|\hat{p}) = \sum_{j=1}^{n_1}\left[p\log\frac{p}{\hat{p}_j} + (1-p)\log\frac{1}{1-\hat{p}_j}\frac{p}{}\right] \tag{6.37}$$

To prevent overweighting, we need to reduce the number of connections between neurons or the square of the weights. To achieve this, we can penalize the following measure:

$$C_W = \| W^{(1)} \|_F^2 + \| W^{(2)} \|_F^2 \qquad (6.38)$$

Summarizing Equations 6.35, 6.37, and 6.38, we obtain the total cost function for learning a sparse autoencoder (SAE):

$$C_{SAE}(W, b) = C_E(W, b) + \mu C_{KL}(p \| \hat{p}) + \lambda C_W, \qquad (6.39)$$

where $\mu$ and $\lambda$ are penalty parameters.

### 6.2.1.2.2 Deep Autoencoders

The basic autoencoders consist of two parts: the encoder and decoder. Deep autoencoder is a stacked autoencoder, which is a multi-multiple layer neural network consisting of sparse autoencoders (Figure 6.3). The outputs of each layer in the stacked autoencoders is sent to the inputs of the successive layer (Zhou et al. 2014). Now we investigate how the data are encoded and decoded in the deep autoencoders. Consider L simple autoencoders, each autoencoder consisting of a hidden layer and reconstruction layer.

For the $l^{th}$ autoencoder, the weight for the connection (encoding parameters) from the $k^{th}$ neuron in the output of the $(l-1)^{th}$ layer to the $j^{th}$ neuron in the hidden layer is denoted by $w_{jk}^{(l,1)}$. The bias of the $j^{th}$ neuron in the hidden layer is denoted by $b_j^{(l,1)}$. The weighted input to the $j^{th}$ neuron and activation of the $j^{th}$ neuron in the hidden layer are denoted by $z_j^{(l,1)}$ and $a_j^{(l,1)}$, respectively. Then, the weighted input $z_j^{(l,1)}$ from the $(l-1)^{th}$ layer to the hidden layer and

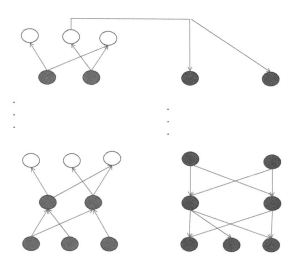

**FIGURE 6.3**
Stacked autoencoders and decoders.

activation $a_j^{(l,1)}$ of the $j^{th}$ neuron in the hidden layer are given by

$$z_j^{(l,1)} = \sum_{k=1}^{n_{l-1}} w_{jk}^{(l,1)} a_k^{(l-1)} + b_j^{(l,1)}, j = 1, \ldots n_l, \tag{6.40}$$

and

$$a_j^{(l,1)} = \sigma\left(z_j^{(l,1)}\right). \tag{6.41}$$

Let $W^{(l,1)} = (w_{jk}^{(l,1)})_{m_{l-1} \times n_l}$ be a weight matrix connecting the output of the $(l - 1)^{th}$ autoencoder to the hidden neurons of the $l^{th}$ encoder, $b^{(l,1)} = [b_1^{(l,1)} \quad \cdots \quad b_{n_1}^{(l,1)}]$ be a vector of biases in the hidden layer of the $l^{th}$ autoencoder. In a matrix form, Equations 6.40 and 6.41 can be rewritten as

$$z^{(l,1)} = W^{(l,1)} a^{(l-1,1)} + b^{(l,1)}, \tag{6.42}$$

and

$$a^{(l,1)} = \sigma\left(z^{(l,1)}\right), \tag{6.43}$$

where $z^{(l,1)} = [z_1^{(l,1)} \quad \cdots \quad z_{n_1}^{(l,1)}]^T$, $a^{(l,1)} = [a_1^{(l,1)} \quad \cdots \quad a_{n_1}^{(l,1)}]^T$, and $\sigma(z)$ denotes an element-wise operation of nonlinear function including logistic sigmoid function. Similarly, for the output layer of the $l^{th}$ autoencoder, we denote the weight for the connection (decoding parameters) from the $k^{th}$ neuron in the hidden layer of the $l^{th}$ autoencoder to the $j^{th}$ neuron in the output layer of the $l^{th}$ autoencoder by $w_{jk}^{(l,2)}$, the bias of the $j^{th}$ neuron in the output layer by $b_j^{(l,2)}$. The weighted input to the $j^{th}$ neuron and activation of the $j^{th}$ neuron in the output layer of the $l^{th}$ autoencoder are denoted by $z_j^{(l,2)}$ and $a_j^{(l,2)}$, respectively. Similar to Equations 6.42 and 6.43, we obtain

$$z_j^{(l,2)} = \sum_{k=1}^{n_1} w_{jk}^{(l,2)} a_k^{(l,1)} + b_j^{(l,2)}, j = 1, \ldots, m_1, \tag{6.44}$$

$$a_j^{(l,2)} = \sigma\left(z_j^{(l,2)}\right) \tag{6.45}$$

Again, in a matrix form, Equations 6.44 and 6.45 can be written as

$$z^{(l,2)} = W^{(l,2)} x + b^{(l,2)}, \tag{6.46}$$

and

$$a^{(l,2)} = \sigma\left(z^{(l,2)}\right), \tag{6.47}$$

where $z^{(l,2)} = [z_1^{(l,2)} \quad \cdots \quad z_{m_l}^{(l,2)}]^T$, $a^{(l,2)} = [a_1^{(l,2)} \quad \cdots \quad a_{m_l}^{(l,2)}]^T$, $W^{(l,2)} = (w_{jk}^{(l,2)})_{n_l \times m_l}$, and $b^{(l,2)} = [b_1^{(l,2)} \quad \cdots \quad b_{m_l}^{(l,2)}]^T$.

### 6.2.1.3 Parameter Estimation of Autoencoders

#### 6.2.1.3.1 Learning Nonnegativity Constrained Autoencoder

Now we introduce the backpropagation algorithm for parameter estimation of deep autoencoders.

Enforcing non-negativity constraints can learn a sparse, part-based representation of imaging data. To incorporate the non-negativity constraints in the weight matrix $W$, the function $C_W$ is changed to the following quadratic constraint (Hosseini-Asl 2016):

$$C_W = \frac{1}{2} \sum_{l=1}^{2} \sum_{i=1}^{s_l} \sum_{j=1}^{s_{l+1}} f\left(w_{ij}^{(l)}\right), \tag{6.48}$$

where

$$f\left(w_{ij}\right) = \begin{cases} w_{ij}^2 & w_{ij} < 0 \\ 0 & w_{ij} \geq 0 \end{cases}$$

Learning non-negativity constrained autoencoder can be formulated as solving the following optimization problem:

$$\min_{W,b} \quad C_{SAE}(W,b) = C_E(W,b) + \mu C_{KL}(p\|\hat{p}) + \lambda C_W \tag{6.49}$$

The backpropagation algorithm (Ng 2011) can be used to solve the minimization problem (6.49). The backpropagation algorithm comes from stochastic gradient methods. Let $\theta = [W,b]$ denote the parameters in the autoencoders and $C_{SAE}(\theta) = C_{SAE}(W,b)$. The standard gradient descent algorithm for updating the parameters $\theta$ of the objective $C_{SAE}(\theta)$ is given by

$$\theta = \theta - \alpha \frac{\partial E[C_{SAE}(\theta)]}{\partial \theta}, \tag{6.50}$$

or its sampling formula:

$$\theta = \theta - \alpha \frac{\partial C_{SAE}(\theta)}{\partial \theta} \tag{6.51}$$

Specifically, the update rules for the weights and biases are, respectively, given by

$$w_{jk}^{(l)} = w_{jk}^{(l)} - \alpha \frac{\partial C_{SAE}(W,b)}{\partial w_{jk}^{(l)}}, \tag{6.52}$$

and

$$b_j^{(l)} = b_j^{(l)} - \alpha \frac{\partial C_{SAE}(W,b)}{\partial b_j^{(l)}}, \tag{6.53}$$

where $\alpha > 0$ is a learning rate.

The derivative of the objective function with respect to the weights can be further decomposed to

$$\frac{\partial C_{SAE}(W,b)}{\partial w_{jk}^{(l)}} = \frac{\partial C_E(W,b)}{\partial w_{jk}^{(l)}} + \mu \frac{\partial C_{KL}(p\|\hat{p})}{\partial w_{jk}^{(l)}} + \lambda \gamma \left(w_{jk}^{(l)}\right), \qquad (6.54)$$

where

$$\gamma(x) = \begin{cases} w_{jk} & w_{jk} < 0 \\ 0 & w_{jk} \geq 0 \end{cases}$$

For convenience, Equation (6.35) can be rewritten as

$$C_E(W,b) = \frac{1}{2m} \sum_{u=1}^{m} C_E\left(W, b, x^{(u)}\right), \qquad (6.55)$$

where

$$C_E\left(W, b, x^{(u)}\right) = \| y^{(u)}(W,b) - x^{(u)} \|^2, \qquad (6.56)$$

and

$$y^{(u)}(W,b) = a_j^{(2)}\left(W, b, x^{(u)}\right) = \sigma\left(z_j^{(2)}(W, b, x^{(u)})\right) \qquad (6.57)$$

Using Equations 6.54 and 6.55, we obtain

$$\frac{\partial C_E(W,b)}{\partial w_{jk}^{(l)}} = \frac{1}{2m} \sum_{u=1}^{m} \frac{\partial C_E\left(W, b, x^{(u)}\right)}{\partial w_{jk}^{(l)}} \qquad (6.58)$$

Using Equations 6.30, 6.31, 6.56, 6.57, and the chain rule, we obtain

$$\frac{\partial C_E\left(W, b, x^{(u)}\right)}{\partial w_{jk}^{(2)}} = -\left(y_j^u(W,b) - x_j^{(u)}\right)\sigma'\left(z_j^{(2)}\right)a_k^{(1)}\left(x^{(u)}\right), \qquad (6.59)$$

where

$$\sigma'(x) = \sigma(x)(1 - \sigma(x)) \qquad (6.60)$$

Let

$$\delta_j^{(2)}(x^{(u)}) = \frac{\partial C_E(W, b, x^{(u)})}{\partial z_j^{(2)}} \qquad (6.61)$$

$$= -(y_j^u(W,b) - x_j^{(u)})\sigma'(z_j^{(2)})$$

Substituting Equation 6.61 into Equation 6.59 gives

$$\frac{\partial C_E(W, b, x^{(u)})}{\partial w_{jk}^{(2)}} = \delta_j^{(2)}(x^{(u)}) a_k^{(1)}(x^{(u)}), \ j = 1,...,n, \ k = 1,...,n_1 \qquad (6.62)$$

The next step for calculation of the cost function gradient is to move the error backward from the last layer to the hidden layer through the network. Note that cost $C_E(W,b,x^{(u)})$ is a function of activations $z_1^{(2)},...,z_n^{(2)}$, that is,

$$C_E\left(W, b, x^{(u)}\right) = C_E\left(W, b, x^{(u)}, z_1^{(2)}, ..., z_n^{(2)}\right) \qquad (6.63)$$

Using the chain rule, we have

$$
\begin{aligned}
\delta_j^{(1)}\left(x^{(u)}\right) &= \frac{\partial C_E(W, b, x^{(u)})}{\partial z_j^{(1)}} \\
&= \sum_{v=1}^{n} \frac{\partial C_E\left(W, b, x^{(u)}, z_1^{(2)}, ..., z_n^{(2)}\right)}{\partial z_v^{(2)}} \frac{\partial z_v^{(2)}}{\partial z_j^{(1)}} \\
&= \sum_{v=1}^{n} \delta_v^{(2)} \frac{\partial z_v^{(2)}}{\partial z_j^{(1)}}
\end{aligned}
\qquad (6.64)
$$

From Equations 6.27 and 6.30 it follows that

$$\frac{\partial z_v^{(2)}}{\partial z_j^{(1)}} = w_{vj}^{(2)} \sigma'\left(z_j^{(1)}\right) \qquad (6.65)$$

Substituting Equation 6.65 into Equation 6.64 yields

$$\delta_j^{(1)}\left(x^{(u)}\right) = \sum_{v=1}^{n} \delta_v^{(2)} w_{vj}^{(2)} \sigma'\left(z_j^{(1)}\right) \qquad (6.66)$$

Now we calculate the rate of change of the cost with respect to any weight connected to the hidden layer: $\frac{\partial C_E(W, b, x^{(u)})}{\partial w_{jk}^{(1)}}$.

Again, using chain rule, Equations 6.26 and 6.64, we obtain

$$
\begin{aligned}
\frac{\partial C_E(W, b, x^{(u)})}{\partial w_{jk}^{(1)}} &= \frac{\partial C_E(W, b, x^{(u)})}{\partial z_j^{(1)}} \frac{\partial z_j^{(1)}}{\partial w_{jk}^{(1)}} \\
&= \delta_j^{(1)} x_k^{(u)}
\end{aligned}
\qquad (6.67)
$$

Now we calculate the rate of change $\dfrac{\partial C_E(W,b,x^{(u)})}{\partial b_j^{(l)}}$ of the cost with respect to any bias in the network. Using Equations 6.61 and 6.64, we obtain

$$\frac{\partial C_E\left(W,b,x^{(u)}\right)}{\partial b_j^{(2)}} = \frac{\partial C_E\left(W,b,x^{(u)}\right)}{\partial z_j^{(2)}}\frac{\partial z_j^{(2)}}{\partial b_j^{(2)}} = \delta_j^{(2)}, \tag{6.68}$$

and

$$\frac{\partial C_E\left(W,b,x^{(u)}\right)}{\partial b_j^{(1)}} = \frac{\partial C_E\left(W,b,x^{(u)}\right)}{\partial z_j^{(1)}}\frac{\partial z_j^{(1)}}{\partial b_j^{(1)}} = \delta_j^{(1)}. \tag{6.69}$$

Next calculate $\dfrac{\partial C_{KL}(p\|\hat{p})}{\partial w_{jk}^{(l)}}$ and $\dfrac{\partial C_{KL}(p\|\hat{p})}{\partial b_j^{(l)}}$. Using Equation 6.37, we obtain

$$\begin{aligned}\frac{\partial C_{KL}(p\|\hat{p})}{\partial w_{jk}^{(l)}} &= -\left[\frac{p}{\hat{p}_j} - \frac{1-p}{1-\hat{p}_j}\right]\frac{\partial \hat{p}_j}{\partial w_{jk}^{(l)}}\\[2mm] &= \frac{\hat{p}_j - p}{\hat{p}_j\left(1-\hat{p}_j\right)}\frac{\partial \hat{p}_j}{\partial w_{jk}^{(l)}}.\end{aligned} \tag{6.70}$$

From Equation 6.36 it follows that

$$\frac{\partial \hat{p}_j\left(x^{(u)}\right)}{\partial w_{jk}^{(l)}} = \sigma'\left(z_j^{(1)}\right)x_k^{(u)}. \tag{6.71}$$

### 6.2.1.3.2 Learning Deep Nonnegativity Constrained Autoencoder

Learning deep autoencoders, we need to extend the cost function from a single autoencoder to multiple autoencoders. The average sum-of-square error $C_{SAE}(W,b,x)$ will not be changed. However, we need to change the KL-distance measure for sparsity. Let

$$\hat{p}_j^{(l)} = \frac{1}{m}\sum_{u=1}^{m}a_j^{(l)}\left(x^{(u)}\right), \; j = 1,\ldots,n_l, \; l = 1,\ldots,L, \tag{6.72}$$

$$a_j^{(l)}\left(x^{(u)}\right) = \sigma\left(\sum_{k=1}^{n_l}w_{jk}^{(l)}x_k^{(u)} + b_j^{(l)}\right). \tag{6.73}$$

Define the KL-distance measure for the whole multiple autoencoders as

$$C_{SKL}(P\|\hat{P}) = \sum_{l=1}^{L}\sum_{j=1}^{n_l}\left[p^{(l)}\log\frac{p^{(l)}}{\hat{p}_j} + \left(1-p^{(l)}\right)\log\frac{1-p^{(l)}}{1-\hat{p}_j^{(l)}}\right]. \tag{6.74}$$

First, we calculate $\dfrac{\partial \hat{p}_j^{(v)}}{\partial w_{kg}^{(l)}}$.

For a fixed sample $x^{(u)}$, consider three cases: $l > v$, $l = v$ and $l < v$.

1. $l > v$

If $l > v$, $\hat{p}_j^{(v)}$ is not a function of $w_{kg}^{(l)}$. Thus

$$\frac{\partial \hat{p}_j^{(v)}}{\partial w_{kg}^{(l)}} = 0 \tag{6.75}$$

2. $l = v$

Using Equation 6.40, we obtain

$$\frac{\partial \hat{p}_j^{(v)}}{\partial w_{jg}^{(v)}} = \sigma'\left(z_j^{(v)}\right) \frac{\partial z_j^{(v)}}{\partial w_{jg}^{(v)}}$$

$$= \sigma'\left(z_j^{(v)}\right) a_g^{(v-1)}, \; g = 1, \ldots, n_{v-1} \tag{6.76}$$

3. $l < v$

Again, using Equation 6.40, we have

$$\frac{\partial \hat{p}_j^{(v)}}{\partial w_{qg}^{(l)}} = \sigma'\left(z_j^{(v)}\right) \sum_{k=1}^{n_{v-1}} w_{jk}^{(v)} \frac{\partial \hat{p}_k^{(v-1)}}{\partial w_{qg}^{(l)}}, \; q = 1, \ldots, n_{v-1}, q \neq j,$$

$$g = 1, \ldots, n_{v-2} \tag{6.77}$$

and

$$\frac{\partial \hat{p}_j^{(v)}}{\partial w_{jg}^{(l)}} = \sigma'\left(z_j^{(v)}\right) w_{jg}^{(v)} \frac{\partial \hat{p}_g^{(v-1)}}{\partial w_{jg}^{(l)}}. \tag{6.78}$$

Similarly, we have the results for $\dfrac{\partial \hat{p}_j^{(v)}}{\partial b_j^{(l)}}$:

$$\frac{\partial \hat{p}_j^{(v)}}{\partial b_g^{(l)}} = \begin{cases} 0 & l > v \\ \sigma'(z_j^{(v)}) & l = v \\ \sigma'(z_j^{(v)}) \sum_{k=1}^{n_{v-1}} w_{jk}^{(v)} \dfrac{\partial \hat{p}_k^{(v-1)}}{\partial b_g^{(l)}}, \; g = 1, \ldots, n_{v-1} & l < v \end{cases} . \tag{6.79}$$

Therefore, summarizing the above discussions, we have

$$\frac{\partial C_{SKL}(p\|\hat{p})}{\partial w_{jk}^{(l)}} = \sum_{v=l}^{L} \frac{\hat{p}_j^{(v)} - p^{(v)}}{\hat{p}_j^{(v)}\left(1 - \hat{p}_j^{(v)}\right)} \frac{\partial \hat{p}_j^{(v)}}{\partial w_{jk}^{(l)}}, \tag{6.80}$$

$$\frac{\partial C_{SKL}(p\|\hat{p})}{\partial b_g^{(l)}} = \sum_{v=l}^{L} \frac{\hat{p}_j^{(v)} - p^{(v)}}{\hat{p}_j^{(v)}\left(1 - \hat{p}_j^{(v)}\right)} \frac{\partial \hat{p}_j^{(v)}}{\partial b_g^{(l)}}. \tag{6.81}$$

Finally, using Equation 6.75, we obtain

$$\frac{\partial C_{SW}}{\partial w_{jk}^{(l)}} = g\left(w_{jk}^{(l)}\right), \; j = 1, \dots, n_l \; , \; k = 1, \dots, n_{l-1}, \tag{6.82}$$

where

$$g\left(w_{jk}^{(l)}\right) = \begin{cases} \left|w_{jk}^{(l)}\right| & w_{jk}^{(l)} < 0 \\ 0 & w_{jk}^{(l)} \geq 0 \end{cases}.$$

Putting the above together, we obtain the desired partial derivatives of the cost function:

$$\frac{\partial C_{SAE}(W,b)}{\partial w_{jk}^{(l)}} = \frac{\partial C_E(W,b)}{\partial w_{jk}^{(l)}} + \mu \frac{\partial C_{SKL}(p\|\hat{p})}{\partial w_{jk}^{(l)}} + \lambda \frac{\partial C_{SW}}{\partial w_{jk}^{(l)}}$$

$$= \frac{1}{2m} \sum_{u=1}^{m} \delta_j^{(l)}\left(x^{(u)}\right) a_k^{(l-1)}\left(x^{(u)}\right) +, \tag{6.83}$$

$$\frac{\mu}{m} \sum_{v=l}^{L} \frac{\hat{p}_j^{(v)}\left(x^{(u)}\right) - p^{(v)}}{\hat{p}_j^{(v)}\left(x^{(u)}\right)\left(1 - \hat{p}_j^{(v)}(x^{(u)})\right)} \frac{\partial \hat{p}_j^{(v)}\left(x^{(u)}\right)}{\partial w_{jk}^{(l)}} + \lambda g\left(w_{jk}^{(l)}\right),$$

$$\frac{\partial C_{SAE}(W,b)}{\partial b_j^{(l)}} = \frac{\partial C_E(W,b)}{\partial b_j^{(l)}} + \mu \frac{\partial C_{SKL}(p\|\hat{p})}{\partial b_j^{(l)}}$$

$$= \frac{1}{2m} \sum_{u=1}^{m} \delta_j^{(l)}\left(x^{(u)}\right) \tag{6.84}$$

$$+ \frac{\mu}{m} \sum_{v=l}^{L} \frac{\hat{p}_j^{(v)}\left(x^{(u)}\right) - p^{(v)}}{\hat{p}_j^{(v)}\left(x^{(u)}\right)\left(1 - \hat{p}_j^{(v)}\left(x^{(u)}\right)\right)} \frac{\partial \hat{p}_j^{(v)}\left(x^{(u)}\right)}{\partial b_g^{(l)}}.$$

The backpropagation algorithm for learning deep autoencoders is summarized as Result 6.2.

### Result 6.2 Backpropagation Algorithm for Learning Deep Autoencoders

Step 1. Initialization. Parameters $w_{jk}^{(l)}$ and $b_j^{(l)}$ in each layer are randomly initialized to small values near to zero using the $N(0,\varepsilon^2)$ distribution with a small prespecified $\varepsilon$.

Step 2. Perform forward activation calculations.

1. Calculate the activation of the neurons in the input layer.

$$z_j^{(1)} = \sum_{k=1}^{n} w_{jk}^{(1)} x_k + b_j^{(1)}, j = 1,...n_1,$$

$$a_j^{(1)} = \sigma\left(z_j^{(1)}\right)$$

2. For $l = 2,....,L$
   For each neuron in layer $l$, set

$$z_j^{(l)} = \sum_{k=1}^{n_{l-1}} w_{jk}^{(l)} a_k^{(l-1)} + b_j^{(l)}, j = 1,...n_l,$$

$$a_j^{(l)} = \sigma\left(z_j^{(l)}\right)$$

Step 3. Update parameters $w_{jk}^{(l)}$ and $b_j^{(l)}$.

1. Set penalty parameters $\mu$ and $\lambda$.
2. For each out neuron $j = 1,...,n$ in the output layer and each neuron $k = 1,...,n_{L-1}$ in the $(L-1)^{th}$ layer, set

$$\delta_j^{(L)}\left(x^{(u)}\right) = \frac{\partial C_E\left(W, b, x^{(u)}\right)}{\partial z_j^{(L)}}$$

$$= \left(x_j^{(u)} - a_j^{(u)}\left(x^{(u)}\right)\right) \sigma'\left(z_j^{(L)}(x^{(u)})\right),$$

$$\frac{\partial C_E\left(W, b, x^{(u)}\right)}{\partial w_{jk}^{(L)}} = \delta_j^{(L)}\left(x^{(u)}\right) a_k^{(L-1)},$$

$$\frac{\partial C_E\left(W, b, x^{(u)}\right)}{\partial b_j^{(L)}} = \delta_j^{(L)}\left(x^{(u)}\right),$$

$$g\left(w_{jk}^{(L)}\right) = \begin{cases} \left|w_{jk}^{(L)}\right| & w_{jk}^{(L)} < 0 \\ 0 & w_{jk}^{(L)} \geq 0 \end{cases},$$

$$\frac{\partial C_{SAE}(W, b)}{\partial w_{jk}^{(L)}} = \frac{1}{2m} \sum_{u=1}^{m} \delta_j^{(L)}\left(x^{(u)}\right) a_k^{(L-1)}\left(x^{(u)}\right) + \lambda g\left(w_{jk}^{(L)}\right),$$

$$\frac{\partial C_{SAE}(W, b)}{\partial b_j^{(L)}} = \frac{1}{2m} \sum_{u=1}^{m} \delta_j^{(l)}\left(x^{(u)}\right),$$

$$w_{jk}^{(L)} = w_{jk}^{(L)} - \alpha \frac{\partial C_{SAE}(W, b)}{\partial w_{jk}^{(L)}},$$

$$b_j^{(L)} = b_j^{(L)} - \alpha \frac{\partial C_{SAE}(W, b)}{\partial b_j^{(L)}}$$

3. For $l = L - 1, \ldots, 1$,

$$\delta_j^{(l)}\left(x^{(u)}\right) = \sum_{k=1}^{n_{l+1}} \delta_k^{(l+1)}\left(x^{(u)}\right) w_{kj}^{(l+1)} \sigma'\left(z_j^{(l)}\right),$$

$$\frac{\partial C_E\left(W, b, x^{(u)}\right)}{\partial w_{jk}^{(l)}} = \delta_j^{(l)}\left(x^{(u)}\right) a_k^{(l-1)}\left(x^{(u)}\right),$$

$$\frac{\partial C_E\left(W, b, x^{(u)}\right)}{\partial b_j^{(l)}} = \delta_j^{(l)}\left(x^{(u)}\right),$$

$$\hat{p}_j^{(l)} = \frac{1}{m} \sum_{u=1}^{m} a_j^{(l)}\left(x^{(u)}\right), j = 1, \ldots, n_l, \ l = 1, \ldots, L$$

$$a_j^{(l)}\left(x^{(u)}\right) = \sigma\left(\sum_{k=1}^{n_l} w_{jk}^{(l)} x_k^{(u)} + b_j^{(l)}\right),$$

$$\frac{\partial \hat{p}_j^{(v)}}{\partial w_{kg}^{(l)}} = 0, l > v,$$

$$\frac{\partial \hat{p}_j^{(v)}}{\partial w_{jg}^{(v)}} = \sigma'\left(z_j^{(v)}\right) a_g^{(v-1)}, g = 1, \ldots, n_{v-1}, l = v,$$

$$\frac{\partial \hat{p}_j^{(v)}}{\partial w_{qg}^{(l)}} = \sigma'\left(z_j^{(v)}\right) \sum_{k=1}^{n_{v-1}} w_{jk}^{(v)} \frac{\partial \hat{p}_k^{(v-1)}}{\partial w_{qg}^{(l)}}, q = 1, \ldots, n_{v-1}, q \neq j,$$

$$g = 1, \ldots, n_{v-2}, l < v,$$

$$\frac{\partial \hat{p}_j^{(v)}}{\partial w_{jg}^{(l)}} = \sigma'\left(z_j^{(v)}\right) w_{jg}^{(v)} \frac{\partial \hat{p}_g^{(v-1)}}{\partial w_{jg}^{(l)}}, l < v,$$

$$\frac{\partial \hat{p}_j^{(v)}}{\partial b_g^{(l)}} = \begin{cases} 0 & l > v \\ \sigma'\left(z_j^{(v)}\right) & l = v \\ \sigma'\left(z_j^{(v)}\right) \sum_{k=1}^{n_{v-1}} w_{jk}^{(v)} \frac{\partial \hat{p}_k^{(v-1)}}{\partial b_g^{(l)}}, g = 1, \ldots, n_{v-1} \ l < v, \end{cases}$$

$$\frac{\partial C_{SKL}(p||\hat{p})}{\partial w_{jk}^{(l)}} = \sum_{v=l}^{L} \frac{\hat{p}_j^{(v)} - p^{(v)}}{\hat{p}_j^{(v)}\left(1 - \hat{p}_j^{(v)}\right)} \frac{\partial \hat{p}_j^{(v)}}{\partial w_{jk}^{(l)}},$$

$$\frac{\partial C_{SKL}(p\|\hat{p})}{\partial b_g^{(l)}} = \sum_{v=l}^{L} \frac{\hat{p}_j^{(v)} - p^{(v)}}{\hat{p}_j^{(v)}\left(1 - \hat{p}_j^{(v)}\right)} \frac{\partial \hat{p}_j^{(v)}}{\partial b_g^{(l)}},$$

$$\frac{\partial C_{SW}}{\partial w_{jk}^{(l)}} = g\left(w_{jk}^{(l)}\right), j = 1, \dots, n_l, k = 1, \dots, n_{l-1},$$

where

$$g(w_{jk}^{(l)}) = \begin{cases} \left|w_{jk}^{(l)}\right| & w_{jk}^{(l)} < 0 \\ 0 & w_j^{(l)} \geq 0 \end{cases},$$

$$\frac{\partial C_E(W,b)}{\partial w_{jk}^{(l)}} = \frac{1}{2m} \sum_{u=1}^{m} \delta_j^{(l)}\left(x^{(u)}\right) a_k^{(l-1)}\left(x^{(u)}\right),$$

$$\frac{\partial C_{SKL}(p\|\hat{p})}{\partial w_{jk}^{(l)}} = \frac{1}{m} \sum_{v=l}^{L} \frac{\hat{p}_j^{(v)}\left(x^{(u)}\right) - p^{(v)}}{\hat{p}_j^{(v)}\left(x^{(u)}\right)\left(1 - \hat{p}_j^{(v)}\left(x^{(u)}\right)\right)} \frac{\partial \hat{p}_j^{(v)}\left(x^{(u)}\right)}{\partial w_{jk}^{(l)}},$$

$$\frac{\partial C_{SAE}(W,b)}{\partial w_{jk}^{(l)}} = \frac{\partial C_E(W,b)}{\partial w_{jk}^{(l)}} + \mu \frac{\partial C_{SKL}(p\|\hat{p})}{\partial w_{jk}^{(l)}} + \lambda \frac{\partial C_{SW}}{\partial w_{jk}^{(l)}},$$

$$\frac{\partial C_E(W,b)}{\partial b_j^{(l)}} = \frac{1}{2m} \sum_{u=1}^{m} \delta_j^{(l)}\left(x^{(u)}\right),$$

$$\frac{\partial C_{SKL}(p\|\hat{p})}{\partial b_j^{(l)}} = \frac{1}{m} \sum_{v=l}^{L} \frac{\hat{p}_j^{(v)}\left(x^{(u)}\right) - p^{(v)}}{\hat{p}_j^{(v)}\left(x^{(u)}\right)\left(1 - \hat{p}_j^{(v)}\left(x^{(u)}\right)\right)} \frac{\partial \hat{p}_j^{(v)}\left(x^{(u)}\right)}{\partial b_g^{(l)}},$$

$$\frac{\partial C_{SAE}(W,b)}{\partial b_j^{(l)}} = \frac{\partial C_E(W,b)}{\partial b_j^{(l)}} + \mu \frac{\partial C_{SKL}(p\|\hat{p})}{\partial b_j^{(l)}},$$

$$w_{jk}^{(l)} = w_{jk}^{(l)} - \alpha \frac{\partial C_{SAE}(W,b)}{\partial w_{jk}^{(l)}},$$

$$b_j^{(l)} = b_j^{(l)} - \alpha \frac{\partial C_{SAE}(W,b)}{\partial b_j^{(l)}}.$$

Step 4. Check for convergence
$\left|C_{SAE}(W^{(t+1)},b^{(t+1)}) - \left|C_{SAE}(W^{(t)},b^{(t)})\right|\right| < \varepsilon$, that is, difference in cost function between the current and previous interactions is less than the prespecified error $\varepsilon$ then stop; otherwise, go to step 1 and repeat the iteration.

### 6.2.1.4 *Convolutional Neural Networks*

In autoencoders, every neuron in the layer will be connected to all the neurons in the previous layer. The number of parameters being estimated will be very large. For example, if we assume that an imaging has $100 \times 100$ pixels, the number of neurons that are connected with the neurons in the input should be 10,000. Therefore, it is necessary to restrict the connections between neurons. Convolutional neural networks (CNN) can achieve this goal by local connectivity (LeCun 1989). For example, as shown in Figure 6.4, the CNN connects each hidden neuron only to a small number of neighboring neurons of the input vector. The deep CNN enforces local connectivity in many layers.

   CNN are originated from application of neural networks to imaging data analysis and are designed to process the multiple array data. A CNN intends to use spatial information across the pixels of an image. The CNN architecture consists of a stack of distinct layers that transform the input volume into an output volume: convolutional layer, pooling layer, rectified linear unit (ReLU) layer, pooling layer, fully connected layer, and loss layer. An essential component of the CNN is convolution. First, we study convolution.

### 6.2.1.4.1 *Convolution or Cross-Correlation Operation*

Convolution is a filter that extracts features and removes noises from the data. Suppose that $f(t)$ is a noisy signal function. To improve accuracy and obtain a less noisy signal, we average several measurements near time $t$. The more recent measurements are more relevant. When we average nearby measurements, we should give more weight to the more recent measurements. Let $g(t)$ be a weight function that is often called a filter or kernel. The signal with less noise can be estimated by the following weighted average operation that is called convolution:

$$s(t) = (f*g)(t) = \int_{-\infty}^{\infty} f(\tau)g(t - \tau)d\tau, \tag{6.85}$$

where * denotes convolution. The first argument is called the input and the second argument is called the kernel. The output is often called the feature map.

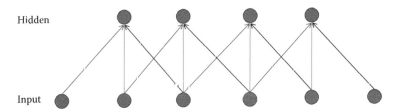

**FIGURE 6.4**
Scheme of locally connected neurons.

In practice, we can only measure signals at regular intervals. If we assume that the signal $f(t)$ and the kernel $g(t)$ can only take values at integer $t$, then the integral in Equation 6.85 can be discretized and the discrete convolution can be defined as

$$s(t) = (f*g)(t) = \sum_{-\infty}^{\infty} f(\tau)g(t - \tau) \tag{6.86}$$

In imaging data analysis, we use a two-dimensional image $A$ as input. A one-dimensional convolution needs to be extended to a two-dimensional convolution. Suppose that image signals (or activations) are denoted as

$$A : \{ a_{i,j}, i = 1, ..., n_1, j = 1, ..., n_2 \}$$

and the kernel is denoted as

$$W = \{ w_{u,v}, u = 0, ..., h_1, v = 0, ..., h_2 \}$$

Discrete convolution of the image or activation $A$ with kernel $W$ is defined as

$$(A*W)_{ij} = \sum_{u=0}^{h_1} \sum_{v=0}^{h_2} w_{u,v} a_{i-u,j-v} \tag{6.87}$$

Let $m = i - u$, $n = j - v$. Then, Equation 6.86 will be changed to

$$(A*W)_{ij} = \sum_{m} \sum_{n} a_{m,n} w_{i-m,j-n} \tag{6.88}$$

In implementation, instead of convolution operation, we often use cross-correlation that is defined as

$$(A*W)_{ij} = \sum_{m} \sum_{n} a_{i+m,j+n} w_{m,n} \tag{6.89}$$

**Example 6.1**

Consider a $4 \times 4$ input matrix $A$ and a $2 \times 2$ kernel matrix $W$ (Figure 6.5). Convolution of $W$ with $A$ generates a $3 \times 3$ output matrix (the feature map).

The structure of the CNN is shown in Figure 6.6. The CNN consists of a set of learnable kernels. Each kernel is a three-dimensional matrix with width, height, and depth. For example, the size $6 \times 6 \times 3$ indicates a pixel with 6 width and height, and depth of the input data. We slide the kernel over the width and height of the data in the previous layer, which generates a two-dimensional activation or feature map that outputs the responses of the kernel at every position of the previous layer.

Depth of the convolutional layer is defined as the number of kernels that is used for the convolution operation. In the network shown in Figure 6.6, three kernels are used for performing convolution of the original image, thus producing three different feature maps as shown. These three feature maps can be thought of as stacked $2d$ matrices. The "depth" of the feature map is three. A depth column that is defined as a set of neurons

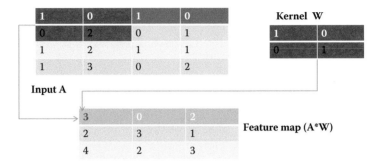

**FIGURE 6.5**
An example of convolution.

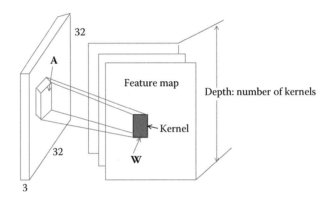

**FIGURE 6.6**
Structure of convolution layer.

with convolution involving the same region of the input is often used in the investigation of parameter sharing.

Stride is defined as the number of pixels with which we slide the kernel over the input matrix. The stride 1 indicates that we move the kernel one pixel at a time. Similarly, stride 2 indicates that the kernel jumps two pixels at a time when the kernel moves. To control the size of output, we often pad the input data with zeros around the border. The convolution with adding zero-padding is referred to as wide convolution, and convolution without zero-padding is called a narrow convolution.

To design the convolution layer, we need to compute the spatial size of the convolutional layer as a function of the input volume size (W), the size (F) of the kernel, the stride (S) with which the kernel moves, and the amount of zero padding used (P) on the border. The size of the convolutional layer in one dimension is calculated by

$$m = \frac{(W - F + 2P)}{S} + 1 \tag{6.90}$$

### Example 6.2

Consider a pace arrangement of convolutional layers as shown in Figure 6.7. We assume $W = 7$, $F = 3$, $P = 1$, and $S = 2$. The size of the output (feature map) is $m = \dfrac{(7 - 3 + 2)}{2} + 1 = 4$. The numbers inside the neurons of the feature maps are the values of the convolutions of the input with the kernel vector $[1,0,-1]$. The elements of the kernel vector are the weights that are also represented by red, green, and purple colors. The weights are shared by four neurons in the feature map.

### Example 6.3

Consider a large, deep CNN with size $227 \times 227 \times 3$ which classifies 1.2 millions of images in the Image Net, a large-scale annotated dataset (Krizhevsky et al. 2012). Each neuron in the convolutional layer uses the kernel with size $11 \times 11$ ($F = 11$), stride $S = 4$, and no zero padding ($P = 0$). Using Equation 6.89, we obtain the size of the output (feature map): $m = \dfrac{(227 - 11 + 0)}{4} + 1 = 55$. In the convolutional layer, $K = 96$ (depth) different kernels were used. Therefore, the convolutional layer consists of 96 feature maps, each feature map having size $55 \times 55$. A total of $55 \times 55 \times 96 = 290{,}400$ neurons in the convolutional layer were used. Each of 290,400 neurons was connected to a region of size $[11 \times 11 \times 3]$ in the previous layer. All 96 neurons in each depth column were connected to the same region of size $[11 \times 11 \times 3]$ in the previous layer, but with different weights. Consequently, each neuron had $11 \times 11 \times 3 = 363$ weights. The total number of weights in the convolutional layer is $290400 \times 363 = 105705600$. It is clear that the number of parameters in the network is too large.

  To control the number of parameters in the convolutional layer, a parameter sharing strategy should be used. In other words, the same parameter should be used for more than one neuron. A feature map is

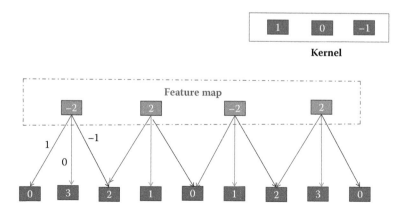

**FIGURE 6.7**
Space arrangement of convolutional layer.

often called a depth slice. In this example, there were 96 depth slices. The kernel in each depth slice used a unique set of weights ($11 \times 11 \times 3 = 363$). The total number of parameters was ($363 \times 96 = 34848$ weights + 96 biases) 34,544. It is clear that parameter sharing dramatically reduces the number of parameters in the convolutional layer.

The convolutional layer is the essential block of the architecture of the CNN. The convolutional layer consists of many feature maps. Each feature map is obtained by convolution of the input layer or previous feature maps with the specific kernel. Assume that the $l^{th}$ layer is a convolutional layer. It consists of $m_1^{(l)}$ feature maps with size $m_2^{(l)} \times m_3^{(l)}$. The input of the convolution layer consists of $m_1^{(l-1)}$ feature maps of size $m_2^{(l-1)} \times m_3^{(l-1)}$. When $l = 1$, the input is an image. Let $A_i^{(l)}$ and $B_i^{(l)}$, respectively, be the activation matrix and bias matrix of the $i^{th}$ feature map in the $l^{th}$ layer:

$$
A_i^{(l)} = \begin{bmatrix} a_{11}^{(l)} & \cdots & a_{1m_3^{(l)}}^{(l)} \\ \vdots & \vdots & \vdots \\ a_{m_2^{(l)}1}^{(l)} & \cdots & a_{m_2^{(l)}m_3^{(l)}}^{(l)} \end{bmatrix}, B_i^{(l)} = \begin{bmatrix} b_{11}^{(l)} & \cdots & b_{1m_3^{(l)}}^{(l)} \\ \vdots & \vdots & \vdots \\ b_{m_2^{(l)}1}^{(l)} & \cdots & b_{m_2^{(l)}m_3^{(l)}}^{(l)} \end{bmatrix}
$$

Define $W_{i,j}^{(l)}$ as the kernel matrix of size $(2h_1^{(l)} + 1) \times (2h_2^{(l)} + 1)$ that connects the $j^{th}$ feature map in the layer $l - 1$ with the $i^{th}$ feature map in the $l^{th}$ layer as follows.

$$
W_{i,j}^{(l)} = \begin{bmatrix} \left(W_{i,j}^{(l)}\right)_{-h_1^{(l)},-h_2^{(l)}} & \cdots & \left(W_{i,j}^{(l)}\right)_{-h_1^{(l)},h_2^{(l)}} \\ \vdots & \vdots & \vdots \\ \left(W_{i,j}^{(l)}\right)_{h_1^{(l)},-h_2^{(l)}} & \cdots & \left(W_{i,j}^{(l)}\right)_{h_1^{(l)},h_2^{(l)}} \end{bmatrix}
$$

The activation matrix $A_i^{(l)}$ is computed by convolution as follows:

$$
\left(A_i^{(l)}\right)_{r,s} = \left(B_i^{(l)}\right)_{r,s}
$$
$$
+ \sum_{j=1}^{m_1^{(l-1)}} \sum_{u=-h_1^{(l)}}^{h_1^{(l)}} \sum_{v=-h_2^{(l)}}^{h_2^{(l)}} \left(W_{i,j}^{(l)}\right)_{u,v} \left(A_j^{(l-1)}\right)_{r+u,s+v} \tag{6.91}
$$

### 6.2.1.4.2 Nonlinear Layer (ReLU)

Since most of the real data are nonlinear and convolution is a linear operation, after every convolution operation, a nonlinear map is often used to capture the nonlinearity of the original data. Assume that the $l^{th}$ layer is a non-linear layer and $(l - 1)^{th}$ layer is its input layer. Let $m_1^{(l)}$ and $m_1^{(l-1)}$ be the number of feature maps in the $l^{th}$ layer and $(l - 1)^{th}$ layer, respectively. We also assume that the size of each feature map is the same, that is, $m_2^{(l-1)} = m_2^{(l)}$ and $m_3^{(l-1)} = m_3^{(l-1)}$. Let $A_i^{(l)}$ and $A_i^{(l-1)}$ be the activation matrix in the $i^{th}$ feature map of the nonlinear layer ($l^{th}$ layer) and input layer (($l - 1)^{th}$ layer), respectively. The nonlinear mapping is given by

$$\left(A_i^{(l)}\right)_{rs} = \sigma\left(\left(A_i^{(l-1)}\right)_{rs}\right), \tag{6.92}$$

where $\sigma$ is a nonlinear activation function, and can be either a tanh function: $\tanh(x) = \dfrac{e^x - e^{-x}}{e^x + e^x}$, or sigmoid function: $sigm(x) = \dfrac{1}{1 + e^{-x}}$.

However, the sigmoid functions often have serious limitations. For the backpropagation process in a neural network, the errors will be squeezed by (at least) a quarter at each layer by activation using the sigmoid function. Therefore, the deeper the network is, more information from the data will be "lost." Even some large errors from the output layer might not be able to affect the weights of a neuron in the previous layers.

To overcome the limitations of the sigmoid function as a nonlinear map function, a rectified linear unit (ReLU) as a nonlinear operation is often used in the recent CNNs. Suppose that the $l^{th}$ layer is a rectified linear layer. Activation in the ReLU is defined as

$$\left(A_i^{(l)}\right)_{r,s} = \max\left(0, \left(A_i^{(l-1)}\right)_{r,s}\right) \tag{6.93}$$

ReLU is the simplest non-linear activation function. It is applied to every neuron in the previous layer and it is an element-wise operation. After the ReLU operation, all negative input signals from the feature map will be replaced by zero. The stage that performs the nonlinear map is often called the detector stage.

### 6.2.1.4.3 Feature Pooling and Subsampling Layer

To reduce the spatial size of the representation and the number of parameters and computation in the network of the feature maps, we often periodically insert a pooling layer in-between successive convolution layers in the CNN. The purpose of spatial pooling is to reduce the dimensionality of each feature map but retain the most essential information. Three types of functions: Max, Average, and Sum are often used for polling operation.

We use the max operation independently on every depth slice of the input feature maps to resize it spatially. Let $l$ be a pooling layer. Assume that the numbers of feature maps in the pooling layer and its input layer are the same, that is, $m_1^{(l)} = m_1^{(l-1)}$. We define a spatial neighborhood, for example, with a window of size $u \times v$. At each window max pooling takes the largest element from the spatial neighborhood. Specifically,

$$\left(A_i^{(l)}\right)_{r,s} = \max_{u,v}\left((A_i^{(l-1)})_{r+u,s+v}\right), \; 1 \leq u \leq p, 1 \leq v \leq q \tag{6.94}$$

Similarly, the average pooling operation is defined as

$$\left(A_i^{(l)}\right)_{r,s} = \frac{1}{pq} \sum_{u=1}^{p} \sum_{v=1}^{q} \left(A_i^{(l-1)}\right)_{r+u,s+v}, \tag{6.95}$$

and the sum pooling operation is defined as

$$\left(A_i^{(l)}\right)_{r,s} = \sum_{u=1}^{p} \sum_{v=1}^{q} \left(A_i^{(l-1)}\right)_{r+u,s+v} \tag{6.96}$$

**Example 6.4**

Consider a rectified feature map of size 6 × 6 and a 3 × 3 kernel (Figure 6.8). Let $S = 3$. The 6 × 6 input matrix is pooled with a kernel of size 3 and stride 3 into a 2 × 2 output matrix. Each max operation is taken over 9 numbers (3 × 3 colored square). Four max pooling operations are performed over four colored squares: red, green, yellow, and purple, which produce a 2 × 2 output matrix.

### 6.2.1.4.4 Normalization Layer

Let the $l^{th}$ layer be a normalization layer. We consider two normalization methods: subtractive normalization and brightness normalization. First, we study subtractive normalization. For each feature map in the input layer $l - 1$, its corresponding output activation in the normalization layer $l$ is defined as

$$A_i^{(l)} = A_i^{(l-1)} - \sum_{j=1}^{m_{l-1}} W_{G(\sigma)} * A_j^{(l-1)}, \tag{6.97}$$

where the Gaussian filter $W_{G(\sigma)}$ is defined as

$$\left(W_{G(\sigma)}\right)_{r,s} = \frac{1}{\sqrt{2\pi\sigma^2}} \exp\left(\frac{r^2 + s^2}{2\sigma^2}\right).$$

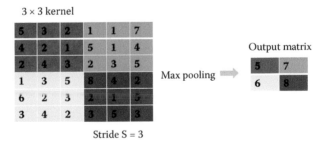

**FIGURE 6.8**
Max pooling operation.

The second popular normalization method is the brightness normalization. The output of the normalization layer for the brightness normalization is defined as

$$\left(A_i^{(l)}\right)_{r,s} = \frac{\left(A_i^{(l-1)}\right)_{r,s}}{\left(\kappa + \mu \sum_{j=1}^{m_1^{(l-1)}} \left(A_j^{(l-1)}\right)_{r,s}^2\right)^\mu}, \tag{6.98}$$

where $\kappa$ and $\mu$ are parameters.

### 6.2.1.4.5 Fully Connected Layer

Finally, after several convolutional layers, max-pooling layers, the task of the CNN will be done via the fully connected layer. Neurons in a fully connected layer have full connections to all activations in the previous layer. Suppose that the $l^{th}$ layer is a fully connected layer and the $(l-1)^{th}$ layer with $m_1^{(l-1)}$ feature maps of size $m_2^{(l-1)} \times m_3^{(l-1)}$ is connected to the $l^{th}$ layer. The activation of the $i^{th}$ unit in the $l^{th}$ layer is given by

$$a_i^{(l)} = f\left(z_i^{(l)}\right), z_i^{(l)} = \sum_{j=1}^{m_1^{(l-1)}} \sum_{u=1}^{m_2^{(l-1)}} \sum_{v=1}^{m_3^{(l-1)}} \left(W_{i,j}^{(l)}\right)_{u,v} \left(A_j^{(l-1)}\right)_{u,v}, \tag{6.99}$$

where $f$ is a nonlinear function and $(W_{i,j}^{(l)})_{u,v}$ is the weight connecting the unit at position $(u,v)$ in the $j^{th}$ feature map of the $(l-1)^{th}$ layer and the $i^{th}$ unit in the $l^{th}$ layer. A softmax is often used as the nonlinear activation function. It transforms the output of each neuron in the $l^{th}$ layer to the interval $[0,1]$. Softmax operation also requires that the total sum of the outputs should be equal to 1. The output of the softmax activation function is equivalent to a class probability density function. Assume that the total number of neurons in the $l^{th}$ layer is $J$. The softmax function is mathematically defined as

$$f\left(z_i^{(l)}\right) = \frac{\exp\left(z_i^{(l)}\right)}{\sum_{j=1}^{J} \exp\left(z_j^{(l)}\right)} \tag{6.100}$$

### 6.2.1.4.6 Parameter Estimation in Convolutional Neural Networks

The computation of CNN is complex. It consumes a large amount of the computational resources (Wei et al. 2017). Learning parameters of CNN consist of two major procedures: the forward procedure and backpropagation procedure. Parameter estimation is iterated between two procedures. In this section, we introduce the backpropagation algorithm for learning CNN which extends the derivation of backpropagation in CNN based on an example with two convolutional layers (Zhang et al. 2016).

The structure of CNN is shown in Figure 6.9. Consider $L$ groups. Each group consists of the pooling layer, convolution layer, and ReLU layer. In the first group, the pooling layer will be replaced by an input layer. In the last

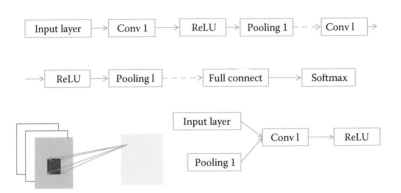

**FIGURE 6.9**
CNN structure.

group, the convolution layer will be replaced by a fully connected layer and ReLU layer will be replaced by the softmax layer. The pooling layer, convolution layer, and ReLU layer in the $l^{th}$ group will be denoted as Sl, Cl, and Rl, respectively. Let $P_l$ be the number of feature maps of the convolution layer Cl. For the $p^{th}$ feature map of the $l^{th}$ group, the output of the $(i,j)^{th}$ neuron in the pooling layer Sl is denoted as $x_p^l(i,j)$, the kernel matrix for the convolution operation is denoted by $W_p^l = (w_p^l(u,v))_{k_1 \times k_2}$, the output of the $(i,j)^{th}$ neuron in the convolution layer Cl is denoted as $z_p^l(i,j)$, the activation output of the $(i,j)^{th}$ neuron in the ReLU layer Rl is denoted by $a_p^l(i,j)$, and a bias is denoted by $b_p^l$.

*6.2.1.4.6.1 Forward Propagation* All parameters should be randomly initialized. They include the kernel matrix $W_p^l = (w_p^l(u,v))_{k_1 \times k_2}$, bias $b_p^l$, the weight matrix $W_{FL}$, and the bias vector $b_{FL}$ in the fully connected layer.

*6.2.1.4.6.1.1 Convolution Layer C1* Let $x_q^1(i,j)$ , $i = 1, ..., m_1$ , $j = 1, ..., n_1$ , $q = 1, ..., q_i$ be the signal of the $(i,j)^{th}$ neuron in the $q^{th}$ input layer and $q_i$ be the number of input layers. Convolution operation of the kernel with the input signals gives the output of the $(i,j)^{th}$ neuron in the convolution layer C1:

$$z_p^1(i,j) = \sum_{q=1}^{q_i} \sum_{u=0}^{k_1-1} \sum_{v=0}^{k_2-1} w_p^1(u,v)x_q^1(i+u,j+u) + b_p^1, \ p = 1, ..., p_1 ,$$

$$i = 1, ..., m_1, \ j = 1, ..., n_1,$$

(6.101)

where $i,j$ are indices for the row and column of the $p^{th}$ feature map. Only the results of convolution without the zero-padding are retained.

*6.2.1.4.6.1.2 ReLU Layer R1* The input to the layer R1 is the output from the convolution layer C1. The activation output of the nonlinear map in layer R1 is given by

$$a_p^1(i,j) = \sigma\left(z_p^1(i,j)\right), \, p = 1, \ldots, p_1 \, , \, i = 1, \ldots, m_1, \, j = 1, \ldots, n_1, \tag{6.102}$$

where $\sigma$ is a nonlinear activation function.

*6.2.1.4.6.1.3 Pooling Layer S1* MAX operation or average operation is often used to pool the activation signals in the ReLU layer. The output $x_p^1(i,j)$, ($i = 1, \ldots, g_1, j = 1, \ldots, h_1$) of the $(i, j)^{th}$ neuron in the pooling layer S1using max operation is given by

$$x_p^1(i,j) = \max\left(a_p^1(i+u, j+v), \, 1 \le u \le r, 1 \le v \le s\right) \tag{6.103}$$

Similarly, the output $x_p^1(i,j)$ of the $(i, j)^{th}$ neuron in the pooling layer S1using average pooling operation is given by

$$x_p^1(i,j) = \frac{1}{rs} \sum_{u=1}^{r} \sum_{v=1}^{s} a_p^1(i+u, j+v), \tag{6.104}$$

*6.2.1.4.6.1.4 Convolution Layer Cl* Now we consider the $l^{th}$ group. Let $x_p^l(i,j)$, $i = 1, \ldots, m_l, j = 1, \ldots, n_l$ be the signal of the $(i, j)^{th}$ neuron in the $l^{th}$ pooling layer Sl. Convolution operation of the kernel with the signals in the pooling layer gives the output of the $(i, j)^{th}$ neuron in the convolution layer Cl:

$$z_p^l(i,j) = \sum_{u=0}^{k_1-1} \sum_{v=0}^{k_2-1} w_p^l(u, v) x_p^{l-1}(i+u, j+u) + b_p^l, \, p = 1, \ldots, p_l,$$
$$i = 1, \ldots, m_l, \, j = 1, \ldots, n_l, \tag{6.105}$$

where $i,j$ are indices for the row and column of the feature map, $m_l$ and $n_l$ are the number of rows and columns of the convolution matrix in the $l^{th}$ layer, respectively, and $p_l$ is the number of feature maps in the $l^{th}$ convolution layer Cl. Only the results of convolution without the zero-padding are retained.

*6.2.1.4.6.1.5 ReLU Layer Rl* The input to the layer Rl is the output from the convolution layer Cl. The activation output of the nonlinear map in layer Rl is given by

$$a_p^l(i,j) = \sigma\left(z_p^l(i,j)\right), \, p = 1, \ldots, p_l, \, i = 1, \ldots, m_l, \, j = 1, \ldots, n_l, \tag{6.106}$$

where $\sigma$ is a nonlinear activation function.

*6.2.1.4.6.1.6 Pooling Layer Sl* MAX operation or average operation is often used to pool the activation signals in the ReLU layer. Again, extension of pooling operation in the first layer to the general $l^{th}$ layer is straightforward.

Let $p_l$ be the number of feature maps, $g_l$ and $h_l$ be the number of rows and columns, respectively, in the $l^{th}$ pooling layer SL. The output $x_p^l(i, j)$, ($i = 1, ..., g_l$, $j = 1, ..., h_l$, $p = 1, ..., p_l$) of the $(i, j)^{th}$ neuron in the pooling layer Sl using the max operation is given by

$$x_p^l(i, j) = \max\left(a_p^l(i + u, j + v), \ 1 \leq u \leq r, 1 \leq v \leq s\right) \tag{6.107}$$

Similarly, the average operation is to average the activation values of the neurons in the region of size $r \times s$ in the layer Rl. The output $x_p^l(i, j)$ of the $(i, j)^{th}$ neuron in the pooling layer S1using average pooling operation is given by

$$x_p^l(i, j) = \frac{1}{rs} \sum_{u=1}^{r} \sum_{v=1}^{s} a_p^l(i + u, j + v) \tag{6.108}$$

*6.2.1.4.6.1.7 Fully Connected Layer*  The neurons in the fully connected layer are arranged as the same as that in the input image. Assume that there are $q_i$ maps in the fully connected layer. Neurons in a fully connected layer have full connections to all activations in the previous $(L - 1)^{th}$ pooling layer. The signals in the $(L - 1)^{th}$ pooling layer are $x_p^{L-1}(i, j)$, ($i = 1, ..., $ g1, $j = 1, ..., h_l$, $p = 1, ..., p_l$).

Let $y_q(i, j)$ be the signal input to the $(i, j)^{th}$ neuron of the $q^{th}$ map of the fully connected layer. Then, $y_q(i, j)$ is given by

$$y_q(i, j) = \sum_{p=1}^{p_{L-1}} \sum_{u=1}^{g_{L-1}} \sum_{v=1}^{h_{L-1}} w_{i,j}^{q,p}(u, v) x_p^{L-1}(u, v) + b_q(i, j), \tag{6.109}$$

where $w_{i,j}^{q,p}$ is the weight of the $(i, j)^{th}$ neuron in the $q^{th}$ map of the fully connected layer corresponding to the $(u,v)^{th}$ neuron of the $p^{th}$ feature map in the $(L - 1)^{th}$ pooling layer, and $b_q(i, j)$ is the bias. The output of the $(i, j)^{th}$ neuron of the $q^{th}$ map of the fully connected layer is

$$a_q(i, j) = \sigma\left(y_q(i, j)\right), \tag{6.110}$$

where $\sigma(.)$ is an activation function.

*6.2.1.4.6.2 Loss Function*  Let $x_q^1(i, j)$, $i = 1, ..., m_1$, $j = 1, ..., n_1$, $q = 1, ..., q_i$ be the signal of the $(i, j)^{th}$ neuron in the $q^{th}$ input layer and $q_i$ be the number of input layers. The CNN is used to reconstruct the input images. The mean square error between the input images and the output of the fully connected layer for assessing the accuracy of the reconstruction is defined as

$$E = \frac{1}{2} \sum_{q=1}^{q_i} \sum_{i=1}^{m1} \sum_{j=1}^{n_1} \left(a_q(i, j) - x_q^1(i, j)\right)^2 \tag{6.111}$$

The mean square error will be used as a loss function. Reconstructing the input images is to adjust the weights such that the output $a_q(i, j)$ of the final fully connected layer is as close as possible to the original images $x_q^1(i, j)$. The backpropagation algorithm is to change the weights according to the gradient descent direction of the square errors.

*6.2.1.4.6.3 Backpropagation*  The backpropagation is a gradient method. It calculates the partial derivatives of the loss function with respect to the weights, biases, and parameters in the kernels from the back to start.

*6.2.1.4.6.3.1  Fully Connected Layer*  We first calculate the partial derivatives $\dfrac{\partial E}{\partial w_{i,j}^{q,p}(u, v)}$. Define

$$e_q(i, j) = \left(a_q(i, j) - x_q^1(i, j)\right) \tag{6.112}$$

Using Equation 6.111 and the chain rule, we obtain

$$\frac{\partial E}{\partial w_{i,j}^{q,p}(u, v)} = e_q(i, j) \frac{\partial a_q(i, j)}{\partial w_{i,j}^{q,p}(u, v)}$$

$$= e_q(i, j) \frac{\partial a_q(i, j)}{\partial y_q(i, j)} \frac{\partial y_q(i, j)}{\partial w_{i,j}^{q,p}(u, v)} \tag{6.113}$$

$$= e_q(i, j)\sigma'\left(y_q(i, j)\right) \frac{\partial y_q(i, j)}{\partial w_{i,j}^{q,p}(u, v)}$$

Recall that

$$\sigma'\left(y_q(i, j)\right) = \sigma\left(y_q(i, j)\right)\left(1 - \sigma(y_q(i, j))\right)$$

$$= a_q(i, j)\left(1 - a_q(i, j)\right) \tag{6.114}$$

Using Equation 6.109, we obtain

$$\frac{\partial y_q(i, j)}{\partial w_{i,j}^{q,p}(u, v)} = x_{p(u,v)}^{L-1} \tag{6.115}$$

Substituting Equations 6.114 and 6.115 into Equation 6.113 gives

$$\frac{\partial E}{\partial w_{i,j}^{q,p}(u, v)} = e_q(i, j)a_q(i, j)\left(1 - a_q(i, j)\right)x_{p(u,v)}^{L-1} \tag{6.116}$$

Similarly, we can derive

$$\frac{\partial E}{\partial b_q(i, j)} = e_q(i, j)a_q(i, j)\left(1 - a_q(i, j)\right) \tag{6.117}$$

Define

$$\delta_p^{L-1}(u,v) = \frac{\partial E}{\partial x_p^{L-1}(u,v)} \tag{6.118}$$

Using Equations 6.109–6.112, 6.114, and 6.118, we obtain

$$\delta_p^{L-1}(u,v) = \sum_{q=1}^{q_i} \sum_{i=1}^{m_1} \sum_{j=1}^{n_1} e_q(i,j) \frac{\partial a_q(i,j)}{\partial x_p^{L-1}(u,v)}$$

$$= \sum_{q=1}^{q_i} \sum_{i=1}^{m_1} \sum_{j=1}^{n_1} e_q(i,j) a_q(i,j) \left(1 - a_q(i,j)\right) w_{i,j}^{q,p}(u,v) \tag{6.119}$$

*6.2.1.4.6.3.2  Pooling Layer Sl*   Since no parameters in the pooling layers need to be estimated, there are no learning tasks to be performed on the pooling layers. To keep track of the pooling, the index that is used during the forward pass is also used for the gradient routing during backpropagation.

*6.2.1.4.6.3.3  Convolution Layer Cl*   Convolution between the $p^{th}$ feature map of dimension $m_l \times n_l$ and the weight kernel of dimension $k_1 \times k_2$ generates an output map.
Define

$$\delta_p^l(i,j) = \frac{\partial E}{\partial z_p^l(i,j)} \ , i = 1, ..., m_l, j = 1, ..., n_l, p = 1, ..., p_l \tag{6.120}$$

Using chain rule and Equations 6.105, 6.106, and 6.120, we obtain

$$\frac{\partial E}{\partial w_p^l(u,v)} = \sum_{i=1}^{m_l} \sum_{j=1}^{n_l} \frac{\partial E}{\partial z_p^l(i,j)} \frac{\partial z_p^l(i,j)}{\partial w_p^l(u,v)}$$

$$= \sum_{i=1}^{m_l} \sum_{j=1}^{n_l} \delta_p^l(i,j) x_p^l(i+u, j+v) \tag{6.121}$$

In Equation 6.121, to transform cross-correlation to convolution, we can flip the matrix $\delta_p^l(i,j)$, which leads to the following equation:

$$\frac{\partial E}{\partial w_p^l(u,v)} = \text{rot}_{180^\circ} \left\{ \delta_p^l(i,j) \right\} * x_p^l(u,v), \tag{6.122}$$

where

$$\text{rot}_{180^\circ} \left\{ \delta_p^l(i,j) \right\} * x_p^l(u,v) = \sum_{i=1}^{m_l} \sum_{j=1}^{n_l} \delta_p^l(i,j) x_p^l(u,v)$$

Now we derive the recursive formula for $\delta_p^l(i,j)$ which measures how the change in a single pixel $z_p^l(i,j)$ in the feature map influences the loss

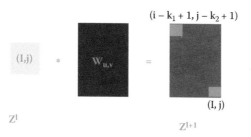

**FIGURE 6.10**
Output region of the (1 + 1)th convolution layer affected by convolution of the kernel with pixel (i,j) in the lth convolution layer.

function $E$. From Equation 6.105, we can see that the value of the variable $z_p^l(i, j)$ in the $l^{th}$ convolution layer affects the output of the variables in the region $[i - k_1 + 1, i] \times [j - k_2 + 1, j]$ (Figure 6.10). Using chain rule and Equation 6.120, we obtain

$$
\delta_{i,j}^l = \frac{\partial E}{\partial z_p^l(i, j)}
$$

$$
= \sum_{u=0}^{k_1-1} \sum_{v=0}^{k_2-1} \frac{\partial E}{\partial z_p^{l+1}(i - u, j - v)} \frac{\partial z_p^{l+1}(i - u, j - v)}{\partial z_p^l(i, j)} \tag{6.123}
$$

$$
= \sum_{u=0}^{k_1-1} \sum_{v=0}^{k_2-1} \delta_p^{l+1}(i - u, j - v) \frac{\partial z_p^{l+1}(i - u, j - v)}{\partial z_p^l(i, j)}
$$

It follows from Equation 6.102 and 6.105 that

$$
z_p^{l+1}(i - u, j - v) = \sum_{m'=0}^{k_1-1} \sum_{n'=0}^{k_2-1} w_p^{l+1}(m', n') \sigma\left( z_p^l(i - u + m', j - v + n') \right)
$$

$$
+ b_p^{l+1} \tag{6.124}
$$

Note that

$$
\frac{\partial \sigma\left( z_p^l(i - u + m', j - v + n') \right)}{\partial z_p^l(i, j)} = \begin{cases} \dfrac{\partial \sigma\left( z_p^l(i, j) \right)}{\partial z_p^l(i, j)} & m' = u, n' = v \\ \\ 0 & \text{otherwise} \end{cases} \tag{6.125}
$$

Substituting Equation 6.125 into Equation 6.124 and using the chain rule, we obtain

$$
\frac{\partial z_p^{l+1}(i - u, j - v)}{\partial z_p^l(i, j)} = w_p^{l+1}(u, v) \sigma'\left( z_p^l(i, j) \right) \tag{6.126}
$$

Substituting Equation 6.126 into Equation 6.123 gives

$$\delta_{i,j}^l = \sum_{u=0}^{k_1-1} \sum_{v=0}^{k_2-1} \delta_p^{l+1}(i-u, j-v) w_p^{l+1}(u, v) \sigma'\left(z_p^l(i, j)\right) \tag{6.127}$$

If we use the flipped kernel, then Equation 6.127 can be rewritten as

$$
\begin{aligned}
\delta_{i,j}^l &= \mathrm{rot}_{180^0}\left\{\sum_{u=0}^{k_1-1} \sum_{v=0}^{k_2-1} \delta_p^{l+1}(i+u, j+v) w_p^{l+1}(u, v)\right\} \sigma'\left(z_p^l(i, j)\right) \\
&= \delta_p^{l+1}(i, j) * \mathrm{rot}_{180^0}\left\{w_p^{l+1}(u, v)\right\} \sigma'\left(z_p^l(i, j)\right),
\end{aligned}
\tag{6.128}
$$

where * denotes the convolution operation.

## 6.2.2 Supervised Deep Learning Methods for Image Segmentation

Segmentation is an important step in imaging data analysis pipelines (Litjens et al. 2017). Image segmentation attempts to identify the set of voxels with specific structures, image space variation, and characteristics. Image segmentation includes pixel segmentation, instance segmentation, and part-based segmentation (Garcia-Garcia et al. 2017). Two types of deep neural networks: CNNs and recurrent neural networks (RNNs) are widely used in medical image segmentation (Ronneberger et al. 2015; Xie et al. 2016). Methods for pixel level segmentation are the basis for all types of segmentation, in this book we focus on pixel level segmentation.

### 6.2.2.1 Pixel-Level Image Segmentation

Pixel level image segmentation is used to assign class labels to each pixel based on image features. CNNs are often used to learn appropriate feature representations for the image segmentation problems and achieve great success (Ciresan et al. 2012; Farabet et al. 2013). A popular CNN is fully convolutional neural networks (FCNNs) that can effectively generate features and use end-to-end training (Lin et al. 2016; Long et al. 2015). The limitation of FCNNs for sematic segmentation is their low-resolution. Contextual relationships widely existed but have not been explored for semantic segmentation by the classical CNNs. Recently, sematic correlations between image regions are incorporated into FCNNs to predict the labels of the image pixel using conditional random fields (CRFs) (Lin et al. 2016).

#### 6.2.2.1.1 CRF for Modeling Semantic Pair-Wise Relations

A CRF is a discriminative undirected probabilistic graphic model. Consider a graph $G = (V, E)$ where $V$ denotes a set of nodes and $E$ denotes a set of edges. Let $X$ be a set of observed feature variables and $Y$ be a set of output variables for which we predict given observed feature variables $X$ (Sutton and McCallum, 2011). Before using CRF to model semantic pair-wise relations, we apply CNNs to generate a feature map from the image data (Figure 6.11a).

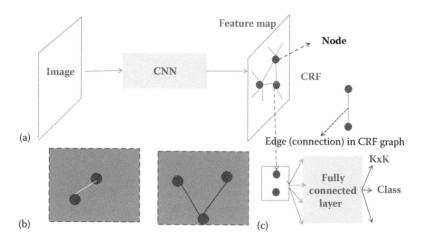

**FIGURE 6.11**
CRF model. (a) Generate a feature map from the image data using CNN. (b) Spatial range box.
(c) Pairwise net.

Each neuron (location) in the feature map is denoted by a node in the CRF graph and it is also connected with all other nodes which are located within the same spatial range box (the pink dashed box in Figure 6.11b) (Lin et al. 2016). If the spatial relations are different, then the different spatial range box will be defined, and different edges will be used to model the different spatial relations (yellow edges and purple edges in Figure 6.11b). To reduce the dimensions of the data in the model, we can define feature vectors that are associated with each node and edge.

Next, we define the distributions over the graphs which take the compatibility of the input-output pair into account (Lin et al. 2016). A given input image is denoted by $x$ and its pixel-wise label is denoted by $y$. Let $X = \{x\}$ be a collection of imaging data and $Y = \{y\}$ be a collection of labels that correspond to the label configuration of each node in the CRF. Let $V = X \cup Y$. We first define a conditional distribution $P(y \mid x)$ of $y$, given $x$. Let $E(y,x,\theta)$ be an energy function, which measures the compatibility of the input-output pair. All the parameters in the CRF are denoted by $\theta$. The conditional distribution for one image is defined as

$$p(y|x) = \frac{1}{Z(x)} e^{-E(y,x)} \tag{6.129}$$

Since $\sum_y p(y \mid x) = 1$, it requires $Z(x) = \sum_y e^{-E(y,x)}$. The function $Z(x)$ is referred to as the partition function.

Next, we define the energy function. Since the CRF is a factor graph $G$, the conditional distribution should be factorized according to $G$:

$$p(y|x) = \frac{1}{Z(x)} \prod_{\alpha=1}^{A} \Psi_\alpha(y_\alpha, x_\alpha), \tag{6.130}$$

where $\Psi_\alpha(y_\alpha, x_\alpha)$ is often called the potential function and has the exponential form:

$$\Psi_\alpha(y_\alpha, x_\alpha) = e^{-E(y_\alpha, x_\alpha)}, \tag{6.131}$$

and the set $A$ of potential functions can be divided into two subsets: the set of unary potential functions and pair-wise potential functions (Lin et al. 2016).

To calculate the unary potential functions, we first stack the feature maps and a shallow fully connected network. The fully connected network is also called the unary-net. The output of the unary note can be taken as a unary potential function. Let $K$ be the number of classes. The output of the node in the unary-net is a $K$-dimensional vector. The unary potential function for the $p^{th}$ node in the unary-net and $y_p$-th class is defined as

$$U\left(y_p, x_p; \theta_U\right) = -\beta_{p,y_p}(x; \theta_U), \tag{6.132}$$

where $\beta_{p,y_p}$ is the output value of the $p^{th}$ node in the unary-net and $y_p$-th class, and $\theta_U$ is the set of parameters in the unary-net $U$. Let $A_U$ be a set of all unary potentials and $N_U$ be a set of nodes for the potential $U$. Then the energy function for the unary potential is defined as

$$E_U(y, x, \theta) = \sum_{U \in A_U} \sum_{p \in N_U} U\left(y_p, x_p; \theta_U\right) \tag{6.133}$$

Next, we investigate the pair-wise potential functions. Consider two connected nodes $p$ and $q$ in the CRF graph. Let $x_p$ and $x_q$ be a feature vector of the nodes $p$ and $q$, respectively. The feature vectors $x_p$ and $x_q$ are from the feature map. Let $x_{pq} = [x_p^T, x_q^T]^T$ be the edge feature vector. The edge feature vector $x_{pg}$ is then imputed to the fully connected network that is called pairwise-net (Figure 6.11c). The total number of classes is $K \times K$, the number of all possible label combinations for a pair of nodes. The pairwise-net predicts the class, denoted by $(y_p, y_q)$. Let $\beta_{p,q,y_p,y_q}(x; \theta_V)$ be the output value of the pairwise-net corresponding to $x_{pq}$, which quantifies the compatibility of the classes $(y_p, y_q)$ under the input image data $x$. The pairwise potential function for the pair of $p$ and $q$ is defined as

$$V\left(y_p, y_q, x_{pq}, \theta_V\right) = -\beta_{p,q,y_p,y_q}(x; \theta_V), \tag{6.134}$$

where $\theta_V$ are the parameters of CNN in defining potential $V$. Let $A_V$ be the set of all types of pairwise potentials and $\varepsilon_V$ be the set of edges for the calculation of pairwise potential $V$. The energy function for a set of pairwise potential is defined as

$$E_V(y, x, \theta) = \sum_{V \in A_V} \sum_{(p,q) \in \varepsilon_V} V\left(y_p, y_q, x_{pq}, \theta_V\right) \tag{6.135}$$

The total energy function is then defined by a set of unary and pairwise potentials:

$$E(y, x, \theta) = E_U(y, x, \theta) + E_V(y, x, \theta) \tag{6.136}$$

### 6.2.2.1.2 Parameter Estimation for CRF Models

Negative log-likelihood can be used to estimate the parameters in CRF models. From Equation 6.129 it follows that the negative log-likelihood for one image is given by

$$- \log p(y|x, \theta) = E(y, x, \theta) + \log Z(x, \theta) \tag{6.137}$$

Assume that $N$ images in the training set are sampled. Let $x_i$ and $y_i$ be the imaging signals and labels of the pixels in the $i^{th}$ image. The negative log-likelihood for the $N$ images is then given by

$$- \sum_{i=1}^{N} \log P\left(y^{(i)} | x^{(i)}, \theta\right) = \sum_{i=1}^{N}\left[E\left(y^{(i)}, x^{(i)}, \theta\right) + \log Z\left(x^{(i)}, \theta\right)\right] \tag{6.138}$$

To improve the efficiency and prediction accuracy, the parameters in the model should be reduced. Therefore, the penalized terms for the parameters should be incorporated into the negative likelihood in Equation 6.137. Finally, the parameter estimation problem is reduced to

$$\min_{\theta}\ l(y, x, \theta) = \sum_{i=1}^{N}\left[E\left(y^{(i)}, x^{(i)}, \theta\right) + \log Z\left(x^{(i)}, \theta\right)\right] + \frac{\lambda}{2} \|\theta\|_2^2 \tag{6.139}$$

Since exact maximum-likelihood estimation that requires repeatedly direct calculations of $Z(x, \theta)$ and its partial derivatives is very computationally expansive, approximate CRF learning methods should be developed. Popular CRF approximate learning methods include pseudo-likelihood learning and piecewise learning (Besag 1977; Sutton and McCallum 2005; Lin et al. 2016). In this section, we will focus on introducing piecewise learning. The basic idea underlying piecewise learning is to divide the whole model (graph) into pieces (subgraphs) which are learned independently, and finally combining the learned weights from each submodel at test time. Assume that the set of nodes that define the unary potential and the set of edges (nodes) that define pairwise potential form pieces. For each piece, we define the likelihood. These likelihoods are independent. The conditional likelihood for the whole model can be approximated by a product of the pieces of independent likelihoods (Lin et al. 2016):

$$P(y|x) = \prod_{U \in A_U} \prod_{p \in N_U} P_U\left(y_p | x\right) \prod_{V \in A_V} \prod_{(p,q) \in r_V} P_V\left(y_p, y_q | x\right), \tag{6.140}$$

where the likelihood $P_U(y_p\,|\,x)$ is defined in terms of the unary potential:

$$P_U\left(y_p\,|\,x\right) = \frac{\exp\left\{-U\left(y_p, x_p\right)\right\}}{\sum_{y_h}\exp\{-U(y_h, x_h)\}},\tag{6.141}$$

and the likelihood $P_V(y_p,y_q\,|\,x)$ is defined in terms of the pairwise potential:

$$P_V\left(y_p, y_q\,|\,x\right) = \frac{\exp\left\{-V\left(y_p, y_q, x_{pq}\right)\right\}}{\sum_{y_g, y_h}\exp\left\{-V\left(y_g, y_h, x_{gh}\right)\right\}}\tag{6.142}$$

After piecewise approximating the likelihood by using Equation 6.140, the optimization problem (6.139) for parameter estimation is reduced to (Lin et al. 2016)

$$\begin{aligned}\min_{\theta}\ l_a(y, x, \theta) = &-\sum_{i=1}^{N}\left[\sum_{U\in A_U}\sum_{p\in N_U^{(i)}}\log P_U\left(y_p\,|\,x^{(i)}, \theta_U\right)\right.\\ &\left.+\sum_{V\in A_V}\sum_{(p,q)\in\varepsilon_V^{(i)}}\log P_V\left(y_p, y_q\,|\,x^{(i)}, \theta_V\right)\right]\\ &+\frac{\lambda}{2}\|\theta\|_2^2\end{aligned}\tag{6.143}$$

The objective function in (6.143) is a summation of the independent log-likelihood. Consequently, the optimization problem (6.143) can be easily solved in parallel.

### 6.2.2.1.3 Prediction

The class label of a pixel can be predicted either by maximizing a posterior distribution: $y^* = \arg\max_{y} P(y\,|\,x)$ or by calculating the label marginal distribution of each variable, that is, for all $p \in N$, we calculate the marginal distribution:

$$P\left(y_p\,|\,x\right) = \sum_{y\backslash y_p}P(y\,|\,x),\tag{6.144}$$

where $y\backslash y_p$ indicates the output class variables $y$ excluding $y_p$. Direct calculation of the marginal distribution is computationally intractable. Approximation methods should be developed. Popular methods for approximate probability inference are mean field methods that search for the distribution that best approximates distribution $P(y\,|\,x,\theta)$ within a tractable subset of distributions (Nowozin and Lampert, 2011).

The Kullback–Leibler (K–L) divergence is a widely used quantity that measures the distance between two distributions. Let $\Omega$ be a family of tractable distributions $q \in \Omega$ on $Y$. Define the K–L divergence $D^{KL}$ between two distributions $q$ and $P(y \mid x, \theta)$ as

$$D_{KL}(q(y) \| P(y \mid x, \theta)) = \sum_{y \in Y} q(y) \log \frac{q(y)}{P(y, x, \theta)} \tag{6.145}$$

Our goal is to obtain the best approximate distribution $q(y)$ by solving the optimization problem:

$$\min_{q \in \Omega} \ D_{KL}(q(y) \| P(y \mid x, \theta)) \tag{6.146}$$

Recall that the marginal distribution $P(y \mid x, \theta)$ can be expressed as

$$P(y \mid x, \theta) = \frac{1}{Z(x, \theta)} e^{-E(y, x, \theta)}, \tag{6.147}$$

where $E(y, x, \theta)$ is an energy function. Using a graphic model, a distribution over a large number of random variables can be represented as a product of local functions that each depends on only a small number of variables. Assume that the set $\Omega$ consists of all factorial distribution, that is

$$q(y) = \prod_{i \in V} q_i(y_i)$$

Using naïve mean field methods, we obtain the best approximate distribution within the set of distributions $\Omega$ (Appendix 6.A):

$$\hat{q}_i(y_i) = \exp \left( \lambda - 1 - \sum_{a \in F, i \in N(a)} \sum_{y_{N(a)} \in Y_{N(a)}, [y_{N(a)}] = y_i} \right.$$
$$\left. \left( \prod_{j \in N(a) \setminus \{i\}} \hat{q}_j(y_j) \right) E_a(y_{N(a)}, x_{N(a)}) \right), \tag{6.148}$$

where

$$\lambda = -\log \left( \sum_{y_i \in Y_i} \exp \left( -1 - \sum_{a \in F, i \in N(a)} \sum_{y_{N(a)} \in Y_{N(a)}, [y_{N(a)}] = y_i} \right. \right.$$
$$\left. \left. \left( \prod_{j \in N(a) \setminus \{i\}} \hat{q}_j(y_j) \right) E_a(y_{N(a)}, x_{N(a)}) \right) \right) \tag{6.149}$$

Updating $q_i(y_i)$ for each $i \in V$ will converge to the solutions to the optimization problem (6.146).

### 6.2.2.2 Deconvolution Network for Semantic Segmentation

The fully convolutional network (FCN) for sematic segmentation has some limitation (Noh et al. 2015). First, label prediction of the pixel is performed using only location information even for large objects, which leads to the inconsistent label predictions of the pixels that belong to the same object. Second, since the deconvolution procedure is too simple, the space variation information and semantic segment structures of an object are often lost. To overcome these limitations, the deconvolution network has been developed (Noh et al. 2015; Zeiler et al. 2011). The deconvolution network consists of deconvolution, unpooling, and rectification. Instead of reducing the size of activations through feedforwarding in the convolution network, the combination of unpooling and deconvolution in the deconvolution network increases the size of activation (Zeiler et al. 2011). The task of the deconvolution network is to produce semantic segmentation from the features generated by the convolution network. The final output of the deconvolution network is a map of the label probability of each pixel with the same size as that in the input image.

### 6.2.2.2.1 Unpooling

The function of the pooling layer in the convolution network is to filter noisy activation by maximizing activation with a single value in a receptive field, which leads to losing spatial variation information (Noh et al. 2015).

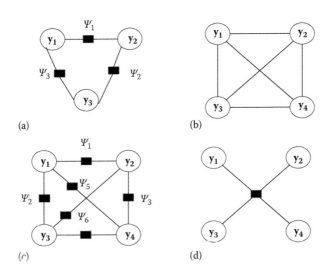

(a)           (b)

(c)           (d)

**FIGURE 6.12**
Factor graphs. (a) A factor graph over three variables (citation was presented in Example A1). (b) A Markov random field with a completely connected graph (citation was presented in Example A2). (c) A possible factorization of the Markov random field (citation was presented in Example A2). (d) A factor graph (citation was presented in Exercise 5).

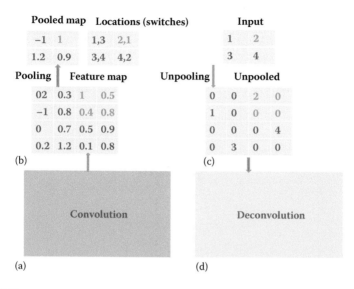

**FIGURE 6.13**
Unpooling and deconvolution. (a) Convolution for producing a feature map. (b) Performing the pooling operation on the feature map. (c) Unpooled feature map. (d) Deconvolution of unpooled features.

Unpooling is a reverse operation to recover the lost spatial variation information. After convolution to produce a feature map (Figure 6.13a), performing the pooling operation on the feature map that maximizes the features in the receptive field gives a pooled map and location (switches) recording the location of the maximum of the absolute feature (Figure 6.13b). In the reverse direction, given corresponding location information of the pooling operation, unpooling operation inserts the feature values in the location that is recorded by the switches in the pooling operation to form the unpooled feature map (Figure 6.13c). The unpooled features are then convolved with the kernels (Figure 6.13d).

### 6.2.2.2 Deconvolution

The deconvolution layer convolves unpooled features with multiple learned kernels to produce multiple outputs. Let $K_l$ be the number of kernels in the $l^{th}$ deconvolution layer, $f^l_{k,m}$ be the $k^{th}$ kernel in the $l^{th}$ layer convolved with the $m^{th}$ feature map in the previous layer, and $z_{k,l}$ be the output of the convolution of the unpooling features in the $(l-1)^{th}$ deconvolution layer with the kernel $f_{k,l}$. Then $z_{k,l}$ is given by

$$z_{k,l} = z_{m,l-1} * f^l_{k,m}, \quad k = 1, 2, \ldots, K_l, \tag{6.150}$$

where $z_{l-1}$ are the unpooling features in the $(l-1)^{th}$ deconvolution layer. Therefore, the output of the deconvolution layer is the dense activation map with the enlarged size.

*6.2.2.2.3 Model and Parameter Estimation*

To develop a cost function for inferring parameters, we consider the $l^{th}$ deconvolutional network layer (Zeiler et al. 2010). Assume a set of images $y = \{y^1,...,y^N\}$ where $N$ is the number of images. Consider the $i^{th}$ image $y^i$. Let $z^i_{k,l}$ be the feature maps of the $l^{th}$ deconvolution layer as input for the $(l - 1)^{th}$ deconvolution layer with the kernel $f^l_{k,m}$, and $z^i_{m,l-1}$ be the future maps from the previous $(l - 1)^{th}$ deconvolution layer. Let $K_{l-1}$ and $K_l$ be the number of feature maps of the $l^{th}$ and $(l - 1)^{th}$ deconvolution layers, respectively. Define the cost function $C_l(y)$ for the $l^{th}$ layer as

$$C_l(y) = \frac{1}{2}\sum_{i=1}^{N}\sum_{m=1}^{K_{l-1}} \| \sum_{k=1}^{K_l} h^l_{k,m}\left(z^i_{k,l} * f^l_{k,m}\right) - z^i_{m,l-1} \|^2_2$$

$$+ \lambda\sum_{i=1}^{N}\sum_{k=1}^{K_l} \| z^1_{k,l} \|_1, \tag{6.151}$$

where $\lambda$ is a penalty parameter, $h^l_{k,m}$ are elements of an indicator matrix defined as

$$h^l_{k,m} = \begin{cases} 1 & \text{if } z^i_{k,l} \text{ is connected to } z^i_{m,l-1} \\ 0 & \text{otherwise} \end{cases}$$

In the first layer, $h^l_{k,m} = 1$ for all $k,m$ are assumed.

The cost function $C_l(y)$ consists of two terms: (1) reconstruction error and (2) regularization term. The learning goal is to make the reconstruction error as small as possible. However, the regularization term penalizes the estimated feature values. These two goals are conflictive. The penalty parameter $\lambda$ balances two goals.

The parameters in the deconvolution network are estimated by minimization of the cost function $C_l(y)$ in Equation 6.150 with respect to the activation $z^l_{k,m}$ in the feature maps and kernel function $f^l_{k,m}$. The alternating direction method of multipliers (ADMM) and proximal methods (Boyd et al. 2011; Parikh and Boyd 2014) can be used to solve the problem.

---

## 6.3 Two- or Three-Dimensional Functional Principal Component Analysis for Image Data Reduction

Dimensionality reduction of image signals plays an important role in image classification and imaging genetics analysis as well (Su et al. 2017). The widely used dimension reduction methods include principle component analysis (PCA), decision boundary feature extraction (DBFE), non-negative matrix factorization, and discriminative analysis feature extraction (DAFE) (Diwaker and Dutta 2015). However, these methods do not explore the spatial information within the image. To overcome the limitations of these methods and fully utilize both the spatial and spectral information, one-dimensional functional principal component analysis (FPCA) is extended to two-dimensional FPCA (Lin et al. 2015). In this section, we primarily introduce three-dimensional FPCA. Two-dimensional FPCA can be viewed as a special case of three-dimensional FPCA.

## 6.3.1 Formulation

Similar to PCA, we consider a linear combination of functional values:

$$f = \iiint_{S\,T\,U} \beta(s,t,u)x(s,t,u)dsdtdu, \tag{6.152}$$

where $\beta(s,t,u)$ is a weight function and $x(s,t,u)$ is a centered random function, for example, intensity function of the image. To capture the variations in the random functions, we chose weight function $\beta(s,t,u)$ to maximize the variance of $f$. By the formula for the variance of the stochastic integral (Henderson and Plaschko, 2006), we have

$$\mathrm{var}(f) = \iiint_{S\,T\,U}\iiint_{S\,T\,U}, \beta(s_1,t_1,u_1)R(s_1,t_1,u_1,s_2,t_2,u_2)$$

$$\beta(s_2,t_2,u_2)ds_1dt_1du_1ds_2t_2du_2, \tag{6.153}$$

where $R(s_1,t_1,u_1,s_2,t_2,u_2) = cov(x(s_1,t_1,u_1),x(s_2,t_2,u_2))$ is the covariance function of $x(s,t,u)$. We define an extended inner product as:

$$(f,g)_\mu = \iiint_{S\,T\,U} f(s,t,u)g(s,t,u)dsdtdu + \mu \iiint_{S\,T\,U} \frac{\partial^6 f(s,t,u)}{\partial s^2\,\partial t^2\,\partial u^2} \frac{\partial^6 g(s,t,u)}{\partial s^2\,\partial t^2\,\partial u^2} dsdtdu$$

Since multiplying $\beta(s,t,u)$ by a constant will not change the maximizer of the variance $Var(f)$, we impose a constraint to make the solution unique:

$$\iiint_{S\,T\,U} \beta^2(s,t,u)dsdt + \mu \iiint_{S\,T\,U}\left[\frac{\partial^6\beta(s,t,u)}{\partial s^2\,\partial t^2\,\partial u^2}\right]^2 dsdtdu = 1 \tag{6.154}$$

Therefore, to find the weight function, we seek to solve the following optimization problem:

$$\mathrm{Max} \iiint_{S\,T\,U}\iiint_{S\,T\,U} \beta(s_1,t_1,u_1)R(s_1,t_1,u_1,s_2,t_2,u_2)\beta(s_2,t_2,u_2)ds_1dt_1du_1ds_2t_2du_2$$

$$\text{s. t. } \iiint_{S\,T\,U} \beta^2(s,t,u)dsdt + \mu \iiint_{S\,T\,U}\left[\frac{\partial^6\beta(s,t,u)}{\partial s^2\,\partial t^2\,\partial u^2}\right]^2 dsdtdu = 1 \tag{6.155}$$

## 6.3.2 Integral Equation and Eigenfunctions

By the Lagrange multiplier, we reformulate the constrained optimization problem (6.155) into the following non-constrained optimization problem:

$$
\max_{\beta} \quad \frac{1}{2} \int_S \int_T \int_U \int_S \int_T \int_U \beta(s_1,t_1,u_1) R(s_1,t_1,u_1,s_2,t_2,u_2) \beta(s_2. \, t_2,u_2) ds_1 dt_1 ds_2 t_2 +
$$

$$
\frac{1}{2}\lambda\left(1 - \int_S \int_T \int_U \beta^2(s_1,t_1,u_1) ds_1 dt_1 du_1 - \mu \int_S \int_T \int_U \left[\frac{\partial^6 \beta(s_1,t_1,u_1)}{\partial s_1^2 \partial t_1^2 \partial u_1^2}\right]^2 ds_1 dt_1 du_1\right),
$$

$$\tag{6.156}$$

where $\lambda$ is a parameter.

By variation calculus (Sagan, 2012), we define the functional

$$
J[\beta] = \frac{1}{2} \int_S \int_T \int_U \int_S \int_T \int_U \beta(s_1,t_1,u_1) R(s_1,t_1,u_1,s_2,t_2,u_2) \beta(s_2. \, t_2) ds_1 dt_1 ds_2 t_2
$$

$$
+\frac{1}{2}\lambda\left(1 - \int_S \int_T \int_U \beta^2(s_1,t_1,u_1) ds_1 dt_1 du_1\right) - \mu \int_S \int_T \int_U \left[\frac{\partial^6 \beta(s_1,t_1,u_1)}{\partial s_1^2 \partial t_1^2 \partial u_1^2}\right]^2 ds_1 dt_1 du_1)
$$

Its first variation is given by

$$
\delta J[h] = \frac{d}{d\varepsilon} J[\beta(s,t,u) + \varepsilon h(s,t,u)]
$$

$$
= \int_S \int_T \int_U \left[\int_S \int_T \int_U \left[R(s_1,t_1,u_1,s_2,t_2,u_2)\beta(s_2,t_2,u_2) ds_2 t_2 du_2\right.\right.
$$

$$
\left.-\lambda\left(\beta(s_1,t_1,u_1) + \mu \frac{\partial^{12}\beta(s_1,t_1,u_1)}{\partial s_1^4 \partial t_1^4 \partial u_1^4}\right)\right] h(s_1,t_1,u_1) ds_1 dt_1 du_1
$$

$$
= \int_S \int_T \int_U \left[\int_S \int_T \int_U R(s_1,t_1,u_1,s_2,t_2,u_2)\beta(s_2,t_2,u_2) ds_2 dt_2 du_2\right.
$$

$$
\left.-\lambda\left(\beta\left(s_1,t_1,u_1 + \mu \frac{\partial^{12}\beta(s_1,t_1,u_1)}{\partial s_1^4 \partial t_1^4 \partial u_1^4}\right)\right)\right]^2 ds_1 dt_1 du_1 = 0,
$$

which implies the following integral equation

$$
\int_S \int_T \int_U R(s_1,t_1,u_1,s_2,t_2,u_2)\beta(s_2,t_2,u_2) ds_2 dt_2 du_2
$$

$$
= \lambda\left[\beta(s_1,t_1,u_1) + \mu \frac{\partial^{12}\beta(s_1,t_1,u_1)}{\partial s_1^4 \partial t_1^4 \partial u_1^4}\right] \tag{6.157}
$$

for an appropriate eigenvalue $\lambda$. The left side of the integral equation (6.157) defines a three-dimensional integral transform $R$ of the weight function $\beta$. Therefore, the integral transform of the covariance function $R(s_1,t_1,u_1,s_2,t_2,u_2)$ is referred to as the covariance operator $R$. The integral equation (6.157) can be rewritten as

$$R\beta = \lambda\beta, \tag{6.158}$$

where $\beta(s_1,t_1,u_1,s_2,t_2,u_2)$ is an eigenfunction and referred to as a principal component function. Equation 6.158 is also referred to as a three-dimensional eigenequation. Clearly, the eigenequation (6.158) looks the same as the eigenequation for the multivariate PCA if the covariance operator and eigenfunction are replaced by a covariance matrix and eigenvector.

Since the number of function values is theoretically infinity, we may have an infinite number of eigenvalues. Provided the functions $X_i$ and $Y_i$ are not linearly dependent, there will be only $N - 1$ nonzero eigenvalues, where $N$ is the total number of sampled individuals ($N = n_A + n_G$). Eigenfunctions satisfying the eigenequation are orthonormal (Ramsay and Silverman, 2005). In other words, Equation 6.158 generates a set of principal component functions

$$R\beta_k = \lambda_k\beta_k, \qquad \text{with } \lambda_1 \geq \lambda_2 \geq \cdots$$

These principal component functions satisfy

1. $\displaystyle\iiint\limits_{S\,T\,U} \beta_k^2(s,t,u)\,dsdtdu + \mu \iiint\limits_{S\,T\,U} [\frac{\partial^6 \beta_k(s,t,u)}{\partial s^2\,\partial t^2\,\partial u^2}]^2 dsdtdu = 1$

2. $\displaystyle\iiint\limits_{S\,T\,U} \beta_k(s,t,u)\beta_m(s,t,u)\,dsdtdu + \mu \iiint\limits_{S\,T\,U} \frac{\partial^6 \beta_k(s,t,u)}{\partial s^2\,\partial t^2\,\partial u^2} \frac{\partial^6 \beta_m(s,t,u)}{\partial s^2\,\partial t^2\,\partial u^2} dsdtd$
$u = 0$, for all $m < k$.

The principal component function $\beta_1$ with the largest eigenvalue is referred to as the first principal component function, and the principal component function $\beta_2$ with the second largest eigenvalue is referred to as the second principal component function.

### 6.3.3 Computations for the Function Principal Component Function and the Function Principal Component Score

The eigenfunction is an integral function and difficult to solve in the closed form. A general strategy for solving the eigenfunction problem is to convert the continuous eigen-analysis problem to an appropriate discrete eigen-analysis task (Ramsay and Silverman, 2005). In this report, we use basis function expansion methods to achieve this conversion.

Let $\{\phi_j(t)\}$ be the series of Fourier functions. For each $j$, define $\omega_{2j-1} = \omega_{2j} = 2\pi j$. We expand each function $x_i(s,t,u)$ as a linear combination of the basis function $\phi_j$:

$$x_i(s,t,u) = \sum_{j=1}^{K}\sum_{k=1}^{K}\sum_{l=1}^{K} c_{jkl}^{(i)}\phi_j(s)\phi_k(t)\phi_l(u) \qquad (6.159)$$

Let $tC_i = [c_{111}^{(i)}, \ldots, c_{11K}^{(i)}, c_{121}^{(i)}, \ldots, c_{12K}^{(i)}, \ldots, c_{KK1}^{(i)}, \ldots, c_{KKK}^{(i)}]^T$ and $\phi(t) = [\phi_1(t), \cdots, \phi_K(t)]^T$. Then, Equation 6.159 can be rewritten as

$$x_i(s,t,u) = C_i^T(\phi(s) \otimes \phi(t) \otimes \phi(u)), \qquad (6.160)$$

where $\otimes$ denotes the Kronecker product of two matrices. Define the vector-valued function $X(s,t,u) = [x_1(s,t,u), \cdots, x_N(s,t,u)]^T$. The joint expansion of all N random functions can be expressed as

$$X(s,t,u) = C(\phi(s) \otimes \phi(t) \otimes \phi(u)), \qquad (6.161)$$

where the matrix C is given by

$$C = \begin{bmatrix} C_1^T \\ \vdots \\ C_N^T \end{bmatrix}$$

In matrix form we can express the variance-covariance function of $x_i(s,t,u)$ as

$$R(s_1,t_1,u_1,s_2,t_2,u_2) = \frac{1}{N}X^T(s_1,t_1,u_1)X(s_2,t_2,u_2)$$

$$= \frac{1}{N}[\phi^T(s_1) \otimes \phi^T(t_1) \otimes \phi^T(u_1)C^TC[\phi(s_2) \otimes \phi(t_2) \otimes \phi(u_2)]$$

$$(6.162)$$

Similarly, the eigenfunction $\beta(s,t,u)$ can be expanded as

$$\beta(s,t,u) = \sum_{j=1}^{K}\sum_{k=1}^{K}\sum_{l=1}^{K} b_{jkl}\phi_j(s)\phi_k(t)\phi_l(u) \quad \text{or}$$

$$\beta(s,t,u) - [\phi^T(s) \otimes \phi^T(t) \otimes \phi(u)]b,$$

$$\frac{\partial^{12}\beta(s,t,u)}{\partial s^4 \partial t^4 \partial u^4} = \sum_{j=1}^{K}\sum_{k=1}^{K}\sum_{l=1}^{K}\omega_j^4\omega_k^4\omega_l^4 b_{jkl}\phi_j(s)\phi_k(t)\phi_l(u),$$

where $b = [b_{111}, ..., b_{11K}, ..., b_{KK1}, ..., b_{KKK}]^T$.

Let $S_0 = \text{diag}(\omega_1^4 \omega_1^4 \omega_1^4, ..., \omega_1^4 \omega_1^4 \omega_K^4, ..., \omega_K^4 \omega_K^4 \omega_1^4, ..., \omega_K^4 \omega_K^4 \omega_K^4)$ and $S = \text{diag}((1 + \mu\omega_1^4 \omega_1^4 \omega_1^4)^{-1/2}, ..., (1 + \mu\omega_1^4 \omega_1^4 \omega_K^4)^{-1/2}, ..., (1 + \mu\omega_K^4 \omega_K^4 \omega_1^4)^{-1/2}, ..., (1 + \mu\omega_K^4 \omega_K^4 \omega_K^4)^{-1/2})$. Then, we have

$$\beta(s_1, t_1, u_1) + \mu \frac{\partial^{12}\beta(s_1, t_1, u_1)}{\partial s_1^4 \, \partial t_1^4 \, \partial u_1^4} = \phi^T(s_1) \otimes \phi^T(t_1) \otimes \phi^T(u_1) S^{-2} b \qquad (6.163)$$

Substituting expansions (6.162) and (6.163) of the variance-covariance $R(s_1, t_1, u_1, s_2, t_2, u_2)$ and eigenfunction $\beta(s,t,u)$ into the functional eigenequation (6.157), we obtain

$$[\phi^T(s_1) \otimes \phi^T(t_1) \otimes \phi^T(u_1)] \frac{1}{N} C^T C b = \lambda \phi^T(s_1) \otimes \phi^T(t_1) \otimes \phi^T(u_1) S^{-2} b \quad (6.164)$$

Since Equation 6.164 must hold for all $t$, we obtain the following eigenequation:

$$\frac{1}{N} C^T C b = \lambda S^{-2} b, \qquad (6.165)$$

which can be rewritten as

$$\left[ S\left( \frac{1}{N} C^T C \right) S \right] [S^{-1} b] = \lambda [S^{-1} b], \text{ or}$$

$$S\left( \frac{1}{N} C^T C \right) S u = \lambda u,$$

where $u = S^{-1} b$. Solving the above eigenequation yields the eigenvector $u$. Thus, $b = Su$, we obtain a set of orthonormal eigenvectors $b_j$. A set of orthonormal eigenfunctions is given by

$$\beta_j(s, t, u) = [\phi^T(s) \otimes \phi^T(t) \otimes \phi^T(u)] b_j, j = 1, ..., J \qquad (6.166)$$

The random functions $x_i(s,t,u)$ can be expanded in terms of eigenfunctions as

$$x_i(s, t, u) = \sum_{j=1}^{J} \xi_{ij} \beta_j(s, t, u), i = 1, ..., N, \qquad (6.167)$$

where

$$\xi_{ij} = \left\langle x_i, \beta_j \right\rangle_{\mu} = C_i^T S^{-2} b_j,$$

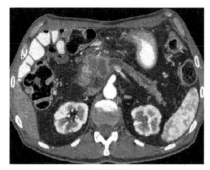 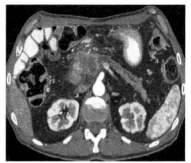

Original CT image          Reconstructed image from FPCA

**FIGURE 6.14**
Original image and reconstructed image of advanced pancreatic adenocarcinoma.

$$S^{-2} = \text{diag}\left(1 + \mu\omega_1^4\omega_1^4\omega_1^4, ..., 1 + \mu\omega_1^4\omega_1^4\omega_k^4, ..., 1 + \mu\omega_K^4\omega_K^4\omega_1^4, ..., 1 + \mu\omega_K^4\omega_K^4\omega_K^4\right).$$

**Example 6.5**

To intuitively illustrate the performance of FPCs in the dimension reduction of image data, we presented Figure 6.14 which showed the original and reconstructed the CT images of advanced pancreatic adenocarcinoma. Figure 6.14 strongly demonstrated that using the FPC score we can accurately approximate the original images.

## 6.4 Association Analysis of Imaging-Genomic Data

The widely used statistical methods for association analysis of imaging-genomic data include univariate and voxel-wise analysis, multivariate regression analysis, and canonical correlation analysis (CCA) (Nathoo et al. 2017; Richfield et al. 2016). The voxel-wise approach is to test association of an SNP with each voxel (Stein et al. 2010). In the multivariate approach, whole image is either reduced to several summary statistics or segmented into several regions of interest, each region of interest is measured by a summary statistic. Then, multivariate regression is used to test the association of an SNP with multiple summary measures of images signals (Wang et al. 2012). CCA seeks the maximum correlation between a combination of summary measures of image and an SNP or combination of SNPs and test for association of SNPs with image (Lin et al. 2014). In this section, we briefly introduce functional data analysis-based association and interaction analysis.

### 6.4.1 Multivariate Functional Regression Models for Imaging-Genomic Data Analysis

#### 6.4.1.1 Model

Consider general multivariate regression models (Jiang et al. 2015). Assume that $n$ individuals are sampled. Let $y_{ik}, k = 1,2,\ldots,K$ be $K$ summary image measures of the $i^{th}$ individual, including image FPC scores. The summary image measures are referred to as trait values. Consider a genomic region $[a, b]$. Let $x_i(t)$ be a genotype or RNA-seq profile of the $i^{th}$ individual defined in the region $[a,b]$. Recall that a regression model for QTL analysis with the $k$-th trait and SNP data is defined as

$$y_{ik} = \mu_k + \sum_{j=1}^{J_1} x_{ij}\alpha_{kj} + \varepsilon_{ik} \tag{6.168}$$

where $\mu_k$ is an overall mean of the $k$-th trait, $\alpha_{kj}$ is the main genetic additive effect of the $j$-th SNP in the genomic region for the $k$-th trait, $x_{ij}$ is an indicator variable for the genotypes at the $j$-th SNP, $\varepsilon_{ik}, k = 1,\ldots,K$ are independent and identically distributed normal variables with mean of zero and covariance matrix $\Sigma$.

Similar to the multiple regression models for QTL analysis with SNP data and multiple quantitative traits, the functional regression model for a quantitative trait with a genotype profile or RNA-seq data can be defined as

$$y_{ik} = \alpha_{0k} + \int_T \alpha_k(t)x_i(t)dt + \varepsilon_{ik}, \tag{6.169}$$

where $\alpha_{0k}$ is an overall mean, $\alpha_k(t)$ is a genetic additive effect of putative QTLs located at the genomic positions $t$ for the $k$-th trait, $k = 1,\ldots,K$, $x_i(t)$ is a genotype profile or RNA-seq data, $\varepsilon_{ik}$ are independent and identically distributed normal variables with a mean of zero and a covariance matrix $\Sigma$.

#### 6.4.1.2 Estimation of Additive Effects

We assume that both trait values and genotype profiles or RNA-seq data are centered. The genotype profiles or RNA-seq data $x_i(t)$ are expanded in terms of the orthonormal basis function as:

$$x_i(t) = \sum_{j=1}^{\infty} \xi_{ij}\phi_j(t) \tag{6.170}$$

where $\phi_j(t)$ are sequences of the orthonormal basis functions. The expansion coefficients $\xi_{ij}$ are estimated by

$$\xi_{ij} = \int_T x_i(t)\phi_j(t)dt \tag{6.171}$$

In practice, numerical methods for the integral will be used to calculate the expansion coefficients. Substituting Equation 6.170 into Equation 6.169, we obtain

$$
\begin{aligned}
y_{ik} &= \int_T \alpha_k(t) \sum_{j=1}^{\infty} \xi_{ij}\phi_j(t)dt + \varepsilon_i \\
&= \sum_{j=1}^{\infty} \xi_{ij} \int_T \alpha_k(t)\phi_j(t)dt + \varepsilon_{ik} \\
&= \sum_{j=1}^{\infty} \xi_{ij}\alpha_{kj} + \varepsilon_{ik}, i = 1, \dots, n, k = 1, \dots, K,
\end{aligned}
\tag{6.172}
$$

where $\alpha_k(t) = \int_T \alpha_k(t)\phi_j(t)dt$. The parameters $\alpha_{kj}$ are referred to as genetic additive effect scores for the $k$-th trait. These scores can also be viewed as the expansion coefficients of the genetic effect functions with respect to orthonormal basis functions:

$$\alpha_k(t) = \sum_j \alpha_{kj}\phi_j(t) \tag{6.173}$$

Let

$$
Y = [Y_1, \dots, Y_K] = \begin{bmatrix} Y_{11} & \cdots & Y_{1K} \\ \vdots & \ddots & \vdots \\ Y_{n1} & \cdots & Y_{nK} \end{bmatrix}, \xi = \begin{bmatrix} \xi_{11} & \cdots & \xi_{1J} \\ \vdots & \ddots & \vdots \\ \xi_{n1} & \cdots & \xi_{nJ} \end{bmatrix}, \quad \xi_i = \begin{bmatrix} \xi_{i1} \\ \vdots \\ \xi_{iJ} \end{bmatrix},
$$

$$
\alpha_k = \begin{bmatrix} \alpha_{k1} \\ \vdots \\ \alpha_{kJ} \end{bmatrix}, \alpha = [\alpha_1, \dots, \alpha_K], \varepsilon = \begin{bmatrix} \varepsilon_{11} & \cdots & \varepsilon_{1K} \\ \cdots & \cdots & \cdots \\ \varepsilon_{n1} & \cdots & \varepsilon_{nK} \end{bmatrix}
$$

Then, Equation 6.171 can be approximated by

$$Y = \xi\alpha + \varepsilon \tag{6.174}$$

The standard least square estimators of $\alpha$ and the variance covariance matrix $\Sigma$ are given by

$$\hat{\alpha} = \left(\xi^T \xi\right)^{-1} \xi^T \left(Y - \bar{Y}\right), \tag{6.175}$$

$$\hat{\Sigma} = \frac{1}{n}\left(Y = \xi\hat{\alpha}\right)^T \left(Y - \xi\hat{\alpha}\right) \tag{6.176}$$

Denote the matrix $(\xi^T \xi)^{-1} \xi^T$ by $A$. Then, the estimator of the parameter $\alpha$ is given by

$$\hat{\alpha} = A\left(Y - \bar{Y}\right) \tag{6.177}$$

The vector of the matrix $\alpha$ can be written as

$$vec(\hat{\alpha}) = (A \otimes I)vec\left(Y - \bar{Y}\right) \tag{6.178}$$

By the assumption of the variance matrix of $Y$, we obtain the variance matrix of $vec(Y)$:

$$var(vec(Y)) = \Sigma \otimes I \tag{6.179}$$

Thus, it follows from Equations 6.178 and 6.179 that

$$\Lambda = var(vec(\hat{\alpha})) = (I_k \otimes A)(\Sigma \otimes I_n)\left(I_k \otimes A^T\right)$$
$$= \Sigma \otimes \left(AA^T\right) \tag{6.180}$$

### 6.4.1.3 Test Statistics

An essential problem in QTL analysis or in integrative analysis of imaging and genomic (genetic or RNA-seq) data is to test the association of the genomic region (or gene) with images. Formally, we investigate the problem of testing the following hypothesis:

$$\alpha_k(t) = 0, \forall\, t \in [a, b], k = 1, \ldots, K,,$$

which is equivalent to testing the hypothesis:

$$H_0 : \alpha = 0$$

Define the test statistic for testing the association of a genomic region (genotype or RNA-seq in the region) with $K$ summary image measures as

$$T = \hat{\alpha}^T \Lambda^{-1} \hat{\alpha} \tag{6.181}$$

Let $r$ = rank $(\Lambda)$.

Then, under the null hypothesis $H_0: \alpha = 0$, $T$ is asymptotically distributed as a central $\chi^2_{(KJ)}$ or $\chi^2_{(r)}$ distribution if $J$ components are taken in the expansion equation (6.170).

### 6.4.2 Multivariate Functional Regression Models for Longitudinal Imaging Genetics Analysis

Let $y_i^{(k)}(t), k = 1, 2, ..., K$, be the $k^{th}$ image summary measure of the $i^{th}$ individual at time $t$. Let $x_i(s)$ be a genotype profile of the gene $G$ which is located in the genomic region $S = [a,b]$. We study the association of the gene $G$ with the multiple image summary measures. The multivariate functional linear model (MFLM) for longitudinal imaging genetics analysis can be defined as

$$y_i^{(k)}(t_m) = \int_S x_i(s)\beta_k(s, t_m)ds + \varepsilon_{ik}(t_m), \quad k = 1, ..., K, i = 1, ..., n, \ m = 1, ..., T \quad (6.182)$$

$\beta_k(s,t_m)$ is a genetic additive effect function of the genotype located at the genomic position $s$ on the $k^{th}$ image summary measure at time $t_m$, and $\varepsilon_{ik}(t_m)$ is the residual function of the noise and unexplained effect for the $i^{th}$ individual. Let $\eta(t_l) = [\eta_1(t_m), ..., \eta_J(t_m)]^T$ be a vector of basis functions. To transform the functional linear model (6.182) into the standard multivariate linear mode, we consider the following eigen functional expansions for the image summary measure $y_{ik}(t_m)$, effect functions $\beta_k(s,t_m)$, and genotype function $x_i(s)$:

$$y_i^{(k)}(t_m) = \sum_{l=1}^{J_y} y_{il}^{(k)} \eta_l(t_m) = y_i^{(k)T} \eta(t_m), \ \eta(t_m) = [\eta_1(t_m), ..., \eta_{J_y}(t_m)]^T, \ x_i(s) = x_i^T \theta(s).$$

$$\beta_k(s, t_m) = \sum_{j=1}^{J_\beta} \sum_{l=1}^{J_y} b_{jl}^{(k)} \theta_j(s) \eta_l(t_m) = \theta^T(s) B^{(k)} \eta(t_m), \ \theta(s) = [\theta_1(s), ..., \theta_{J_\beta}(s)]^T, \ \text{where}$$

$$y_i^{(k)} = \begin{bmatrix} y_{i1}^{(k)} \\ \vdots \\ y_{iJ_y}^{(k)} \end{bmatrix}, B_k = \begin{bmatrix} b_{11}^{(k)} & \cdots & b_{1J_y}^{(k)} \\ \vdots & \ddots & \vdots \\ b_{J_\beta 1}^{(k)} & \cdots & b_{J_\beta J_y}^{(k)} \end{bmatrix}.$$

$B^{(k)} = (b_{jl}^{(k)})_{J_\beta \times J_y}$ is a matrix of expansion coefficients of the genetic additive effect function for the $k^{th}$ image summary measure, $\varepsilon_{ik}(t_m) = \sum_{l=1}^{J_y} \varepsilon_{ikl} \eta_l(t_m) = \varepsilon_{ik}^T \eta(t_m), \ \varepsilon_{ik} = [F_{ik1}, ..., \varepsilon_{ikJ_y}]^T$. The integral $\int_S x_i(s)\beta(s,t_l)ds$ can be expanded as

$$\int_S x_i(s)\beta_k(s, t_m)ds = \int_S x_i(s)\theta^T(s)ds B^{(k)} \eta(t_m) = x_i^T B^{(k)} \eta(t_m),$$

where

$$x_i = \left[ \int_T x_i(s)\theta_1(s)ds, \ldots, \int_T x_i(s)\theta_{J_\beta}(s)ds \right]^T = \left[ x_{i1}, \ldots, x_{iJ_\beta} \right]^T$$

Substituting these expansions into Equation 6.182, we obtain

$$y_i^{(k)T}\eta(t_m) = x_i^T B^{(k)}\eta(t_m) + \varepsilon_i^{(k)T}\eta(t_m), k = 1, \ldots, K, \ m = 1, \ldots, T, \ i = 1, \ldots, n \quad (6.183)$$

Since Equation 6.183 should hold for all $t_m$, we must have

$$y_i^{(k)T} = x_i^T B^{(k)} + \varepsilon_i^{(k)T}, i = 1, \ldots, n, k = 1, \ldots, K \quad (6.184)$$

The model (6.184) is a standard linear model. Instead of using the observed data as the values of the response and predictor variables, we use their expansion coefficients as the values of the response and predictor variables in the linear model (6.184). Equation 6.184 can be further written in a matrix form.

Let

$$Y^{(k)} = [Y_1^{(k)}, \ldots, Y_{J_y}^{(k)}] = \begin{bmatrix} Y_{11}^{(k)} & \cdots & Y_{1J_y}^{(k)} \\ \vdots & \ddots & \vdots \\ Y_{n1}^{(k)} & \cdots & Y_{nJ_y}^{(k)} \end{bmatrix}, \ X = \begin{bmatrix} x_{11} & \cdots & x_{1J_\beta} \\ \vdots & \ddots & \vdots \\ x_{n1} & \cdots & x_{nJ_\beta} \end{bmatrix},$$

$$B^{(k)} = \begin{bmatrix} b_{11}^{(k)} & \cdots & b_{1J_y}^{(k)} \\ \vdots & \ddots & \vdots \\ b_{J_\beta 1}^{(k)} & \cdots & b_{J_\beta J_y}^{(k)} \end{bmatrix}, \ \varepsilon^{(k)} = \begin{bmatrix} \varepsilon_{11}^{(k)} & \cdots & \varepsilon_{1J_y}^{(k)} \\ \vdots & \ddots & \vdots \\ \varepsilon_{n1}^{(k)} & \cdots & \varepsilon_{nJ_y}^{(k)} \end{bmatrix}$$

Then, Equation 6.182 can be approximated by

$$Y^{(k)} = XB^{(k)} + \varepsilon^{(k)} \quad (6.185)$$

Let

$$Y = \begin{bmatrix} Y^{(1)} & \cdots & Y^{(K)} \end{bmatrix} = X \begin{bmatrix} B^{(1)} & \cdots & B^{(k)} \end{bmatrix} + \begin{bmatrix} \varepsilon^{(1)} & \cdots & \varepsilon^{(K)} \end{bmatrix}$$

$$= XB + \varepsilon \quad (6.186)$$

$$\text{Let } \Sigma = \begin{bmatrix} \Sigma^{(11)} & \Sigma^{(12)} & \cdots & \Sigma^{(1K)} \\ \Sigma^{(21)} & \Sigma^{(22)} & \cdots & \Sigma^{(2K)} \\ \vdots & \vdots & \ddots & \vdots \\ \Sigma^{(K1)} & \Sigma^{(K2)} & \cdots & \Sigma^{(KK)} \end{bmatrix}, \Sigma^{(kl)} = \begin{bmatrix} \sigma_{11}^{(kl)} & 0 & \cdots & 0 \\ 0 & \sigma_{22}^{(kl)} & \cdots & 0 \\ \vdots & \vdots & \ddots & \vdots \\ 0 & 0 & \cdots & \sigma_{J_y J_y}^{(kl)} \end{bmatrix} \text{ and}$$

$$Cov(vec(\varepsilon)) = \Sigma \otimes I_n, cov\left(vec\left(\varepsilon^{(k)}\right)\right) = \Sigma^{(kk)} \otimes I_n \qquad (6.187)$$

It follows from Equation 6.186 that

$$Vec(Y) = (I \otimes X)vec(B) + Vec(\varepsilon) \qquad (6.188)$$

First estimate $\Sigma$.

$$\hat{\Sigma} = \frac{1}{nJ_y - J_\beta} Y^T \left(I_{nJ_y} - X(X^T X)^{-1} X^T\right) Y \qquad (6.189)$$

$$\Lambda = cov\left(Vec\left(\hat{B}\right)\right) = \Sigma \otimes (X^T X)^{-1} \text{ and } \Lambda_k = cov\left(vec\left(\hat{B}^{(k)}\right)\right) = \Sigma^{(kk)} \otimes (X^T X)^{-1}$$

$$\Lambda^{-1} = \Sigma^{-1} \otimes (X^T X) \text{ and } \Lambda_k^{-1} = \left(\Sigma^{(kk)}\right)^{-1} \otimes (X^T X)$$

Minimizing the following objective function:

$$F = Tr\{[vec(Y) - (I \otimes X)vec(B)]^T \Lambda^{-1}[vec(Y) - (I \otimes X)vec(B)],$$

gives the estimation of the parameters.
Define the test statistics:

$$T = [vec(B)]^T \Lambda^{-1} vec(B) \text{ and } T_k = \left[vec\left(\hat{B}^{(k)}\right)\right]^T \Lambda_k^{-1} vec\left(\hat{B}^{(k)}\right)$$

$T$ is distributed as $\chi^2_{(KJ_Y J_\beta)}$ under the null hypothesis of no association of gene or genomic region with image signal variation.
Recall that $\Lambda_k = cov(vec(\hat{B}^{(k)})) = \Sigma^{(kk)} \otimes (X^T X)^{-1}$, where $\Sigma^{(kk)}$ are obtained from Equation 6.188. Thus, we have

$$\Lambda_k^{-1} = \left(\hat{\Sigma}^{(kk)}\right)^{-1} \otimes (X^T X)$$

Define the test statistic for testing association of the gene with the $k^{th}$ image summary measure ($k^{th}$ image region):

$$T_k = \left[vec\left(\hat{B}^{(k)}\right)\right]^T \Lambda_k^{-1} vec\left(\hat{B}^{(k)}\right)$$

$T_k \sim \chi^2_{(J_Y J_\beta)}$) under the null hypothesis of no association of the gene with the $k^{th}$ image region.

## 6.4.3 Quadratically Regularized Functional Canonical Correlation Analysis for Gene–Gene Interaction Detection in Imaging-Genetic Studies

Consider two genomic regions. There are $p$ SNPs and $q$ SNPs in the first and second genomic regions, respectively. Assume that $n$ individuals are sampled. For the $i^{th}$ individual, two vectors of the genetic variation data (genotypes or functional principal component scores) for the first and second genomic regions are denoted by $x_i = [x_{i1},...,x_{ip}]$ and $z_i = [z_{i1},...,z_{iq}]$, respectively. Let $\xi_i = [x_{i1}z_{i1},...x_{i1}z_{iq},...,x_{ip}z_{i1},...,x_{ip}z_{iq}]$. The single variate regression (for a single image summary measure), and multivariate regression (for multiple image summary measures) will be used to pre-process the genotype data for removing the genetic main effects of two genomic regions before taking their residuals for interaction analysis. We start with interaction analysis of a single summary measure.

### 6.4.3.1 Single Image Summary Measure

Before performing interaction analysis, we first regress the single image summary measure on the genotypes of two genomic regions. Let $y_i$ be a single image summary measure of the $i^{th}$ individual. Consider a regression model:

$$y_i = \mu + \sum_{m=1}^{M} \tau_{im}\theta_m + \sum_{j=1}^{p} x_{ij}\alpha_j + \sum_{l=1}^{q} z_{il}\beta_l + \varepsilon_i \tag{6.190}$$

where $\mu$ is an overall mean, $\tau_{im}$ are covariates such as age, sex, and principal component (PC) scores for removing the impact of population structure, $\theta_m$ are their corresponding regression coefficients, $\alpha_j$ and $\beta_l$ are genetic main effects for the first and second genomic regions, respectively. After the model fits the data, we calculate the residual for each individual:

$$\eta_i = y_i - \hat{\mu} - \sum_{m=1}^{M} \tau_{im}\hat{\theta}_m - \sum_{j=1}^{p} x_{ij}\hat{\alpha}_j - \sum_{l=1}^{q} z_{il}\hat{\beta}_l \tag{6.191}$$

Define a vector of residual and a genetic data matrix:

$$\eta = \begin{bmatrix} \eta_1 \\ \vdots \\ \eta_m \end{bmatrix} \text{ and } \xi = \begin{bmatrix} \xi_{11} & \cdots & \xi_{1pq} \\ \vdots & \vdots & \vdots \\ \xi_{n1} & \vdots & \xi_{npq} \end{bmatrix} \tag{6.192}$$

### 6.4.3.2 Multiple Image Summary Measures

Consider $K$ image summary measures. Let $y_{ik}$, be the $k^{th}$ image summary measure of the $i^{th}$ individual. The multivariate regression model is

$$y_{ik} = \mu_k + \sum_{m=1}^{M} \tau_{im}\theta_{mk} + \sum_{j=1}^{J_1} x_{ij}\alpha_{kj} + \sum_{l=1}^{J_2} z_{il}\beta_{kl} + \varepsilon_{ik}, \qquad (6.193)$$

where $\mu_k$ is an overall mean of the $k^{th}$ image summary measure, $\theta_{mk}$ is the regression coefficient associated with the covariate $\tau_{im}$, $\alpha_{kj}$ is the main genetic additive effect of the $j^{th}$ genetic variant in the first genomic region for the $k^{th}$ image summary measure, and $\beta_{kl}$ is the main genetic additive effect of the $l^{th}$ genetic variant in the second genomic region for the $k^{th}$ image summary measure.

Let $\hat{y}_{ik}$ be the predicted value by the fitted model:

$$\hat{y}_{ik} = \hat{\mu}_k + \sum_{m=1}^{M} \tau_{im}\hat{\theta}_{mk} + \sum_{j=1}^{J_1} x_{ij}\hat{\alpha}_{kj} + \sum_{l=1}^{J_2} z_{il}\hat{\beta}_{kl}$$

The residual is defined as

$$\eta_{ik} = y_{ik} - \hat{y}_{ik}$$

Define residual matrix:

$$\eta = \begin{bmatrix} \eta_{11} & \cdots & \eta_{1K} \\ \vdots & \vdots \\ \eta_{n1} & \cdots & \eta_{nK} \end{bmatrix} \qquad (6.194)$$

Data matrix $\xi$ is defined as before.

### 6.4.3.3 CCA and Functional CCA for Interaction Analysis

For convenience, we assume that $K \le pq$. Define the covariance matrix:

$$\Sigma = \begin{bmatrix} \Sigma_{\xi\xi} & \Sigma_{\xi\eta} \\ \Sigma_{\eta\xi} & \Sigma_{\eta\eta} \end{bmatrix}$$

The matrix $\Sigma$ will be estimated by

$$\hat{\Sigma} = \begin{bmatrix} \hat{\Sigma}_{\xi\xi} & \hat{\Sigma}_{\xi\eta} \\ \hat{\Sigma}_{\eta\xi} & \hat{\Sigma}_{\eta\eta} \end{bmatrix} = \frac{1}{n} \begin{bmatrix} \xi^T\xi & \xi^T\eta \\ \eta^T\xi & \eta^T\eta \end{bmatrix} \qquad (6.195)$$

The solution to CCA starts with defining the $R^2$ matrix:

$$R^2 = \hat{\Sigma}_{\eta\eta}^{-1/2}\hat{\Sigma}_{\eta\xi}\hat{\Sigma}_{\xi\xi}^{-1}\hat{\Sigma}_{\xi\eta}\hat{\Sigma}_{\eta\eta}^{-1/2} \tag{6.196}$$

Let

$$W = \hat{\Sigma}_{\xi\xi}^{-1/2}\hat{\Sigma}_{\xi\eta}\hat{\Sigma}_{\eta\eta}^{-1/2} \tag{6.197}$$

Suppose that the singular value decomposition (SVD) of the matrix is given by

$$W = U\Lambda V^T, \tag{6.198}$$

where $\Lambda = diag\ (\lambda_1,...,\lambda_d)$ and $d = min(K,pq)$. It is clear that

$$W^TW = R^2 = V\Lambda^2 V^T \tag{6.199}$$

The matrices of canonical covariates are defined as

$$A = \Sigma_{\xi\xi}^{-1/2}U,$$
$$B = \Sigma_{\eta\eta}^{-1/2}V \tag{6.200}$$

The vector of canonical correlations is

$$CC = [\lambda_1, ..., \lambda_d]^T \tag{6.201}$$

Canonical correlations between the interaction terms and image summary measures the strength of the interaction. The CCA produces multiple canonical correlations. However, we wish to use a single number to quantify the interaction. We propose to use the summation of the square of the singular values as a measure to quantify the interaction:

$$r = \sum_{i=1}^{d}\lambda_i^2 = \text{Tr}(\Lambda^2) = \text{Tr}(R^2) \tag{6.202}$$

To test the interaction between two genomic regions is equivalent to testing independence between $\xi$ and $\eta$ or to test the hypothesis that each variable in the set $\xi$ is uncorrelated with each variable in the set $\eta$. The null hypothesis of no interaction can be formulated as

$$H_0 : \Sigma_{\xi\eta} = 0$$

The likelihood ratio for testing $H_0 : \Sigma_{\xi\eta} = 0$ is

$$\Lambda_r = \frac{|\Sigma|}{|\Sigma_{\xi\xi} \| \Sigma_{\eta\eta}|} = \prod_{i=1}^{d}(1 - \lambda_i^2), \tag{6.203}$$

which is equal to the Wilks' lambda $\Lambda$. This demonstrates that testing for interaction using multivariate linear regression can be treated as special case of CCA.

We usually define the likelihood ratio test statistic for testing the interaction as:

$$T_{CCA} = -N\sum_{i=1}^{d} \log(1 - \lambda_i^2) \qquad (6.204)$$

For small $\lambda_i^2$, $T_{CCA}$ can be approximated by $N\sum_{i=1}^{d} \lambda_i^2 = Nr$, where $r$ is the measure of interaction between two genomic regions. The stronger the interaction, the higher the power that the test statistic can test the interaction.

Under the null hypothesis $H_0 : \Sigma_{\xi\eta} = 0$, $T_{CCA}$ is asymptotically distributed as a central $\chi^2_{Kpq}$. When sample size is large, Bartlett (1939) suggests using the following statistic for hypothesis testing:

$$T_{CCA} = -\left[N - \frac{(d+3)}{2}\right]\sum_{i=1}^{d} \log(1 - \lambda_i^2) \qquad (6.205)$$

If the functional principal component scores are taken as genetic variants in the matrix $\xi$, then the multivariate CCA becomes the functional CCA. All previous discussion for the multivariate CCA can be applied to the functional CCA.

## 6.5 Causal Analysis of Imaging-Genomic Data

Imaging data can serve as an endophenotype that is close to the actions of the genes and offers a much higher power to discover the risk genes than the final clinical outcome of the disease (Medland et al. 2014). New imaging techniques provide fine-grained measures of the brain and other tissues structural connectivity and rich physiological information such as cellularity and metabolism. The complex structure of images of the brain and other tissues is strongly shaped by genetic and expression variation. Despite their success in imaging-genomic association analysis, the various multivariate regression and data reduction methods have several serious limitations. First, the results are the lack of direct biological interpretation. Second, lack of methods for imaging-genomic data analysis explicitly consider the connection structure of images. Third, none of methods can properly address the pleiotropic effects of the genomic variants on the imaging measures. Sparse Gaussian graphical models and structural equation models (SEMs) for the integrated causal analysis of structural imaging and genomic data will be useful.

### 6.5.1 Sparse SEMs for Joint Causal Analysis of Structural Imaging and Genomic Data

The SEMs can be used for the integrated causal analysis of structural imaging and genomic data. Specifically, we use super pixel and deep learning techniques to segment the whole image into a number of homogeneous regions as discussed in Section 6.2. We treat the measures of each imaging region as phenotypes and hence transform the problems of imaging-genomic causal analysis into causal inference for genotypes and multiple complex phenotypes. The methods for causal inference of genotype–phenotype discussed in previous chapters can be adapted to joint causal analysis of imaging and genomic data. In this section, we briefly introduce the SEMs for the causal analysis of imaging and genomic data. The estimation methods and test statistics are referred to in the previous chapters.

Assume that $n$ individuals are sampled. We consider $M$ image summary measures for the $M$ image regions which are referred to as endogenous variables. We denote the $n$ observations on the $M$ endogenous variables by the matrix $y = [y_1,...,y_M]$. Covariates, genetic variants that are defined as exogenous or predetermined variables are denoted by $X = [x_1,...,x_K]$. Similarly, random errors are denoted by $E = [e_1,...,e_M]$. The linear structural equations for modeling relationships among phenotypes and genotypes can be written as

$$y_1\gamma_{11} + y_2\gamma_{21} + \cdots + y_M\gamma_{M1} + x_1\beta_{11} + x_2\beta_{21} + \cdots + x_K\beta_{K1} + e_1 = 0$$

$$\vdots \qquad\qquad \vdots \qquad\qquad (6.206)$$

$$y_1\gamma_{1M} + y_2\gamma_{2M} + \cdots + y_M\gamma_{MM} + x_1\beta_{1M} + x_2\beta_{2M} + \cdots + x_K\beta_{KM} + e_M = 0,$$

where the $\gamma$'s and $\beta$'s are the structural parameters of the system that are unknown. The parameter $\gamma$'s characterize the relationships between the image regions. The parameter $\beta$'s measure genetic contribution of the SNPs to the image intensity variation. In matrix notation, the SEMs in Equation 6.206 can be rewritten as

$$Y\Gamma + XB + E = 0, \qquad (6.207)$$

where $\Gamma, B, E$ are corresponding matrices.

In general, genotype-image networks are sparse. Therefore, $\Gamma$ and $B$ are sparse matrices. To obtain sparse estimates of $\Gamma$ and $B$, the natural approach is the $l_1$-norm penalized regression of Equation 6.206. Let $y_i$ be the vector of observations of the variable $i$. Let $Y_{-i}$ be the observation matrix $Y$ after removing $y_i$ from it and $\gamma_{-i}$ be the parameter vector $\Gamma_i$ after removing the parameter $\gamma_{ii}$. The $i^{th}$ equation can be written as

$$y_i = W_i\Delta_i + e_i, \qquad (6.208)$$

where $W_i = [Y_{-i}, X], \Delta_i = [\gamma_{-i}, B_i]$. Using the $l_1$-norm penalization, we can form the following optimization problem:

$$\min \ f(\Delta_i) + \lambda \| \Delta_i \|_1, \tag{6.209}$$

where $f(\Delta_i) = (X^T y_i - X^T W_i \Delta_i)^T (X^T X)^{-1} (X^T y_i - X^T W_i \Delta_i)$.

The size of the genotype-image network may be large. An efficient alternating direction method of multipliers (ADMM) (Section 1.3.2) (Boyd et al. 2011) is used to solve the optimization problem (6.209).

### 6.5.2 Sparse Functional Structural Equation Models for Phenotype and Genotype Networks

To utilize multi-locus genetic information, we propose to use a gene as a unit and to model genotypes across the loci within a gene as a function of genomic position. Functional data analysis techniques as a tool for reducing dimension are used to develop sparse functional structural equation models (FSEMs) for inferring the image and genotype (next-generation sequencing data) networks and for defining cost function in causal analysis. Let $t$ be a genomic position and $x_i(t)$ be a genotype profile of the $i^{th}$ individual. Suppose that we are interested in $G$ genes with genomic regions $[a_j, b_j]$, denoted as $T_j, j = 1, \ldots, k$. Consider the FSEMs

$$y_1 \gamma_{11} + y_2 \gamma_{21} + \cdots + y_M \gamma_{M1} + \int_{T_1} x_1(t) \beta_{11}(t) dt + \cdots + \int_{T_k} x_k(t) \beta_{k1}(t) dt + e_1 = 0$$

$$y_1 \gamma_{12} + y_2 \gamma_{22} + \cdots + y_M \gamma_{M2} + \int_{T_1} x_1(t) \beta_{12}(t) dt + \cdots + \int_{T_k} x_k(t) \beta_{k2}(t) dt + e_2 = 0$$

$$\vdots \qquad\qquad \vdots \qquad\qquad \vdots$$

$$y_1 \gamma_{1M} + y_2 \gamma_{2M} + \cdots + y_M \gamma_{MM} + \int_{T_1} x_1(t) \beta_{1M}(t) dt + \cdots + \int_{T_k} x_k(t) \beta_{kM}(t) dt + e_M = 0, \tag{6.210}$$

where $\beta_{ij}(t)$ are genetic effect functions. We expand $x_{nj}(t), n = 1, \ldots, N, j = 1, 2, \ldots, k$ in each genomic region in terms of orthogonal principal component functions:

$$x_{nj}(t) = \sum_{l=1}^{L_j} \eta_{njl} \phi_{jl}(t), j = 1, \ldots, k, \tag{6.211}$$

where $\phi_{jl}(t), j = 1, \ldots, k, l = 1, \ldots, L_j$ are the $l^{th}$ principal component function in the $j^{th}$ genomic region or gene and $\eta_{njl}$ are the functional principal component scores of the $n^{th}$ individual.

Using the functional principal component expansion of $x_{nj}(t)$, we obtain

$$\int_{T_j} x_{nj}(t)\beta_{ji}(t)dt = \int_{T_j} \sum_{l=1}^{L_j} \eta_{njl}\phi_{jl}(t)\beta_{ji}(t)dt = \sum_{l=1}^{L_j} \eta_{njl}b_{jl}^{(i)}, n = 1, \dots, N, j$$

$$= 1, \dots, k, i = 1, \dots, M, \tag{6.212}$$

where $b_{jl}^{(i)} = \int_{T_j} \phi_{jl}(t)\beta_{ji}(t)dt$.

Let $x_j(t) = [x_{1j}(t), \dots, x_{nj}(t)]^T$, $\eta_{jl} = [\eta_{1jl}, \dots, \eta_{Njl}]^T$. Substituting Equation 6.212 into Equation 6.210, we obtain

$$y_1 r_{11} + y_2 r_{21} + \cdots + y_M r_{M1} + \sum_{l=1}^{L_1} \eta_{1l}b_{1l}^{(1)} + \cdots + \sum_{l=1}^{L_k} \eta_{kl}b_{kl}^{(1)} + e_1 = 0$$

$$\vdots \qquad \vdots \qquad \vdots \tag{6.213}$$

$$y_1 r_{1M} + y_2 r_{2M} + \cdots + y_M r_{MM} + \sum_{l=1}^{L_1} \eta_{1l}b_{1l}^{(M)} + \cdots + \sum_{l=1}^{L_k} \eta_{kl}b_{kl}^{(M)} + e_M = 0$$

Therefore, the FSEMs are transformed to the multivariate SEMs. We then use techniques that were developed in previous sections to solve this problem.

### 6.5.3 Conditional Gaussian Graphical Models (CGGMs) for Structural Imaging and Genomic Data Analysis

After segmenting the image into $M$ subregions, let $y = [y_1, \dots, y_M]^T$ be a vector of image summary measures of $M$ subregions and $x = [x_1, \dots, x_K]^T$ be a vector of variables for genomic variables such as indicator variables for the common SNPs, overall gene expression levels. Let $\Sigma$ be a covariance matrix and $\Theta = \Sigma^{-1}$ be its inverse matrix. The elements in the matrix $\Theta$ represent the presence (or absence) of edges (connection between image subregions) in the network. To infer the genomic-image network, we minimize

$$\frac{1}{2}\text{Tr}\left(Y\Theta_{yy}Y^T + \text{Tr}\left(X\Theta_{xy}Y^T\right) + \lambda_1\|\Theta_{xy}\|_1 + \lambda_2\|\Theta_{yy}\|_1, \tag{6.214}\right.$$

where $Y$ is a matrix of the imaging summary measures, $\Sigma_{yy}$ is covariance matrix of $y$, $\Theta_{yy} = \Sigma_{yy}^{-1}$, $X$ is a matrix of the genomic variables, $\Sigma_{xy}$ is covariance matrix between $x$ and $y$, $\lambda_1$ and $\lambda_2$ are penalty parameters. We use ADMM algorithms (Boyd et al. 2011) to learn the sparse CGGM (Exercise 8) which will finally lead to the genome-imaging network.

## 6.6 Time Series SEMs for Integrated Causal Analysis of fMRI and Genomic Data

Functional magnetic resonance imaging (fMRI) can be used for inferring brain effective connectivity networks that combine structural and effectivity into a directed graph to capture causal influence among brain regions in which neuronal activity in one region can predict activity in another region (Yu et al. 2015). The Granger causality, dynamic causal modeling, graphic theory-based methods, and model independent methods are widely used methods for connectivity analysis of fMRI (Bressler and Seth 2011; Friston et al. 2003). Advances in imaging and genomic technologies have generated extremely high dimensional fMRI, RNA-seq, and NGS genotype data, which raise great challenges for their analysis. To meet these challenges, we introduce some recently developed methods for integrated casual analysis of fMRI and genomics. The methods consist of three major components: (1) sparse SEMs with both functional endogenous and exogenous variables for vector time series, (2) directed graphic models for accurate estimation of effective connectivity among brain regions and causal relations between brain regions and genomic variations including NGS genotype and RNA-seq data, and (3) IP and DP algorithms for searching for the best causal graph.

### 6.6.1 Models

Suppose that a brain image is segmented into $M$ subregions. The summary measures of the $M$ brain subregions are referred to as endogenous variables. We denote the $n$ observations on the $M$ endogenous variables at the time $t,t = 1,...,T$, by the matrix $Y(t) = [y_1(t),y_2(t),...,y_M(t)]$, where $y_i(t) = [y_{1i}(t), ..., y_{ni}(t)]^T$ is a vector of collecting $n$ observation of the endogenous variable $i$, the image summary measure of the $i^{th}$ brain subregion at time $t$. Covariates, genetic variants defined as exogenous or predetermined variables are denoted by $X(0) = [x_1,...,x_K]$ where $x_i = [x_{1i}, ..., x_{ni}]^T$. Similarly, random errors are denoted by $e_i(t) = [e_{1i}(t), ..., e_{ni}(t)]^T, e(t) = [e_1(t), ..., e_M(t)]^T, E = [e(1), ..., e(T)]$ where we assume $E[e_i(t)] = 0, E[e_i(t)e(t)_i^T] = \sigma_i^2 I_n$ for $i = 1,...,M$, and $E[e_i(t)e_i^T(s)] = 0_{n \times n}, t \neq s$.

Let $X(1) = Y(t-1),...,X(p) = Y(t-p)$. Define $X = [X(0) \quad X(1) \quad ... \quad X(p)]$. We develop the following structural equation model for the joint causal analysis of the fMRI and genomics:

$$Y(t)\Gamma + X(0)B(0) + X(1)B(1) + ... + X(p)B(p) + e(t) = 0, \qquad (6.215)$$

where

$$\Gamma = [\Gamma_1, ..., \Gamma_M], \Gamma =^i [\gamma_{1i}, ..., \gamma_{Mi}]^T,$$

$$B(0) = \begin{bmatrix} \beta_{11} & \cdots & \beta_{1M} \\ \vdots & \ddots & \vdots \\ \beta_{K1} & \cdots & \beta_{KM} \end{bmatrix} = [\,\beta_1 \;\cdots\; \beta_M\,] = [\,B_1(0) \;\cdots\; B_M(0)\,],$$

$$e(t) = \begin{bmatrix} e_{11}(t) & \cdots & e_{1M}(t) \\ \vdots & \ddots & \vdots \\ e_{n1}(t) & \cdots & e_{nM}(t) \end{bmatrix} = [\,e_1(t) \;\cdots\; e_M(t)\,] \text{ and } B(j) = \begin{bmatrix} \alpha_{11}^{(j)} & \cdots & \alpha_{1M}^{(j)} \\ \vdots & \ddots & \vdots \\ \alpha_{M1}^{(j)} & \cdots & \alpha_{MM}^{(j)} \end{bmatrix}, \; j$$

$$= 1, ..., p$$

Let

$$B = \begin{bmatrix} B(0) \\ B(1) \\ \vdots \\ B(p) \end{bmatrix} = \begin{bmatrix} B_1(0) & \cdots & B_i(0) & \cdots & B_M(0) \\ B_1(1) & \cdots & B_i(1) & \cdots & B_M(1) \\ \vdots & \vdots & \vdots & \vdots & \vdots \\ B_1(p) & \cdots & B_i(p) & \cdots & B_M(p) \end{bmatrix} = [\,B_1 \;\cdots\; B_i \;\cdots\; B_M\,]$$

Equation 6.215 can be rewritten as

$$Y(t)\Gamma + XB + e(t) = 0 \tag{6.216}$$

Consider the $i^{th}$ equation in (6.216), we obtain

$$Y(t)\Gamma_i + XB_i + e_i(t) = 0, \tag{6.217}$$

where $B_i = \begin{bmatrix} \beta_i \\ B_i(1) \\ \vdots \\ B_i(p) \end{bmatrix}$.

We set $\gamma_{ii} = -1$ and define

$$\Gamma_i = \begin{bmatrix} -1 \\ \gamma_i \end{bmatrix}$$

Let $Y_{-i}(t)$ be the observation matrix $Y(t)$ after removing $y_i(t)$ from it and $\gamma_{-i}$ be the parameter vector $\Gamma_i$ after removing the parameter $\gamma_{ii}$.

With rearrangement, Equation 6.217 can be rewritten as

$$y_i(t) = Y_{-i}(t)\gamma_i + XB_i + e_i(t)$$
$$= Z_i(t)\delta_i + e_i(t), \tag{6.218}$$

where

$$Z_i(t) = \begin{bmatrix} Y_{-i}(t) \ X \end{bmatrix} \text{ and } \delta_i = \begin{bmatrix} \gamma_i \\ \beta_i \\ B_i(1) \\ \vdots \\ B_i(p) \end{bmatrix}$$

### 6.6.2 Reduced Form Equations

We assume that the matrix $\Gamma$ is nonsingular. We can express the endogenous variables (image summary measures) $Y(t)$ at the time $t$ as a function of the predetermined variables $X$ and the lagged endogenous variables (image summary measures) $Y(t - j)$, $j = 1,\ldots,p$:

$$\begin{aligned} Y(t) &= -XB\Gamma^{-1} - e(t)\Gamma^{-1} \\ &= X\Pi + V(t), \end{aligned} \tag{6.219}$$

where

$$\begin{aligned} \Pi &= -B\Gamma^{-1} \\ &= \begin{bmatrix} -B_1\Gamma^{-1} & \cdots & -B_M\Gamma^{-1} \end{bmatrix} \\ &= \begin{bmatrix} \Pi_1 & \cdots & \Pi_M \end{bmatrix}, \end{aligned}$$

$$\Pi_i = -B_i\Gamma^{-1} = \begin{bmatrix} -\beta_i\Gamma^{-1} \\ B_i(1)\Gamma^{-1} \\ \vdots \\ B_i(p)\Gamma^{-1} \end{bmatrix}, \tag{6.220}$$

$$V(t) = -e(t)\Gamma^{-1} = \begin{bmatrix} V_1(t) & \cdots & V_M(t) \end{bmatrix}$$

Therefore, the $i^{th}$ equation in reduced form is given by

$$y_i(t) = X\Pi_i + V_i \tag{6.221}$$

The least square estimators of the regression coefficients of the $i^{th}$ equation and the whole system are

$$\hat{\Pi}_i = \left(X^T X\right)^{-1} X^T y_i(t) \quad \text{and}$$

$$\hat{\Pi} = \left(X^T X\right)^{-1} X^T Y(t),\tag{6.222}$$

respectively.

## 6.6.3 Single Equation and Generalized Least Square Estimator

Substituting Equation 6.222 into Equation 6.120, we obtain

$$\left(X^T X\right)^{-1} X^T Y(t) \begin{pmatrix} -1 \\ \gamma_i \end{pmatrix} = -B_i \tag{6.223}$$

Multiplying Equation 6.223 by $X^T X$, we obtain

$$X^T Y(t) \begin{pmatrix} -1 \\ \gamma_i \end{pmatrix} = -\left(X^T X\right) B_i, \tag{6.224}$$

which implies

$$- X^T y_i(t) + X_T Y_{-i}(t)\hat{\gamma}_i = -X^T X \hat{B}_i$$

or

$$X^T y_i(t) = X^T Y_{-i}(t)\hat{\gamma}_i + X^T X \hat{B}_i \tag{6.225}$$

It follows from Equations 6.218 and 6.225 that

$$X^T y_i(t) = X^T Z_i(t)\delta_i + X^T e_i(t) \tag{6.226}$$

To efficiently estimate $\delta_i$ we need to explore the weighted least square estimation methods. The variance of $X^T e_i(t)$ is given by

$$Var\left(X^T e_i(t)\right) = \sigma_i^2 X^T X \tag{6.227}$$

The generalized least square estimator of $\delta_i$ is estimated by minimizing the sum of square errors:

$$F = \sum_{t=p+1}^{T} \left[X^T y_i(t) - X^T Z_i(t)\delta_i\right]^T \left(\sigma_i^2 X^T X\right)^{-1} \left[X^T y_i(t) - X^T Z_i(t)\delta_i\right] \tag{6.228}$$

Setting the partial derivative of the function $F$

$$\frac{\partial F}{\partial \delta_i} = -2\sum_{t=p+1}^{T} Z_i^T(t) X \left(\sigma_i^2 X^T X\right)^{-1} \left[X^T y_i(t) - X^T Z_i(t)\delta_i\right]$$

to be zero, we obtain

$$\hat{\delta}_i = \left[\sum_{t=p+1}^{T} Z_i^T(t)X(X^TX)^{-1}X^TZ_i(t)\right]^{-1}\sum_{t=p+1}^{T}Z_i^T(t)X(X^TX)^{-1}X^Ty_i(t) \quad (6.229)$$

The variance-covariance matrix of the estimator is given by

$$Var(\hat{\delta}_i) = \sigma_i^2\left[\sum_{t=p+1}^{T}Z_i^T(t)X(X^TX)^{-1}X^TZ_i(t)\right]^{-1} \quad (6.230)$$

The variance $\sigma_i^2$ is estimated by

$$\hat{\sigma}_i^2 = \frac{\sum_{t=p+1}^{T}\left[y_i(t) - Z_i(t)\hat{\delta}_i\right]^T\left[y_i(t) - Z_i(t)\hat{\delta}_i\right]}{(T-p-1)(n-m_i-k_i-)}, \quad (6.231)$$

where $m_i$ is the number of nonzero elements in $\Gamma_i$ and $k_i$ is the number of nonzero elements in $B_i$.

### 6.6.4 Sparse SEMs and Alternating Direction Method of Multipliers

In general, the networks are sparse. To obtain sparse estimates of the matrices $\Gamma$ and $B$, the natural approach is the $l_1$-norm penalized regression of Equation 6.226. Let $W_i(t) = Z_i(t), \Delta_i = \delta_i$. $Z_i(t) = [Y_{-i}(t)\ X], X(0) = [x_1,...,x_K]$ where $x_i = [x_{1i},...,x_{ni}]^T$, $X = [X(0)\ X(1)\ \cdots\ X(p)]$ and $X^Ty_i(t) = X^TW_i(t)\Delta_i + X^Te_i(t)$.

Using weighted least square and $l_1$-norm penalization, we can form the following optimization problem:

$$\min_{\Delta_i}\ f(\Delta_i) + \lambda\|\Delta_i\|_1$$

where $f(\Delta_i) = \sum_{t=p+1}^{T}\left[(X^Ty_i(t) - X^TW_i(t)\Delta_i)^T(X^TX)^{-1}(X^Ty_i(t) - X^TW_i(t)\Delta_i)\right]$

$$(6.232)$$

The optimization problem (6.232) can be further reduced to

$$\min\quad f(\Delta_i) + \lambda\|Z_i\|_1$$
$$\text{subject to}\quad \Delta_i - Z_i = 0, \quad (6.233)$$

where $f(\Delta_i) = \sum_{t=p+1}^{T}[(X^Ty_i(t) - X^TW_i(t)\Delta_i)^T(X^TX)^{-1}(X^Ty_i(t) - X^TW_i(t)\Delta_i)]$.

To solve the optimization problem (6.233), we form the augmented Lagrangian

$$L_\rho(\Delta_i, Z_i, \mu) = f(\Delta_i) + \lambda\|Z_i\|_1 + \mu^T(\Delta_i - Z_i) + \frac{\rho}{2}\|\Delta_i - Z_i\|_2^2 \quad (6.234)$$

The alternating direction method of multipliers (ADMM) consists of the iterations:

$$\Delta_i^{(k+1)} :== \arg\min_{\Delta_i} L_\rho\left(\Delta_i, Z_i^{(k)}, \mu^{(k)}\right) \tag{6.235}$$

$$Z_i^{(k+1)} :== \arg\min_{Z_i} L_\rho\left(\Delta_i^{(k+1)}, Z_i, \mu^{(k)}\right) \tag{6.236}$$

$$\mu^{(k+1)} :== \mu^{(k+1)} + \rho\left(\Delta_i^{(k+1)} - Z_i^{(k+1)}\right), \tag{6.237}$$

where $\rho > 0$.

Let $u = \dfrac{\mu}{\rho}$. Equations 6.235–6.237 an be reduced to

$$\Delta_i^{(k+1)} :== \arg\min_{\Delta_i} \left(f(\Delta_i) + \frac{\rho}{2}\|\Delta_i - Z_i^{(k)} + u^{(k)}\|_2^2\right) \tag{6.238}$$

$$Z_i^{(k+1)} :== \arg\min_{Z_i} \left(\lambda\|Z_i\|_1 + \frac{\rho}{2}\|\Delta_i^{(k+1)} - Z_i + u^{(k)}\|_2^2\right) \tag{6.239}$$

$$u^{(k+)} :== u^{(k)} + \left(\Delta_i^{(k+1)} - Z_i^{(k+1)}\right) \tag{6.240}$$

Solving the minimization problem (6.238), we obtain

$$\Delta_i^{(k+1)} = \left\{\left[\sum_{t=p+1}^{T} W_i^T(t)X(X^TX)^{-1}X^TW(t)\right] + \rho I\right\}^{-1}$$

$$\left\{\left[\sum_{t=p+1}^{T} W_i^T(t)X(X^TX)^{-1}X^Ty_i(t)\right] + \rho\left(Z_i^k - u^k\right)\right\}$$

The optimization problem (6.239) is non-differentiable. Although the first term in (6.239) is not differentiable, we still can obtain a simple closed-form solution to the problem (6.239) using subdifferential calculus. Let $\Gamma_j$ be a generalized derivative of the $j^{th}$ component $Z_i^j$ of the vector $Z_i$ and $\Gamma = [\Gamma_1, ..., \Gamma_{M+K-1}]^T$ where

$$\Gamma_j = \begin{cases} 1 & Z_i^j > 0 \\ [-1, 1] & Z_i^j = 0 \\ -1 & Z_i^j < 0 \end{cases}$$

Then, we have

$$\frac{\lambda}{\rho}\Gamma + Z_i = \Delta_i^{k+1} + u^k,$$

which implies that

$$Z_i^{(k+1)} = \text{sgn}\left(\Delta_i^{k+1} + u^k\right)\left(\left|\Delta_i^{k+1} + u^k\right| - \frac{\lambda}{\rho}\right)_+,\tag{6.241}$$

where

$$|x|_+ = \begin{cases} x & x \geq 0 \\ 0 & x < 0 \end{cases}$$

The algorithm is summarized as the Result 6.3.

**Result 6.3 Algorithm for Learning Sparse fMRI**

For $i = 1,...,M$
Step 1. Initialization

$$u^0 := 0$$
$$\Delta_i^0 := \left[\sum_{t=p+1}^{T} W_i^T(t)X(X^TX)^{-1}X^TW_i(t)\right]^{-1}\sum_{t=p+1}^{T}W_i^T(t)X(X^TX)^{-1}X^Ty_i(t)$$
$$Z_i^0 := \Delta_i^0$$

Carry out steps 2, 3 and 4 until convergence
Step 2.

$$\Delta_i^{(k+1)} = \left\{\left[\sum_{t=p+1}^{T}W_i^T(t)X(X^TX)^{-1}X^TW(t)\right] + \rho I\right\}^{-1}$$
$$\left\{\left[\sum_{t=p+1}^{T}W_i^T(t)X(X^TX)^{-1}X^Ty_i(t)\right] + \rho\left(Z_i^k - u^k\right)\right\}$$

Step 3.

$$Z_i^{(k+1)} := \text{sgn}\left(\Delta_i^{k+1} + u^k\right)\left(\left|\Delta_i^{k+1} + u^k\right| - \frac{\lambda}{\rho}\right)_+,$$

where

$$|x|_+ = \begin{cases} x & x \geq 0 \\ 0 & x < 0 \end{cases}$$

Step 4.

$$u^{(k+)} := = u^{(k)} + \left(\Delta_i^{(k+1)} - Z_i^{(k+1)}\right)$$

Check convergence:
If $\|\Delta_i^{(k+1)} - \Delta_i^{(k)}\| > \varepsilon$ then $\Delta_i^{(k)} \leftarrow \Delta_i^{(k+1)}$, $Z_i^{(k+1)} \leftarrow Z_i^{(k)}$, $u^{(k+1)} \leftarrow u^{(k)}$, $k \leftarrow k+1$, go to step 2;
Otherwise stop.

**Example 6.6**

To evaluate the feasibility of the proposed models for the effective connectivity analysis of fMRI, we applied the integer programming (IP) with the sparse time series SEMs as a tool for defining score function and super structure for causal graph learning to the rest of the fMRI data with six individuals downloaded from the public available data warehouse OpenfMRI (https://openfmri.org/). The fMRI data were acquired by the balloon analog risk-taking in an event-related design. The duration time for the fMRI measurement is 200 seconds. The image dimension of a single frame is 64*64*10. We used simple linear iterative clustering (SLIC) supervoxels algorithms to segment the brain into 34 subregions. The total volume of each subregion was used as a measure of the imaging signals. The results of effective connectivity for the rest of the fMRI was shown in Figure 6.15 in which we showed connectivity between the regions in the cortex (other regions were not shown). We presented the varying effective connectivity patterns of the brain at the times $t - 1$ and $t$. The directions "Superior Frontal→ Orbital Frontal," "Middle Temporal → Inferior Frontal," "Superior Temporal→ Middle Temporal" at the time $t - 1$ remain at the time $t$ and directions: "Super Frontal→ Central→ Superior Temporal" at the time $t - 1$ was changed to directions: "Super Frontal→ Superior Temporal." The direction "Occipital"→ Superior Temporal" at the time $t - 1$ was lost at the time $t$. When we transit from the time $t - 1$ to time $t$, two directions "Super Frontal→Central" and "Superior→Middle Temporal" came from the directions at the time $t - 1$, but the directions: "Inferior Frontal→Superior Temporal" and "Orbital Frontal→ Central" were newly created.

## 6.7 Causal Machine Learning

The current paradigm for disease risk prediction and association analysis is to use unstructured variables for classification and testing (Sun et al. 2014). Less attention has been paid to structured data (Wu et al. 2015). However, causal networks provide a holistic view of genetic, epigenetic, and imaging structures of complex diseases. The traditional machine learning methods often attempt to explore statistical associations between variables for making predictions and considers less underlying the data generating process (Gershman and Daw 2017). The real-world prediction systems are constantly under changes. It will often happen that the distributions of training and test data differ significantly. Imaging data provide imperfect information about the underlying state of the system. The widely used models for imaging-genomic data analysis do not consider the data generating mechanism. We often observed that predictions using imaging data are accurate in the training dataset, but not satisfactory in the test dataset. Without causal

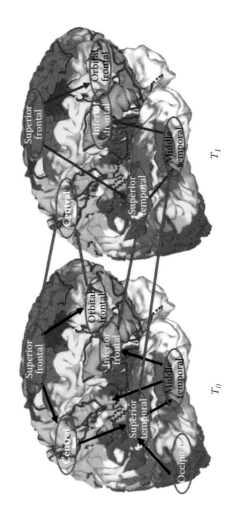

**FIGURE 6.15**
Effective connectivity map of fMRI.

knowledge, we often select features that do not generalize to the test setting, which will result in inaccurate predictions (Magliacane et al. 2017).

The goal of causal machine learning is to explore the causal structure underlying the data generating process for feature selection and classification. To employ the causal structure underlying the data generating process, we first develop a clever feature representation for causal graphs and then develop sparse sufficient dimension reduction (SDR) for feature selection in causal graphs (Mairal and Yu 2013). The SDR is a supervised dimension reduction method in which both genetic information and disease information will be used. The SDR will select a minimal set of linear combinations of the original predictors without loss of disease information for classification (Chen et al. 2010).

Consider a DAG $G = (V,E)$ with $V = \{x_1,...,x_k\}$ and $E = \{E_j, j = 1,...,m\}$, $E_j = \{(x_{a_j}, x_{b_j}) | x_{a_j}, x_{b_j} \in V\}$ which can be inferred by the methods for causal inference discussed in Section 6.5 and other chapters in the book. Order the nodes in the DAG. Taking all node and edge information in the DAG, we define a novel feature-vector for the DAG as $X = [x_1, ..., x_k, x_{a_1}x_{b_1}, ..., x_{a_m}x_{b_m}]^T$. Let $Y$ be a class label of the individual. SDR aims to find a linear subspace S such that the response $Y$ depends on $X$ only through vectors in the subspace S, denoted by $S_{Y|X}$. Let $B = [\beta_1,...,\beta_d]$ be the matrix of the basis forming $S_{Y|X}$. To maximally employ information in the reduced space, we solve the optimization problem (Wang T and Zhu 2013):

$$\min_\beta E\left[\| T_i(Y) - E(T_i(Y)) - (X - E(X))^T \beta \|_2^2\right], \tag{6.242}$$

where $T_i(y)$ are orthogonal and are a set of transformation functions of the functional $y$. Using the basis function expansion of the transformation function, the above problem can be transformed to an optimal score problem (Clemmensen et al. 2011). Optimal score formulation of the SDR can work well for a small problem but is problematic when the number of features is large. To solve this problem, we further transform the optimal score problem to

$$|Y\theta_i - X\beta_i \|_2^2 + \lambda \sum_l |\beta_l^* \|_2^{1-r} \text{st} \theta_i^T D\theta_i = 1, \theta_i^T D\theta_j = 0, \tag{6.243}$$

where $B = [\beta_1, ..., \beta_i] = [\beta_1^*, ..., \beta_p^*]^T$ and use ADMM (Boyd et al. 2011) to solve it.

### Example 6.7

As a pilot study to validate the feasibility of the causal machine learning for risk prediction, the proposed causal machine learning (CML) was applied to the schizophrenia (SCZ) imaging and genomic dataset with 142 series of DTI and 14,412 genes (746,575 SNPs) typed in 64 SCZ patients and 78 healthy controls. A five-fold cross validation was used to evaluate

**TABLE 6.1**

The Accuracy of Prediction of SCZ Using CML and Other Methods

| Methods | Training | | | Test | | |
|---------|----------|-------------|-------------|----------|-------------|-------------|
|         | Accuracy | Sensitivity | Specificity | Accuracy | Sensitivity | Specificity |
| CML      | 1     | 1     | 1     | 0.951 | 0.962 | 0.939 |
| MPCA Only | 0.989 | 0.994 | 0.984 | 0.542 | 0.48  | 0.619 |
| MPCA SDR | 0.977 | 0.959 | 0.99  | 0.611 | 0.307 | 0.845 |
| FPC Only | 0.984 | 0.975 | 0.983 | 0.574 | 0.089 | 0.952 |
| FPC SDR  | 0.986 | 0.983 | 0.986 | 0.662 | 0.515 | 0.793 |

the accuracy. Table 6.1 presents average results over five folds. A total of 11 brain subregions were selected by all five folds. Seven of them were subregions that were significantly connected by genes with P-values < 3.5E-06. We found that the gene ALDOA that regulated the differentially expressed miRNA and was associated with the SCZ was connected with the brain regions that had high accuracy of prediction of SCZ. Table 6.1 showed that the average prediction accuracy using the CML in the test datasets was much higher than that of other methods and that the CML has great generalization ability. The CML analysis can calculate the classification accuracy for each region. For example, the brain subregions can be ranked by the classification accuracy as: MOG (left) (0.672), SCC (0.651), MOG (right) (0.648), PG (left) (0.647), SG (0.634), and DT (0.621), where the number in the bracket is the average classification accuracy over five folds in the test datasets. This shows that using a single brain region may obtain higher accuracy than using all brain imaging information by some methods. It is reported that abnormal function and structure of MOG (Takahashi et al. 2011), SCC (Balevich et al. 2015), PG (Brauns et al. 2015), and DT (Pergola et al. 2015) contributes to the pathophysiology of SCZ.

---

## Software Package

Software for U-Net: Convolutional Networks for Biomedical Image Segmentation can be downloaded from https://lmb.informatik.uni-freiburg.de /people/ronneber/u-net/.

All models are trained and tested with Caffe on a single NVIDIA Tesla K40c. Our models and code are publicly available at http://fcn .berkeleyvision.org.

The code for semantic segmentation based on the conditional random file can be downloaded from https://github.com/Vaan5/piecewisecrf. Software DeconvNet for learning deconvolution network for semantic segmentation can be downloaded from the website https://github.com/HyeonwooNoh/DeconvNet.

---

## Appendix 6.A   Factor Graphs and Mean Field Methods for Prediction of Marginal Distribution

Factor graphs and mean filed methods are a powerful tool for marginal distribution estimation. Factor graphs are undirected graphical models that factorize a distribution over a large number of random variables as a product of local functions that each depends on only a small number of variables (Nowozin and Lampert 2011; Sutton and McCallum 2011). Let $V$ be a set of variable nodes, $F$ be a set of factor nodes, and $E$ be a set of edges that connect a variable node and a factor node. A factor graph is a bipartite graph $G = (V, F, E)$ that consists of a set of variable nodes $V = 1, 2, \ldots, |V|$, a set of factor nodes $F = 1, 2, \ldots, A$, and a set of edges $E = 1, 2, \ldots, |E|$. We often use $i$ to index nodes in $V$ and $a$ to index factors in $F$. The neighbors of the factor with index $a \in F$ is defined as

$$N(a) = (i \in V : (i, a) \in E). \tag{6.A.1}$$

A factor or a local function is a non-negative function and is denoted by $\Psi_a, a \in F$. A distribution $P(y)$ can be factorized according to a factor graph $G$:

$$P(y) = \frac{1}{Z} \prod_{a \in F} \Psi_a \left( y_{N(a)} \right), \tag{6.A.2}$$

where $\psi_a(y_{N(a)})$ is a local function of $y_{N(a)}$ and $Z$ is a normalization constant:

$$Z = \sum_{y \in Y} \prod_{a \in F} \Psi_a \left( y_{N(a)} \right). \tag{6.A.3}$$

In the factor graph, the variable nodes are often drawn as $\circ$ and the factor nodes are often drawn as

### Example 6.A.1

Figure 6.12a shows a factor graph over three variables. Three circles are nodes of three variables $y_1$, $y_2$, and $y_3$, and the black boxes are three factor nodes $\Psi_1$, $\Psi_2$, and $\Psi_3$. The neighbors of factor $\Psi_1$ is $N(1) = \{1, 2\}$. Thus, $y_{N(1)} = \{y_1, y_2\}$ and $\Psi_1(y_1, y_2) = \Psi_1(y_1, y_2)$. Similarly, we have $\Psi_2(y_{N(2)}) = \Psi_2(y_2, y_3)$ and $\Psi_3(y_{(N)3}) = \Psi_3(y_1, y_3)$. Using Equation 6.A.2, the distribution $p(y)$ can be factorized as

$$p(y) = p(y_1, y_2, y_3) \propto \Psi_1(y_1, y_2) \Psi_2(y_2, y_3) \Psi_3(y_1, y_3).$$

**Example 6.A.2**

Consider a Markov random field with a completely connected graph as shown in Figure 6.12b. A possible factorization of the Markov random field shown in Figure 6.12b is shown in Figure 6.12c where there are four variable nodes: $V = \{y_1, y_2, y_3, y_4\}$ and six factor nodes: $F = \{\Psi_1, \Psi_2, \Psi_3, \Psi_4, \Psi_5, \Psi_6\}$. By the similar arguments for Example 6.A.1, we obtain $y_{N(1)} = \{y_1, y_2\}$ and $\Psi_1(y_{N(1)}) = \Psi_1(y_1, y_2)$. Similarly, we have $\Psi_2(y_{N(2)}) = \Psi_2(y_1, y_3)$, $\Psi_3(y_{N(3)}) = \Psi_3(y_2, y_3)$, $\Psi_4(y_{N(4)}) = \Psi_4(y_3, y_4)$, $\Psi_5(y_{N(5)}) = \Psi_5(y_1, y_4)$, and $\Psi_6(y_{N(6)}) = \Psi_6(y_2, y_3)$. The distribution $p(y)$ can be factorized as

$$P(y) = \frac{1}{Z} \prod_{a \in F} \Psi_a(y_{N(a)})$$

$$= \frac{1}{Z} \Psi_1(y_1, y_2) \Psi_2(y_1, y_3) \Psi_3(y_2, y_3) \Psi_4(y_3, y_4) \Psi_5(y_1, y_4) \Psi_6(y_2, y_3).$$

Consider a local function $\Psi_a(y_{N(a)})$. Let $E_a(y_{N(a)})$ be an energy function. The energy function is defined as

$$E_a(y_{N(a)}) = -\log(\Psi_a(y_{N(a)})). \tag{6.A.4}$$

The factorization formula (6.A.2) can be rewritten as a function of the energy function:

$$P(y) = -\frac{1}{Z} \prod_{a \in F} \Psi_a(y_{N(a)})$$

$$= \frac{1}{Z} \prod_{a \in F} \exp\{-E_a(y_{N(a)})\} \tag{6.A.5}$$

$$= \frac{1}{Z} \exp\left\{-\sum_{a \in F} E_a(y_{N(a)})\right\},$$

where the normalizing constant is

$$Z = \sum_{y \in Y} \exp\left\{-\sum_{a \in F} E_a(y_{N(a)})\right\}. \tag{6.A.6}$$

The mean field methods that search for the best approximation distribution are to minimize the K–L distance:

$$\min_{q \in \Omega} \; D_{KL}(q(y) \| P(y|x, \theta)), \tag{6.A.7}$$

where $\Omega$ is a family of tractable distributions.

By definition, the K–L distance can be reduced to

$$D_{KL}(q(y) \parallel P(y|x, \theta)) = \sum_{y \in Y} q(y) \log \frac{q(y)}{P(y|x, \theta)}$$

$$= \sum_{y \in Y} a(y) \log q(y) - \sum_{y \in Y} q(y) \log P(y|x, \theta) \quad (6.A.8)$$

$$= -H(q) - \sum_{y \in Y} q(y) \log P(y|x, \theta).$$

Substituting Equation 6.A.5 into Equation 6.A.8 gives

$$D_{KL}(q(y) \parallel P(y|x, \theta)) = -H(q) + \sum_{y \in Y} q(y) \sum_{a \in F} E_a \left( y_{N(a)}, x_{N(a)}, \theta \right)$$

$$+ \log Z(x, \theta)$$

$$= -H(q) + \sum_{a \in F} \sum_{y_{N(a)} \in Y_{N(a)}} \left[ \sum_{(y \in Y) \setminus y_{N(a)}} q(y) \right]$$

$$E_a \left( y_{N(a)}, x_{N(a)}, \theta \right) + \log Z(x, \theta)$$

$$= -H(q) + \sum_{a \in F} \sum_{y_{N(a)} \in Y_{N(a)}} q \left( y_{N(a)} \right) E_a \left( y_{N(a)}, x_{N(a)}, \theta \right)$$

$$+ \log Z(x, \theta),$$

$$(6.A.9)$$

where $q(y_{N(a)}) = \sum_{(y \in Y) \setminus y_{N(a)}} q(y)$ is the marginal distribution of $q(y)$ on the variables $N(a)$.

To reduce complexity of the computation, we assume that the set $\Omega$ consists of all factorial distributions, that is,

$$q(y) = \prod_{i \in V} q_i(y_i). \quad (6.A.10)$$

The mean field that searches for the best approximate distribution among all factorial distributions is called naïve mean field method (Nowozin and Lampert, 2011). Under factorial distribution, the entropy is reduced to

$$H(q) = -\sum_{y \in Y} \prod_{i \in V} q_i(y_i) \sum_{i \in V} \log(q_i(y_i))$$

$$= \sum_{i \in V} \left[ -\sum_{y_i \in Y_i} q_i(y_i) \log(q_i(y_i)) \right] \quad (6.A.11)$$

$$= \sum_{i \in V} H_i(q_i).$$

The marginal distribution of $q(y)$ on the variables $N(a)$ can also be reduced to

$$
\begin{aligned}
q\left(y_{N(a)}\right) &= \sum_{(y \in Y) \setminus y_{N(a)}} q(y) \\
&= \prod_{i \in N(a)} q_i(y_i).
\end{aligned}
\tag{6.A.12}
$$

Substituting Equations 6.A.11 and 6.A.12 into Equation 6.A.9, we obtain

$$
\begin{aligned}
D_{KL}(q(y) \| P(y|x, \boldsymbol{\theta})) &= \sum_{i \in V} \sum_{y_i \in Y_i} q_i(y_i) \log(q_i(y_i)) \\
&+ \sum_{a \in F} \sum_{y_{N(a)} \in Y(a)} \prod_{i \in N(a)} q_i(y_i) E_a\left(y_{N(a)}, x_{N(a)}, \boldsymbol{\theta}\right) \\
&+ \log Z(x, \boldsymbol{\theta}).
\end{aligned}
\tag{6.A.13}
$$

Note that the probability distribution $q_i(y_i)$ should satisfy the following constraints:

$$
\begin{aligned}
&\sum_{y_i \in Y_i} q_i(y_i) = 1 \\
&q_i(y_i) \geq 0, i \in N(a).
\end{aligned}
\tag{6.A.14}
$$

The first term in (6.A.13) is entropy and hence is convex. However, the second term in (6.A.13) involves products $\prod_{i \in N(a)} q_i(y_i)$. that are non-convex. To make the second term in (6.A.13) convex, we only consider optimization with respect to a specific term $q_i(y_i)$ and hold all other variables $q_j(y_j), j \in N(a) \setminus \{i\}$ fixed. Let $\hat{q}_j(y_j) = q_j(y_j)$ be held fixed. After all constant terms not affected by $q_i(y_i)$ are dropped, Equation 6.A.13 is reduced to

$$
\begin{aligned}
D_{KL}(q(y) \| P(y|x, \boldsymbol{\theta})) &\approx \sum_{i \in V} \sum_{y_i \in Y_i} q_i(y_i) \log(q_i(y_i)) \\
&+ \sum_{a \in F} \sum_{y_{N(a)} \in Y_{N(a)}} \left( \prod_{j \in N(a) \setminus \{i\}} \hat{q}_j(y_i) \right) \\
&q_i(y_i) E_a\left(y_{N(a)}, x_{N(a)}, \boldsymbol{\theta}\right).
\end{aligned}
\tag{6.A.15}
$$

Now the optimization problem for searching for the best approximate distribution is reduced to

$$\min \quad D_{KL}(q(y) \| P(y|x, \theta))$$
$$\text{s.t} \sum_{y_i \in Y_i} q_i(y_i) = 1. \tag{6.A.16}$$

Using the Lagrange multiplier method, the constraint optimization problem (6.A.16) can be reduced to

$$\min \quad G(q_i(y_i)) = D_{KL}(q(y) \| P(y|x, \theta)) + \lambda \left(1 - \sum_{y_i \in Y_i} q_i(y_i)\right). \tag{6.A.17}$$

Taking partial derivatives of $G(q_i(y_i))$ with respect to $q_i(y_i)$ and setting it to zero, we obtain

$$\log(q_i(y_i)) + 1 + \sum_{a \in F} \sum_{y_{N(a)} \in Y_{N(a)}} \left(\prod_{j \in N(a) \setminus \{i\}} \hat{q}_j \left(y_j\right)\right)$$
$$E_a \left(y_{N(a)}, x_{N(a)}, \theta\right) - \lambda) = 0. \tag{6.A.18}$$

Solving Equation 6.A.18 for $q_i(y_i)$ gives

$$\hat{q}_i(y_i) = \exp\left(\lambda - 1 - \sum_{a \in F, i \in N(a)} \sum_{y_{N(a)} \in Y_{N(a)}, [y_{N(a)}] = y_i}\right.$$
$$\left.\left(\prod_{j \in N(a) \setminus \{i\}} \hat{q}_j \left(y_j\right)\right) E_a \left(y_{N(a)}, x_{N(a)}\right)\right) \tag{6.A.19}$$

Imposing the constraint $\sum_{y_i \in Y_i} \hat{q}_i(y_i) = 1$ in Equation 6.A.19, we obtain

$$\lambda = -\log\left(\sum_{y_i \in Y_i} \exp\left(-1 - \sum_{a \in F, i \in N(a)} \sum_{y_{N(a)} \in Y_{N(a)}, [y_{N(a)}] = y_i}\right.\right.$$
$$\left.\left.\left(\prod_{j \in N(a) \setminus \{i\}} \hat{q}_j(y_j)\right) E_a \left(y_{N(a)}, x_{N(a)}\right)\right)\right), \tag{6.A.20}$$

where the empty product is defined as 1.

## Exercises

**Exercise 1.** Show

$$\frac{\partial C_E\left(W,b,x^{(u)}\right)}{\partial w_{jk}^{(l)}} = -\left(y_j^u(W,b) - x^{(u)}\right)\sigma'\left(z_j^{(2)}\right)a_k^{(1)},$$

where

$$\sigma'(x) = \sigma(x)(1 - \sigma(x)).$$

**Exercise 2.** Show that
for $l = L - 1,\ldots,1$, we have

$$\delta_j^{(l)}\left(x^{(u)}\right) = \sum_{k=1}^{n_{l+1}} \delta_k^{(l+1)}\left(x^{(u)}\right) w_{kj}^{(l+1)} \sigma'\left(z_j^{(l)}\right),$$

$$\frac{\partial C_E\left(W,b,x^{(u)}\right)}{\partial b_j^{(l)}} = \delta_j^{(l)}\left(x^{(u)}\right),$$

$$\frac{\partial C_E\left(W,b,x^{(u)}\right)}{\partial w_{jk}^{(l)}} = \frac{\partial C_E\left(W,b,x^{(u)}\right)}{\partial z_j^{(l)}} \frac{\partial z_j^{(l)}}{\partial w_{jk}^{(l)}} = \delta_j^{(l)}\left(x^{(u)}\right) a_k^{(l-1)}\left(x^{(u)}\right).$$

**Exercise 3.** Show

$$\frac{\partial \hat{p}_j^{(v)}}{\partial b_g^{(l)}} = \begin{cases} 0 & l > v \\ \sigma'\left(z_j^{(v)}\right) & l = v \\ \sigma'\left(z_j^{(v)}\right)\sum_{k=1}^{n_{v-1}} w_{jk}^{(v)} \dfrac{\partial \hat{p}_k^{(v-1)}}{\partial b_g^{(l)}} & l < v. \end{cases}$$

**Exercise 4.** Consider the data in Example 6.2. If we assume $S = 1$, generate the feature map.

**Exercise 5.** Consider a factor graph shown in Figure 6.12d. Write the factorization of distribution $p(y)$.

**Exercise 6.** Show

$$\delta J[h] = \frac{d}{d\varepsilon} J[\beta(s,t,u) + \varepsilon h(s,t,u)]$$

$$= \int_S \int_T \int_U \left[ \int_S \int_T \int_U \left[ R(s_1,t_1,u_1,s_2,t_2,u_2)\beta(s_2,t_2,u_2)ds_2t_2du_2 \right. \right.$$

$$\left. -\lambda \left( \beta(s_1,t_1,u_1) + \mu \frac{\partial^{12}\beta(s_1,t_1,u_1)}{\partial s_1^4 \partial t_1^4 \partial u_1^4} \right] h(s_1,t_1,u_1)ds_1dt_1\mathbf{du}_1 \right.$$

$$= \int_S \int_T \int_U \left[ \int_S \int_T \int_U R(s_1,t_1,u_1,s_2,t_2,u_2)\beta(s_2,t_2,u_2)ds_2dt_2du_2 \right.$$

$$\left. -\lambda \left( \beta \left( s_1,t_1,u_1 + \mu \frac{\partial^{12}\beta(s_1,t_1,u_1)}{\partial s_1^4 \partial t_1^4 \partial u_1^4} \right) \right]^2 ds_1dt_1du_1 = 0.$$

**Exercise 7.** Rewrite the following equation:

$$y_1 r_{11} + y_2 r_{21} + \cdots + y_M r_{M1} + \sum_{l=1}^{L_1} \eta_{1l} b_{1l}^{(1)} + \cdots + \sum_{l=1}^{L_k} \eta_{kl} b_{kl}^{(1)} + e_1 = 0$$

$$\vdots \qquad\qquad\qquad \vdots \qquad\qquad\qquad \vdots$$

$$y_1 r_{1M} + y_2 r_{2M} + \cdots + y_M r_{MM} + \sum_{l=1}^{L_1} \eta_{1l} b_{1l}^{(M)} + \cdots + \sum_{l=1}^{L_k} \eta_{kl} b_{kl}^{(M)} + e_M = 0$$

in a matrix form.

**Exercise 8.** Develop ADMM algorithms to solve the following optimization problem:

$$\min_{\Theta_{YY},\Theta_{xy}} \quad \frac{1}{2}\mathrm{Tr}\left( Y\Theta_{yy}Y^T + \mathrm{Tr}\left( X\Theta_{xy}Y^T \right) + \lambda_1 \| \Theta_{xy} \|_1 + \lambda_2 \| \Theta_{yy} \|_1.$$

# 7

# From Association Analysis to Integrated Causal Inference

Next generation genomic, epigenomic, sensing, and image technologies produce ever deeper multiple omic, physiological, imaging, environmental, and phenotypic data with millions of features. Analysis of increasingly larger and deep omic and phenotype data provides invaluable information for the holistic discovery of the genetic structure of disease and precision medicine. The current approach to genomic analysis lacks breadth (number of variables analyzed at a time) and depth (the number of steps that are taken by the genetic variants to reach the clinical outcomes, across genomic and molecular levels) and its paradigm of analysis is association and correlation analysis. Despite significant progress in dissecting the genetic architecture of complex diseases by association analysis, understanding the etiology and mechanism of complex diseases remains elusive. Using association analysis as a major analytic platform for genetic studies of complex diseases is a key issue that hampers the theoretic development of genomic science and its application in practice.

Causal inference is an essential component for the discovery of mechanism of diseases. Many researchers suggest making the transition from association to causation (Clyde 2017). Although causal inference may have great potential to improve prevention, management, and therapy of complex diseases (Peters et al. 2017; Orho-Melander 2015), most genomic, epigenomic, and image data are observational data. Many confounding variables are not or cannot be measured. The unmeasured confounding variables will invalidate the most traditional causal inference analysis. The gold-standard for causal inference is to perform a randomized controlled trial, which can control the confounding effects (Statnikov et al. 2012). Unfortunately, in many cases, performing experiments is unethical or infeasible. In most genetic studies inferring causal relations must be from observational data alone. Despite its fundamental role in science, engineering, and biomedicine (Granger 1969; Sims 1972; The Prize in Economic Sciences, 2011), the traditional causal inference from observational data alone is unable to identify unique causal-effect relations among variables. These "ununique" causal solutions seriously limit their translational application (Nowzohour and Bühlmann 2016).

In the past decade, causal inference theory is undergoing exciting and profound changes from discovering only up to the Markov equivalent class to identify unique causal structure. This chapter will introduce the assumptions for learning causal-effect models, and additive noise models for causal discovery of both qualitative and quantitative traits (Peters et al. 2012; Mooij et al. 2016; Peters and Bühlman 2014; Peters et al. 2017).

In causal analysis, the variables that are not measured or recorded are called hidden variables or confounding factors. It is known that the hidden variables often cause severe complications (Spirtes 1999). Therefore, causal analysis also demands deep genetic and molecular analysis. However, the causal models for deep genetic and molecular analysis have not been well developed. The current causal inference for unraveling mechanisms of complex diseases faces two big challenges. The first challenge is to develop an innovative and paradigm-shifting analytic platform for effectively integrating multi-level genetic, molecular, and phenotype datasets into multilevel omics networks through integrated analyses of WGS, other omics, environmental, imaging, and clinical data to reveal the deep causal chain of mechanisms underlying the disease. Well-founded and informative causal models that cross multiple levels of analysis are rare. Developing methods for joint causal analysis across multiple levels using multiple types of molecules are urgently needed. The second challenge is to develop search algorithms for construction of DAGs with large number of variables. This chapter will address the issues for integrating heterogeneous genomic, epigenomic, environmental, imaging, and phenotypic data into multilayer networks underlying disease and health. It is time to shift the current paradigm of genetic analysis from shallow association analysis in homogeneous populations to deep causal inference in heterogeneous populations.

## 7.1 Genome-Wide Causal Studies

In genomic and epigenomic data analysis, we usually consider four types of associations: association of discrete variables (DNA variation) with continuous variables (phenotypes, gene expressions, methylations, imaging signals, and physiological traits), association of continuous variables (expressions, methylations, and imaging signals) with continuous variables (gene expressions, imaging signals, phenotypes, and physiological traits), association of discrete variables (DNA variation) with binary trait (disease status), and association of continuous variables (gene expressions, methylations, phenotypes, and imaging signals) with binary trait (disease status). We will extend these four types of associations to four types of causations in this section.

### 7.1.1 Mathematical Formulation of Causal Analysis

Both association and causation characterize the dependence between two variables. The association between two variables $X$ and $Y$ includes (1) $X$ causes $Y(X \rightarrow Y)$, (2) $Y$ causes $X(Y \rightarrow X)$, and (3) both $X$ and $Y$ is caused by a third variable $X \leftarrow Z \rightarrow Y$. For simplicity, for the time being we assume that there is no confounding. In other words, the third case is excluded from study in this section. We will release this condition in a later section. For the linear relations, if both variables $X$ and $Y$ are continuous, the association between $X$ and $Y$ is studied by regression:

$$Y = \alpha X + \varepsilon \quad \text{or}$$

$$X = \beta Y + e$$

The parameters in the regression can be estimated by

$$\alpha = r\sqrt{\frac{\text{var}(Y)}{\text{var}(X)}} \text{ and } \beta = r\sqrt{\frac{\text{var}(X)}{\text{var}(Y)}}, \text{ where}$$

$$r = \frac{\text{cov}(X, Y)}{\sqrt{\text{var}(X)\text{var}(Y)}} \text{ is correlation coefficient.}$$

Two regressions can be inverted and their regression coefficients depend on the correlation coefficient. Causation implies association, but association may not indicate causation. Association has two potential causation directions. In causal inference, we should first test association between two variables and then remove one causal direction.

In this section we formalize the basic concepts and mathematic models of causal-effect for only two variables. In Section 2.5.1.1 we discussed an additive noise model for bivariate causal discovery with two continuous variables. Now we give a more general model for causal-effect definition. We consider two random variables $X$ and $Y$. For example, $X$ represents a gene expression and $Y$ represents blood pressure. Suppose that the gene expression causes variation of blood pressure. The following functional structural equation model gives the formal definition of causation which encodes the data-generating process (Peters et al. 2017).

#### Definition 7.1: Cause-Effect

Cause-effect is defined by the following functional structural equation model (SEM):

$$X = f_x(\varepsilon_x)$$
$$Y = f_y(X, \varepsilon_y),$$

(7.1)

where $\varepsilon_x, \varepsilon_y$ are noise variables, $\varepsilon_y$, and $X$ are independent. The random variable $X$ is called cause and the random variable $Y$ is called the effect, which is denoted as $X \rightarrow Y$.

A SEM quantifies the relationships between the marginal distribution of each variable and the distribution of its direct effects. The function $f_y$ represents the causal mechanism, which makes the relationships between variables $X$ and $Y$ asymmetric (Ernest 2016).

### Example 7.1

Assume that $\varepsilon_x, \varepsilon_y$ are normally distributed as $N(0,1)$. Consider the SEM:

$$X = \varepsilon_x$$
$$Y = 3X + \varepsilon_y.$$

Then, the distributions of $X$ and $Y$ are $N(0,1)$ and $N(0,10)$, respectively. If variable $X$ is set to the value 3, then the distribution of the effect variable $Y$ is $N(1,9)$. If $Y$ is set to the value 5, then the SEM in Equation 7.1 becomes

$$X = f_x(\varepsilon_x)$$
$$Y = 5.$$

### 7.1.2 Basic Causal Assumptions

The SEM in Equation 7.1 is too general to find unique causal solutions (Ernest 2016). To achieve better causal solutions, the SEMs should be restricted, and assumptions need to be introduced (Peters et al. 2017). Given joint distribution of two random variables $X$ and $Y$, the SEMs cannot determine whether $X \rightarrow Y$ or $Y \rightarrow X$. It was shown (Peters 2012; Peters et al. 2017) that for every joint distribution $P(x,y)$ of two random variables $X$ and $Y$ we can always construct SEMs consistent with the distribution $P(x,y)$:

$$Y = f_Y(X, \varepsilon_Y), \text{ where } X \text{ and } \varepsilon_Y \text{ are independent}$$

or

$$X = f_X(Y, \varepsilon_X), \text{ where } Y \text{ and } \varepsilon_X \text{ are independent.}$$

Indeed, define the SEM:

$$Y = \frac{\varepsilon_Y}{P_{Y|X}(y)}, \tag{7.2}$$

where $\varepsilon_Y$ is assumed to be uniformly distributed on $[0,1]$ and $P_{Y|X}(y)$ is the conditional density function of $Y$, given $X$. Make transformation:

$$X = X,$$
$$Y = \frac{\varepsilon_Y}{P_{Y|X}(y)}.$$

The determinant of its Jacobian matrix is

$$J(X, \varepsilon_Y) = \frac{1}{P_{Y|X}(y)}.$$

Then, the joint density function of variables $X$ and $Y$ is (Ross 2014)

$$P(x, y) = P_X(x)P_{Y|X}(y).$$

Similarly, we can prove that the SEM

$X = f_X(Y, \varepsilon_X)$ entails the joint distribution $P(x, y) = P_Y(y)P_{X|Y}(x)$.

This shows that using only SEMs cannot identify the causal structures between two variables. To distinguish cause from effect using observational data and joint distribution, the SEMs should be restricted and additional assumptions for SEM are needed. If the functions and noises in the defining Equation 7.2 are restricted to make it impossible that two causal directions $X \rightarrow Y$ and $Y \rightarrow X$ induce the same joint distributions $P(X,Y)$. Let $C$, for example, gene expression, be a casual variable and $P_C$ be its distribution. The effect variable, for example, blood pressure, is denoted by $E$ and its marginal distribution is denoted by $P_E$. Intuitively, if gene expression causes changes in blood pressure, then the conditional distribution $P_{E|C}$ of effect $E$ (blood pressure), given the cause $C$ (gene expression) will not change when the cause $C$ changes. In other words, the conditional distribution $P_{E|C}$ does not contain information about causal marginal distribution $P_C$. If we think the conditional distribution $P_{E|C}$ as a cause mechanism, then the above statement indicates that the cause is independent of mechanism.

In the SEM (7.1), the function $f_Y$ and noise distribution $\varepsilon_Y$ can be viewed as mechanism. The function $f_Y$ and distribution $P_{\varepsilon_Y}$ contain no information about the distribution $P_X$. In terms of the SEM, the independence of cause and mechanism implies that the distribution of the cause ($X$) should be independent of the function in the model and the noise distribution ($\varepsilon_Y$) (Peters et al. 2017). The additive noise models can implement this assumption. The linear SEMs with non-Gaussian distribution and functional additive SEMs can be used for causal inference.

### 7.1.3 Linear Additive SEMs with Non-Gaussian Noise

Let $X$ and $Y$ be two observed random variables, and $u$ be a covariate variable. The models can be extended to multiple covariate variables. For the simplicity of presentation, we only consider a single covariate variable. We assume that both $X$ and $Y$ are standardized to have zero mean and unit variance. The popular methods for causal inference between two variables $X$ and $Y$ under the linear non-Gaussian acyclic model (LiNGAM) are independent component analysis (ICA)-based methods (Shimizu et al. 2006, 2011; Moneta et al. 2013) and likelihood ration-based methods (Hyvärinen and Smith 2013). The first potential causal model, denoted by $X \rightarrow Y$, is defined as

$$X = \alpha u + \varepsilon_x$$
$$Y = \rho X + \beta u + \varepsilon_y,$$
(7.3)

where residuals $\varepsilon_x$ and $\varepsilon_y$ are independent and $\varepsilon_y$ is independent of $X$. The second potential causal model, denoted by $Y \rightarrow X$, is defined as

$$Y = \beta u + \varepsilon_y$$
$$X = \rho Y + \alpha u + \varepsilon_x,$$
(7.4)

where residuals $\varepsilon_x$ and $\varepsilon_y$ are independent and $\varepsilon_x$ is independent of $Y$. In regression equation (7.3), the regression coefficient is estimated by

$$\rho = \frac{\text{cov}(Y, X)}{\text{var}(X)} = \text{cov}(Y, X) = corr(Y, X).$$
(7.5)

Similarly, in regression equation (7.4), the regression coefficient is also estimated by the correlation coefficient $\rho$. The regression coefficients in two models are the same and equal to the correlation coefficient $\rho$.

The likelihood ratio can be used to distinguish two models from the observed data (Hyvärinen and Smith 2013). Let $L(X \rightarrow Y)$ be the likelihood of the LiNGAM in which $X \rightarrow Y$ and $L(Y \rightarrow X)$ be the the likelihood of the LiNGAM in which $Y \rightarrow X$. The average log-likelihood ratio is defined as

$$R = \frac{1}{n} \log L(X \rightarrow Y) - \frac{1}{n} \log L(Y \rightarrow X).$$
(7.6)

Result 7.1 can be derived (Hyvärinen and Smith 2013; Appendix 7.A).

**Result 7.1: Likelihood Ratio Test for Causal Direction**

The average of the log likelihood ratio is given by

$$R = \frac{1}{n} l(X \rightarrow Y) - \frac{1}{n} l(Y \rightarrow X)$$
$$= \frac{1}{n} \sum_{i=1}^{n} \left[ G_x(x_i) + G_d \left( \frac{y_i - \rho x_i - \beta u_i}{\sqrt{1 - \rho^2}} \right) - G_y(y_i) - G_e \left( \frac{x_i - \rho y_i - \alpha u_i}{\sqrt{1 - \rho^2}} \right) \right].$$
(7.7)

If $R$ is positive we infer that the causal direction is $X \rightarrow Y$, and if it is negative then the causal direction is $Y \rightarrow X$. We can further show (Appendix 7.A) that the log-likelihood ratio can be approximated by

$$R \approx R_{NC} + R_C,$$
(7.8)

where $R_{NC} = \rho\hat{E}[x\tanh(y) - y\tanh(x)]$, $R_C = \beta\hat{E}[u\tanh(y_i)] - \alpha\hat{E}[u\tanh(x)]$, $\hat{E}$ denotes the sampling mean and $\tanh(t) = \dfrac{e^t - e^{-t}}{e^t + e^{-t}}$. When there are no covariates (or confounders), Equation 7.8 becomes

$$R \approx \rho\hat{E}[x\tanh(y) - y\tanh(x)]. \tag{7.9}$$

Equation 7.8 shows that the loglikelihood ratio consists of two parts: $R_{NC}$ and $R_C$. If we only consider two variables $X$ and $Y$ then $R \approx R_{NC}$. The log-likelihood ratio $R_C$ due to confounders can be either positive or negative and hence will affect the sign of $R$ and conclusion of causal direction.

The asymptotic limit of the log-likelihood ratio (Hyvärinen and Smith 2013; Appendix 7.A) is

$$R \rightarrow -H(x) - H\left(\frac{\hat{\varepsilon}_y}{\sigma_{\varepsilon_y}}\right) + H(y) + H\left(\frac{\hat{\varepsilon}_x}{\sigma_e}\right), \tag{7.10}$$

where $H$ denotes differential entropy.

Using Taylor expansion of the log-likelihood ratio, we can obtain Result 7.2.

**Result 7.2: Differential Entropy Approximation of Log-Likelihood Ration**

$R$ in Equation 7.10 can be approximated by

$$R \approx R_0 + \left[\frac{\beta}{\sigma_{\varepsilon_y}}h\left(\frac{\varepsilon_y^0}{\sigma_{\varepsilon_y}}\right) - \frac{\alpha}{\sigma_{\varepsilon_x}}h\left(\frac{\varepsilon_x^0}{\sigma_{\varepsilon_x}}\right)\right]E[u], \tag{7.11}$$

where

$$R_0 = -H(x) - H\left(\frac{\varepsilon_y^0}{\sigma_{\varepsilon_y}}\right) + H(y) + H\left(\frac{\varepsilon_x^0}{\sigma_e}\right), \quad \varepsilon_x^0 = X - \rho Y, \quad \varepsilon_y^0 = Y - \rho X,$$

$$h(z) = -2k_1\{E[\log\cosh(z)] - \gamma\}E[\tanh(z)] - 2k_2 E[ze^{-\frac{z^2}{2}}]E[(1 - z^2)e^{-\frac{z^2}{2}}] \quad \text{and}$$

differential entropy can be approximated by

$$\tilde{H}(z) = H(u) - k_1\{E[\log\cosh(z)] - \gamma\}^2 - k_2\left\{E\left[ze^{-z^2/2}\right]\right\}^2,$$

$$H(u) = \frac{1}{2}(1 + 2\pi), k_1 \approx 79.047, k_2 \approx 7.4129, \gamma \approx 0.37457.$$

Again, the entropy approximation of the log-likelihood ratio consists of two parts. One part is the entropy approximation of the log-likelihood ratio without considering confounding variables and the second part is due to the confounding variables. The second part may change the sign of the first part $R_0$ and affect the conclusion of causal direction. Considering confounding variables in the log-likelihood ratio will improve the accuracy of causal direction inference.

### 7.1.4 Information Geometry Approach

In Section 2.5, we discussed the additive noise model for distinguishing cause from effect. In this section we introduce the information geometric approach for inferring causal direction that exploits the asymmetry between cause and effect (Janzing et al. 2010, 2012; Mooij et al. 2016). Consider two variables $X$ and $Y$. To assess whether $X$ causes $Y$ is to assess the independence between the distribution of the cause $P(X)$ and the conditional distribution mapping cause to effect $P(Y \mid X)$. Clearly, the traditional correlation coefficient between two variables cannot be used to measure the dependence relationship between two distributions. The natural extension of the correlation coefficient is the Kullback–Leibler (K–L) distance between two distributions $P(X)$ and $P(Y \mid X)$. A nice property of K–L distance is that independence between two distributions corresponds to 0 K–L distance between two distributions. It is clear that the K–L distance-based information geometry can be used for causal inference.

#### 7.1.4.1 Basics of Information Geometry

We first introduce several concepts of information theory (Cover and Thomas, 1991). In 7.10, we mentioned differential entropy, but did not provide a definition. Now we give its strict mathematic definition.

**Definition 7.2: Differential Entropy**

Let $f(x)$ be a density function of a continuous variable $X$. The differential entropy $S(f)$ of a continuous variable $X$ with s support set $S$ is defined as

$$S(f) = -\int_S f(x) \log f(x) dx, \tag{7.12}$$

where the log is a natural logarithm.

**Example 7.2: Uniform Distribution**

Consider a uniform distribution with its density function $u(x) = \dfrac{1}{b-a}$ of a random variable with the support $S = [a,b]$. Then, differential entropy of the uniform distribution is

$$S(u) = -\int_a^b \frac{1}{b-a} \log \frac{1}{b-a} dx = \log(b-a). \tag{7.13}$$

### Example 7.3: Exponential Distribution

Consider an exponential distribution $f(x) = \lambda e^{-\lambda x}$ for $x \geq 0$. Its differential entropy is

$$S(e) = -\int_0^\infty \lambda e^{-\lambda x} \log(\lambda e^{-\lambda x}) dx$$

$$= -\log \lambda \int_0^\infty \lambda e^{-\lambda x} + \lambda \int_0^\infty x \lambda e^{-\lambda x} dx \qquad (7.14)$$

$$= -\log \lambda + 1.$$

### Example 7.4: Normal Distribution

Consider a normal distribution: $\phi(x) = \dfrac{1}{\sqrt{2\pi\sigma^2}} \exp\{-\dfrac{(x-\mu)^2}{2\sigma^2}\}$. Its differential entropy can be calculated as follows.

$$S(\phi) = -\int_{-\infty}^\infty \frac{1}{\sqrt{2\pi\sigma^2}} \exp\left\{-\frac{(x-\mu)^2}{2\sigma^2}\right\} \log \frac{1}{\sqrt{2\pi\sigma^2}} \exp\left\{-\frac{(x-\mu)^2}{2\sigma^2}\right\} dx$$

$$= \frac{1}{2} \log(2\pi\sigma^2) \int_{-\infty}^\infty \frac{1}{\sqrt{2\pi\sigma^2}} \exp\left\{-\frac{(x-\mu)^2}{2\sigma^2}\right\} dx + E\left[\frac{(x-\mu)^2}{2\sigma^2}\right]$$

$$= \frac{1}{2} \log(2\pi\sigma^2) + \frac{1}{2}$$

$$= \frac{1}{2} \log(2\pi e\sigma^2). \qquad (7.15)$$

### Example 7.5: Gamma Distribution

Let $f(x) = \dfrac{\beta^\alpha x^{\alpha-1} e^{-\beta x}}{\Gamma(\alpha)}, x > 0$ be the density function of the gamma distribution. We start with calculating the integral:

$$J(\alpha, \beta) = \frac{d}{d\alpha} \int_0^\infty x^{\alpha-1} e^{-\beta x} dx = \int_0^\infty x^{\alpha-1} e^{-\beta x} \log x\, dx. \qquad (7.16)$$

It is clear that

$$\int_0^\infty x^{\alpha-1} e^{-\beta x} dx = \int_0^\infty \frac{t^{\alpha-1}}{\beta^{\alpha-1}} e^{-y} \frac{dy}{\beta}$$

$$= \int_0^\infty \frac{t^{\alpha-1}}{\beta^\alpha} e^{-y} dy = \frac{\Gamma(\alpha)}{\beta^\alpha}. \qquad (7.17)$$

Substituting Equation 7.17 into Equation 7.16 gives

$$J(\alpha, \beta) = \frac{\Gamma(\alpha)}{\beta^\alpha} \frac{d}{d\alpha} \log\left(\frac{\Gamma(\alpha)}{\beta^\alpha}\right)$$

$$= \frac{\Gamma(\alpha)}{\beta^\alpha} \frac{d}{d\alpha} \left(\log\left(\Gamma(\alpha)\right) - \alpha \log \beta\right)$$

$$= \frac{\Gamma(\alpha)}{\beta^\alpha} \left(\frac{\Gamma'(\alpha)}{\Gamma(\alpha)} - \log \beta\right)$$

$$= \frac{\Gamma(\alpha)}{\beta^\alpha} \left(\psi(\alpha) - \log(\beta)\right),$$

(7.18)

where $\psi(\alpha) = \dfrac{\dfrac{d\Gamma(\alpha)}{d\alpha}}{\Gamma(\alpha)}$ is a digamma function.

Now we calculate the differential entropy of the gamma distribution. By definition, the differential entropy of the gamma distribution is

$$S(g) = -\int_0^\infty \frac{\beta^\alpha x^{\alpha-1} e^{-\beta x}}{\Gamma(\alpha)} \log \frac{\beta^\alpha x^{\alpha-1} e^{-\beta x}}{\Gamma(\alpha)} dx$$

$$= -\int_0^\infty \frac{\beta^\alpha x^{\alpha-1} e^{-\beta x}}{\Gamma(\alpha)} \left[\alpha \log \beta + (\alpha - 1) \log x - \beta x - \log \Gamma(\alpha)\right] dx$$

$$= -\alpha \log \beta + \log \Gamma(\alpha) + (1 - \alpha) \int_0^\infty \frac{\beta^\alpha x^{\alpha-1} e^{-\beta x}}{\Gamma(\alpha)} \log(x) dx$$

(7.19)

$$+ \int_0^\infty \frac{\beta^{\alpha+1} x^\alpha e^{-\beta x}}{\Gamma(\alpha)} dx$$

$$= -\alpha \log \beta + \log \Gamma(\alpha) + (1 - \alpha) \frac{\beta^\alpha}{\Gamma(\alpha)} J(\alpha, \beta) + \alpha \int_0^\infty \frac{\beta^{\alpha+1} x^\alpha e^{-\beta x}}{\Gamma(\alpha+1)} dx.$$

Substituting Equation 7.18 into Equation 7.19 yields the differential entropy of the gamma function:

$$S(g) = -\alpha \log \beta + \log \Gamma(\alpha) + (1 - \alpha) \frac{\beta^\alpha}{\Gamma(\alpha)} \frac{\Gamma(\alpha)}{\beta^\alpha} \left(\psi(\alpha) - \log \beta\right) + \alpha$$

$$= \log \frac{\Gamma(\alpha)}{\beta} + (1 - \alpha) \psi(\alpha) + \alpha.$$

(7.20)

### Definition 7.3: Joint Differential Entropy

Let $f(x_1,...,x_k)$ be the joint density function of a set of continuous variables $x_1,...,x_k$. Their joint differential entropy is defined as

$$S(X_1,...,X_k) = -\int f(x_1,...,x_k)\log f(x_1,...,x_k)dx_1 dx_2...dx_n. \qquad (7.21)$$

### Definition 7.4: Conditional Differential Entropy

Let $X = (X_1,...,X_k)$ and $Y = (Y_1,...,Y_m)$. If they have joint density function $f(x,y)$, the conditional differential entropy of $X$, given $Y$ is

$$S(X|Y) = -\int f(x,y)\log f(x|y)dxdy. \qquad (7.22)$$

### Example 7.6: Multivariate Normal Distribution

Let $X$ be distributed as a multivariate normal distribution $N(\mu,\Sigma)$. Its density function is

$$f(x) = \frac{1}{(\sqrt{2\pi})^n |\Sigma|^{1/2}} \exp\left\{-\frac{1}{2}(x-\mu)^T \Sigma^{-1}(x-\mu)\right\}. \qquad (7.23)$$

The differential entropy of the multivariate normal distribution is

$$
\begin{aligned}
S(mn) &= -\int f(x)\log\left[\frac{1}{(\sqrt{2\pi})^n |\Sigma|^{1/2}} \exp\left\{-\frac{1}{2}(x-\mu)^T \Sigma^{-1}(x-\mu)\right\}\right]dx \\
&= -\int f(x)\left[-\frac{1}{2}\log(2\pi)^n|\Sigma| - \frac{1}{2}(x-\mu)^T \Sigma^{-1}(x-\mu)\right] \\
&= \frac{1}{2}\log(2\pi)^n|\Sigma| + \frac{n}{2} \qquad (7.24) \\
&= \frac{1}{2}\log(2\pi e)^n|\Sigma|.
\end{aligned}
$$

### Example 7.7: Multivariate Uniform Distribution

The density function of the multivariate uniform distribution is

$$f(x) = \frac{1}{a^k}, \ 0 \le x_1 \le a,...,0 \le x_k \le a. \qquad (7.25)$$

Its differential entropy is

$$
\begin{aligned}
S(mu) &= -\int_0^a ...\int_0^a \frac{1}{a^k}\log\frac{1}{a^k}dx_1...dx_k \\
&= k\log a.
\end{aligned}
\qquad (7.26)
$$

Next, we introduce K–L distance to measure the similarity between two distributions.

**Definition 7.5: K–L Distance**

Let $f(x)$ and $g(x)$ be two density functions. The K–L distance between density functions $f(x)$ and $g(x)$ is defined as

$$D(f\|g) = \int f(x) \log \frac{f(x)}{g(x)} dx, \qquad (7.27)$$

where the support set of $f(x)$ is contained in the support set of $g(x)$ and we set

$$0 \log \frac{0}{0} = 0.$$

**Example 7.8: K–L Distance Between Two Normal Distribution**

Suppose that we have two multivariate normal distributions $N(\mu_0, \Sigma_0)$ and $N(\mu_1, \Sigma_1)$. Assume that two normal distributions have the same dimension $n$. The K–L distance between two distributions is

$$
\begin{aligned}
D(N_0\|N_1) &= \int f_0(x) \log \frac{\dfrac{1}{(\sqrt{2\pi})^n |\Sigma_0|^{1/2}} \exp\left\{ -\dfrac{1}{2}(x-\mu_0)^T \Sigma_0^{-1}(x-\mu_0) \right\}}{\dfrac{1}{(\sqrt{2\pi})^n |\Sigma_1|^{1/2}} \exp\left\{ -\dfrac{1}{2}(x-\mu_1)^T \Sigma_1^{-1}(x-\mu_1) \right\}} dx \\
&= \frac{1}{2} \log \frac{|\Sigma_1|}{|\Sigma_0|} - \frac{1}{2} E_0 \left[ \operatorname{Tr}\left(\Sigma_0^{-1}(x-\mu_0)(x-\mu_0)^T\right) \right. \\
&\quad \left. -\operatorname{Tr}\left(\Sigma_1^{-1}(x-\mu_1)(x-\mu_1)^T\right) \right] \qquad (7.28) \\
&= \frac{1}{2} \log \frac{|\Sigma_1|}{|\Sigma_0|} - \frac{1}{2}n + \frac{1}{2} E_0 \left[ \operatorname{Tr}\left( \Sigma_1^{-1}(x-\mu_0+\mu_0-\mu_1) \right. \right. \\
&\quad \left. \left. (x-\mu_0+\mu_0-\mu_1)^T\right) \right] \\
&= \frac{1}{2} \left( \log \frac{|\Sigma_1|}{|\Sigma_0|} - n + \operatorname{Tr}\left(\Sigma_1^{-1}\Sigma_0\right) + (\mu_0-\mu_1)^T \Sigma_1^{-1}(\mu_0-\mu_1) \right).
\end{aligned}
$$

**Result 7.3: Properties of Entropy and K–L Distance**

K–L distance is always non-negative

$$D(f\|g) \geq 0 \qquad (7.29)$$

With equality if and only if $f = g$.

In fact,

$$-D(f\|g) = \int f \log \frac{g}{f} dx$$

$$\leq \log \int f \frac{g}{f} dx$$

$$= \log \int g dx$$

$$= \log 1 = 0,$$

which implies $D(f \mid\mid g) \geq 0$.

## 7.1.4.2 Formulation of Causal Inference in Information Geometry

We start with distribution theory for transformation of random variables which is useful for deriving distributions of many statistics in information geometry (Ross 2014).

### Theorem 7.1: Distributions of Functions of the Random Variables

Let $P_X(x_1,...,x_n)$ be the joint density function of the $n$ random variables $X_1,...,X_n$ and $Y_1,...,Y_n$ be the functions of the variables $X_1,...,X_n$ defined as

$$Y_1 = f_1(X_1, ..., X_n)$$

$$Y_2 = f_2(X_1, ..., X_n)$$

$$\cdot \quad \cdot \quad \cdot \quad \cdot \quad \cdot \quad \cdot$$

$$Y_n = f_n(X_1, ..., X_n)$$

(7.30)

Assume that the functions $f_i$ have continuous partial derivatives and that the Jacobian determinant $J(x_1,...,x_n) \neq 0$ at all points $(x_1,...,x_n)$ where

$$J(x_1, ..., x_n) = \begin{vmatrix} \dfrac{\partial f_1}{\partial x_1} & \dfrac{\partial f_1}{\partial x_2} & \cdots & \dfrac{\partial f_1}{\partial x_n} \\[2mm] \dfrac{\partial f_2}{\partial x_1} & \dfrac{\partial f_2}{\partial x_2} & \cdots & \dfrac{\partial f_2}{\partial x_n} \\[2mm] \vdots & \vdots & \vdots & \vdots \\[2mm] \dfrac{\partial f_n}{\partial x_1} & \dfrac{\partial f_n}{\partial x_2} & \cdots & \dfrac{\partial f_n}{\partial x_n} \end{vmatrix}.$$

Assume that all functions $f_i$ are invertible and the solutions to Equation 7.30 are unique:

$$x_1 = g_1(y_1, ..., y_n)$$

$$x_2 = g_2(y_1, ..., y_n)$$

$$\cdots \cdots \cdots$$   (7.31)

$$x_n = g_n(y_1, ..., y_n).$$

Then, the joint density function of the random variables $Y_1, Y_2, ..., Y_n$ is given by

$$P_Y(y_1, ..., y_n) = \frac{P_X(x_1, ...., x_n)}{|J(x_1, ..., x_n)|}.$$   (7.32)

### Example 7.9

Let the density function $p_X(x) = 1$ of the random variable $X$ with uniform distribution on $[0,1]$ and $Y = e^X$. Then, the density function $P_Y(y)$ of the random variable $Y$ is

$$P_Y(y) = \frac{1}{e^x} = \frac{1}{y}.$$

Next we discuss how to use information geometric theory for assessing causation (Daniusis et al. 2010; Janzing et al. 2012, 2014). Consider two variables $X$ and $Y$. The task is to assess whether $X$ causes $Y$ or $Y$ causes $X$, assuming that there is no common cause. The asymmetry between cause and effect that if $X$ causes $Y$ then the mechanisms of generating distribution $P(X)$ and conditional distribution $P(Y|X)$ in nature are independent and that $P(X)$ and $P(Y|X)$ contain no information about each other is the fundamental postulate for inferring causal relation between just two observed variables (Janzing and Schölkopf 2010; Lemeire and Janzing 2012; Janzing et al. 2014). In this section, we only consider deterministic relationships between two variables:

$$Y = f(X).$$   (7.33)

Furthermore, we assume that $f$ is monotonically increasing. The fundamental principle in causal inference between two variables is the following postulate (independent mechanism of input and function):

### Postulate 7.1

If $X \rightarrow Y$, then the distribution of $X$ and function $f$ that maps $X$ to $Y$ are independent.

This postulate is informal. The question is how to formally measure dependency between a distribution and a function. In information geometry, $f$-divergence is a non-negative measure of dissimilarity between two distributions or functions (Gibbs and Su 2002). A host of metrics available to quantify the dissimilarity or distance between distributions include the total variation distance $|P - Q|$, K–L distance $D(P \parallel Q)$, $\chi^2$-divergence $\chi^2(P \parallel Q)$, Hellinger divergence $H_a(P \parallel Q)$, and Rényi divergence $D_a(P \parallel Q)$ (Sason and Verdú 2016). A widely used measure of distance between two distributions is the K–L distance. The K–L distance will be used to assess causal relations between two variables.

To illustrate the basic idea behind the information geometry approach to causal inference, we first further study information geometry of the exponential family (Amari 2011). Consider the density function of the exponential family:

$$P(x, \theta) = \exp\left\{\sum_i \theta_i T_i(x) - \psi(\theta)\right\}. \tag{7.34}$$

Define

$$\eta_i = E[T_i(x)]. \tag{7.35}$$

The Legendre transformation and inverse transformation are (Exercise 5)

$$\eta_i = \frac{\partial \psi(\theta)}{\partial \theta_i} \tag{7.36}$$

and

$$\theta_i = \frac{\partial \phi(\eta)}{\partial \eta_i}, \tag{7.37}$$

where

$$\phi(\eta) = E[\log p(x, \theta)] \tag{7.38}$$

is the negative entropy.

It follows from Equations 7.34 and 7.38 that

$$\begin{aligned}
\phi(\eta) &= E\left[\sum_i \theta_i T_i(x) - \psi(\theta)\right] \\
&= \sum_i \theta_i E[T_i(x)] - \psi(\theta) \\
&= \sum_i \theta_i \eta_i - \psi(\theta).
\end{aligned} \tag{7.39}$$

Entropy measures the degree of randomness. The maximum entropy principle makes statistical inference that maximizes entropy or uncertainty (Stein et al. 2015). In other words, the estimated distribution should be a uniform distribution that best models all that is known and assumes nothing about what is unknown.

From Equation 7.37 maximizing entropy implies that

$$\theta_i = \frac{\partial \varphi(\eta)}{\partial \eta_i} = 0. \tag{7.40}$$

Equations 7.34 and 7.40 state that if the distribution is in the exponential family the distribution determined by maximum entropy is

$$P(x,0) = \exp[-\psi(0)], \tag{7.41}$$

which is a uniform distribution. The K–L distance between distribution $P(x,\theta)$ and uniform distribution $P(x,0)$ is

$$D(P(x,\theta)||P(x,0)) = -H(\theta) + \psi(0). \tag{7.42}$$

Suppose that the density function of cause $X$ is $P(x,\theta)$. Without any additional information, the distribution that is the closest distribution of $P(x,\theta)$ can be inferred by maximum entropy of $P(x,\theta)$. Equation 7.41 shows that the inferred distribution is the uniform distribution $P(x,0)$. For the convenience of presentation, let $u(x)$ be the uniform distribution on $[0,1]$ for cause $X$. Let $Y = f(x)$, $x = f^{-1}(y) = g(y)$, and $v(y)$ be the uniform distribution on $[0,1]$ for $Y$. It is clear that $\dfrac{df^{-1}(y)}{dy} = g'(y) = \dfrac{1}{f'(x)}$. Furthermore, assume that $u_f$ is the image of $X$ under $f$ and $v_g$ is the image of $Y$ under $g$ (Figure 7.1b). Using Theorem 7.1, the density functions of $u_f$ and $v_g$ are, respectively, given by

$$u_f(y) = \frac{u(x)}{f'(x)} = u(f^{-1}(y))g'(y) = g'(y) \tag{7.43}$$

and

$$v_g(x) = \frac{v(y)}{g'(y)} = \frac{v(f(x))}{1/f'(x)} = f'(x). \tag{7.44}$$

Postulate 7.1 states that the distribution $P(x,\theta)$ is independent of the function $f$. Therefore, by information geometry, the line connecting the

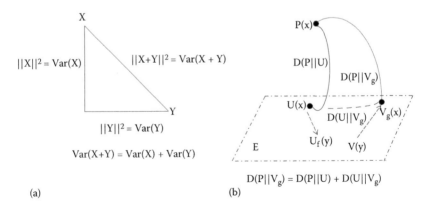

**(a)**                                                                  **(b)**

**FIGURE 7.1**
Pythagorean theorem. (a) Pythagorean theorem and covariance. (b) Pythagorean theorem and independence.

points $P(X)$ and $U(x)$, and the line connecting $U(x)$ and image $V_g(x)$ under $f$ is orthogonal (Figure 7.1b). By the Pythagorean theorem, we obtain Result 7.4.

### Result 7.4: Causal Formulation in Information Geometry

If $X \rightarrow Y$, then the following Pythagorean theorem:

$$D\Big(P(X) \| V_g(X)\Big) = D(P(X) \| U(X)) + D\Big(U(x) \| V_g(X)\Big) \tag{7.45}$$

holds where $P(x)$ is the density function of the cause $X$, $U(x)$ is the density function of uniform distribution for $X$ and $V_g(x)$ is the density function of the images of $Y$ under mapping $g$. We can show that Result 7.4 implies Condition (2) in the paper (Daniušis et al. 2010).

### Result 7.5

The Pythagorean theorem in Equation 7.45 requires

$$\int_0^1 \log f'(x) P(x) dx = \int_0^1 \log f'(x) dx. \tag{7.46}$$

Proof
By definition, we have

$$D(P(x) \| V_g(x)) = \int_0^1 P(x) \log \frac{P(x)}{f'(x)} dx = \int_0^1 P(x) \log P(x) dx$$
$$- \int_0^1 P(x) \log f'(x) dx$$

$$D(P(x) \| U(x)) = \int P(x) \log P(x) dx = \int_0^1 P(x) \log P(x) dx$$

$$D(U(x) \| V_g(x)) = \int_0^1 \log \frac{1}{f'(x)} dx = -\int_0^1 \log f'(x) dx.$$

The above equation clearly shows that Results 7.4 and 7.5 are equivalent.

### Example 7.10

If $P(x)$ is a uniform distribution, then for any function $f(x)$ Equations 7.45 and 7.46 will hold.

### Example 7.11

Suppose that $P(x) = N(0, \sigma^2)$ and $f(x) = e^x$. Then,

$$V_g(x) = e^x,$$

$$D(P(X) \| V_g(X)) = \int P(x) \log P(x) dx - E[X]$$

$$= \int P(x) \log P(x) dx - 0 = \int P(x) \log P(x) dx$$

$$D(P(X) \| U(X)) = \int P(x) \log P(x) dx$$

$$D(U(X) \| V_g(X)) = \int \log \frac{1}{e^x} dx = -\int x dx.$$

It is clear that

$$D(P(X) \| V_g(X)) \neq D(P(X) \| U(X)) + D(U(X) \| V_g(X)).$$

This shows that Results 7.4 and 7.5 have no power to detect causation when a nonlinear map is the exponential function. Similar to the Pythagorean theorem in probability space (Figure 7.1a) we can use Result 7.5 to show Result 7.6.

**Result 7.6**

If $X \rightarrow Y$ and $Y = f(X)$ then

$$Cov_U\left(\log \frac{P(X)}{U(X)}, \log \frac{U(X)}{V_g(X)}\right) = 0. \qquad (7.47)$$

In fact, by definition we have

$$Cov_U\left(\frac{P(X)}{U(X)}, \log \frac{V_g(X)}{U(X)}\right) = \int P(x) \log f'(x) dx - \int P(x) dx \int \log f'(x) dx$$

$$= \int P(x) \log f'(x) dx - \int \log f'(x) dx = 0.$$

Next, we examine whether the Pythagorean theorem for the distributions $P(Y), V(Y)$, and $U_f(Y)$ holds. Using Theorem 7.1 we obtain the distribution of $Y = f(X)$, assuming $X$ is uniformly distributed.

$$P(y) = \frac{P(x)}{f'(x)} = \frac{1}{f'(x)} = g'(y). \qquad (7.48)$$

Next calculate the pair-wise K–L distances of three distributions. The K–L distances are calculated as follows.

$$D(P(Y) \| V(Y)) = \int P(y) \log \frac{P(y)}{V(y)} dy$$

$$= \int P(y) \log P(y) dy \qquad (7.49)$$

$$= \int g'(y) \log g'(y) dy.$$

Using Equation 7.43, we have

$$D(V(Y) \parallel U_f(Y)) = \int \log \frac{1}{g'(y)} \, dy, \tag{7.50}$$

and

$$D(P(Y) \parallel U_f(Y)) = \int P(y) \log \frac{P(y)}{g'(y)} \, dy$$

$$= \int g'(y) \log \frac{g'(y)}{g'(y)} \, dy = 0. \tag{7.51}$$

Since $P \neq V$ and $V \neq U_f$ Result 7.3 implies that

$$D(P(Y) \parallel V(Y)) > 0, \tag{7.52}$$

and

$$D(V(Y) \parallel U_f(Y)) > 0. \tag{7.53}$$

Combining Equations 7.51–7.53, we claim that

$$D(P(Y) \parallel V(Y)) + D(V(Y) \parallel U_f(Y)) > D(P(Y) \parallel U_f(Y)). \tag{7.54}$$

This demonstrates that if $Y$ does not cause $X$ the Pythagorean theorem for the distributions $P(Y), V(Y)$, and $U_f(Y)$ does not hold.

### 7.1.4.3 Generalization

Now we generalize uniform distributions to any reference distributions (Janzing et al. 2012). We begin with defining $U_f(Y)$ and $V_g(X)$ in terms of any reference distributions.

**Definition 7.6: Output Distribution**

Let $U(X)$ and $V(Y)$ denote reference densities for $X$ and $Y$, respectively. Define

$$U_{out}(y) = \int P(y|x)U(x)dx \tag{7.55}$$

as the output distribution of the system with the reference input $U(X)$ and

$$V_{out}(x) = \int P(x|y)V(y)dy \tag{7.56}$$

as the output distribution of the system with the reference input $V(Y)$.

When $Y = f(x)$, $U_{out}(y)$ and $V_{out}(x)$ are reduced to

$$U_{out}(y) = u_f(y), V_{out}(x) = V_g(x). \tag{7.57}$$

Next, we define the K–L distances or structure functions among these distributions.

**Definition 7.7: Structure Functions**

$$D_{CY}(x) = D\left(P_{Y|x} \parallel V(y)\right) = \int P(y|x) \log \frac{P(y|x)}{V(y)} dy \tag{7.58}$$

$$D_{CU_o}(x) = D\left(P_{Y|x} \parallel U_{out}(y)\right) = \int P(y|x) \log \frac{P(y|x)}{U_{out}(y)} dy \tag{7.59}$$

$$D_{diff}(x) = D_{CY}(x) - D_{CU_o}(x) = \int P(y|x) \log \frac{U_{out}(y)}{V(y)} dy \tag{7.60}$$

For the convenience of presentation, the Pythagorean theorem in Equation 7.45 can be written in a general form (Janzing et al. 2012):

$$D(P \parallel Q) = D(P \parallel R) + D(R \parallel Q). \tag{7.61}$$

The Pythagorean theorem implies the following orthogonality condition.

**Result 7.7: Orthogonal Condition**

The Pythagorean theorem in Equation 7.61 is equivalent to

$$\int \log \frac{R(x)}{Q(x)} P(x) dx = \int \log \frac{R(x)}{Q(x)} R(x) dx. \tag{7.62}$$

Proof

$$\int \log \frac{R(x)}{Q(x)} P(x) dx = \int \log \frac{P(x)}{Q(x)} \frac{R(x)}{P(x)} P(x) dx$$

$$= \int \log \frac{P(x)}{Q(x)} P(x) dx + \int \log \frac{R(x)}{P(x)} P(x) dx \tag{7.63}$$

$$= D(P \parallel Q) - D(P \parallel R).$$

Combining Equations 7.62 and 7.63 gives

$$D(P \parallel Q) - D(P \parallel R) = D(R \parallel Q),$$

which implies

$$D(P \parallel Q) = D(P \parallel R) + D(R \parallel Q).$$

Orthogonality is related to covariance in statistics. We consider the covariance between $D_{CY}(X)$ and $\dfrac{P(X)}{U(X)}$, the covariance between $D_{CU_o}$ and $\dfrac{P(X)}{U(X)}$, the covariance between $D_{diff}$ and $\dfrac{P(X)}{U(X)}$, with a reference distribution $U(X)$. Specifically, we define

$$Cov_{U_X}\left(D_{CY}, \frac{P(X)}{U(X)}\right) = \int D_{CY}(x)\frac{P(x)}{U(x)}U(x)dx - \int D_{CY}(x)U(x)dx$$

$$\int \frac{P(x)}{U(x)}U(x)dx$$

$$= \int D_{CY}(x)P(x)dx - \int D_{CY}(x)U(x)dx \int P(x)dx$$

$$= \int D_{CY}(x)P(x)dx - \int D_{CY}(x)U(x)dx,$$

(7.64)

$$Cov\left(D_{CU_o}, \frac{P(X)}{U(X)}\right) = \int D_{CU_o}(x)\frac{P(x)}{U(x)}U(x)dx - \int D_{CU_o}(x)U(x)dx$$

$$\int \frac{P(x)}{U(x)}U(x)dx$$

$$= \int D_{CU_o}(x)P(x)dx - \int D_{CU_o}(x)U(x)dx,$$

(7.65)

and

$$Cov\left(D_{diff}, \frac{P(X)}{U(X)}\right) = \int D_{diff}(x)\frac{P(x)}{U(x)}U(x)dx - \int D_{diff}(x)U(x)dx$$

$$\int \frac{P(x)}{U(x)}U(x)dx$$

$$= \int D_{diff}(x)P(x)dx - \int D_{diff}(x)U(x)dx.$$

(7.66)

**Result 7.8: Orthogonality Conditions**

In Appendix 7.B we show that conditions (7.64), (7.65) and (7.66) are, respectively, equivalent to

$$D(P_{Y,X} \| U_X V_Y) = D(P_{Y,X} \| U_X P_{Y|X}) + D(U_X P_{Y|X} \| U_X V_Y), \quad (7.67)$$

$$D(P_{Y,X} \| U_X U_{out}(Y)) = D(P_{Y,X} \| U_X P_{Y|X}) + D(U_X P_{Y|X} \| U_X U_{out}(y)), \quad (7.68)$$

$$D(P_Y \parallel V_Y) = D(P_Y \parallel U_{out}(Y)) + D(U_{out}(Y) \parallel V_Y) \qquad (7.69)$$

Next, we intuitively examine the relations between causation $X \to Y$ and orthogonality. We previously stated that if $X \to Y$ then the conditional distribution $P_{Y|X}$ is independent of distribution $P_X$ of cause $X$. If $Cov_{U_X}(D_{CY}, \frac{P(X)}{U(X)}) = 0$, Equation 7.64 implies that

$$\int D_{CY}(x)P(x)dx = \int D_{CY}(x)U(x)dx. \qquad (7.70)$$

Note that

$$\int D_{CY}(x)P(x)dx = \int\int \log\frac{P_{Y|X}}{V(y)} P_{Y|X} \frac{P_X}{U_X} U_X dydx$$
$$= \int\int \log\frac{P_{Y|X}}{V(y)} P_{Y|X} P_X dydx \qquad (7.71)$$

and

$$\int D_{CY}(x)U(x)dx = \int\int \log\frac{P_{Y|X}}{V(y)} P_{Y|X} U_X dydx. \qquad (7.72)$$

Equations 7.70, 7.71, and 7.72 indicate that

$$\int\int \log\frac{P_{Y|X}}{V(y)} P_{Y|X} P_X dydx = \int\int \log\frac{P_{Y|X}}{V(y)} P_{Y|X} U_X dydx. \qquad (7.73)$$

Since integrals on both sides of Equation 7.73 are equal, the differences $P_X$ and $U_X$ between both integrals should have no effects on the common parts $P_{Y|X}$. In other words, conditional distribution $P_{Y|X}$ and distribution $P_X$ are independent. Therefore, $Cov_{U_X}(D_{CY}, \frac{P(X)}{U(X)}) = 0$ can be taken as a characteristic of $X \to Y$.

To justify Equation 7.67, we first calculate $D(P_{Y,X} \parallel U_X V_Y)$. By the definition of K–L distance, we have

$$D(P_{Y,X} \parallel U_X V_Y) = \int \log\frac{P(x,y)}{U(x)V(y)} U(x)V(y)dxdy$$
$$= \int \log\frac{P(y|x)P(x)}{U(x)V(y)} U(x)V(y)dxdy$$
$$- \int \log\frac{P(x)}{U(x)} U(x)dx + \int \log\frac{P(y|x)}{V(y)} U(x)V(y) \qquad (7.74)$$
$$= f_1(P_X) + f_2(P(y|x)).$$

Then, we calculate $D(P_{Y,X}\|U_XP_{Y|X})$ and $D(U_XP_{Y|X}\|U_XV_Y)$. Again, using the definition of K–L distance, we have

$$D\left(P_{Y,X} \parallel U_XP_{Y|X}\right) = \int \log \frac{P(x,y)}{U(x)P(y|x)}P(x,y)dxdy$$

$$= \int \log \frac{P(x)}{U(x)}P(x,y)dxdy \qquad (7.75)$$

$$= \int \log \frac{P(x)}{U(x)}P(x)dx = g_1(P_X).$$

Similarly,

$$D\left(U_XP_{Y|X} \parallel U_XV_Y\right) = \int \log \frac{U(x)P(y|x)}{U(x)V(y)}U(x)P(y|x)dxdy$$

$$= \int \log \frac{P(y|x)}{V(y)}P(y|x)U(x)dxdy \qquad (7.76)$$

$$= g_2(P(y|x)).$$

Combining Equations 7.67 and 7.74–7.76, we obtain

$$f_1(P_X) + f_2(P(y|x)) = g_1(P_X) + g_2(P(Y|x)). \qquad (7.77)$$

Since $P_X$ and $P_{Y|X}$ are independent, they are orthogonal. Therefore, the Pythagorean theorem (7.67) holds. Equation 7.68 can be similarly justified.

**Example 7.12**

Suppose that $X$ and $Y$ are defined on the interval $[0,1]$ and that $U(X)$ and $V(Y)$ are uniformly distributed on $[0,1]$. Then, $D_{CY}(x)$ is negative entropy of the conditional distribution $P_{Y|X}$:

$$D_{CY}(x) = D\left(P_{Y|x} \parallel V(y)\right) = \int P(y|x) \log P(y|x)dy = -S\left(P_{Y|x}\right). \qquad (7.78)$$

When reference distribution is a uniform distribution, Equation 7.58 is reduced to

$$\int S\left(P_{Y|x}\right)P_X dx = \int S\left(P_{Y|x}\right)dx. \qquad (7.79)$$

For completeness, Result 7.8 is extended to Result 7.9 to cover more equivalent characterization of orthogonality summarized in (Janzing et al. 2012).

**Result 7.9: Equivalent Pythagorean Theorem Formulations**

The following formulations of orthogonality are equivalent

1. Distributions of input and backward mapping are uncorrelated:

$$Cov_{U_X}\left(\log\frac{V_{out}}{U_X}, \frac{P_X}{U_X}\right) = 0. \tag{7.80}$$

2. Pythagorean theorem among $P_Y, V_Y$ and $U_{out}$:

$$D(P_Y\|V_Y) = D(P_Y \| U_{out}(Y)) + D(U_{out}(Y) \| V_Y). \tag{7.81}$$

3. Pythagorean theorem among $P_X, U_X$ and $V_{out}$:

$$D(P_X \| V_{out}) = D(P_X \| U_X) + D(U_X \| V_{out}). \tag{7.82}$$

4. Orthogonality of input distribution and forward mapping:

$$D(P_Y \| V_Y) = D(P_X \| U_X) + D(U_{out}(Y) \| V_Y). \tag{7.83}$$

5. Pythagorean theorem of approximation errors:

$$D(P_X \| V_{out}) = D(P_Y \| U_{out}(Y)) + D(U_{out}(Y) \| V_Y). \tag{7.84}$$

Next, we examine these five conditions in a backward direction which are summarized in Result 7.10.

**Result 7.10: Relations in Backward Direction**

Assume $X \rightarrow Y$ and that the image of the image of reference distribution $U_X$ under mapping $f$ does not coincide with $V_Y$. Corresponding to orthogonality Result 7.9 in forward direction, the following results show that Pythagorean theorem does not hold in backward direction (Appendix 7.D).

1. $\quad D(P_X\|V_{out}(X)) + D(V_{out}(X) \| U_X) > D(P_X\|U_X) \tag{7.85}$

2. $\quad Cov_{V_Y}\left(\log\frac{U_{out}(Y)}{V_Y}, \frac{P_Y}{V_Y}\right) > 0 \tag{7.86}$

3. $\quad D(P_Y\|V_Y) + D(V_Y\|U_{out}(Y)) > D(P_Y\|U_{out}(Y)) \tag{7.87}$

4. $\quad D(P_Y \| V_Y) + D(V_{out}(X) \| U_X) > D(P_X \| U_X) \tag{7.88}$

5. $\quad D(P_X \| V_{out}(X)) + D(V_{out}(X) \| U_X) > D(P_Y \| U_{out}(Y)). \tag{7.89}$

### 7.1.4.4 Information Geometry for Causal Inference

The information geometry results in Section 7.1.4.3 can be used for casual inference. We begin with definition of information projection.

**Definition 7.8: Information Projection**

Let $\varepsilon$ be a set of probability densities. The information projection of a probability distribution $P$ onto the set of distributions $\varepsilon$ is

$$\varepsilon^* = \arg\min_{Q \in \varepsilon} D(P \| Q). \tag{7.90}$$

A set of probability distributions is referred to as a reference manifold. If the set of distributions is from exponential family, the manifold is referred to as an exponential manifold.

Equivalent Pythagorean theorem versions in Result 7.9 are formulated in terms of specific reference distributions. Now we extend Result 7.9 to general exponential manifolds (Postulate 2, Janzing et al. 2012).

**Assumption: Pythagorean Theorem for Reference Manifolds**

Let $\varepsilon_X$ and $\varepsilon_Y$ be reference manifolds for $X$ and $Y$, respectively. If $X$ causes $Y$, then Result 7.9 of several equivalent Pythagorean theorem formulations hold approximately, where $U_X$ and $U_Y$ are the projections of the distributions of $P_X$ and $P_Y$ onto manifolds $\varepsilon_X$ and $\varepsilon_Y$, and are denoted by $\varepsilon_X^*$ and $\varepsilon_Y^*$, respectively.

Under this assumption, Equation 7.83 can be rewritten as

$$D(P_Y \| \varepsilon_Y^*) = D(P_X \| \varepsilon_X^*) + D(U_{out}(Y) \| \varepsilon_Y^*). \tag{7.91}$$

Therefore, if $X \rightarrow Y$ then

$$D(P_X \| \varepsilon_X^*) \leq D(P_Y \| \varepsilon_Y^*). \tag{7.92}$$

Their difference can be used to measure causality:

$$C_{X \rightarrow Y} = D(P_X \| \varepsilon_X^*) - D(P_Y \| \varepsilon_Y^*). \tag{7.93}$$

In other words, the causality measure $C_{X \rightarrow Y}$ is defined as the difference in K–L distances between the cause distribution $P_X$ and its projection $\varepsilon_X^*$ on manifold of reference distribution $\varepsilon_X$, and K–L distance between the effect distribution $P_Y$ and its projection $\varepsilon_Y^*$ on manifold of reference distribution $\varepsilon_Y$. To make its calculation easier, computation of the K–L distance can be reduced difference in entropy (Janzing et al 2012).

In fact, let $U_X^{(0)}$ be a uniform distribution contained in the manifold $\varepsilon_X$. The Pythagorean theorem in information space implies

$$D(P_X \| \varepsilon_X^*) + D\left(\varepsilon_X^* \| U_X^{(0)}\right) = D\left(P_X \| U_X^{(0)}\right). \tag{7.94}$$

Therefore,

$$D(P_X \parallel \varepsilon_X^*) = D\left(P_X \parallel U_X^{(0)}\right) - D\left(\varepsilon_X^* \parallel U_X^{(0)}\right)$$

$$= \int \log \frac{P_X}{U_X^{(0)}} P_X dx - \int \log \frac{\varepsilon_X^*}{U_X^{(0)}} \varepsilon_X^* dx \qquad (7.95)$$

$$= \int P_X \log P_X dx - \int \varepsilon_X^* \log \varepsilon_X^* dx$$

$$= -S(P_X) + S(\varepsilon_X^*).$$

Similarly, we have

$$D(P_Y \parallel \varepsilon_Y^*) = -S(P_Y) + S(\varepsilon_Y^*). \qquad (7.96)$$

Combining Equation 7.93, 7.95, and 7.96, we obtain (Janzing et al. 2012) the following.

### Result 7.11: Cause Measure as Difference of Entropies

Let $P_X$ and $P_Y$ be density functions on $R^d$. Assume that $\varepsilon_X^*$ and $\varepsilon_Y^*$ are the projections of $P_X$ on manifold $\varepsilon_X$ and $P_Y$ on manifold $\varepsilon_Y$. Then, the cause measure can be computed by

$$C_{X \to Y} = S(\varepsilon_X^*) - S(P_X) - (S(\varepsilon_Y^*) - S(P_Y)). \qquad (7.97)$$

Let $Y = f(X)$. Then, from changing variable theory, we have

$$dy = |\nabla f(x)| dx \text{ and } P(y) = \frac{P(x)}{|\nabla f(x)|}.$$

The entropy of $Y = f(X)$ can be calculated in terms of the entropy of $X$ and the Jacobi determinant $|\nabla f(x)|$ of transformation function $f(X)$. By definition of entropy of $Y$ we obtain

$$S(P_Y) = -\int P(y) \log P(y) dy$$

$$= -\int \frac{P(x)}{|\nabla f(x)|} \log \frac{P(x)}{|\nabla f(x)|} |\nabla f(x)| dx \qquad (7.98)$$

$$= S(P_X) + \int P(x) \log |\nabla f(x)| dx.$$

Substituting Equation 7.84 into Equation 7.85 gives Result 7.12 (Janzing et al. 2012).

### Result 7.12: Cause Measure as Mean of Log Jacobi Determinant of Transformation

$$C_{X \to Y} = S(\varepsilon_X^*) - S(\varepsilon_Y^*) + \int P(x) \log |\nabla f(x)| dx. \qquad (7.99)$$

### 7.1.4.5 Information Geometry-Based Causal Inference Methods

Results 7.11 and 7.12 indicate that cause measure depends on entropies and their projections to the reference spaces. Therefore, information geometry-based causal inference methods are classified according to reference measures and estimation methods of entropies.

#### 7.1.4.5.1 Uniform Reference Measure

Consider two datasets: $X = [x_1,...,x_n]$ and $Y = [y_1,...,y_n]$ where the $x$-values and $y$-values are assumed to be ascended, that is, $x_{i+1} \geq x_i$ and $y_{i+1} \geq y_i$. The entropies can be estimated by (Kraskov et al. 2003)

$$S(P_X) \approx \psi(n) - \psi(1) + \frac{1}{n-1} \sum_{i=1}^{n-1} \log|x_{i+1} - x_i|, \tag{7.100}$$

$$S(P_Y) \approx \psi(n) - \psi(1) + \frac{1}{n-1} \sum_{i=1}^{n-1} \log|y_{i+1} - y_i|. \tag{7.101}$$

Substituting Equations 7.100 and 7.101 into Equation 7.97 gives the statistics for testing causation (Janzing et al. 2012, Janzing et al. 2015).

**Result 7.13: Entropy-Based Statistic for Testing Causation**

Let

$$C_{X \to Y} = \frac{1}{n-1} \sum_{i=1}^{n-1} \log\left|\frac{y_{i+1} - y_i}{x_{i+1} - x_i}\right| \quad \text{and} \quad C_{Y \to X} = \frac{1}{n-1} \sum_{i=1}^{n} \log\left|\frac{x_{i+1} - x_i}{y_{i+1} - y_i}\right|.$$

Define the statistic:

$$T_E = C_{X \to Y} - C_{Y \to X}. \tag{7.102}$$

If

1. $T_E = 0$ then no causation,
2. $T_E < 0$ then $X \to Y$,
3. $T_E > 0$ then $Y \to X$.

#### 7.1.4.5.2 Gaussian Reference Measure

Let $X$ and $Y$ be $d$-dimensional random vectors. Assume that both $\varepsilon_X$ and $\varepsilon_Y$ are the manifolds of $d$-dimensional Gaussian distributions. Then, the projections $\varepsilon_X^*$ and $\varepsilon_Y^*$ are the $d$-dimensional Gaussian with the same mean vectors and covariance matrices $\Sigma_X$ and $\Sigma_Y$ as $X$ and $Y$, respectively. Using Equation 7.24, we obtain

$$S(\varepsilon_X^*) = \frac{1}{2} \log (2\pi e)^d |\Sigma_X| \text{ and } S(\varepsilon_Y^*) = \frac{1}{2} \log (2\pi e)^d |\Sigma_Y|. \tag{7.103}$$

Substituting Equation 7.103 into Equation 7.97 gives

$$C_{X \to Y} = \frac{1}{2} \log \frac{|\Sigma_X|}{|\Sigma_Y|} - S(P_X) + S(P_Y)), \tag{7.104}$$

where entropies $S(P_X)$ and $S(P_Y)$ can be estimated from the data.

When, $d = 1$ and both $X$ and $Y$ are rescaled such that both variances of $X$ and $Y$ are equal to 1, then the first term in Equation 7.104 is equal to zero and Equation 7.104 is reduced to

$$C_{X \to Y} = -S(P_X) + S(P_Y)). \tag{7.105}$$

The statistic $C_{X \to Y}$ can be estimated by

$$C_{X \to Y} = \frac{1}{n-1} \sum_{i=1}^{n-1} \log \left| \frac{y_{i+1} - y_i}{x_{i+1} - x_i} \right|. \tag{7.106}$$

Similarly, we have

$$C_{Y \to X} = \frac{1}{2} \log \frac{|\Sigma_Y|}{|\Sigma_X|} + S(P_X) - S(P_Y)), \tag{7.107}$$

$$C_{Y \to X} = S(P_X) - S(P_Y)), \tag{7.108}$$

and

$$C_{Y \to X} = \frac{1}{n-1} \sum_{i=1}^{n} \log \left| \frac{x_{i+1} - x_i}{y_{i+1} - y_i} \right|. \tag{7.109}$$

In summary, we have the results (Janzing et al. 2012):

**Result 7.14: Causation Test for the Gaussian Reference Measure**

Define the statistic:

$$T_E = C_{X \to Y} - C_{Y \to X}. \tag{7.110}$$

If

4. $T_E = 0$ then no causation,
5. $T_E < 0$ then $X \to Y$,
6. $T_E > 0$ then $Y \to X$,

where $C_{X \to Y}$ and $C_{Y \to X}$ are defined in Equations 7.105–7.109.

### 7.1.4.5.3 *Isotropic Gaussian Reference Measure and Trace Method*

Assume that $X$ and $Y$ are $n$ and $m$ dimensional multivariate normal vectors with zero mean and covariance matrices $\Sigma_X$ and $\Sigma_Y$, respectively (Janzing et al. 2010). Further assume that $Y$ is linearly transformed from $X$:

$$Y = AX, \tag{7.111}$$

where $A$ is a $m \times n$ matrix. The renormalized trace is defined as

$$\tau_n(.) = \frac{\text{Tr}(.)}{n} \tag{7.112}$$

Let $O(n)$ be the group of $n \times n$ real orthogonal matrices. In Appendix 7.E we show the following fundamental results for the trace method (Theorem 1 in Janzing et al. 2010).

**Result 7.15: Multiplicativity of Traces**

Assume that $\Sigma$ is a symmetric, positive definite $n \times n$ matrix and $A$ is a $m \times n$ matrix. Let $U$ be an orthogonal matrix randomly chosen from the group of $n \times n$ real orthogonal matrices $O(n)$ according to the Haar measure. Then,

$$\left| \tau_m \left( AU\Sigma U^T A^T \right) - \tau_n(\Sigma)\tau_m \left( AA^T \right) \right| \le 2\varepsilon \, \| \Sigma \| \| AA^T \| \tag{7.113}$$

with probability at least $q = 1 - \exp(- \kappa(n - 1)\varepsilon^2)$ for some constant $\kappa$ that is independent of $\Sigma$, $A$, $n$, $m$ and $\varepsilon$, where $\| . \|$ denotes the norm of a matrix. Result 7.15 implies that the pairs $(A, \Sigma_X)$ satisfy

$$\tau_n \left( A\Sigma_X A^T \right) \approx \tau_n(\Sigma)\tau_m \left( AA^T \right) \tag{7.114}$$

Equation 7.99 can be used to infer causation. Define

$$\Delta_{X \to Y} = \log \tau_n \left( A\Sigma A^T \right) - \log \tau_n(\Sigma) - \log \tau_m \left( AA^T \right) \tag{7.115}$$

Note that Exercise 6 shows that $\Delta$ is equal to zero for dimension one, which implies that the trace method cannot be applied to two one-dimensional variables.

Next we will show that if we assume $n \le m$ and $A$ has rank $n$, then $\Delta_{X \to Y} = 0$ and $\Delta_{Y \to X} \le 0$.

Consider deterministic linear models:

$$Y = AX \text{ and } X = A^-Y,$$

where $A^-$ denotes the pseudo inverse of the matrix $A$.
Define

$$\Delta_{Y \to X} = \log \tau_n \left( A^-\Sigma_{YY} A^{-T} \right) - \log \tau_m(\Sigma_{YY}) - \log \tau_n \left( A^- A^{-T} \right) \tag{7.116}$$

Assume that $Z$ is a real-valued random variable that follows the empirical distribution of eigenvalues of $AA^T$, that is, $\tau_m((AA^T)^k) = E(Z^k)$ for all $k \in Z$. We can show (Exercise 7) that

$$\Delta_{X \to Y} + \Delta_{Y \to X} = -\log(1 - \text{cov}(Z, 1/Z)) + \log \frac{n}{m}. \qquad (7.117)$$

Since roughly $E[Z] \geq 0$, $E[1/Z] \geq 0$, $E[Z]E[1/Z] \geq 1$ which implies

$$Cov(Z, 1/Z) \leq 0 \text{ and } -\log(1 - Cov(Z, 1/Z)) \leq 0.$$

When $n = m$ then we have

$$\Delta_{X \to Y} + \Delta_{Y \to X} \leq 0. \qquad (7.118)$$

Next, we consider isotropic Gaussian as a reference measure and anisotropy of covariance matrix of causal $X$. We first give the anisotropy of the covariance matrix (Janzing et al. 2010).

### Definition 7.9: Anisotropy of the Covariance Matrix

The anisotropy of the covariance matrix $\Sigma_1$ is defined as the smallest K–L distance between the Gaussian $Z_1$ and the isotropic Gaussian with $\Sigma_0 = \lambda I$:

$$D(\Sigma_1) = \min_\lambda \ D\left(P_{\Sigma_1} \| P_{\Sigma_0}\right). \qquad (7.119)$$

In Appendix 7.F, we show

$$D(\Sigma_1) = \frac{1}{2}\left(n \log \tau_n(\Sigma_1) - \log|\Sigma_1|\right). \qquad (7.120)$$

Assume that both $\varepsilon_X$ and $\varepsilon_Y$ are the manifold of isotropic Gaussians, and $\varepsilon_X^*$ and $\varepsilon_Y^*$ are projections of $P_X$ and $P_Y$. In Appendix 7.F we also show Result 7.16 (Janzing et al. 2010).

### Result 7.16: Anisotropy of the Output Covariance Matrix and K–L Distance

Assume that $P_X$ and $P_Y$ have covariance matrices $\Sigma_X$ and $\Sigma_Y = A\Sigma_X A^T$, respectively. Then,

$$D(\Sigma_X) = D(P_X \| \varepsilon_X^*) = \frac{1}{2}\left(n \log \tau_n(\Sigma_X) - \log|\Sigma_X|\right), \qquad (7.121)$$

and

$$D\left(A\Sigma_X A^T\right) = D(P_Y \| \varepsilon_Y^*) = \frac{1}{2}\left(n \log \tau_n(\Sigma_Y) - \log|\Sigma_Y|\right)$$
$$= \frac{n}{2}\Delta_{X \to Y} + D(\Sigma_X) + D\left(AA^T\right). \qquad (7.122)$$

In Appendix 7.F, we also show independence Result 7.17 (Janzing et al. 2012; Zscheischler et al. 2011).

**Result 7.17: Independence of Causal and Linear Transformation Matrix**

If $X \rightarrow Y$ and $Y = AX$ then we have

$$\tau_n\left(A\Sigma_X A^T\right) = \tau_n(\Sigma_X)\tau_n\left(AA^T\right). \tag{7.123}$$

Next, we extend the deterministic relation to stochastic relation between $X$ and $Y$. Consider the general linear model (Janzing et al. 2010):

$$Y = AX + e, \tag{7.124}$$

where $A$ is a $m \times n$ matrix and $e$ is a vector of noise with zero mean and covariance matrix $\Sigma_e$, statistically independent of $X$. Then, we can easily show

$$\Sigma_Y = A\Sigma_X A^T + \Sigma_e, \tag{7.125}$$

$$\hat{A} = \Sigma_{YX}\Sigma_X^{-1}. \tag{7.126}$$

The backward model is given by

$$X = \tilde{A}Y + \tilde{e}, \tag{7.127}$$

where

$$\tilde{A} = \Sigma_{XY}\Sigma_Y^{-1}. \tag{7.128}$$

If we assume that $A$ is an orthogonal transformation and $e$ is isotropic, that is, $\Sigma_e = \lambda I$ then we can show Result 7.18 (Janzing et al. 2010) (Appendix 7.G).

**Result 7.18: Trace Method for Noise Linear Model**

Consider $Y = AX + e$ where $A$ is the orthogonal matrix and $e$ is a vector of noises with $\Sigma_e = \lambda I$, $\lambda > 0$. Then,

$$\Delta_{Y \rightarrow X} = \log \frac{\tau_n\left(\tilde{A}\Sigma_Y \tilde{A}^T\right)}{\tau_n(\Sigma_Y)\tau_n\left(\tilde{A}\tilde{A}^T\right)}$$

$$= \log \frac{1}{n}\sum_{i=1}^{n}\frac{\mu_i^2}{(\mu_i + \lambda)} - \log \frac{1}{n}\sum_{i=1}^{n}(\mu_i + \lambda) - \log \frac{1}{n}\sum_{i=1}^{n}\frac{\mu_i^2}{(\mu_i + \lambda)^2} > 0, \tag{7.129}$$

where $\mu_i$ with $\mu_1 \geq \mu_2 \geq \dots \mu_n \geq 0$ are eigenvalues of the covariance matrix $\Sigma_X$ and $n > 1$.

If we assume that the association of $X$ with $Y$ can be measured by

$$r = \frac{1}{n} \sum_{i=1}^{n} \lambda_i^2 = \frac{1}{n} \text{Tr}(R^2),$$

then under the same model, we show Result 7.19 charactering association (Appendix 7.H).

It is clear that under the same model we have

$$\Delta_{X \rightarrow Y} = \log \tau_n \left( A \Sigma_X A^T \right) - \log \tau_n (\Sigma_X) - \log \tau_n \left( A A^T \right)$$

$$= \log \tau_n (\Sigma_X) - \log \tau_n (\Sigma_X) - \log (1) = 0.$$

**Result 7.19: Association Characterization**

Assume the model:

$$Y = AX + e$$

where $A$ is an invertible matrix and $e$ is a vector of noises with $\Sigma_e = \lambda I, \lambda > 0$. Then, the association measure between $X$ and $Y$ is

$$r = \frac{1}{n} \sum_{i=1}^{n} \frac{\mu_i}{\mu_i + \lambda} \tag{7.130}$$

and

$$r < 1, \log r < 0.$$

Without orthogonal assumption of $A$, the measures for causation and association are calculated as

$$\Delta_{X \rightarrow Y} = \log \tau_m \left( \Sigma_{YX} \Sigma_X^{-1} \Sigma_{XY} \right) - \log \tau_n (\Sigma_X) - \log \tau_m \left( \Sigma_{YX} \Sigma_X^{-2} \Sigma_{XY} \right), \tag{7.131}$$

$$\Delta_{Y \rightarrow X} = \log \tau_n (\Sigma_{XY} \left( \Sigma_{YX} \Sigma_X^{-1} \Sigma_{XY} + \lambda I \right)^{-1} \Sigma_{YX}) - \log \tau_m \left( \Sigma_{YX} \Sigma_Y^{-1} \Sigma_{XY} + \lambda I \right)$$

$$- \log(\tau_n \left( \Sigma_{XY} (\Sigma_{YX} \Sigma_X^{-1} \Sigma_{XY} + \lambda I )^{-2} \Sigma_{YX}, \tag{7.132}$$

and

$$r = \log \tau_n \left( \Sigma_{YX} \Sigma_X^{-1} \Sigma_{XY} \Sigma_Y^{-1} \right). \tag{7.133}$$

In summary, If $X \rightarrow Y$ then $\Delta_{X \rightarrow Y} = 0$ and $\Delta_{Y \rightarrow X} > 0$; if $Y \rightarrow X$, then $\Delta_{X \rightarrow Y} > 0$ and $\Delta_{Y \rightarrow X} = 0$; if $X$ is associated with $Y$, then $r < 0$.

Statistical procedure for testing linear causal models can be summarized in Result 7.20 (Janzing et al. 2010).

**Result 7.20: Trace Algorithm for Linear Causal Testing**

Trace algorithm is summarized as follows.

Step 1: Estimate $\Sigma_X$, $\Sigma_Y$, $\Sigma_{XY}$ and $\Sigma_{YX}$.
Step 2: Calculate $A = \Sigma_{YX}\Sigma_{XX}^{-1}$.
Step 3: Calculate $\tilde{A} = \Sigma_{XY}\Sigma_{YY}^{-1}$.
Step 4: Calculate

$$\Delta_{X \to Y} = \log \frac{\tau_m\left(A\Sigma_X A^T\right)}{\tau_n(\Sigma_X)\,\tau_m(AA^T)} \quad \text{and}$$

$$\Delta_{Y \to X} = \log \frac{\tau_n\left(\tilde{A}\Sigma_Y \tilde{A}^T\right)}{\tau_m(\Sigma Y)\,\tau_n\left(\tilde{A}\tilde{A}^T\right)} \; .$$

Step 5: If $|\Delta_{Y \to X}| > (1 + \varepsilon)\,|\Delta_{X \to Y}|$ then

$$X \to Y,$$

Else
If $|\Delta_{X \to Y}| > (1 + \varepsilon)\,|\Delta_{Y \to X}|$ then

$$Y \to X,$$

Else
No cause and then test association
End if
End if.

### 7.1.4.5.4 *Kernelized Trace Method*

In the previous section, we studied the linear trace method for causal inference. However, in practice, the relationship between $X$ and $Y$ may be nonlinear. The linear trace methods may lead to incorrect causal conclusions if they are applied to data with nonlinear relations. Chen et al. (2013) extend the linear race method to a kernelized trace method for nonlinear causal discovery. In this section, we present such an extension.

*7.1.4.5.4.1 Problem Formulation* Assume that we map the data from the original space to the high-dimensional feature space:

$$\Psi : x \in \chi \mapsto \Psi(x) \in H,$$

where the inner product in the feature space is defined by the kernel function $k(x,x')$ as

$$< \Psi(x), \Psi(x') >= k(x, x').$$

Assume that $\{\psi_1(x), ..., \psi_{n_H}(x)\}$ form a set of orthonormal basis functions. Let $f_i(x) \in H$ be a nonlinear function. Then, function $f_i(x)$ can be expanded in terms of basis functions $\psi_j(x)$:

$$f_i(x) = \sum_{j=1}^{n_H} a_{ij} \psi_j(x). \tag{7.134}$$

Let

$$y = \begin{bmatrix} y_1 \\ \vdots \\ y_m \end{bmatrix} = \begin{bmatrix} f_1(x) \\ \vdots \\ f_m(x) \end{bmatrix} = f(x), \Psi(x) = \begin{bmatrix} \psi_1(x) \\ \vdots \\ \psi_{n_H}(x) \end{bmatrix} \text{ and } A = \begin{bmatrix} a_{11} & \cdots & a_{1n_H} \\ \vdots & \vdots & \vdots \\ a_{m1} & \cdots & a_{mn_H} \end{bmatrix}.$$

It follows from Equation 7.134 that

$$y = A\Psi(x). \tag{7.135}$$

Suppose that $L$ points are sampled. Define

$$Y = \begin{bmatrix} y_{11} & \cdots & y_{1L} \\ \vdots & \vdots & \vdots \\ y_{m1} & \cdots & y_{mL} \end{bmatrix} \text{ and } \Psi = \begin{bmatrix} \psi_1(x_1) & \cdots & \psi_1(x_L) \\ \vdots & \vdots & \vdots \\ \psi_{n_H}(x)^1 & \cdots & \psi_{n_H}(x_L) \end{bmatrix}.$$

Then, output data matrix $Y$ can be expressed as

$$Y = A\Psi. \tag{7.136}$$

The covariance matrix $\Sigma_{\Psi(x)}$ of $\Psi(x)$ can be estimated by

$$\Sigma_{\Psi(x)} = \frac{1}{N} \Psi\Psi^T. \tag{7.137}$$

Define the kernel Gram matrix of $X$:

$$K_x = \begin{bmatrix} k(x_1, x_1) & \cdots & k(x_1, x_L) \\ \vdots & \vdots & \vdots \\ k(x_L, x_1) & \cdots & k(x_L, x_L) \end{bmatrix} = \Psi^T\Psi.$$

Now we consider variables in the feature space and want to assess the causal relations between Y and $\Psi$. Using linear trace method, we can define

$$\Delta_{\Psi(x) \to Y} = \log \frac{\tau_m \left( A\Sigma_{\Psi(x)} A^T \right)}{\tau_m \left( AA^T \right) \tau_L \left( \Sigma_{\Psi(x)} \right)}. \tag{7.138}$$

Similarly, consider expansion:

$$x_i = g_i(x) = \sum_{j=1}^{n_y} b_{ij} \phi_j(y). \tag{7.139}$$

Let

$$X = \begin{bmatrix} x_{11} & \cdots & x_{1L} \\ \vdots & \vdots & \vdots \\ x_{n1} & \cdots & x_{nL} \end{bmatrix}, B = \begin{bmatrix} b_{11} & \cdots & b_{1n_y} \\ \vdots & \vdots & \vdots \\ b_{n1} & \cdots & b_{nn_y} \end{bmatrix} \text{ and } \Phi = \begin{bmatrix} \phi_1(y_1) & \cdots & \phi_1(y_L) \\ \vdots & \vdots & \vdots \\ \phi_{n_y}(y_1) & \cdots & \phi_{n_y}(y_L) \end{bmatrix}.$$

Then, we have

$$X = B\Phi. \tag{7.140}$$

Again, the covariance matrix $\Sigma_{\Phi(y)}$ of $\Phi(y)$ can be estimated by

$$\Sigma_{\Phi(y)} = \frac{1}{N} \Phi\Phi^T. \tag{7.141}$$

Define the kernel Gram matrix of $Y$:

$$K_y = \begin{bmatrix} k(y_1, y_1) & \cdots & k(y_1, y_L) \\ \vdots & \vdots & \vdots \\ k(y_L, y_1) & \cdots & k(y_L, y_L) \end{bmatrix} = \Phi^T\Phi.$$

Now define $\Delta_{\Phi(y) \to x}$ as

$$\Delta_{\Phi(y) \to x} = \log \frac{\tau_n \left( B\Sigma_{\Phi(y)} B^T \right)}{\tau_n (BB^T) \tau_L \left( \Sigma_{\Phi(y)} \right)}. \tag{7.142}$$

**Result 7.21: Kernelized Trace Method**

If $X \to Y$, then

$$\Delta_{\Psi(x) \to Y} = \log \frac{\tau_m \left( A\Sigma_{\Psi(x)} A^T \right)}{\tau_m (AA^T) \tau_L \left( \Sigma_{\Psi(x)} \right)} \approx 0 \tag{7.143}$$

and

$$\Delta_{\Phi(y) \to x} = \log \frac{\tau_n \left( B\Sigma_{\Phi(y)} B^T \right)}{\tau_n (BB^T) \tau_L \left( \Sigma_{\Phi(y)} \right)} < 0. \tag{7.144}$$

*7.1.4.5.4.2 Estimation of Parameters*   Using Equations 7.128 and 7.129 to discover causation, we need to estimate matrices $A$ and $B$ to implement the kernelized trace method. Suppose that matrix $A$ can be expressed as

$$A = R\Psi^T, \tag{7.145}$$

where matrix $R$ will be estimated.

Substituting Equation 7.145 into Equation 7.136, we can re-express $A$ as

$$Y = R\Psi^T\Psi = RK_x. \tag{7.146}$$

Therefore, $R$ can be estimated by minimizing the following regularized loss:

$$F = \text{Tr}\left((Y - RK_x)^T(Y - RK_x)\right) + \frac{\lambda}{2}\text{Tr}\left(RK_xR^T\right). \tag{7.147}$$

Using a matrix derivative formula and setting equation $\dfrac{\partial F}{\partial R} = 0$, we obtain

$$\frac{\partial F}{\partial R} = -(Y - RK_x)K_x^T + \lambda RK_x^T = 0. \tag{7.148}$$

Solving Equation 7.148 for $R$ gives

$$R = Y(K_x + \lambda I)^{-1}. \tag{7.149}$$

Substituting Equation 7.149 into Equation 7.145, we obtain

$$A = Y(K_x + \lambda I)^{-1}\Psi^T. \tag{7.150}$$

Using similar arguments, we have

$$\tau_m\left(AA^T\right) = \tau_m\left(K_xR^TR\right), \tag{7.151}$$

$$
\begin{aligned}
\tau_m\left(A\Sigma_{\Psi(x)}A^T\right) &= \frac{1}{N}\tau_m\left(A\Psi\Psi^TA^T\right) \\
&= \frac{1}{N}\tau_m\left(R\Psi^T\Psi\Psi^T\Psi R^T\right) \\
&= \frac{1}{N}\tau_m\left(RK_x^2R^T\right) \\
&= \frac{1}{N}\tau_m\left(K_x^2R^TR\right),
\end{aligned}
\tag{7.152}
$$

$$\tau_L\left(\Sigma_{\Psi(x)}\right) = \tau_L\left(\Sigma_{\Psi(x)}\right)$$

$$= \frac{1}{N}\tau_L\left(\Psi\Psi^T\right)$$

$$= \frac{1}{N}\tau_L\left(\Psi^T\Psi\right). \tag{7.153}$$

$$= \frac{1}{N}\tau_L\left(K_x\right).$$

Therefore, we have

$$\Delta_{\Psi(x)\to Y} = \log\frac{\tau_m\left(K_x^2 R^T R\right)}{\tau_m\left(K_x R^T R\right)\tau_L\left(K_x\right)}. \tag{7.154}$$

Similarly, we have

$$X = B\Phi, \tag{7.155}$$

$$B = Q\Phi^T, \tag{7.156}$$

and

$$X = QK_y. \tag{7.157}$$

After solving the following optimization problem:

$$F = \text{Tr}\left(X - QK_y\right)^T\left(X - QK_y\right) + \frac{\lambda}{2}\text{Tr}\left(QK_yQ^T\right)$$

We obtain

$$Q = X\left(K_y + \lambda I\right)^{-1},$$

$$\Delta_{\Phi(y)\to x} = \log\frac{\tau_n\left(K_y^2 Q^T Q\right)}{\tau_n\left(K_y Q^T Q\right)\tau_L\left(K_y\right)} < 0. \tag{7.158}$$

Now we summarize algorithms for implementing the kernelized trace method in Result 7.21.

**Result 7.21: Algorithms for Kernelized Trace Method**

Step 1: Select kernel function and calculate $K_x$ and $K_y$. Select the penalty parameter $\lambda$.

Step 2: Calculate $R = Y(K_x + \lambda I)^{-1}$ and $Q = X(K_y + \lambda I)^{-1}$.

Step 3: Calculate

$$\Delta_{\Psi(x) \to Y} = \log \frac{\tau_m (K_x^2 R^T R)}{\tau_m (K_x R^T R) \tau_L (K_x)} \quad \text{and} \quad \Delta_{\Phi(y) \to x} = \log \frac{\tau_n (K_y^2 Q^T Q)}{\tau_n (K_y Q^T Q) \tau_L (K_y)}.$$

Step 5: If $|\Delta_{\Phi(y) \to x}| > (1 + \varepsilon) |\Delta_{\Psi(x) \to Y}|$ then

$$X \to Y$$

> Else
> If $|\Delta_{\Psi(X) \to Y}| > (1 + \varepsilon) |\Delta_{\Phi(y) \to x}|$ then

$$Y \to X$$

> Else
> No cause and then test association
> End if
> End if.

### 7.1.4.5.5 Sparse Trace Method

Trace methods are used to discover causal relations between two sets of variables. The set of effect variables is often called responses and the set of causal variables is often called predictors. Both high-dimensional responses and predictors have natural group structures. For example, when we study causal regulatory relations between two pathways, a gene expression forms a group. Some genes may contribute causal relations, some genes may not. Causal analysis should identify important responses and important predictors that generate cause-effect relations between high-dimensional responses and predictors and remove unimportant response and predictor variables. In this section, we will combine trace methods with multivariate sparse group lasso (Li et al. 2015) to develop sparse trace methods for sparse causal inference.

Consider the linear model:

$$Y = AX + W, \tag{7.159}$$

where

$$Y = \begin{bmatrix} y_{11} & \cdots & y_{1L} \\ \vdots & \vdots & \vdots \\ y_{m1} & \cdots & y_{mL} \end{bmatrix}, A = \begin{bmatrix} a_{11} & \cdots & a_{1n} \\ \vdots & \vdots & \vdots \\ a_{m1} & \cdots & a_{mn} \end{bmatrix}, X = \begin{bmatrix} x_{11} & \cdots & x_{1L} \\ \vdots & \vdots & \vdots \\ x_{n1} & \cdots & x_{nL} \end{bmatrix}$$

$$\text{and } W = \begin{bmatrix} w_{11} & \cdots & w_{1L} \\ \vdots & \vdots & \vdots \\ w_{m1} & \cdots & w_{mL} \end{bmatrix}.$$

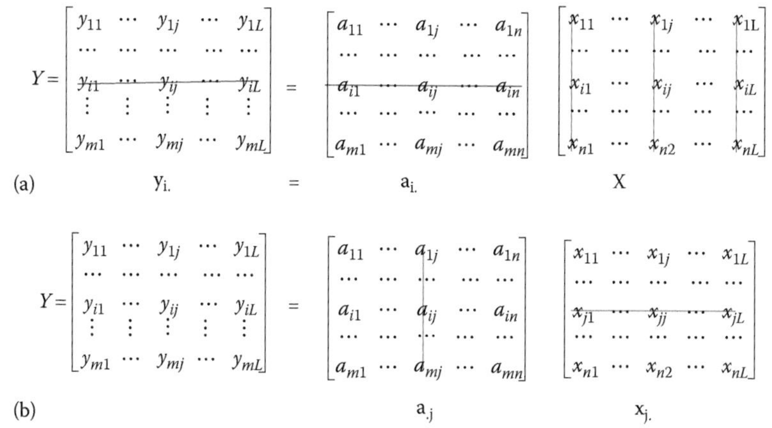

$$
Y = \begin{bmatrix} y_{11} & \cdots & y_{1j} & \cdots & y_{1L} \\ \cdots & \cdots & \cdots & \cdots & \cdots \\ y_{i1} & \cdots & y_{ij} & \cdots & y_{iL} \\ \vdots & \vdots & \vdots & \vdots & \vdots \\ y_{m1} & \cdots & y_{mj} & \cdots & y_{mL} \end{bmatrix} = \begin{bmatrix} a_{11} & \cdots & a_{1j} & \cdots & a_{1n} \\ \cdots & \cdots & \cdots & \cdots & \cdots \\ a_{i1} & \cdots & a_{ij} & \cdots & a_{in} \\ \cdots & \cdots & \cdots & \cdots & \cdots \\ a_{m1} & \cdots & a_{mj} & \cdots & a_{mn} \end{bmatrix} \begin{bmatrix} x_{11} & \cdots & x_{1j} & \cdots & x_{1L} \\ \cdots & \cdots & \cdots & \cdots & \cdots \\ x_{i1} & \cdots & x_{ij} & \cdots & x_{iL} \\ \cdots & \cdots & \cdots & \cdots & \cdots \\ x_{n1} & \cdots & x_{n2} & \cdots & x_{nL} \end{bmatrix}
$$

(a)             $y_{i\cdot}$             $=$             $a_{i\cdot}$                  $X$

$$
Y = \begin{bmatrix} y_{11} & \cdots & y_{1j} & \cdots & y_{1L} \\ \cdots & \cdots & \cdots & \cdots & \cdots \\ y_{i1} & \cdots & y_{ij} & \cdots & y_{iL} \\ \vdots & \vdots & \vdots & \vdots & \vdots \\ y_{m1} & \cdots & y_{mj} & \cdots & y_{mL} \end{bmatrix} = \begin{bmatrix} a_{11} & \cdots & a_{1j} & \cdots & a_{1n} \\ \cdots & \cdots & \cdots & \cdots & \cdots \\ a_{i1} & \cdots & a_{ij} & \cdots & a_{in} \\ \cdots & \cdots & \cdots & \cdots & \cdots \\ a_{m1} & \cdots & a_{mj} & \cdots & a_{mn} \end{bmatrix} \begin{bmatrix} x_{11} & \cdots & x_{1j} & \cdots & x_{1L} \\ \cdots & \cdots & \cdots & \cdots & \cdots \\ x_{j1} & \cdots & x_{jj} & \cdots & x_{jL} \\ \cdots & \cdots & \cdots & \cdots & \cdots \\ x_{n1} & \cdots & x_{n2} & \cdots & x_{nL} \end{bmatrix}
$$

(b)                         $a_{\cdot j}$                     $x_{j\cdot}$

**FIGURE 7.2**
Group structure. (a) Penalty for variable $y_i$, (b) penalty for variable $x_j$.

Consider two group structures of the matrix $A$ for penalty (Figure 7.2). Group $a_{i\cdot} = [a_{i1},...,a_{in}]$ is used to remove the response $y_i$ and $a_{\cdot j} = [a_{1j},...,a_{mj}]$ is used to remove predictor $x_j$. Define

$$
\| a_{i\cdot} \|_2 = \sqrt{a_{i\cdot}^T a_{i\cdot}} \text{ and } \| a_{\cdot j} \|_2 = \sqrt{a_{\cdot j}^T a_{\cdot j}}.
$$

The multivariate sparse group lasso for the multivariate linear model (7.159) is defined as solving the following nonsmooth optimization problem:

$$
\min_{a_i, a_j} \quad F = \mathrm{Tr}\left((Y - AX)^T(Y - AX)\right) + \lambda \sum_{j=1}^{n} m\| a_{\cdot j} \|_2 + \lambda \sum_{i=1}^{m} n\| a_{i\cdot} \|_2 \quad (7.160)
$$

We can use $a_g$ to denote $a_{i\cdot}$ and $a_{\cdot j}$, and let $G = n + m$. The optimization problem (7.159) can be rewritten as

$$
\min_{a_g} \quad F = \mathrm{Tr}\left((Y - AX)^T(Y - AX)\right) + \lambda \sum_{g=1}^{G} \| a_g \|_2. \quad (7.161)
$$

The traditional group lasso problems consider only penalization on either predictors or responses. However, the current group lasso problems need to simultaneously consider penalization on both predictors and responses. The coordinate descent method will be used to solve the problem. The algorithm is summarized in Result 7.22 (Appendix 7.I).

### Result 7.22: Algorithms for Sparse Trace Method

Step 1. Initialization. Select penalty parameter $\lambda$. Let $A^{(0)}$ be an initial estimator of the matrix $A$. Let $A^{(-j)}$ be $A$ with the elements of the $j^{th}$ column vector replaced by zeros. Let

$$X = \begin{bmatrix} X_1 \\ \vdots \\ X_i \\ \vdots \\ X_n \end{bmatrix} = \begin{bmatrix} x_{11} & \cdots & x_{1j} & \cdots & x_{1L} \\ \vdots & \vdots & \vdots & & \vdots & \vdots \\ x_{i1} & \cdots & x_{ij} & \cdots & x_{iL} \\ \vdots & \vdots & \vdots & \vdots & \vdots \\ x_{n1} & \cdots & x_{nj} & \cdots & x_{nL} \end{bmatrix}, X_j = \begin{bmatrix} x_{1j} \\ \vdots \\ x_{ij} \\ \vdots \\ x_{nj} \end{bmatrix} \text{ and } X_i$$

$$= \lfloor x_{i1} \cdots x_{ij} \cdots x_{iL} \rfloor.$$

Define $S_j = (Y - A^{(-j)}X)X_j^T$. Then,

$$a_j^{(0)} = \left( 1 - \frac{m\lambda}{\| S_j \|_2} \right)_+ \frac{S_j}{\| x_j \|_2^2}, j = 1, .., n, \tag{7.162}$$

$$A^{(0)} = \begin{bmatrix} a_{.1}^{(0)} & \cdots & a_{.j}^{(0)} & \cdots & a_{.n}^{(0)} \end{bmatrix}$$

Step 2. Repeat

$$t \leftarrow t + 1$$

For $i = 1,...,m, j = 1,...,n,$
Calculate

$$S_{kj}^{(t)} = \left( Y_k - \left( a_k^{-(kj)} \right)^{(t)} X \right) X_j^T \tag{7.163}$$

If $|S_{kj}^{(t)}| \le \lambda(m + n)$, then set $a_{kj}^{(t+1)} = 0$; otherwise, update

$$a_{kj}^{(t+1)} = \frac{S_{kj}^{(t)}}{\| X_j \|_2^2 + \lambda \left( \dfrac{n}{\| a_{.k}^{(t)} \|_2} + \dfrac{m}{\| a_{.j}^{(t)} \|_2} \right)}, \tag{7.164}$$

$S_{kj}^{(t)}, a_{k.}^{(t)}$ and $a_{.j}^{(t)}$ are calculated using the matrix $A^{(t)}$.

Step 3. Check convergence
If $\| A^{(t+1)} - A^{(t)} \|_F \le \varepsilon$ where $\| A \|_F = \sqrt{\text{Tr}(A^T A)}$ is the Frobenius norm of the matrix and $\varepsilon$ is pre-specified error, then go to step 4; otherwise

$$A^{(t)} \leftarrow A^{(t+1)}, \text{ go to Step 2.}$$

Step 4. Initialization. Select penalty parameter $\lambda$. Let $\tilde{A}^{(0)}$ be an initial estimator of the matrix $\tilde{A}$. Let $\tilde{A}^{(-j)}$ be $\tilde{A}$ with the elements of the $j^{th}$ column vector replaced by zeros. Let

$$
Y = \begin{bmatrix} Y_{1.} \\ \vdots \\ Y_{i.} \\ \vdots \\ Y_{m.} \end{bmatrix} = \begin{bmatrix} y_{11} & \cdots & y_{1j} & \cdots & y_{1L} \\ \vdots & \vdots & \vdots & & \vdots \vdots \\ y_{i1} & \cdots & y_{ij} & \cdots & y_{iL} \\ \vdots & \vdots & \vdots & \vdots & \vdots \\ y_{m1} & \cdots & y_{mj} & \cdots & y_{mL} \end{bmatrix}, Y_{.j} = \begin{bmatrix} y_{1j} \\ \vdots \\ y_{ij} \\ \vdots \\ y_{mj} \end{bmatrix} \text{ and } Y_i
$$

$$
= \lfloor y_{i1} \cdots y_{ij} \cdots y_{iL} \rfloor
$$

Define $\tilde{S}_j = (X - \tilde{A}^{(-j)}Y)Y_j^T$. Then,

$$
\tilde{a}_j^{(0)} = \left(1 - \frac{n\lambda}{\|\tilde{S}_j\|_2}\right)_+ \frac{\tilde{S}_j}{\|Y_j\|_2^2} \, , j = 1, \ldots, m, \tag{7.165}
$$

$$
\tilde{A}^{(0)} = \begin{bmatrix} \tilde{a}_{.1}^{(0)} & \cdots & \tilde{a}_{.j}^{(0)} & \cdots & \tilde{a}_m^{(0)} \end{bmatrix}.
$$

Step 5. Repeat

$$
t \leftarrow t + 1
$$

For $i = 1, \ldots, n$, $j = 1, \ldots, m$,
Calculate

$$
\tilde{S}_{kj}^{(t)} = \left(X_{k.} - \left(\tilde{a}_{k.}^{-(kj)}\right)^{(t)} Y\right)^{(t)} Y_{.j}^T \tag{7.166}
$$

If $|S_{kj}^{(t)}| \le \lambda(m + n)$, then set $\tilde{a}_{kj}^{(t+1)} = 0$; otherwise, update

$$
\tilde{a}_{kj}^{(t+1)} = \frac{\tilde{S}_{kj}^{(t)}}{\|X_{j.}\|_2^2 + \lambda \left(\dfrac{m}{\|\tilde{a}_{k.}^{(t)}\|_2} + \dfrac{n}{\|\tilde{a}_{.j}^{(t)}\|_2}\right)}, \tag{7.167}
$$

$\tilde{S}_{kj}^{(t)}, \tilde{a}_{k.}^{(t)}$ and $\tilde{a}_{.j}^{(t)}$ are calculated using the matrix $\tilde{A}^{(t)}$.

Step 6. Check convergence
If $\|\tilde{A}^{(t+1)} - A^{(t)}\|_F \le \varepsilon$ then go to step 7; otherwise

$A^{(t)} \leftarrow A^{(t+1)}$, go to Step 5.

Step 7. Estimate $\Sigma_X$, $\Sigma_Y$, $\Sigma_{XY}$ and $\Sigma_{YX}$.

Step 8. Calculate

$$\Delta_{X \to Y} = \log \frac{\tau_m\left(A\Sigma_X A^T\right)}{\tau_n(\Sigma_X)\,\tau_m(AA^T)} \quad \text{and}$$

$$\Delta_{Y \to X} = \log \frac{\tau_n\left(\tilde{A}\Sigma_Y \tilde{A}^T\right)}{\tau_m(\Sigma Y)\,\tau_n\left(\tilde{A}\tilde{A}^T\right)}$$

Step 9. If $|\Delta_{Y \to X}| > (1 + \varepsilon)|\Delta_{X \to Y}|$ then

$$X \to Y$$

Else

If $|\Delta_{X \to Y}| > (1 + \varepsilon)\,|\Delta_{Y \to X}|$ then

$$Y \to X$$

Else

No cause and then test association

End if

End if


### 7.1.5 Causal Inference on Discrete Data

In genetic studies of complex diseases, the disease status is a discrete variable and the genotype variable is also a discrete variable. It is necessary to extend causal inference from continuous variables to discrete variables. In this section, we introduce the distance correlation-based method for causal inference on discrete variables (Liu and Chan 2016).

In previous sections, we introduce the basis principal for assessing causation $X \to Y$ that the distribution $P(X)$ of causal $X$ is independent of the causal mechanism or conditional distribution $P(Y|X)$ of the effect $Y$, given causal $X$. Now the question is how to assess their independence. Recently, distance correlation is proposed to measure dependence between random vectors which allows for both linear and nonlinear dependence (Székely et al. 2007; Székely and Rizzo 2009). Distance correlation extends the traditional Pearson correlation in two remarkable directions:

1. Distance correlation extends the Pearson correlation defined between two random variables to the correlation between two sets of variables with arbitrary numbers;

2. Zero of distance correlation indicates independence of two random vectors.

Discretizing distributions $P(X)$ and $P(Y|X)$, and viewing their discretized distributions as two vectors $P(X)$ and $P(Y|X)$, the distance correlation between $P(X)$ and $P(Y|X)$ can be used to assess causation between $X$ and $Y$.

### 7.1.5.1 Distance Correlation

Consider two vectors of random variables: $p$- dimensional vector $X$ and $q$- dimensional vector $Y$. Let $P(x)$ and $P(y)$ be density functions of the vectors $X$ and $Y$, respectively. Let $P(x,y)$ be the joint density function of $X$ and $Y$. There are two ways to define independence between two vectors of variables: (1) density definition and (2) characteristic function definition. In other words, if $X$ and $Y$ are independent then either

1. $P(x,y) = P(x)P(y)$ or
2. $f_{X,Y}(t,s) = f_X(t)f_Y(s)$,

where $f_{X,Y}(t,s) = E[e^{i(t^T x + s^T y)}]$, $f_X(t) = E[e^{it^T x}]$, and $f_Y(s) = E[e^{is^T y}]$ are the characteristic functions of $(X,Y)$, $X$, and $Y$, respectively. Therefore, we can use both distances $\|P(x,y) - P(x)P(y)\|$ and $\|f_{X,Y}(t,s) - f_X(t)f_Y(s)\|$ to measure dependence between two vectors $X$ and $Y$. Distance correlation (Székely et al. 2007) uses distance between characteristic function to define the dependence measure. We first define a squared distance covariance between two vectors as

$$V^2(X, Y) = \| f_{X,Y}(t,s) - f_X(t)f_Y(s) \|_w^2$$

$$= \int_{R^{p+q}} |f_{X,Y}(t,s) - f_X(t)f_Y(s)|^2 w(t,s)dtds, \tag{7.168}$$

where

$$w(t,s) = \frac{1}{c_p c_q \| t \|_p^{1+p} \| s \|_q^{1+q}}, \; c_p = \frac{\pi^{(1+p)/2}}{\Gamma((1+p)/2)}, \; c_q = \frac{\pi^{(1+q)/2}}{\Gamma((1+q)/2)}$$

and $\| . \|_p$ is the Euclidean norm of $p$- dimensional vector. Since $t$ and $s$ represent the frequencies of the Fourier expansion of the probability density function, the weight function will be small for high-frequency components of the characteristic function (Cowley and Vinci 2014).

Similarly, we define a squared distance variance as

$$V^2(X) = V^2(X, X) = \int_{R^{2p}} \frac{|f_{X,X}(t,s) - f_X(t)f_X(s)|^2}{c_p^2 \| t \|_p^{1+p} \| s \|_p^{1+p}} dtds \tag{1.769}$$

Then, we can define the distance correlation between random vectors:

$$R^2(X, Y) = \begin{cases} \dfrac{V^2(X, Y)}{\sqrt{V^2(X)V^2(Y)}} & V^2(X)V^2(Y) > 0 \\[2ex] 0 & V^?(X)V^2(Y) = 0. \end{cases} \tag{7.170}$$

Theorem 3 in the paper (Székely et al. 2007) states that $0 \le R \le 1$ and $R(X,Y) = 0$ if and only if $X$ and $Y$ are independent.

Now we introduce the sampling formula for calculation of distance correlation (Székely et al. 2007; Székely and Rizzo 2009). Assume that pairs of $(X_k, Y_K)$, $k = 1,...,n$ are sampled. Calculate the Euclidean distances:

$$a_{kl} = \| X_K - X_l \|_p, b_{kl} = \| Y_k - Y_l \|_q, k = 1,...,n, l = 1,...,n.$$

Define

$$\bar{a}_{k.} = \frac{1}{n}\sum_{l=1}^{n}a_{kl}, \quad \bar{a}_{.l} = \frac{1}{n}\sum_{k=1}^{n}a_{kl}, \quad \bar{a}_{..} = \frac{1}{n^2}\sum_{k=1}^{n}\sum_{l=1}^{n}a_{kl},$$

$$\bar{b}_{k.} = \frac{1}{n}\sum_{l=1}^{n}b_{kl}, \quad \bar{b}_{.l} = \frac{1}{n}\sum_{k=1}^{n}b_{kl} \text{ and } \bar{b}_{..} = \frac{1}{n^2}\sum_{k=1}^{n}\sum_{l=1}^{n}b_{kl}.$$

Define two matrices:

$$A = (A_{kl})_{n\times n} \text{ and } B = (B_{kl})_{n\times n},$$

where

$$A_{kl} = a_{kl} - \bar{a}_{k.} - \bar{a}_{.l} + \bar{a}_{..}, \tag{7.171}$$

$$B_{kl} = b_{kl} - \bar{b}_{k.} - \bar{b}_{.l} + \bar{b}_{..}, \ k,l = 1,...,n \tag{7.172}$$

Finally, the sampling distance covariance $V_n(X,Y)$, variance $V_n(X)$, and correlation $R_n(X,Y)$ are defined as

$$V_n^2(X, Y) = \frac{1}{n^2}\sum_{k=1}^{n}\sum_{l=1}^{n}A_{kl}B_{kl}, \tag{7.173}$$

$$V_n^2(X) = V_n^2(X, X) = \frac{1}{n^2}\sum_{k=1}^{n}\sum_{l=1}^{n}A_{kl}^2, \ V_n^2(Y) = \sum_{k=1}^{n}\sum_{l=1}^{n}B_{kl}^2, \tag{7.174}$$

$$R_n^2(X, Y) = \begin{cases} \dfrac{V_n^2(X, Y)}{\sqrt{V_n^2(X)V_n^2(Y)}}, & V_n^2(X)V_n^2(Y) > 0 \\ 0 & V_n^2(X)V_n^2(Y) = 0, \end{cases} \tag{7.175}$$

respectively.

### 7.1.5.2 Properties of Distance Correlation and Test Statistics

Properties of distance covariance and correlation that have been investigated (Székely et al. 2007; Székely and Rizzo 2009) are summarized in Result 7.23.

### Result 7.23: Properties of Distance Covariance and Correlation

1. Almost surely convergence.
   The sample distance covariance converges almost surely to the distance covariance $V_n(X, Y) \xrightarrow{a.s} V(X, Y)$ and the sample distance correlation converges almost surely to the distance correlation $R_n^2(X, Y) \xrightarrow{a.s} R^2(X, Y)$.
2. Distance correlation and independence
   a. $0 \leq R(X,Y) \leq 1$ and $R(X,Y) = 0$ if and only if $X$ and $Y$ are independent.
   b. Consider two random vectors $X = (X_1, X_2)$ and $Y = (Y_1, Y_2)$. If $X$ and $Y$ are independent, then

$$V(X_1 + Y_1, X_2 + Y_2) \leq V(X_1, X_2) + V(Y_1, Y_2),$$

   equality holds if $X_1, X_2, Y_1, Y_2$ are mutually independent.
   c. Let $a \in R^P$ be a constant vector and $b$ be a scalar and $C$ be a $p \times p$ dimensional orthonormal matrix. Then

$$V(a + bCX) = |b| V(X).$$

   d. $V(X) = 0$ indicates that $X = E[X]$.
   e. Assume that $X$ and $Y$ are independent. Then, we have

$$V(X + Y) \leq V(X) + V(Y).$$

3. Distribution
   a. If $X$ and $Y$ are independent and $E(\|X\|_p + \|Y\|_p) < \infty$, Then,

$$nV_n^2 \xrightarrow{D} \| \zeta(t,s) \|^2, \qquad (7.176)$$

   where $\zeta(.)$ is a complex-valued zero mean Gaussian random process with covariance function

$$R(u, u_0) = \left( f_X(t - t_0) - f_X(t)\overline{f_X(t_0)} \right) \left( f_Y(s - s_0) - f_Y(s)\overline{f_Y(s_0)} \right), u$$

$$= (t, s), u_0 = (t_0, s_0),$$

   b. If $X$ and $Y$ are independent and $E(\|X\|_p + \|Y\|_p) < \infty$, then

$$n\frac{V_n^2}{a \cdot b} \xrightarrow{D} Q, \qquad (7.177)$$

   where

$$Q = \sum_{j=1}^{\infty} \lambda_j Z_j^2, \qquad (7.178)$$

$Z_j$ are independent normal random variables, $\lambda_j$ non-negative constants and are determined by the distribution of $(X,Y)$ and $E[Q] = 1$.

c. If $X$ and $Y$ are dependent, then

$$nV_n^2 \xrightarrow{P} \infty \text{ and } n\frac{V_n^2}{\bar{a}\ \bar{b}} \xrightarrow{P} \infty \qquad (7.179)$$

Independence can be formally tested. The null hypothesis is defined as $H_0 : X$ and $Y$ are independent.

We can use Equation 7.177 to define a test statistic:

$$T_{IND} = n\frac{V_n^2}{\bar{a}\ \bar{b}} \qquad (7.180)$$

Although Equation 7.178 can be used to compute an asymptotic distribution, $\lambda_j$ are difficult to calculate. We often use permutations to calculate P-values. Specifically, we can permutate $X$ and $Y$ millions of times. For each permutation, we compute test statistic $T_{IND}$. Therefore, via permutations we can calculate the empirical distribution of $T_{IND}$. Using an empirical distribution, we can calculate the P-value as

$$P - value = P(T_{IND} > T_{IND0}),$$

where $T_{IND0}$ is the observed value of $T_{IND}$ in real data.

### 7.1.5.3 Distance Correlation for Causal Inference

Distance correlation can be used to test independence between causal and causal generating mechanisms. Consider $p$-dimensional random vector $X$ and $q$-dimensional random vector $Y$. Let $P(X,Y)$ be their joint distribution. Let $P(X)$ and $P(Y|X)$ be the density function of $X$ and conditional density function of $Y$, given $X$, respectively. Similarly, we can define $P(Y)$ and $P(X|Y)$. Unlike association analysis where dependence is measured between two random vectors, in causal analysis, dependence is measured between two distributions.

**Definition 7.10: Distance Correlation Measure between Causal and Effect**

The distance correlation dependence measures between two distributions are defined as

$$\Delta_{X \to Y} = R(P(X), P(Y|X)), \qquad (7.181)$$

$$\Delta_{Y \to X} = R(P(Y), P(X|Y)). \tag{7.182}$$

Suppose that $X$ and $Y$ are discretized (or divided) into $m$ and $k$ groups, respectively. Let $m_i$ be the number of points $X$ in the $i^{th}$ group and $k_{ij}$ be the number of points $(X,Y)$ where $X$ is in the $i^{th}$ group and $Y$ is in the $j^{th}$ group. Then, $n = \sum_{i=1}^{m} m_i$, $m_i = \sum_{j=1}^{k} k_{ij}$. Let $X^{(i)}$ be the collection of all points $X$ in the $i^{th}$ group and $Y^{(j)}$ be the collection of all points $Y$ in the $j^{th}$ group. Then, the estimated density function $P(X^{(i)})$ is $P(X^{(i)}) = \dfrac{m_i}{n}$ and the conditional density function $P(Y^{(j)}|X^{(i)}) = \dfrac{k_{ij}}{m_i}$.

### Example 7.13

Consider two vectors $X$ and $Y$. Each component of the vectors is discretized (or divided) into two categories. Therefore, $X$ and $Y$ can be expressed as (Figure 7.3)

$$X = [X_1 \ X_2] = \begin{bmatrix} x_{11} & x_{12} \\ x_{21} & x_{22} \end{bmatrix} \text{ and } Y = [Y_1 \ Y_2] = \begin{bmatrix} y_{11} & y_{12} \\ y_{21} & y_{22} \end{bmatrix}.$$

Both $X$ and $Y$ are divided into four groups:

$$X^{(1)} = (x_{11}, x_{12}), X^{(2)} = (x_{11}, x_{22}), X^{(3)} = (x_{21}, x_{12}), X^{(4)} = (x_{21}, x_{22}) \text{ and } Y^{(1)}$$

$$= (y_{11}, y_{12}), Y^{(2)} = (y_{11}, y_{22}), Y^{(3)} = (y_{21}, y_{12}), Y^{(4)} = (y_{21}, y_{22}).$$

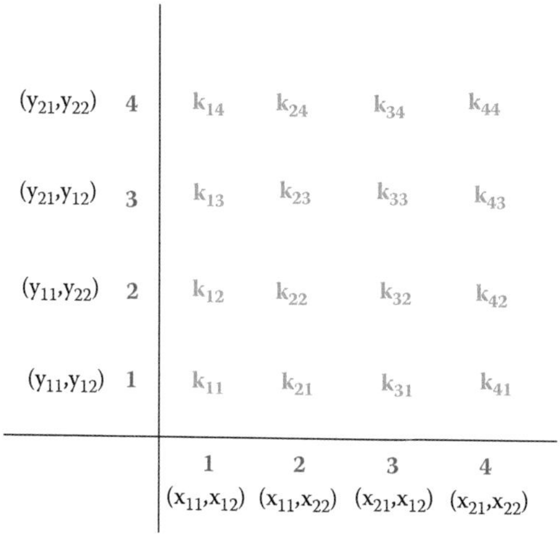

| | | | | |
|---|---|---|---|---|
| $(y_{21}, y_{22})$ 4 | $k_{14}$ | $k_{24}$ | $k_{34}$ | $k_{44}$ |
| $(y_{21}, y_{12})$ 3 | $k_{13}$ | $k_{23}$ | $k_{33}$ | $k_{43}$ |
| $(y_{11}, y_{22})$ 2 | $k_{12}$ | $k_{22}$ | $k_{32}$ | $k_{42}$ |
| $(y_{11}, y_{12})$ 1 | $k_{11}$ | $k_{21}$ | $k_{31}$ | $k_{41}$ |
| | 1 | 2 | 3 | 4 |
| | $(x_{11}, x_{12})$ | $(x_{11}, x_{22})$ | $(x_{21}, x_{12})$ | $(x_{21}, x_{22})$ |

**FIGURE 7.3**
Data are organized into groups.

Figure 7.3 shows that $X$ and $Y$ jointly form 16 groups. The vectors of probability density and conditional probability density functions are defined as

$$P\left(X^{(i)}\right) = \frac{m_i}{n}, i = 1, 2, 3, 4, \text{ and}$$

$$P\left(Y^{(j)}\middle|X^{(i)}\right) = \frac{k_{ij}}{m_i}, i = 1, ..., 4, j = 1, ..., 4$$

Similarly, we can calculate $P(Y)$ and $P(X|Y)$.
Next we study how to calculate distance correlation. Define

$$\alpha_i = P\left(X^{(i)}\right), \ i = 1, ..., m, \tag{7.183}$$

$$\beta_i = \left[P\left(Y^{(1)}\middle|X^{(i)}\right), ..., P\left(Y^{(k)}\middle|X^{(i)}\right)\right]^T \tag{7.184}$$

$$= [\beta_{i1}, ..., \beta_{ik}]^T, i = 1, ..., m.$$

Define

$$a_{ij} = |\alpha_i - \alpha_j|, \ a_{i.} = \frac{1}{m}\sum_{j=1}^m a_{ij}, \ a_{.j} = \frac{1}{m}\sum_{i=1}^m a_{ij}, \ a_{..}$$

$$= \frac{1}{m^2}\sum_{i=1}^m\sum_{j=1}^m a_{ij}, \tag{7.185}$$

$$b_{ij} = \|\beta_i - \beta_j\|_2 = \sqrt{\left(\beta_{i1} - \beta_{j1}\right)^2 + ... + \left(\beta_{ik} - \beta_{jk}\right)^2},$$

$$b_{i.} = \frac{1}{m}\sum_{j=1}^m b_{ij}, b_{.j} = \frac{1}{m}\sum_{i=1}^m b_{ij}, b_{..} = \frac{1}{m^2}\sum_{i=1}^m\sum_{j=1}^m b_{ij},$$

$$A_{ij} = a_{ij} - a_{i.} - a_{.j} + a_{..}, \tag{7.186}$$

$$B_{ij} = b_{ij} - b_{i.} - b_{.j} + b_{..}, \ i = 1, ..., m, j = 1, ..., m \tag{7.187}$$

Let $S_{X \to Y} = a_{..}b_{..}$.
Distance covariance is defined as

$$V_m^2(P(X), P(Y|X)) = \frac{1}{m^2}\sum_{i=1}^m\sum_{j=1}^m A_{ij}B_{ij}. \tag{7.188}$$

Similarly, $V_m^2(P(Y), P(X|Y))$ and $S_{Y \to X}$ can be similarly defined.
Define

$$\Delta_{X \to Y} = \frac{mV_m^2(P(X), P(Y|X))}{S_{X \to Y}}, \tag{7.189}$$

$$\Delta_{Y \to X} = \frac{mV_m^2(P(Y), P(X|Y))}{S_{Y \to X}}. \tag{7.190}$$

The null hypothesis for testing is $H_0$ : no causation between two vectors $X$ and $Y$.

The statistic for testing the causation between two vectors $X$ and $Y$ is defined as

$$T_C = |\Delta_{X \to Y} - \Delta_{Y \to X}| \qquad (7.191)$$

When $T_C$ is large, either $\Delta_{X \to Y} > \Delta_{Y \to X}$ which implies $Y$ causes $X$, or $\Delta_{Y \to X} > \Delta_{X \to Y}$ which implies that $X$ causes $Y$. When $T_C \approx 0$, this indicates that no causal decision can be made.

To more accurately assess the causation and calculate the P-value of the test, we need to compute the sampling distribution of the test statistic when the null hypothesis is true. However, the analytic formula for computing the distribution of the test statistic $T_C$ is unknown. A permutation test that gives a simple way to compute the sampling distribution for the test statistic $T_C$ under the null hypothesis of no causation decision can be used to calculate the P-value of the test. By randomly shuffling $X$ and $Y$, we can generate many data sets. If the null hypothesis is true, the shuffled data sets should look like the real data. For each permutated dataset, we calculate the value of $T_C$ and hence generate the empirical distribution of the test statistics. Assume that $K$ permutations are carried out and we observe $T_{C_0}$ from the real data. Also assume that the number of simulations showing $T_C > T_{C_0}$ is $l$. Then, the p-value is $P < \dfrac{l}{K}$.

When the presence of causation is inferred, we then identify causal direction. Again, permutation tests can be used to discover causal direction. In other words, we can also use permutations to calculate the P-values of the statistics $\Delta_{Y \to X}$ for testing the null hypothesis $H_0:Y \to X$ and the P-values of the statistics $\Delta_{X \to Y}$ for testing the null hypothesis $H_0 : X \to Y$.

The above discussion can be summarized in the following algorithm.

**Result 7.24: Algorithm for Causal Inference Using Distance Variances**

1. Encode the vector for the distribution $P(X)$ and the matrix for the conditional distribution $P(Y|X)$. Use Equations 7.168–7.175 to calculate $\Delta_{X \to Y}$.
2. Similarly, encode the vector for the distribution $P(Y)$ and the matrix for the conditional distribution $P(X|Y)$. Use Equations similar to 7.166–7.175 to calculate $\Delta_{Y \to X}$.
3. Use Equation 7.177 to calculate the test statistic $T_C$.
4. Use permutations to calculate the P-values $P_C$, $P_{X \to Y}$ and $P_{Y \to X}$ for $T_C$, $\Delta_{X \to Y}$ and $\Delta_{Y \to X}$, respectively.
5. If $P_C$ is small ($P_C$ < pre-specified value) and either $P_{Y \to X}$ or $P_{X \to Y}$ is large, then causation between $X$ and $Y$ exists.
6. If $P_C$ is small ($P_C$ < pre-specified value) and $P_{Y \to X}$ is small ($P_{Y \to X}$ < pre-specified value) and $P_{Y \to X} < P_{X \to Y}$ (large), then $Y \to X$.
7. If $P_C$ is small ($P_C$ < pre-specified value) and $P_{X \to Y}$ is small ($P_{X \to Y}$ < pre-specified value) and $P_{X \to Y} < P_{Y \to X}$ (large), then $Y \to X$.

**TABLE 7.1**

Data for Example 7.14

| Code for disease status | 0 | 1 | 2 | 3 |
|---|---|---|---|---|
| Disease status | Normal | Bipolar | Schizophrenia | Depress |
| Number of individuals | 47 | 43 | 45 | 14 |
| Code for genotype | 0 | 1 | 2 | |
| Genotype | TT | CT | CC | |
| Number of individuals | 137 | 11 | 1 | |
| $P(X)$ | 0.9195 | 0.0738 | 0.0067 | |
| $P(0 \mid X)$ | 0.3066 | 0.3636 | 1 | |
| $P(1 \mid X)$ | 0.2847 | 0.3636 | 0 | |
| $P(2 \mid X)$ | 0.3217 | 0.0909 | 0 | |
| $P(3 \mid X)$ | 0.0876 | 0.1818 | 0 | |

**Example 7.14**

Consider 149 individuals who suffered from one of the schizophrenia, bipolar, depression diseases or are healthy. Consider one SNP that is denoted by $X$. Let $Y$ denote disease status. Data, density function $P(X)$, and conditional density function $P(Y \mid X)$ are summarized in Table 7.1. The distance correlation between $P(X)$ and $P(Y \mid X)$ is 0.7311 and the distance correlation between $P(Y)$ and $P(X \mid Y)$ is 0.8032. The P-values of $P_C, P_{X \to Y}$, and $P_{Y \to X}$ are 0.8995, 0.4197, and 0.3990, respectively.

### 7.1.5.4 Additive Noise Models for Causal Inference on Discrete Data

In this section we introduce the additive noise model approach (Shimizu et al. 2006; Peters et al. 2011) to discover causal directions: $X \to Y$ or $Y \to X$ on discrete data. Additive noise models for casual inference on discrete data can be applied to genome-wide causal studies of complex diseases where causation of two discrete variables—disease variable and genotype variable of SNP—is investigated.

#### 7.1.5.4.1 Integer Models

We first consider an integer additive noise model:

$$Y = f(X) + N_1 \text{ and } N_1 \perp X, \tag{7.192}$$

where $X$ and $Y$ are two integer random variables, $f : Z \to Z$ is a function mapping value in integer space to values in integer space, $N_1$ is additive noise that takes integer values and is independent of the hypothetical cause $X$. The model in Equation 7.192 is called an additive noise model from $X$ to $Y$ in the forward direction. We also can define an additive noise model from $Y$ to $X$ in the backward direction:

$$X = g(Y) + N_2 \text{ and } N_2 \perp Y, \tag{7.193}$$

where $g : Z \rightarrow Z$ is a function mapping value in integer space to values in integer space, $N_2$ is additive noise that takes integer values and is independent of the hypothetical cause $Y$.

Can both the additive noise models in the forward and backward directions simultaneously fit the data well? This is the basic question about using the additive noise model to assess causal direction. Consider a trivial example: $P(X = 0) = 1, f(0) = 0, P(N_1 = 0) = 1, P(Y = 0) = 1, g(0) = 0$ and $P(N_2 = 0) = 1$. Then, it is clear that these distributions satisfy both additive noise models (7.192) and (7.193) in both directions. However, these cases are very rare (Peters et al. 2011). In most cases, either additive disease models in forward direction or additive disease models in backward direction fit the data, but both models do not fit the data. If both models fit the data, we are unable to make causal direction selection. Figure 7.4 shows that the joint distribution allows an additive noise model (ANM) only from $X$ to $Y$, but not from $Y$ to $X$. However, Figure 7.5 demonstrates that the joint distribution in this example allows for an ANM in both directions: from $X$ to $Y$ and from $Y$ to $X$.

### 7.1.5.4.2 Cyclic Models

First we introduce the concepts of the modular operator and finite ring $Z/mZ$. When $a$ and $b$ are integers we divide two integers:

$$\frac{a}{b} = k \text{ remainder } r.$$

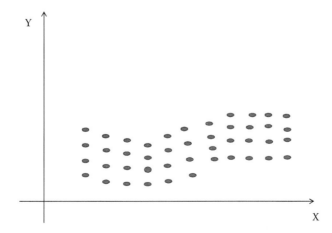

**FIGURE 7.4**
Joint distribution allows ANM in the X -> Y direction, but not in the Y -> X direction.

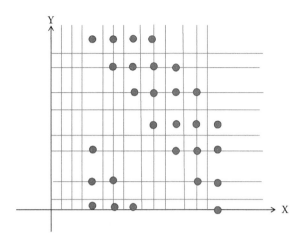

**FIGURE 7.5**
Joint distribution allows ANM in both directions.

We are interested in the remainder $r$. We then define $r$ as $a$ modulo $b$ denote it as $r = a \bmod b$. For example, $14 \bmod 3 = 2$. The finite ring $Z/mZ$ is defined as the set of all possible remainders in the mod by $n$, that is,

$$Z/mZ = \{Z \bmod m\} = \{0,1,...,m-1\}.$$

Now we define cyclic additive noise models (Peters et al. 2011). Let $X$ and $Y$ be $m$ and $\tilde{m}$ cyclic random variables, respectively. Let $f : Z/mZ \to Z/\tilde{m}Z$ be a function and $N$ be a $\tilde{m}$ cyclic noise. Define an additive noise model from $X$ to $Y$ as

$$Y = f(X) + N \ and \ N \perp X, \tag{7.194}$$

where $N$ is independent of $X$ and $P(N = 0) > P(N = j), j \neq 0$. If this model is reversible, then there is a function $g : Z/\tilde{m}Z \to Z/mZ$ and an $m$ cycle noise $\tilde{N}$ such that

$$X = g(Y) + \tilde{N} \ and \ \tilde{N} \perp Y. \tag{7.195}$$

There are some examples that show that joint distributions allow ANMs in both forward and backward directions. However, in generic cases, the ANMs are identifiable (Peters et al. 2011). Some conditions for the identifiability of the ANMs have been explored (Peters et al. 2011) but will not be presented here due to their complexities and space limitations of the book.

### 7.1.5.4.3 Statistical Test for ANM with Discrete Variables

In practice, a key issue for causal inference using ANMs with discrete variables is to develop statistical tests for the ANMs. Peters et al. (2011) presented a general procedure for causal inference using ANMs that is restated in Result 7.25.

### Result 7.25: General Procedure for Causal Inference Using ANMs

A general procedure for causal inference using ANMs is given as follows.

**Step 1.** Input $n$ samples of data $\{(x_1,y_1),...,(x_n,y_n)\}$.

**Step 2.** Perform nonlinear regression of the forward model: $Y = f(X) + N$ and calculate the residuals $\hat{N}_i = y_i - f(x_i), i = 1, ..., n$.

**Step 3.** Perform nonlinear regression of the backward model: $X = g(Y) + \tilde{N}$ and calculate the residuals $\tilde{N}_i = x_i - g(y_i), i = 1, ..., n$.

**Step 4.** If the residuals $\hat{N}$ is independent of $X$ and $\tilde{N}$ is not independent of $Y$ then $X$ is causing $Y$ ($X \rightarrow Y$); If the residuals $\hat{N}$ is not independent of $X$ and $\tilde{N}$ is independent of $Y$ then $Y$ is causing $X$ ($Y \rightarrow X$).

**Step 5.** If both residuals $\hat{N}$ is independent of $X$ and $\tilde{N}$ is independent of $Y$ or if both residuals $\hat{N}$ is not independent of $X$ and $\tilde{N}$ is not independent of $Y$ then, we are unable to make decisions. In these cases, causation is unknown.

To implement procedures, we need to develop statistical methods for nonlinear regressions with discrete variables and statistics for testing the independence between two random variables (Peters et al. 2011). First, we introduce regressions with discrete variables. Since our goal is to select a regression function that makes the residuals as independent of the regressor (the potential cause) as possible, taking a dependence measure between the residuals and regressor as a loss function for the regression is an appropriate choice. This measure is denoted as $DM(\hat{N}, X)$ for the ANM: $Y = f(X) + N$. Let $\hat{P}(X, Y)$ be the sample distribution of the joint distribution $P(X,Y)$. Here, we present distance correlation-based regression that can cover multivariate cases and can be viewed as a simple extension of algorithm 1 in the paper (Peters et al. 2011). In Result 7.26, we extend the one variate ANMs to multivariate ANMs and assume that $W$ is an $m$-dimensional vector of variables and $Z$ is a $q$-dimensional vector of variables. A multivariate ANM is given as follows:

$$Z = F(W) + N,$$

where $Z = [Z_1,...,Z_m]$, $W = [W_1,...,W_q]$, $F(W) = [f_1(W),...f_m(W)]$, $N = [N_1,...,N_m]$ and $N$ is independent of $W$.

### Result 7.26: Distance Regression with Distance Correlation Dependence Measure

**Step 1:** Calculate the sampling distribution $\hat{P}(W, Z)$.

**Step 2:** Initialization.

$$F^{(0)}\left(w^i\right) = \arg\max_{Z} \hat{P}\left(W = w^i, Z = z\right), \ t = 0,$$

where $w^i = [w_1^i, ..., w_q^i]$.

**Step 3:** Repeat

$$t = t + 1;$$

**Step 4: for** $i = 1, \ldots, n$ **do**
**Step 5:** $F^{(t)}(w^i) = \arg\min_{Z} DM(w^i, Z - F^{(t-1)}(w^i))$ **end for**
**Step 6: until** $\| F^{(t)} - F^{(t)} \|_2 < \varepsilon$ or $t = T$,
     where $\varepsilon$ and $T$ are pre-specified.
   If the model $Y = F(X) + N$ is considered, then $Z = Y$ and $W = X$. In the backward direction: $X = G(Y) + \tilde{N}$ we set $Z = Y$ and $W = Y$. For inferring causation involving at least one vector of variables, distance correlation will be used as $DM$. If we infer the causal structure between two variables, the P-values for testing independence between residuals and regressor (causal variable) based on Pearson's $\chi^2$ test and Fisher's exact test (Peters et al. 2011) will be used as $DM$.

### Example 7.15

Let $X$ and $Y$ be $m$ and $\tilde{m}$ cyclic random variables, respectively. Simulate 1000 different models where four combinations $(m, \tilde{m}) \in \{(4,4), (4,6), (6,4), (6,6)\}$ are considered. A total of 10,000 data points was sampled with the distributions $P(X)$, $P(N)$ and nonconstant function $f$. A significance level of $\alpha = 0.05$ was assumed. The ANMs with cyclic models where P-values based on the $\chi^2$ test were used as dependence measure were applied to fit the simulated data. The results were summarized in Table 7.2. We observe that the proportions of the ANMs that cannot fit the data in both directions are close to the significance level of $\alpha = 0.05$.

## 7.2 Multivariate Causal Inference and Causal Networks

In the previous sections, we studied genome-wide single trait and multiple trait causal studies that involve causal inference between two variables or two vectors of variables. In this section, we extend causal inference from two variables or two vectors to multiple variables that often form causal networks.

**TABLE 7.2**

The Results of ANM with Cyclic Models

| $(m, \tilde{m})$ | (4,4) | (4,6) | (6,4) | (6,6) |
|---|---|---|---|---|
| Correct direction | 0.941 | 0.952 | 0.941 | 0.944 |
| Wrong direction | 0 | 0 | 0 | 0 |
| Both directions | 0 | 0 | 0 | 0 |
| Unfitted in both directions | 0.059 | 0.048 | 0.059 | 0.056 |

In Chapters 1 and 2, we introduced graph concepts and structural equation models. Although causal inference may have substantial potential to improve prevention, management, and therapy of complex diseases, most genomic, epigenomic, and image data are observational data. Many confounding variables are not or cannot be measured. The unmeasured confounding variables will invalidate the most traditional causal inference analysis. The gold-standard for causal inference is to perform a randomized controlled trial, which can control the confounding effects. Unfortunately, in many cases, performing experiments is unethical or infeasible. In most genetic studies inferring causal relations must be from observational data alone. Despite its fundamental role in science, engineering, and biomedicine, the traditional causal inference from observational data alone is unable to identify unique causal-effect relations among variables. Un-uniqueness of the inferred causal relations seriously limits the application of the traditional causal inference to genetic studies and biomedical research.

To overcome limitations of the traditional causal inference, in this section, we introduce concepts of Markov property, faithfulness, causal sufficiency, and minimality. We will show that under some assumptions the additive noise models are identifiable. Then, we review and develop new causal models and learning methods to uniquely discover the causal relationships (Peters et al. 2017).

### 7.2.1 Markov Condition, Markov Equivalence, Faithfulness, and Minimality

One of basic assumptions in the graphical model is the Markov condition (Peters et al. 2017; Ernest 2016). Markov condition encodes conditional independence in the distribution. In other words, the graph is Markov if all conditional independences imposed by the graph structure (*d*-separation) can be identified in the distribution (Peters et al. 2011). Definition 7.11 gives a formal definition of the Markov condition.

#### Definition 7.11: Markov Condition

The Markov condition is defined as the property that every variable is independent of its non-descendants in the directed acyclic graph (DAG) (excluding the parents), given its parents in the DAG. The Markov condition implies the Markov factorization property:

$$P(x) = P\left(x_1, \ldots, x_p\right) = \prod_{j=1}^{p} P\left(x_j \mid pa_j\right) \qquad (7.196)$$

#### Example 7.16

A DAG for Example 7.16 is shown in Figure 7.6. The conditional independences that are encoded by the causal Markov condition for the DAG in Figure 7.6 are $X_2 \perp X_4 \mid X_1$ and $X_5 \perp (X_1, X_2, X_4) \mid X_3$. Clearly, knowing

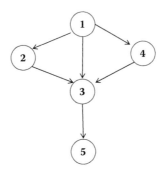

**FIGURE 7.6**
Illustration of Markov condition.

$X_1$, $X_2$ will not provide any new information about $X_4$. Similarly, if we already know $X_3$, then $X_5$ does not contain any new information about $(X_1, X_2, X_4)$. This example shows that the DAG encodes the conditional independence distribution. However, there is no one to one correspondence between the DAG and conditional independence distribution. Definition 7.12 states that different DAG may encode the exact same conditional independence distribution (Peters et al. 2017). Before introducing Definition 7.12, we briefly describe one concept in the graphs. A graph is called a partially directed acyclic graph (PDAG) if there is no directed cycle in the graph.

**Definition 7.12: Markov Equivalence Classes**

Markov equivalence class is defined as the set of DAGs that satisfy the same set of conditional independence distributions or the same Markov property with respect to the graph.

Consider three simple DAGs: $x \rightarrow y \rightarrow z$, $x \leftarrow y \leftarrow z$, and $x \rightarrow y \rightarrow z$. Three variables $x$, $y$, and $z$ in all three DAGs satisfy the causal Markov condition: $x$ and $z$ are independent, given $y$. This indicates that these three DAGs form a Markov equivalence class. However, these three DAGs represent three different causal relationships among variables $x$, $y$, and $z$, which prohibit unique causal identification.

All DAGs are clustered in a Markov equivalence class. Before discussing characterization of Markov equivalence class, we introduce several concepts of graphs (Kalisch and Bühlmann 2007). A graph is called a partially directed acyclic graph (PDAG) if there is no directed cycle in the graph. A PDAG is called completed PDAG (CPDAG) if (1) every directed edge in the CPDAG also exists in all DAGs that belong to the Markov equivalence class and (2) for every undirected edge $i - j$ in the CPDAG there exists a DAG with $i \rightarrow j$ and a DAG with $i \leftarrow j$ in the Markov equivalence class. The Markov equivalence class can be represented as CPDAG. Consider an ordered triple of nodes $a,b$, and $c$. The structure with $a \rightarrow b \leftarrow c$ and, $a$ and $c$ not directly connected is called a **v-structure**.

The **skeleton** of a DAG is defined as the undirected graph with all directed edges in the DAG replaced by undirected edges.

The Markov equivalent class can be represented by CPDAG. All DAGs in a Markov equivalence class share the same skeleton and v-structures (Verma and Pearl 1991; Ernest 2006).

### Example 7.17

Figure 7.7 presents a Markov equivalence class (a,b) and its representation CPDAG (c). The graphs (a) and (b) in the Markov equivalence class share the same skeleton and v-structures: $x_1 \rightarrow x_4 \leftarrow x_5$ and $x_2 \rightarrow x_3 \leftarrow x_4$. Graph (c) is the CPDAG corresponding to the Markov equivalence class (a,b). We observe that the CPDAG keeps all directed edges shared in graphs (a) and (b). However, directed edge $x_8 \rightarrow x_7$ in (a) and directed edge $x_8 \leftarrow x_7$ in (b) that have opposite directions are replaced by undirected edge $x_7 - x_8$ in the CPDAG (c).

This example shows that different graphs may satisfy the same set of conditional independence. Therefore, causal inference based on the conditional independences may not have unique solutions. To enforce the unique causal inference solution, we introduce the concept of faithfulness condition (Ernest 2016).

### Definition 7.13: Faithfulness Condition and Causal Minimality

A distribution $P(X)$ is called faithful with respect to the DAG $G$ if there exists a one-to-one correspondence between conditional independence distributions and $d$ separation structures in the DAG $G$. In other words, faithfulness condition requires that every conditional independence in the distribution must correspond to the Markov condition that is applied to the DAG. If a distribution is Markovian with respect to the DAG $G$, but not to any proper subgraph, then we say that this distribution satisfies causal minimality with respect to $G$.

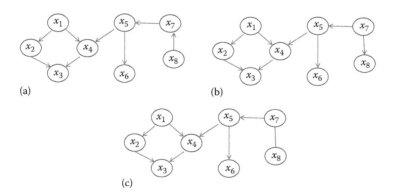

(a)          (b)          (c)

**FIGURE 7.7**
Markov equivalence class (a,b) and its representation CPDAG (c).

**Example 7.18**

Consider a DAG in Figure 7.8a. Its linear structural equations are given as follows.

$$X = N_X,$$
$$Y = aX + N_Y,$$
$$Z = bY + N_Z,$$
$$W = cZ + dX + N_W,$$

(7.197)

where $N_X \sim N(0, \sigma_X^2)$, $N_Y \sim N(0, \sigma_Y^2)$, $N_Z \sim N(0, \sigma_Z^2)$ and $N_W \sim N(0, \sigma_W^2)$ are jointly independent.

Equation 7.197 can be reduced to

$$X = N_X,$$
$$Y = aX + N_Y,$$
$$Z = abX + bN_Y + N_Z,$$
$$W = (abc + d)X + cN_Z + N_W$$

(7.198)

If we assume

$$abc + d = 0,$$

(7.199)

then Equation 7.198 is reduced to

$$X = N_X,$$
$$Y = \tilde{a}X + N_Y,$$
$$Z = \tilde{b}X + \tilde{N}_Z,$$
$$W = \tilde{N}_W,$$

(7.200)

where

$$\tilde{b} = ab,$$
$$\tilde{N}_Z = bN_Y + N_Z,$$
$$\tilde{N}_W = CN_Z + N_W$$

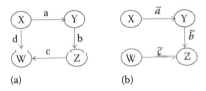

(a)                              (b)

**FIGURE 7.8**
Illustrate faithfulness. (a) A DAG for structural Equation 7.197. (b) A DAG for structural Equation 7.200.

Since $Z_Z$ and $N_W$ are independent of $N_X$, $W$ and $X$ are independent which is not implied by Figure 7.8a. Therefore, distributions defined by Equation 7.197 are not faithful with respect to Figure 7.8a. Equation 7.200 corresponds to Figure 7.8b.

## 7.2.2 Multilevel Causal Networks for Integrative Omics and Imaging Data Analysis

### 7.2.2.1 Introduction

Deep integrative omic analysis will shift the current paradigms of genomic, epigenomic, and image studies of complex diseases by opening new avenues for unravelling disease mechanisms and designing therapeutic interventions. Common diseases result from the interplay of DNA sequence variation and nongenetic factors acting through molecular networks (Zhang et al. 2013; Jiang et al. 2015). Their etiology is complex with multiple steps between genes and phenotypes (Delude 2015; Köhler et al. 2017). Each step is influenced by genomic and epigenomic variation and can obscure the causal mechanism of the phenotype. Efficient genetic analysis consists of two major parts: (1) breadth (the number of variables/phenotypes that are connected) and (2) depth (the number of steps influenced by genetic variants on the way to clinical outcomes). Only by broadly and deeply searching the enormous path space connecting genetic variants to clinical outcomes will we uncover the mechanisms of disease. Precision medicine demands deep, systematic, comprehensive, and precise analysis of omics data — "and the deeper you go, the more you know" (Huan et al. 2015).

In this section we will use causal inference theory to infer multilevel, causal omic, and imaging networks which integrates genotype subnetworks, environment subnetworks, methylation subnetworks, gene expression subnetworks, micro RNA subnetworks, metabolic subnetworks, image subnetworks, the intermediate phenotype subnetworks, and multiple disease subnetworks into a single connected multilevel genotype–disease network to reveal the deep causal chain of mechanisms underlying the diseases (Figure 7.9).

In Chapter 2, we used integer programming (IP) as a general framework for estimation of a single causal network (DAG). Again, in this section we will extend IP from a single causal network estimation to joint multiple causal network estimations to integrate genomic, epigenomic, and imaging data.

### 7.2.2.2 Additive Noise Models for Multiple Causal Networks

#### 7.2.2.2.1 Models

The classical causal inference assumes both Markov conditions and faithfulness. It is now clear that Markov conditions and faithfulness ensure only to identify up to the Markov equivalence class, which cannot distinguish

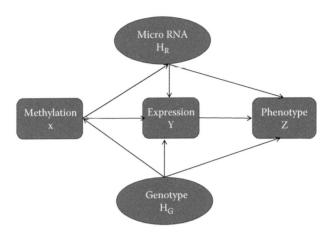

**FIGURE 7.9**
Mediation model with confounding.

between two Markov equivalent graphs. Recently, it has been shown that additive noise models with continuous variables or discrete variables, linear Gaussian models with equal error variables, and linear non-Gaussian acyclic models are identifiable (Peters et al. 2017). In Chapter 2, we discussed non-linear structural equations for causal discovery. In this section, we study ANMs for causal networks with several different types of data.

For the convenience of discussion, consider $M$ gene expression variables $Y_1,...,Y_M$, $Q$ methylation variables $Z_1,...,Z_Q$, and $K$ genotype variables $X_1,..., X_K$. Let $pa_D(d)$ be the parent set of the node $d$ including gene expression, methylation, and genotype variables. Consider three types of ANMs. First, we consider a general ANM model for the gene expression:

$$Y_d = f_d\Big(Y_i \in pa_D(d), Z_q \in pa_D(d), X_j \in pa_D(d)\Big) + \varepsilon_d, \quad d = 1,...,M, \quad (7.201)$$

$$Z_q = f_q\big(Z_l \in pa_Q(q), X_m \in pa_Q(q)\big) + \varepsilon_q, \quad q = 1,..,Q, \quad (7.202)$$

where $f_d$ and $f_q$ are nonlinear functions from $R^{|pa_D|} \rightarrow R$ and $R^{|pa_Q|} \rightarrow R$, respectively, and the errors $\varepsilon_d$ and $\varepsilon_q$ are independent following distributions $P_{\varepsilon_d}$ and $P_{\varepsilon_q}$, respectively. Equation 7.201 defines a causal network that connects gene expressions, methylations, and genotypes. Equation 7.202 defines a causal network that connects methylations and genotypes. Let $k_i$, $k_q$, and $k_j$ be the number of gene expression variables, the number of methylation variables, and the number of genotype variables that belong to the set of parents of node $d$, respectively. Let $k_l$ and $k_m$ be the number of methylation variables and the number of genotype variables that belong to the set of parents of node $q$. Then, we consider the ANMs for $Y_d$ that connect three subnetworks (gene expression subnetwork, methylation subnetwork, and genotype subnetwork):

$$Y_d = F_{di}(Y_i \in pa_D(d)) + F_{dq}\left(Z_q \in pa_D(d)\right) + F_{dj}\left(X_j \in pa_D(d)\right) + \varepsilon_d \quad (7.203)$$

and the ANMs for $Z_q$ that connect two subnetworks (methylation subnetwork and genotype subnetwork):

$$Z_q = F_{ql}(Z_l \in pa_Q(q)) + F_{qm}\left(X_m \in pa_Q(q)\right) + \varepsilon_q, \quad (7.204)$$

where $F_{di}: R^{k_i} \to R$, $F_{dq}: R^{k_q} \to R$, $F_{dj}: R^{k_j} \to R$, $F_{ql}: R^{k_l} \to R$, and $F_{qm}: R^{k_m} \to R$ are nonlinear functions whose forms are unknown, $\varepsilon_d$ and $\varepsilon_q$ are defined as before.

Finally, we consider a special case of the ANMs:

$$Y_d = \sum_{i \in pa_D(d)} f_{di}(Y_i) + \sum_{q \in pa_D(d)} f_{dq}\left(Z_q\right) + \sum_{j \in pa_D(d)} f_{dj}\left(X_j\right) + \varepsilon_d, \quad (7.205)$$

and

$$Z_q = \sum_{l \in pa_Q(q)} f_{ql}(Z_l) + \sum_{m \in pa_Q(q)} f_{qm}(X_m) + \varepsilon_q, \quad (7.206)$$

where $\varepsilon_d$ and $\varepsilon_q$ are independent with

$$\varepsilon_d \sim N\left(0, \sigma_d^2\right), \ \sigma_d^2 > 0, \ \varepsilon_q \sim N\left(0, \sigma_q^2\right) > 0, \ d = 1,...,M, q = 1,..,Q.$$

### 7.2.2.2.2 Loglikelihood Functions and Penalized Methods for the ANMs with Continuous Variables

The nodes in the network consist of two types of variables: continuous and discrete variables. The estimation methods for the ANMs with continuous variables are, in general, different from that for the ANMs with discrete variables. The penalized methods are often used for continuous variables but are difficult to be applied to discrete variables. In this section, we study the penalized maximum likelihood for the ANMs. For the convenience of presentation, we mainly discuss the models in Equations 7.205 and 7.206.

Assume that $n$ individuals are sampled. Assume that the parent nodes can be found by encouraging sparsity in the full models (Bühlmann et al. 2014; Meier et al. 2009). Consider the full model for Equation 7.206:

$$Z_{qu} = \sum_{l=1, l \neq q}^Q f_{ql}\left(Z_l^{(u)}\right) + \sum_{m=1}^K f_{qm}\left(X_m^{(u)}\right) + \varepsilon_{qu}, \ u = 1,...,n \quad (7.207)$$

and the full model for Equation 7.205:

$$Y_{du} = \sum_{i=1, i \neq d}^M f_{di}\left(Y_i^{(u)}\right) + \sum_{q=1}^Q f_{dq}\left(Z_q^{(u)}\right) + \sum_{j=1}^K f_{dj}\left(X_j^{(u)}\right)$$
$$+ \varepsilon_{du}, u = 1,...,n \quad (7.208)$$

where variables $Y$, $Z$, and $X$ are centered to zero.

Define

$$
f_{ql} = \begin{bmatrix} f_{ql}\left(z_l^{(1)}\right) \\ \vdots \\ f_{ql}\left(z_l^{(n)}\right) \end{bmatrix}, \; f_{qm} = \begin{bmatrix} f_{qm}\left(x_m^{(1)}\right) \\ \vdots \\ f_{qm}\left(x_m^{(n)}\right) \end{bmatrix}, \; f_{d_i} = \begin{bmatrix} f_{di}\left(y_i^{(1)}\right) \\ \vdots \\ f_{di}\left(y_i^{(n)}\right) \end{bmatrix},
$$

$$
f_{dq} = \begin{bmatrix} f_{dq}\left(z_q^{(1)}\right) \\ \vdots \\ f_{dq}\left(z_q^{(n)}\right) \end{bmatrix}, \; f_{dj} = \begin{bmatrix} f_{dj}\left(x_j^{(1)}\right) \\ \vdots \\ f_{dj}\left(x_j^{(n)}\right) \end{bmatrix},
$$

$$
Y_d = \begin{bmatrix} Y_{d1} \\ \vdots \\ Y_{dn} \end{bmatrix}, \; Z_q = \begin{bmatrix} Z_{q_1} \\ \vdots \\ Z_{qn} \end{bmatrix}, \text{ and Euclidean norm of a vector } f \in R^n : \|f\|_2^2 =
$$
$$
\frac{1}{n}\sum_{u=1}^{n} f_u^2.
$$

Nonlinear functions can be expressed in terms of basis functions including smoothing splines and B-splines. In Chapter 2, we introduced smoothing splines for approximating nonlinear functions. In this section, we use B-spline as basis functions for expansion of nonlinear functions. To obtain sparse and smooth function estimators, we define the penalty that penalizes both sparsity and smoothness:

$$
J\left(f_{ql}\right) = \sqrt{\|f_{ql}\|_2^2 + \mu S^2\left(f_{ql}\right)}, \; J\left(f_{qm}\right) = \sqrt{\|f_{qm}\|_2^2 + \mu S^2\left(f_{qm}\right)}, \; J(f_{di})
$$

$$
= \sqrt{\|f_{di}\|_2^2 + \mu S^2(f_{di})},
$$

$$
J\left(f_{dq}\right) = \sqrt{\|f_{dq}\|_2^2 + \mu S^2\left(f_{dq}\right)} \text{ and } J(f_{dj}) = \sqrt{\|f_{dj}\|_2^2 + \mu S^2\left(f_{dj}\right)},
$$

where

$$
S^2\left(f_{ql}\right) = \int \left(f''_{ql}(z)\right)^2 dz, \; S^2\left(f_{qm}\right) = \int \left(f''_{qm}(x)\right)^2 dx, \; S^2(f_{di}) = \int \left(f''_{di}(y)\right)^2 dy,
$$

$$
S^2\left(f_{dq}\right) = \int \left(f''_{dq}(z)\right)^2 dz \text{ and } S^2\left(f_{dj}\right) = \int \left(f''_{dj}(x)\right)^2 dx.
$$

The parameter $\mu$ control the smoothness of the functions.

Define the total penalty as

$$J(f) = \sum_{l=1,\neq q}^{Q} J(f_{ql}) + \sum_{m=1}^{Q} J(f_{qm}) + \sum_{i=1,\neq d}^{M} J(f_{di}) + \sum_{q=1}^{Q} J(f_{dq}) + \sum_{j=1}^{K} J(f_{dj}).$$
(7.209)

To estimate the functional and network structure, we define the penalized least square estimates as

$$F = \| Y_d - \sum_{i=1,i\neq d}^{M} f_{di} - \sum_{q=1}^{Q} f_{dq} - \sum_{j=1}^{K} f_{dj} \|_2^2$$

$$+ \| Z_q - \sum_{l=1,l\neq q}^{Q} f_{ql} - \sum_{m=1}^{K} f_{qm} \|_2^2 + J(f).$$
(7.210)

Assume that $Y$ and $Z$ are centered. The minimization of the objective function $F$ automatically enforces all nonlinear function terms in Equation 7.210 to be centered (Meier et al. 2009).

To calculate $F$, we need to derive a formula for computing the penalty function. Cubic B-spline basis functions are used to approximate each function $f$. Specifically, the expansions of the functions in terms of basis functions are given by

$$f_{di}(y) = \sum_{v=1}^{V} \beta_{di,v} b_{di,v}(y), \; f_{dq}(z) = \sum_{v=1}^{V} \beta_{dq,v} b_{dq,v}(z), \; f_{dj}(x) = \sum_{v=1}^{V} \beta_{dj,v} b_{dj,v}(x),$$

$$f_{ql}(z) = \sum_{v=1}^{V} \beta_{ql,v} b_{ql,v}(z) \text{ and } f_{qm}(x) = \sum_{v=1}^{V} \beta_{qm,v} b_{qm,v}(x)$$
(7.211)

Now using these expansions, we can compute the Euclidean norm of the functions at the observed values. For example,

$$f_{d_i} = \begin{bmatrix} f_{di}\left(y_i^{(1)}\right) \\ \vdots \\ f_{di}\left(y_i^{(n)}\right) \end{bmatrix} = \begin{bmatrix} b_{di,1}\left(y_i^{(1)}\right) & \cdots & b_{di,V}\left(y_i^{(1)}\right) \\ \vdots & \vdots & \vdots \\ b_{di,1}\left(y_i^{(n)}\right) & \cdots & b_{di,V}\left(y_i^{(n)}\right) \end{bmatrix} \begin{bmatrix} \beta_{di,1} \\ \vdots \\ \beta_{di,V} \end{bmatrix} = A_{di}\alpha_{di},$$
(7.212)

where

$$A_{di} = \begin{bmatrix} b_{di,1}\left(y_i^{(1)}\right) & \cdots & b_{di,V}\left(y_i^{(1)}\right) \\ \vdots & \vdots & \vdots \\ b_{di,1}\left(y_i^{(n)}\right) & \cdots & b_{di,V}\left(y_i^{(n)}\right) \end{bmatrix}, \; \alpha_{di} = \begin{bmatrix} \beta_{di,1} \\ \vdots \\ \beta_{di,V} \end{bmatrix}.$$

Similarly, we obtain

$$f_{dq} = B_{dq}\beta_{dq}, \; f_{dj} = C_{dj}\gamma_{dj}, \; f_{ql} = H_{ql}\delta_{ql} \text{ and } f_{qm} = G_{qm}\theta_{qm},$$
(7.213)

where

$$
B_{dq} = \begin{bmatrix} b_{dq,1}\left(z_q^{(1)}\right) & \cdots & b_{dq,V}\left(z_q^{(1)}\right) \\ \vdots & \vdots & \vdots \\ b_{dq,1}\left(z_q^{(n)}\right) & \cdots & b_{dq,V}\left(z_q^{(n)}\right) \end{bmatrix}, \ \beta_{dq} = \begin{bmatrix} \beta_{dq,1} \\ \vdots \\ \beta_{dq,V} \end{bmatrix}, \ C_{dj}
$$

$$
= \begin{bmatrix} b_{dj,1}\left(x_j^{(1)}\right) & \cdots & b_{dj,V}\left(x_j^{(1)}\right) \\ \vdots & \vdots & \vdots \\ b_{dj,1}\left(x_j^{(n)}\right) & \cdots & b_{dj,V}\left(x_j^{(n)}\right) \end{bmatrix}, \ \gamma_{dj} = \begin{bmatrix} \beta_{dj,1} \\ \vdots \\ \beta_{dj,V} \end{bmatrix}
$$

$$
H_{ql} = \begin{bmatrix} b_{ql,1}\left(z_l^{(1)}\right) & \cdots & b_{dl,V}\left(z_l^{(1)}\right) \\ \vdots & \vdots & \vdots \\ b_{ql,1}\left(z_l^{(n)}\right) & \cdots & b_{ql,V}\left(z_l^{(n)}\right) \end{bmatrix}, \ \delta_{ql} = \begin{bmatrix} \beta_{ql,1} \\ \vdots \\ \beta_{ql,V} \end{bmatrix}, \ G_{qm}
$$

$$
= \begin{bmatrix} b_{qm,1}\left(x_m^{(1)}\right) & \cdots & b_{qm,V}\left(x_m^{(1)}\right) \\ \vdots & \vdots & \vdots \\ b_{qm,1}\left(x_m^{(n)}\right) & \cdots & b_{qm,V}\left(x_m^{(n)}\right) \end{bmatrix}, \ \theta_{qm} = \begin{bmatrix} \beta_{qm,1} \\ \vdots \\ \beta_{qm,V} \end{bmatrix}
$$

Therefore, we have

$$
\| f_{ql} \|_2^2 = \frac{1}{n} \delta_{ql}^T H_{ql}^T H_{ql} \delta_{ql}, \ \| f_{dq} \|_2^2 = \frac{1}{n} \beta_{dq}^T B_{dq}^T B_{dq} \beta_{dq}, \ \| f_{dj} \|_2^2 = \frac{1}{n} \gamma_{dj}^T C_{dj}^T C_{dj} \gamma_{dj},
$$

$$
\| f_{di} \|_2^2 = \frac{1}{n} \alpha_{di}^T A_{di}^T A_{di} \alpha_{di} \text{ and } \| f_{qm} \|_2^2 = \frac{1}{n} \theta_{qm}^T G_{qm}^T G_{qm} \theta_{qm}. \tag{7.214}
$$

Next, we calculate the measures of the smoothness of functions. For example, function $f_{ql}(z)$ in Equation 7.211 can be re-expressed as

$$
f_{ql}(z) = \delta_{ql}^T b_{ql}(z), \tag{7.215}
$$

where $\delta_{ql} = [\beta_{ql,1}, \dots, \beta_{ql,V}]^T$ and $b_{ql}(z) = [b_{ql,1}(z), \dots, b_{ql,V}(z)]^T$.
Using Equation 7.215, we obtain

$$
\begin{aligned}
S^2\left(f_{ql}\right) &= \int \left(f''_{ql}(z)\right)^2 dz \\
&= \delta_{ql}^T \int b''_{ql}(z)\left(b''_{ql}(z)\right)^T dz \delta_{ql} \\
&= \delta_{ql}^T \Omega_{ql} \delta_{ql},
\end{aligned} \tag{7.216}
$$

where $\Omega_{ql} = (\Omega_{ql,uv})_{V \times V}$, $\Omega_{ql,uv} = \int b''_{qi,u}(z) b''_{ql,v}(z) dz$.
Similarly, we have

$$S^2\left(f_{qm}\right) = \theta_{qm}^T \Omega_{qm} \theta_{qm}, \; S^2(f_{di}) = \alpha_{di}^T \Omega_{di} \alpha_{di}, \; S^2\left(f_{dq}\right) = \beta_{dq}^T \Omega_{dq} \beta_{dq}, \; S^2\left(f_{dj}\right)$$

$$= \gamma_{dj}^T \Omega_{dj} \gamma_{dj}.$$

Combining Equations 7.214 and 7.216, we obtain

$$J\left(f_{ql}\right) = \sqrt{\frac{1}{n} \delta_{ql}^T H_{ql}^T H_{ql} \delta_{ql} + \mu \delta_{ql}^T \Omega_{ql} \delta_{ql}}$$

$$= \sqrt{\delta_{ql}^T \left(\frac{1}{n} H_{ql}^T H_{ql} + \mu \Omega_{ql}\right) \delta_{ql}}. \tag{7.217}$$

Similarly, we have

$$J\left(f_{qm}\right) = \sqrt{\theta_{qm}^T \left(\frac{1}{n} G_{qm}^T G_{qm} + \mu \Omega_{qm}\right) \theta_{qm}}, \; J(f_{di}) = \sqrt{\alpha_{di}^T \left(\frac{1}{n} A_{di}^T A_{di} + \mu \Omega_{di}\right) \alpha_{di}},$$

$$J\left(f_{dq}\right) = \sqrt{\beta_{dq}^T \left(\frac{1}{n} B_{dq}^T B_{dq} + \mu \Omega_{dq}\right) \beta_{dq}}, \; and \; J\left(f_{dj}\right)$$

$$= \sqrt{\gamma_{dj}^T \left(\frac{1}{n} C_{dj}^T C_{dj} + \mu \Omega_{dj}\right) \gamma_{dj}}. \tag{7.218}$$

For notation convenience, we define

$$\Psi_{ql} = \left(\frac{1}{n} H_{ql}^T H_{ql} + \mu \Omega_{ql}\right), \; \Psi_{qm} = \left(\frac{1}{n} G_{qm}^T G_{qm} + \mu \Omega_{qm}\right), \; \Psi_{di}$$

$$= \left(\frac{1}{n} A_{di}^T A_{di} + \mu \Omega_{di}\right),$$

$$\Psi_{dq} = \left(\frac{1}{n} B_{dq}^T B_{dq} + \mu \Omega_{dq}\right) \; and \; \Psi_{dj} = \left(\frac{1}{n} C_{dj}^T C_{dj} + \mu \Omega_{dj}\right). \tag{7.219}$$

Define $A = [A_{d1} \; \cdots \; A_{dM}]$ in which the matrix with index $di$ is removed, $B = [B_{d1} \; \cdots \; B_{dQ}]$, $C = [C_{d1} \; \cdots \; C_{dK}]$, $H = [H_{q1} \; \cdots \; H_{qQ}]$ in which matrix with index $qq$ is removed, $G = [G_{q1} \; \cdots \; G_{qK}]$ and matrices $D_Y = [A \; B \; C]$ and $D_Z = [H \; G]$. Define their corresponding vectors

$$\alpha = [\alpha_{d1} \; \cdots \; \alpha_{dM}]^T, \; \beta = [\beta_{d1} \; \cdots \; \beta_{dQ}]^T, \; \gamma = [\gamma_{d1} \; \cdots \; \gamma_{dK}], \; \delta = \lfloor \delta_{q1} \; \cdots \; \delta_{qQ} \rfloor,$$

$$\theta = \begin{bmatrix} \theta_{q1} & \cdots & \theta_{qK} \end{bmatrix}, \ \eta_Y = \begin{bmatrix} \alpha^T & \beta^T & \gamma^T \end{bmatrix}^T \text{ and } \eta_Z = \begin{bmatrix} \delta^T & \theta^T \end{bmatrix}^T.$$

Combining equations 7.208–7.219, we obtain

$$
\begin{aligned}
F = \| Y_d - D_Y \eta_Y \|_2^2 + \| Z_q - D_Z \eta_Z \|_2^2 + \lambda \sqrt{V} \\
\left( \sum_{i=1, \neq d}^{M} \sqrt{\alpha_{d_i}^T \Psi_{di} \alpha_{di}} + \sum_{q=1}^{Q} \sqrt{\beta_{dq}^T \Psi_{dq} \beta_{dq}} \right. \\
\left. + \sum_{j=1}^{K} \sqrt{\gamma_{dj}^T \Psi_{dj} \gamma_{dj}} + \sum_{l=1, \neq q}^{Q} \sqrt{\delta_{ql}^T \Psi_{ql} \delta_{ql}} + \sum_{m=1}^{Q} \sqrt{\theta_{qm}^T \Psi_{qm} \theta_{qm}} \right).
\end{aligned}
\tag{7.220}
$$

Our goal is to find the minimum of $F$. This is a typical group lasso problem. Using group lasso methods, we can solve the problem. The minimum of $F$ that can be taken as a node score and non-zero components of $\eta_Y$ and $\eta_Z$ indicates what nodes are the parents of the node $d$. Extension of Equation 7.220 to multiple methylated genes in $Z_q$ is straightforward but involves many complicated notations.

### 7.2.2.3 Integer Programming as a General Framework for Joint Estimation of Multiple Causal Networks

We collect multiple types of data: genotype, gene expression, microRNA expression, methylation, metabolite, image, phenotype, and disease data (Figure 7.9). We want to estimate multiple causal networks with different types of data. For example, consider $M$ gene expression variables $Y_1, \ldots, Y_M$, $Q$ methylation variables $Z_1, \ldots, Z_Q$, and $K$ genotype variables $X_1, \ldots, X_K$. Let $pa_D(d)$ be the parent set of the node $d$ including gene expression, methylation, and genotype variables. Consider an ANM model for the gene expression:

$$Y_d = f_d \left( Y_i \in pa_D(d), Z_q \in pa_D(d), X_j \in pa_D(d) \right) + \varepsilon_d, \ d = 1, \ldots, M, \tag{7.221}$$

and an ANM for the methylation:

$$Z_q = g_q \left( Z_l \in pa_Q(q), X_k \in pa_Q(q) \right) + \varepsilon_q, \ q = 1, \ldots, Q. \tag{7.222}$$

Assume that nonlinear functions are approximated by B-splines. After the sets $W_{di}$ and $W_{ql}$ of parents are specified, using techniques introduced in Section 7.2.2.2.2 we can calculate matrices $D_Y^i$ and $D_Z^l$ which correspond to the parent sets $W_{di}$ and $W_{ql}$ in Equations 7.221 and 7.222, respectively. The scores of the nodes $Y_d$ and $Z_q$ are, respectively, given by

$$C(Y_d, W_{di}) = Y_d^T \left( I - D_Y^i \left( \left( D_Y^i \right)^T D_Y^i \right)^{-1} \left( D_Y^i \right)^T \right) Y_d \tag{7.223}$$

and

$$C\left(Z_q, W_{ql}\right) = Z_q^T \left(I - D_Z^l \left(\left(D_Z^l\right)^T D_Z^l\right)^{-1} \left(D_Z^l\right)^T\right) Z_q. \qquad (7.224)$$

Let $V_E$ be the set of nodes in the gene expression network and $V_M$ be the set of nodes in the methylation network. Let $C_E$ be a subset of nodes in $V_E$ and $C_M$ be a subset of nodes in $V_M$. Similar to Section 2.6.2, a joint expression and methylation causal network can be formulated as the following 0-1 integer linear programming:

$$\text{Min} \qquad \sum_{d=1}^M \sum_{i \in pa_D(d)} C(d, W_{di}) \chi(W_{di} \to d)$$

$$+ \sum_{q=1}^Q \sum_{l \in pa_Q(q)} C\left(q, W_{ql}\right) \chi\left(W_{lq} \to q\right)$$

$$\text{s.t.} \qquad \sum_{i \in pa_D(d)} \chi(W_{di} \to d) = 1, \ d = 1, \dots, M,$$

$$\sum_{l \in pa_Q(q)} \chi\left(W_{ql} \to q\right) = 1, \quad q = 1, \dots, Q,$$

$$\forall \ C_E \subseteq V_E : \sum_{d \in C_E} \sum_{W_d \,:\, W_d \cap C_E = \phi} \chi(W_d \to d) \geq 1, \qquad (7.225)$$

$$\forall \ C_M \subseteq V_M \ : \sum_{q \in C_M} \sum_{W_q \,:\, W_q \cap C_M = \phi} \chi\left(W_q \to q\right) \geq 1.$$

Using branch and bound and other methods for solving the IP discussed in Section 2.6, we can solve the IP problem to obtain the best joint causal genotype-methylation-expression and genotype-methylation network fitting the data.

## 7.3 Causal Inference with Confounders

In the previous discussions, we assumed that all variables are measured. Although integrated genomic, epigenomic, and imaging data analysis take a huge number of variables into consideration in causal inference, there are still unmeasured causally relevant variables. A major challenge to the validity of causal inference from observation studies is the presence of unmeasured confounding that affects both the intervention and the outcome (Baiocchi et al. 2014; Louizos et al. 2017). The task of causal inference is to discover causal structures (1) among the observed variables, (2) among the observed variables and unobserved variables, and (3) among unobserved variables only (Wolfe et al. 2016). If the confounding variables are observed, the common approach is to include these variable into models and adjust for them. If the confounding variables are unobserved, these variables are often overlooked. Ignoring unobserved confounding may make incorrect causal structure inference and biased causal effect estimation.

Developing effective statistical methods and computational algorithms for causal inference with hidden confounding variables is urgently needed. In this section we will introduce several approaches to causal inference with hidden confounding variables. First, we will introduce the concept of causal sufficiency. Then, we study proxy variables or instrument variables that can be measured to replace the unmeasured confounding variables. Several graphical representations for structural causal models with hidden confounding variables will be discussed.

## 7.3.1 Causal Sufficiency

In the previous discussion, we implicitly assume that the available set of variables is causally sufficient. However, we have not given its precise definition. Intuitively, the system in which we are interested is closed on causal operation. In other words, the system includes all variables that do not share common causal variables outside the system. We have observed all the common causes of observed variables.

**Definition 7.14: Causal Sufficiency**

A set of variables is called causally sufficient, if there are no hidden common causes of any of the observed variables. Common causes are often called confounders. The assumption of causal sufficiency requires that all potential causal variables are observed (Ernest 2016). However, in genetic studies of complex diseases we cannot include all genetic and environmental factors into a study. We must assume that some common causes are hidden. The methods for causal inference with hidden variables must be developed.

## 7.3.2 Instrumental Variables

Consider a methylation-genotype-micro RNA-expression-phenotype network (Figure 7.9). Assume that the methylation, gene expression, and phenotype are observed, and the genotype and micro RNA variables are hidden variables. Figure 7.9 represents a mediation model with confounding variables. The methylation affects phenotype via mediation of gene expression. Assume that mediation is an endogenous variable. The observed correlation between methylation and gene expression may be due to variation of the hidden genotype variable. Genotype and micro RNA may create correlation between gene expression and phenotype which may be interpreted as the causal effect of the gene expression on the phenotype. This implies that the mediation model represented in Figure 7.9 is unidentified.

Instrument variable methods are often used to solve identification problems (Dippel et al. 2017; Peters et al. 2017). Consider three observed variables $X$ (methylation), $Y$ (gene expression), and $Z$ (phenotype). In Figure 7.9, the variable $H_G$ that is an unobserved exogenous variable is called a general confounder and the variable $H_R$ that is caused by $X$ and causes observed mediator $Y$, is called the unobserved mediator. Consider five independent error variables $\varepsilon_X, \varepsilon_Y, \varepsilon_Z, \varepsilon_{H_G}$, and $\varepsilon_{H_R}$. Assume that a mediation model is given by

$$H_G = f_{H_G}(\varepsilon_{H_G}),$$
$$X = f_X(H_G, \varepsilon_X),$$
$$H_R = f_{H_R}(X, \varepsilon_{H_R}),$$
$$Y = f_Y(X, H_R, H_G, \varepsilon_Y),$$
$$Z = f_Z(X, Y, H_R, H_G, \varepsilon_Z)$$

(7.226)

Clearly, unobserved confounder $H_G$ creates correlations between the variable $X$ and mediator $Y$, and outcome $Z$. Unobserved confounder $H_G$ and the unobserved mediator $H_R$ generates correlation between the mediator $Y$ and the outcome $Z$.

For the convenience of presentation, we first introduce a standard instrument variable (IV) model as shown in Figure 7.10a where $I$ denotes IV, $X$ denotes a causal variable, $Z$ denotes effect variable, and $H_G$ denotes hidden confounder. The conditions for a variable to be IV are (Peters et al. 2017)

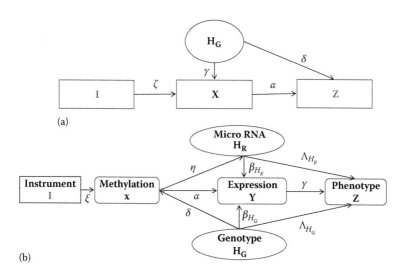

(a)

(b)

**FIGURE 7.10**

Instrument variable and mediation model. (a) A standard instrument variable (IV) model. (b) An instrument variable for the mediation causal model with unobserved confounder and unobserved mediator.

1. $I$ is independent of $H_G$,
2. $I$ is correlated with $X$,
3. $I$ affects $Y$ only through $X$.

The structural equation model for Figure 7.10a is given by

$$X = \xi I + \gamma H_G + N_X, \tag{7.227}$$

$$Z = \alpha X + \delta H_G + N_Z. \tag{7.228}$$

To estimate the direct effect of $X$ on $Z$, using the assumption of uncorrelation between IV and confounder $H_G$, we take covariance with IV $I$ on both sides of Equation 7.228 gives

$$\text{cov}(Z, I) = \alpha \text{cov}(X, I),$$

which implies that

$$\hat{\alpha} = \frac{\text{cov}(Z, I)}{\text{cov}(X, I)}. \tag{7.229}$$

Next, we introduce IV to the mediation causal model with unobserved confounder and unobserved mediator (Figure 7.10b). For simplicity, we consider the following linear mediation model with IV, confounding and unobserved mediator:

$$H_G = N_G, \tag{7.230}$$

$$X = \xi I + \delta H_G + N_X, \tag{7.231}$$

$$H_R = \eta X + N_{H_R}, \tag{7.332}$$

$$Y = \alpha X + \beta_{H_R} H_R + \beta_{H_G} H_G + N_Y, \tag{7.333}$$

$$Z = \pi X + \gamma Y + \Lambda_{H_G} H_G + \Lambda_{H_R} H_R + N_Z, \tag{7.334}$$

where the IV $I$ are uncorrelated with confounder $H_G$, unmeasured mediator $H_R$, the error variables $N_G, N_X, N_{H_R}, N_Y,$ and $N_Z$ are independent.

Similar to the approach in Dippel et al. (2017), we can estimate the effect of methylation $X$ on phenotype $Z$ using the two-stage-least-square estimation method and IV. The first and second stage regressions are defined as

$$X = C_x + \beta I + \varepsilon_x, \tag{7.335}$$

$$Z = C_Z + \alpha X + \varepsilon_Z. \tag{7.336}$$

Taking the covariance of both sides with $I$ gives

$$\hat{\alpha} = \frac{\text{cov}(Z, I)}{\text{cov}(X, I)}. \tag{7.337}$$

In Appendix 7.J, we show that the regression coefficient $\hat{\alpha}$ is equal to the effect of $X$ on $Z$ where

$$\hat{\alpha} = \pi + \gamma(\alpha + \beta_{HR}\eta) + \Lambda_{H_R}\eta. \tag{7.338}$$

Next the two-stage least-square estimator of the effect of methylation $X$ on gene expression $Y$ is defined by the following regression equations:

$$X = C_x + \beta I + \varepsilon_x, \tag{7.339}$$

$$Y = C_y + \delta_Y X + \varepsilon_Y. \tag{7.340}$$

Using similar arguments as before, we obtain the estimator $\hat{\delta}_Y$ from regression (7.340):

$$\hat{\delta}_Y = \frac{\text{cov}(Y, I)}{\text{cov}(X, I)}. \tag{7.341}$$

In Appendix 7.J, we show that it follows from Equation 7.332 that the estimator $\hat{\delta}_Y$ can be expressed as

$$\hat{\delta}_Y = \alpha + \beta_{H_R}\eta, \tag{7.342}$$

which is equal to the effect of $X$ on $Y$.

Finally, we estimate the effect of $Y$ (gene expression) on $Z$ (phenotype) conditional on $X$ (methylation). The two-stage least-square estimator is defined by the following two equations:

$$Y = C_{Y|X} + \beta_1 I + \beta_2 X + \varepsilon_{Y|X}, \tag{7.343}$$

$$Z = C_{Z|X} + \alpha_1 Y + \alpha_2 X + \varepsilon_{Z|X}. \tag{7.344}$$

The two-stage least-square estimators $\alpha_1$ and $\alpha_2$ are, respectively, given by

$$\alpha_1 = \frac{\text{cov}(Z, I)\text{cov}(X, X) - \text{cov}(Z, X)\text{cov}(X, Z)}{\Delta}, \tag{7.345}$$

$$\alpha_2 = \frac{\text{cov}(Y, I)\text{cov}(Z, X) - \text{cov}(Y, X)\text{cov}(Z, I)}{\Delta}, \tag{7.346}$$

where $\Delta = \text{cov}(Y, I)\text{cov}(X, X) - \text{cov}(X, I)\text{cov}(Y, X)$.

In Appendix 7.J, we show that the two-stage least-square estimators $\alpha_1$ and $\alpha_2$ can be, respectively, expressed as

$$\hat{\alpha}_1 = \frac{\gamma \beta_{H_G} + \Lambda_{H_G}}{\beta_{H_G}},$$ (7.347)

and

$$\hat{\alpha}_2 = \pi + \eta \Lambda_{H_R} - \frac{\left(\alpha + \eta \beta_{H_R}\right) \Lambda_{H_G}}{\beta_{H_G}}.$$ (7.348)

Clearly, if $\Lambda_{H_G} = 0$, then $\hat{\alpha}_1 = \alpha_1^*$ and $\hat{\alpha}_2 = \alpha_2^*$, and the two-stage least-square-estimators converge to the effect of $Y$ (gene expression) on $Z$ (phenotype) conditional on $X$ (methylation).

## 7.3.3 Confounders with Additive Noise Models

### 7.3.3.1 Models

In the previous sections, we consider additive noise models

$$Y = f(X) + N_X \text{ or}$$
$$X = g(Y) + N_Y$$

for inferring whether $X \rightarrow Y$ or $Y \rightarrow X$, assuming no confounders. Now we consider the third class of causal models in the presence of confounder: $X \leftarrow Z \rightarrow Y$. Janzing et al. (2009) proposed the following additive noise models for confounders (ANMC):

$$X = f(Z) + N_X$$
$$Y = g(Z) + N_Y,$$ (7.349)

where $X, Y$, and $Z$ are random variables taking real values, $f$ and $g$ are continuously differentiable functions, $N_X$, $N_Y$ are real-valued random variables, and $N_X$, $N_Y$, $Z$ are jointly independent.

Now we consider several special cases of the ANMC (Equation 7.349).

Case 1: Assume that $N_X = 0$ and function $f$ is invertible.

In this case, we have

$$Z = f^{-1}(X) = h(x).$$ (7.350)

Substituting Equation 7.350 into the second equation in Equation 7.349, we obtain

$$Y = g(h(X)) + N_Y$$
$$= r(X) + N_Y. \tag{7.351}$$

In the ANMC, we assume that $X$ and $Y$ are independent. Therefore, we infer $X \rightarrow Y$.

**Case 2:** Assume that $N_Y = 0$ and function $g$ is invertible.

By the similar argument, we can infer $Y \rightarrow X$.

Next we introduce the algorithm for fitting the ANMC to the data and identifying confounder proposed by Janzing et al. (2009). The algorithm for fitting the ANMC to the data consists of two parts. The first part is to find the common confounder $T$ such that $N_X$ and $N_Y$, $N_X$ and $Z$, $N_Y$ and $Z$ are as independent as possible, assuming that the functions $f$ and $g$ are known. The second part is to find functions $f$ and $g$ via nonlinear regressions of $X$ on $Z$ and $Y$ on $Z$, respectively.

### 7.3.3.2 Methods for Searching Common Confounder

Let DEP be a dependence measure, $f$ and $g$ be known functions. For a common confounder $\hat{Z}$, the residuals are given by

$$\hat{N}_{X,k} = X_k - \hat{f}(Z), \tag{7.352}$$

$$\hat{N}_{Y,k} = Y_k - \hat{g}(Z). \tag{7.353}$$

The first part for finding the common confounder $\hat{Z}$ can be formulated as solving the following optimization problem:

$$\min_Z \text{Dep}\left(\hat{N}_X, \hat{N}_Y\right) + \text{Dep}\left(\hat{N}_X, Z\right) + \text{Dep}\left(\hat{N}_Y, Z\right). \tag{7.354}$$

There are a number of dependence measures that can be used for solving problem (7.354). Similar to the ANMs, we can use the Hilbert–Schmidt independence criterion (HSIC) as a dependence measure (Janzing et al. 2009) and transform problem (7.354) as the following optimization problem:

$$\min_Z \text{HSIC}\left(\hat{N}_X, \hat{N}_Y\right) + \text{HSIC}\left(\hat{N}_X, \hat{Z}\right) + \text{HSIC}\left(\hat{N}_Y, \hat{Z}\right). \tag{7.355}$$

We can use a biased estimator of HSIC (for details, please see Section 5.3.2.3) to calculate HSIC in problem (7.355).

Define Gaussian radial basis function kernels for $\hat{N}_X, \hat{N}_Y$ and $\hat{Z}$:

$$K_{\hat{N}_X}\left(\hat{N}_{X,k}, \hat{N}_{X,l}\right) = \exp\left(-\frac{\|\hat{N}_{X,k} - \hat{N}_{X,l}\|_2^2}{2\sigma_{\hat{N}_X}^2}\right), \tag{7.356}$$

$$K_{\hat{N}_Y}\left(\hat{N}_{Y,k}, \hat{N}_{Y,l}\right) = \exp\left(-\frac{\|\hat{N}_{Y,k} - \hat{N}_{Y,l}\|_2^2}{2\sigma_{\hat{N}_Y}^2}\right), \tag{7.357}$$

$$K_{\hat{Z}}\left(\hat{Z}_K, \hat{Z}_L\right) = \exp\left(-\frac{\|\hat{Z}_k - \hat{Z}_l\|_2^2}{2\sigma_{\hat{Z}}^2}\right). \tag{7.358}$$

The kernel matrices are given by

$$K_{\hat{N}_X} = \left(K_{\hat{N}_X}\left(\hat{N}_{X,k}, \hat{N}_{X,l}\right)\right)_{n \times n}, \tag{7.359}$$

$$K_{\hat{N}_Y} = \left(K_{\hat{N}_Y}\left(\hat{N}_{Y,k}, \hat{N}_{Y,l}\right)\right)_{n \times n}, \tag{7.360}$$

$$K_{\hat{Z}} = \left(K_{\hat{Z}}\left(\hat{Z}_K, \hat{Z}_l\right)\right)_{n \times n}. \tag{7.361}$$

Let $H = I_n - \frac{1}{n}\mathbf{1}\mathbf{1}^T$ be a centering matrix that centers the rows or columns, where $\mathbf{1}$ is a vector of 1s.

Define

$$\text{HSIC}\left(\hat{N}_X, \hat{N}_Y\right) = \frac{1}{n^2}\,\text{tr}\left(K_{\hat{N}_X} H K_{\hat{N}_Y} H\right), \tag{7.362}$$

$$\text{HSIC}\left(\hat{N}_X, \hat{Z}\right) = \frac{1}{n^2}\,\text{tr}\left(K_{\hat{N}_X} H K_{\hat{Z}} H\right), \tag{7.363}$$

$$\text{HSIC}\left(\hat{N}_Y, \hat{Z}\right) = \frac{1}{n^2}\,\text{tr}\left(K_{\hat{N}_Y} H K_{\hat{Z}} H\right). \tag{7.364}$$

Substituting Equations 7.362–7.364 into Equation 7.355, we obtain the following optimization problem:

$$\min_Z \frac{1}{n^2}\,\text{tr}\left(K_{\hat{N}_X} H K_{\hat{N}_Y} H + K_{\hat{N}_X} H K_{\hat{Z}} H + K_{\hat{N}_Y} H K_{\hat{Z}} H\right). \tag{7.365}$$

### 7.3.3.3 Gaussian Process Regression

A key issue of the ANMC is to learn unknown functions in nonlinear regression (Rasmussen and Williams 2006; Janzing et al. 2009). A powerful tool for nonlinear regression is a Gaussian process regression. A Gaussian process is a distribution over functions and is defined as a collection of random variables, any finite number of which have joint Gaussian distributions. We begin Gaussian process regression with reviewing the Bayesian analysis of the linear regression model. Consider the linear regression model with Gaussian noise (Rasmussen and Williams 2006):

$$y = f(x) + e,$$
$$f(x) = x^T w, \tag{7.366}$$

where $x$ is a$q$-dimensional input vector, $w$ is a $q$-dimensional vector of weights, $f$ is the function value, $y$ is the observed scalar response value, and the noise $e$ is independently and identically distributed Gaussian distribution $N(0, \sigma_n^2)$. Consider a training dataset $D = \{(x_i, y_i), i = 1, \ldots, n\}$.

The likelihood function for the dataset $D$ is

$$l(y|X, w) = \frac{1}{(2\pi\sigma_n^2)^{n/2}} \exp\left\{ -\frac{1}{2\sigma_n^2} (y - X^T w)^T (y - X^T w) \right\}, \tag{7.367}$$

where $y = [y_1, \ldots, y_n]^T$, $X = [x_1, \ldots, x_n]$ and $w = [w_1, \ldots, w_q]$.

We assume that a prior over the weights is a Gaussian distribution with a zero mean and covariance matrix $\Sigma_q$, i.e.,

$$w \sim N\left(0, \Sigma_q\right). \tag{7.368}$$

The posterior distribution over the weights is proportional to the product of the likelihood and the prior:

$$p(w|X, y) \propto l(y|X, w)p(w)$$

$$\propto \exp\left\{ -\frac{1}{2\sigma_n^2} (y - X^T w)^T (y - X^T w) - \frac{1}{2} w^T \Sigma_q^{-1} w \right\}$$

$$\propto \exp\left\{ -\frac{1}{2} \left[ w^T \left( \frac{XX^T}{\sigma_n^2} + \Sigma_q^{-1} \right) w - \frac{1}{\sigma_n^2} (y^T X^T w + w^T Xy) + \frac{y^T y}{\sigma_n^2} \right] \right\}. \tag{7.369}$$

Consider the equation

$$(w - \mu)^T \left( \frac{XX^T}{\sigma_n^2} + \Sigma_q^{-1} \right)(w - \mu) = w^T \left( \frac{XX^T}{\sigma_n^2} + \Sigma_q^{-1} \right) w - \frac{1}{\sigma_n^2} \left( y^T X^T w + w^T Xy \right)$$

$$+ \mu^T \left( \frac{XX^T}{\sigma_n^2} + \Sigma_q^{-1} \right) \mu.$$

(7.370)

Solving Equation 7.370 for $\mu$, we obtain

$$\mu = \frac{1}{\sigma_n^2} \left( \frac{XX^T}{\sigma_n^2} + \Sigma_q^{-1} \right)^{-1} Xy.$$

(7.371)

Substituting Equation 7.371 into Equation 7.369 gives the posterior distribution

$$p(w|X,y) \propto \exp \left\{ -\frac{1}{2}(w - \mu)^T \left( \frac{XX^T}{\sigma_n^2} + \Sigma_q^{-1} \right)(w - \mu) \right\}.$$

(7.372)

Equation 7.372 can be written as

$$p(w|X,y) \sim N\left(\mu, \Lambda^{-1}\right),$$

(7.373)

where

$$\Lambda = \frac{XX^T}{\sigma_n^2} + \Sigma_q^{-1}.$$

(7.374)

Now we investigate the distribution of the prediction. Suppose that $x_{new}$ is a new input dataset. The distribution of the prediction $y_{new}$ can be obtained by averaging all possible output values of the linear models over the Gaussian posterior using Equations 7.367 and 7.373 (Appendix 7.J):

$$p(y_{new}|x_{new}, X, y) = \int p(y_{new}|x_{new}, w)p(w|X,y)dw$$

$$= N\left( \frac{1}{\sigma_n^2} x_{new}^T \Lambda^{-1} Xy, x_{new}^T \Lambda^{-1} x_{new} \right)$$

(7.375)

To expand input space, the data can be mapped to high-dimensional feature space. Let $\phi(x)$ be a feature map from a $q$-dimensional input space to $D$-dimensional feature space. Let $\Phi(x) = [\varphi(x_1),\ldots,\varphi(x_n)]$.

In the feature space, the linear model is given by

$$f(x) = \varphi(x)^T w.$$

(7.376)

In the feature space, similar to Equation 7.375, the distribution of prediction is given by

$$p(y_{new}|x_{new}, X, y) = N\left(\frac{1}{\sigma_n^2} \varphi(x_{new})^T \Lambda^{-1} \Phi(x) y, \varphi(x_{new})^T \Lambda^{-1} \varphi(x_{new})\right), \quad (7.377)$$

where

$$\Lambda = \frac{\Phi(x)\Phi(x)^T}{\sigma_n^2} + \Sigma_q^{-1}. \quad (7.378)$$

Multiplying both sides of Equation 7.378 by $\Sigma_q$ from the right, we obtain

$$\Lambda\Sigma_q = \frac{\Phi(x)\Phi(x)^T \Sigma_q}{\sigma_n^2} + I. \quad (7.379)$$

Let $K = \Phi(x)^T \Sigma_q \Phi(x)$. Again, multiplying both sides of Equation 7.379 by $\Phi(x)$ from the right yields

$$\Lambda\Sigma_q\Phi(x) = \frac{\Phi(x)\Phi(x)^T \Sigma_q \Phi(x)}{\sigma_n^2} + \Phi(x)$$

$$= \frac{1}{\sigma_n^2} \Phi(x)(K + \sigma_n^2 I). \quad (7.380)$$

Multiplying both sides of Equation 7.381 by $(K + \sigma_n^2 I)^{-1}$ from the right gives

$$\Lambda\Sigma_q\Phi(x)(K + \sigma_n^2 I)^{-1} = \frac{1}{\sigma_n^2} \Phi(x). \quad (7.381)$$

Finally, multiplying both sides of Equation 7.381 by $\Lambda^{-1}$ from the left, we obtain

$$\Sigma_q\Phi(x)(K + \sigma_n^2 I)^{-1} = \frac{1}{\sigma_n^2} \Lambda^{-1}\Phi(x). \quad (7.382)$$

Substituting Equation 7.382 into Equation 7.377, we obtain the new expression of the mean of the distribution of precision:

$$\frac{1}{\sigma_n^2} \varphi(x_{new})^T \Lambda^{-1}\Phi(x) y = \varphi(x_{new})^T \Sigma_q (K + \sigma_n^2 I)^{-1} y. \quad (7.383)$$

The inverse of matrix $\Lambda$ is

$$\Lambda^{-1} - \Sigma_q - \Sigma_q\Phi(x)\left(\sigma_n^2 I + \Phi(x)^T \Sigma_q \Phi(x)\right)^{-1}\Phi(x)^T \Sigma_q. \quad (7.384)$$

Substituting Equation 7.384 into Equation 7.377, we obtain the new expression of the covariance of the distribution of prediction:

$$\varphi(x_{new})^T \Lambda^{-1} \varphi(x_{new})$$

$$= \varphi(x_{new})^T \Sigma_q \varphi(x_{new}) - \varphi(x_{new})^T \Sigma_q \Phi(x)(K + \sigma_n^2 I) \Phi(x)^T \Sigma_q \varphi(x_{new}). \qquad (7.385)$$

Summarizing Equations 7.383 and 7.385, we can rewrite the distribution of prediction as

$$p(y_{new}|x_{new}, X, y) = N(\varphi(x_{new})^T \Sigma_q (K + \sigma_n^2 I)^{-1} y,$$

$$\varphi(x_{new})^T \Sigma_q \varphi(x_{new}) - \varphi(x_{new})^T \Sigma_q \Phi(x)(K + \sigma_n^2 I)\Phi(x)^T \Sigma_q \varphi(x_{new}))$$

$$(7.386)$$

It is well known that the inner products in the feature space are often computed by the kernel function. We define kernel matrices

$$K(X, X) = \Phi(x)^T \Sigma_q \Phi(x), \qquad (7.387)$$

$$K(X, X_{new}) = \Phi(x)^T \Sigma_q \Phi(x_{new}), \qquad (7.388)$$

$$K(X_{new}, X) = \Phi(x_{new})^T \Sigma_q \Phi(x), \qquad (7.389)$$

$$K(X_{new}, X_{new}) = \Phi(x_{new})^T \Sigma_q \Phi(x_{new}), \qquad (7.390)$$

where $\Phi(x_{new}) = [\varphi(x_{new1}), \dots, \varphi(x_{newl})]$.

Now we establish the relationships between a Bayesian linear model and a Gaussian process (Rasmussen and Williams 2006). Consider the Bayesian linear regression model $f(x) = \varphi(x)^T w$ with prior $w \sim N(0, \Sigma_q)$. The value of the function defines a Gaussian process with mean and covariance

$$E[f(x)] = \varphi^T e[w] = 0,$$

$$\text{cov}(f(x), f(x')) = E[f(x)f(x')] = \varphi(x)^T E[ww^T]\varphi(x') = \varphi(x)^T \Sigma_q \varphi(x') = k(x, x').$$

$$(7.391)$$

Consider the train data set $(X, f(X))$ and $(X_{new}, f(X_{new}))$. Assume that the observations are noise free. The joint distribution of the training outputs and test outputs under the prior is

$$\begin{bmatrix} f(X) \\ f(X_{new}) \end{bmatrix} \sim N\left( 0, \begin{bmatrix} K(X, X) & K(X, X_{new}) \\ K(X_{new}, X) & K(X_{new}, X_{new}) \end{bmatrix} \right). \qquad (7.392)$$

The distribution of prediction $f(X_{new})$ can be viewed as the conditional distribution of prediction $f(X_{new})$, given the training and test datasets:

$$p(f(X_{new})|X_{new}, X, f(X)) =$$

$$N\left( K(X_{new}, X)K(X, X)^{-1}f(X),\ K(X_{new}, X_{new}) - K(X_{new}, X)K(X, X)^{-1}K(X, X_{new}) \right).$$

$$(7.393)$$

Now consider the noise observations that are modeled by $y = f(x) + \varepsilon$ where $\varepsilon$ is distributed as $N(0, \sigma_n^2 I)$. Then, we have

$$\text{cov}(y, y) = K(X, X) + \sigma_n^2 I. \qquad (7.394)$$

The joint distribution of the training outputs and test outputs under the prior given in Equation 7.392 is changed to

$$\begin{bmatrix} f(X) \\ f(X_{new}) \end{bmatrix} \sim N\left( 0, \begin{bmatrix} K(X, X) + \sigma_n^2 I & K(X, X_{new}) \\ K(X_{new}, X) & K(X_{new}, X_{new}) \end{bmatrix} \right). \qquad (7.395)$$

Similarly, the conditional distribution of prediction $f(X_{new})$, given the noise training and test datasets, is changed to

$$p(f(X_{new})|X_{new}, X, f(X)) = N(K(X_{new}, X)[K(X, X) + \sigma_n^2 I]^{-1}f(X),$$

$$K(X_{new}, X_{new}) - K(X_{new}, X)K(X, X)^{-1}K(X, X_{new})) \qquad (7.396)$$

The kernel functions are defined in Equations 7.387–7.390. The mean and variance of the estimator of prediction, given $x_{new}$ are, respectively, given by

$$\bar{y}_{new} = k(x_{new}, X)^T \left[ K(X, X) + \sigma_n^2 I \right]^{-1} y, \qquad (7.397)$$

and

$$\text{var}(\bar{y}_{new}) = k(x_{new}, x_{new}) - k(x_{new}, X)^T \left[ K(X, X) + \sigma_n^2 I \right]^{-1} k(x_{new}, X). \qquad (7.398)$$

Let $\alpha = [K(X, X) + \sigma_n^2 I]^{-1} y$. Then, Equation 7.397 can be rewritten as

$$\bar{y}_{new} = k(x_{new}, X)^T \alpha = \sum_{i=1}^{n} \alpha_i k(x_{new}, x_i), \qquad (7.399)$$

where $a_i$ can be taken as parameters of the Gaussian process regression and $k$ $(x_{new}, x_i)$ as the elements of the design matrix. Recall the model

$$y = f(X) + \varepsilon. \qquad (7.400)$$

From Equations 7.392 and 7.395, we obtain the marginal likelihood functions for $f \mid X$ and $y \mid X$:

$$p(f|X) = \frac{1}{(2\pi)^{n/2}|K(X,X)|^{1/2}} \exp\left\{-\frac{1}{2}f^T K(X,X)^{-1}f\right\}, \tag{7.401}$$

$$p(y|X) = \frac{1}{(2\pi)^{n/2}|K(X,X)+\sigma_n^2 I|^{1/2}} \exp\left\{-\frac{1}{2}f^T \left[K(X,X)+\sigma_n^2 I\right]^{-1}f\right\}. \tag{7.402}$$

Summarizing Equations 7.399 and 7.402, we obtain the following result.

### Result 7.27: Gaussian Process Regression

**Step 1.** Define Gaussian kernel matrix $K(X, X)$ and calculate initial variance of noise:

$$(\hat{\sigma}_n^2)^{(0)} = \frac{1}{n-q}\sum_{i=1}^{n}(y_i - \bar{y})^2, \ \ \bar{y} = \frac{1}{n}\sum_{i=1}^{n}y_i. \text{ Let } t = 0.$$

**Step 2.** Estimate regression coefficients: $\hat{\alpha} = (K(X,X) + (\hat{\sigma}_n^2)^{(t)}I)^{-1}y$.

**Step 3.** Calculate the prediction value, given $x_{new}$: $\bar{y}_{new} = k(x_{new}, X)^T\hat{\alpha}$ and noise $\varepsilon_{new} = y_{new} - \bar{y}_{new}$.

**Step 4.** Calculate the covariance matrix of the estimator of prediction value and predictive distribution for the test data $y = y_{new}$:

$$V(\hat{y}_{new}) = K(X_{new}, X_{new}) - K(X_{new}, X)K(X,X)^{-1}K(X, X_{new}),$$

$$p(y|X) = \frac{1}{(2\pi)^{n/2}|K(X,X)+\sigma_n^2 I|^{1/2}} \exp\left\{-\frac{1}{2}y^T \left[K(X,X)+\sigma_n^2 I\right]^{-1}y\right\}.$$

**Step 5.** Calculate the noise variance for the test dataset.

$$\left(\hat{\sigma}_n^2\right)^{(t)} = \frac{1}{m-q}\sum_{i=1}^{m}\left(y_{new_i} - k(x_{new_i}, X)^T\hat{\alpha})^2\right).$$

**Step 6.** Check convergence.

Let $\varepsilon$ be a prespecified error.

If $|(\hat{\sigma}_n^2)^{(t)} - (\hat{\sigma}_n^2)^{(t-1)}| < \varepsilon$ then stop. Otherwise, go to Step 2.

If cholesky decomposition is used for calculation of inverse of matrix, then Result 7.27 is reduced to Result 7.28 (Rasmussen and Williams 2006).

### Result 7.28: Gaussian Process Regression with Cholesky Decomposition

**Step 1.** Define Gaussian kernel matrix $K(X, X)$ and calculate initial variance of noise:

$$(\hat{\sigma}_n^2)^{(0)} = \frac{1}{n-q}\sum_{i=1}^{n}(y_i - \bar{y})^2, \ \bar{y} = \frac{1}{n}\sum_{i=1}^{n}y_i. \text{ Let } t = 0.$$

**Step 2.** Estimate regression coefficients.

Cholesky decomposition: $(K(X, X) + (\hat{\sigma}_n^2)^{(t)} I) = LL^T$.

Solve the equation $Lz = y$ for $z$.

Solve the equation $L^T a = z$ for $a$.

**Step 3.** Calculate the prediction value, given $x_{new}$: $\bar{y}_{new} = k(x_{new}, X)^T \hat{\alpha}$ and noise $\varepsilon_{new} = y_{new} - \bar{y}_{new}$.

**Step 4.** Calculate the covariance matrix of the estimator of the prediction value and marginal likelihood.

Solve the equation $L\gamma = K(X, X_{new})$ for $\gamma$.

Calculate the variance:

$$V(\hat{y}_{new}) = K(X_{new}, X_{new}) - \gamma^T \gamma.$$

Calculate the predictive distribution for noise test data $y = y_{new}$.

$$p(y|X) = \frac{1}{(2\pi)^{n/2}|L|} \exp\left\{-\frac{1}{2} y^T \hat{\alpha}\right\}.$$

**Step 5.** Calculate noise variance for the test dataset.

$$\left(\hat{\sigma}_n^2\right)^{(t)} = \frac{1}{m-q} \sum_{i=1}^{m} \left(y_{new_i} - k(x_{new_i}, X)^T \hat{\alpha}\right)^2.$$

**Step 6.** Check convergence.

Let $\varepsilon$ be a prespecified error.

If $|(\hat{\sigma}_n^2)^{(t)} - (\hat{\sigma}_n^2)^{(t-1)}| < \varepsilon$ then stop. Otherwise, go to Step 2.


### 7.3.3.4 Algorithm for Confounder Identification Using Additive Noise Models for Confounder

Now we are ready to introduce algorithm for confounder identification using ANMC (Janzing et al. 2009). Suppose that the data are divided into the training dataset $D_{train} = \{x_i, y_i, i = 1, \dots, n\}$ and test dataset $D_{test} = \{x_{new_j}, y_{new_j}, j = 1, \dots, m\}$. Before we study the algorithm for confounder discovery, we introduce the algorithms for initialization.

#### Result 7.29: Initialization

**Step 1.** Using the Isomap algorithm (Izenman 2008), we found the initial guest values $z_k^{(0)}, k = 1, \dots, n + m$. Let $Z_{train}^{(0)} = [z_1^{(0)}, \dots, z_n^{(0)}]$ and $Z_{test}^{(0)} = [z_{n+1}^{(0)}, \dots, z_{n+m}^{(0)}]$. Set $t = 0$.

**Step 2.** For fixed $z_k^{(t)}$, using Result 7.27 (or Result 7.28), we regress $x$ on $z$ and $y$ on $z$ to obtain

$\hat{x} = f(z_{test}^{(t)}) = K(z_{test}^{(t)}, Z_{Train}^{(t)})^T \hat{\alpha}^{(t)}$, where $\hat{\alpha}^{(t)} = (K(Z_{train}^{(t)}, Z_{train}^{(t)}) + (\hat{\sigma}_n^2)^{(t)} I)^{-1} x_{train}$

and

$\hat{y} = g(z_{test}^{(t)}) = K(z_{test}^{(t)}, Z_{Train}^{(t)})^T \hat{\alpha}^{(t)}$, where $\hat{\beta}^{(t)} = (K(Z_{train}^{(t)}, Z_{train}^{(t)}) + (\hat{\sigma}_n^2)^{(t)} I)^{-1}$

$y_{train}$.

**Step 3.** The curves $f(z)$ and $g(z)$ are fixed. Each data point $(x_k, y_k), k = 1, \ldots,$ $n+m$ is projected to the nearest point of the curve to find $z_k^{(t+1)}$ such that

$$z_k^{(t+1)} = \arg\min_{z^t} \|(f\left(z_{test}^{(t)}\right), g\left(z_{test}^{(t)}\right) - (x_k, y_k)\|_2 .$$

**Step 4.** Check convergence.

If $|z_k^{(t+1)} - z_k^{(t)}| < \varepsilon$ then stop and $z_k \leftarrow z_k^{(t+1)}$; otherwise, $z_k^{(t)} \leftarrow z_k^{(t+1)}$, go to Step 2.

The algorithm for confounder identification using the ANMC is summarized in the following result.

### Result 7.30: Algorithm for Confounder Identification Using the ANMC

**Step 1.** Initialization. The algorithm for initialization is specified in Result 7.29.

The set of estimated values $\{z_i, i = 1, \ldots, n + m\}$ will be used as the initial values for minimization in the projection below.

**Step 2.** Repeat

**Step 3.** Projection.

Use direct search methods such as the Nelder–Mead method (Lagarias et al. 1998) to solve the unconstrained optimization problem:

$$\min_{Z} \frac{1}{n^2} \text{tr}(K_{\hat{N}_X} H K_{\hat{N}_Y} H + K_{\hat{N}_X} H K_{\hat{Z}} H + K_{\hat{N}_Y} H K_{\hat{Z}} H),$$

where kernels are defined in Section 7.3.3.2, the noises $N_X$ and $N_Y$ are calculated by $\hat{N}_X = x - f(z)$ and $\hat{N}_Y = y - g(z)$, and $f(z)$ and $g(z)$ are calculated by the Gaussian process regression (Result 7.27 or Result 7.28).

**Step 4.** Test independence.

If $\hat{N}_X, \hat{N}_Y$, and $\hat{Z}$ are pair-wise independent, then stop; the confounder is found. Output the values of the confounder at the test data points: $\hat{z}_{n+1}$, $\ldots, \hat{z}_{n+m}$ and functions $f(\hat{z}_j)$ and $g(\hat{z}_j)$; otherwise, go to Step 5.

**Step 5.** Gaussian process regression.

Using Result 7.27 or 7.28 and the currently estimated confounder $\hat{Z}_*$, we perform regression:

$$x = f(z) + N_X \quad \text{and} \quad y = g(z) + N_Y.$$

Set $\hat{Z}^{(0)} = \hat{Z}_*$ and go to Step 3.

until $T$ (a prespecified number) iterations. Stop. This demonstrates that the data cannot be fitted by the additive noise models for the confounder.

## Software Package

A complete MATLAB® code package for implementing causal discovery algorithm based on linear non-Gaussian acyclic models can be downloaded from https://sites.google.com/site/sshimizu06/lingam.

A code package implementing the algorithm for testing linear causal models can be downloaded from http://webdav.tuebingen.mpg.de/causality/.

Package "CAM" that fits a causal additive model (CAM) for estimating the causal structure of the underlying process can be downloaded from https://cran.r project.org/web/packages/CAM/CAM.pdf. The pcalg package for R that can be used for the causal structure learning and estimation of causal effects from observational and/or interventional data can be downloaded from https://cran.r project.org/web/packages/pcalg/index.html.

## Appendix 7.A    Approximation of Log-Likelihood Ratio for the LiNGAM

Derivation of approximation follows the approach of Hyvärinen and Smith (2013). Consider the causal model $X \rightarrow Y$ in (7.3)

$$
\begin{aligned}
X &= \alpha u + \varepsilon_x \\
Y &= \rho X + \beta u + \varepsilon_y.
\end{aligned}
\tag{7.A.1}
$$

First, we standardize the variable $\varepsilon_Y$. It follows from Equation 7.A.1 that

$$
\varepsilon_Y = Y - \rho X - \beta u,
\tag{7.A.2}
$$

which implies

$$
\mathrm{Var}(\varepsilon_Y) = \mathrm{Var}(Y) + \mathrm{Var}(\rho X) - 2\mathrm{cov}(Y, \rho X).
\tag{7.A.3}
$$

By assumption that the variables $X$ and $Y$ are standardized, we have

$$
\begin{aligned}
\mathrm{Var}(X) &= 1 \\
\mathrm{Var}(Y) &= 1 \\
\mathrm{Cov}(X, Y) &= \rho\,\mathrm{Var}(X) = \rho.
\end{aligned}
\tag{7.A.4}
$$

Substituting Equation 7.A.4 into Equation 7.A.3 gives

$$
\begin{aligned}
\mathrm{Var}(\varepsilon_Y) &= \mathrm{Var}(Y) + \rho^2 \mathrm{Var}(X) - 2\rho\,\mathrm{cov}(Y, X) \\
&= 1 + \rho^2 - 2\rho^2 = 1 - \rho^2.
\end{aligned}
\tag{7.A.5}
$$

Define the standardized $\varepsilon_Y$ as

$$
d = \frac{\varepsilon_Y}{\sqrt{1 - \rho^2}}.
\tag{7.A.6}
$$

Let $P_x(u)$ be the density function of the random variable $X$ and $P_d(u)$ be the density function of the variable $d$. It follows from Equation 7.A.1 that the density function $\varepsilon_X$ is equal to the density function $P_x(u)$ of the variable $X$. It is clear from Equation 7.A.6 that

$$P(\varepsilon_Y) = \frac{P_d(u)}{\sqrt{1-\rho^2}}. \qquad (7.A.7)$$

Define $G_x(u) = \log P_x(u)$ and $G_d(u) = \log P_d(u)$. The joint density function of the random variables $X$ and $Y$ is

$$\begin{aligned} P(x,y) &= P(\varepsilon_X, \varepsilon_Y) \\ &= P(\varepsilon_X)P(\varepsilon_Y). \end{aligned} \qquad (7.A.8)$$

Substituting Equation 7.A.7 into Equation 7.A.8 yields

$$P(x,y) = P_x(x) \frac{P_d\left(\dfrac{y - \rho x - \beta u}{\sqrt{1-\rho^2}}\right)}{\sqrt{1-\rho^2}}. \qquad (7.A.9)$$

Taking the logarithm on both sides of Equation 7.A.9, we obtain the log-likelihood of the LiNGAM with $X \to Y$ in Equation 7.A.1:

$$l(X \to Y) = \sum_{i=1}^{n}\left[ G_x(x_i) + G_d\left(\frac{y_i - \rho x_i - \beta u_i}{\sqrt{1-\rho^2}}\right)\right] - \frac{n}{2}\log(1-\rho^2). \qquad (7.A.10)$$

Recall that the second potential causal model $Y \to X$ in Equation 7.4 is

$$\begin{aligned} Y &= \beta u + \varepsilon_y \\ X &= \rho Y + \alpha u + \varepsilon_x. \end{aligned} \qquad (7.A.11)$$

Define

$$e = \frac{X - \rho Y - \alpha u}{\sqrt{1-\rho^2}}.$$

Let $P_y(u)$ be the density function of the random variable $Y$ and $P_e(u)$ be the density function of the variable $e$. Define $G_y(u) = \log P_y(u)$ and $G_e(u) = \log P_e(u)$. Similarly, we can obtain the log-likelihood of the LiNGAM with $X \leftarrow Y$ in Equation 7.A.11:

$$l(X \leftarrow Y) = \sum_{i=1}^{n}\left[ G_y(y_i) + G_e\left(\frac{x_i - \rho y_i - \alpha u_i}{\sqrt{1-\rho^2}}\right)\right] - \frac{n}{2}\log(1-\rho^2). \qquad (7.A.12)$$

Therefore, the average of the log likelihood ratio is given by

$$R = \frac{1}{n} l(X \to Y) - \frac{1}{n} l(Y \to X)$$

$$= \frac{1}{n} \sum_{i=1}^{n} \left[ G_x(x_i) + G_d\left( \frac{y_i - \rho x_i - \beta u_i}{\sqrt{1 - \rho^2}} \right) - G_y(y_i) G_e\left( \frac{x_i - \rho y_i - \alpha u_i}{\sqrt{1 - \rho^2}} \right) \right].$$

(7.A.13)

Taylor expansion can be used to approximate the log density function. A first-order approximation of $G_d(\frac{y_i - \rho x_i - \beta u_i}{\sqrt{1 - \rho^2}})$ is

$$G_d\left( \frac{y_i - \rho x_i - \beta u_i}{\sqrt{1 - \rho^2}} \right) = G(y) - (\rho x + \beta u) g(y) + O(\rho^2),$$

(7.A.14)

where $g$ is the derivative of $G$.

Similarly, we have

$$G_e\left( \frac{x_i - \rho y_i - \alpha u_i}{\sqrt{1 - \rho^2}} \right) = G(x) - (\rho y + \alpha u) g(x) + O(\rho^2).$$

(7.A.15)

Substituting Equations 7.A.14 and 7.A.15 into Equation 7.A.13 yields

$$R \approx \frac{1}{n} \sum_{i=1}^{n} [G(x_i) + G(y_i) - (\rho x_i + \beta u_i) g(y_i) - G(y_i) + (\rho y_i + \alpha u_i) g(x_i)]$$

$$= \frac{1}{n} \sum_{i=1}^{n} [-(\rho x_i - \beta u_i) g(y_i) + (\rho y_i + \alpha u_i) g(x_i)].$$

(7.A.16)

Hyvärinen and Smith (2013) used the following logistic density function to approximate the log pdf $G$:

$$G(z) = -2 \log \cosh \left( \frac{\pi}{2\sqrt{3}} z \right) + \text{const},$$

(7.A.17)

where

$$\cosh(z) = \frac{e^z + e^{-z}}{2}$$

If we assume that $\frac{\pi}{2\sqrt{3}} \approx 1$ and ignore the constant 2 in Equation 7.A.17, then the derivative of $G(z)$ can be approximated by

$$g(z) = \frac{dG(z)}{dz} \approx -\frac{e^z - e^{-z}}{e^z + e^{-z}} = -\tanh(z),$$

(7.A.18)

where $\tanh(z) = \dfrac{e^z - e^{-z}}{e^z + e^{-z}}$ is a hyperbolic tangent function.

Substituting Equation 7.A.18 into Equation 7.A.16, we obtain

$$
\begin{aligned}
R &\approx \frac{1}{n}\sum\nolimits_{i=1}^{n}[-(\rho x_i + \beta u_i)g(y_i) + (\rho y_i + \alpha u_i)g(x_i)] \\
&= \frac{1}{n}\sum\nolimits_{i=1}^{n}[(\rho x_i + \beta u_i)\tanh(y_i) - (\rho y_i + \alpha u_i)\tanh(x_i)] \\
&= \frac{1}{n}\sum\nolimits_{i=1}^{n}[\rho(x_i\tanh(y_i) - y_i\tanh(x_i)) + \beta u_i\tanh(y_i) - \alpha u_i\tanh(x_i)] \\
&= \rho\hat{E}[x\tanh(y) - y\tanh(x)] + \beta\hat{E}[u\tanh(y_i)] - \alpha\hat{E}[u\tanh(x)].
\end{aligned}
\tag{7.A.19}
$$

Define

$$
R_{NC} = \rho\hat{E}[x\tanh(y) - y\tanh(x)]
\tag{7.A.20}
$$

and

$$
R_C = \beta\hat{E}[u\tanh(y_i)] - \alpha\hat{E}[u\tanh(x)].
\tag{7.A.21}
$$

Then, the log-likelihood ratio can be approximated by

$$
R \approx R_{NC} + R_C.
\tag{7.A.22}
$$

Next, we discuss how to approximate the four log-pdfs $G_x$, $G_y$, $G_d$, and $G_e$. By central limit theorem, we have

$$
\frac{1}{n}\sum\nolimits_{i=1}^{n}G_x(x_i) \rightarrow E[G_x(x)] = \int P_x(x)\log P_x(x)dx = -H(x),
\tag{7.A.23}
$$

where $H$ denotes differential entropy.

Similarly, we have

$$
\frac{1}{n}\sum\nolimits_{i=1}^{n}G_d\left(\frac{y_i - \rho x_i - \beta u_i}{\sqrt{1 - \rho^2}}\right) \rightarrow -H\left(\frac{\hat{\varepsilon}_y}{\sigma_{\varepsilon_y}}\right),
\tag{7.A.24}
$$

where

$$
\varepsilon_Y = Y - \rho X - \beta u \text{ and } \sigma_{\varepsilon_y} = \sqrt{\operatorname{var}(\varepsilon_Y)}
$$

$$
\frac{1}{n}\sum\nolimits_{i=1}^{n}G_y(y_i) \rightarrow -H(y),
\tag{7.A.25}
$$

$$
\frac{1}{n}\sum\nolimits_{i=1}^{n}G_e\left(\frac{x_i - \rho y_i - \alpha u_i}{\sqrt{1 - \rho^2}}\right) \rightarrow -H\left(\frac{\hat{\varepsilon}_x}{\sigma_e}\right),
\tag{7.A.26}
$$

where

$$
\varepsilon_x = X - \rho Y - \alpha u \text{ and } \sigma_{\varepsilon_x} = \sqrt{\operatorname{var}(\varepsilon_x)}.
$$

Therefore, combining Equations 7.A.13 and 7.A.23–7.A.26 gives

$$R \rightarrow -H(x) - H\left(\frac{\hat{\varepsilon}_y}{\sigma_{\varepsilon_y}}\right) + H(y) + H\left(\frac{\hat{\varepsilon}_x}{\sigma_e}\right). \tag{7.A.27}$$

Define

$$\varepsilon_x^0 = X - \rho Y \text{ and } \varepsilon_y^0 = Y - \rho X.$$

Then, we have

$$\varepsilon_x = \varepsilon_x^0 - \alpha u \tag{7.A.28}$$

and

$$\varepsilon_y = \varepsilon_y^0 - \beta u. \tag{7.A.29}$$

Since the density functions of random variables $X$ and $Y$ are unknown, computing differential entropy is not easy in practice. One way to approximate the differential entropy (Hyvärinen 1998; Hyvärinen and Smith 2013) is

$$\tilde{H}(z) = H(u) - k_1 \left\{ E\left[\log \cosh(z)\right] - \gamma \right\}^2 - k_2 \left\{ E\left[ze^{-z^2/2}\right] \right\}^2, \tag{7.A.30}$$

where $H(u) = \dfrac{1}{2}(1 + 2\pi)$, $k_1 \approx 79.047$, $k_2 \approx 7.4129$, $\gamma \approx 0.37457$.

Taking the derivative of $\tilde{H}(z)$, we obtain

$$h(z) = \frac{d\tilde{H}(z)}{dz} = -2k_1 \left\{ E\left[\log \cosh(z)\right] - \gamma \right\} E[\tanh(z)]$$

$$- 2k_2 E\left[ze^{-\frac{z^2}{2}}\right] E\left[(1 - z^2)e^{-\frac{z^2}{2}}\right]. \tag{7.A.31}$$

Now we use Taylor expansion to approximate $R$ in Equation 7.A.27. The first order Taylor expansion approximation of $H\left(\dfrac{\hat{\varepsilon}_y}{\sigma_{\varepsilon_y}}\right)$ is

$$H\left(\frac{\hat{\varepsilon}_y}{\sigma_{\varepsilon_y}}\right) = H\left(\frac{\varepsilon_y^0}{\sigma_{\varepsilon_y}} - \frac{\beta}{\sigma_{\varepsilon_y}}u\right)$$

$$\approx H\left(\frac{\varepsilon_y^0}{\sigma_{\varepsilon_y}}\right) - \frac{\beta}{\sigma_{\varepsilon_y}}h\left(\frac{\varepsilon_y^0}{\sigma_{\varepsilon_y}}\right)E[u]. \tag{7.A.32}$$

Similarly, we have

$$H\left(\frac{\hat{\varepsilon}_x}{\sigma_e}\right) \approx H\left(\frac{\varepsilon_x^0}{\sigma_{\varepsilon_x}}\right) - \frac{\alpha}{\sigma_{\varepsilon_x}}h\left(\frac{\varepsilon_x^0}{\sigma_{\varepsilon_x}}\right)E[u]. \tag{7.A.33}$$

Substituting Equations 7.A.32 and 7.A.33 into Equation 7.A.27, we obtain

$$R \approx -H(X) - H\left(\frac{\varepsilon_y^0}{\sigma_{\varepsilon_y}}\right) + \frac{\beta}{\sigma_{\varepsilon_y}} h\left(\frac{\varepsilon_y^0}{\sigma_{\varepsilon_y}}\right) E[u]$$

$$+ H(y) + H\left(\frac{\varepsilon_x^0}{\sigma_{\varepsilon_x}}\right) - \frac{\alpha}{\sigma_{\varepsilon_x}} h\left(\frac{\varepsilon_x^0}{\sigma_{\varepsilon_x}}\right) E[u] \qquad (7.A.34)$$

$$= R_0 + \left[\frac{\beta}{\sigma_{\varepsilon_y}} h\left(\frac{\varepsilon_y^0}{\sigma_{\varepsilon_y}}\right) - \frac{\alpha}{\sigma_{\varepsilon_x}} h\left(\frac{\varepsilon_x^0}{\sigma_{\varepsilon_x}}\right)\right] E[u],$$

where

$$R_0 = -H(x) - H\left(\frac{\varepsilon_y^0}{\sigma_{\varepsilon_y}}\right) + H(y) + H\left(\frac{\varepsilon_x^0}{\sigma_e}\right). \qquad (7.A.35)$$

---

## Appendix 7.B  Orthogonality Conditions and Covariance

Let

$$g_1(x) = D\left(P_{Y|x} \| U_Y\right) = \int \log\frac{P(y|x)}{U(y)} P(y|x) dy. \qquad (7.B.1)$$

Now we calculate $\text{cov}_{U_x}\left(g_1, \frac{P_x}{U_x}\right)$. By definition, we have

$$\text{cov}_{U_x}\left(g_1, \frac{P_X}{U_X}\right) = \int\int \log\frac{P(y|x)}{U(y)} P(y|x) dy \frac{P_X}{U_X} U_X dx$$

$$- \int\int \log\frac{P(y|x)}{U(y)} P(y|x) dy U_X dx \int \frac{P_X}{U_X} U_X dx$$

$$= \int\int \log\frac{P(y|x)}{U(y)} P(y|x) dy P_X dx \qquad (7.B.2)$$

$$- \int\int \log\frac{P(y|x)}{U(y)} P(y|x) dy U_X dx.$$

Note that the first term in Equation 7.B.2 can be expressed as

$$\iint \log \frac{P(y|x)}{U(y)} P(y|x) dy P_X dx = \iint \log \frac{P(y|x)P(x)}{U(y)P(x)} P(y|x) P_X dy dx$$

$$= \iint \log \frac{P(x,y)U(x)}{U(y)U(x)P(x)} P(x,y) dy dx$$

$$= \iint \log \frac{P(x,y)}{U(y)U(x)} P(x,y) dy dx$$

$$+ \iint \log \frac{U(x)}{P(x)} P(x,y) dy dx$$

$$= D(P_{Y,X} \| U_X U_Y) + \iint \log \frac{U(x)}{P(x)} P(x,y) dy dx.$$

The second term in Equation 7.B.3 can be reduced to $\qquad$ (7.B.3)

$$\iint \log \frac{U(x)}{P(x)} P(x,y) dy dx = -\iint \log \frac{P(x)}{U(x)} P(x,y) dx dy$$

$$= -\iint \log \frac{P(y|x)P(x)}{P(y|x)U(x)} P(x,y) dx dy \qquad (7.B.4)$$

$$= -\iint \log \frac{P(x,y)}{P(y|x)U(x)} P(x,y) dx dy$$

$$= -D(P_{Y,X} \| U_X P_{Y|X}).$$

Substituting Equation 7.B.4 into Equation 7.B.3 gives

$$\iint \log \frac{P(y|x)}{U(y)} P(y|x) dy P_X dx = D(P_{Y,X} \| U_X U_Y) - D(P_{Y\,X} \| U_X P_{Y|X}). \quad (7.B.5)$$

Now we consider the second term in Equation (7.B.2) which can be expressed as

$$\iint \log \frac{P(y|x)}{U(y)} P(y|x) dy U(x) dx = \iint \log \frac{P(y|x)U(x)}{U(y)U(x)} P(y|x) U(x) dx dy \qquad (7.B.6)$$

$$= D(P(y|x)U_X \| U_Y).$$

Substituting Equations 7.B.5 and 7.B.6 into Equation 7.B.2, we obtain

$$\text{cov}_{U_x}\left(g_1, \frac{P_X}{U_X}\right) = D(P_{Y,X} \| U_X U_Y) - D(P_{Y\,X} \| U_X U_{Y|X})$$

$$- D(P(y|x)U_X \| U_Y). \qquad (7.B.7)$$

Therefore, $\mathrm{cov}_{U_x}\left(g_1, \dfrac{P_X}{U_X}\right) = 0$ is equivalent to

$$D(P_{Y,X} \| U_X U_Y) = D(P_{Y\,X} \| U_X P_{Y|X}) + D(P(y|x)U_X \| U_Y). \qquad (7.B.8)$$

Next consider

$$g_3 = \int \log \frac{U_{out}(y)}{U(y)} P(y|x) dy. \qquad (7.B.9)$$

By definition, we obtain

$$\mathrm{cov}_{U_x}\left(g_3, \frac{P_X}{U_X}\right) = \int\int \log \frac{U_{out}(y)}{U(y)} P(y|x) dy \frac{P_X}{U_X} U_X dx$$

$$- \int\int \log \frac{U_{out}(y)}{U(y)} P(y|x) dy U_X dx \int \frac{P_X}{U_X} U_X dx$$

$$= \int\int \log \frac{U_{out}(y)}{U(y)} P(y|x) dy P_X dx - \int\int \log \frac{U_{out}(y)}{U(y)} P(y|x) dy U_X dx.$$

$$(7.B.10)$$

The first term in Equation 7.B.10 can be reduced to

$$\int\int \log \frac{U_{out}(y)}{U(y)} P(y|x) dy P_X dx = \int\left[\int \log \frac{U_{out}(y)}{U(y)} P(x,y) dx\right] dy$$

$$= \int \log \frac{U_{out}(y)}{U(y)} \left(\int P(x,y) dx\right) dy$$

$$= \int \log \frac{U_{out}(y)}{U(y)} P(y) dy \qquad (7.B.11)$$

$$= -\int \log \frac{U(y)}{U_{out}(y)} P(y) dy.$$

Next to express Equation 7.B.11 in terms of K-L distance, we transfer the final term in Equation 7.B.11 to

$$\int \log \frac{U(y)}{U_{out}(y)} P(y) dy = \int \log \frac{P(y)}{U_{out}(y)} \frac{U(y)}{P(y)} P(y) dy$$

$$= \int \log \frac{P(y)}{U_{out}(y)} P(y) dy + \int \log \frac{U(y)}{P(y)} P(y) dy \qquad (7.B.12)$$

$$= D\left(P_Y \| \vec{P}_Y\right) - D(P_Y \| U_Y).$$

Again, the second term in Equation 7.B.10 can be further reduced to

$$\iint \log \frac{U_{out}(y)}{U(y)} P(y|x) dy U(x) dx = \int \log \frac{U_{out}(y)}{U(y)} \left( \int P(y|x) U(x) dx \right) dy$$

$$= \int \log \frac{U_{out}(y)}{U(y)} U_{out}(y) dy \qquad (7.B.13)$$

$$= D(U_{out}(y) \| U_Y).$$

Substituting Equations 7.B.11–7.B.13 into Equation 7.B.10 gives

$$\text{cov}_{U_X} \left( g_3, \frac{P_X}{U_X} \right) = -D((P_Y) \| U_{out}(Y)) + D(P_Y \| U_Y)$$

$$- D(U_{out}(Y) \| U_Y) \qquad (7.B.14)$$

Equation 7.B.14 implies that $\text{cov}_{U_X} \left( g_3, \dfrac{P_X}{U_X} \right) = 0$ is equivalent to

$$D(P_Y \| U_Y) = D(P_Y | U_{out}(Y)) + D(U_{out}(Y) \| U_Y) \qquad (7.B.15)$$

Similarly, we can prove Equation 7.52.

---

## Appendix 7.C    Equivalent Formulations Orthogonality Conditions

We first show that Equation 7.64 implies Equation 7.66. By definition of covariance, we have

$$\text{cov}_{U_X} \left( \log \frac{V_{out}}{U_X}, \frac{P_X}{U_X} \right) = \int \log \frac{V_{out}}{U_X} \frac{P_X}{U_X} U_X dx - \int \log \frac{V_{out}}{U_X} U_X dx \int \frac{P_X}{U_X} U_X dx$$

$$= \int \log \frac{V_{out}}{U_X} P_X dx - \int \log \frac{V_{out}}{U_X} U_X dx . \qquad (7.C.1)$$

The first term in Equation 7.C.1 can be reduced to

$$\int \log \frac{V_{out}}{U_X} P_X dx = \int \log \frac{P_X}{U_X} \frac{V_{out}}{P_X} P_X dx$$

$$= \int \log \frac{P_X}{U_X} P_X dx + \int \log \frac{V_{out}}{P_X} P_X dx \qquad (7.C.2)$$

$$= D(P_X \| U_X) - D(P_X \| V_{out})$$

The second term in Equation 7.C.1 can be rewritten as

$$\int \log \frac{V_{out}}{U_X} U_X dx = -D(U_X \| V_{out}).$$

(7.C.3)

Substituting Equations 7.C.2 and 7.C.3 into Equation 7.C.1 gives

$$\text{cov}_{U_X}\left(\log \frac{V_{out}}{U_X}, \frac{P_X}{U_X}\right) = D(P_X \| U_X) - D(P_X \| V_{out}) + D(U_X \| V_{out}).$$

(7.C.4)

Therefore, Equation 7.64 implies Equation 7.66:

$$D(P_X \| V_{out}) = D(P_X \| U_X) + D(U_X \| V_{out}).$$

(7.C.5)

Next, we show that Equations 7.65 and 7.66 are equivalent. Recall that

$$P_Y = \frac{P_X}{f'(x)} \text{ and } V_{out} = f'(x)V_Y.$$

(7.C.6)

Note that

$$
\begin{aligned}
D(P_Y \| V_Y) &= \int \log \frac{P_Y}{V_Y} P_Y dy \\
&= \int \log \frac{P_X}{f'(x)V_Y} \frac{P_X}{f'(x)} f'(x) dx \\
&= \int \log \frac{P_X}{V_{out}} P_X dy \\
&= D(P_X \| V_{out}),
\end{aligned}
$$

(7.C.7)

$$
\begin{aligned}
D(P_Y \| U_{out}(Y)) &= \int \log \frac{P_Y}{U_{out}(Y)} P_Y dy \\
&= \int \log \frac{P_X f'(x)}{f'(x)U_X f'(x)} \frac{P_X}{f'(x)} f'(x) dx \\
&= \int \log \frac{P_X}{U_X} P_X dx \\
&= D(\bar{P}_X \| U_X),
\end{aligned}
$$

(7.C.8)

$$D(U_{out}(Y) \| V_Y) = \int \log \frac{U_{out}(Y)}{V_Y} U_{out}(Y)dy$$

$$= \int \log \frac{U_X}{f'(x)V_Y} \frac{U_X}{f'(x)} f'(x)dx \qquad (7.C.9)$$

$$= \int \log \frac{U_X}{V_{out}} U_X dx$$

$$= D(U_X \| V_{out}).$$

Combining Equations 7.C.7–7.C.9 shows that

$$D(P_Y \| V_Y) = D(P_Y \| U_{out}(Y)) + D(U_{out}(Y) \| V_Y)$$

is equivalent to

$$D(P_X \| V_{out}) = D(P_X \| U_X) + D(U_X \| V_{out}).$$

By the similar arguments, we can show other equivalence.

---

## Appendix 7.D  M–L Distance in Backward Direction

Using Equation 7.66 we obtain

$$D(P_X \| U_X) = D(P_X \| V_{out}(X)) - D(U_X \| V_{out}(X)). \qquad (7.D.1)$$

Note that

$$-D(U_X \| V_{out}) < D(V_{out}(X) \| U_X). \qquad (7.D.2)$$

Substituting Equation 7.D.2 into Equation 7.D.1 gives

$$D(P_X \| V_{out}(X)) + D(V_{out}(X) \| U_X) > D(P_X \| U_X). \qquad (7.D.3)$$

Recall that

$$V_{out}(x) = V_Y f'(x), \quad U_{out}(Y) = \frac{U_X}{f'(x)}. \qquad (7.D.4)$$

Using Equation 7.D.4, we obtain

$$
\begin{aligned}
D(V_Y \| U_{out}(Y)) &= \int \log \frac{V_Y}{U_{out}(Y)} V_Y dy \\
&= \int \log \frac{V_Y f'(x)}{U_X} V_Y f'(x) dx \\
&= \int \log \frac{V_{out}(X)}{U_X} V_{out}(X) dx \\
&= D(V_{out}(X) \| U_X).
\end{aligned}
\tag{7.D.5}
$$

Using Equations 7.C.7 and 7.D.5, we obtain

$$
D(P_Y \| V_Y) + D(V_Y \| U_{out}(Y)) = D(P_X \| V_{out}(X)) + D(V_{out}(X) \| U_X). \tag{7.D.6}
$$

Using Equation 7.C.8, 7.D.6, and Equation 7.70 gives

$$
D(P_Y \| V_Y) + D(V_Y \| U_{out}(Y)) > D(P_Y \| U_{out}(Y)), \tag{7.D.7}
$$

which proves inequality 7.71.

Now we calculate $\mathrm{cov}_{V_Y}\left( \log \dfrac{U_{out}(Y)}{V_Y}, \dfrac{P_Y}{V_Y} \right)$. By definition of covariance, we have

$$
\begin{aligned}
\mathrm{cov}_{V_Y}\left( \log \frac{U_{out}(Y)}{V_Y}, \frac{P_Y}{V_Y} \right) &= \int \log \frac{U_{out}(Y)}{V_Y} \frac{P_Y}{V_Y} V_Y dy \\
&\quad - \int \log \frac{U_{out}(Y)}{V_Y} V_Y dy \int \frac{P_Y}{V_Y} V_Y dy \\
&= \int \log \frac{U_{out}(Y)}{V_Y} P_Y dy - \int \log \frac{U_{out}(Y)}{V_Y} V_Y dy \\
&= \int \log \frac{U_{out}(Y)}{P_Y} \frac{P_Y}{V_Y} P_Y dy + D(V_Y \| U_{out}(Y)) \\
&= D(P_Y \| V_Y) - D(P_Y \| U_{out}(Y)) + D(V_Y \| U_{out}(Y)).
\end{aligned}
\tag{7.D.8}
$$

Using inequality (7.D.7), Equation 7.D.8 shows

$$
\mathrm{cov}_{V_Y}\left( \log \frac{U_{out}(Y)}{V_Y}, \frac{P_Y}{V_Y} \right) > 0.
$$

Other inequalities in Result 7.10 can be similarly proved.

## Appendix 7.E    Multiplicativity of Traces

For the completeness, in this appendix we follow the approach of Janzing et al. (2010) to give detailed proof of the multiplicativity of traces.

### Definition 7.E.1: Lipschitz Continuous Function

A real-valued function $g: R^d \to R$ is called Lipschitz continuous if there exists a positive real constant K such that, for all points $x_1 \in R^d$ and $x_2 \in R^d$,

$$|g(x_1) - g(x_2)| \leq \|x_1 - x_2\|, \tag{7.E.1}$$

where $\|.\|$ is a norm of the vector.

Next, we introduce a lemma that is useful in proving the main result.

### Levy's Lemma

Given a Lipschitz continuous function $g: S_d \to R$ defined on the $d$-dimensional hypersphere $S_d$ with Lipschitz constant

$$L = \max_{x \neq x'} \frac{|g(x) - g(x')|}{\|x - x'\|}. \tag{7.E.2}$$

If a point $x \in S_d$ is selected at random with respect to the uniform measure on the sphere, then we have

$$P\{|g(x) - \bar{g}| \leq \varepsilon\} \geq 1 - \exp\left\{-\frac{\kappa(d-1)\varepsilon^2}{L^2}\right\} \tag{7.E.3}$$

for some constant $\kappa$, where $\bar{g}$ is the mean or median of $g(x)$.

Now we prove the results about multiplicativity of traces.

### Definition 7.E.2: Group

A group is a set $G$, combined with an operation *, such that:

1. The group contains an identity. In other words, there exists an element $e$ in the set $G$, such that $a^* e = e^* a = a$ for all elements $a$ in $G$.
2. The group contains inverses, that is, for all elements $a$ in $G$, there exists an element $b$ in $G$, such that $a^* b = e$ and $b^* a = e$ where $e$ is the identity element.
3. The operation is associative, that is, $(a^* b)^* c = a^* (b^* c)$ for all $a,b,c \in G$.
4. The group is closed under the operation, that is, for all elements $a,b \in G, a^* b \in G$.

### Definition 7.E.3: Topology

A collection T of subsets of a nonempty set $G$ where subsets are referred to as open sets is called a topology if

1. The empty set $\phi \in T$ and the set $G \in T$;
2. The union of a collection of open sets $G_\alpha \in T$ for $\alpha \in A$ is in the collection $T(\cup_{\alpha \in A} G_\alpha \in T)$; and
3. The intersection of a finite number of open sets $G_i \in T$ for $i = 1,\ldots,$ n is in the collection $T (\cap_{i=1}^{n} G_i \in T)$.

The pair $(G, T)$ is referred to as a topological space.

### Example 7.E.1

Assume that $G$ is a nonempty set. The collection of the nonempty set and the whole set $\{\phi, G\}$ which satisfies the above three conditions is a topology on $G$, and often called indiscrete topology. The power set $P(G)$ of $G$ that consists of all subsets of $G$, is a topology on $G$ and is often referred to as the discrete topology. For example, consider a set $G = \{a,b,c\}$. The empty set $\{\}$ is a subset of $G = \{a,b,c\}$. The set $G$ also includes $\{a\},\{b\},\{c\},\{a,b\},$ $\{a,c\},\{b,c\}$ and whole set $\{a,b,c\}$. Let T denote a topology on $G$. T contains all collections of subsets in $G$. The power set $P(G)$ satisfies

1. The empty set $\{\phi\} \in T$ and the whole set $G \in T$;
2. The union of the power sets is in T, for example, $\{a\} \cup \{b,c\} \in \{\{a\},\{b, c\}\} \in T$; and
3. The power set with the intersection in T, for example $\{a\} \cap \{b,c\} = \{\phi\} \in T$.

Therefore, the power set $P(G)$ is a topology.

### Example 7.E.2: Euclidean Examples (Strickland, 2017)

To define a metric topology, we first define open balls and open sets. Let $x$ be a point in a space $G$, the open ball of radius $r$ about $x$ is defined as the set:

$$B(x,r) = \{y \in G | d(x,y) < r\}.$$

A subset $U$ of $G$ is called open if and only if for every $x \in U$ here exists an open ball $B(x,r) \subset U$. A collection of open sets induced by a metric is called metric topology. The metric in $R$ is defined as $d(x,y) = |x - y|$ and the metric in $R^n$ is defined as $d(x,y) = \|x - y\|$ where $\|x\|$ is a norm of a vector and can be, for example,

$$\|x\| = \|x\|_1,$$

$$\|x\| = \|x\|_2, \text{ and}$$

$$\|x\| = \|x\|_\infty.$$

### Definition 7.E.4: Hausdorff Topology

A set $V \subset G$ is a neighborhood of a point $x \in G$ if there exists an open set $E \subset V$ with $x \in E$. A topology T on $G$ is called Hausdorff if every pair of distinct points $x, y \in G$, $x \neq y$ has a pair of neighborhoods $V_x$ of $x$ and $V_y$ of $y$ such that $V_x \cap V_y = \phi$.

### Definition 7.E.5: Topology Group

A topological group $G$ is a group with a Hausdorff topology such that the group's binary operation (multiplication map $G \times G \to G$, $(\alpha, \beta) \mapsto \alpha\beta$ and the group's inverse function (inverse map $G \to G$, $\alpha \mapsto \alpha^{-1}$) are continuous. A topological group that is a compact topological space is called a compact group.

### Example 7.E.3

Define $U(1) = \{z \in C \mid |z| = 1\} = \{e^{i\theta} \mid 0 \leq \theta < 2\pi\}$ with the ordinary multiplication in complex number $C$ and the topology in $C$. In other words, the binary operation multiplication is defined as

$$(\alpha = e^{i\theta}, \beta = e^{i\varphi}) = \alpha\beta = e^{i(\theta+\varphi)} \in U(1)$$

and the inverse function is defined as

$$\alpha^{-1} = e^{-i\theta}.$$

Define the identity as $e^{i0}$ and distance matric as

$$d\left(e^{i\theta}, e^{i\varphi}\right) = \sqrt{(\cos\theta - \cos\varphi)^2 + (\sin\theta - \sin\varphi)^2}.$$

### Example 7.E.4: Matrix Group

Let $GL(n)$ be the set of real invertible $n \times n$ matrices, $SL(n)$ be the set of $n \times n$ real matrices with determinant 1, and $O(n)$ be the set of $n \times n$ real orthogonal matrices. Clearly, the sets $GL(n)$, $SL(n)$, and $O(n)$ form groups under matrix multiplication. The topology of matrices can be defined via the inner product of matrices. Let $A$ and $B$ be two matrices. Their inner product is defined as

$$<A, B> = \text{Tr}\left(AB^T\right) = \sum_{i=1}^{n} \sum_{j=1}^{n} a_{ij} b_{ij},$$

where $a_{ij}$ and $b_{ij}$ are elements of the matrices $A$ and $B$, respectively.

A matrix norm induced by the inner product is defined as

$$\| A \| = <A, A>^{1/2}.$$

A metric $\rho$ on the set of matrices $GL(n)$, $SL(n)$, and $O(n)$ is defined as

$$\rho(A, B) = \| A - B \|.$$

Using this metric, we can define a topology on $GL(n)$, $SL(n)$, or $O(n)$. In fact, we can define an open ball of radius $r$ and open sets. For example, let $A \in GL(n)$. Define

$$N_{GL(n)}(A, r) = \{B \in GL(n)|\rho(A, B) < r\}.$$

Similarly, if $Y \subset GL(n)$ and $A \in Y$, we can define an open ball in $Y$ as

$$N_Y(A, r) = \{B \in Y| \|B - A\| < r\} = N_{GL(n)}(A, r) \cap Y.$$

Now we can define an open set in $Y$. A subset $V \subseteq Y$ is open in $Y$ if and only if for every set in $V(A \in V)$ there exists a $\delta > 0$ such that $N_Y(A, \delta) \subseteq V$. Next, we define a continuous map.

### Definition 7.E.6: Continuous Map

Let $Y$ be in any one of $Gl(n)$, $SL(n)$, and $O(n)$ and $(X, T)$ be a topological space. A continuous function or a continuous map $f:Y \rightarrow X$ is defined as for very $A \in Y$ and $U \in T$ such that $f(A) \in U$ there is a $\delta > 0$ for which

$$B \in N_Y(A, \delta) \text{ implies } f(B) \in U.$$

Or equivalently, $f$ is continuous can be defined as if and only if for every $U \in T, f^{-1}(U) \subseteq Y$ is open in $Y$.

It is clear that by definition, $Gl(n)$, $SL(n)$ and $O(n)$ form a topological group.

Next, we will define an invariant measure on a topological group.

### Definition 7.E.7: Haar Measure

A Haar measure on a topological group $G$ is defined as a measure $\mu$ such that

1. $\mu(G) = 1$
2. $\mu(\alpha S) = \mu(S)$ for all $\alpha \in G$, where $\alpha S = \{\alpha\beta \mid \beta \in S\}$.

The Haar measure can also be defined as a bounded linear functional $E$

$$E(f) = \int_G f(\alpha) d\mu(\alpha) \tag{7.E.4}$$

Define indicator functions:

$$I_S(\gamma) = \begin{cases} 1 & \gamma \in S \\ 0 & \gamma \notin S \end{cases} \text{ and } I_{\alpha S}(\gamma) = \begin{cases} 1 & \gamma \in \alpha S \\ 0 & \gamma \notin \alpha S \end{cases}.$$

In terms of linear functional $E$, the conditions (1) and (2) of definition 7.E.7 are reduced to

$$E[1] = \int_G d\mu(\alpha) = \mu(G) = 1 \tag{7.E.5}$$

and

$$E[I_{\alpha S}] = \int_G I_{\alpha S}(\gamma)d\mu(\gamma) = \mu(\alpha S) = \mu(S) = \int_G I_S(\gamma)d\mu(\gamma) = E[I_S], \quad (7.E.6)$$

respectively.

### Example 7.E.5: Haar Measure for a Group of Real Numbers

Let $G$ be the group of real numbers with multiplication as an operation. Consider

$$\int_G u(y)f(y)dy.$$

To find Haar measure for the group of real numbers, we need to select $u$ $(y)$ such that

$$\int_G u(y)f(\alpha^{-1}y)dy = \int_G u(y)f(y)dy. \quad (7.E.7)$$

Changing of variables $x = \alpha^{-1}y$, we obtain

$$\int_G u(y)f(\alpha^{-1}y)dy = \int_G u(\alpha x)f(x)|\alpha|dx. \quad (7.E.8)$$

Substituting $\alpha = x^{-1}$ into Equation 7.E.8 gives

$$\begin{aligned}
\int_G u(\alpha x)f(x)|\alpha|dx &= \int_G u(1)\frac{f(x)}{|x|}dx \\
&= \int_G u(1)\frac{f(y)}{|y|}dy.
\end{aligned} \quad (7.E.9)$$

Since $u(1)$ is a constant, taking $u(y) = \dfrac{1}{|y|}$ gives Equation (7.E). The Haar measure and integral is given by

$$\int_G \frac{1}{|y|}dy \quad (7.E.10)$$

and

$$\int_G \frac{f(y)}{|y|}dy, \quad (7.E.11)$$

respectively.

### Example 7.E.6

Define $U_n(1) = \{z \in C^n | |z| = 1\} = \{(e^{i\theta_1}, ..., e^{i\theta_n} | 0 \le \theta_1 < 2\pi, ..., 0 \le \theta_n \le 2\pi\}$ with the ordinary multiplication in $n$ dimensional space of the complex numbers. Define a map $r{:}R^n \to U_n$, $r(\theta_1, ..., \theta_n) = (e^{i\theta_1}, ..., e^{i\theta_n})$.

Any function $f$ defined on $U_n(1)$ determines the function $f(r(\theta_1,...,\theta_n))$. Then, the Haar integral is

$$E[f] = \int_P f(r(\theta_1, ..., \theta_n))d\theta_1 ... d\theta_n, \qquad (7.E.12)$$

where $P = \{(\theta_1,...,\theta_n) \mid 0 \le \theta_i \le 2\pi, i = 1,...,n\}$.

Now we check its left invariance. Let $y = (e^{is_1}, ..., e^{is_n}) \in U_n(1)$. Then, $L_y f$ $(\theta) = f(y^{-1}\theta)$. We have the following equation:

$$\begin{aligned}
E\left[L_y f\right] &= \int_P f(y^{-1}r(\theta_1, ..., \theta_n)d\theta_1 ... d\theta_n) \\
&= \int_P f(r(\theta_1 - s_1, ..., \theta_n - s_n))d\theta_1 ... d\theta_n \qquad (7.E.13) \\
&= \int_{P-y} f(r(\theta_1, ..., \theta_n))d\theta_1 ... d\theta_n,
\end{aligned}$$

where $P - y = \{(\theta_1,...,\theta_n) \mid -s_i \le \theta_i \le 2\pi - s_i\}$.

Since $e^{i(\theta + 2\pi)} = e^{i\theta}$, we have

$$\int_{P-y} f(r(\theta_1, ..., \theta_n))d\theta_1 ... d\theta_n = \int_P f(r(\theta_1, ..., \theta_n))d\theta_1 ... d\theta_n,$$

which implies

$$E\left[L_y f\right] = E[f].$$

### Example 7.E.7: Discrete Group

Let $G$ be a discrete group. Define

$$C_c(G) = \{f \mid f : G \to C \text{ is continuous, } f$$

$$= 0 \text{ except for finite numbers of points}\}.$$

The Haar measure for the discrete group is

$$E[f] = \sum_{x \in G} f(x).$$

Since $\gamma x \in G$, we have $E[L_y f] = E[f]$.
Define

$$L^1(G) = \{f \mid f \text{ with countable support set } S, \sum_{x \in S} |f(x)| < +\infty\}.$$

When $G = Z$, we set $f(n) = a_n$. Thus, the Haar measure is

$$E[f] = \sum_{n=-\infty}^{\infty} a_n. \qquad (7.E.14)$$

**Example 7.E.8: Haar Measure of $GL_n(R)$**

Consider $A, X \in GL_n(R)$ and a linear transformation: $L_A: X \rightarrow AX$. Then, in terms of vector space, this transformation can be expressed as

$$vec(X) = \begin{bmatrix} A & 0 & \cdots & 0 \\ 0 & A & \cdots & 0 \\ \vdots & \vdots & \vdots & \vdots \\ 0 & 0 & \cdots & A \end{bmatrix} vec(X). \qquad (7.E.15)$$

Let $f \in C_c(G)$ and $L_\gamma f(x) = f(\gamma^{-1} x)$. The Haar measure should satisfy

$$E[L_\gamma f] = E[f]. \qquad (7.E.16)$$

Note that

$$E[L_\gamma f] = \int_S f(\gamma^{-1} x) dx_1 \ldots dx_{n^2}. \qquad (7.E.17)$$

Making change of variables: $y = \gamma^{-1} x$ or

$$vec(y) = \begin{bmatrix} \gamma^{-1} & 0 & \cdots & 0 \\ 0 & \gamma^{-1} & \cdots & 0 \\ \vdots & \vdots & \vdots & \vdots \\ 0 & 0 & \cdots & \gamma^{-1} \end{bmatrix} vec(x), \qquad (7.E.18)$$

Equation 7.E.17 is reduced to

$$E[L_\gamma f] = \int_S f(y) |\gamma|^n dy_1 \ldots dy_{n^2}. \qquad (7.E.19)$$

Let $dy_1 \ldots dy_n = \dfrac{dz_1 \ldots dz_{n^2}}{|\gamma|^n}$ and. Then, Equation 7.E.19 is reduced to

$$E[L_\gamma f] = E[f].$$

Let $\gamma = x$. Then, the Haar measure is

$$\mu(S) = \int_S \frac{dx}{|x|^n} \qquad (7.E.20)$$

The Haar integral is

$$E[f] = \int_S \frac{f(x)}{|x|^n} dx. \qquad (7.E.21)$$

For the self-contained, we prove results along the approach of Janzing et al. 2010.

Consider a set of orthonormal bases $(\varphi_j), j = 1, \ldots, m$. Expanding the matrix $AU\Sigma U^T A^T$ in terms of orthonormal bases, we obtain

$$AU\Sigma U^T A^T = \sum_{j=1}^{m} \xi_j \varphi_j^T. \qquad (7.E.22)$$

Multiplying both sides of Equation 7.E.22 by $\varphi_j$ gives

$$AU\Sigma U^T A^T \varphi_j = \sum_{k=1}^{m} \xi_k \varphi_k^T \varphi_j = \xi_j. \tag{7.E.23}$$

Taking trace on both sides of Equation 7.E.22 and using Equation 7.E.23 yields

$$\begin{aligned}
\text{Tr}(AU\Sigma U^T A^T) &= \sum_{j=1}^{m} \text{Tr}\left(AU\Sigma U^T A^T \varphi_j \varphi_j^T\right) \\
&= \sum_{j=1}^{m} \varphi_j^T AU\Sigma U^T A^T \varphi_j \\
&= \sum_{j=1}^{m} \left(U^T A^T \varphi_j\right)^T \Sigma \left(U^T A^T \varphi_j\right).
\end{aligned} \tag{7.E.24}$$

To apply Levy's lemma, we need to normalize the items in Equation (7.E.24) Note that

$$\| U^T A^T \varphi_j \|_2^2 = \varphi_j^T A A^T \varphi_j. \tag{7.E.25}$$

Then, using Equation 7.E.25 we obtain

$$\begin{aligned}
\left(U^T A^T \varphi_j\right)^T \Sigma \left(U^T A^T \varphi_j\right) &= \frac{\left(U^T A^T \varphi_j\right)^T \Sigma \left(U^T A^T \varphi_j\right)}{\| U^T A^T \varphi_j \|_2^2} \| U^T A^T \varphi_j \|_2^2 \\
&= \alpha_j^T \Sigma \alpha_j \left(\varphi_j^T A A^T \varphi_j\right),
\end{aligned} \tag{7.E.26}$$

where $\alpha_j = \dfrac{U^T A^T \varphi_j}{\| U^T A^T \varphi_j \|}$ and $\|\alpha_j\| = 1$.

Define

$$f\left(\alpha_j\right) = \alpha_j^T \Sigma \alpha_j \text{ and}$$

$$\begin{aligned}
\bar{f}\left(\alpha_j\right) &= \frac{1}{n} \alpha_j^T \Sigma \alpha_j \\
&= \frac{1}{n} \text{Tr}\left(\alpha_j^T \Sigma \alpha_j\right) \\
&= \frac{1}{n} \text{Tr}(\Sigma).
\end{aligned} \tag{7.E.27}$$

Before applying Levy's lemma, we calculate Lipschitz constant $L$. Note that

$$\begin{aligned}
\left| f\left(\alpha_j\right) - f\left(\alpha_j^*\right)\right| &= \left| \alpha_j^T \Sigma \alpha_j - \left(\alpha_j^*\right)^T \Sigma \alpha_j^* \right| \\
&\leq \left| \left(\alpha_j - \alpha_j^*\right)^T \Sigma \alpha_j \right| + \left| \left(\alpha_j^*\right)^T \Sigma \left(\alpha_j - \alpha_j^*\right) \right| \\
&\leq \| \Sigma \| \, \| \alpha_j \| \, \| \alpha_j - \alpha_j^* \| + \| \Sigma \| \, \| \alpha_j^* \| \, \| \alpha_j - \alpha_j^* \| \\
&= 2 \| \Sigma \| \, \| \alpha_j - \alpha_j^* \|.
\end{aligned}$$

Thus,

$$\frac{\left| f\left( \alpha_j \right) - f\left( \alpha_j^* \right) \right|}{\| \alpha_j - \alpha_j^* \|} \le 2 \| \Sigma \|,$$

which implies

$$L = \max_{\alpha_j \neq \alpha_j^*} \frac{\left| f\left( \alpha_j \right) - f\left( \alpha_j^* \right) \right|}{\| \alpha_j - \alpha_j^* \|} = 2 \| \Sigma \|.$$

Let

$$\varepsilon = \frac{\varepsilon'}{L} \text{ or } \varepsilon' = L\varepsilon.$$

Applying Levy's lemma, we have

$$\left| f\left( \alpha_j \right) - \bar{f} \right| = \left| \alpha_j^T \Sigma \alpha_j - \frac{1}{n} \text{Tr}(\Sigma) \right| \tag{7.E.28}$$

$$\le 2\varepsilon \| \Sigma \|$$

with probability $1 - exp\{- k(n - 1)\varepsilon^2\}$.

Recall that

$$\tau_m\left( A U \Sigma U^T A^T \right) = \frac{1}{m} \text{Tr}\left( A U \Sigma U^T A^T \right).$$

Using Equations 7.E.24 and 7.E.26, we obtain

$$\tau_m\left( A U \Sigma U^T A^T \right) = \frac{1}{m} \sum_{j=1}^m \alpha_j^T \Sigma \alpha_j \left( \varphi_j^T A A^T \varphi_j \right) \tag{7.E.29}$$

It is clear that

$$\tau_m\left( A A^T \right) = \frac{1}{m} \sum_{j=1}^m \text{Tr}\left( A A^T \right)$$

$$= \frac{1}{m} \sum_{j=1}^m \text{Tr}\left( \varphi_j^T A A^T \varphi_j \right) \tag{7.E.30}$$

$$= \frac{1}{m} \sum_{j=1}^m \varphi_j^T A A^T \varphi_j.$$

Using Equations 7.E.28–7.E.30, we obtain

$$\left| \tau_m\left( A U \Sigma U^T A^T \right) - \tau_n(\Sigma) \tau_m\left( A A^T \right) \right| = \left| \frac{1}{m} \sum_{j=1}^m \alpha_j^T \Sigma \alpha_j \left( \varphi_j^T A A^T \varphi_j \right) \right.$$

$$\left. - \frac{1}{n} \text{Tr}(\Sigma) \frac{1}{m} \sum_{j=1}^m \varphi_j^T A A^T \varphi_j \right|$$

$$= \left| \frac{1}{m} \sum_{j=1}^m \left( \varphi_j^T A A^T \varphi_j \right) \left( \alpha_j^T \Sigma \alpha_j - \frac{1}{n} \text{Tr}(\Sigma) \right) \right|$$

$$\le \frac{1}{m} \sum_{j=1}^m \left( \varphi_j^T A A^T \varphi_j \right) \left| \left( \alpha_j^T \Sigma \alpha_j - \frac{1}{n} \text{Tr}(\Sigma) \right) \right|$$

$$\le \frac{2\varepsilon}{m} \sum_{j=1}^m \left( \varphi_j^T A A^T \varphi_j \right) \| \Sigma \|$$

$$= 2\varepsilon \| \Sigma \| \| A A^T \|.$$

## Appendix 7.F   Anisotropy and K–L Distance

Consider two Gaussians $Z_1$ and $Z_0$ with equal mean and covariance matrices $\Sigma_1$ and $\Sigma_0$. The K–L distance between two Gaussians is defined as

$$D\left(P_{\Sigma_1} \| P_{\Sigma_0}\right) = E_{Z_1}\left[\log \frac{\dfrac{1}{(2\pi)^{n/2}|\Sigma_1|^{1/2}} \exp\left\{-1/2\operatorname{Tr}\left(\Sigma_1^{-1}Z_1Z_1^T\right)\right\}}{\dfrac{1}{(2\pi)^{n/2}|\Sigma_0|^{1/2}} \exp\left\{-1/2\operatorname{Tr}\left(\Sigma_0^{-1}Z_0Z_0^T\right)\right\}}\right]$$

$$= \frac{1}{2}\log\frac{|\Sigma_0|}{|\Sigma_1|} - \frac{1}{2}E_{Z_1}\left[\operatorname{Tr}\left(\Sigma_1^{-1}Z_1Z_1^T\right)\right] + \frac{1}{2}E_{Z_1}\left[\operatorname{Tr}\left(\Sigma_0^{-1}Z_0Z_0^T\right)\right] \quad (7.F.1)$$

$$= \frac{1}{2}\log\frac{|\Sigma_0|}{|\Sigma_1|} - \frac{1}{2}n + \frac{1}{2}\operatorname{Tr}\left(\Sigma_0^{-1}\Sigma_1\right)$$

$$= \frac{1}{2}\left(\log\frac{|\Sigma_0|}{|\Sigma_1|} + n\left[\tau_n\left(\Sigma_0^{-1}\Sigma_1\right) - 1\right]\right).$$

Let $\Sigma_0 = \lambda I$. Define the anisotropy of $\Sigma$ as the smallest K–L distance between the Gaussian $Z_1$ and the isotropic Gaussian with $\Sigma_0 = \lambda I$:

$$D(\Sigma_1) = \min_{\lambda}\ D\left(P_{\Sigma_1} \| P_{\Sigma_0}\right) \qquad (7.F.2)$$

Note that

$$D\left(P_{\Sigma_1} \| P_{\Sigma_0}\right) = \frac{1}{2}\left(n\log\lambda - \log|\Sigma_1| + \frac{1}{\lambda}\operatorname{Tr}(\Sigma_1) - n\right). \qquad (7.F.3)$$

The minimum of $D(P_{\Sigma_1} \| P_{\Sigma_0})$ is obtained by setting the derivative of the right hand of Equation 7.F.3 equal to zero:

$$\frac{n}{\lambda} - \frac{\operatorname{Tr}(\Sigma_1)}{\lambda^2} = 0. \qquad (7.F.4)$$

Solving Equation 7.F.4 for $\lambda$ gives

$$\lambda = \tau_n(\Sigma_1). \qquad (7.F.5)$$

Substituting Equation 7.F.5 into Equation 7.F.3, we obtain

$$D\left(P_{\Sigma_1} \| P_{\Sigma_0}\right) = \frac{1}{2}\left(n\log\tau_n(\Sigma_1) - \log|\Sigma_1| + n\frac{\operatorname{Tr}(\Sigma_1)}{\operatorname{Tr}(\Sigma_1)} - n\right)$$

$$\qquad (7.F.6)$$

$$= \frac{1}{2}\left(n\log\tau_n(\Sigma_1) - \log|\Sigma_1|\right).$$

Substituting Equation 7.F.6 into Equation 7.F.2 yields

$$D(\Sigma_1) = \frac{1}{2}\left(n\log\tau_n(\Sigma_1) - \log|\Sigma_1|\right). \qquad (7.F.7)$$

Assume that both $\varepsilon_X$ and $\varepsilon_Y$ are the manifold of isotropic Gaussians, and $\varepsilon_X^*$ and $\varepsilon_Y^*$ are projections of $P_X$ and $P_Y$. Recall that $P_X$ and $P_Y$ have covariance matrices $\Sigma_X$ and $A\Sigma_X A^T$, respectively. Then, we have

$$D(\Sigma_X) = D(P_X \parallel \varepsilon_X^*) = \frac{1}{2}\left(n\log\tau_n(\Sigma_X) - \log|\Sigma_X|\right) \tag{7.F.8}$$

and

$$D(\Sigma_Y) = D(P_Y \parallel \varepsilon_Y^*) = \frac{1}{2}\left(n\log\tau_n(\Sigma_Y) - \log|\Sigma_Y|\right). \tag{7.F.9}$$

Note that

$$\tau_n(\Sigma_Y) = \tau_n\left(A\Sigma_X A^T\right) \tag{7.F.10}$$

and

$$\log|\Sigma_Y| = \log|\Sigma_X| + \log|AA^T|. \tag{7.F.11}$$

Recall that

$$\Delta_{X\to Y} = \log\tau_n\left(A\Sigma A^T\right) - \log\tau_n(\Sigma) - \log\tau_m\left(AA^T\right) \text{ or}$$

$$\log\tau_n\left(A\Sigma A^T\right) = \Delta_{X\to Y} + \log\tau_n(\Sigma) + \log\tau_m\left(AA^T\right). \tag{7.F.12}$$

Substituting Equations 7.F.10–7.F.12 gives

$$D\left(A\Sigma_X A^T\right) = \frac{n}{2}\Delta_{X\to Y} + \frac{n}{2}\log\tau_n(\Sigma_X) + \frac{n}{2}\log\tau_m\left(AA^T\right) - \frac{1}{2}\log|\Sigma_X| - \frac{1}{2}\log|AA^T|$$

$$= \frac{n}{2}\Delta_{X\to Y} + \frac{n}{2}\log\tau_n(\Sigma_X) - \frac{1}{2}\log|\Sigma_X|\frac{n}{2}\log\tau_m\left(AA^T\right) - \frac{1}{2}\log|AA^T|$$

$$= \frac{n}{2}\Delta_{X\to Y} + D(\Sigma_X) + D\left(AA^T\right). \tag{7.F.13}$$

Recall that $U_{out}(Y)$ denotes the distribution of $Y = AX$ and $\varepsilon_Y^*$ is the distribution of the projection of the variable $Y$ onto the manifold of isotropic Gaussian $\varepsilon_Y$. The covariance matrix of $U_{out}(Y)$ is $\Sigma_1 = A\tau_n(\Sigma_X)IA^T = \tau_n(\Sigma_X)AA^T$ and the covariance matrix of $\varepsilon_Y^*$ is $\Sigma_0 = \tau_n(\Sigma_Y)I$. Then, using Equation 7.F.1, we obtain the K–L distance between $U_{out}(Y)$ and $\varepsilon_Y^*$ (Janzing et al. 2012):

$$D(U_{out}(Y) \parallel \varepsilon_Y^*) = \frac{1}{2}\left(\log\frac{(\tau_n(\Sigma_Y))^n}{(\tau_n(\Sigma_X))^n|AA^T|} + n\left[\frac{\tau_n(\Sigma_X)\tau_n\left(AA^T\right)}{\tau_n(\Sigma_Y)} - 1\right]\right) \tag{7.F.14}$$

Note that

$$|\Sigma_Y| = |\Sigma_X||AA^T|). \tag{7.F.15}$$

Substituting Equations 7.F.8, 7.F.9, and 7.F.15 into Equation 7.F.14 gives

$$D(P_Y \| \varepsilon_Y^*) = D(P_X \| \varepsilon_X^*) + D(U_{out}(Y) \| \varepsilon_Y^*) + \frac{n}{2}\left[1 - \frac{\tau_n(\Sigma_X)\tau_n(AA^T)}{\tau_n(\Sigma_Y)}\right] \quad (7.F.16)$$

Comparing Equation 7.F.16 with Equation 7.68, we conclude that $X \to Y$ must imply

$$\tau_n(\Sigma_Y) = \tau_n(\Sigma_X)\tau_n(AA^T) \quad (7.F.17)$$

in order to make two equations to be equal.

---

## Appendix 7.G  Trace Method for Noise Linear Model

Recall that

$$\Sigma_{XY} = \Sigma_X A^T \quad (7.G.1)$$

and

$$\Sigma_Y = A\Sigma_X A^T + \lambda I, \quad (7.G.2)$$

where $A$ is an orthogonal matrix.

It follows from Equation 7.112 that

$$\tilde{A} = \Sigma_X A^T (A\Sigma_X A^T + \lambda I)^{-1}$$
$$= \Sigma_X(\Sigma_X + \lambda I)^{-1}A^T. \quad (7.G.3)$$

Assume that the eigenvalue decomposition of the covariance matrix $\Sigma_X$ is

$$\Sigma_X = U\Lambda U^T, \quad (7.G.4)$$

where $U^T U = I$ and $\Lambda = diag(\mu_1,\ldots, \mu_n)$ with $\mu_1 \geq \mu_2 \geq \ldots\mu_n \geq 0$.

Then, using Equations 7.G.3 and 7.G.4 gives

$$\tilde{A} = U\Lambda U^T (U\Lambda U^T + \lambda I)^{-1}A^T$$
$$= U\Lambda(\Lambda + \lambda I)^{-1}U^T A^T \quad (7.G.5)$$

Thus,

$$\tau_n(\tilde{A}\tilde{A}^T) = \tau_n\left(U\Lambda(\Lambda + \lambda I)^{-1}U^T A^T A U(\Lambda + \lambda I)^{-1}\Lambda U^T\right)$$
$$= \tau_n\left(\left(\Lambda(\Lambda + \lambda I)^{-1}\right)^2\right). \quad (7.G.6)$$

Using Equations 7.G.2 and 7.G.4, we obtain

$$\Sigma_Y = AU\Delta U^T A^T + \lambda I, \tag{7.G.7}$$

which implies

$$\tau_n(\Sigma_Y) = \tau_n(\Delta) + \lambda. \tag{7.G.8}$$

Now consider the model:

$$X = \tilde{A}Y + \tilde{e}$$

and calculate $\tau_n(\tilde{A}\Sigma_Y\tilde{A}^T)$:

$$
\begin{aligned}
\tau_n\left(\tilde{A}\Sigma_Y\tilde{A}^T\right) &= \tau_n\left(U\Delta(\Delta + \lambda I)^{-1}U^T A^T (AU\Delta U^T A^T + \lambda I)AU(\Delta + \lambda I)^{-1}\Delta U^T\right) \\
&= \tau_n\left(U\Delta(\Delta + \lambda I)^{-1}(\Delta + \lambda I)(\Delta + \lambda I)^{-1}\Delta U^T\right) \\
&= \tau_n\left(U\Delta(\Delta + \lambda I)^{-1}\Delta U^T\right) \\
&= \tau_n\left(\Delta^2(\Delta + \lambda I)^{-1}\right).
\end{aligned}
\tag{7.G.9}
$$

Then, using Equations 7.G.6, 7.G.8, and 7.G.9, we obtain the statistic for assessing causal $Y \to X$:

$$
\begin{aligned}
\Delta_{Y\to X} &= \log\frac{\tau_n\left(\tilde{A}\Sigma_Y\tilde{A}^T\right)}{\tau_n(\Sigma_Y)\tau_n\left(\tilde{A}\tilde{A}^T\right)} \\
&= \log\frac{\tau_n\left(\Delta^2(\Delta + \lambda I)^{-1}\right)}{(\tau_n(\Delta) + \lambda)\tau_n\left(\left(\Delta(\Delta + \lambda I)^{-1}\right)^2\right)} \\
&= \log\frac{1}{n}\sum_{i=1}^{n}\frac{\mu_i^2}{(\mu_i + \lambda)} - \log\frac{1}{n}\sum_{i=1}^{n}(\mu_i + \lambda) - \log\frac{1}{n}\sum_{i=1}^{n}\frac{\mu_i^2}{(\mu_i + \lambda)^2}.
\end{aligned}
\tag{7.G.10}
$$

Now we show that $\Delta_{Y\to X} > 0$ by induction.
Let $n = 2$. We show

$$\frac{1}{2}\left(\frac{\mu_1^2}{\mu_1 + \lambda} + \frac{\mu_2^2}{\mu_2 + \lambda}\right) > \frac{1}{2}(\mu_1 + \lambda + \mu_2 + \lambda)\frac{1}{2}\left(\frac{\mu_1^2}{(\mu_1 + \lambda)^2} + \frac{\mu_2^2}{(\mu_2 + \lambda)^2}\right) \tag{7.G.11}$$

In fact, Equation 7.G.11 implies that

$$2\left(\frac{\mu_1^2}{\mu_1 + \lambda} + \frac{\mu_2^2}{\mu_2 + \lambda}\right) > (\mu_1 + \lambda + \mu_2 + \lambda)\left(\frac{\mu_1^2}{(\mu_1 + \lambda)^2} + \frac{\mu_2^2}{(\mu_2 + \lambda)^2}\right). \tag{7.G.12}$$

Expanding the right side of Equation 7.G.12, we obtain

$$(\mu_1 + \lambda + \mu_2 + \lambda) \left( \frac{\mu_1^2}{(\mu_1 + \lambda)^2} + \frac{\mu_2^2}{(\mu_2 + \lambda)^2} \right)$$

$$= \frac{\mu_1^2}{\mu_1 + \lambda} + \frac{\mu_2^2}{\mu_2 + \lambda} + (\mu_1 + \lambda) \frac{\mu_2^2}{(\mu_2 + \lambda)^2} + (\mu_2 + \lambda) \frac{\mu_1^2}{(\mu_1 + \lambda)^2}. \tag{7.G.13}$$

Combining Equations 7.G.12 and 7.G.13, we need to prove

$$\left( \frac{\mu_1^2}{\mu_1 + \lambda} + \frac{\mu_2^2}{\mu_2 + \lambda} \right) > (\mu_1 + \lambda) \frac{\mu_2^2}{(\mu_2 + \lambda)^2} + (\mu_2 + \lambda) \frac{\mu_1^2}{(\mu_1 + \lambda)^2}. \tag{7.G.14}$$

Note that

$$\frac{\mu_1^2}{\mu_1 + \lambda} - (\mu_2 + \lambda) \frac{\mu_1^2}{(\mu_1 + \lambda)^2} = \left( 1 - \frac{\mu_2 + \lambda}{\mu_1 + \lambda} \right) \frac{\mu_1^2}{\mu_1 + \lambda}$$

$$= \frac{\mu_1 - \mu_2}{\mu_1 + \lambda} \frac{\mu_1^2}{\mu_1 + \lambda} \tag{7.G.15}$$

and

$$\frac{\mu_2^2}{\mu_2 + \lambda} - (\mu_1 + \lambda) \frac{\mu_2^2}{(\mu_2 + \lambda)^2} = \left( 1 - \frac{\mu_1 + \lambda}{\mu_2 + \lambda} \right) \frac{\mu_2^2}{\mu_2 + \lambda}$$

$$= \frac{\mu_2 - \mu_1}{\mu_2 + \lambda} \frac{\mu_2^2}{\mu_2 + \lambda}. \tag{7.G.16}$$

Using Equations 7.G.15 and 7.G.16, we obtain

$$\left( \frac{\mu_1^2}{\mu_1 + \lambda} + \frac{\mu_2^2}{\mu_2 + \lambda} \right) - (\mu_1 + \lambda) \frac{\mu_2^2}{(\mu_2 + \lambda)^2} - (\mu_2 + \lambda) \frac{\mu_1^2}{(\mu_1 + \lambda)^2}$$

$$= \frac{\mu_1 - \mu_2}{\mu_1 + \lambda} \frac{\mu_1^2}{\mu_1 + \lambda} + \frac{\mu_2 - \mu_1}{\mu_2 + \lambda} \frac{\mu_2^2}{\mu_2 + \lambda}$$

$$= (\mu_1 - \mu_2) \left( \frac{\mu_1^2}{(\mu_1 + \lambda)^2} - \frac{\mu_2^2}{(\mu_2 + \lambda)^2} \right) \tag{7.G.17}$$

$$= (\mu_1 - \mu_2) \left( \frac{\mu_1}{\mu_1 + \lambda} - \frac{\mu_2}{\mu_2 + \lambda} \right) \left( \frac{\mu_1}{\mu_1 + \lambda} + \frac{\mu_2}{\mu_2 + \lambda} \right)$$

$$= \frac{\lambda(\mu_1 - \mu_2)^2}{(\mu_1 + \lambda)(\mu_2 + \lambda)} \left( \frac{\mu_1}{\mu_1 + \lambda} + \frac{\mu_2}{\mu_2 + \lambda} \right) > 0.$$

This proves inequality (7.G.11). Now suppose that when $n = k$ the following inequality

$$\frac{1}{n}\sum_{i=1}^{n}\frac{\mu_i^2}{(\mu_i + \lambda)} > \left(\frac{1}{n}\sum_{i-1}^{n}(\mu_i + \lambda)\right)\left(\frac{1}{n}\sum_{i=1}^{n}\frac{\mu_i^2}{(\mu_i + \lambda)^2}\right) \tag{7.G.18}$$

holds. Then, we show that when $n = k + 1$ inequality (7.G.18) still holds. Let

$$S_k^1 = \sum_{i=1}^{k}\frac{\mu_i^2}{(\mu_i + \lambda)}, \;\; S_k^2 = \sum_{i=1}^{k}(\mu_i + \lambda) \text{ and } S_k^3 = \sum_{i=1}^{k}\frac{\mu_i^2}{(\mu_i + \lambda)^2}.$$

Then, inequality (7.G.18) shows that

$$kS_k^1 > S_k^2 S_k^3 \tag{7.G.19}$$

holds.

Note that

$$\sum_{i=1}^{k+1}\frac{\mu_i^2}{(\mu_i + \lambda)} = S_k^1 + \frac{\mu_{k+1}^2}{\mu_{k+1} + \lambda}, \;\; \sum_{i=1}^{k+1}(\mu_i + \lambda) = S_k^2 + \mu_{k+1} + \lambda \text{ and}$$

$$\sum_{i=1}^{k+1}\frac{\mu_i^2}{(\mu_i + \lambda)^2} = S_K^3 + \frac{\mu_{k+1}^2}{(\mu_{k+1} + \lambda)^2}.$$

To show inequality (7.G.18) holds we must show

$$\frac{1}{k+1}\sum_{i=1}^{k+1}\frac{\mu_i^2}{(\mu_i + \lambda)} > \left(\frac{1}{k+1}\sum_{i=1}^{k+1}(\mu_i + \lambda)\right)\left(\frac{1}{k+1}\sum_{i=1}^{k+1}\frac{\mu_i^2}{(\mu_i + \lambda)^2}\right)$$

or

$$(k+1)\sum_{i=1}^{k+1}\frac{\mu_i^2}{(\mu_i + \lambda)} > \left(\sum_{i=1}^{k+1}(\mu_i + \lambda)\right)\left(\sum_{i=1}^{k+1}\frac{\mu_i^2}{(\mu_i + \lambda)^2}\right). \tag{7.G.20}$$

Note that

$$\sum_{i=1}^{k+1}\frac{\mu_i^2}{(\mu_i + \lambda)} = S_k^1 + \frac{\mu_{k+1}^2}{\mu_{k+1} + \lambda}, \tag{7.G.21}$$

$$\sum_{i=1}^{k+1}(\mu_i + \lambda) = S_k^2 + \mu_{k+1} + \lambda, \tag{7.G.22}$$

$$\sum_{i=1}^{k+1} \frac{\mu_i^2}{(\mu_i + \lambda)^2} = S_k^3 + \frac{\mu_{k+1}^2}{(\mu_{k+1} + \lambda)^2}. \tag{7.G.23}$$

Substituting Equations 7.G.21–7.G.23 into Equation 7.G.20 gives

$$(k+1)S_k^1 + (k+1)\frac{\mu_{k+1}^2}{\mu_{k+1} + \lambda} > \left(S_k^2 + \mu_{k+1} + \lambda\right)\left(S_k^3 + \frac{\mu_{k+1}^2}{(\mu_{k+1} + \lambda)^2}\right). \tag{7.G.24}$$

Using Equation 7.G.19, we can reduce Equation 7.G.24 to

$$S_k^1 + k\frac{\mu_{k+1}^2}{\mu_{k+1} + \lambda} > (\mu_{k+1} + \lambda)S_k^3 + \frac{\mu_{k+1}^2}{(\mu_{k+1} + \lambda)^2}S_k^2. \tag{7.G.25}$$

Moving the right side of Equation 7.G.25 to the left side, we obtain

$$\sum_{i=1}^{k} \frac{\mu_i^2}{\mu_i + \lambda}\left(1 - \frac{\mu_{k+1} + \lambda}{\mu_i + \lambda}\right) + \sum_{i=1}^{k} \frac{\mu_{k+1}^2}{\mu_{k+1} + \lambda}\left(1 - \frac{\mu_i + \lambda}{\mu_{k+1} + \lambda}\right)$$

$$= \sum_{i=1}^{k} \frac{\mu_i^2}{\mu_i + \lambda}\frac{\mu_i - \mu_{k+1}}{\mu_i + \lambda} + \sum_{i=1}^{k} \frac{\mu_{k+1}^2}{\mu_{k+1} + \lambda}\frac{\mu_{k+1} - \mu_i}{\mu_{k+1} + \lambda}$$

$$= \sum_{i=1}^{k} (\mu_i - \mu_{k+1})\left(\frac{\mu_i^2}{(\mu_i + \lambda)^2} - \frac{\mu_{k+1}^2}{(\mu_{k+1} + \lambda)^2}\right) \tag{7.G.26}$$

$$= \sum_{i=1}^{k} (\mu_i - \mu_{k+1})\left(\frac{\mu_i}{\mu_i + \lambda} - \frac{\mu_{k+1}}{\mu_{k+1} + \lambda}\right)\left(\frac{\mu_i}{\mu_i + \lambda} + \frac{\mu_{k+1}}{\mu_{k+1} + \lambda}\right)$$

$$= \sum_{i=1}^{k} (\mu_i - \mu_{k+1})^2 \lambda \frac{1}{(\mu_{i+1} + \lambda)(\mu_{k+1} + \lambda)}\left(\frac{\mu_i}{\mu_i + \lambda} + \frac{\mu_{k+1}}{\mu_{k+1} + \lambda}\right) \geq 0$$

Combining Equations 7.G.20, 7.G.24–7.G.26, we show that

$$\frac{1}{n}\sum_{i=1}^{n} \frac{\mu_i^2}{(\mu_i + \lambda)} > \left(\frac{1}{n}\sum_{i=1}^{n} (\mu_i + \lambda)\right)\left(\frac{1}{n}\sum_{i=1}^{n} \frac{\mu_i^2}{(\mu_i + \lambda)^2}\right) \tag{7.G.27}$$

and

$$\Delta_{Y \to X} = \log\frac{1}{n}\sum_{i=1}^{n} \frac{\mu_i^2}{(\mu_i + \lambda)} - \log\frac{1}{n}\sum_{i=1}^{n} (\mu_i + \lambda)$$

$$- \log\frac{1}{n}\sum_{i=1}^{n} \frac{\mu_i^2}{(\mu_i + \lambda)^2} > 0. \tag{7.G.28}$$

## Appendix 7.H  Characterization of Association

It is well known that the measure of association between $Y$ and $X$ is

where

$$r = \frac{1}{n}\sum_{i=1}^{n}\lambda_i^2 = \frac{1}{n}\mathrm{Tr}(R^2),$$ (7.H.1)

$$R^2 = \Sigma_Y^{-1/2}\Sigma_{YX}\Sigma_X^{-1}\Sigma_{XY}\Sigma_Y^{-1/2}.$$

Using Equations 7.G.1, 7.G.2, and 7.G.4, we obtain

$$
\begin{aligned}
r &= \frac{1}{n}\mathrm{Tr}\left(\Sigma_{YX}\Sigma_X^{-1}\Sigma_{XY}\Sigma_Y^{-1}\right) \\
&= \frac{1}{n}\mathrm{Tr}\left(A\Sigma_X\Sigma_X^{-1}\Sigma_X A^T\left(A\Sigma_X A^T + \lambda I\right)^{-1}\right) \\
&= \frac{1}{n}\mathrm{Tr}\left(A\Sigma_X A^T\left(A\Sigma_X A^T + \lambda I\right)^{-1}\right) \\
&= \frac{1}{n}\mathrm{Tr}\left(\Sigma_X(\Sigma_X + \lambda I)^{-1}\right).
\end{aligned}
$$ (7.H.2)

Substituting Equation 7.G.4 gives

$$
\begin{aligned}
r &= \frac{1}{n}\mathrm{Tr}\left(\Delta(\Delta + \lambda I)^{-1}\right) \\
&= \frac{1}{n}\sum_{i=1}^{n}\frac{\mu_i}{\mu_i + \lambda}.
\end{aligned}
$$ (7.H3)

If $\lambda > 0$, then $\mu_i < \mu_i + \lambda$. Therefore, we obtain

$$r < 1 \text{ and } \log r < 0.$$

## Appendix 7.I  Algorithm for Sparse Trace Method

First, we apply the traditional group lasso to estimate the initial values of the matrix $A$. An optimization problem for implementing the traditional group lasso method can be defined as

$$\min_{a_{\cdot j}} \quad F = \mathrm{Tr}\left((Y - AX)^T(Y - AX)\right) + \lambda\sum_{j=1}^{n} m\| a_{\cdot j}\|_2.$$ (7.I.1)

Let

$$\rho(A) = \text{Tr}\big((Y - AX)^T (Y - AX)\big).$$

Using matrix calculus, we obtain

$$\frac{\partial \rho(A)}{\partial A} = -(Y - AX)X^T, \tag{7.I.2}$$

which can be further reduced to

$$\frac{\partial \rho(A)}{\partial A} = -\left(Y - A^{(-j)}X\right)X^T + a_{.j}\| X_{j.} \|_2^2, \tag{7.I.3}$$

where

$$A^{(-j)}, X_{j.}$$

are defined in Result 7.22.

It follows from Equation 7.I.3 that

$$\frac{\partial \rho(A)}{\partial a_{.j}} = -\left(Y - A^{(-j)}X\right)X_{j.}^T + a_{.j}\| X_{j.} \|_2^2. \tag{7.I.4}$$

It is well known that the subgradient of the norm $\| a_{.j} \|_2$ is

$$\partial \| a_{.j} \|_2 = \begin{cases} \dfrac{a_{.j}}{\| a_{.j} \|_2} & \| a_{.j} \|_2 \neq 0 \\[2mm] s, \| s \|_2 \leq 1 & \| a_{.j} \|_2 = 0. \end{cases} \tag{7.I.5}$$

A point $a_{.j}^*$ is the minimum of the objective function $F$ in (7.I.1) if and only if

$$0 \in \partial F\left(a_{.j}^*\right) = \frac{\partial \rho(A)}{\partial a_{.j}} + \lambda m \, \partial \| a_{.j} \|_2. \tag{7.I.6}$$

It follows from Equations 7.I.5 and 7.I.6 that when $\| a_{.j} \|_2 \neq 0$ then we have

$$-\left(Y - A^{(-j)}X\right)X_{j.}^T + a_{.j}\| X_{j.} \|_2^2 + \frac{\lambda m a_{.j}}{\| a_{.j} \|_2} = 0. \tag{7.I.7}$$

Let

$$S_j = \left(Y - A^{(-j)}X\right)X_{j.}^T. \tag{7.I.8}$$

Substituting Equation 7.I.8 into Equation 7.I.7 gives

$$-S_j + a_{.j}\| X_j \|_2^2 + \frac{\lambda m a_{.j}}{\| a_{.j} \|_2} = 0, \tag{7.I.9}$$

which implies

$$\left( \| X_{j.} \|_2^2 + \frac{\lambda m}{\| a_{.j} \|_2} \right) a_{.j} = S_{j.}$$

(7.I.10)

Taking norm $\| . \|_2$ on both sides of Equation 7.I.10, we obtain

$$\left( \| X_{j.} \|_2^2 + \frac{\lambda m}{\| a_{.j} \|_2} \right) \| a_{.j} \|_2 = \| S_j \|_2$$

or

$$\| X_{j.} \|_2^2 \| a_{j.} \|_2 + \lambda m = \| S_j \|_2.$$

(7.I.11)

Solving Equation 7.I.11 for $\| a_{j.} \|_2$, we obtain

$$\| a_{j.} \|_2 = \frac{\| S_j \|_2 - \lambda m}{\| X_{j.} \|_2^2}.$$

(7.I.12)

Substituting Equation 7.I.12 into Equation 7.I.10, we obtain

$$a_{.j} = \left( 1 - \frac{\lambda m}{\| S_j \|_2} \right) \frac{S_j}{\| X_{j.} \|_2^2}.$$

(7.I.13)

Now we consider case where $\| a_{.j} \|_2 = 0$. When $\| a_{.j} \|_2 = 0$ Equation 7.I9 becomes

$$- S_j + \lambda m \delta = 0,$$

(7.I.14)

where $\| \delta \|_2 \leq 1$ or

$$\| S_j \|_2 = \lambda m \| \delta \|_2 \leq \lambda m,$$

which implies

$$\frac{\lambda m}{\| S_j \|_2} \geq 1.$$

(7.I.15)

In other words, if $\frac{\lambda m}{\| S_j \|_2} \geq 1$ we must have

$$a_{.j} = 0, \text{ for } \frac{\lambda m}{\| S_j \|_2} \geq 1.$$

(7.I.16)

Combining Equations 7.I.13 and 7.I.16 gives the final solution:

$$a_{.j} = \left( 1 - \frac{\lambda m}{\| S_j \|_2} \right)_+ \frac{S_j}{\| X_{j.} \|_2^2},$$

(7.I.17)

where

$$(b)_+ = \begin{cases} b & \text{if } b > 0 \\ 0 & b \leq 0. \end{cases}$$

Next, we study the major step in the algorithm. Similar to Equation 7.I.4, we can have

$$\frac{\partial \rho(A)}{\partial a_{kj}} = -\left(Y_k - a_k^{-(kj)} X\right) X_j^T + a_{kj} \| X_j \|_2^2, \tag{7.I.18}$$

where

$$Y_{k.} = [y_{k1} \quad \cdots \quad y_{kL}], \ a_{k.} = \lfloor a_{k1} \quad \cdots \quad a_{kj} \quad \cdots a_{kn} \rfloor,$$

and $a_{k.}^{(kj)}$ is $a_{k.}$ with element $kj$ replaced by zero.

Let $S_{kj} = (Y_{k.} - a_{k.}^{-(kj)} X) X_j^T$. Equation 7.I.18 can be rewritten as

$$\frac{\partial \rho(A)}{\partial a_{kj}} = -S_{kj} + a_{kj} \| X_{j.} \|_2^2. \tag{7.I.19}$$

Again, subdifferential of $\| a_{.j} \|_2$ at the point $a_{kj}$ is

$$\frac{\partial \| a_{.j} \|_2}{\partial a_{kj}} = \begin{cases} \dfrac{a_{kj}}{\| a_{.j} \|_2} & a_{kj} \neq 0 \\ |\alpha| \leq 1 & a_{kj} = 0. \end{cases} \tag{7.I.20}$$

Now consider $a_{kj} \neq 0$. In this case, the optimal condition for solving problem (7.144) is

$$\frac{\partial F}{\partial a_{kj}} = -S_{kj} + a_{kj} \| X_{j.} \|_2^2 + \lambda \left( \frac{n}{\| a_{k.} \|_2} + \frac{m}{\| a_{.j} \|_2} \right) a_{kj} = 0. \tag{7.I.21}$$

Solving Equation 7.I.21 for $a_{kj}$ is

$$a_{kj}^{(t+1)} = \frac{S_{kj}^{(t)}}{\| X_{j.} \|_2^2 + \lambda \left( \dfrac{n}{\| a_{k.}^{(t)} \|_2} + \dfrac{m}{\| a_{.j}^{(t)} \|_2} \right)}, \tag{7.I.22}$$

where $S_{kj}^{(t)}$, $a_{k.}^{(t)}$ and $a_{.j}^{(t)}$ are calculated using the matrix $A^{(t)}$.

Next consider $a_{kj} = 0$. When $a_{kj} = 0$ Equation 7.I.21 is changed to

$$\frac{\partial F}{\partial a_{kj}} = -S_{kj} + \lambda(m + n)\alpha = 0, \tag{7.I.23}$$

where

$$|\alpha| \leq 1$$

Therefore, if $|S_{kj}| \leq \lambda(m + n)$, then set $a_{kj} = 0$.

---

## Appendix 7.J    Derivation of the Distribution of the Prediction in the Bayesian Linear Models

For the self-contained, in this appendix, we briefly give the derivation of the predictive distribution in the Bayesian linear models following the approach by Rasmussen and Williams (2016).

Using Equations 7.367 and 7.373, we obtain

$$p(y_{new}|x_{new}, w)p(w|X, y)$$

$$\propto \exp\left\{-\frac{1}{2}\left[\frac{1}{\sigma_n^2}\left(y_{new} - x_{new}^T w\right)^T \left(y_{new} - x_{new}^T w\right) + (w - \mu)^T \Lambda(w - \mu)\right]\right\}. \quad (7.\text{J}.1)$$

Note that

$$\frac{1}{\sigma_n^2}(y_{new} - x_{new}^T w)T(y_{new} - x_{new}^T w)$$

$$= \frac{1}{\sigma_n^2} y_{new}^2 - \frac{2}{\sigma_n^2} w^T x_{new} y_{new} + \frac{w^T x_{new} x_{new}^T}{\sigma_n^2} w, \quad (7.\text{J}.2)$$

$$(w - \mu)^T \Lambda(w - \mu) = w^T \Lambda w - 2w^T \mu + \mu^T \Lambda\mu. \quad (7.\text{J}.3)$$

Substituting Equations 7.J.2 and 7.J.3 into Equation 7.J.1, we obtain

$$p(y_{new}|x_{new}, w)p(w|X, y)$$

$$\propto \exp\left\{-\frac{1}{2}\left[w^T\left(\Lambda + \frac{x_{new} x_{new}^T}{\sigma_n^2}\right)w - 2w^T(\Lambda\mu + \frac{x_{new} y_{new}}{\sigma_n^2}) + \frac{y_{new}^2}{\sigma_n^2}\right]\right\}. \quad (7.\text{J}.4)$$

Let

$$V = \Lambda + \frac{x_{new} x_{new}^T}{\sigma_n^2}. \quad (7.\text{J}.5)$$

Then,

$$V^{-1} = \Lambda^{-1} - \frac{1}{\sigma_n^2} \frac{\Lambda^{-1} x_{new} x_{new}^T \Lambda^{-1}}{1 + \frac{1}{\sigma_n^2} x_{new}^T \Lambda^{-1} x_{new}}. \tag{7.J.6}$$

Define

$$(w - m)^T V (w - m) = w^T V w - 2w^T V m + m^T V m. \tag{7.J.7}$$

To find the mean $m$, we set the equation

$$Vm = \Lambda\mu + \frac{x_{new} y_{new}}{\sigma_n^2}. \tag{7.J.8}$$

Solving Equation 7.J.8 for $m$ gives

$$m = V^{-1} \left( \Lambda\mu + \frac{x_{new} y_{new}}{\sigma_n^2} \right), \tag{7.J.9}$$

which implies

$$m^T V m = \left( \Lambda\mu + \frac{x_{new} y_{new}}{\sigma_n^2} \right)^T V^{-1} \left( \Lambda\mu + \frac{x_{new} y_{new}}{\sigma_n^2} \right)$$

$$= \frac{x_{new}^T V^{-1} x_{new}}{\sigma_n^4} y_{new}^2 + 2 \frac{\mu^T \Lambda V^{-1} x_{new} y_{new}}{\sigma_n^2} + const. \tag{7.J.10}$$

Substituting Equation 7.J.10 into Equation 7.J.4, we obtain

$$p(y_{new}|x_{new}, w) p(w|X, y) \propto \exp\left\{ -\frac{1}{2} \left[ (w - m)^T V (w - m) - m^T V m + \frac{y_{new}^2}{\sigma_n^2} \right] \right\}$$

$$\propto \exp\left\{ -\frac{1}{2} (w - m)^T V (w - m) \right.$$

$$\left. -\frac{1}{2} \left[ \frac{y_{new}^2}{\sigma_n^2} - \frac{x_{new}^T V^{-1} x_{new}}{\sigma_n^4} y_{new}^2 - 2 \frac{\mu T \Lambda V^{-1} x_{new} y_{new}}{\sigma_n^2} + const. \right] \right\} \tag{7.J.11}$$

Substituting Equation 7.J.11 into Equation 7.375, we obtain

$$p(y_{new}|x_{new}, X, y) = \int p(y_{new}|x_{new}, w) p(w|X, y) dw$$

$$\propto \exp\left\{ -\frac{1}{2} \left[ \frac{1}{\sigma_n^2} \left( 1 - \frac{1}{\sigma_n^2} x_{new}^T V^{-1} x_{now} \right) y_{new}^2 - 2 \frac{\mu^T \Lambda V - 1 x_{new} y_{new}}{\sigma_n^2} \right] \right\}. \tag{7.J.12}$$

Setting

$$\lambda(y_{new} - \bar{y})^2 = \frac{1}{\sigma_n^2} \left( 1 - \frac{1}{\sigma_n^2} x_{new}^T V^{-1} x_{new} \right) y_{new}^2 - 2 \frac{\mu T \Lambda V^{-1} x_{new} y_{new}}{\sigma_n^2} + \bar{y}^2,$$

we obtain

$$\lambda = \frac{1}{\sigma_n^2}\left(1 - \frac{1}{\sigma_n^2}x_{new}^T V^{-1}x_{new}\right) \tag{7.J.13}$$

and

$$\bar{y} = \frac{1}{\lambda}\frac{1}{\sigma_n^2}x_{new}^T V^{-1}\Lambda\mu, \tag{7.J.14}$$

where

$$\mu = \frac{1}{\sigma_n^2}\Lambda^{-1}Xy.$$

Note that

$$x_{new}^T V^{-1}x_{new} = x_{new}^T \Lambda^{-1}x_{new} - \frac{1}{\sigma_n^2}\frac{\left(x_{new}^T\Lambda^{-1}x_{new}\right)^2}{1 + \frac{1}{\sigma_n^2}x_{new}^T\Lambda^{-1}x_{new}}$$

$$= x_{new}^T\Lambda^{-1}x_{new}\left[1 - \frac{1}{\sigma_n^2}\frac{x_{new}^T\Lambda^{-1}x_{new}}{1 + \frac{1}{\sigma_n^2}x_{new}^T\Lambda^{-1}x_{new}}\right] \tag{7.J.15}$$

$$= \frac{x_{new}^T\Lambda^{-1}x_{new}}{1 + \frac{1}{\sigma_n^2}x_{new}^T\Lambda^{-1}x_{new}}$$

Substituting Equation 7.J.15 into Equation 7.J.13, we obtain

$$\lambda = \frac{1}{\sigma_n^2\left(1 + \frac{1}{\sigma_n^2}x_{new}^T\Lambda^{-1}x_{new}\right)} \tag{7.J.16}$$

or

$$\frac{1}{\lambda} = \sigma_n^2\left(1 + \frac{1}{\sigma_n^2}x_{new}^T\Lambda^{-1}x_{new}\right). \tag{7.J.17}$$

Recall that

$$\bar{y} = \frac{1}{\lambda}\frac{1}{\sigma_n^2}x_{new}^T V^{-1}\Lambda\mu$$

$$= \mu^T\frac{\Lambda V^{-1}x_{new}}{\lambda^T\lambda\sigma_n^2} \tag{7.J.18}$$

Next we show

$$\frac{\Lambda V^{-1}x_{new}}{\lambda\sigma_n^2} = x_{new}.$$

To prove this, we make the following transformation:

$$V\Lambda^{-1}x_{new} = \left(\Lambda + \frac{x_{new}x_{new}^T}{\sigma_n^2}\right)\Lambda^{-1}x_{new}$$

$$= x_{new} + \frac{x_{new}x_{new}^T\Lambda^{-1}x_{new}}{\sigma_n^2} \qquad (7.J.19)$$

$$= \left(1 + \frac{x_{new}^T\Lambda^{-1}x_{new}}{\sigma_n^2}\right)x_{new}.$$

It follows from Equation 7.J.17 that

$$\frac{1}{\lambda\sigma_n^2} = \frac{1}{\lambda\sigma_n^2}x_{new}^T\Lambda^{-1}x_{new}. \qquad (7.J.20)$$

Substituting Equation 7.J.20 into Equation 7.J.19 gives

$$V\Lambda^{-1}x_{new} = \frac{1}{\lambda\sigma_n^2}x_{new}. \qquad (7.J.21)$$

Multiplying both sides of Equation 7.J.21 by $\Lambda V^{-1}$, we obtain

$$x_{new} = \frac{\Lambda V^{-1}x_{new}}{\lambda\sigma_n^2} \qquad (7.3A22)$$

Substituting Equation 7.J.22 into Equation 7.J.18 yields

$$\bar{y} = \mu^T x_{new}$$

$$= x_{new}^T\mu. \qquad (7.J.23)$$

Combining Equations 7.J.14 and 7.J.23, we obtain

$$\bar{y} = x_{new}^T\frac{1}{\sigma_n^2}\Lambda^{-1}Xy. \qquad (7.J.24)$$

$$p(y_{new}|x_{new}, X, y) \propto exp\left\{-\frac{\left(y_{new} - \frac{1}{\sigma_n^2}x_{new}^T\Lambda^{-1}Xy\right)}{2x_{new}^T\Lambda^{-1}x_{new}}\right\} \qquad (7.J.25)$$

$$= N\left(\frac{1}{\sigma_n^2}x_{new}^T\Lambda^{-1}Xy, \Lambda^{-1}x_{new}\right).$$

## Exercises

Exercise 1. Consider the SEM

$$X = \varepsilon_x$$

$$Y = X^2 + \varepsilon_y,$$

where $\varepsilon_x, \varepsilon_y$ are normally distributed as $N(0,1)$. Find the distribution of effect $Y$ if $X$ is set to the value 3.

Exercise 2. Let $f(x) = \dfrac{1}{2b} \exp\left\{-\dfrac{|x - \mu|}{b}\right\}$ be the density function of the Laplace distribution. Show that its differential entropy is

$$S(l) = 1 + \log(2b).$$

Exercise 3. Let $f(x) = \dfrac{x^{\alpha-1}(1 - x)^{\beta-1}}{B(\alpha, \beta)}$ be the density function of the Beta distribution. Show that its differential entropy is

$$S(b) = \log B(\alpha, \beta) - (\alpha - 1)[\psi(\alpha) - \psi(\alpha + \beta)] - (\beta - 1)$$

$$\times [\psi(\beta) - \psi(\alpha + \beta)],$$

where $\psi$ is a digamma function.

Exercise 4. Let $f(x)$ be a beta distribution and $g(x)$ a uniform distribution on $[0,1]$. Calculate K-L distance $D(f \parallel g)$.

Exercise 5. Let

$$g_2(x) = D\left(P_{Y|x} \| \vec{P}(y)\right) = \int \log \frac{P(y|x)}{\vec{P}(y)} P(y|x) dy.$$

Discuss the implication of $\operatorname{cov}_{U_x}\left(g_2, \dfrac{P_X}{U_X}\right) = 0.$

Exercise 6. Show that if $X$ and $Y$ are dimension one then

$$\Delta_{X \to Y} = 0 \text{ and } \Delta_{Y \to X} = 0.$$

Exercise 7. Show that

$$\Delta_{X \to Y} + \Delta_{Y \to X} = -\log(1 - \operatorname{cov}(Z, 1/Z)) + \log \frac{n}{m}$$

# References

Ajami, S. and Teimouri, F. (2015). Features and application of wearable biosensors in medical care. *Journal of Research in Medical Sciences: The Official Journal of Isfahan University of Medical Sciences* **20**:1208–1215.

Akutekwe, A. and Seker, H. (2015). Inference of nonlinear gene regulatory networks through optimized ensemble of support vector regression and dynamic Bayesian networks. *Engineering in Medicine and Biology Society (EMBC), 2015 37th Annual International Conference of the IEEE* 8177–8180.

Albarqouni, S., Baur, C., Achilles, F., Belagiannis, V., Demirci, S. and Navab, N. (2016). Aggnet: Deep learning from crowds for mitosis detection in breast cancer histology images. *IEEE Transactions on Medical Imaging* **35**:1313–1321.

Alonso, A. M., Casado, D. and Romo, J. (2012). Supervised classification for functional data: A weighted distance approach. *Computational Statistics & Data Analysis* **56**: 2334–2346.

Anders, S. and Huber, W. (2010). Differential expression analysis for sequence count data. *Genome Biol* **11**:R106.

Anderson, T. W. (1984). *An introduction to multivariate statistical analysis. 2nd ed.*, John Wiley & Sons, New York.

Andrew, N. (2011). Sparse autoencoder. *CS294A Lecture Notes* **72**.

Aschard, H., Vilhjálmsson, B. J., Greliche, N., Morange, P.-E., Trégouët, D.-A. and Kraft, P. (2014). Maximizing the power of principal-component analysis of correlated phenotypes in genome-wide association studies. *The American Journal of Human Genetics* **94**:662–676.

Bühlmann, P., Peters, J. and Ernest, J. (2014). CAM: Causal additive models, high-dimensional order search and penalized regression. *The Annals of Statistics* **42**: 2526–2556.

Bailey, D. L., Townsend, D. W., Valk, P. E. and Maisey, M. N. (2005). *Positron emission tomography.* Springer.

Baiocchi, M., Cheng, J. and Small, D. S. (2014). Instrumental variable methods for causal inference. *Statistics in Medicine* **33**:2297–2340.

Balevich, E. C., Haznedar, M. M., Wang, E. et al. (2015). Corpus callosum size and diffusion tensor anisotropy in adolescents and adults with schizophrenia. *Psychiatry Research: Neuroimaging* **231**:244–251.

Bartlett, M. and Cussens, J. (2013). Advances in Bayesian network learning using integer programming. *arXiv:1309.6825.*

Bartlett, M. S. (1939). A note on tests of significance in multivariate analysis. *Mathematical Proceedings of the Cambridge Philosophical Society* 180–185.

Barzel, B. and Barabási, A.-L. (2013). Network link prediction by global silencing of indirect correlations. *Nature Biotechnology* **31**:720–725.

Belilovsky, E., Varoquaux, G. and Blaschko, M. B. (2016). Testing for differences in Gaussian graphical models: Applications to brain connectivity. *arXiv:1512.08643.*

Besag, J. (1977). Efficiency of pseudolikelihood estimation for simple Gaussian fields. *Biometrika* 616–618.

697

Bollen, K. A. (1989). *Structural equations with latent variables.* John Wiley & Sons, New York.

Bollen, K. A. (2012). Instrumental variables in sociology and the social sciences. *Annual Review of Sociology* **38**:37–72.

Bolstad, B. M., Irizarry, R. A., Åstrand, M. and Speed, T. P. (2003). A comparison of normalization methods for high density oligonucleotide array data based on variance and bias. *Bioinformatics* **19**:185–193.

Bouvrie, J. (2006). Notes on convolutional neural networks. http://cogprints.org /5869/1/cnn_tutorial.pdf.

Boyd, S., Parikh, N., Chu, E., Peleato, B. and Eckstein, J. (2011). Distributed optimization and statistical learning via the alternating direction method of multipliers. *Foundations and Trends® in Machine Learning* **3**:1–122.

Boyle, E. A., Li, Y. I. and Pritchard, J. K. (2017). An Expanded View of Complex Traits: From Polygenic to Omnigenic. *Cell* **169**:1177–1186.

Brauns, S., Gollub, R. L., Walton, E. et al. (2013). Genetic variation in GAD1 is associated with cortical thickness in the parahippocampal gyrus. *Journal of Psychiatric Research* **47**:872–879.

Bressler, S. L. and Seth, A. K. (2011). Wiener–Granger causality: A well established methodology. *Neuroimage* **58**:323–329.

Brito, C. and Pearl, J. (2002). Generalized instrumental variables. *Proceedings of the Eighteenth Conference on Uncertainty in Artificial Intelligence* 85–93.

Cai, X., Bazerque, J. A. and Giannakis, G. B. (2013). Inference of gene regulatory networks with sparse structural equation models exploiting genetic perturbations. *PLoS Comput Biol* **9**:e1003068.

Callaway, E. (2017). Genome studies attract criticism. *NATURE* **546**:463.

Cancer Genome Atlas Research Network. (2011). Integrated genomic analyses of ovarian carcinoma. *NATURE* **474**:609–615.

Candès, E. J., Li, X., Ma, Y. and Wright, J. (2011). Robust principal component analysis? *Journal of the ACM (JACM)* **58**:11.

Cascio, D., Magro, R., Fauci, F., Iacomi, M. and Raso, G. (2012). Automatic detection of lung nodules in CT datasets based on stable 3D mass–spring models. *Computers in Biology and Medicine* **42**:1098–1109.

Chang, C. C. and Lin, C. J. (2011). LIBSVM: A library for support vector machines. *ACM Transactions on Intelligent Systems and Technology* **2**:1–27.

Chapter 2: Variational Bayesian theory. www.cse.buffalo.edu/faculty/mbeal/thesis /beal03_2.pdf.

Chen, B. and Pearl, J. (2014). Graphical tools for linear structural equation modeling. *UCLA Cognitive Systems Laboratory, Technical Report (R-432).*

Chen, X., Zou, C. and Cook, R. D. (2010). Coordinate-independent sparse sufficient dimension reduction and variable selection. *The Annals of Statistics* 3696–3723.

Chen, Z., Zhang, K. and Chan, L. (2013). Nonlinear causal discovery for high dimensional data: A kernelized trace method. *Data Mining (ICDM), 2013 IEEE 13th International Conference on* 1003–1008.

Cheng, J., Grainer, G., Kelly, J., Bell, D. and Liu, W. (2002). Learning bayesian networks from data: An information-theory based approach. *Artificial Intelligence* **137**:43–90.

Cheng, S., Guo, M., Wang, C., Liu, X., Liu, Y. and Wu, X. (2016). MiRTDL: A deep learning approach for miRNA target prediction. *IEEE/ACM Transactions on Computational Biology and Bioinformatics* **13**:1161–1169.

Ciresan, D., Giusti, A., Gambardella, L. M. and Schmidhuber, J. (2012). Deep neural networks segment neuronal membranes in electron microscopy images. *Advances in Neural Information Processing Systems* 2843–2851.

Clemmensen, L., Hastie, T., Witten, D. and Ersbøll, B. (2011). Sparse discriminant analysis. *Technometrics* **53**:406–413.

Clyde, D. (2017). Disease genomics: Transitioning from association to causation with eQTLs. *Nature Reviews Genetics* **18**:271.

Cover, T. M. and Thomas, J. A. (2012). *Elements of information theory*. John Wiley & Sons, New York.

Cowley, B. and Vinci, G. (2014). Summary and discussion of: "Brownian Distance Covariance". http://www.stat.cmu.edu/~ryantibs/journalclub/dcov.pdf.

Cussens, J. (2012). Bayesian network learning with cutting planes. *arXiv:1202.3713*.

Cussens, J. (2014). Integer programming for Bayesian network structure learning. *Quality Technology & Quantitative Management* **11**:99–110.

Danaher, P., Wang, P. and Witten, D. M. (2014). The joint graphical lasso for inverse covariance estimation across multiple classes. *Journal of the Royal Statistical Society: Series B (Statistical Methodology)* **76**:373–397.

Daniusis, P., Janzing, D., Mooij, J. et al. (2012). Inferring deterministic causal relations. *Proceedings of the 26th Conference on Uncertainty in Artificial Intelligence (UAI)* pp. 1–8.

Das, K., Li, J., Wang, Z. et al. (2011). A dynamic model for genome-wide association studies. *Human genetics* **129**:629–639.

De Campos, L. M. and Huete, J. F. (2000). A new approach for learning belief networks using independence criteria. *International Journal of Approximate Reasoning* **24**:11–37.

Delude, C. M. (2015). Deep phenotyping: The details of disease. *NATURE* **527**:S14–S15.

Dhawan, A. P. (2011). *Medical image analysis*. Vol. 31. John Wiley & Sons.

Dippel, C., Gold, R., Heblich, S. and Pinto, R. (2017). Instrumental Variables and Causal Mechanisms: Unpacking The Effect of Trade on Workers and Voters. *National Bureau of Economic Research*.

Dutta, M. (2015). Assessment of feature extraction techniques for hyperspectral image classification. *Computer Engineering and Applications (ICACEA), 2015 International Conference on Advances in* 499–502.

Ernest, J. (2016). *Causal inference in semiparametric and nonparametric structural equation models*, Ph.D. Thesis, ETH Zurich.

Fan, X., Malone, B. and Yuan, C. (2014). Finding optimal Bayesian network structures with constraints learned from data. *Proceedings of the Thirtieth Conference on Uncertainty in Artificial Intelligence* 200–209.

Farabet, C., Couprie, C., Najman, L. and LeCun, Y. (2013). Learning hierarchical features for scene labeling. *IEEE Transactions on Pattern Analysis and Machine Intelligence* **35**:1915–1929.

Fortin, J. P., Triche, T. J., Jr. and Hansen, K. D. (2017). Preprocessing, normalization and integration of the Illumina Human Methylation EPIC array with minfi. *Bioinformatics* **33**:558–560.

Friedman, J., Hastie, T. and Tibshirani, R. (2008). Sparse inverse covariance estimation with the graphical lasso. *Biostatistics* **9**:432–441.

Friedman, N. and Nachman, I. 2000. Gaussian process networks. In *Proceedings of the Sixteenth Conference on Uncertainty in Artificial Intelligence*. Stanford, CA: Morgan Kaufmann Publishers Inc.

Friston, K. J., Harrison, L. and Penny, W. (2003). Dynamic causal modelling. *Neuroimage* **19**:1273–1302.

Fügenschuh, A. and Martin, A. (2005). Computational integer programming and cutting planes. *Handbooks in Operations Research and Management Science* **12**:69–121.

Fusi, N. and Listgarten, J. (2016). Flexible modelling of genetic effects on function-valued traits. *Journal of Computational Biology* **24**:524–535.

Gamboa, J. C. B. (2017). Deep learning for time-series analysis. *arXiv:1701.01887*.

Gao, W., Emaminejad, S., Nyein, H. Y. Y. et al. (2016). Fully integrated wearable sensor arrays for multiplexed in situ perspiration analysis. *NATURE* **529**:509–514.

Garcia-Garcia, A., Orts-Escolano, S., Oprea, S., Villena-Martinez, V. and Garcia-Rodriguez, J. (2017). A review on deep learning techniques applied to semantic segmentation. *arXiv:1704.06857*.

Garin-Muga, A. and Borro, D. (2014). Review and challenges of brain analysis through DTI measurements. *Studies in Health Technology and Informatics* **207**:27–36.

*Gene expression.* https://www2.stat.duke.edu/courses/Spring04/sta278/refinfo/Gene_Expression.pdf.

Gershman, S. J. and Daw, N. D. (2017). Reinforcement learning and episodic memory in humans and animals: An integrative framework. *Annual Review of Psychology* **68**:101–128.

Gianola, D. and Sorensen, D. (2004). Quantitative genetic models for describing simultaneous and recursive relationships between phenotypes. *Genetics* **167**:1407–1424.

Gibbs, A. L. and Su, F. E. (2002). On choosing and bounding probability metrics. *International Statistical Review* **70**:419–435.

Gibson, E., Hu, Y., Huisman, H. J. and Barratt, D. C. (2017). Designing image segmentation studies: Statistical power, sample size and reference standard quality. *Med Image Anal* **42**:44–59.

Glover, G. H. (2011). Overview of functional magnetic resonance imaging. *Neurosurgery Clinics of North America* **22**:133–139.

González, I. (2014). Tutorial: Statistical analysis of RNA-Seq data. www.nathalievilla.org/doc/pdf/tutorial-rnaseq.pdf.

Granger, C. W. (1969). Investigating causal relations by econometric models and cross-spectral methods. *Econometrica: Journal of the Econometric Society* 424–438.

Gretton, A. (2015). Notes on mean embeddings and covariance operators. http://www.gatsby.ucl.ac.uk/~gretton/coursefiles/lecture5_covarianceOperator.pdf.

Gretton, A., Borgwardt, K. M., Rasch, M. J., Schölkopf, B. and Smola, A. (2012). A kernel two-sample test. *Journal of Machine Learning Research* **13**:723–773.

Gretton, A., Bousquet, O., Smola, A. and Scholkopf, B. (2005). Measuring statistical dependence with Hilbert-Schmidt norms. *Lecture Notes in Computer Science* 63–78.

Guo, J., Levina, E., Michailidis, G. and Zhu, J. (2011). Joint estimation of multiple graphical models. *Biometrika* **98**:1–15.

Hachiya, T., Furukawa, R., Shiwa, Y. et al. (2017). Genome-wide identification of inter-individually variable DNA methylation sites improves the efficacy of epigenetic association studies. *npj Genomic Medicine* **2**:11.

Hansen, K. D., Wu, Z., Irizarry, R. A. and Leek, J. T. (2011). Sequencing technology does not eliminate biological variability. *Nature Biotechnology* **29**:572–573.

Henderson, D. and Plaschko, P. (2006). *Stochastic differential equations in science and engineering:(With CD-ROM)*. World Scientific.

Hernandez, K. M. (2015). Understanding the genetic architecture of complex traits using the function—Valued approach. *New Phytologist* **208**:1–3.

Hoffmann, A., Ziller, M. and Spengler, D. (2016). The future is the past: Methylation QTLs in schizophrenia. *Genes* **7**:104.

Hong, S., Chen, X., Jin, L. and Xiong, M. (2013). Canonical correlation analysis for RNA-seq co-expression networks. *Nucleic Acids Research* **41**:e95.

Hosseini-Asl, E. (2016). *Sparse feature learning for image analysis in segmentation, classification, and disease diagnosis.* https://doi.org/10.18297/etd/2456.

Hoyer, P. O., Janzing, D., Mooij, J. M., Peters, J. and Schölkopf, B. (2009). Nonlinear causal discovery with additive noise models. *Advances in Neural Information Processing Systems* 689–696.

Hrdlickova, R., Toloue, M. and Tian, B. (2017). RNA-Seq methods for transcriptome analysis. *Wiley Interdiscip Rev RNA* **8**.

Huan, T., Meng, Q., Saleh, M. A. et al. (2015). Integrative network analysis reveals molecular mechanisms of blood pressure regulation. *Molecular Systems Biology* **11**: 799.

Huang, H.-C., Niu, Y. and Qin, L.-X. (2015). Differential expression analysis for RNA-seq: An overview of statistical methods and computational software. *Cancer informatics* **14**:57–67.

Hyvärinen, A. (1998). New approximations of differential entropy for independent component analysis and projection pursuit. *Advances in Neural Information Processing Systems* **10**:273–279.

Hyvärinen, A. and Smith, S. M. (2013). Pairwise likelihood ratios for estimation of non-Gaussian structural equation models. *Journal of Machine Learning Research* **14**: 111–152.

Ideker, T. and Nussinov, R. (2017). Network approaches and applications in biology. *PLoS Comput Biol* **13**:e1005771.

Jaakkola, T., Sontag, D., Globerson, A. and Meila, M. (2010). Learning Bayesian network structure using LP relaxations. *Proceedings of the Thirteenth International Conference on Artificial Intelligence and Statistics* 358–365.

Janzing, D., Hoyer, P. O. and Schölkopf, B. (2009). Telling cause from effect based on high-dimensional observations. In Proceedings of the 27th International Conference on Machine Learning pp. 479–486.

Janzing, D., Mooij, J., Zhang, K. et al. (2012). Information-geometric approach to inferring causal directions. *Artificial Intelligence* **182**:1–31.

Janzing, D. and Scholkopf, B. (2010). Causal inference using the algorithmic Markov condition. *IEEE Transactions on Information Theory* **56**:5168–5194.

Janzing, D., Steudel, B., Shajarisales, N. and Schölkopf, B. (2015). Justifying information-geometric causal inference. In *Measures of complexity,* Springer.

Javery, O., Shyn, P. and Mortele, K. (2013). FDG PET or PET/CT in patients with pancreatic cancer: When does it add to diagnostic CT or MRI? *Clinical Imaging* **37**: 295–301.

Jefkine. (2016). *Backpropagation in convolutional neural networks.* http://www.jefkine.com/general/2016/09/05/backpropagation-in-convolutional-neural-networks.

Jiang, J., Lin, N., Guo, S., Chen, J. and Xiong, M. (2015). Multiple functional linear model for association analysis of RNA-seq with imaging. *Quantitative Biology* **3**:90–102.

Jiang, P., Scarpa, J. R., Fitzpatrick, K. et al. (2015). A systems approach identifies networks and genes linking sleep and stress: Implications for neuropsychiatric disorders. *Cell Reports* **11**:835–848.

Jin, B., Li, Y. and Robertson, K. D. (2011). DNA methylation: Superior or subordinate in the epigenetic hierarchy? *Genes Cancer* **2**:607–617.

Jin, L.-p. and Dong, J. (2016). Ensemble deep learning for biomedical time series classification. *Computational Intelligence and Neuroscience* **2016**:6212684.

Judge, G. G., Hill, R. C., Griffiths, W., Lutkepohl, H. and Lee, T. C. (1982). *Introduction to the theory and practice of econometrics.* John Wiley & Sons, New York.

Köhler, S., Vasilevsky, N. A., Engelstad, M. et al. (2017). The human phenotype ontology in 2017. *Nucleic Acids Research* **45**:D865–D876.

Kalisch, M. and Bühlmann, P. (2007). Estimating high-dimensional directed acyclic graphs with the PC-algorithm. *Journal of Machine Learning Research* **8**:613–636.

Kalyagin, V. A., Koldanov, A. P., Koldanov, P. A. and Pardalos, P. M. (2017). Optimal statistical decision for Gaussian graphical model selection. *arXiv:1701.02071.*

Koivisto, M. and Sood, K. (2004). Exact Bayesian structure discovery in Bayesian networks. *Journal of Machine Learning Research* **5**:549–573.

Kozlov, A. V. and Koller, D. (1997). Nonuniform dynamic discretization in hybrid networks. *Proceedings of the Thirteenth Conference on Uncertainty in Artificial Intelligence* 314–325.

Kpotufe, S., Sgouritsa, E., Janzing, D. and Schölkopf, B. (2014). Consistency of causal inference under the additive noise model. *Proceedings of the 31st International Conference on Machine Learning (ICML-14)* 478–486.

Krämer, N., Schäfer, J. and Boulesteix, A.-L. (2009). Regularized estimation of large-scale gene association networks using graphical Gaussian models. *BMC Bioinformatics* **10**:384.

Kraskov, A., Stögbauer, H. and Grassberger, P. (2004). Estimating mutual information. *Physical Review E* **69**:066138.

Kremling, A. and Saez-Rodriguez, J. (2007). Systems biology—An engineering perspective. *Journal of Biotechnology* **129**:329–351.

Krizhevsky, A., Sutskever, I. and Hinton, G. E. (2012). Imagenet classification with deep convolutional neural networks. *Advances in Neural Information Processing Systems* 1097–1105.

Krueger, F., Kreck, B., Franke, A. and Andrews, S. R. (2012). DNA methylome analysis using short bisulfite sequencing data. *Nature Methods* **9**:145–151.

Kulig, P., Pach, R. and Kulig, J. (2014). Role of abdominal ultrasonography in clinical staging of pancreatic carcinoma: A tertiary center experience. *Pol Arch Med Wewn* **124**:225–232.

Kurdyukov, S. and Bullock, M. (2016). DNA Methylation Analysis: Choosing the Right Method. *Biology (Basel)* **5**:3.

Kwak, I.-Y., Moore, C. R., Spalding, E. P. and Broman, K. W. (2014). A simple regression-based method to map quantitative trait loci underlying function-valued phenotypes. *Genetics* **197**:1409–1416.

Kwak, I.-Y., Moore, C. R., Spalding, E. P. and Broman, K. W. (2016). Mapping quantitative trait loci underlying function-valued traits using functional principal component analysis and multi-trait mapping. *G3: Genes, Genomes, Genetics* **6**:79–86.

Laird, P. W. (2010). Principles and challenges of genome wide DNA methylation analysis. *Nature Reviews Genetics* **11**:191–203.

Lappalainen, T. and Greally, J. M. (2017). Associating cellular epigenetic models with human phenotypes. *Nat Rev Genet* **18**:441–451.

Lappalainen, T., Sammeth, M., Friedländer, M. R. et al. (2013). Transcriptome and genome sequencing uncovers functional variation in humans. *NATURE* **501**:506–511.

LeCun, Y. (1989). Generalization and network design strategies. *Connectionism in perspective* 143–155.

LeCun, Y., Bengio, Y. and Hinton, G. (2015). Deep learning. *NATURE* **521**:436–444.

Lee, D. D. and Seung, H. S. (1999). Learning the parts of objects by non-negative matrix factorization. *NATURE* **401**:788–791.

Lee, D. D. and Seung, H. S. (2001). Algorithms for non-negative matrix factorization. *Advances in Neural Information Processing Systems* 556–562.

Lee, D. Y. (2015). *Nonlinear functional regression model for sequencing-based association and gene-gene interaction analysis of physiological traits and their applications to sleep apnea.* Thesis, The University of Texas School of Public Health.

Lelli, K. M., Slattery, M. and Mann, R. S. (2012). Disentangling the many layers of eukaryotic transcriptional regulation. *Annual Review of Genetics* **46**:43–68.

Lemeire, J. and Janzing, D. (2013). Replacing causal faithfulness with algorithmic independence of conditionals. *Minds and Machines* **23**:227–249.

Li, B. and Dewey, C. N. (2011). RSEM: Accurate transcript quantification from RNA-Seq data with or without a reference genome. *BMC Bioinformatics* **12**:323.

Li, B., Ruotti, V., Stewart, R. M., Thomson, J. A. and Dewey, C. N. (2009). RNA-Seq gene expression estimation with read mapping uncertainty. *Bioinformatics* **26**: 493–500.

Li, D., Xie, Z., Pape, M. L. and Dye, T. (2015). An evaluation of statistical methods for DNA methylation microarray data analysis. *BMC Bioinformatics* **16**:217.

Li, J., Witten, D. M., Johnstone, I. M. and Tibshirani, R. (2012). Normalization, testing, and false discovery rate estimation for RNA-sequencing data. *Biostatistics* **13**: 523–538.

Li, L. and Xiong, M. (2015). Dynamic model for RNA-seq data analysis. *BioMed Research International* **2015**:916352.

Li, P., Piao, Y., Shon, H. S. and Ryu, K. H. (2015). Comparing the normalization methods for the differential analysis of Illumina high-throughput RNA-Seq data. *BMC Bioinformatics* **16**:347.

Li, X., Dunn, J., Salins, D. et al. (2017). Digital health: Tracking physiomes and activity using wearable biosensors reveals useful health-related information. *PLoS Biology* **15**:e2001402.

Li, Y., Nan, B. and Zhu, J. (2015). Multivariate sparse group lasso for the multivariate multiple linear regression with an arbitrary group structure. *Biometrics* **71**:354–363.

Lienart, T. (2015). *RKHS Embeddings.* https://www.stats.ox.ac.uk/~lienart/gml15 _rkhsembeddings.html.

Lin, D., Calhoun, V. D. and Wang, Y.-P. (2014). Correspondence between fMRI and SNP data by group sparse canonical correlation analysis. *Med Image Anal* **18**:891–902.

Lin, G., Shen, C., van den Hengel, A. and Reid, I. (2016). Efficient piecewise training of deep structured models for semantic segmentation. *Proceedings of the IEEE Conference on Computer Vision and Pattern Recognition* 3194–3203.

Lin, N., Jiang, J., Guo, S. and Xiong, M. (2015). Functional principal component analysis and randomized sparse clustering algorithm for medical image analysis. *PLoS One* **10**:e0132945.

Litjens, G., Kooi, T., Bejnordi, B. E. et al. (2017). A survey on deep learning in medical image analysis. *Med Image Anal* **42**:60–88.

Liu, F. and Chan, L. (2016). Causal inference on discrete data via estimating distance correlations. *Neural Computation* **28**:801–814.

Liu, F., Zhang, S.-W., Guo, W.-F., Wei, Z.-G. and Chen, L. (2016). Inference of gene regulatory network based on local bayesian networks. *PLoS Comput Biol* **12**: e1005024.

Liu, G., Lin, Z., Yan, S., Sun, J., Yu, Y. and Ma, Y. (2013). Robust recovery of subspace structures by low-rank representation. *IEEE Transactions on Pattern Analysis and Machine Intelligence* **35**:171–184.

Liu, X., Shi, X., Chen, C. and Zhang, L. (2015). Improving RNA-Seq expression estimation by modeling isoform-and exon-specific read sequencing rate. *BMC Bioinformatics* **16**:332.

Long, J., Shelhamer, E. and Darrell, T. (2015). Fully convolutional networks for semantic segmentation. *Proceedings of the IEEE Conference on Computer Vision and Pattern Recognition* 3431–3440.

Louizos, C., Shalit, U., Mooij, J., Sontag, D., Zemel, R. and Welling, M. (2017). Causal effect inference with deep latent-variable models. *arXiv:1705.08821*.

Luo, L., Zhu, Y. and Xiong, M. (2013). Smoothed functional principal component analysis for testing association of the entire allelic spectrum of genetic variation. *European Journal of Human Genetics* **21**:217–224.

Ma, C.-X., Casella, G. and Wu, R. (2002). Functional mapping of quantitative trait loci underlying the character process: A theoretical framework. *Genetics* **161**:1751–1762.

Magliacane, S., van Ommen, T., Claassen, T., Bongers, S., Versteeg, P. and Mooij, J. M. (2017). Causal transfer learning. *arXiv:1707.06422*.

Mairal, J. and Yu, B. (2013). Supervised feature selection in graphs with path coding penalties and network flows. *The Journal of Machine Learning Research* **14**:2449–2485.

Marblestone, A. H., Wayne, G. and Kording, K. P. (2016). Toward an integration of deep learning and neuroscience. *Frontiers in Computational Neuroscience* **10**:94.

Mazumder, R. and Hastie, T. (2012). The graphical lasso: New insights and alternatives. *Electronic Journal of Statistics* **6**:2125–2149.

McCarthy, D. J., Chen, Y. and Smyth, G. K. (2012). Differential expression analysis of multifactor RNA-Seq experiments with respect to biological variation. *Nucleic Acids Research* **40**:4288–4297.

McRae, A., Marioni, R. E., Shah, S. et al. (2017). Identification of 55,000 Replicated DNA Methylation QTL. *bioRxiv*: 166710.

Medland, S. E., Jahanshad, N., Neale, B. M. and Thompson, P. M. (2014). Whole-genome analyses of whole-brain data: Working within an expanded search space. *Nature Neuroscience* **17**:791–800.

Meier, L., Van de Geer, S. and Bühlmann, P. (2009). High-dimensional additive modeling. *The Annals of Statistics* **37**:3779–3821.

Menéndez, P., Kourmpetis, Y. A., ter Braak, C. J. and van Eeuwijk, F. A. (2010). Gene regulatory networks from multifactorial perturbations using Graphical Lasso: Application to the DREAM4 challenge. *PLoS One* **5**:e14147.

Mi, X., Eskridge, K., Wang, D. et al. (2010). Regression-based multi-trait QTL mapping using a structural equation model. *Statistical Applications in Genetics and Molecular Biology* **9**:Article38.

Moeskops, P., de Bresser, J., Kuijf, H. J. et al. (2018). Evaluation of a deep learning approach for the segmentation of brain tissues and white matter hyperintensities of presumed vascular origin in MRI. *NeuroImage: Clinical* **17**:251–262.

Mohammadi, S., Zuckerman, N., Goldsmith, A. and Grama, A. (2017). A Critical Survey of Deconvolution Methods for Separating Cell Types in Complex Tissues. *Proceedings of the IEEE* **105**:340–366.

Mohan, K., London, P., Fazel, M., Witten, D. and Lee, S.-I. (2014). Node-based learning of multiple gaussian graphical models. *The Journal of Machine Learning Research* **15**:445–488.

Moneta, A., Entner, D., Hoyer, P. O. and Coad, A. (2013). Causal inference by independent component analysis: Theory and applications. *Oxford Bulletin of Economics and Statistics* **75**:705–730.

Mooij, J. and Janzing, D. (2010). Distinguishing between cause and effect. *Causality: Objectives and Assessment* 147–156.

Mooij, J. M., Peters, J., Janzing, D., Zscheischler, J. and Schölkopf, B. (2016). Distinguishing cause from effect using observational data: Methods and benchmarks. *The Journal of Machine Learning Research* **17**:1103–1204.

Nariai, N., Hirose, O., Kojima, K. and Nagasaki, M. (2013). TIGAR: Transcript isoform abundance estimation method with gapped alignment of RNA-Seq data by variational Bayesian inference. *Bioinformatics* **29**:2292–2299.

Nariai, N., Kojima, K., Mimori, T., Kawai, Y. and Nagasaki, M. (2016). A Bayesian approach for estimating allele-specific expression from RNA-Seq data with diploid genomes. *BMC Genomics* **17**:Suppl 1:2.

Nelms, B. D., Waldron, L., Barrera, L. A. et al. (2016). CellMapper: Rapid and accurate inference of gene expression in difficult-to-isolate cell types. *Genome Biol* **17**:201.

Nesterov, Y. and Nemirovskii, A. (1994). Interior-point polynomial algorithms in convex programming. *SIAM*.

Noh, H. and Gunawan, R. (2016). Inferring gene targets of drugs and chemical compounds from gene expression profiles. *Bioinformatics* **32**:2120–2127.

Noh, H., Hong, S. and Han, B. (2015). Learning deconvolution network for semantic segmentation. *Proceedings of the IEEE International Conference on Computer Vision* 1520–1528.

Nowozin, S. and Lampert, C. H. (2011). Structured learning and prediction in computer vision. *Foundations and Trends® in Computer Graphics and Vision* **6**:185–365.

Nowzohour, C. and Bühlmann, P. (2016). Score-based causal learning in additive noise models. *Statistics* **50**:471–485.

Ogata, K. (1998). *System dynamics*. 3rd ed., Prentice Hall, New Jersey.

Orho-Melander, M. (2015). Genetics of coronary heart disease: Towards causal mechanisms, novel drug targets and more personalized prevention. *Journal of Internal Medicine* **278**:433–446.

Parikh, N. and Boyd, S. (2014). Proximal algorithms. *Foundations and Trends® in Optimization* **1**:127–239.

Pearl, J. (2009). *Causality: Models, reasoning, and inference*. 2nd ed., Cambridge University Press, New York.

Pearl, J. (2013). The mathematics of causal inference. *Proceedings of the Joint Statistical Meetings Conference*.

Pergola, G., Selvaggi, P., Trizio, S., Bertolino, A. and Blasi, G. (2015). The role of the thalamus in schizophrenia from a neuroimaging perspective. *Neuroscience & Biobehavioral Reviews* **54**:57–75.

Peters, J. and Bühlmann, P. (2013). Identifiability of Gaussian structural equation models with equal error variances. *Biometrika* **101**:219–228.

Peters, J., Janzing, D. and Schölkopf, B. (2017). *Elements of causal inference: Foundations and learning algorithms*. MIT Press, Boston.

Peters, J., Janzing, D. and Schölkopf, B. (2011). Causal inference on discrete data using additive noise models. *IEEE Transactions on Pattern Analysis and Machine Intelligence* **33**:2436–2450.

Peters, J., Mooij, J., Janzing, D. and Schölkopf, B. (2012). Identifiability of causal graphs using functional models. *arXiv:1202.3757*.

Peters, J. M. (2012). *Restricted structural equation models for causal inference*, Ph.D. Thesis, ETH Zurich.

Pettit, J. B., Tomer, R., Achim, K., Richardson, S., Azizi, L. and Marioni, J. (2014). Identifying cell types from spatially referenced single-cell expression datasets. *PLoS Comput Biol* **10**:e1003824.

Pidsley, R., Zotenko, E., Peters, T. J. et al. (2016). Critical evaluation of the Illumina MethylationEPIC BeadChip microarray for whole-genome DNA methylation profiling. *Genome Biol* **17**:208.

Poyton, A., Varziri, M. S., McAuley, K. B., McLellan, P. and Ramsay, J. O. (2006). Parameter estimation in continuous-time dynamic models using principal differential analysis. *Computers & Chemical Engineering* **30**:698–708.

Rakyan, V. K., Down, T. A., Balding, D. J. and Beck, S. (2011). Epigenome-wide association studies for common human diseases. *Nat Rev Genet* **12**:529–541.

Ramsay, J. and Silverman, B. W. (2005). *Functional data analysis*. 2nd ed., Springer-Verlag, New York.

Rapin, J., Bobin, J., Larue, A. and Starck, J. L. (2012). Robust non-negative matrix factorization for multispectral data with sparse prior. *Proceedings of ADA7*.

Ravì, D., Wong, C., Lo, B. and Yang, G.-Z. (2017). A deep learning approach to on-node sensor data analytics for mobile or wearable devices. *IEEE Journal of Biomedical and Health Informatics* **21**:56–64.

Ravier, P., Leclerc, F., Dumez-Viou, C. and Lamarque, G. (2007). Redefining performance evaluation tools for real-time QRS complex classification systems. *IEEE Transactions on Biomedical Engineering* **54**:1706–1710.

Richfield, O., Alam, M. A., Calhoun, V. and Wang, Y.-P. (2016). Learning Schizophrenia Imaging Genetics Data Via Multiple Kernel Canonical Correlation Analysis. *Bioinformatics and Biomedicine (BIBM), 2016 IEEE International Conference on* 507–511.

Robinson, M. D. and Oshlack, A. (2010). A scaling normalization method for differential expression analysis of RNA-seq data. *Genome Biol* **11**:R25.

Ronneberger, O., Fischer, P. and Brox, T. (2015). U-net: Convolutional networks for biomedical image segmentation. *International Conference on Medical Image Computing and Computer-Assisted Intervention* 234–241.

Rosa, G. J., Valente, B. D., de los Campos, G., Wu, X.-L., Gianola, D. and Silva, M. A. (2011). Inferring causal phenotype networks using structural equation models. *Genetics Selection Evolution* **43**:6.

Ross, S. M. (2014) *Introduction to probability models*. 11th ed., Academic Press, San Diego.

Sagan, H. (2012). *Introduction to the calculus of variations*. Courier Corporation.

Sandler, R. (2010). *Nonnegative matrix factorization for segmentation analysis*. Technion-Israel Institute of Technology, Faculty of Computer Science.

Sason, I. and Verdú, S. (2016). $f$-Divergence Inequalities. *IEEE Transactions on Information Theory* **62**:5973–6006.

Sathyanarayana, A., Joty, S., Fernandez-Luque, L. et al. (2016). Sleep quality prediction from wearable data using deep learning. *JMIR mHealth and uHealth* **4**:e125.

Sauwen, N., Acou, M., Sima, D. M. et al. (2017). Semi-automated brain tumor segmentation on multi-parametric MRI using regularized non-negative matrix factorization. *BMC Medical Imaging* **17**:29.

Scanagatta, M., De Campos, C. P. and Zaffalon, M. (2014). Min-BDeu and max-BDeu scores for learning Bayesian networks. *European Workshop on Probabilistic Graphical Models* 426–441.

Scutari, M. (2017). *Package 'bnlearn'*. http://www.bnlearn.com.

Shen-Orr, S. S. and Gaujoux, R. (2013). Computational deconvolution: Extracting cell type-specific information from heterogeneous samples. *Current Opinion in Immunology* **25**:571–578.

Shim, H. and Stephens, M. (2015). Wavelet-based genetic association analysis of functional phenotypes arising from high-throughput sequencing assays. *The Annals of Applied Statistics* **9**:665–686.

Shimizu, S., Hoyer, P. O., Hyvärinen, A. and Kerminen, A. (2006). A linear non-Gaussian acyclic model for causal discovery. *Journal of Machine Learning Research* **7**:2003–2030.

Shimizu, S., Inazumi, T., Sogawa, Y. et al. (2011). DirectLiNGAM: A direct method for learning a linear non-Gaussian structural equation model. *Journal of Machine Learning Research* **12**:1225–1248.

Shrout, P., Keyes, K. and Ornstein, K. (2011). Causality and psychopathology: *Finding the determinants of disorders and their cures*. Oxford University Press.

Sims, C. A. (1972). Money, income, and causality. *The American Economic Review* **62**: 540–552.

Skelly, D. A., Johansson, M., Madeoy, J., Wakefield, J. and Akey, J. M. (2011). A powerful and flexible statistical framework for testing hypotheses of allele-specific gene expression from RNA-seq data. *Genome Research* **21**:1728–1737.

Spirtes, P., Meek, C., Richardson, T. and Meek, C. (1999). An algorithm for causal inference in the presence of latent variables and selection bias.

Statnikov, A., Henaff, M., Lytkin, N. I. and Aliferis, C. F. (2012). New methods for separating causes from effects in genomics data. *BMC Genomics* **13**:S22.

Stegle, O., Teichmann, S. A. and Marioni, J. C. (2015). Computational and analytical challenges in single-cell transcriptomics. *Nature Reviews Genetics* **16**:133–145.

Stein, J. L., Hua, X., Lee, S. et al. (2010). Voxelwise genome-wide association study (vGWAS). *neuroimage* **53**:1160–1174.

Stephens, M. (2013). A unified framework for association analysis with multiple related phenotypes. *PLoS One* **8**:e65245.

Su, J., Yi, D., Liu, C., Guo, L. and Chen, W.-H. (2017). Dimension reduction aided hyperspectral image classification with a small-sized training dataset: Experimental comparisons. *Sensors* **17**:2726.

Sun, A., Venkatesh, A. and Hall, D. A. (2016). A multi-technique reconfigurable electrochemical biosensor: Enabling personal health monitoring in mobile devices. *IEEE Transactions on Biomedical Circuits and Systems* **10**:945–954.

Sun, K., Gonçalves, J. P., Larminie, C. and Pržulj, N. (2014). Predicting disease associations via biological network analysis. *BMC Bioinformatics* **15**:304.

Sun, Z. and Zhu, Y. (2012). Systematic comparison of RNA-Seq normalization methods using measurement error models. *Bioinformatics* **28**:2584–2591.

Sutton, C. A. and McCallum, A. (2005). Piecewise training for undirected models. Proceeding UAI'05 Proceedings of the Twenty-First Conference on Uncertainty in Artificial Intelligence. Pp. 568–575. Edinburgh, Scotland — July 26–29, 2005.

Sutton, C. and McCallum, A. (2011). An introduction to conditional random fields. *Foundations and Trends® in Machine Learning* **4**:267–373.

Székely, G. J. and Rizzo, M. L. (2009). Brownian distance covariance. *The Annals of Applied Statistics* **3**:1236–1265.

Székely, G. J., Rizzo, M. L. and Bakirov, N. K. (2007). Measuring and testing dependence by correlation of distances. *The Annals of Statistics* **35**:2769–2794.

Takahashi, T., Zhou, S.-Y., Nakamura, K. et al. (2011). A follow-up MRI study of the fusiform gyrus and middle and inferior temporal gyri in schizophrenia spectrum. *Progress in Neuro-Psychopharmacology and Biological Psychiatry* **35**:1957–1964.

Tian, Y., Morris, T. J., Webster, A. P. et al. (2017). ChAMP: Updated methylation analysis pipeline for Illumina BeadChips. *Bioinformatics* **33**:3982–3984.

Tibshirani, R., Saunders, M., Rosset, S., Zhu, J. and Knight, K. (2005). Sparsity and smoothness via the fused lasso. *Journal of the Royal Statistical Society: Series B (Statistical Methodology)* **67**:91–108.

Trapnell, C., Hendrickson, D. G., Sauvageau, M., Goff, L., Rinn, J. L. and Pachter, L. (2013). Differential analysis of gene regulation at transcript resolution with RNA-seq. *Nature Biotechnology* **31**:46–53.

Udell, M., Horn, C., Zadeh, R. and Boyd, S. (2016). Generalized low rank models. *Foundations and Trends® in Machine Learning* **9**:1–118.

Valente, B. D. and de Magalhaes Rosa, G. J. (2013). Mixed effects structural equation models and phenotypic causal networks. *Genome-Wide Association Studies and Genomic Prediction* 449–464.

Varvarigou, V., Dahabreh, I. J., Malhotra, A. and Kales, S. N. (2011). A review of genetic association studies of obstructive sleep apnea: Field synopsis and meta-analysis. *Sleep* **34**:1461–1468.

Vavasis, S. A. (2009). On the complexity of nonnegative matrix factorization. *SIAM Journal on Optimization* **20**:1364–1377.

Verma, T. and Pearl, J. (1991). Equivalence and synthesis of causal models. In Proceedings of the 6th Annual Conference on Uncertainty in Artificial Intelligence (UAI) 255–270.

Vincent, M., Mundbjerg, K., Skou Pedersen, J. et al. (2017). epiG: Statistical inference and profiling of DNA methylation from whole-genome bisulfite sequencing data. *Genome Biol* **18**:38.

Wang, B., Huang, L., Zhu, Y., Kundaje, A., Batzoglou, S. and Goldenberg, A. (2017). Vicus: Exploiting local structures to improve network-based analysis of biological data. *PLoS Comput Biol* **13**:e1005621.

Wang, H., Nie, F., Huang, H. et al. (2011). Identifying quantitative trait loci via group-sparse multitask regression and feature selection: An imaging genetics study of the ADNI cohort. *Bioinformatics* **28**:229–237.

Wang, J., Ding, H., Azamian, F. et al. (2017). Detecting cardiovascular disease from mammograms with deep learning. *IEEE Transactions on Medical Imaging* **36**:1172–1181.

Wang, K., Zhao, Y., Xiong, Q. et al. (2016). Research on healthy anomaly detection model based on deep learning from multiple time-series physiological signals. *Scientific Programming* **2016**:5642856.

Wang, P. (2016). *Causal genetic network analysis for multiple phenotypes from complex diseases*. Thesis, Fudan University.

Wang, P., Rahman, M., Jin, L. and Xiong, M. (2016). A new statistical framework for genetic pleiotropic analysis of high dimensional phenotype data. *BMC genomics* **17**:881.

Wang, T. and Zhu, L. (2013). Sparse sufficient dimension reduction using optimal scoring. *Computational Statistics & Data Analysis* **57**:223–232.

Wang, Y. (2011). *Smoothing splines: Methods and applications*. CRC Press, New York.

Wang, Z., Fang, H., Tang, N. L.-S. and Deng, M. (2017). VCNet: Vector-based gene co-expression network construction and its application to RNA-seq data. *Bioinformatics* **33**:2173–2181.

Wei, B., Sun, X., Ren, X. and Xu, J. (2017). Minimal effort back propagation for convolutional neural networks. *arXiv:1709.05804*.

Wolfe, E., Spekkens, R. W. and Fritz, T. (2016). The inflation technique for causal inference with latent variables. *arXiv:1609.00672*.

Wright, S. (1921). Correlation and causation. *Journal of Agricultural Research* **20**:557–585.

Wu, J., Pan, S., Zhu, X. and Cai, Z. (2015). Boosting for multi-graph classification. *IEEE Transactions on Cybernetics* **45**:416–429.

Wu, M. C., Lee, S., Cai, T., Li, Y., Boehnke, M. and Lin, X. (2011). Rare-variant association testing for sequencing data with the sequence kernel association test. *Am J Hum Genet* **89**:82–93.

Xie, Y., Zhang, Z., Sapkota, M. and Yang, L. (2016). Spatial clockwork recurrent neural network for muscle perimysium segmentation. *International Conference on Medical Image Computing and Computer-Assisted Intervention* 185–193.

Xiong, H., Brown, J. B., Boley, N., Bickel, P. J. and Huang, H. (2014). DE-FPCA: Testing gene differential expression and exon usage through functional principal component analysis. In *Statistical analysis of next generation sequencing data*: Springer.

Xiong, M. M. (2018). Big data in omics and imaging: Association analysis. Chapman and Hall/CRC.

Xu, K., Jin, L. and Xiong, M. (2017). Functional regression method for whole genome eQTL epistasis analysis with sequencing data. *BMC Genomics* **18**:385.

Yang, J., Wu, R. and Casella, G. (2009). Nonparametric functional mapping of quantitative trait loci. *Biometrics* **65**:30–39.

Yong, W. S., Hsu, F. M. and Chen, P. Y. (2016). Profiling genome-wide DNA methylation. *Epigenetics Chromatin* **9**:26.

Yu, Q., Erhardt, E. B., Sui, J. et al. (2015). Assessing dynamic brain graphs of time-varying connectivity in fMRI data: Application to healthy controls and patients with schizophrenia. *Neuroimage* **107**:345–355.

Yuan, M. and Lin, Y. (2006). Model selection and estimation in regression with grouped variables. *Journal of the Royal Statistical Society: Series B (Statistical Methodology)* **68**:49–67.

Zeiler, M. D., Krishnan, D., Taylor, G. W. and Fergus, R. (2010). Deconvolutional networks. Computer Vision and Pattern Recognition (CVPR), 2010 IEEE Conference on 2528–2535.

Zeiler, M. D., Taylor, G. W. and Fergus, R. (2011). Adaptive deconvolutional networks for mid and high level feature learning. *Computer Vision (ICCV), 2011 IEEE International Conference on* 2018–2025.

Zhang, B., Gaiteri, C., Bodea, L.-G. et al. (2013). Integrated systems approach identifies genetic nodes and networks in late-onset Alzheimer's disease. *Cell* **153**: 707–720.

Zhang, K., Wang, Z., Zhang, J. and Schölkopf, B. (2016). On estimation of functional causal models: General results and application to the post-nonlinear causal model. *ACM Transactions on Intelligent Systems and Technology (TIST)* **7**:1:22.

Zhang, Q., Filippi, S., Gretton, A. and Sejdinovic, D. (2017). Large-scale kernel methods for independence testing. *Statistics and Computing* pp. 1–18.

Zhang, Z. (2016). *Derivation of Backpropagation in Convolutional Neural Network (CNN)*. http://web.eecs.utk.edu/~zzhang61.

Zhang, Z. H., Jhaveri, D. J., Marshall, V. M. et al. (2014). A comparative study of techniques for differential expression analysis on RNA-Seq data. *PLoS One* **9**: e103207.

Zhao, S., Zhang, Y., Gordon, W. et al. (2015). Comparison of stranded and non-stranded RNA-seq transcriptome profiling and investigation of gene overlap. *BMC Genomics* **16**:675.

Zheng, Y., Liu, Q., Chen, E., Ge, Y. and Zhao, J. L. (2016). Exploiting multi-channels deep convolutional neural networks for multivariate time series classification. *Frontiers of Computer Science* **10**:96–112.

Zhou, S., Rütimann, P., Xu, M. and Bühlmann, P. (2011). High-dimensional covariance estimation based on Gaussian graphical models. *Journal of Machine Learning Research* **12**:2975–3026.

Zhou, X. and Stephens, M. (2014). Efficient multivariate linear mixed model algorithms for genome-wide association studies. *Nature Methods* **11**:407–409.

Zhou, Y., Arpit, D., Nwogu, I. and Govindaraju, V. (2014). Is Joint Training Better for Deep Auto-Encoders? *arXiv:1405.1380*.

Zhu, H., Rao, R. S. P., Zeng, T. and Chen, L. (2012). Reconstructing dynamic gene regulatory networks from sample-based transcriptional data. *Nucleic Acids Research* **40**:10657–10667.

Zscheischler, J., Janzing, D. and Zhang, K. (2012). Testing whether linear equations are causal: A free probability theory approach. *arXiv:1202.3779*.

Zyprych-Walczak, J., Szabelska, A., Handschuh, L. et al. (2015). The impact of normalization methods on RNA-Seq data analysis. *BioMed Research International* **2015**:621690.

# *Index*

Page numbers followed by f and t indicate figures and tables, respectively.